Autodesk® Inventor® Essentials Plus: 2013 and beyond

Autodesk® Inventor® Essentials Plus: 2013 and beyond

DANIEL T. BANACH
TRAVIS JONES

DELMAR
CENGAGE Learning®

Australia • Brazil • Japan • Korea • Mexico • Singapore • Spain • United Kingdom • United States

Autodesk® Inventor® Essentials Plus: 2013 and beyond

Daniel T. Banach, Travis Jones

Vice President, Editorial: Dave Garza

Director of Learning Solutions: Sandy Clark

Acquisitions Editor: Stacy Masucci

Managing Editor: Larry Main

Senior Product Manager: John Fisher

Editorial Assistant: Kaitlin Murphy

Vice President, Marketing: Jennifer Baker

Marketing Director: Deborah Yarnell

Marketing Manager: Katie Hall

Associate Marketing Manager: Jillian Borden

Production Director: Wendy Troeger

Art Direction and Content Project Management: PreMediaGlobal

Technology Project Manager: Joe Pliss

Interior Graphics: Dave Gink/Shadowland Studios

Library of Congress Control Number: 2012935640

ISBN-13: 978-1-133-94222-1

ISBN-10: 1-133-94222-9

Delmar
5 Maxwell Drive
Clifton Park, NY 12065-2919
USA

Cengage Learning is a leading provider of customized learning solutions with office locations around the globe, including Singapore, the United Kingdom, Australia, Mexico, Brazil, and Japan. Locate your local office at: **international.cengage.com/region**

Cengage Learning products are represented in Canada by Nelson Education, Ltd.

To learn more about Delmar, visit **www.cengage.com/delmar**

Purchase any of our products at your local college store or at our preferred online store **www.cengagebrain.com**

Printed in the United States of America
1 2 3 4 5 6 7 16 15 14 13 12

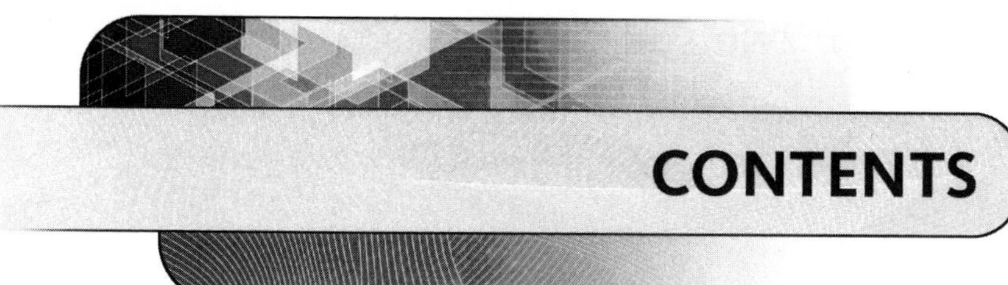

CONTENTS

CHAPTER 4 CREATING PLACED FEATURES 135

CHAPTER 5 CREATING AND EDITING DRAWING VIEWS 198

CHAPTER 6 CREATING AND DOCUMENTING ASSEMBLIES 296

CHAPTER 7 ADVANCED MODELING TECHNIQUES 403

CHAPTER 8 iCOMPONENTS AND PARAMETERS 503

CHAPTER 9 ADVANCED ASSEMBLY MODELING TECHNIQUES 586

CHAPTER 10 SHEET METAL DESIGN 664

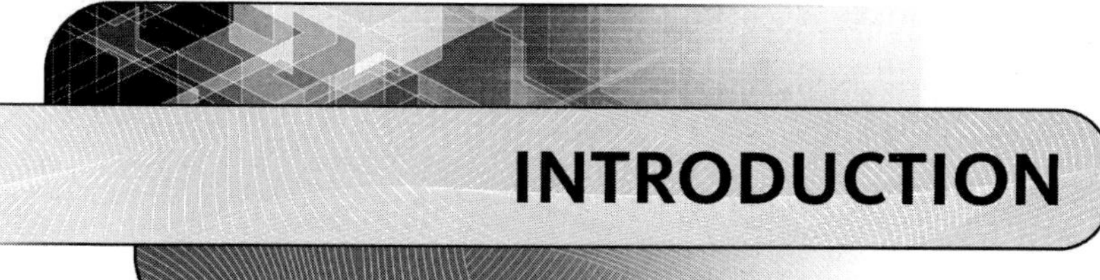

INTRODUCTION

Welcome to the *Autodesk Inventor 2013 Essentials Plus* manual. This manual provides a thorough coverage of the features and functionalities offered in Autodesk Inventor.

Each chapter in this manual is organized with the following elements:

Objectives. Describes the content and learning objectives.

Topic Coverage. Presents a concise, thorough review of the topic.

Exercises. Presents the workflow for a specific command or process through illustrated, step-by-step instructions.

Checking Your Skills. Tests your understanding of the material using True/False and multiple-choice questions.

NOTE TO THE LEARNER

Autodesk Inventor is designed for easy learning. Autodesk Inventor's help system provides you with ongoing support as well as access to online documentation.

As described above, each chapter in this manual has the same instructional design, making it easy to follow and understand. Each exercise is task-oriented and based on real-world mechanical engineering examples.

WHO SHOULD USE THIS MANUAL?

The manual is designed to be used in instructor-led courses, although you may also find it helpful as a self-paced learning tool.

Recommended Course Duration

Four days (32 hours) to seven days (56 hours) are recommended, although you may use the manual for specific Autodesk Inventor topics that may last only a few hours.

User Prerequisites

It is recommended that you have a working knowledge of Microsoft® Windows XP Professional® or Windows 7™ as well as a working knowledge of mechanical design principles.

Manual Objectives

The primary objective of this manual is to provide instruction on how to create part and assembly models, document those designs with drawing views, and automate the design process.

Upon completion of all chapters in this manual, you will be proficient in the following tasks:

- Basic and advanced part modeling techniques
- Drawing view creation techniques
- Assembly modeling techniques
- Sheet metal design

While working through these materials, we encourage you to make use of the Autodesk Inventor help system, where you may find solutions to additional design problems that are not addressed specifically in this manual.

Manual Description

This manual provides the foundation for a hands-on course that covers basic and advanced Autodesk Inventor features used to create, edit, document, and print parts and assemblies. You learn about the part and assembly modeling tools through online and print documentation and through the real-world exercises in this manual.

ESSENTIALS EXERCISE FILES

The exercise files for each chapter can be downloaded from http://www.cengagebrain.com.

Accessing a Student Companion site from CengageBrain:

1. Go to: http://www.cengagebrain.com.
2. TYPE Banach, Autodesk® Inventor® 2013 Essentials Plus or 1133942229 in the **Search** window.
3. LOCATE the desired product and click on the title.
4. When you arrive at the Product Page, CLICK on the **Free Stuff** tab.
5. Use the "**Click Here**" link to be brought to the Companion site.
    - Note: you will only see the Click Here link if there is a companion product available.

CAD Connect Website

We also include a CAD Connect website with this text. This website contains datasets for use in the exercises. It will also contain future content addressing features of subsequent releases of the software. The access code included with this text enables you to access this website. Instructions for redeeming your access code and accessing the website are below.

Redeeming an Access Code:

1. Go to HYPERLINK http://www.cengagebrain.com.
2. Enter the Access code in the Prepaid Code or Access Key field, **Register**.
3. Register as a new user or log in as an existing user if you already have an account with Cengage Learning or visit http://cengagebrain.com/.
4. Open the product from the My Account page.

Accessing CAD Connect from CengageBrain- My Account:

1. Sign in to your account at: http://www.CengageBrain.com.
2. Go to My Account to view purchases.
3. Locate the desired product.
4. Click on the **Open** button next to the CAD Connect site entry, or Premium Website for Autodesk Inventor Essentials Plus 2013 and beyond.

Projects

Most designers and engineers work on several projects at a time, with each project consisting of a number of files. To accommodate this, Autodesk Inventor uses projects to help organize related files and maintain links between files.

Each project has a *project file* that stores the paths to all files related to the project. When you attempt to open a file, Autodesk Inventor uses the paths in the current project file to locate other necessary files.

For convenience, a project file is provided for the *exercises.*

Using the Project File

Before starting the exercise, you must complete the following steps:

1. Start Autodesk Inventor.
2. On the Get Started tab > Launch panel, click Projects.
3. In the Projects window, select Browse. Navigate to the folder, or click the Application Menu > Manage > Projects, where you placed the Essentials Exercises, and double-click the file "Inv 2013 Ess Plus.ipj".
4. The "Inv 2013 Ess Plus" project will become the current project.
5. You can now start doing the exercises.

Projects are reviewed in more detail in Chapter 1.

AUTODESK INVENTOR 2013 PROFESSIONAL EXAM OBJECTIVES

The Autodesk Inventor 2013 Professional Certification exam assesses your knowledge of the commands, features, and common tasks that you use in Autodesk Inventor 2013. Following are the possible exam objectives, and the section that each exam objective is covered in the book. For more information about the Autodesk Certification Exams please visit: www.autodesk.com/certification.

- Control a project file
  Chapter 1—Projects in Autodesk Inventor
- Create dynamic input dimensions
  Chapter 2—Dynamic Input and Adding Dimensions Manually
- Use sketch constraints
  Chapter 2—Constraining the Sketch
- Create extrude features
  Chapter 3—Extruding a Sketch
- Create revolve features
  Chapter 3—Create Revolve Features
- Use the Project Geometry and Project Cut Edges commands
  Chapter 3—Projecting Part Edges

- Identify how to use visual style to control the appearance of a model
  Chapter 3—Visual Style
- Create fillet features
  Chapter 4—Fillets
- Create hole features
  Chapter 4—Holes
- Create a shell feature
  Chapter 4—Shelling
- Create a pattern of features
  Chapter 4—Patterns
- Create work features
  Chapter 4—Creating Work Planes, UCS—User Coordinate System
- Edit a section view
  Chapter 5—Creating Section Views
- Modify a style in a drawing
  Chapter 5—Drawing Standards and Styles
- Edit base and projected views
  Chapter 5—Editing Drawing Views
- Create and edit dimensions in a drawing
  Chapter 5—Adding Dimensions to a View
- Create and edit a hole table
  Chapter 5—Creating Hole Tables
- Create a part in the context of an assembly
  Chapter 6—Creating Parts in Place, Exercise 6-2: Designing Parts in the Assembly Context
- Apply and use assembly constraints
  Chapter 6—Assembly Constraints
- Find minimum distance between parts and components
  Chapter 6—The Minimum Distance Command
- Animate a presentation file
  Chapter 6—Creating Presentation Files
- Modify a bill of materials
  Chapter 6—Modify a bill of materials
- Modify a parts list
  Chapter 6—Parts List
- Emboss text and a profile
  Chapter 7—Embossed Text and Closed Profiles
- Create a sweep feature
  Chapter 7—Sweep Features
- Create a 3D path using the Intersection Curve and the Project to Surface commands
  Chapter 7—3D Sketch from Intersection Geometry, Project to Surface, and Project to 3D Sketch
- Create a loft feature
  Chapter 7—Loft Features
- Create a multi-body part
  Chapter 7—Multi-Body Parts
- Create a part using surfaces
  Consult the Help System
- Describe and use Shrinkwrap
  Chapter 7—Shrinkwrap

- Create and constrain sketch blocks
  Consult the Help System
- Create an iPart
  Chapter 8—Creating iParts
- Use iLogic
  Consult the Help System
- Create a level of detail
  Chapter 9—Levels of Detail Representations
- Create a positional representation
  Chapter 9—Positional Representations
- Use the Frame Generator commands
  Chapter 9—The Frame Generator
- Create components using the Design Accelerator commands
  Chapter 9—Design Accelerator
- Create sheet metal features
  Chapter 10—Contour Flange, Contour Roll, Flange, Lofted Flange
- Describe sheet metal features
  Chapter 10—Contour Flange, Contour Roll, Flange, Lofted Flange, Hems
- Create and edit a sheet metal flat pattern
  Chapter 10—Flat Pattern
- Annotate a sheet metal part in a drawing
  Chapter 10—Detailing Sheet Metal Designs

ACKNOWLEDGEMENTS

Copyedit

The authors would like to thank PreMediaGlobal for the comprehensive and attentive copyediting. Their expertise, knowledge, and attention to detail have added a great deal to this book.

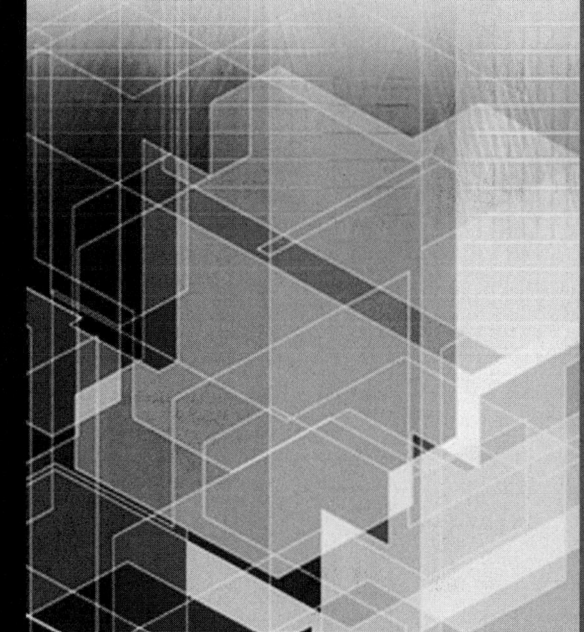

CHAPTER 1

Getting Started

INTRODUCTION

In this chapter you learn about the following: Inventor's user interface, application options that control how Inventor looks and acts, how to start commands, create and control projects, and how to change your viewpoint to view parts from different perspectives. The knowledge that you learn in this chapter will lay a strong foundation for you to master Inventor.

OBJECTIVES

In this chapter, you will be able to:

- List the main areas of Inventor's user interface
- Open files
- Create new files
- List the file types used in Autodesk Inventor
- Explain how to use the three save commands
- Use the Application Options to control how Inventor looks
- Start commands
- Describe how to use Inventor's help system
- Describe the purposes of a project file
- Create a project file for a single user
- Use navigation commands to change how you view a part

GETTING STARTED WITH AUTODESK INVENTOR

The welcome to Autodesk Inventor screen looks similar to Figure 1.1. From here you can open existing files, create new files, see a list of recent files, configure the unit and drawing standard of the default templates, start tutorials, watch skills videos, learn what's new in this release, and browse for applications that run with Autodesk Inventor.

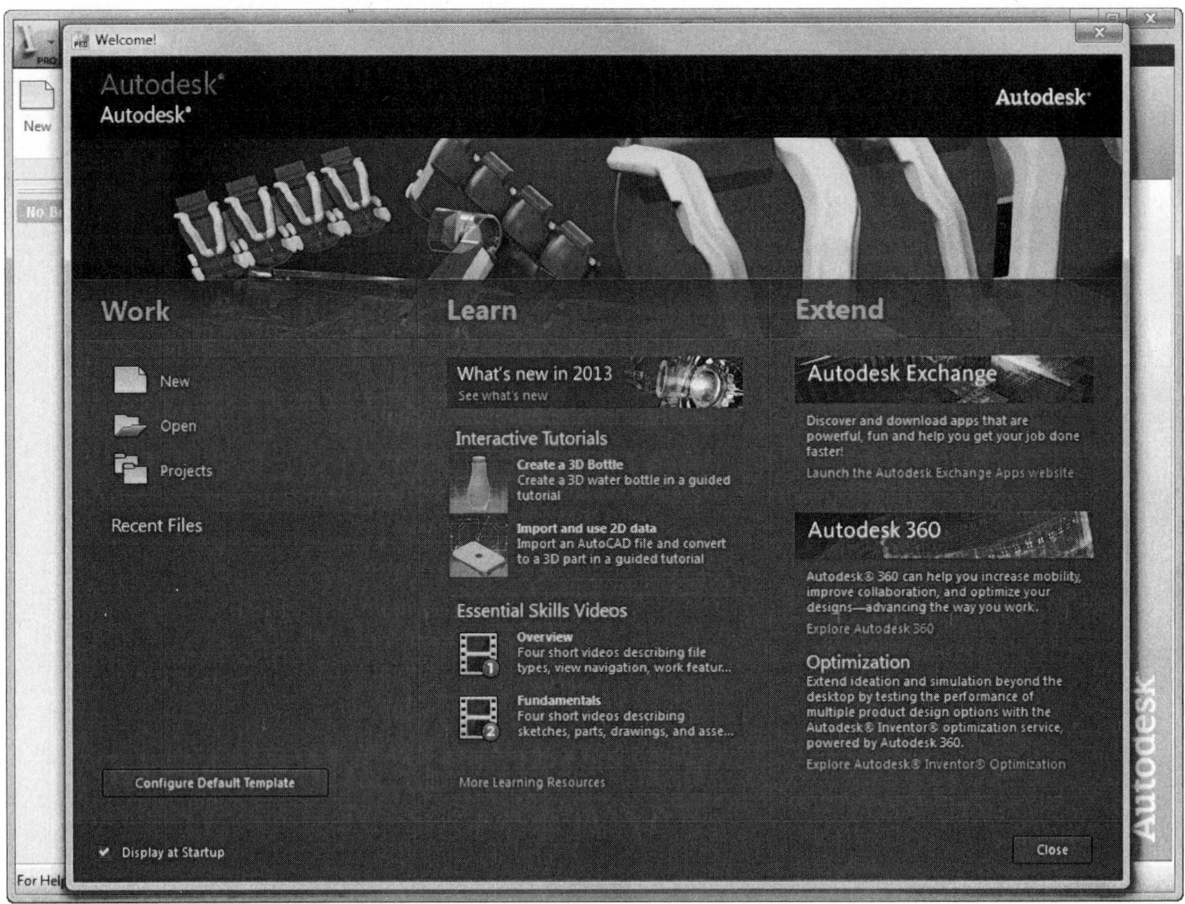

FIGURE 1.1

USER INTERFACE

The default sketch environment of a part (*.ipt*) in the Autodesk Inventor application window is shown in Figure 1.2.

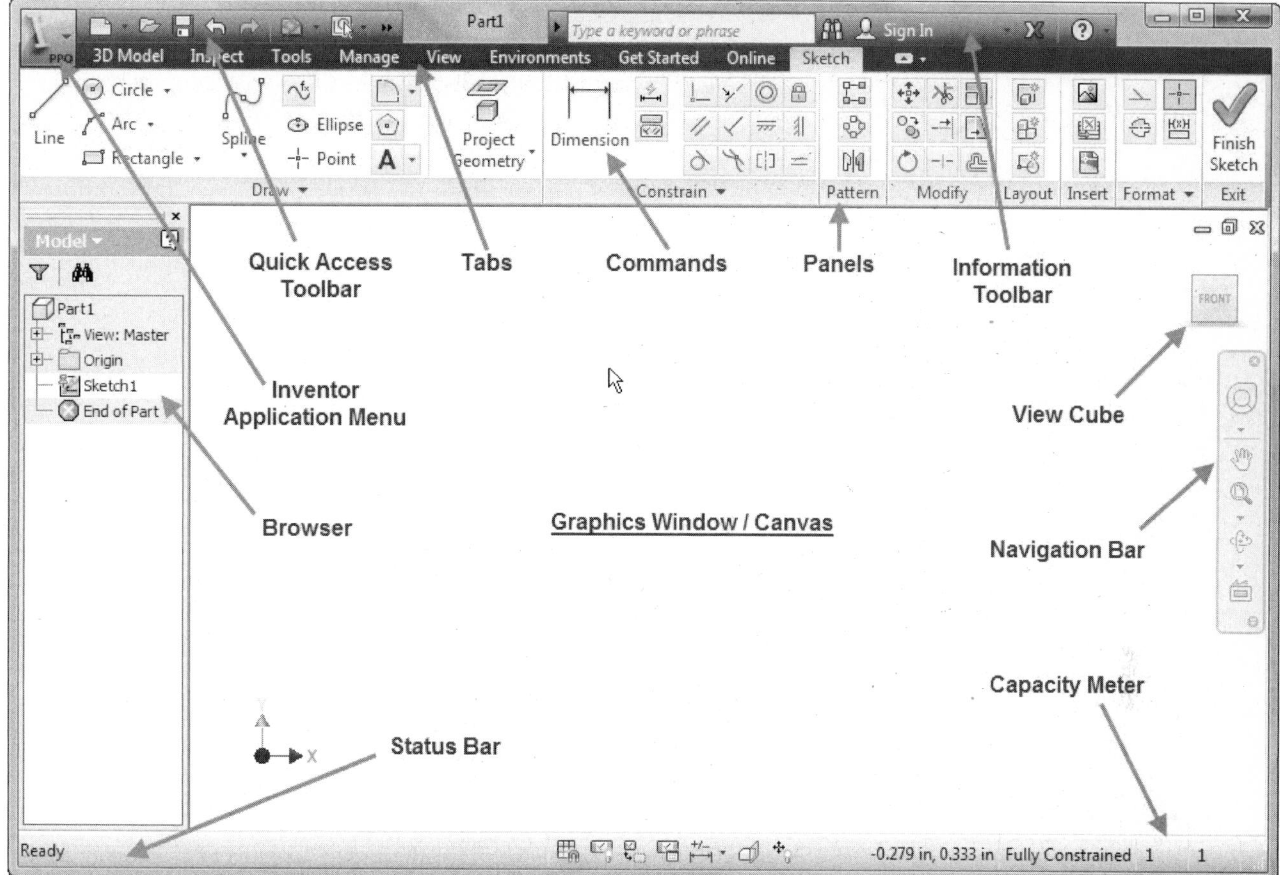

FIGURE 1.2

The screen is divided into the following areas:

Inventor Application Menu

Contains common commands for working with files.

Quick Access Toolbar

Accesses common commands as well as commands that can be added or removed.

Tabs

Changes available commands by clicking on a tab.

Commands

Accesses basic Windows and Autodesk Inventor commands. The set of commands in the changes to reflect the environment in which you are working.

Panels

Changes to show available commands for the current tab, click on a tab to display a set of new panels and commands.

Information Toolbar

Displays common help commands as well as Subscription services.

ViewCube

Displays current viewpoint and allows you to change the orientation of the view.

Navigation Bar

Displays common viewing commands. Viewing commands can be added by clicking the bottom drop arrow.

Capacity Meter

Displays how many occurrences (parts) are in the active document, the number of open documents in the current session, and how much memory is being used. Note: The capacity meter that shows memory usage is available only on 32-bit computers.

Status Bar

Views text messages about the current process.

Browser

Shows the history of how the contents in the file were created. The browser can also be used to edit features and components.

Graphics Window/Canvas

Displays the graphics of the current file.

INVENTOR APPLICATION MENU

Besides selecting commands for working with files you can control how the recent or open documents are listed in the menu. Figure 1.3 shows the functionality available from the Inventor Application Menu.

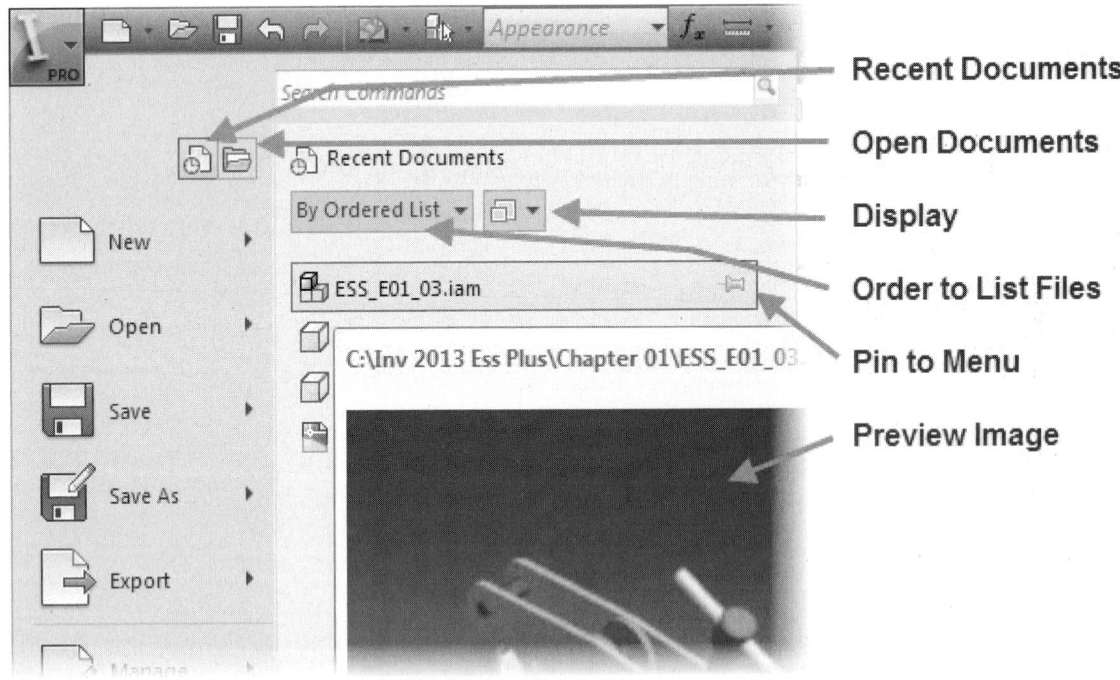

FIGURE 1.3

Recent Documents

Displays documents that were previously opened.

Open Documents
Displays documents that are opened.

Display
Controls what is displayed in the list: icons or images and what size.

Order to List Files
Controls the order that the files are listed: by Ordered List, by Access Date, by Size, by Type opened.

Pin to Menu
Click the push pin to keep the file in the list no matter when it was last opened.

Preview Image
Hovers the cursor over a file to see a larger image of the file when it was last saved.

NOTE

Double-click on the Inventor Application button, and a dialog box will appear asking you to save each unsaved document and then Inventor will close.

RIBBON
The ribbon displays commands that are relevant to the selected tab. The current tab is highlighted in green and is also green if it supports the current environment. The commands are arranged by panels. The most common commands are larger in size while less used commands are smaller and positioned to the right of the larger command. Tools may also be available in the drop list of a command or in the name of the panel. For example there is a drop list available under the Circle icon and the Draw panel as shown in Figure 1.4 that shows the Sketch tab active in a part file.

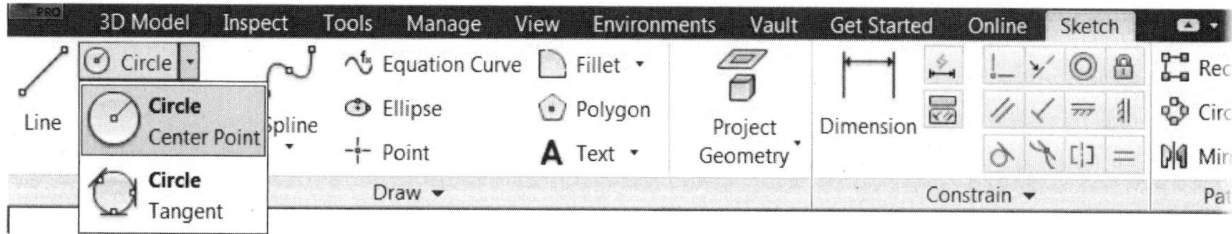

FIGURE 1.4

The ribbon can be modified by right clicking on the Panel; Figure 1.5 shows the available options.

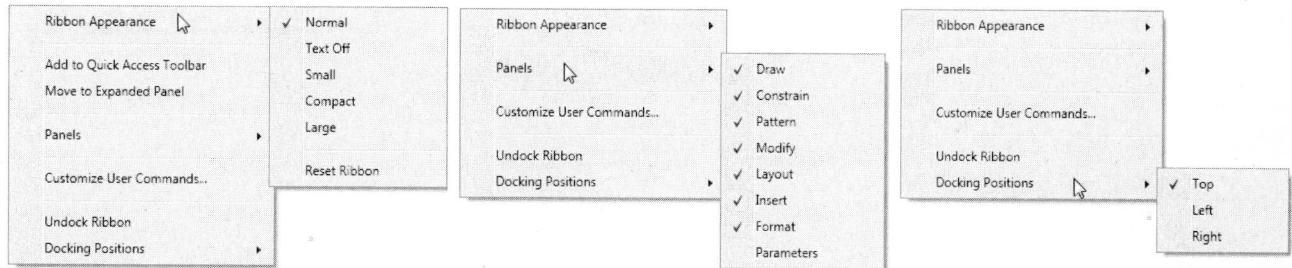

FIGURE 1.5

- Ribbon Appearance – Changes how the Ribbon looks, turns text off, and changes the size of the commands
- Panels – Adds and removes Panels
- Customize User Commands – Adds commands to a User Panel
- Undock Ribbon – Allows the ribbon to be freely moved
- Docking Position – Changes the location of the Ribbon

QUICK ACCESS TOOLBAR

Tools can be removed or added from the Quick Access Toolbar.

To add a command follow these steps

1. Move the cursor over a command to add and right-click.
2. Click Add to Quick Access Toolbar as shown in Figure 1.6.

To remove a command follow these steps

3. Move the cursor over the command to remove and right-click.
4. Click Remove from Quick Access Toolbar as shown in Figure 1.6.

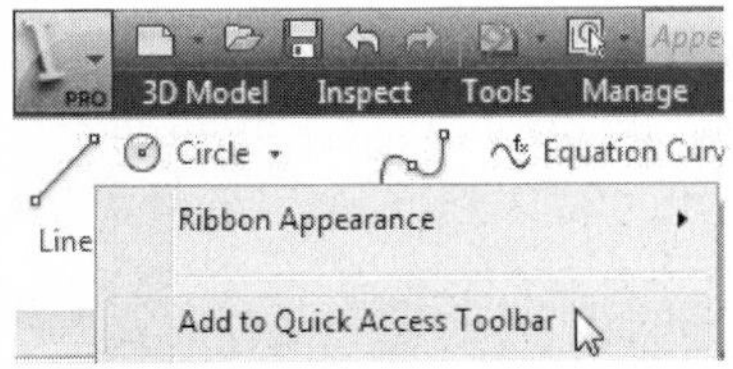

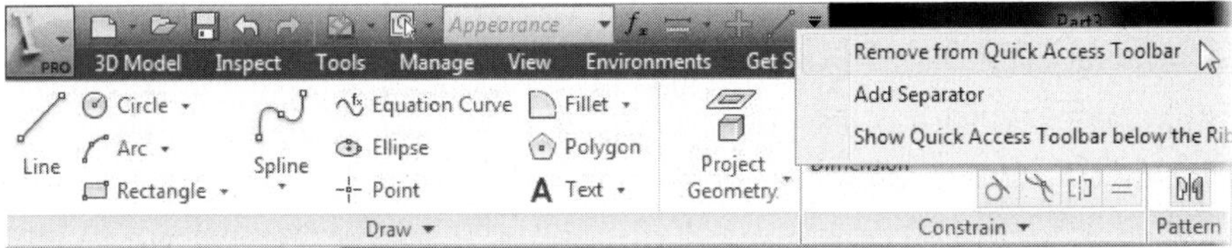

FIGURE 1.6

Open

To open files follow one of these techniques:

- Click the Open command in the Get Started tab in the Launch Panel as shown on the left of Figure 1.7.
- Click the Open command in the Quick Access Toolbar as shown on the left of Figure 1.7.
- Click the Inventor Application menu in the top-left corner and click Open as shown on the right of Figure 1.7.
- Press CTRL + O.

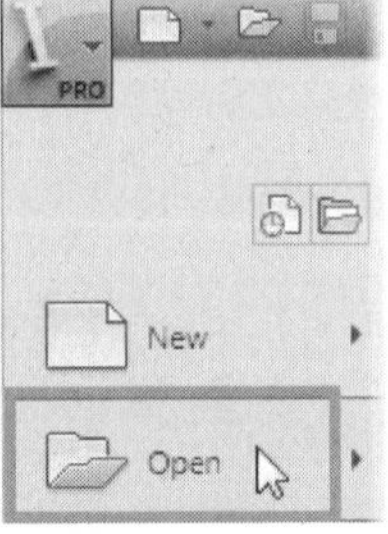

FIGURE 1.7

The Open dialog box will appear as shown in the Figure 1.8. The directory that opens by default is set in the current project file. You can open files from other directories that are not defined in the current project file, but this is not recommended. Part, drawing, and assembly relationships may not be resolved when you reopen an assembly that contains components outside the locations defined in the current project file. Projects are covered later in this chapter.

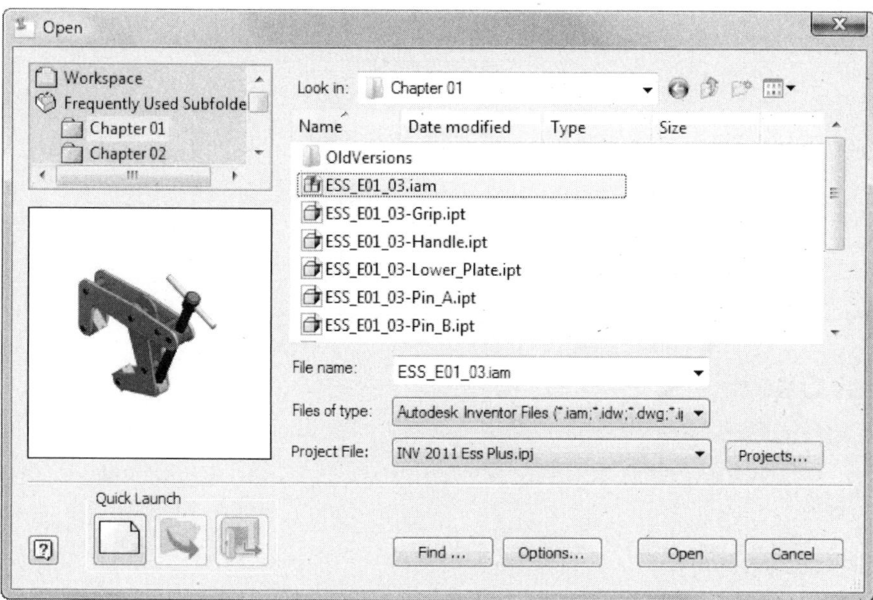

FIGURE 1.8

Opening Multiple Documents

You can open multiple Autodesk Inventor files at the same time by holding down the CTRL key and selecting the files to open as shown in the Figure 1.9. Each file will be opened in its own window in a single Autodesk Inventor session. To switch between the open documents, click the file on the Windows menu. The files can also be arranged to fit the screen or to appear cascaded. If the files are arranged or cascaded, click a file to activate it. Only one file can be active at a time.

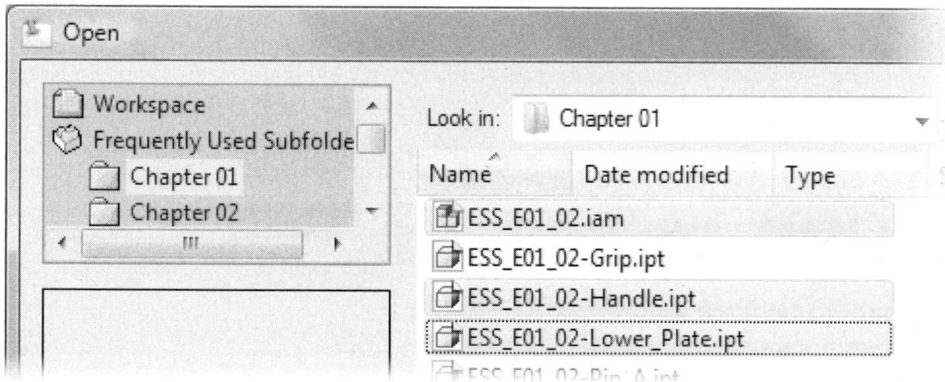

FIGURE 1.9

Document Tabs

When multiple documents are open in Inventor, each document appears in a tab in the lower left corner of the graphics window. The current document is represented with an "x" to the right of the file name and the tab's background is white. You can see a preview of an open document by hovering the cursor over the tab. In the same area, you can cascade, arrange, or list the open documents as shown in the Figure 1.10.

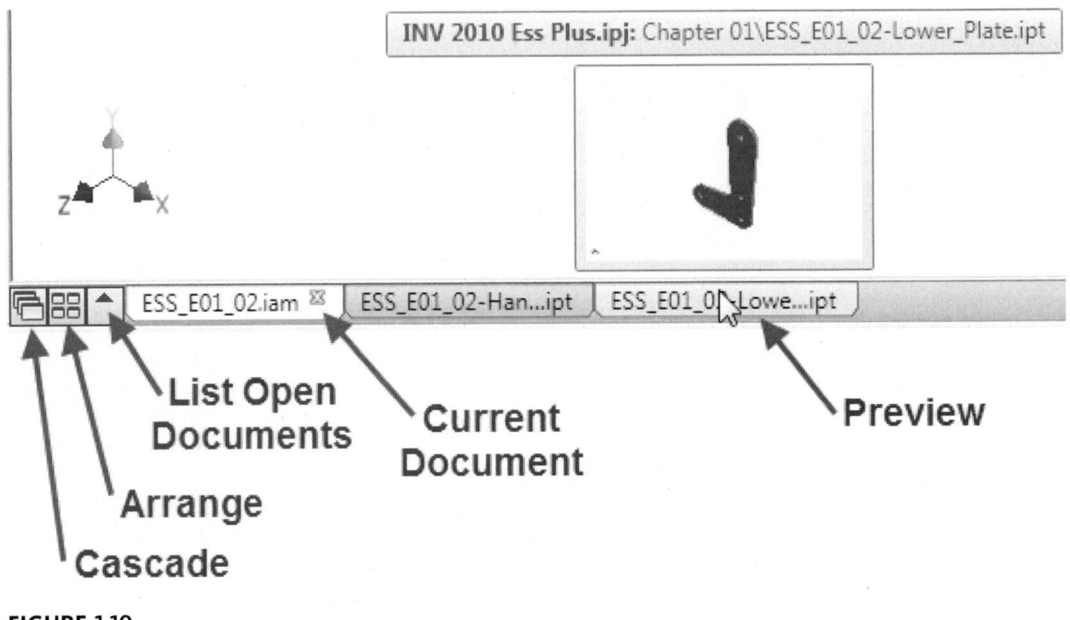

FIGURE 1.10

New Files

Like the Open command there are many ways to create a new file.

To create a new Inventor file, follow one of these techniques:

- Click the New command in the Get Started tab in the Launch Panel as shown on the left of Figure 1.11.
- Click the New command in the Quick Access Toolbar as shown on the left of Figure 1.11.
- Click the Inventor Application button in the top-left corner and click New as shown in the middle of Figure 1.11.
- Press CTRL + N.
- To start a new file based on one of the default templates click the down arrow next to the New icon in the Quick Access Toolbar as shown on the right of Figure 1.11.

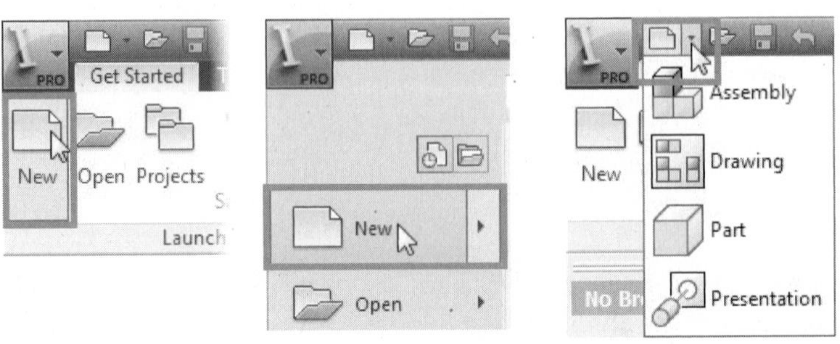

FIGURE 1.11

The New dialog box will appear as shown in Figure 1.12. Begin by selecting the type of file to create or one of the drafting standards named on the folders on the left column, and then click on a template from the Part, Assembly, Drawing or Presentation section and a preview and a description will appear on the right side of the dialog box. To start a new file double-click on a template or after a template is selected click Create on the bottom-right of the dialog box. If Autodesk Inventor Professional is installed, a Mold Design folder will exist in the list of template folders.

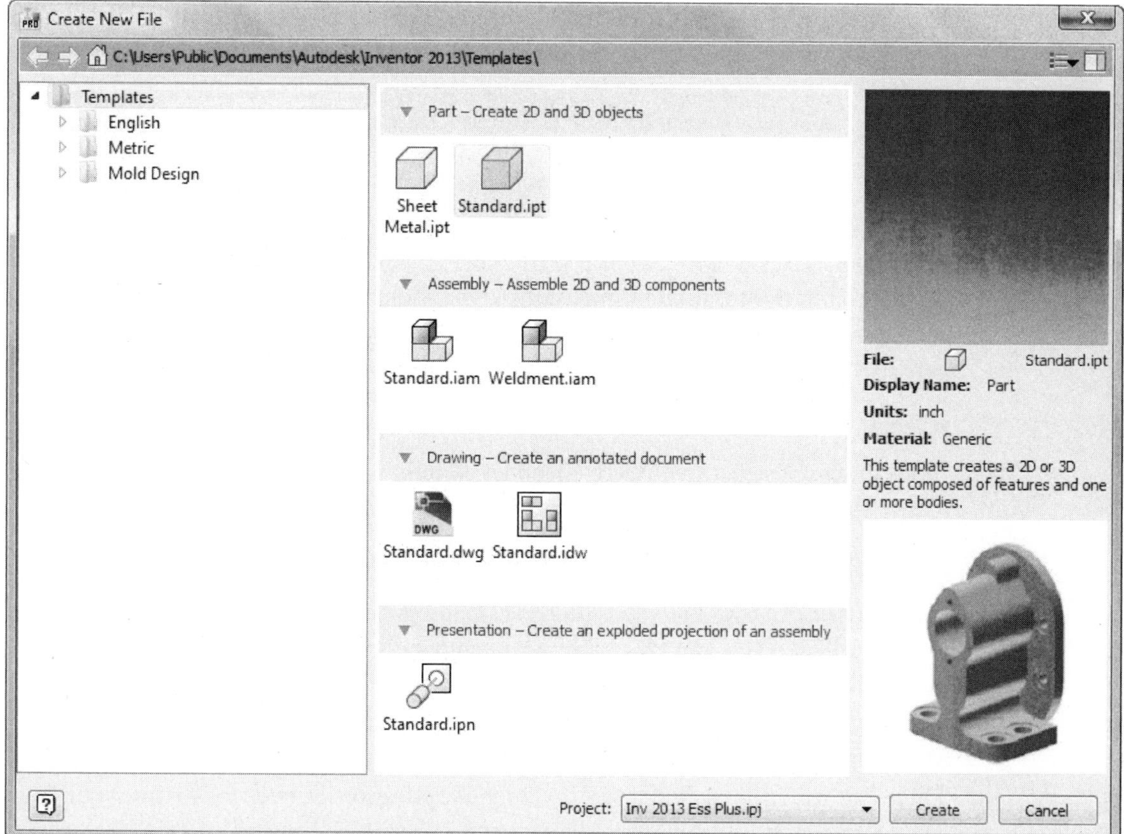

FIGURE 1.12

FILE INFORMATION

While creating parts, assemblies, presentation files, and drawing views, data is stored in separate files with different file extensions. This section describes the different file types and the options for creating them.

File Types

The following section describes the main file types that you can create in Autodesk Inventor, their file extensions, and descriptions of their uses.

Part (.ipt)

Part files contain only one part, which can be either 2D or 3D.

Assembly (.iam)

Assembly files can consist of a single part, multiple parts, or subassemblies. The parts themselves are saved to their own part file and are referenced (linked) in the assembly file. See Chapter 6 for more information about assemblies.

Presentation (.ipn)

Presentation files show parts of an assembly exploded in different states. A presentation file is associated with an assembly, and any changes made to the assembly will be updated in the presentation file. A presentation file can be animated, showing how parts are assembled or disassembled. The presentation file extension is *ipn*, but you save animations as an AVI or WMV file. See Chapter 6 for more information about presentation files.

Sheet Metal (.ipt)

Sheet metal files are part files that have the sheet metal environment loaded. In the sheet metal environment, you can create sheet metal parts and flat patterns. You can create a sheet metal part while in a regular part. This requires that you load the sheet metal environment manually. See Chapter 10 for more information about creating sheet metal parts.

Drawing (.dwg and .idw)

Drawing files can contain 2D projected drawing views of parts, assemblies, and/or presentation files. You can add dimensions and annotations to drawing views. The parts and assemblies in drawing files are linked, like the parts and assemblies in assembly and presentation files. See Chapter 5 for more information about drawing views.

Project (.ipj)

Project files are structured XML files that contain search paths to locations of all the files in the project. The search paths are used to find the files in a project.

iFeature (.ide)

iFeature files can contain one or more 3D features or 2D sketches that can be inserted into a part file. You can place size limits and ranges on iFeatures to enhance their functionality. See Chapter 8 for more information about creating iFeatures.

Save Options

There are three options on the Inventor Application button for saving your files: Save, Save Copy As, and Save All as shown in Figure 1.13.

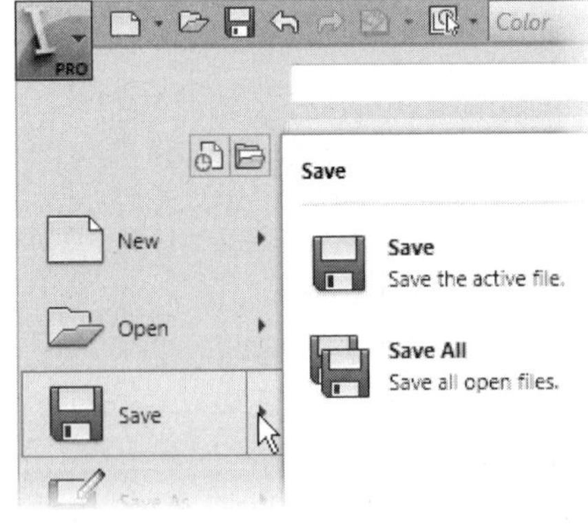

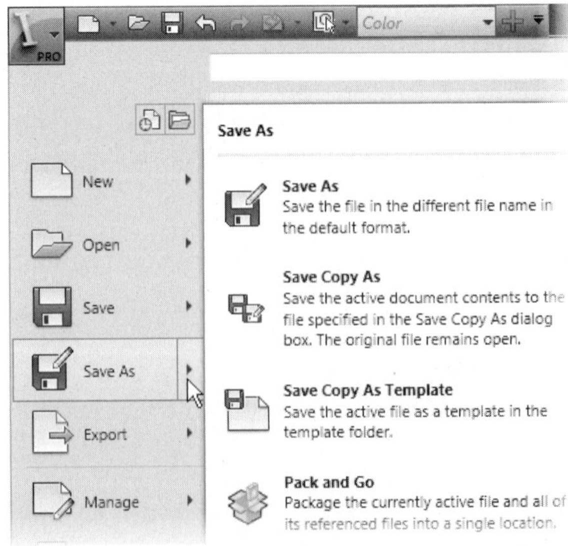

FIGURE 1.13

Save

The Save command saves the current document with the same name and to the location where you created it. If this is the first time that a new file is saved, you are prompted for a file name and file location.

TIP

To run the Save command, click the Save icon on the Quick Access Toolbar, use the shortcut keys CTRL + S, or click Save on the File menu.

Save All

Use the Save All command to save all open documents and their dependents. The files are saved with the same name to the location where you created them. The first time that a new file is saved, you will be prompted for a file name and file location.

Save As

Use the Save As command to save the active document with a new name and location, if required. A new file is created and is made active.

Save Copy As

Use the Save Copy As command to save the active document with a new name and location, if required. A new file is created but is not made active. You can also save the current file as different file formats that other CAD systems can open.

Save Copy As Template

Use the Save Copy As Template command to save the current file to the template folder. New files can be based on the template file. Templates can be saved in the existing folders, or you can create a subdirectory in the *Autodesk\Inventor (version number)\templates* directory, and add a file to it. A new template tab, with the same name as the subdirectory, is created automatically when a file is added to the new folder.

Pack and Go

Use the Pack and Go command to copy all the files that are used to create the current file to a specified location.

Save Reminder

You can have Inventor remind you to save a file. Inventor will NOT automatically save the file. After a predetermined amount of time has expired without saving the file, a notification bubble appears in the upper right corner of the screen as shown on the left of Figure 1.14. The time can be adjusted via the Application Options > Save tab as shown on the right of Figure 1.14. The time can be adjusted from 1 minute to 9999 minutes, or uncheck this option to turn off the notification.

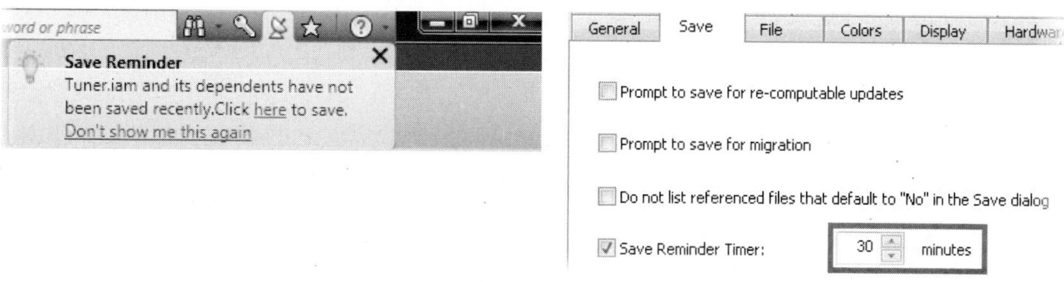

FIGURE 1.14

APPLICATION OPTIONS

Autodesk Inventor can be customized to your preferences. On the Inventor Application Menu, click Options, or from the Tools tab click Application Options, to open the Options dialog box as shown in Figure 1.15. You set options on each of the tabs to control specific actions in the Autodesk Inventor software. The application options affect all Inventor documents that are open or will be created. Each section is covered in more detail in the pertinent sections throughout this book. For more information about application options, see the Help system.

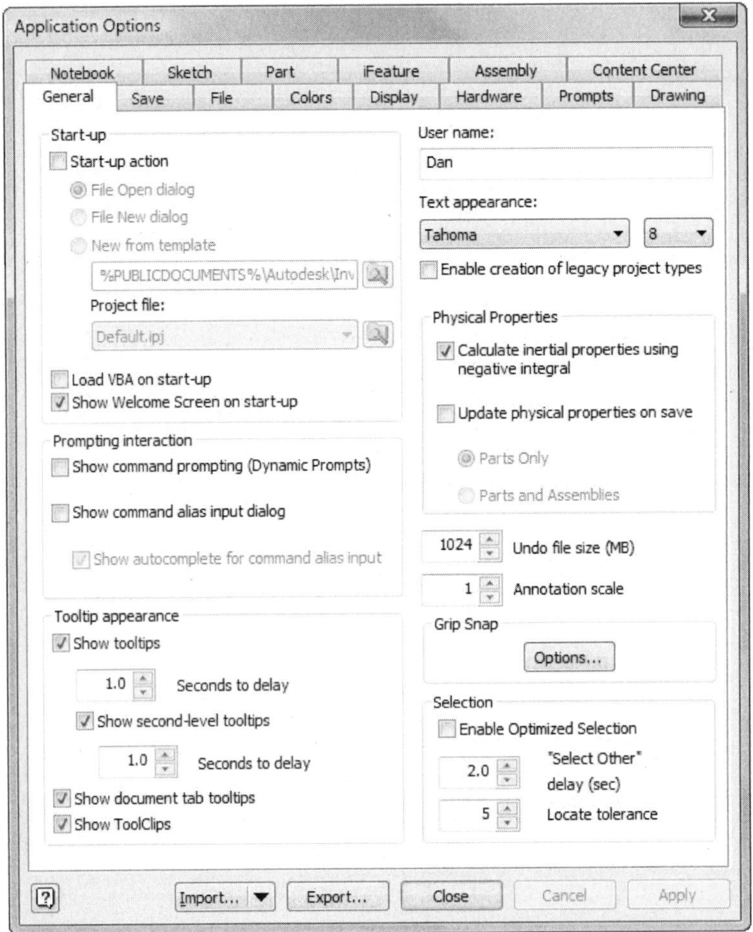

FIGURE 1.15

General

Set general options for how Autodesk Inventor operates.

Save

Set how files are saved.

File

Set where files are located.

Colors

Change the color scheme and color of the background on your screen. Determine if reflections and textures will be displayed.

Display

Adjust how parts look. Your video card and your requirements affect the appearance of parts on your screen. Experiment with different settings to achieve maximum video performance.

Hardware

Adjust the interaction between your video card and the Autodesk Inventor software. The software is dependent upon your video card. Take time to make sure that you are running a supported video card and the recommended video drivers. If you experience video-related issues, experiment with the options on the Hardware tab. For more information about video drivers, click Graphics Drivers on the Help menu.

Prompts

Modify the response given to messages that are displayed.

Drawing

Specify the way that drawings are created and displayed.

Notebook

Specify how the Engineer's Notebook is displayed.

Sketch

Modify how sketch data is created and displayed.

Part

Change how parts are created.

iFeature

Adjust where iFeatures data is stored.

Assembly

Specify how assemblies are controlled and behave.

Content Center

Specify the preferences for using the Content Center.

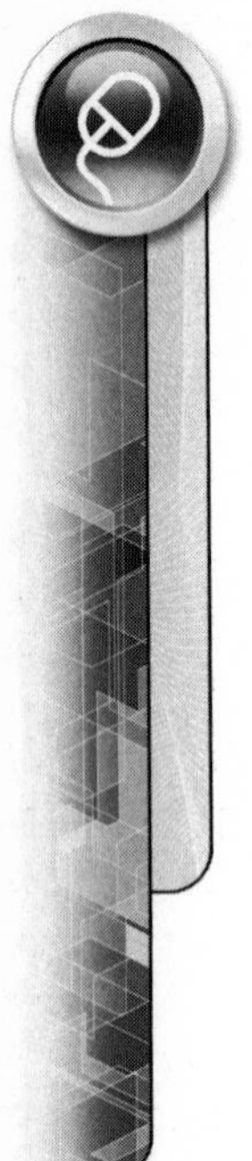

EXERCISE 1-1: USER INTERFACE

In this exercise, you change the user interface by moving the Ribbon, Quick Access Toolbar, and add and remove commands to the Quick Access Toolbar.

1. Click the New command from the Quick Access Toolbar, in the Create New File dialog box click the English folder on the left column, and in the Part section double-click Standard (in).ipt as shown in Figure 1.16.

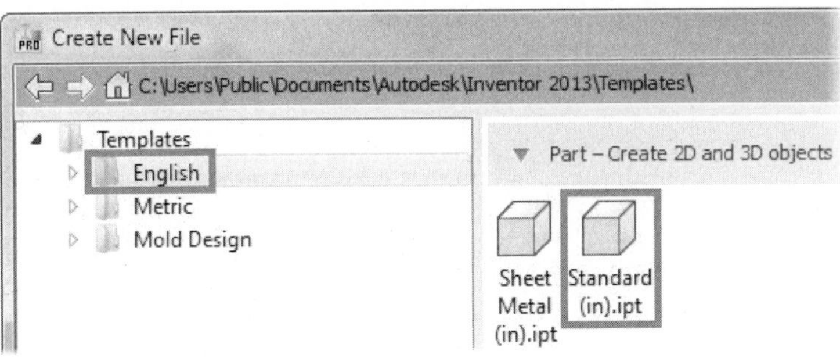

FIGURE 1.16

2. Move the Ribbon to different locations. Move the cursor anywhere over the Ribbon and right-click, and from the menu click Docking Positions and click Left and then repeat the process to move the Ribbon to the Right side.
3. Undock the Ribbon, right-click on the Ribbon, and from the menu click Undock Ribbon. Move the Ribbon to different locations.
4. Move the Ribbon back to its original top position, right-click on the Ribbon, and from the menu click Docking Positions > Top as shown in Figure 1.17.

FIGURE 1.17

5. Change the appearance of the Ribbon, right-click on the Ribbon, and from the menu click Ribbon Appearance. Try the different options to change the Ribbon's appearance.
6. Reset the Ribbon back to its original state by clicking Reset Ribbon from the same menu.

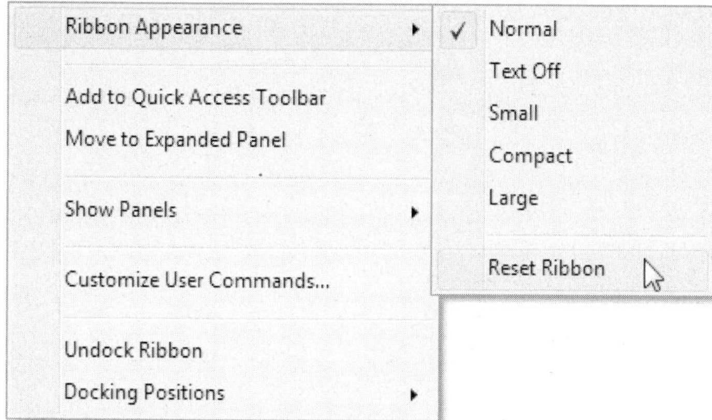

FIGURE 1.18

7. Add a command to the Quick Access Toolbar. Click the Tools tab > Options panel and right-click on the Application Options command right-click, and click Add to Quick Access Toolbar.

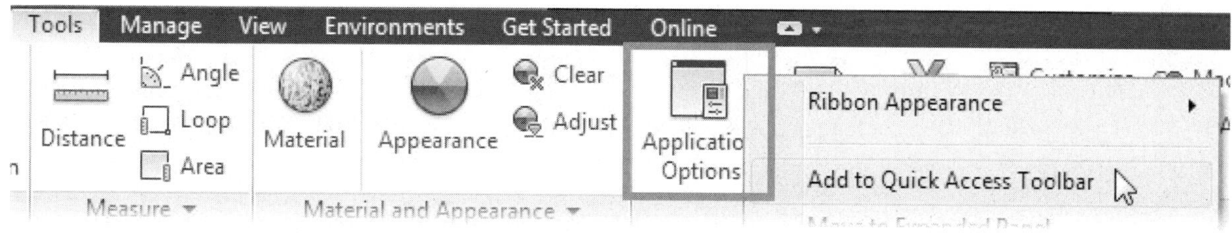

FIGURE 1.19

8. If desired remove the Application Options command from the Quick Access Toolbar. Move the cursor over the Application Options command in the Quick Access Toolbar, right-click, and click Remove from Quick Access Toolbar.

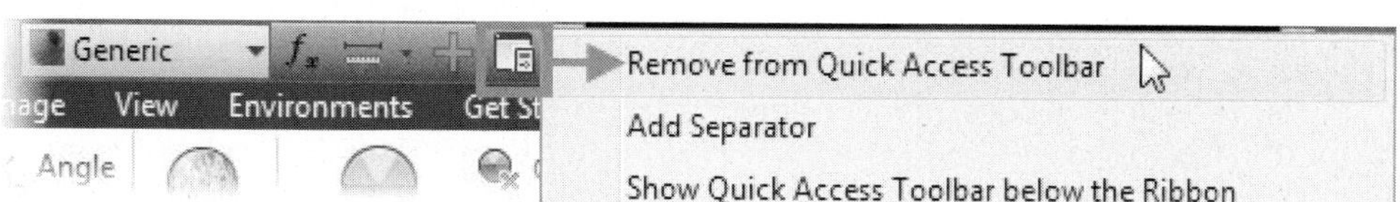

FIGURE 1.20

9. Change the background color of the graphics screen. Click the Application Options command that you just added to the Quick Access Toolbar.
10. Click the Colors tab, and from the Color scheme area select an option and click Apply to see the change.
11. Experiment with the background options.
12. Experiment changing the colors of the icons. In the Color Theme area, click the Amber option and click Apply. Notice the color of the icons change.

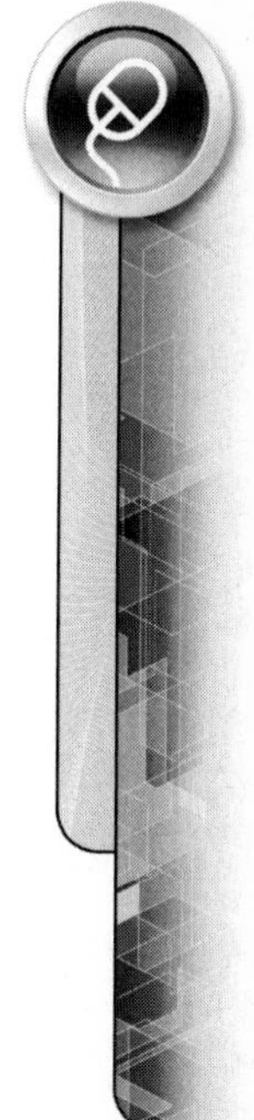

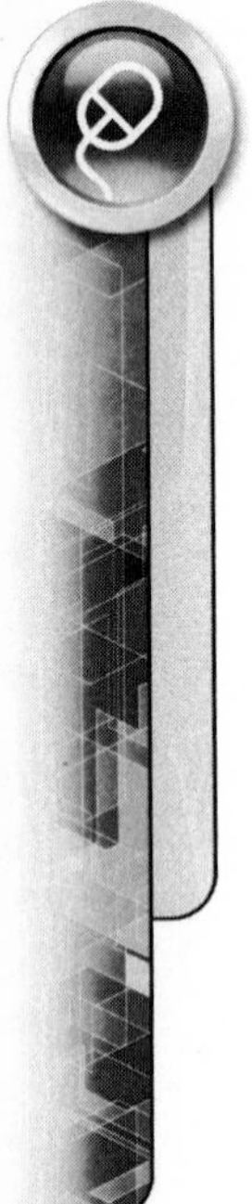

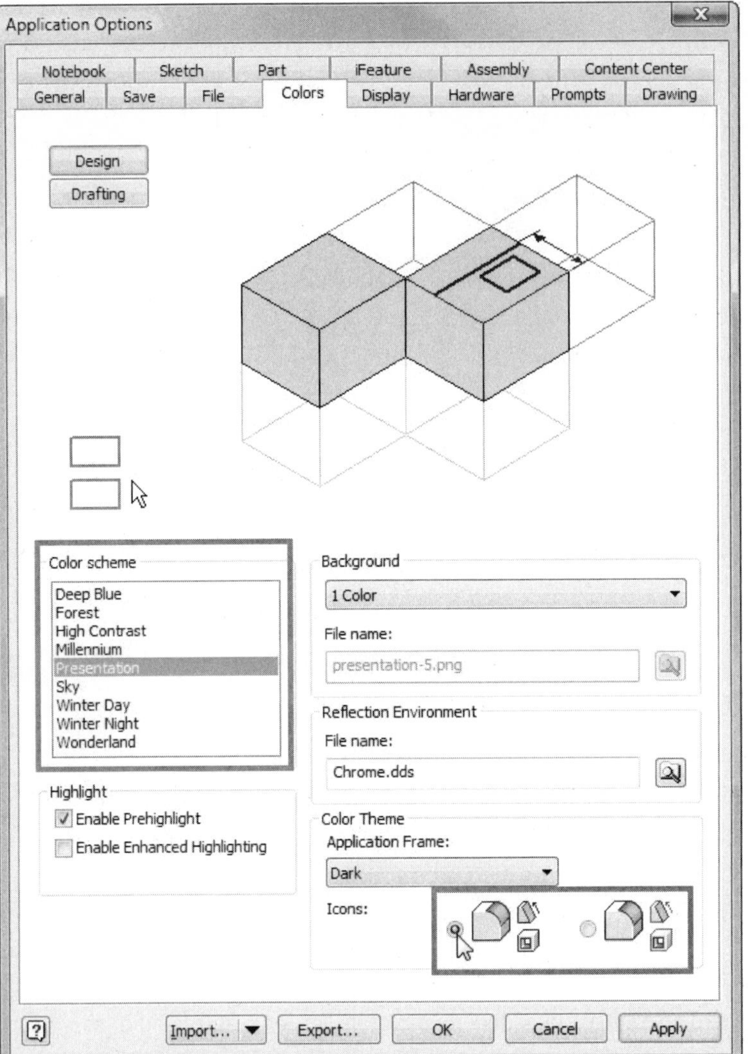

FIGURE 1.21

13. If desired change the icons color back to Cobalt as shown in Figure 1.21 and click OK.
14. As you work with Inventor, adjust the user interface to meet your requirements.
15. Close the file. Do not save changes. End of exercise.

Command Entry

There are several methods to issue commands in Autodesk Inventor. In the following sections, you will learn how to start a command. There are no right or wrong methods for starting a command, and with experience, you will develop your own preference.

To stop a command, either press the Esc key, right-click, and click Done from the menu or select another icon.

Panels and Tooltips

In the last section, you learned how to control the appearance of the ribbon. The main function of the Ribbon is to hold the commands in a logical fashion, which is done by dividing the commands into panels. To start a command from a panel, move the cursor over the desired icon, and a command tip appears with the name of the command. You can control the tooltip from the Application Options under the General tab as shown on the left of Figure 1.22. The first level tooltip displays an abbreviated command description as shown in the middle of Figure 1.22. If the cursor hovers over the icon longer, a more detailed command description will appear as shown on the right of Figure 1.22.

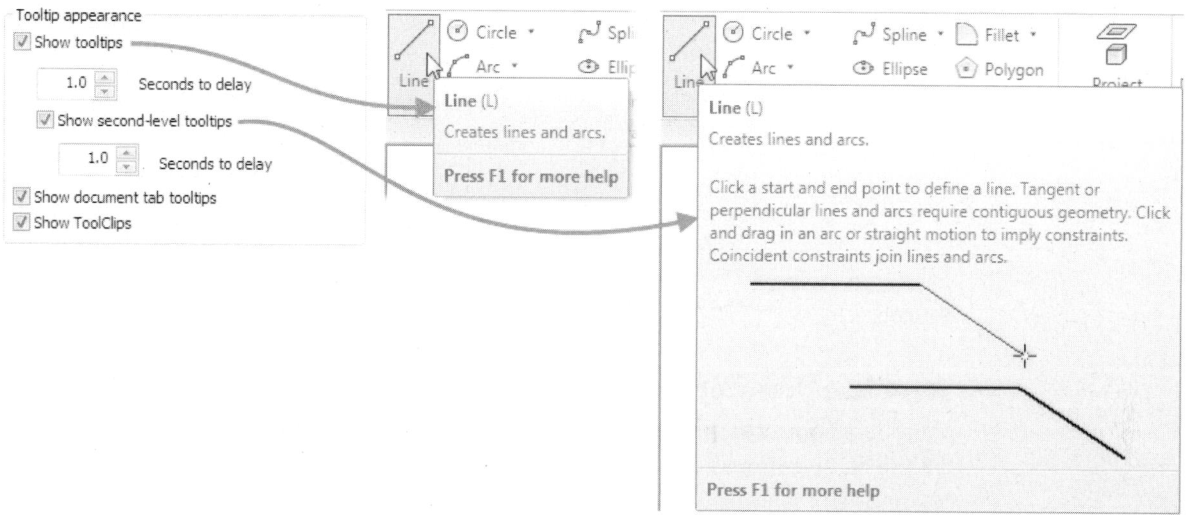

FIGURE 1.22

Some of the icons in the Panel have a small down arrow in the right side. Select the arrow to see additional commands. To activate a command, move the cursor over a command icon and click. The command that is selected will appear first in the list replacing the previous command.

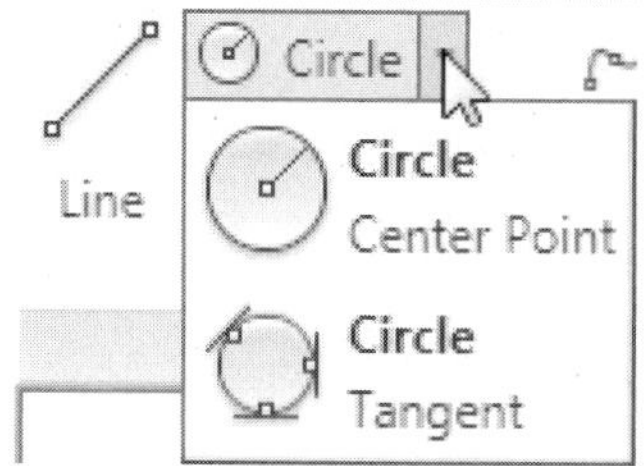

FIGURE 1.23

Marking Menus and Context (Overflow) Menus

Autodesk Inventor also uses marking menus and context menus also referred to as overflow menus. These menus appear when you press the right mouse button. The marking menu consists of commands that appear around the center of the cursor and consist of commands that are commonly performed for the current environment. The context (overflow) menus appear below the marking menu and contain options that are relevant to the current task. A marking menu only appears when you right-click in the graphics window. Figure 1.24 on the left shows the marking menu that appears while in the Line command. As you gain experience with Inventor you can start a command from the top portion of the marking menu by right-clicking and moving the cursor (before the menu appears within 250 milliseconds) in the direction of the command and release the mouse button as shown on the right of Figure 1.24. This technique is referred to as gesture behavior.

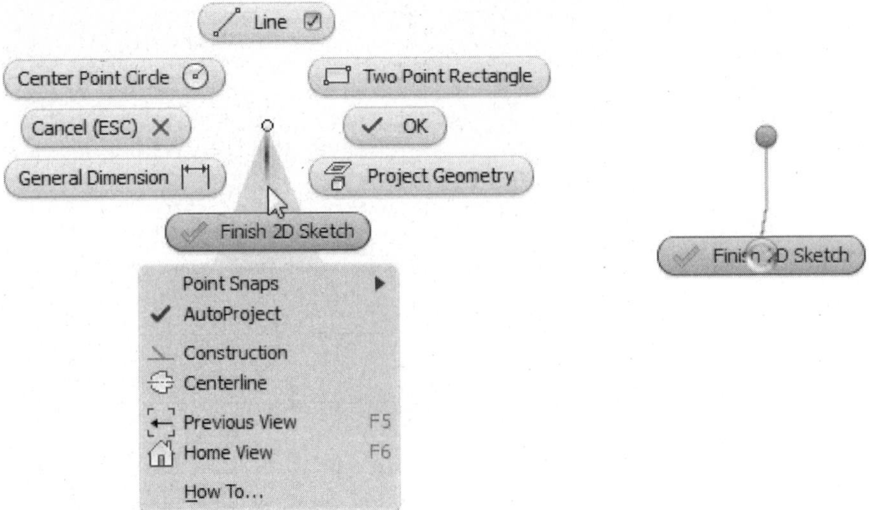

FIGURE 1.24

Autodesk Inventor Shortcut Keys

Autodesk Inventor has keystrokes called shortcut keys that are preprogrammed. While in a command, the tooltip displays the shortcut key in parenthesis if a shortcut key is available. To start a command via a shortcut key, press the desired preprogrammed key(s). The keys can be reprogrammed by clicking the Tools tab and click Customize.

REPEAT LAST COMMAND

To restart a command without reselecting the command in a panel bar, either press ENTER or the spacebar, or right-click and click the top entry in the menu, Repeat "the last command." Figure 1.25 shows the Line command being restarted.

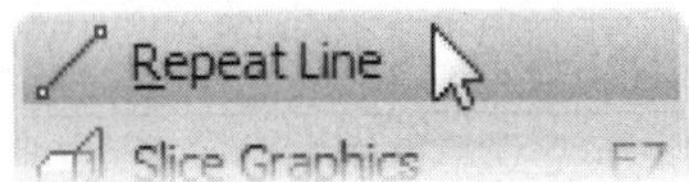

FIGURE 1.25

Undo and Redo

You may want to undo an action that you just performed, or undo an undo. The Undo command backs up Autodesk Inventor one function at a time. If you undo too far, you can use the Redo command to move forward one step at a time. The Zoom, Orbit, and Pan commands do not affect the Undo and Redo commands. To start the commands, select the command from the Quick Access Toolbar as shown in Figure 1.26. The Undo command is to the left, and the Redo command is to the right. The shortcut keys are CTRL + Z for Undo and CTRL + Y for Redo.

NOTE

To set the Undo file size allocation, click Tools tab > Application Options. On the General tab of the Options dialog box, change the Maximum size of Undo file (MB).

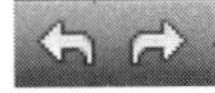

FIGURE 1.26

HELP SYSTEM

The Help system in Autodesk Inventor goes beyond basic command definition by offering assistance while you design. The commands in the Information Toolbar on the top-right corner of the screen will assist you while you design. To get help on a topic, enter a keyword in the area entitled "Type a keyword or phrase." To see the other help mechanisms that make up the Help System, click the drop arrow next to the right of the question mark as shown in Figure 1.27.

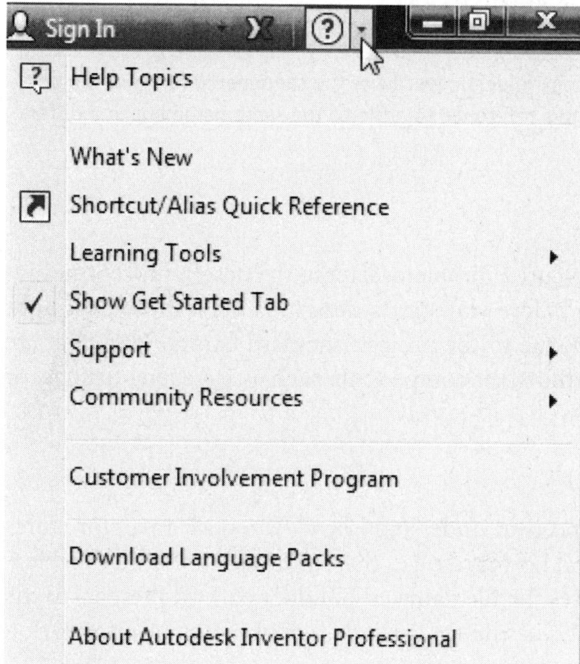

FIGURE 1.27

Other options to access the Help system include the following methods:

- Press the F1 key, and the Help system assists you with the active operation.
- Click an option on the Help menu.
- Click a Help option on the Information Toolbar.
- In any dialog box, click the ? icon.
- Click an option on the Get Started tab > Videos and Tutorials panel.
- Click on How To on a context menu when you are in a command.

PROJECTS IN AUTODESK INVENTOR

Almost every design that you create in Autodesk Inventor involves more than a single file. Each part, assembly, presentation, and drawing created is stored in a separate file. Each of these files has its own unique file extension. There are many times when a design will reference other files. An assembly file, for example, will reference a number of individual part files and/or additional subassemblies. When you open the parent or top-level assembly, it must contain information that allows Autodesk Inventor to locate each of the referenced files. Autodesk Inventor uses a project file to organize and manage these file-location relationships. There is no limit to the number of projects you can create, but only one project can be active at any given time.

You can structure the file locations for a design project in many ways. A single-person design shop has different needs from a large manufacturing company or a design team with multiple designers working on the same project. In addition to project files, Autodesk Inventor includes a program called Autodesk Vault on the DVD that controls basic check-out and check-in file-reservation mechanisms; these control file access for multiuser design teams. Autodesk Inventor always has a project named Default. Specifically, if all the files defining a design are located in a single folder, or in a folder tree where each referenced part is located with its parent or in a subfolder underneath the parent, the Default project may be all that is required.

NOTE

It is recommended that files in different folders never have the same name to avoid the possibility of Autodesk Inventor resolving a reference to a file of the same name but in a different folder.

Project Setup

To reduce the possibility of file resolution problems later in the design process, always plan your project folder structure before you start a design. A typical project might consist of parts and assemblies unique to the project; standard components that are unique to your company; and off-the-shelf components such as fasteners, fittings, or electrical components.

Project File Search Options

Before you create a project, you need to understand how Autodesk Inventor stores cross-file reference information and how it resolves that information to find the referenced file. Autodesk Inventor stores the file name, a subfolder path (if present) to the file, and a library name (optionally) as the three fundamental pieces of information about the referenced file.

When you use the Default project file, the subfolder path is located relative to the folder containing the referencing file. It may be empty or may go deeper in the subfolder hierarchy, but it can never be located at a level above the parent folder.

When you create a project file, you do not need to add subfolder as search paths. The subfolder(s) are automatically searched and do not need to be added to the project file.

Creating Projects

To create a new project or edit an existing project, use the Autodesk Inventor Project File Editor. The Project File Editor displays a list of shortcuts to previously active projects. A project file has an *.ipj* file extension and typically is stored in the home folder for the design-specific documents, while a shortcut to the project file is stored in the Projects Folder. The Projects Folder is specified on the Files tab of the Options dialog box as shown in Figure 1.28. All projects with a shortcut in the Projects Folder are listed in the top pane of the Project File Editor.

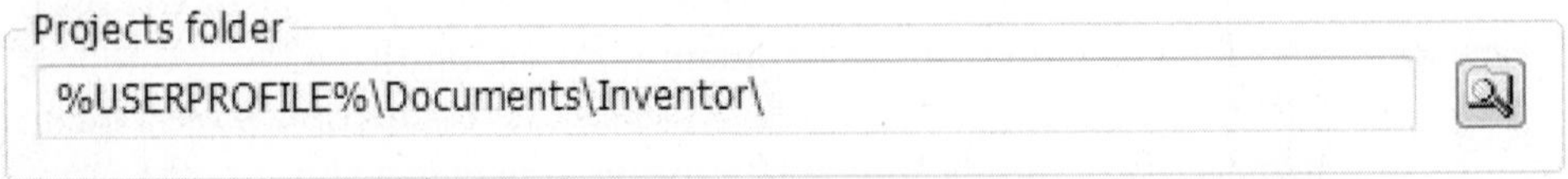

Projects folder
%USERPROFILE%\Documents\Inventor\

FIGURE 1.28

You create or edit a project file by clicking the Project button in the New or Open dialog box or by clicking Get Started tab > Launch panel > Projects as shown in Figure 1.29 on the left, or from the Inventor Application Menu click Manage and click Projects as shown on the right of Figure 1.29.

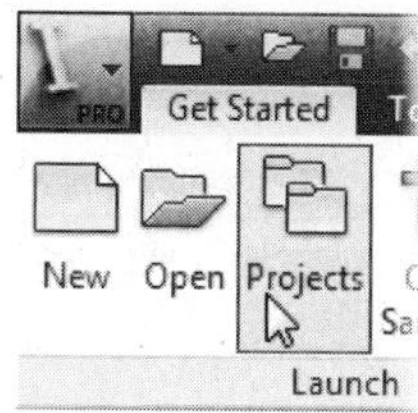

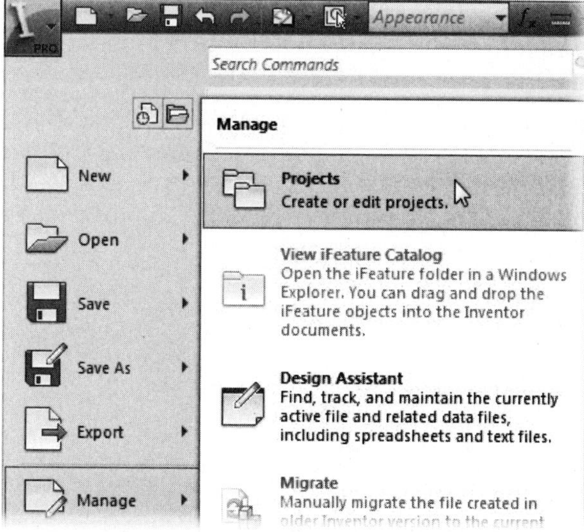

FIGURE 1.29

The Projects dialog box will appear as shown in Figure 1.30. The Projects dialog box is divided into two panes. The top pane lists shortcuts to the project files that have been active previously. Double-click on a project's name to make it the active project. All Inventor files must be closed before making a project current. Only one project file can be active in Autodesk Inventor at a time. The bottom portion reflects information about the project selected in the top pane. If a project file already exists, click on the Browse button on the bottom of the dialog box, then navigate to, and select the project file. The bottom pane of the dialog box lists information about the highlighted project.

NOTE

When defining a path to a folder on a network, it is recommended to define a Universal Naming Convention (UNC) path starting with the server name (\\Server\...) and not to use shared (mapped) network drives.

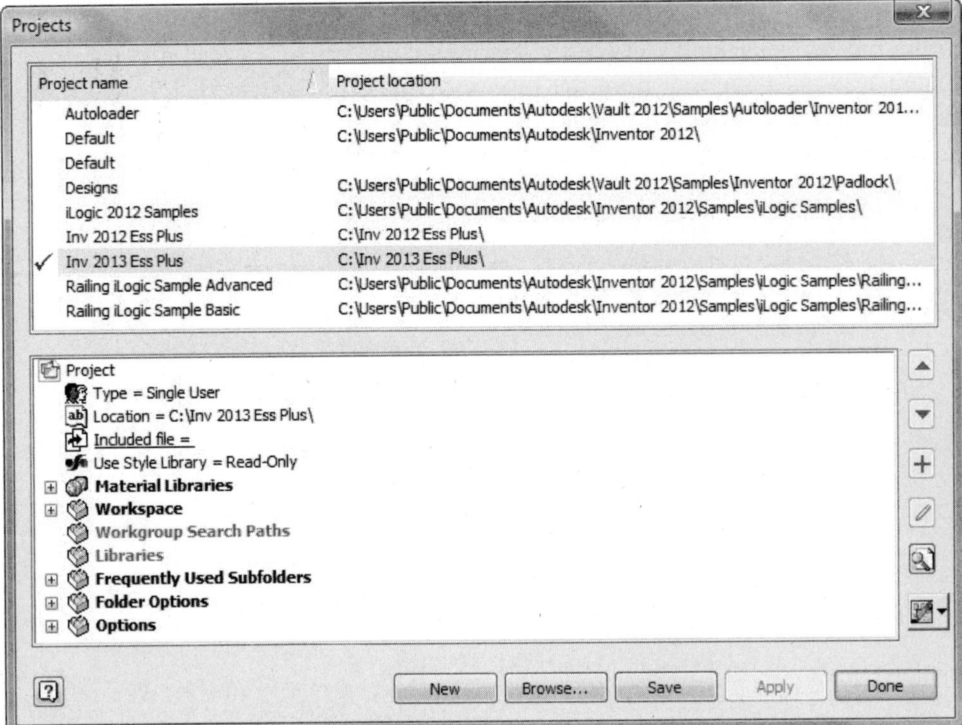

FIGURE 1.30

To create a new project, follow these steps:

1. In the Projects dialog box, click the New button at the bottom to initiate the Inventor project wizard.
2. In the Inventor project wizard, follow the prompts to the following questions.

What Type of Project Are You Creating?

If Autodesk Vault is installed, you will be prompted to create a New Vault project or a New Single-User project. If Autodesk Vault is not installed, only a New Single-User project type will appear in the list.

New Vault Project

This project type is used with Autodesk Vault and is not available until you install Autodesk Vault.

It creates a project with one workspace and any needed library location(s), and it sets the multiuser mode to Vault. More information about Autodesk Vault appears later in this section.

New Single-User Project

This is the default project type, which is used when only one user will reference Autodesk Inventor files. It creates one workspace where Autodesk Inventor files are stored and any needed library location(s), and it sets the Project Type to Single User. No workgroup is defined but can be defined later.

The next section covers the steps for creating a new single-user project. For more information on projects, consult the online Help system.

Creating a New Single-User Project

Click on New Single User Project as shown in Figure 1.31. If Autodesk Vault is not installed, New Single User Project will be the only available option.

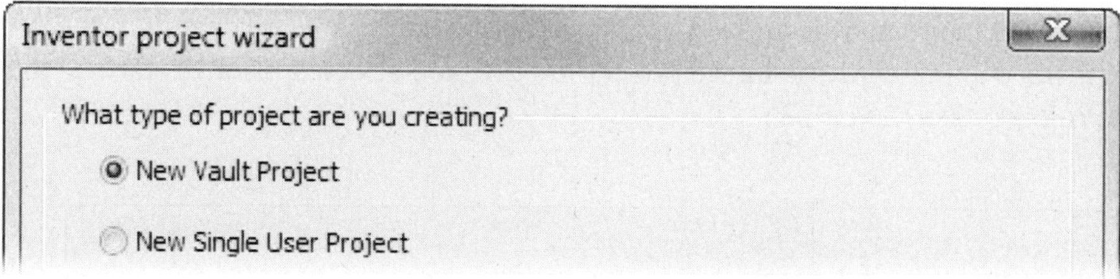

FIGURE 1.31

Click the Next button, and specify the project file name and location on the second page of the Inventor project wizard as shown in Figure 1.32.

Name

Enter a descriptive project file name in the Name field. The project file will use this name with an *.ipj* file extension.

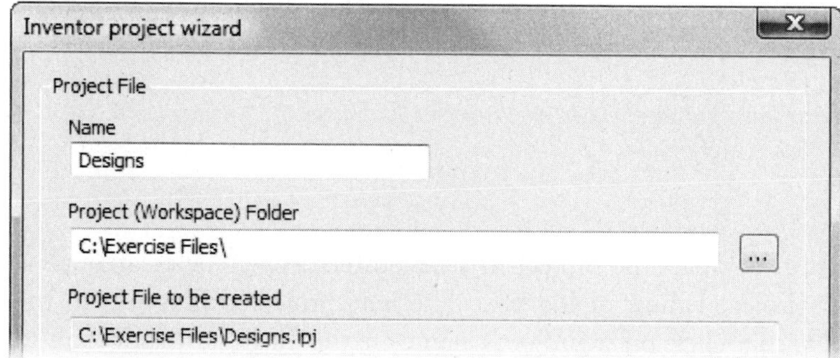

FIGURE 1.32

Project (Workspace) Folder

This specifies the path to the home or top-level folder for the project. You can accept the suggested path, enter a path, or click the Browse button (...) to manually locate the path. The default home folder is a subfolder under My Documents, or wherever you last browsed, that is named to match the project file name.

Project File to Be Created

The full path name of the project file is displayed below the Location field.

You can specify the home or top-level folder as the location of your workspace, but it is preferable for the project file (*.ipj*) to be the only Autodesk Inventor file stored in the home folder. This makes it easier to create other project locations, such as library and workspace folders, as subfolders without creating nesting situations that can lead to confusion.

Click the Next button at the bottom of the Inventor project wizard, and specify the project library search paths.

You can add library search paths from existing project files to this new project file. The library search paths from every project with a shortcut in your Projects Folder are listed on the left in the Inventor project wizard dialog box as shown in Figure 1.33. You can add and remove libraries from the New Project area by clicking on their names in either the All Projects area or the New Project area and then clicking the arrows in the middle of the dialog box. The libraries listed by default in the All Projects list will match those in the project file that you selected prior to starting the New Project process.

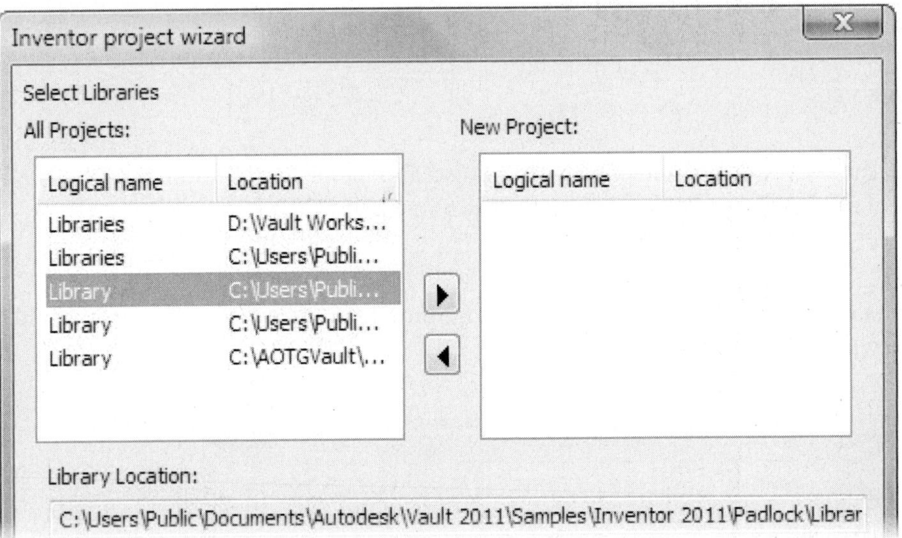

FIGURE 1.33

Click the Finish button to create the project. If a new directory will be created, click OK in the Inventor Project Editor dialog box. The new project will appear in the Open dialog box. Double-click on a project's name in the Projects dialog box to make it the active project. A checkmark will appear to the left of the active project as shown in Figure 1.34.

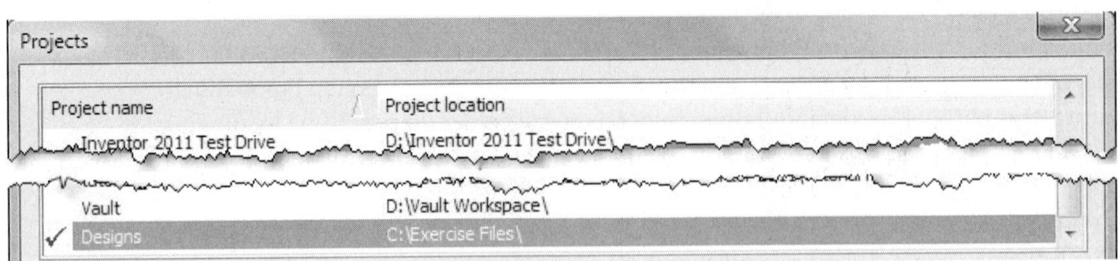

FIGURE 1.34

Autodesk Vault

Autodesk Vault is available when Inventor is installed. The Autodesk Vault enhances the data management process by managing more than just Autodesk Inventor files and by tracking file versions as well as team member access. Controlling access to data, tracking modifications, and communicating the design history are important aspects of managing collaborative data. When working with a vault project, your data files are stored in a central repository that records the entire development history of the design. The vault manages Inventor and non-Inventor files alike. In order to modify a file, it first must be checked out of the vault. When the file is checked back into the vault, the modifications are stored as the most recent version for the project, and the previous version is sequentially indexed as part of the living history of the design.

EXERCISE 1-2: PROJECTS

In this exercise, you create a project file for a single-user project, open existing files, delete the project file, and make an existing project file current.

1. Prior to creating a new project file, close all Inventor files and ensure that the exercise files have been installed. To download the exercise, see the Essentials Exercise Files section in the Introduction section in the front of the book.
2. From the Get Started tab > Launch panel click Projects.
3. You now create a new project file. Click the New button at the bottom of the Projects dialog box.
4. If Vault is installed click New Single User Project, and click Next.
5. In the next dialog box of the Inventor project wizard, enter **ABC Machine** in the Name field.
6. In the Project (Workspace) Folder field type C:\My First Project as shown in the following image.

TIP

It is a good practice to place files such as parts, assemblies, and drawings in folders below the top-level folder, but not in the top-level folder where the project file is located.

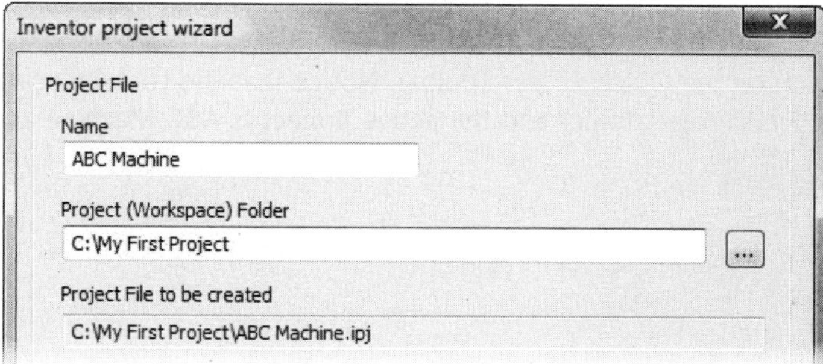

FIGURE 1.35

7. Click Next to see a list of Libraries that are used in other projects a library search path. The New Project list on the right side should be blank. For this project no libraries will be added.
8. Click Finish and then click OK to create the new project path. The new project file is highlighted in the upper pane of the Project File Editor. The options for the active project are listed in the lower pane. The new project should be the current project.

9. Expand Workspace and notice that the Workspace search path is listed as a period "." as shown in Figure 1.36. The "." denotes that the workspace location is relative to the location where the project file is saved.

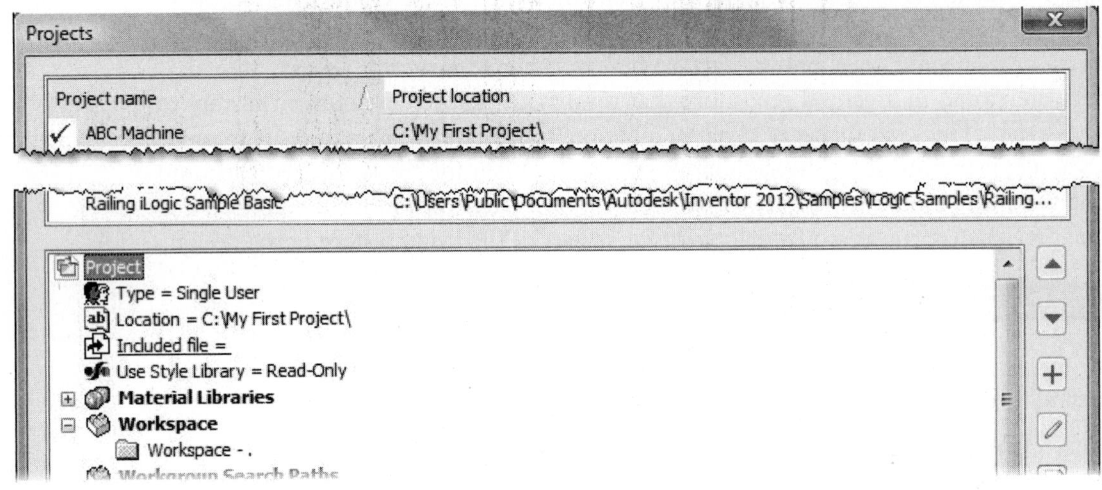

FIGURE 1.36

NOTE

To a make it easier to use subfolders below the workspace, you can add them to the Frequently Used Subfolders by right-clicking on the Frequently Used Subfolders option, and click Add Paths from Directory as shown in Figure 1.37. The project file that you will use for the remaining exercises has the frequently used subfolders already added.

FIGURE 1.37

10. Click Done.
11. Click the Open icon on the Quick Access Toolbar. Notice that the Look in location is in the My First Project folder and the active project is ABC Machine as shown in Figure 1.38.

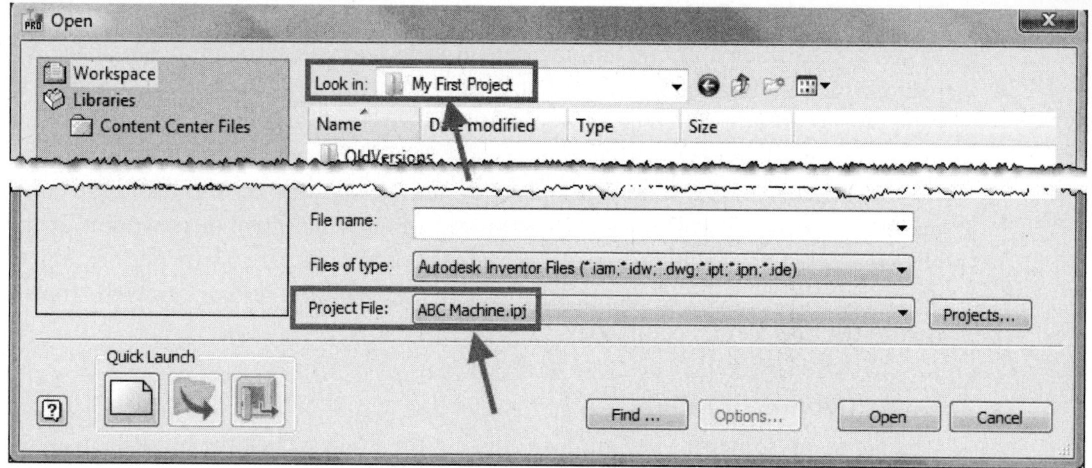

FIGURE 1.38

12. **You must change the active project file to successfully complete the remaining exercises. Close all files currently open in Autodesk Inventor.**
13. Click Inventor Application Menu then click Manage > Projects.
14. To add an existing project file to the current list, click the Browse ... button at the bottom of the dialog box. Navigate to and select *C:\Inv 2013 Ess Plus\Inv 2013 ESS Plus.ipj*. This project should now be the current project. This project will be used for the remaining exercises. Expand the Frequently Used Subfolders section to verify that each chapter has its own subfolder.
15. Right-click on the project **ABC Machine** in the upper pane and select delete as shown in Figure 1.39.

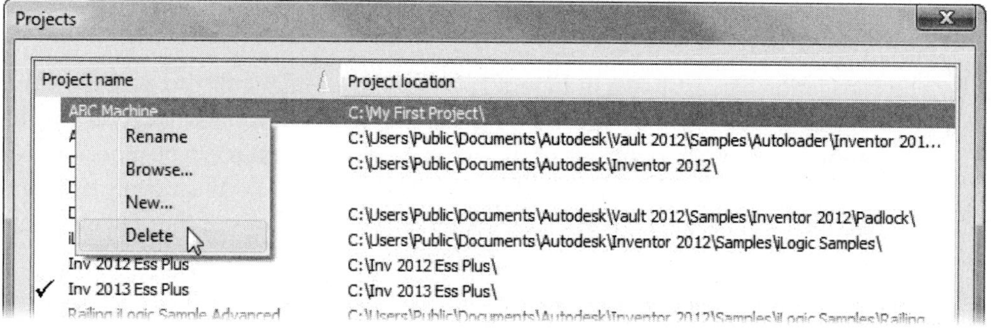

FIGURE 1.39

16. End of exercise.

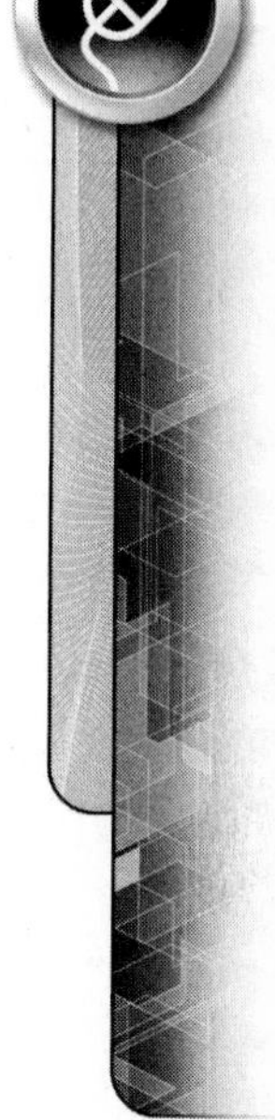

VIEWPOINT OPTIONS

When you work on a 2D sketch, the default view is looking straight down at the XY plane, which is often referred to as a plan view. When you work in 3D, it is helpful to view objects from a different viewpoint and to zoom in and out or pan the objects on the screen. The next section guides you through the most common methods for viewing objects from different perspectives and viewpoints. As you use these commands, the physical objects remain unmoved. Your perspective or viewpoint of the objects is what creates the perceived movement of the part. If you are performing an operation while a viewing command is issued, the operation resumes after the transition to the new view is completed.

Home (Isometric) View

Change to an isometric viewpoint by pressing the F6 key, click the home icon above the ViewCube as shown on the left of Figure 1.40, or by right-clicking in the graphics window and then selecting Home View from the menu as shown on the right of Figure 1.40. The view on the screen transitions to a predetermined home view. You can redefine the home view with a ViewCube option. ViewCube is explained later in this section.

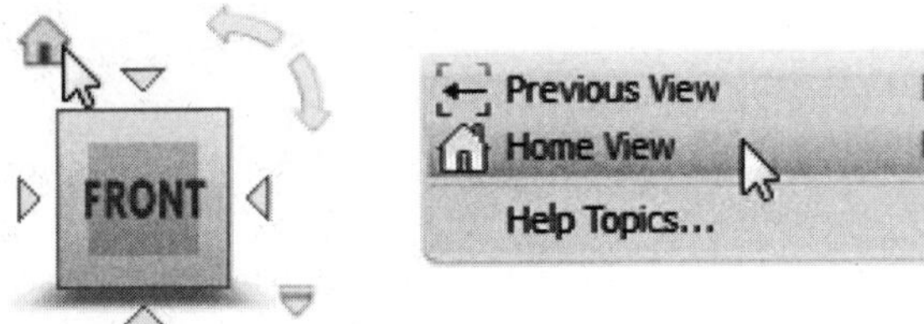

FIGURE 1.40

Navigation Bar

The Navigation bar, as shown in Figure 1.41, contains commands that will allow you to zoom, pan, and rotate the geometry on the screen. The default location for the Navigation bar is on the right side of the graphics window, but it can be repositioned. Commands are also available below the displayed commands. When commands are selected, they will become the top level command. Descriptions of the viewing commands follow.

NOTE

The navigation commands are also available on the View tab > Navigate panel.

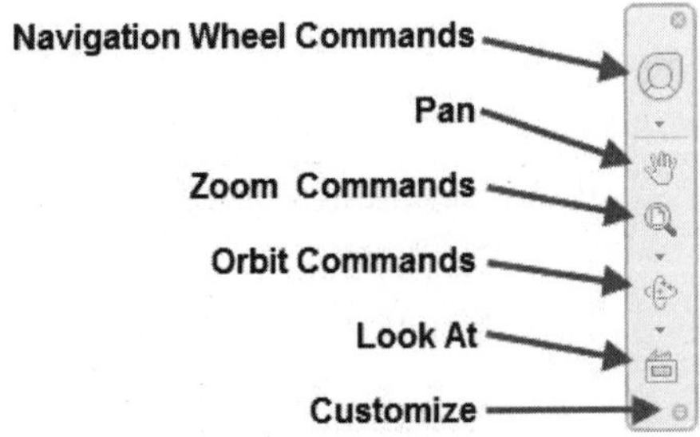

FIGURE 1.41

Navigation Wheel

Click the Navigation Wheel icon to turn on the steering wheel. Click one of the eight options and that option will be active.

Orbit – Rotates the viewpoint.

Zoom – Zooms in and zooms out.

Rewind – Changes to a previous view. A series of images appears representing the previous views. Move the cursor over the image that represents the view to restore.

Pan – Moves the view to a new location.

Center – Centers the view based on the position of the cursor over the wheel.

Walk – Swivels the viewpoint.

Up/Down – Changes the viewpoint vertically.

Look – Moves the view without rotating the viewpoint.

Click the down arrow below the Navigation Wheel icon to change the look of the Navigation wheel.

Pan

Moves the view to a new location. Issue the Pan View command, or select the F2 key. Press and hold the left mouse button, and the screen moves in the same direction that the cursor moves. If you have a mouse with a wheel, hold down the wheel, and the screen moves in the same direction that the cursor moves.

Zoom All

Maximizes the screen with all parts that are in the current file. The screen transitions to the new view. Other commands under Zoom All command are Zoom, Zoom Window, and Zoom Selected.

Zoom. Zooms in or out from the parts. Issue the Zoom In-Out command, or press the F3 key. Then, in the graphics window, press and hold the left mouse key. Move the mouse toward you to make the parts appear larger, and away from you to make the parts appear smaller. If you have a mouse with a wheel, roll the wheel toward you, and the parts appear larger; roll the wheel away from you, and the parts appear smaller.

Zoom Window. Zooms in on an area that is designated by two points. Issue the Zoom Window command and select the first point. With the mouse button depressed, move the cursor to the second point. A rectangle representing the window appears. When the correct window is displayed on the screen, release the mouse button, and the view transitions to it.

Zoom Selected. Fills the screen with the maximum size of a selected face, faces, or a part. Either select the face or faces and then issue the Zoom Selected command, or press the END key, or launch the Zoom Selected command, or press the END key. Then select the face or faces to which you wish to zoom. Use the Select Other command to select a part to zoom to.

Orbit

Rotates your viewpoint dynamically. Issue the Orbit command; a circular image with lines at the quadrants and center appears. To rotate your viewpoint, click a point inside the circle, and keep the mouse button pressed as you move the cursor. Your viewpoint rotates in the direction of the cursor movement. When you release the mouse button, the viewpoint stops rotating. To accept the view orientation, either press the ESC key or right-click and select Done from the menu. Click the outside of the circle to rotate the viewpoint about the center of the circle. To rotate the viewpoint about the vertical axis, click one of the horizontal lines on the circle and, with the mouse button pressed, move the cursor sideways. To rotate the viewpoint about the horizontal axis, click one of the vertical lines on the circle and, with the mouse button pressed, move the cursor upward or downward. The Constrained Orbit command is available below the Orbit command.

Shortcut Options for Orbit

Press and hold down the F4 key and a circular image appears with lines at the quadrants and center. With the F4 key depressed, rotate the viewpoint. When you finish rotating the viewpoint, release the F4 key. If you are performing an operation while the F4 key is pressed, that operation will resume after you release the F4 key.

Another option to quickly rotate your viewpoint is to hold down the Shift key and press down the wheel on the mouse. Once the rotate glyph appears on the screen, you can release the Shift key. While holding down the wheel, move the mouse, and your viewpoint will rotate. With this option, no other options are available. Release the wheel to complete the operation.

Constrained Orbit. Use the Constrained Orbit command to rotate the model about the horizontal screen axes like a turntable. Click one of the horizontal lines on the circle and, with the mouse button pressed, move the cursor sideways. The model is rotated about the model space center point set in the Navigation Wheel CENTER command.

Look At

Changes your viewpoint so that you are looking perpendicular to a plane or rotates the screen viewpoint to be horizontal to an edge. Issue the Look At command or press the PAGE UP key; then select a plane or edge. The Look At command can also be issued by selecting a plane or edge, right-clicking while the cursor is in the graphics window, and selecting Look At from the menu.

Customize the Navigation Bar

To modify the Navigation bar, click the down arrow on the bottom of the Navigation Toolbar as shown in Figure 1.42. Click a command to add or remove it from the Navigation bar. You can also reposition the toolbar by clicking the options under Docking position.

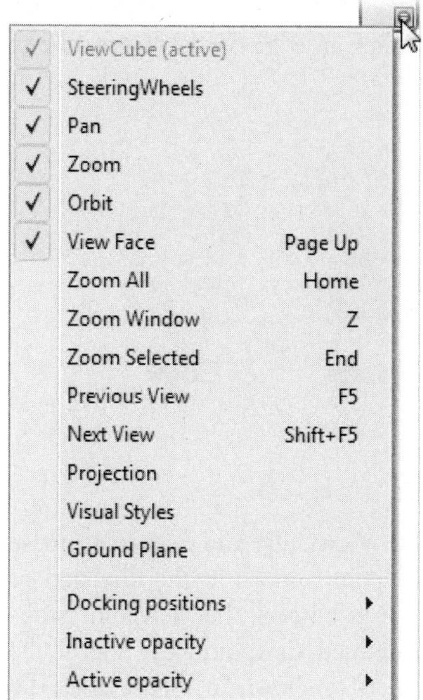

FIGURE 1.42

ViewCube

The ViewCube allows you to quickly change the viewpoint of the screen. With the ViewCube turned on, move the cursor over the ViewCube. Move the cursor over the home in the ViewCube to return to the default home (isometric) view as shown on the left of Figure 1.43.

FIGURE 1.43

Change the viewpoint by using one of the following techniques.

Isometric

Change to a different isometric view by clicking a corner on the ViewCube as shown in Figure 1.44 second from the left.

Face

Change the viewpoint so it looks directly at a plane by clicking a plane on the ViewCube as shown in the middle of Figure 1.44.

Rotate in 90 Degree Increments

When looking at the plane of a ViewCube, you can rotate the view 90 degrees by clicking one of the arcs with arrows or one of the four inside facing triangles as shown second from the right in Figure 1.44.

Edge

You can also orientate the viewpoint to an edge. Click an edge on the ViewCube as shown on the right of Figure 1.44.

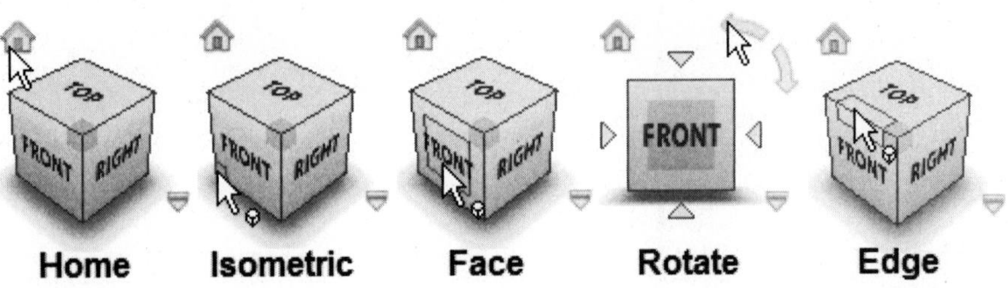

FIGURE 1.44

Dynamic Rotate

To dynamically rotate your viewpoint, click on the ViewCube and keep the mouse button pressed as you move the cursor. Your viewpoint rotates in the direction of the cursor movement. When you release the mouse button, the viewpoint stops rotating. When the viewpoint does not match a defined viewpoint, the ViewCube edges appear in dashed lines from one of the corners as shown in Figure 1.45. To smooth the rotation, right-click on the ViewCube and click Options > and uncheck Snap to closest view.

FIGURE 1.45

Dynamic Rotation

Rotates viewpoint dynamically. While working, press and hold down the F4 key. The circular image appears with lines at the quadrants and center. With the F4 key still depressed, rotate the viewpoint. When you finish rotating the viewpoint, release the F4 key. If you are performing an operation while the F4 key is pressed, that operation will resume after you release the F4 key.

Another option to quickly rotate your viewpoint is to hold down the Shift + Wheel (middle mouse button). Once the rotate glyph appears on the screen, you can release the Shift key and, while holding down the wheel, move the mouse, and your viewpoint will rotate. With this option, no other options are available. The model (s) are rotated about the center of the model(s). Release the wheel to complete the operation.

EXERCISE 1-3: VIEWING A MODEL

In this exercise, you use Navigation commands that make it easier to work on your designs.

1. Open C:\Inv 2013 Ess Plus\Chapter 01\Motor Housing.ipt.

Click the Chapter 01 subfolder from the Frequently Used Subfolders area, and then click the file in the file area as shown in Figure 1.46.

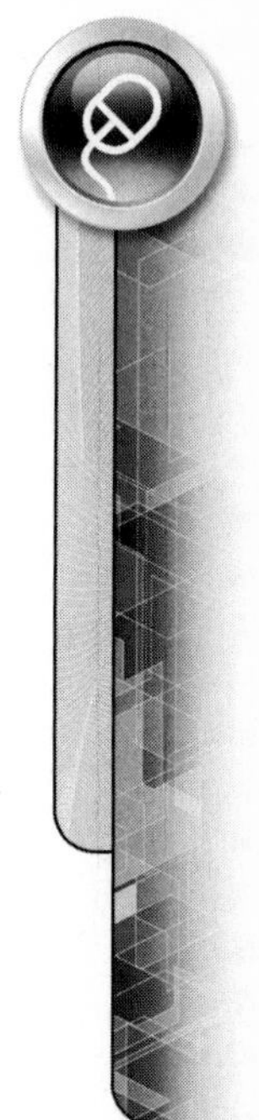

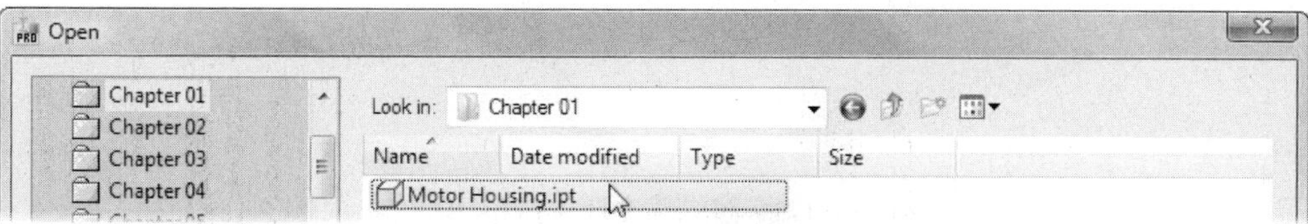

FIGURE 1.46

2. Move the cursor over the lower left foot of the motor housing and spin the wheel toward you. Inventor will zoom in/out to the location of the cursor. This will fill the screen as shown in Figure 1.47.

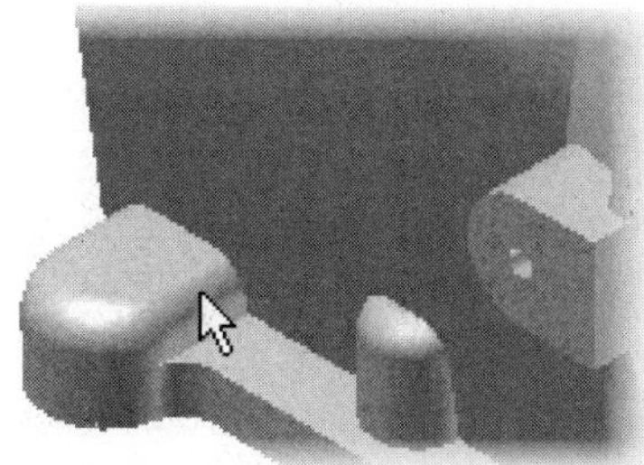

FIGURE 1.47

3. Press F5 a few times to change the viewpoint to the previous view.
4. Press the F6 key to change the the viewpoint to the home view.
5. Use the Orbit command to rotate the viewpoint. From the Navigation bar, click the Orbit command. The 3D Rotate symbol appears in the window as shown in Figure 1.48.

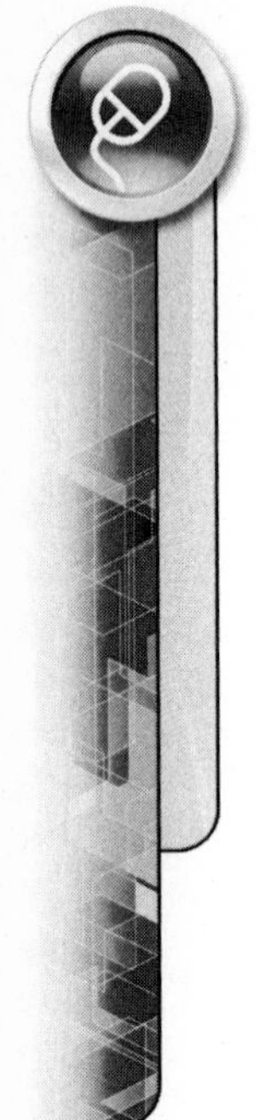

FIGURE 1.48

6. Move the cursor inside the circle of the 3D Rotate symbol; notice how the cursor (glyph) changes as the cursor moves inside the orbit's circle.
7. Click and drag the cursor to rotate the model.
8. To return to the Home View, press F6.
9. Place the cursor on one of the horizontal handles of the 3D Rotate symbol, noting the cursor display.
10. Drag the cursor right or left to rotate the viewpoint about the Y axis. As you rotate the viewpoint, you will see the bottom of the assembly.
11. Press and hold down the F4 key, move the cursor inside the orbit's circle and click and hold down the left mouse button, and move the mouse to rotate the viewpoint.
12. Press and hold down the Shift key, move the cursor inside the orbit's circle and click and hold down the wheel on the mouse, and move the mouse to rotate the viewpoint.
13. Return to the home view by clicking the Home symbol above the ViewCube.
14. To change the viewpoint to predetermined isometric views, or directly at a plane, use the ViewCube. Click the Top-Front left corner of the ViewCube as shown on the left of Figure 1.49. Continue clicking the other corners of the ViewCube as shown in middle two images of Figure 1.49.
15. Change to the front view by clicking the Front plane on the ViewCube as shown on the right of Figure 1.49.

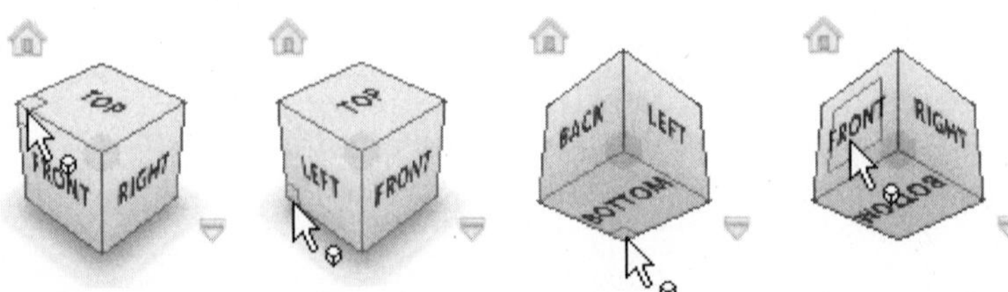

FIGURE 1.49

16. Rotate the view 90 degrees by clicking the arc with arrow in the ViewCube as shown on the left of Figure 1.50.
17. To look at the Top view that is also rotated by 90 degrees, click the left arrow as shown on the right of Figure 1.50.

FIGURE 1.50

18. Practice changing views by clicking the corners, faces, and edges of the ViewCube.
19. Click the Look At command from the Navigation bar as shown in the following image on the left. Select the planar face on the part as shown in the following middle image. The viewpoint will change so you are looking perpendicular to the selected plane as shown in Figure 1.51 on the right.

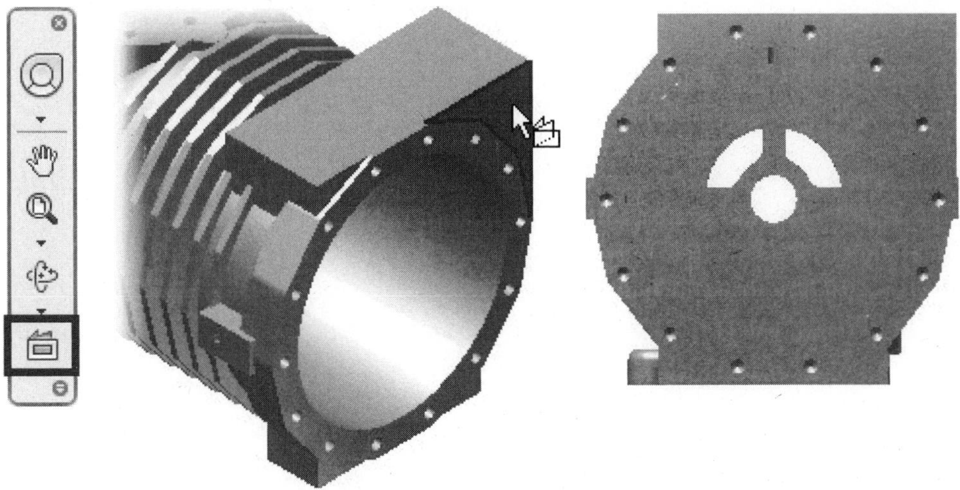

FIGURE 1.51

20. Continue to practice using the navigation commands.
21. Close the file. Do not save changes model. End of exercise.

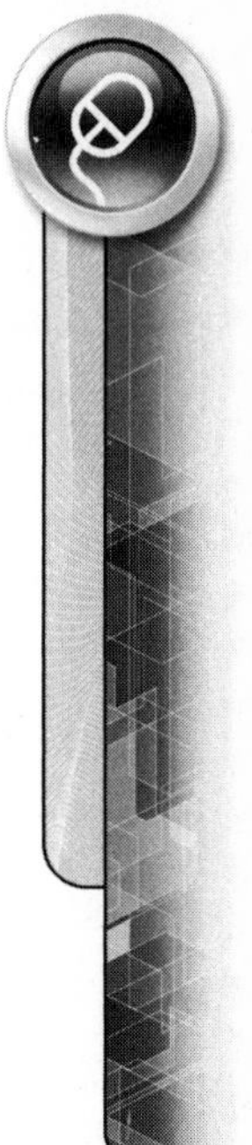

CHECKING YOUR SKILLS

Use these questions to test your knowledge of the material covered in this chapter.

1. Explain the reasons why a project file is used.
2. True __ False __ Only one project can be active at any time.
3. True __ False __ Subfolders need to be added to the project file in order for the files to be found.
4. True __ False __ Autodesk Inventor stores the part, assembly information, and related drawing views in the same file.

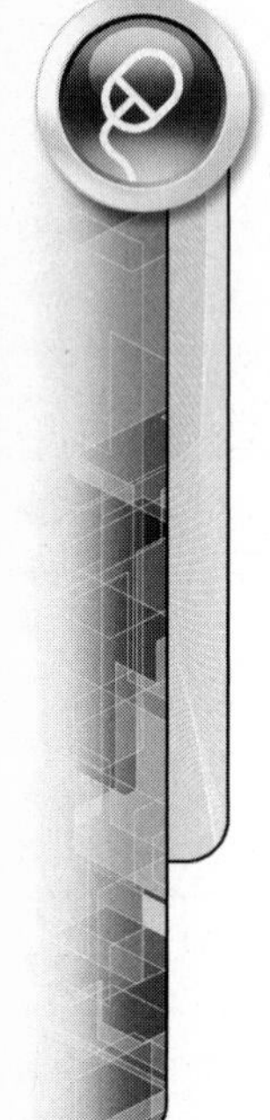

5. True ___ False ___ Press and hold down the F4 key to dynamically rotate the viewpoint.
6. True ___ False ___ The Save Copy As command saves the active document with a new name, and then makes it current.
7. List four ways to access the Help system.
8. Explain how to change the location of the Ribbon.
9. True ___ False ___ Commands can be added to the Quick Access toolbar.
10. True ___ False ___ The ViewCube is used to rotate the active part or assembly file about the X axis.
11. True ___ False ___ Marking menus are used to add annotations to a part file.
12. True ___ False ___ The settings in the Application Options are global and affect all open and new Inventor files.
13. List three methods to repeat the last command.
14. Explain how to start a command from the marking menu without clicking a on a specific command.
15. True ___ False ___ By default Inventor will automatically save the current file every 10 minutes.

CHAPTER 2

Sketching, Constraining, and Dimensioning

INTRODUCTION

Most 3D parts in Autodesk Inventor start from a 2D sketch. This chapter first provides a look at the application options for creating a part file and sketching. It then covers the three steps in creating a 2D parametric sketch: sketching a rough 2D outline of a part, applying geometric constraints, and then adding parametric dimensions. Lastly you learn how to use 2D AutoCAD data in a sketch.

OBJECTIVES

After completing this chapter, you will be able to do the following:

- Change the part and sketch Application Options to meet your needs
- Sketch an outline of a part
- Create geometric constraints to a sketch to control design intent
- Use construction geometry to help constrain a sketch
- Dimension a sketch
- Change a dimension's value in a sketch
- Insert AutoCAD DWG data into a part's sketch

PART AND SKETCH APPLICATION OPTIONS

Before you start a new part, examine the part and sketch options in Autodesk Inventor that will affect how the part file will be created and how the sketching environment will look and act. While learning Autodesk Inventor, refer back to these option settings to determine which ones work best for you—there are no right or wrong settings.

Part Options

You can customize Autodesk Inventor Part options to your preferences. Click Inventor Application Menu > Options button, and click on the Part tab, as shown in Figure 2.1. Descriptions of a couple of the most common Part options follow. For more information about the Application Options consult the help system. These settings are global—they will affect all active and new Autodesk Inventor documents.

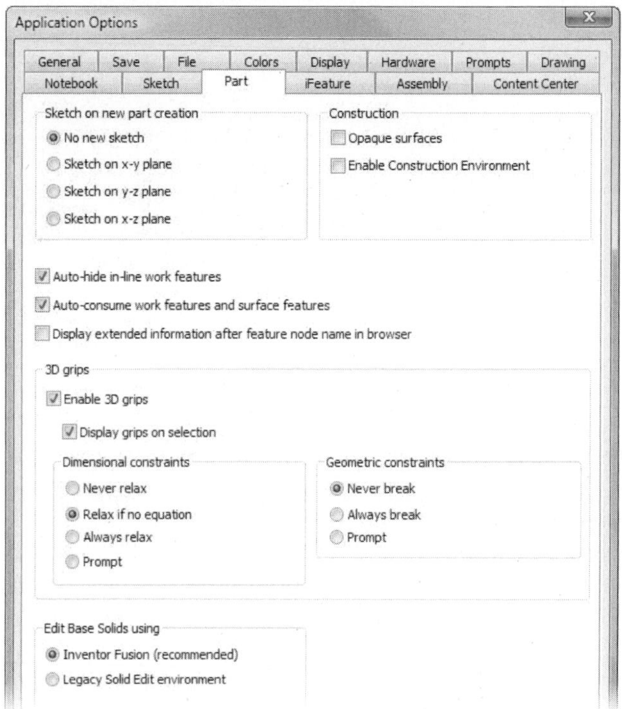

FIGURE 2.1

Sketch on New Part Creation

- No new sketch (is the default setting)

When checked, Inventor does not set a sketch plane when you create a new part.

- Sketch on x-y plane

When checked, Inventor sets the x-y plane as the current sketch plane when you create a new part.

- Sketch on y-z plane

When checked, Inventor sets the y-z plane as the current sketch plane when you create a new part.

- Sketch on x-z plane

When checked, Inventor sets the x-z plane as the current sketch plane when you create a new part.

Edit Base Solids Using

- Inventor Fusion

When checked, Inventor Fusion will be used to edit base solids (solids that have no features).

- Legacy Solid Edit environment

When checked, Inventor commands from the Edit Base Solid tab are used to edit base solids (solids that have no features).

Sketch Options

Autodesk Inventor sketching options can be customized to your preferences. Click Inventor Application Menu > Options, and then click on the Sketch tab, as displayed in Figure 2.2. Descriptions of the most common Sketch options follow. For more information about the Application Options consult the help system. These settings are global, and all of them affect currently active Autodesk documents and Autodesk Inventor documents you open in the future.

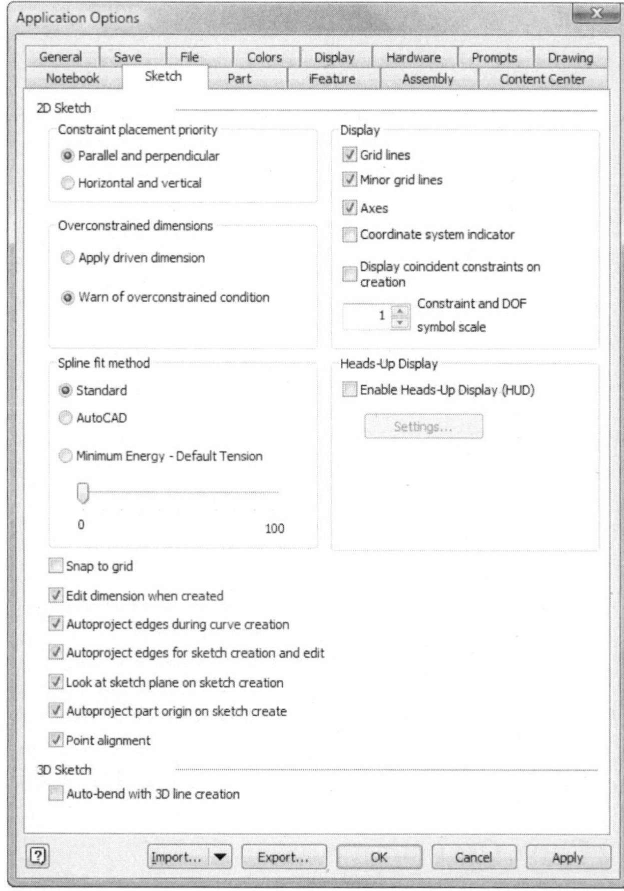

FIGURE 2.2

Display

- Grid lines

Toggles both minor and major grid lines on the screen on and off. To set the grid distance, click the Tools tab > Options panel > Document Settings, and on the Sketch tab of the Document Settings dialog box, change the Snap Spacing and Grid Display.

- Minor grid lines

Toggles the minor grid lines on the screen on and off.

- Axes

Toggles the lines that represent the X- and Y-axis of the current sketch on and off.

- Coordinate system indicator

Toggles the icon on and off that represents the X-, Y-, and Z-axes at the 0, 0, 0 coordinates of the current sketch.

- Display coincident constraints on creation

When checked, coincident constraints are represented with a dot after the constraint is created.

- Constraint and DOF Symbol Scale

Adjusts the scale of the sketch constraint toolbars and the sketch degree of freedom symbols to make them larger or smaller.

Over-Constrain Dimension

- Apply driven dimensions

When checked, and when you add dimensions that would over-constrain the sketch, adds the dimension as a driven (reference) dimension.

- Warn of over-constrained condition

When checked, and when you add dimensions that would over-constrain the sketch, a dialog box appears warning of the condition. This allows you to accept the placement of a driven dimension or cancel the dimension placement.

Snap to Grid

When checked, endpoints of sketched objects snap to the intersections of the grid as the cursor moves over them.

Edit Dimension When Created

When checked, edits the values of a dimension in the Edit Dimension dialog box immediately after you position the dimension. It is recommended to turn this setting on.

Autoproject Edges During Curve Creation

When checked, and while sketching, place the cursor over an object and it will be projected onto the current sketch. You can also toggle Autoproject on and off while sketching by right-clicking and selecting Autoproject from the menu.

Autoproject Edges for Sketch Creation and Edit

When checked, automatically projects all of the edges that define that plane onto the sketch plane as reference geometry when you create a new sketch.

Look at Sketch Plane on Sketch Creation

When checked, automatically changes the view orientation to look directly at the new or active sketch.

Autoproject Part Origin on Sketch Create

When checked, the part's origin point will automatically be projected when a new sketch is created. It is recommended to keep this setting on.

Point Alignment On

When checked, automatically infers alignment (horizontal and vertical) between end-points of newly created geometry. No sketch constraint is applied. If this option is not checked, points can still be inferred; this technique is covered later in this chapter in the Inferred Points section.

Enable Heads-Up Display (HUD)

When checked Dynamic Input is turned on, which allows numeric and angular values to be entered into the value input boxes when creating sketch geometry. Click the Settings button to change the Heads-Up Display settings.

UNITS

Autodesk Inventor uses a default unit of measurement for every part and assembly file. The default unit is set from the template file from which you created the part or assembly file. When specifying numbers in dialog boxes with no unit, the default unit will be used. You can change the default unit in the active part or assembly document by clicking Tools tab > Options panel > Document Settings button and click the Units tab as shown in Figure 2.3. The unit system values change for all of the existing values in that file.

In a *drawing* file, the appearance of dimensions is controlled by dimension styles. Drawing settings will be covered in Chapter 5.

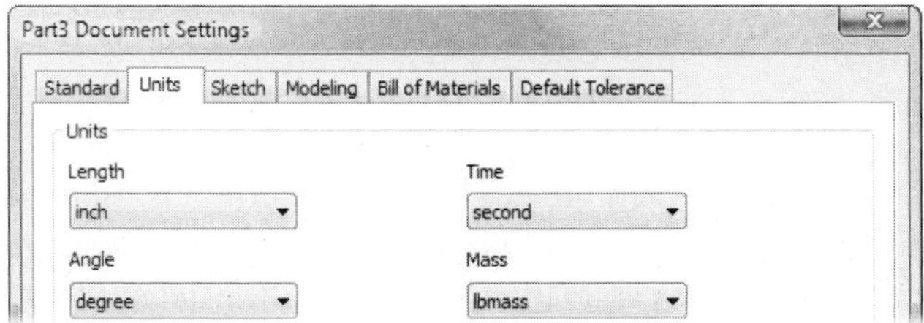

FIGURE 2.3

You can override the default unit for any value by entering the desired unit. If you were working in a metric file whose unit is set to mm, for example, and you placed a horizontal dimension whose default value was 50.8 mm, you could enter 2 in. Dimensions appear on the screen in the default units.

For the previous example, 50.8 mm would appear on the screen. When you edit a dimension, the overridden unit appears in the Edit Dimension dialog box as shown in Figure 2.4.

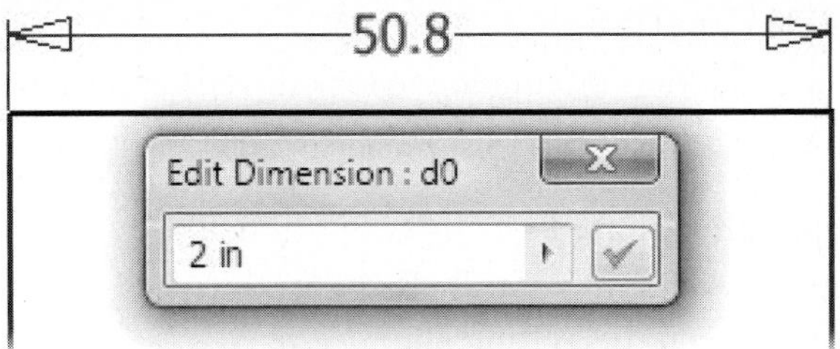

FIGURE 2.4

TEMPLATES

Each new file is created from a template. You can modify existing templates or add your own templates. As you work, make note of the changes that you make to each file. You then create a new template file or modify an existing file that contains all of the changes and save that file to your template directory, which by default in Windows Vista and Windows 7 is *C:\Users\Public\Public Documents\ Autodesk\Inventor 2013\Templates* directory. You can also create a new subdirectory under the templates folder, and place any Autodesk Inventor file in this new directory. After adding an Inventor file the new tab will appear, and it will be available as a template.

You can use one of two methods to share template files among many users. You can modify the location of templates by clicking the Inventor Application Menu > Options button, clicking on the File tab, and modifying the Templates location as shown in Figure 2.5. The Templates location will need to be modified for each user who needs access to templates that are not stored in the local location. You can also change the default unit of measurement (inches or millimeters) and the default drawing standard (ANSI, DIN, ISO, etc.) by clicking the Configure Default Template button above the file location.

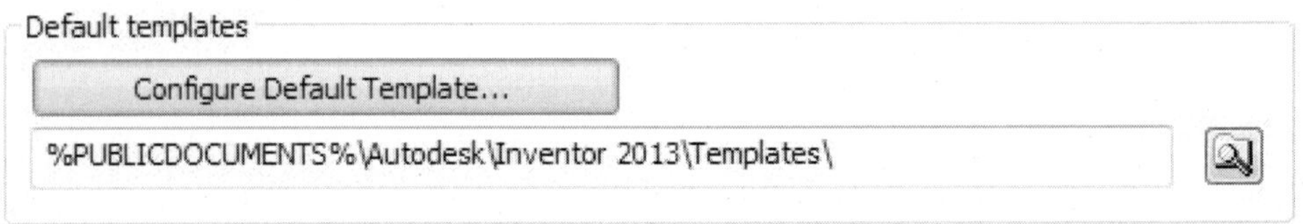

FIGURE 2.5

You can also set the Templates location in each project file. This method is useful if you need different templates files for each project. While editing a project file, change the Templates location in the Folder Options area. Figure 2.6 shows the default location in Windows Vista and Windows 7. The Template location in the project file takes precedence over the Templates option in the Application Options, File tab.

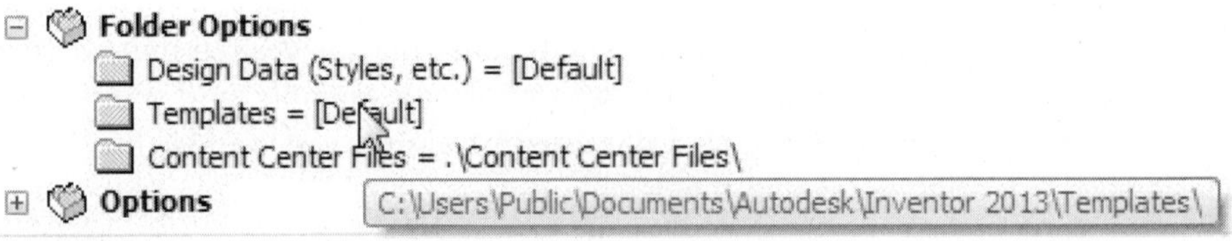

Folder Options
Design Data (Styles, etc.) = [Default]
Templates = [Default]
Content Center Files = .\Content Center Files\
Options C:\Users\Public\Documents\Autodesk\Inventor 2013\Templates\

FIGURE 2.6

Template files have file extensions that are identical to other files of the same type, but they are located in the template directory. Template files should not be used as production files.

NOTE

CREATING A PART

The first step in creating a part is to start or create a new part file in an assembly. The first four chapters in this book deal with creating parts in a new part file, and Chapter 6 covers creating and documenting assemblies. You can use the following methods to create a new part file:

- Click New on the Inventor Application menu, and then click the *Standard.ipt* icon as shown on the left of Figure 2.7. You can also click New - Create a file from the list of templates, to select a template file that is not in the top level template folder.
- Click the down arrow of the New icon, and select Part from the left side of the Quick Access toolbar as shown in the middle of Figure 2.7. This creates a new part file based on the units that were selected when Autodesk Inventor was installed or on the *Standard.ipt* that exists in the Templates folder.
- Click the New icon in the Application menu, in the Get Started tab > Launch panel, or in Quick Launch area of the Open dialog box, and then click the desired templates folder on the left side of the Create New File dialog box and then from the Part section on the right side of the dialog box click on the desired part template file, as shown in Figure 2.7 on the right, or click *Standard (unit).ipt* on one of the other tabs.

After starting a new part file using one of the previous methods, Autodesk Inventor's screen will change to reflect the part environment.

The units for the files located in the top level folder are based upon the units you selected when you installed Autodesk Inventor.

NOTE

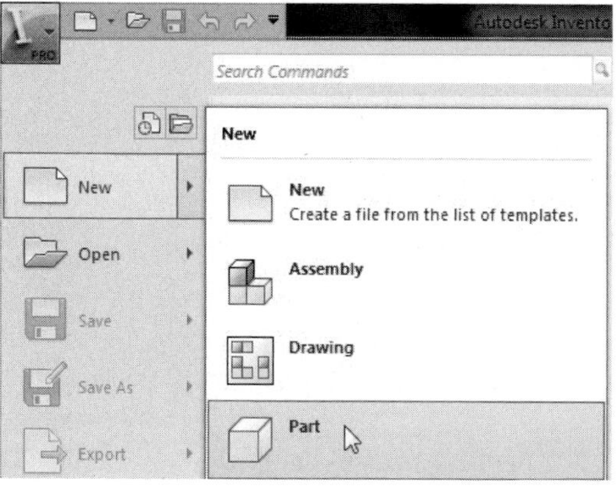

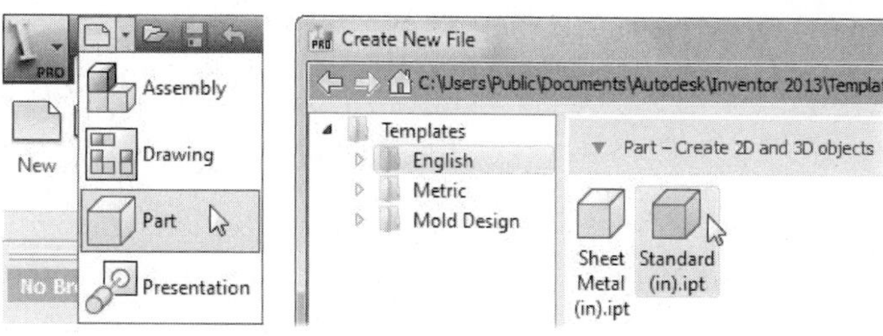

FIGURE 2.7

Sketches and Default Planes

Before you start sketching, you select a plane on which to draw. A sketch is a plane on which 2D objects are sketched. You can use any planar part face or work plane to create a sketch. By default, when you create a new part file no sketch is created, and you will select an origin plane to sketch on. You can change the default plane on which you will create the sketch by selecting the Inventor Application Menu > Options and clicking on the Part tab. Choose the sketch plane to which new parts should default.

Each time you create a new Autodesk Inventor part or assembly file, there are three planes (XY, YZ, and XZ), three axes (X, Y, and Z), and the center (origin) point at the intersection of the three planes. You can use these default planes to create an active sketch. To see the planes, axes, or center point, expand the Origin entry in the browser by clicking on the + to the left side of the text. You can then move the cursor over the names, and they will appear in the graphics area. Figure 2.8 illustrates the default planes, axes, and to leave the visibility of the planes or axes on, right-click in the browser while the cursor is over the name, right-click and click Visibility from the menu.

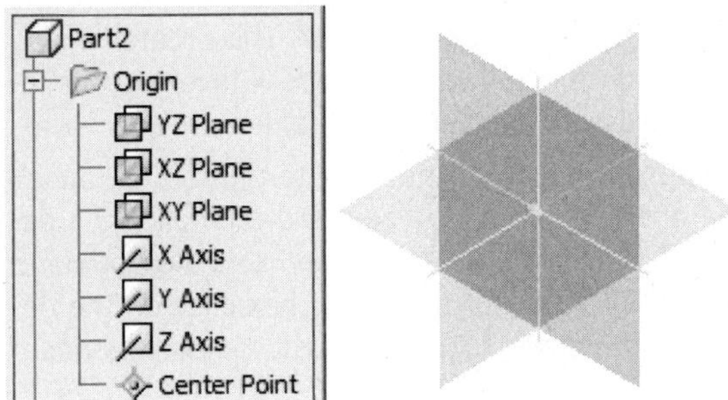

FIGURE 2.8

Origin 3D Indicator

When working in 3D, it is common to get your orientation turned around. By default in the lower left corner of the graphics screen, there is an XYZ axis indicator that shows the default (world) coordinate system as shown in Figure 2.9 on the left. The direction of these planes and axes cannot be changed. The arrows are color coded:

Red arrow = X axis

Green arrow = Y axis

Blue arrow = Z axis

In the Application Options dialog box under the Display tab, you can turn the axis indicator and the axis labels on and off as shown in Figure 2.9 on the right.

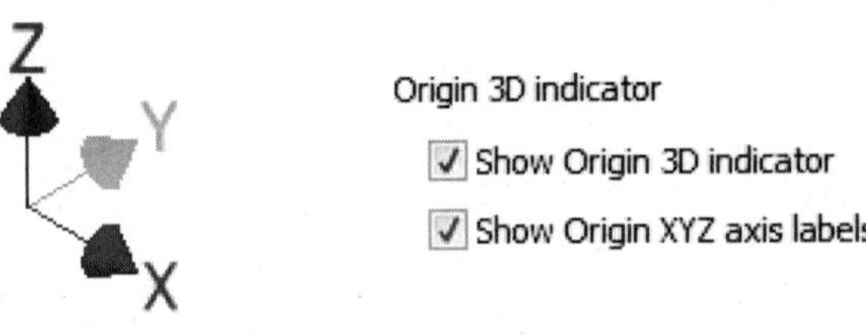

FIGURE 2.9

By default, Inventor will automatically project the origin point (0,0) when a new sketch is created in a part file. The origin point can be used to constrain a sketch to the 0, 0 point of the sketch. If desired, you can turn this option off by clicking Tools tab Application Options or the Application Menu, click on the Sketch tab, and then uncheck Autoproject part origin on sketch create as displayed in the Figure 2.10.

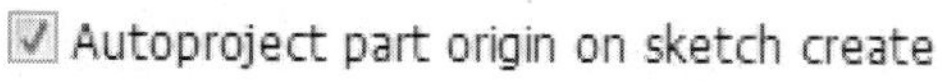

☑ Autoproject part origin on sketch create

FIGURE 2.10

New Sketch

By default, when you create a new part file no sketch is active. You can define a plane from the origin folder to be the default by selecting a default plane from the Application Menu > Options > Part tab. Issue the Create Sketch command to create a new sketch on a planar part face or a work plane or to activate a nonactive sketch in the part. When you are in a part file that does not have a sketch defined and when you start the create sketch command, the origin planes will be displayed in the graphics window, and you can select one of these planes to create the sketch on. To create a new sketch or make an existing sketch active, use one of these methods:

- Click the 3D Model tab > Sketch panel > Create 2D Sketch as shown in Figure 2.11 on the left. Then click a planar face, a work plane, or an existing sketch in the browser.
- Press the S key (a keyboard shortcut) and click a planar face of a part, a work plane, or an existing sketch in the browser.
- While not in the middle of an operation, right-click in the graphics window, and select New Sketch from the marking menu as shown in the middle. Then click a planar face, a work plane, or an existing sketch in the browser.
- While not in the middle of an operation, click a planar face of a part, a work plane, or an existing sketch in the browser. Then right-click in the graphics area, and click Create Sketch from the mini-toolbar as shown in Figure 2.11 on the right.

TIP

You can either start the command first and then select a plane or you can select a plane first and then start the command.

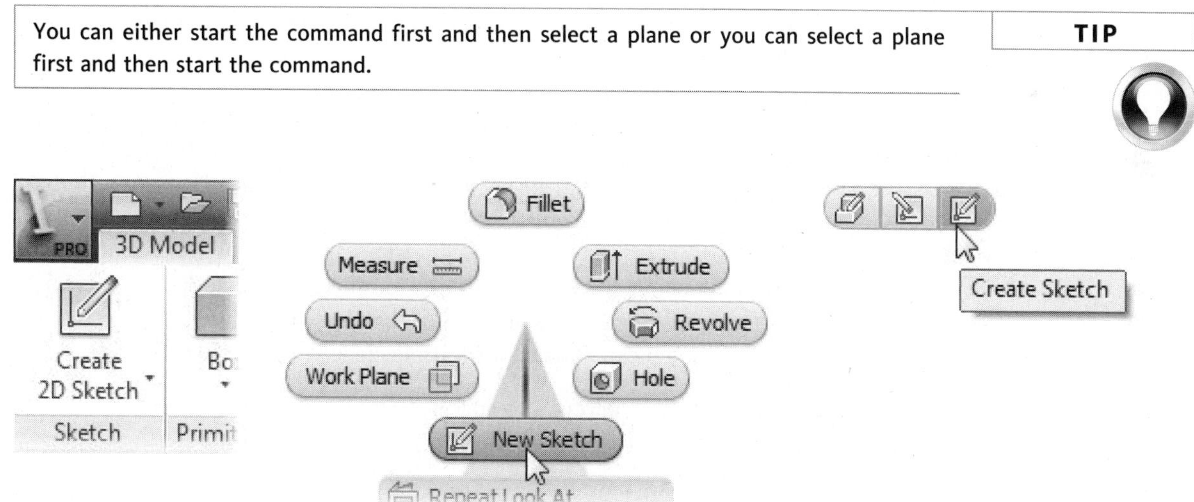

FIGURE 2.11

After defining a sketch, a Sketch entry will appear in the browser as shown and a Sketch tab will appear in the ribbon. By default after you have defined a sketch, the X- and Y-axes will align automatically to this plane, and you can begin to sketch.

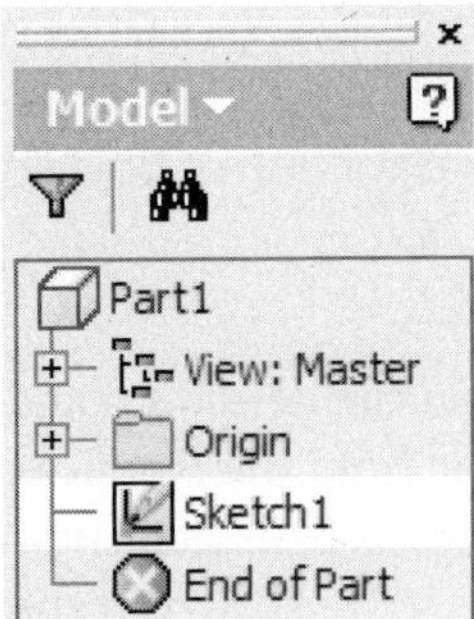

FIGURE 2.12

STEP 1—SKETCH THE OUTLINE OF THE PART

As stated at the beginning of this chapter, 3D parts usually start with a 2D sketch of the outline shape of the part. You can create a sketch with lines, arcs, circles, splines, or any combination of these elements. The next section will cover sketching strategies, commands, and techniques.

Sketching Overview

When deciding what outline to start with, analyze how the finished shape will look. Look for the 2-dimensional shape that best describes the part. When looking for this outline, try to look for a flat 2-dimensional shape that can be extruded or revolved to create a shape that other features can be added to, to create the finished part. It is usually easier to sketch a 2-dimensional geometry than 3-dimensional geometry. As you gain modeling experience, you can reflect on how you created the model and think about other ways that you could have built it. There is usually more than one way to generate a given part.

When sketching, draw the geometry so that it is close to the desired shape and size—you do not need to be concerned about exact dimensional values. Even though Autodesk Inventor allows islands in the sketch (closed objects that lie within another closed object) it is NOT recommended to sketch islands (when you extrude a sketch, island(s) may become voids in the solid). A better method is to place features, which make editing a part easier. For example, instead of sketching a circle inside a rectangle, place a hole feature. A sketch can consist of multiple closed objects that are coincident.

The following guidelines will help you to successfully generate sketches:

- Select a 2-dimensional outline that best represents the part. The 2D outline will be used to create the base feature.
- Draw the geometry close to the finished size. If you want a 20 inch square, for example, do not draw a 200 inch square. Use dynamic input to define the size of the geometry. Dynamic input is covered in a later section in this chapter.
- Create the sketch proportional in size to the finished shape. When drawing the first object verify its size in the lower-right corner of status bar. Use this information as a guide.
- Draw the sketch so that it does not have geometry over geometry, that is, line on top of another line.
- Do not allow the sketch to have a gap; the geometry should start and end at a single point, just as the start and end points of a rectangle share the same point.

- Keep the sketches simple. Leave out fillets and chamfers when possible. You can easily place them as features after the sketch turns into a solid. The simpler the sketch, the fewer the number of constraints and dimensions that will be required to constrain the model.

NOTE

If you want to create a solid, the sketch must form a closed shape. If it is open, it can only be turned into a surface.

Sketching Tools

Before you start sketching the outline of the part, examine the 2D sketching commands that are available. By default, the 2D sketch commands appear in the browser. The most frequently used commands will be explained throughout this chapter. Consult the help system for information about the remaining commands.

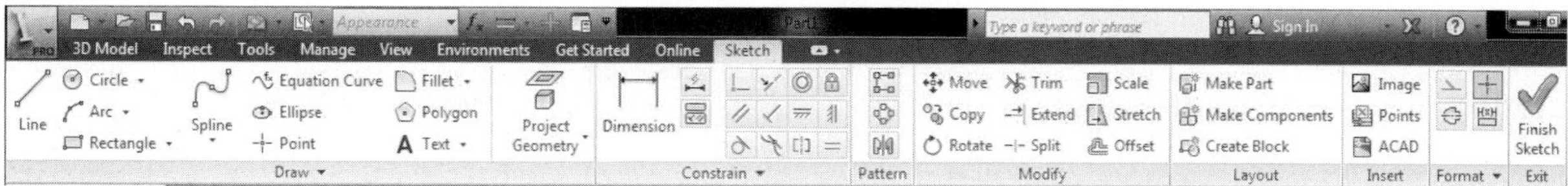

FIGURE 2.13

Using the Sketch Tools

After starting a new part, a sketch will automatically be active so that you can now use the sketch commands to draw the shape of the part. To start sketching, issue the sketch command that you need, click a point in the graphics area, and follow the prompt on the status bar. The sections that follow will introduce techniques that you can use to create a sketch.

Dynamic Input

Dynamic input in the sketch environment makes a Heads-Up Display (HUD), which displays information near the cursor for many sketching commands that helps you keep your eyes on the screen. While using the Line, Circle, Arc, Rectangle, or Point commands, you can enter values in the input fields. You can toggle between the value input fields by pressing the TAB key. Figure 2.14 shows examples of entering Polar and Cartesian Coordinates.

NOTE

If no data is entered in the input fields and you click in the graphics window to locate geometry, dimensions will NOT automatically be placed. You can manually place dimensions and constraints after the geometry is sketched.

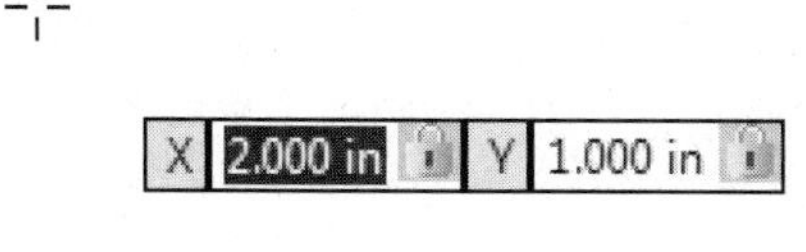

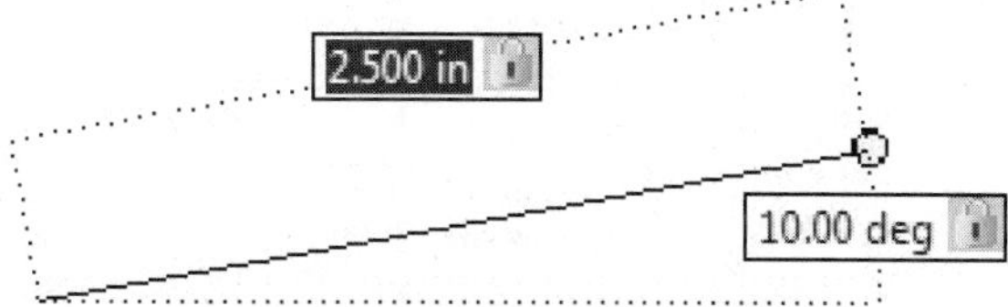

Cartesian Coordinates **Polar Coordinates**

FIGURE 2.14

Coordinate Types

While sketching you can right-click and change the coordinate types shown in Figure 2.15. You must select one option from the top and bottom of the menu.

Absolute Coordinate #: Values are relative to the 0,0 coordinates on the sketch.

Relative Coordinate @: Values are relative to the last selected point.

Polar Coordinate <: #: When defining a second or subsequent point, enter X value and angle.

Cartesian Coordinate >: #: Enter X and Y coordinates.

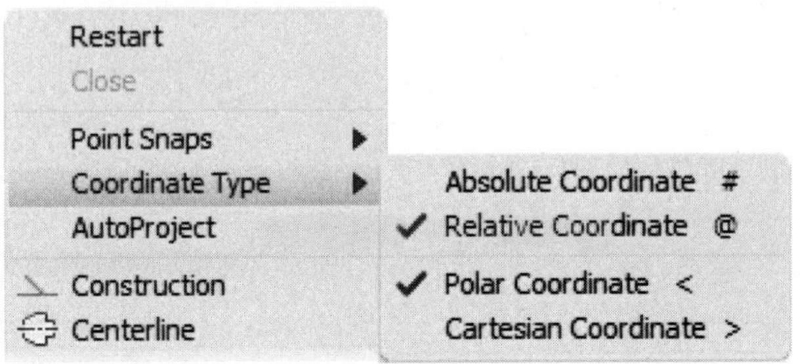

FIGURE 2.15

Polar Coordinates. By default you enter values using Polar coordinates by specifying the distance and angle of sketch elements. After entering the first value, press TAB to display a lock icon in the field, which constrains the cursor by the value that you entered and moves the cursor to the next field.

Cartesian Coordinates. The Rectangle and Point commands use Cartesian coordinates, where you specify X and Y coordinates. After entering the first value, press TAB to display a lock icon in the field, which constrains the cursor by the value that you entered and moves the cursor to the next field.

Dimension Input. When dimension input is on, the value input fields display polar coordinates when defining lengths and angles for a second point. The dimensional values change as you move the cursor. Press TAB to move to the next input field or click in another cell. After pressing the Tab key, the value will be locked and a lock icon will appear to the right of the value as shown in Figure 2.16. After a dimension's value is locked, the parametric dimension will be created after clicking a point or pressing the Enter key. You can change the value in an input filed by either clicking in the field or pressing the Tab key until the field is highlighted and then typing in a new value.

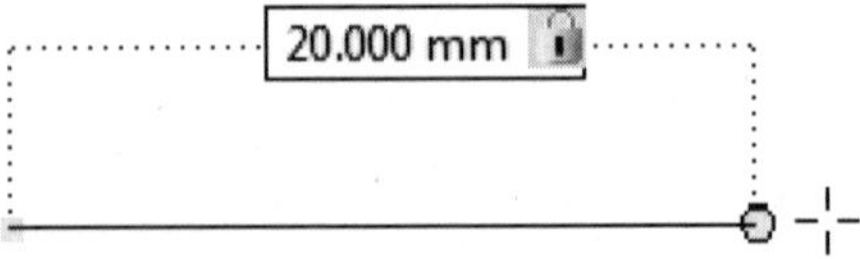

FIGURE 2.16

Line Command

The Line command, is one of the most powerful commands that you will use to sketch. Not only can you draw lines with it, but also you can draw an arc from the endpoint of a line segment. To start sketching lines, click the Line command from the Sketch tab > Draw panel as shown in Figure 2.17 on the left, or right-click in a blank area in the graphics window and click Create Line from the marking menu as shown in Figure 2.17 in the middle, or press the L key on the keyboard. After starting the Line command you will be prompted to click a first point, select a point in the graphics window, and then click a second point. Figure 2.17 on the right shows the line being created with the dynamic input as well as the horizontal constraint.

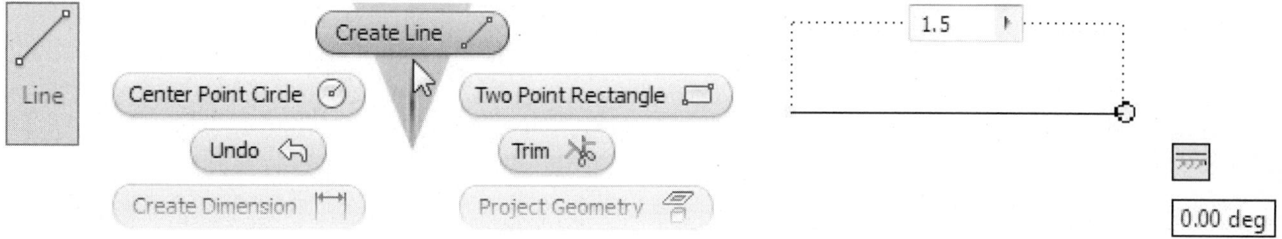

FIGURE 2.17

You can continue drawing line segments, or you can sketch an arc from the endpoint. Move the cursor over the endpoint of a line segment or arc, and a small gray circle will appear at that endpoint as shown in Figure 2.18 on the left. Click on the small circle, and with the left mouse button pressed down, move the cursor in the direction that you want the arc to go. Up to eight different arcs can be drawn, depending upon how you move the cursor. The arc will be tangent to the horizontal or vertical edges that are displayed from the selected endpoint. Figure 2.18 on the right shows an arc that is normal to the sketched line being drawn.

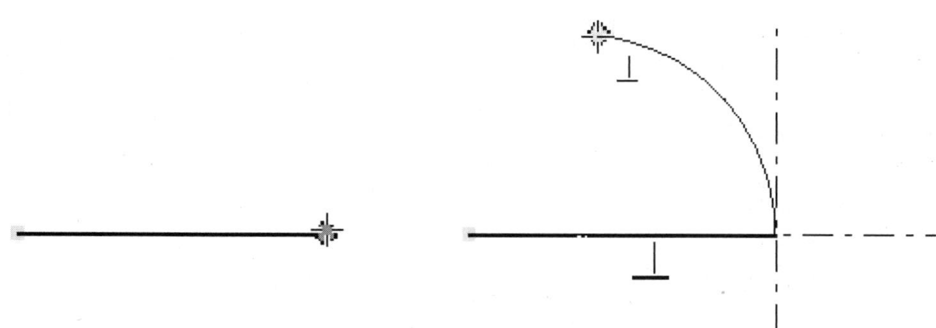

FIGURE 2.18

> **TIP**
>
> **When sketching, look at the bottom-right corner of the status bar (bottom of the screen) to see the coordinates, length, and angle of the objects that you are drawing. Figure 2.19 shows the status bar when a line is being drawn.**

0.650 in, 0.375 in x=0.650 in y=0.375 in

FIGURE 2.19

Object Tracking – Inferred Points

If the Point Alignment On option is checked from the Sketch tab of the Application Options, as you sketch dashed lines will appear on the screen. These dotted lines represent the endpoints; midpoints; and theoretical intersections of lines, arcs, and center points of arcs and circles that represent their horizontal, vertical, or perpendicular positions. As the cursor gets close to these inferred points, it will snap to that location. If that is the point that you want, click that point; otherwise, continue to move the cursor until it reaches the desired location. When you select inferred points, no constraints (geometric rules such as horizontal, vertical, collinear, and so on) are applied from them. Using inferred points helps create more accurate sketches. Figure 2.20 shows the inferred points from two midpoints that represent their horizontal and vertical position.

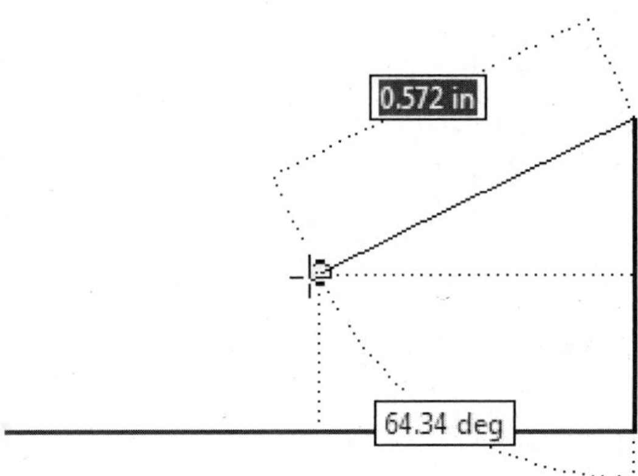

FIGURE 2.20

Automatic Constraints

As you sketch, a small constraint symbol appears that represents geometric constraint(s) that will be applied to the object. If you do not want a constraint to be applied, uncheck the Constraint Persistent option under the Constrain panel drop arrow or hold down the CTRL key when you select the point. Figure 2.21 shows a line being drawn from the arc, tangent to the arc, and parallel to the angled line, and the dynamic input is also displayed. The symbol appears near the object from which the constraint is coming. Constraints will be covered in the next section.

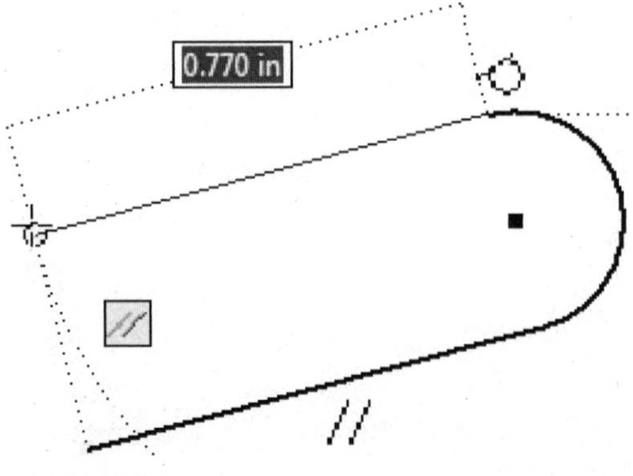

FIGURE 2.21

Scrubbing

As you sketch, you may prefer to apply a constraint different from the one that automatically appears on the screen. You may want a line to be perpendicular to a given line, for example, instead of being parallel to a different line. The technique to change the constraint is called *scrubbing*. To place a different constraint while sketching, move the cursor so it touches (scrubs) the other object to which the constraint should be related. Move the cursor back to its original location, and the constraint symbol changes to reflect the new constraint. The same constraint symbol will also appear near the scrubbed object, representing that it is the object to which the constraint is matched. Continue sketching as normal. Figure 2.22 shows the top horizontal line being drawn with a perpendicular constraint that was scrubbed from the left vertical line. Without scrubbing the left vertical line, the applied constraint would have been parallel to the bottom line.

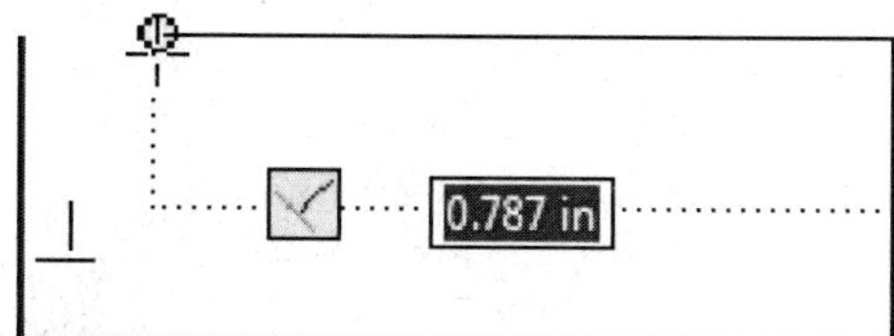

FIGURE 2.22

Common Sketch Tools

The following chart lists common 2D sketch commands that are not covered elsewhere in this chapter. Some commands are available by clicking the down arrow in the lower-right corner of the top command. Consult the help system for more information about these commands.

Tool	Function
Center-point Circle	Creates a circle by clicking a center point for the circle and then a point on the circumference of the circle.
Tangent Circle	Creates a circle that will be tangent to three lines or edges by clicking the lines or edges.
Three-Point Arc	Creates an arc by clicking a start and endpoint and then a point that will lie on the arc.
Tangent Arc	Creates an arc that is tangent to an existing line or arc by clicking the endpoint of a line or arc and then clicking a point for the other endpoint of the arc.
Center-Point Arc	Creates an arc by clicking a center point for the arc and then clicking a start and endpoint.
Two-Point Rectangle	Creates a rectangle by defining a point and then clicking another point to define the opposite side of the rectangle. The edges of the rectangle will be horizontal and vertical. If values were entered, dimensions will be placed on the rectangle.
Three-Point Rectangle	Creates a rectangle by clicking two points that will define an edge and then clicking a point to define the third corner. You can also type values to define the three points of the rectangle, and dimensions will be created that define the size of the rectangle.

(Continued)

Tool	Function
Two-Point Center Rectangle	Creates a rectangle by defining a center point and another point to define the rectangle's size or type values for the center point and its X and Y values of the rectangle. The edges of the rectangle will be horizontal and vertical, and if values were entered, dimensions will be created.
Three-Point Center Rectangle	Creates a rectangle by defining a center point, a point to define the rectangle's starting point and its angle and another point size or type values for the center point, and size of the rectangle. The edges of the rectangle will be horizontal and vertical, and if values were entered, dimensions will be created.
Fillet	Creates a fillet between two nonparallel lines, two arcs, or a line and an arc at a specified radius. If you select two parallel lines, a fillet is created between them without specifying a radius. When the first fillet is created, a dimension will be created. If many fillets are placed in the same operation, you choose to either apply or not apply an equal constraint.
Chamfer	Creates a chamfer between lines. There are three options to create a chamfer: both sides equal distances, two defined distances, or a distance and an angle.
Polygon	Creates an inscribed or a circumscribed polygon with the number of faces that you specify. The polygon's shape is maintained as dimensions are added.
Mirror	Mirrors the selected objects about a centerline. A symmetry constraint will be applied to the mirrored objects.
Rectangular Pattern	Creates a rectangular array of a sketch with a number of rows and columns that you specify.
Circular Pattern	Creates a circular array of a sketch with a number of copies and spacing that you specify.
Offset	Creates a duplicate of the selected objects that are a given distance away. By default, an equal-distance constraint is applied to the offset objects.
Trim	Trims the selected object to the next object it finds. Click near the end of the object that you want trimmed. While using the Trim command, hold down the SHIFT key to extend objects. If desired, hold down the CTRL key to select boundary objects. While in the Trim command you can also hold down the left mouse button and move the cursor to dynamically trim geometry. While in the Dynamic mode you can hold down the Shift key to dynamically extend geometry.
Extend	Extends the selected object to the next object it finds. Click near the end of the object that you want extended. While using the Extend command, hold down the SHIFT key to trim objects. If desired, hold down the CTRL key to select boundary objects. While in the Extend command you can also click and hold down the left mouse button and move the cursor to dynamically extend geometry. While in the Dynamic mode you can hold down the Shift key to dynamically trim geometry.

Selecting Objects

After sketching objects, you may need to move, rotate, or delete some or all of the objects. To edit an object, it must be part of a selection set. There are two methods that you can use to place objects into a selection set.

- You can select objects individually by clicking on them. To select multiple individual objects, hold down the CTRL or SHIFT key while clicking the objects. You can remove selected objects from a selection set by holding down the CTRL or SHIFT key and reselecting them. As you select objects, their color will change to show that they have been selected.
- You can select multiple objects by defining a selection window. Not all commands allow you to use the selection window technique and only allow single selections. To define the window, click a starting point. With the left mouse button depressed, move the cursor to define the box. If you draw the window from left to right (solid lines), as shown in Figure 2.23 on the left, only the objects that are fully enclosed in the window will be selected. If you draw the window from right to left (dashed lines), as shown in Figure 2.23 on the right, all of the objects that are fully enclosed in the window *and* the objects that are touched by the window will be selected.
- You can use a combination of the methods to create a selection set.

When you select an object, its color will change according to the color style that you are using. To remove all of the objects from the selection set, click in a blank section of the graphics area.

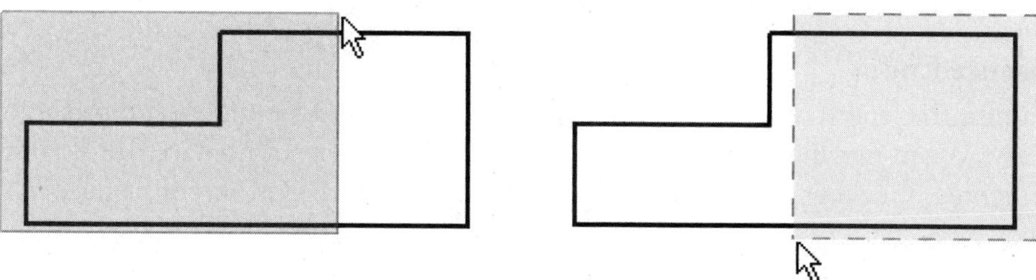

FIGURE 2.23

Deleting Objects

To delete objects first cancel the command that you are in by pressing the Esc key. Then select objects to delete, and then either press the DELETE key or right-click and select Delete from the menu as shown in Figure 2.24.

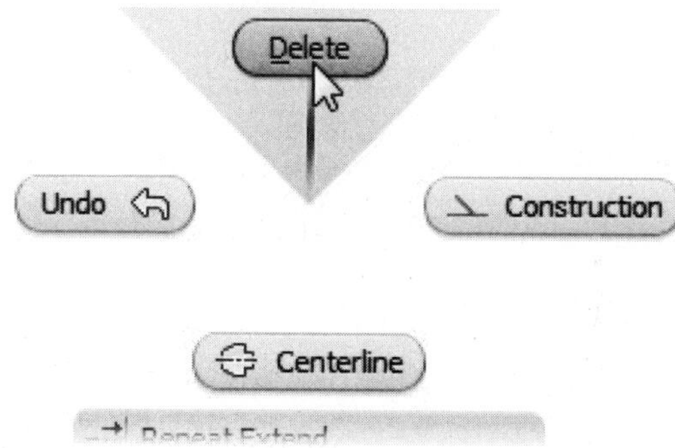

FIGURE 2.24

Measure Commands

Measure commands can assist in analyzing sketch, part, and assembly models. The measure command is not a replacement for dimensions; they are additional tools to give you more information. You can measure distances, angles, and loops, and you can perform area calculations. You can start the measure command first and then select the geometry, or select the geometry first and then start a measure command.

The Measure commands are located on the Quick Access toolbar as shown in Figure 2.25 or on the Tools tab > Measure panel as shown in Figure 2.25. Below is a description of common measuring tools that you could do with 2D geometry. The measure command will also be covered in Chapter 6.

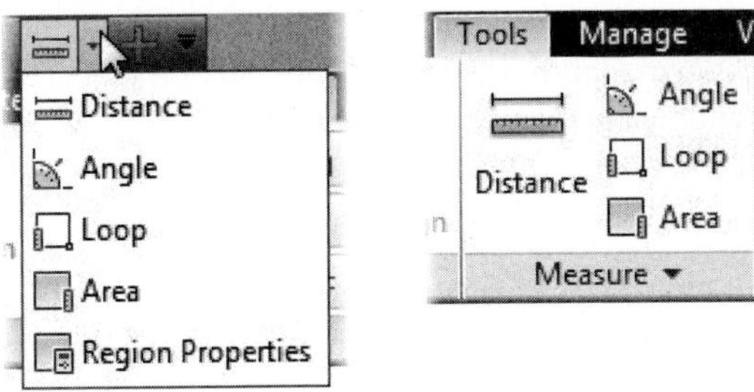

FIGURE 2.25

Measure Loop

Measures the length of closed or open loops defined by face boundaries or other geometry. When moving your cursor over a part face, all edges of the face will become highlighted. Clicking on this face will calculate the closed loop or perimeter of the shape.

Measure Area

Measures the area of enclosed regions or faces. Moving your cursor inside the closed outer shape will cause the outer shape and all holes (referred to as "islands") to also become highlighted. Clicking inside this shape will calculate the area of the shape.

When you click the arrow beside the measurement box, a menu similar to Figure 2.26 will appear.

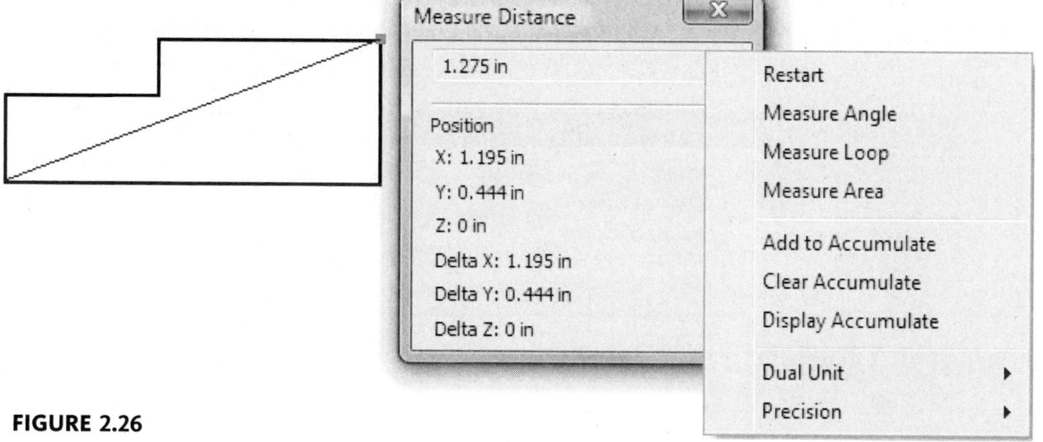

FIGURE 2.26

Region Properties

Available by clicking the drop arrow on the bottom of the Measure panel. While in a sketch you can determine the properties such as the area, perimeter, and Moment of Inertia properties of a closed 2D sketch. All measurements are taken from the sketch coordinate system. The properties can be displayed in dual units, the default unit of the document, or a unit of your choice. Figure 2.27 shows the region properties of two circles.

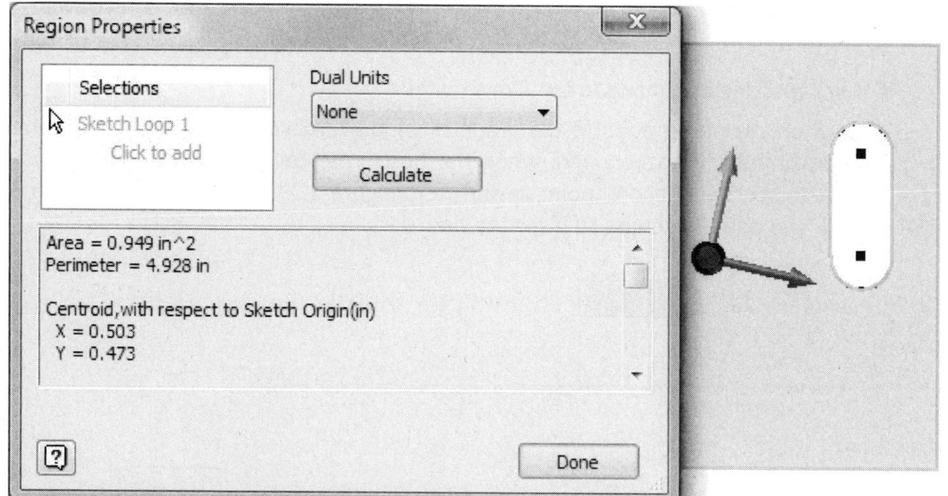

FIGURE 2.27

EXERCISE 2-1: CREATING A SKETCH WITH LINES

In this exercise, you create a new part file, and then create sketch geometry using basic construction techniques. In this exercise no dimensions will be created.

1. Click the New command on the Quick Access toolbar, click the English folder, and then double-click *Standard (in).ipt,* or if inch is the default unit; from the left side of the Quick Access toolbar you can click the down arrow on the New icon, and select Part.

2. Click the Create 2D Sketch command on the 3D Model tab > Sketch panel and then select the XY origin plane in the graphics window as shown in Figure 2.28.

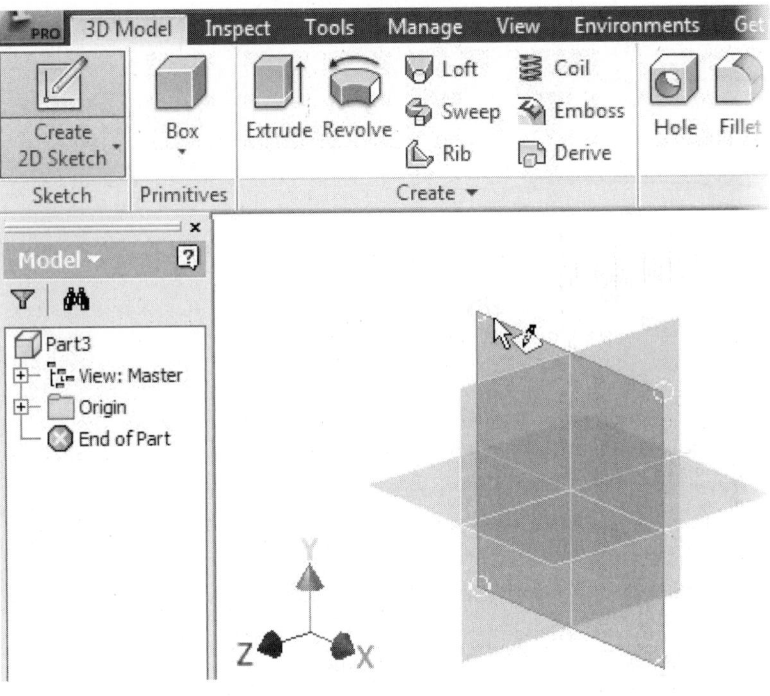

FIGURE 2.28

3. Click the Line command in the Draw panel.
4. Click on the origin point in the graphics window, move the cursor to the right approximately 4 inches, and, when the horizontal constraint symbol displays, click to specify a second point as shown in Figure 2.29. You may need to zoom back and pan the screen to draw the line.

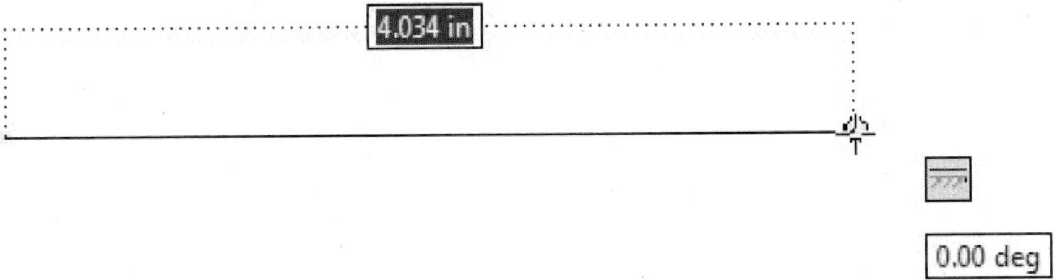

FIGURE 2.29

NOTE

Symbols indicate the geometric constraint. In the figure above, the symbol indicates that the line is horizontal. When you create the first entity in a sketch, make it close to final size. The length and angle of the line are displayed in the lower-right corner of the window to assist you.

5. Move the cursor up until the perpendicular constraint symbol displays beside the first line and then click to create a perpendicular line that is approximately 2 inches as shown in Figure 2.30 on the left.

6. Move the cursor to the left and create a horizontal line approximately 1 inch, that is, parallel to the first horizontal line. The parallel constraint symbol is displayed as shown in Figure 2.30 on the right.

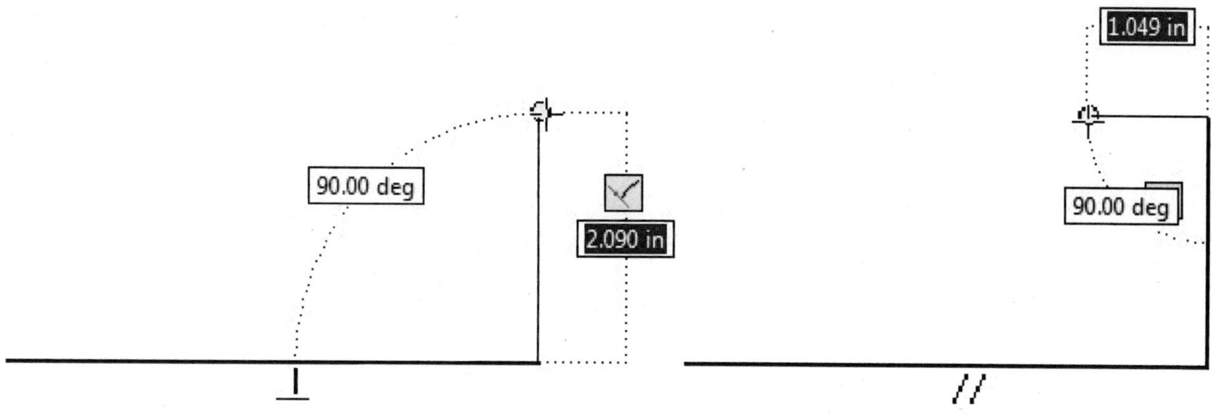

FIGURE 2.30

7. Move the cursor down, and create a line that is parallel to the first vertical line and is approximately 1 inch.
8. Move the cursor left to create a horizontal line of approximately 2 inches.
9. Move the cursor up until the parallel constraint symbol is displayed, and a dotted alignment line appears as shown in Figure 2.31 on the left. If the parallel constraint does not appear, move (scrub) the cursor over the inside vertical line to create a relationship to it. Click to locate the point.
10. Move the cursor to the left until the parallel constraint symbol is displayed, and a dotted alignment line appears as shown in Figure 2.31 on the right. Then click to locate the point.

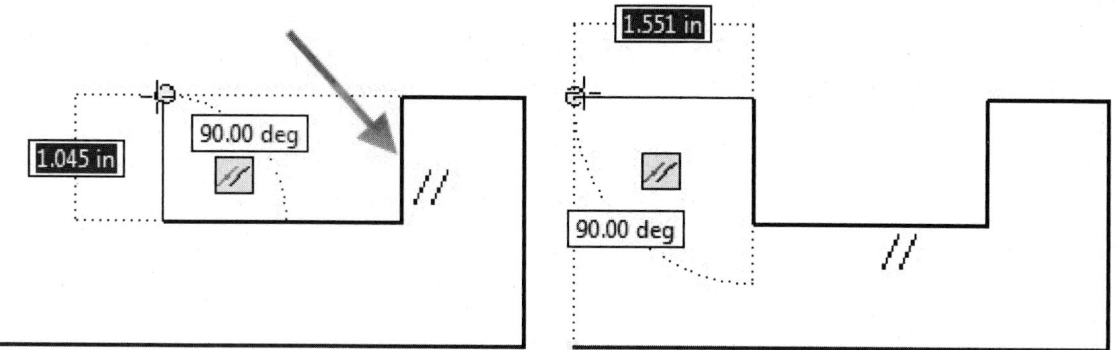

FIGURE 2.31

11. To close the profile right-click and click Close from the menu.
12. Your screen should resemble Figure 2.32.
13. Right-click in the graphics screen, and click Done to end the Line command.
14. Right-click in the graphics screen again, and click Finish Sketch.
15. Close the file. Do not save changes. End of exercise.

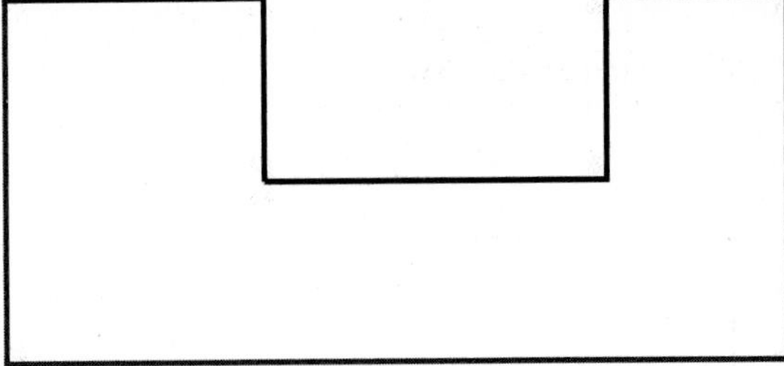

FIGURE 2.32

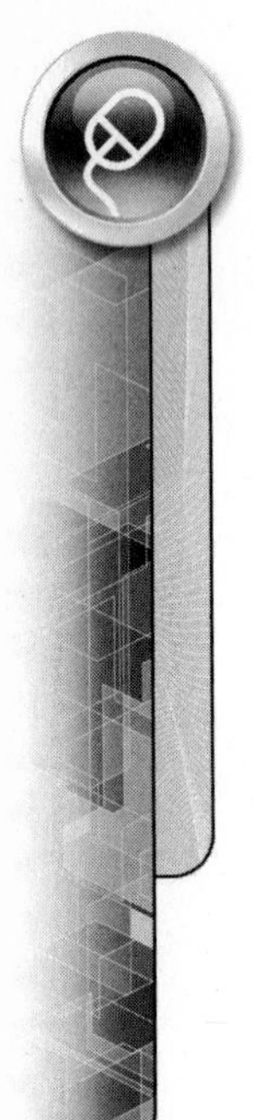

EXERCISE 2-2: CREATING A SKETCH WITH TANGENCIES

In this exercise, you create a new part file, and then you create a profile consisting of lines and tangent arcs.

1. Click the New command, and then double-click *Standard (inch).ipt*; or if inch is the default unit, from the left side of the Quick Access toolbar you can click the down arrow of the New icon, and select Part.
2. Click the Create 2D Sketch command on the 3D Model tab > Sketch panel and then select the XY origin plane as shown in Figure 2.33.

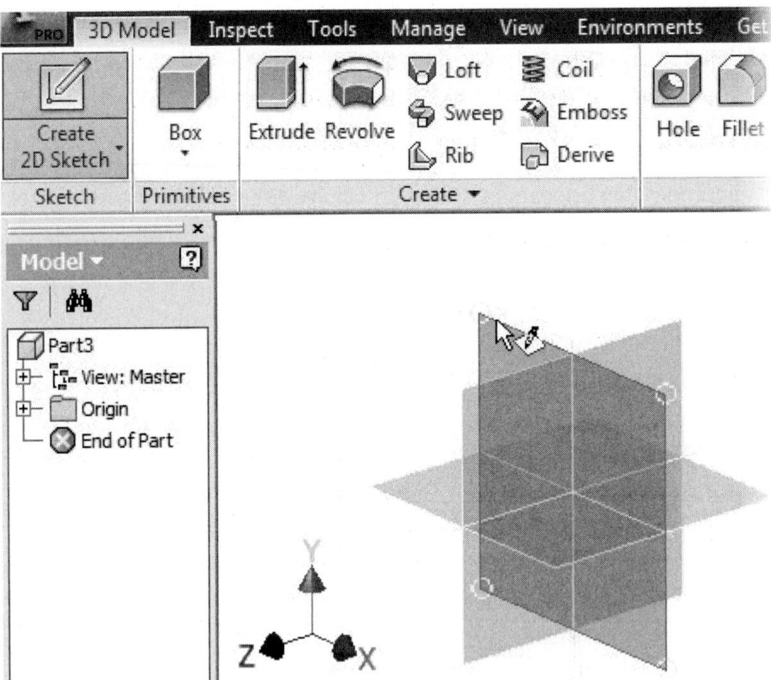

FIGURE 2.33

3. Click the Line command in the Draw panel.
4. Click on the projected origin point in the middle of the graphics window, and create a horizontal line to the right of the origin point and type **3** (inches will be assumed as the unit because the part file is based on the unit of inch) in the input field and then press enter.

NOTE

If the second point of the line lies off the screen, roll the mouse wheel away from you to zoom out, hold down the mouse wheel, and drag to pan the view.

5. Create a perpendicular line, move the cursor up, type **2** in the input field as shown in Figure 2.34 on the left, and press enter.
6. In this step, you infer points, meaning that no sketch constraint is applied. Move the cursor to the intersection of the midpoints of the right-vertical line and bottom horizontal line. Dashed lines (inferred points) appear as shown in Figure 2.34 on the right, and then click to create the line. No dimension was created since a value was not entered.

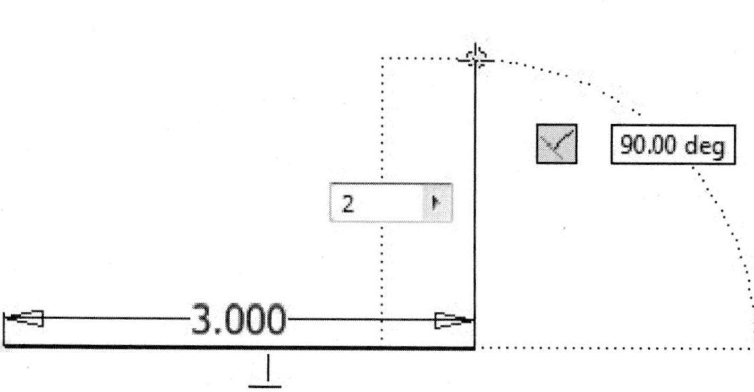

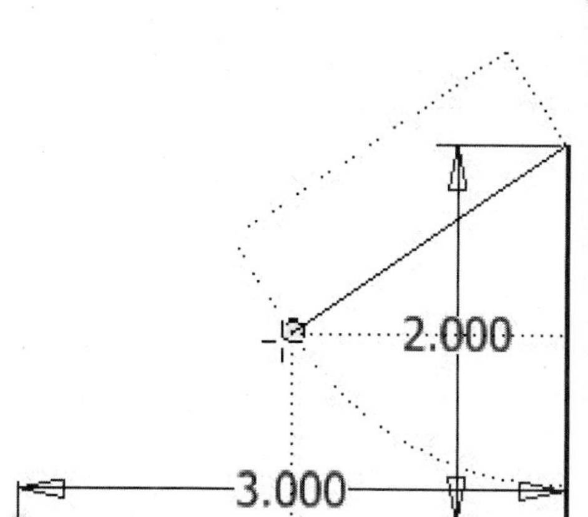

FIGURE 2.34

7. Next you create a line that is parallel to the bottom line. You scrub the bottom line by moving the cursor over the bottom line (do NOT click), and then move the cursor up and to the left until the vertical inferred line and the constraints are displayed as shown in Figure 2.35 on the left, and then click to place the line.
8. Next you sketch an arc while in the line command. Click on the gray dot at the left end of the line, and hold and drag the endpoint to create a tangent arc. **Do not release the mouse button.**
9. Move the cursor over the endpoint of the first line segment until a coincident constraint (green circle) and the two tangency constraints at start and end points of the arc are displayed as shown in Figure 2.35 on the right.

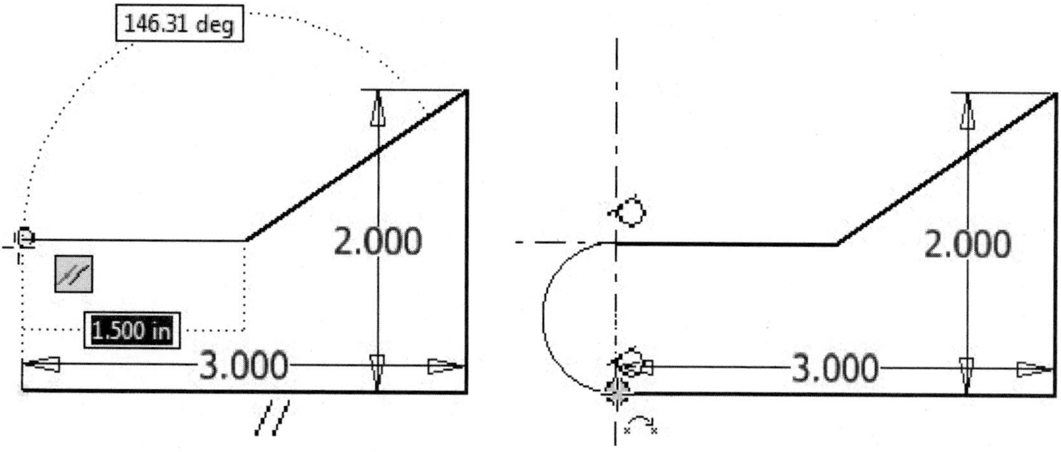

FIGURE 2.35

10. Release the mouse button to place the arc.
11. Right-click in the graphics window, and then click Done. Later in this chapter you will learn how to create dimensions.
12. Close the file. Do not save changes. End of exercise.

STEP 2—CONSTRAINING THE SKETCH

After you draw the sketch, you may want to add geometric constraints to it to add design intent. Geometric constraints apply behavior to a specific object or create a relationship between two objects. An example of using a constraint is applying a vertical constraint to a line so that it will always be vertical. You could apply a parallel constraint between two lines to make them parallel to one another; then, as one of the line's angles changes, so will the other's. You can apply a tangent constraint to a line and an arc or to two arcs.

When you add a constraint, the number of constraints or dimensions that are required to fully constrain the sketch will decrease. On the bottom-right corner of Autodesk Inventor, the number of constraints or dimensions will be displayed similar to what is shown in Figure 2.36. A fully constrained sketch is a sketch whose objects cannot move or stretch.

FIGURE 2.36

To help you see which objects are constrained, Autodesk Inventor will change the color of constrained objects if a point on the sketch has a coincident constraint applied to the origin point or another point of an existing sketch. If you have not sketched a point coincident to the projected origin point or another point of an existing sketch, the color of the sketch will not change. You could apply a fix constraint instead of using the origin point, but it is not recommended. If you have not sketched a point coincident to the origin point or applied a fix constraint to the sketch, objects are free to move in their sketch plane.

NOTE

Autodesk Inventor does not force you to fully constrain a sketch. However, it is recommended that you fully constrain a sketch, as this will allow you to better predict how a part will react when you change dimensions values.

Constraint Types

Autodesk Inventor has 12 geometric constraints that you can apply to a sketch. Figure 2.37 shows the constraint types that can be applied from the Sketch tab > Constrain panel and the symbols that represent them when the constrains are displayed in a sketch. Descriptions of the constraints follow.

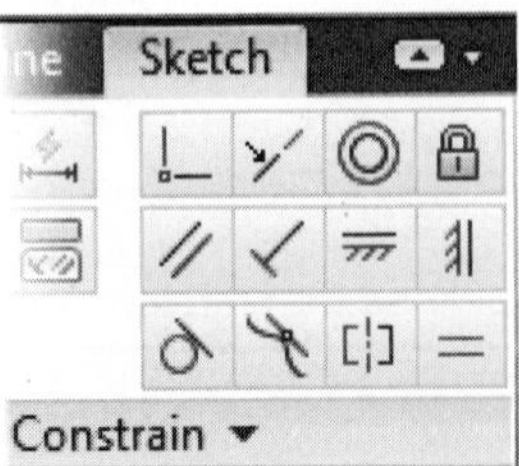

FIGURE 2.37

Button	Tool	Function
	Perpendicular	Lines will be repositioned at 90° angles to one another. The first line sketched will stay in its position, and the second will rotate until the angle between them is 90°.
	Parallel	Lines will be repositioned so that they are parallel to one another. The first line sketched will stay in its position, and the second will move to become parallel to the first.
	Tangent	An arc, circle, or line will become tangent to another arc or circle.
	Smooth (G2)	A spline and another spline, line, or arc that connect at an endpoint with a coincident constraint will represent a smooth G2 (continuous curvature) condition.
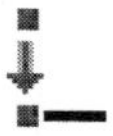	Coincident	A point is constrained to lie on another point or curve (line, arc, etc.).
	Concentric	Arcs and/or circles will share the same center point.
	Collinear	Two selected lines will line up along a single line; if the first line moves, so will the second. The two lines do not have to be touching.
	Equal	If two arcs or circles are selected, they will have the same radius or diameter. If two lines are selected, they will become the same length. If one of the objects changes, so will the other object to which the Equal constraint command has been applied. If the Equal constraint command is applied after one of the arcs, circles, or lines has been dimensioned, the second arc, circle, or line will take on the size of the first one. If you select multiple similar objects (lines, arcs, etc.) before selecting this command, the constraint is applied to all of them. This rule applies to some of the other sketch constraints as well.
	Horizontal	Lines are positioned parallel to the X-axis, or a horizontal constraint can be applied between any two points in the sketch. The selected points will be aligned such that a line drawn between them will be parallel to the X-axis.
	Vertical	Lines are positioned parallel to the Y-axis, or a vertical constraint can be applied between any two points in the sketch. The selected points will be aligned such that a line drawn between them will be parallel to the Y-axis.
	Fix	Applying a fixed point or points will prevent the selected point from moving. The Fix constraint command overrides any other constraint. Any endpoint or segment of a line, arc, circle, spline segment, or ellipse can be fixed. Multiple points in a sketch can be fixed. If you select near the endpoint of an object, the endpoint will be locked from moving. If you select a line segment, the angle of the line will be fixed and only its length will be able to change. You can remove a fix constraint as needed. Deleting constraints will be covered later in this section.

(*Continued*)

Button	Tool	Function
L|J	Symmetry	Selected points defining the selected geometry are made symmetric about the selected line.

NOTE

To fully constrain a base (first) sketch, constrain or dimension a point in the sketch to the center point or apply a fix constraint.

Adding Constraints

As stated previously in this chapter, you can apply constraints while you sketch objects. You can also apply additional constraints after the sketch is drawn. However, Autodesk Inventor will not allow you to over-constrain the sketch or add duplicate constraints. If you add a constraint that would conflict with another, you will be warned with the message, "Adding this constraint will over-constrain the sketch." For example, if you try to add a vertical constraint to a line that already has a horizontal constraint, you will be alerted. To apply a constraint, follow these steps:

1. Click a constraint from the Constrain panel, or right-click in the graphics window and click Create Constraint from the menu. Then click the specific constraint from the menu as shown in Figure 2.37 before the chart.
2. Click the object or objects to apply the constraints.

Showing and Deleting Constraints

To see the constraints that are applied to an object, use the Show Constraints command from the Constrain panel as shown on the left side of Figure 2.38. After issuing the Show Constraints command, select an object and a constraint icon and yellow squares that represent coincident constraints will appear with the constraints that are applied to the selected object—similar to what is shown in the middle of Figure 2.38. You can modify the size of the constraint toolbar by clicking Tools tab > Application Options, clicking on the Sketch tab, and modifying the size of the Constraint and DOF Symbol Scale. Figure 2.38 on the right shows the scale increased from 1.0 to 1.5.

FIGURE 2.38

To show and hide all the constraints for all of the objects in the sketch, do the following:

- While not in an command click Show All Constraints on the middle of the Status bar (bottom of the screen) as shown in Figure 2.39 on the left, or press the F8 key.
- To hide all the constraints click Hide All Constraints from the middle of the Status bar (bottom of the screen) as shown in Figure 2.39 on the right.
- If a constraint is added when sketch constraints are displayed you will need to exit out of the active command and then press the F8 key again.

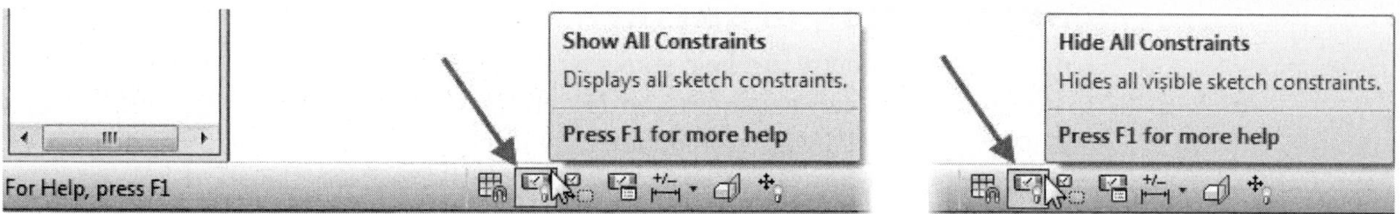

FIGURE 2.39

As you move the cursor over a constraint icon, the matching sketch constraint and the object that is linked to that constraint will be highlighted. The coincident constraint will appear as a yellow square at the point that the constraint exists. The perpendicular constraint will appear once at the location the constraint is applied. Figure 2.40 on the left shows the cursor over the perpendicular constraint on the right vertical edge; the bottom horizontal and right vertical lines are highlighted. Figure 2.40 on the right shows the cursor over the bottom horizontal line, and the constraints that are related to the line are highlighted.

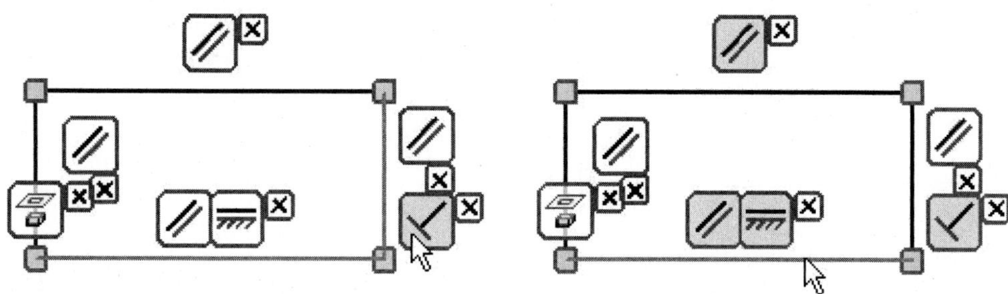

FIGURE 2.40

To delete the constraint, move the cursor over the constraint right-click and click Delete from the marking menu or press the Delete key. Figure 2.41 shows a parallel constraint being deleted.

To hide constraint icons, click the × on the right side of the constraint symbols toolbar or right-click on a constraint and click Hide All Constraints from the marking menu. You can also press the F9 key or click Hide All Constraints on the middle of the Status Bar.

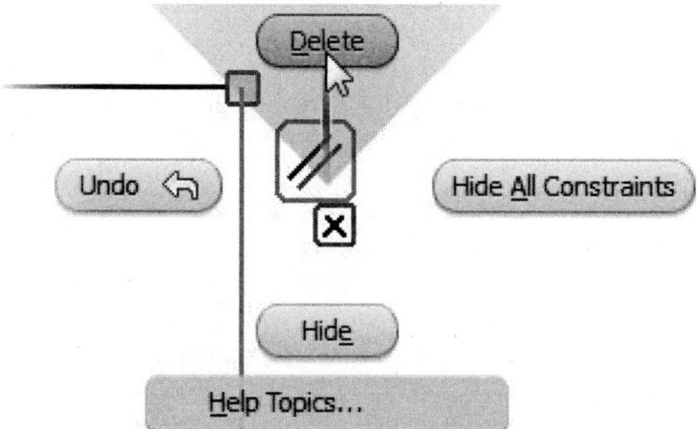

FIGURE 2.41

CONSTRUCTION GEOMETRY

Construction geometry can help you create sketches that would be otherwise difficult to constrain. You can constrain and dimension construction geometry like normal geometry, but the construction geometry will not be recognized as a profile edge in the part when you turn the sketch into a feature. When you sketch, the sketches by default have a normal geometry style, meaning that the sketch geometry is visible in the feature. Construction geometry can reduce the number of constraints and dimensions required to constrain a sketch fully, and it can help to define the sketch. For example, a construction circle that is tangent to the inside a hexagon (drawn with individual lines and not the Polygon command) can drive the size of the hexagon. Without construction geometry, the hexagon would require six constraints and dimensions. With construction geometry, it would require only three constraints and dimensions; the circle would have tangent or coincident constraints applied to it and the hexagon. You create construction geometry by changing the line style before or after you sketch geometry in one of the following two ways:

- After creating the sketch, select the geometry that you want to change and click the Construction icon on the Format panel as shown in Figure 2.42.
- Before sketching, click the Construction icon on the Format panel, as shown in Figure 2.42 on the left. All geometry created will be construction until Construction is unselected.

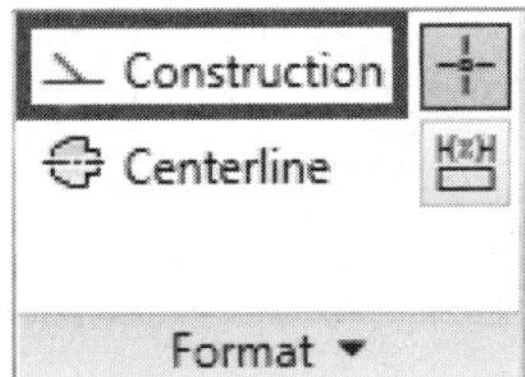

FIGURE 2.42

After turning the sketch into a feature, the construction geometry will be consumed with the sketch and is maintained in the sketch. When you edit a feature's sketch that you created with construction geometry, the construction geometry will reappear during editing and disappear when the part is updated. You can add or delete construction geometry to or from a sketch just like any geometry that has a normal style. In the graphics window, construction geometry will be displayed as a dashed line, lighter in color and thinner in width than normal geometry. Figure 2.43 on the left shows a sketch with a construction line for the angled line. The angled line has a coincident constraint applied to it at every point that touches it. The image on the right shows the sketch after it has been extruded. Note that the construction line was not extruded.

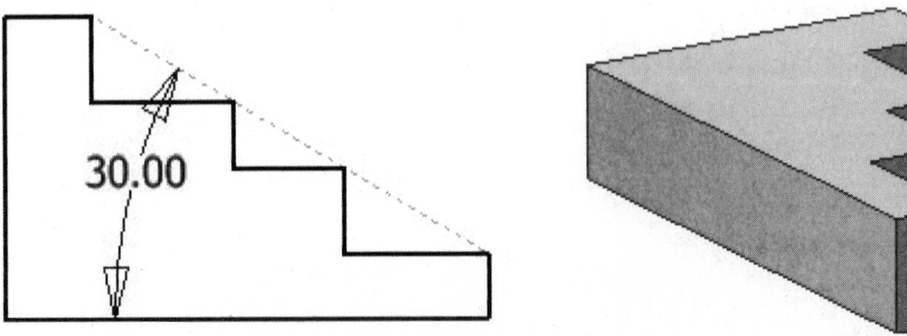

FIGURE 2.43

Snaps

Another method used to place geometry with a coincident constraint is snaps: midpoint, center, and intersection. After using a snap, a coincident constraint will be applied, and it will maintain the relationship that you define. For example, if you use a midpoint snap, the sketched point will always be in the middle of the selected object, even if the selected object's length changes.

To use snaps, follow these steps:

1. Start a sketch command, and right-click in the graphics window.
2. From the menu, select the desired snap.
3. Click on the object to which the sketched object will be constrained. For the intersection snap, select two objects.

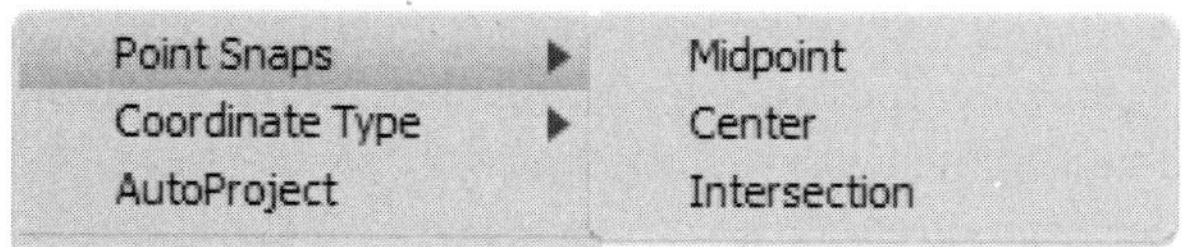

FIGURE 2.44

Sketch Degrees of Freedom

While constraining and dimensioning a sketch there are multiple ways to determine the open degrees of freedom. When you add a constraint or dimension the number of constraints or dimensions that are required to fully constrain, the sketch will decrease.

Status Bar

On the bottom-right corner of Autodesk Inventor, the number of constraints or dimensions will be displayed similar to what is shown in Figure 2.45 on the left. A fully constrained sketch is a sketch whose objects cannot move or stretch and the message Fully Constrained will appear in the bottom-right corner of the screen as shown in Figure 2.45 on the right.

FIGURE 2.45

Degrees of Freedom

To see the areas in the sketch that are NOT constrained, you can display the degrees of freedom. While in a sketch click Show Degree of Freedom on the bottom of the status bar as shown in Figure 2.46 on the left. Lines and arcs with arrows will appear as shown in Figure 2.46 in the middle. As constraints and dimensions are added to the sketch, degrees of freedom will disappear. To remove the degree of freedom symbols from the screen, click Hide All Degrees of Freedom on the bottom of the status bar as shown in Figure 2.46 on the right.

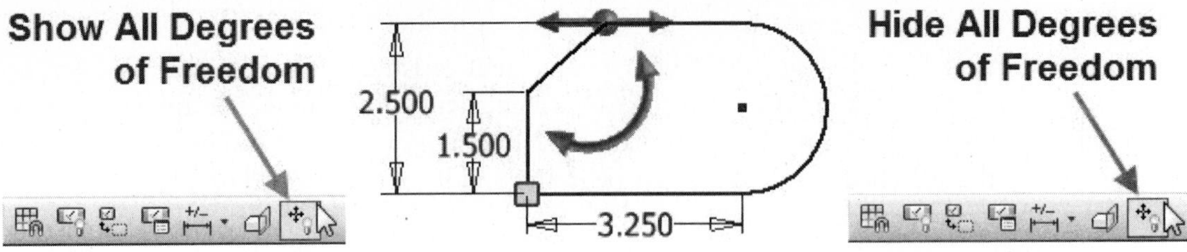

FIGURE 2.46

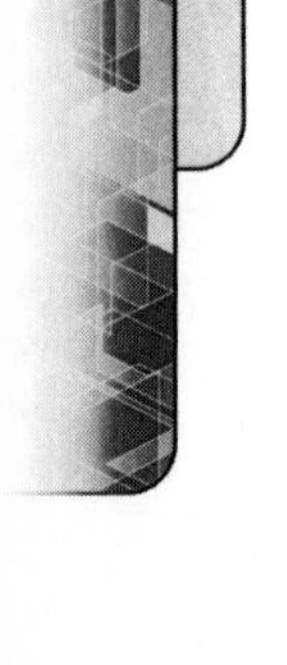

Dragging a Sketch

Another method to determine whether or not an object is constrained is to try to drag it to a new location. While not in a command, click a point or an edge, or select multiple objects on the sketch. With the left mouse button depressed, drag it to a new location. If the geometry stretches, it is underconstrained. For example, if you draw a rectangle that has two horizontal and two vertical constraints applied to it and you drag a point on one of the corners, the size of the rectangle will change, but the lines will maintain their horizontal and vertical behaviors. If dimensions are set on the object, they too will prevent the object from stretching.

EXERCISE 2-3: ADDING AND DISPLAYING CONSTRAINTS

In this exercise, you add geometric constraints to sketch geometry to control the shape of the sketch.

1. Click the New command, click the English folder, and double-click *Standard (in).ipt*.
2. Click the Create 2D Sketch command on the 3D Model tab > Sketch panel and then select the XY origin plane as shown in Figure 2.47.

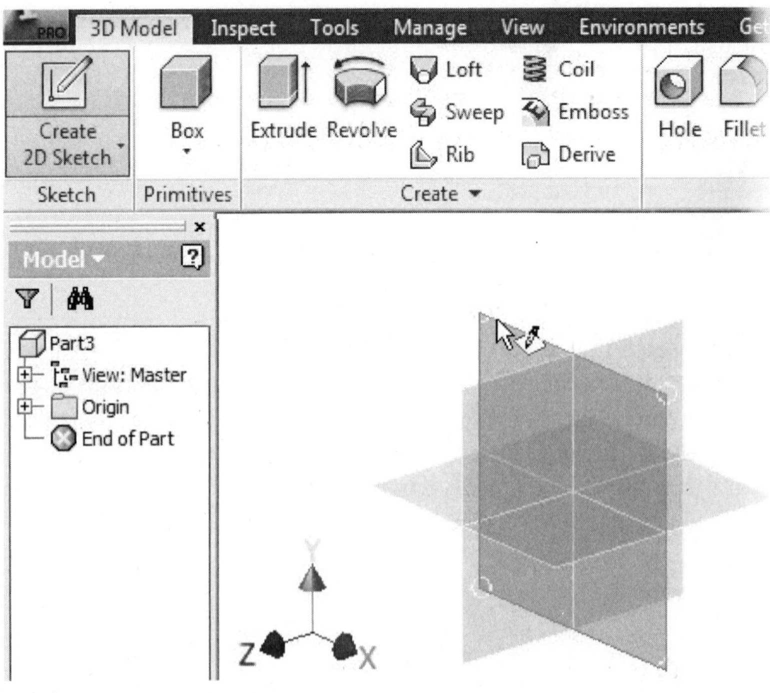

FIGURE 2.47

3. Sketch the geometry as shown in Figure 2.48, with an approximate size of 2 inches in the X (horizontal) and 1.125 inches in the Y (vertical). Do not apply dimensions dynamically. Place the lower-left corner of the sketch on the origin point. Right-click in the graphics window, and then click Done. By starting the line at the origin point, that point is constrained to the origin.
4. Click Show All Constraints on the Status Bar, or press the F8 key. Your screen should resemble Figure 2.48.

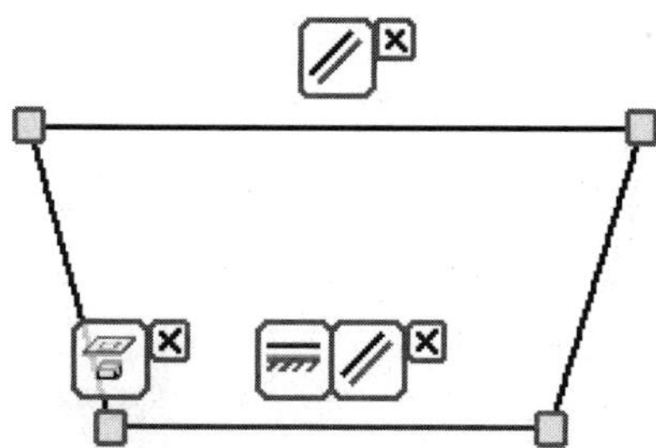

FIGURE 2.48

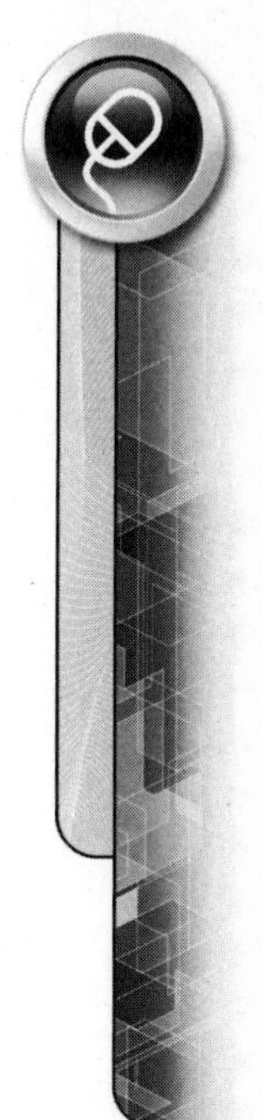

5. If another constraint appears, place the cursor over it, right-click, and then click Delete from the marking menu.
6. On the Constrain panel, click the Parallel constraint icon.
7. Select the two angled lines. Depending upon the order in which you sketched the lines, the angles may be opposite of Figure 2.49 on the left.
8. Press the ESC key twice to stop adding constraints.
9. Press the F8 key to refresh the visible constraints. Your screen should resemble Figure 2.49 on the right.

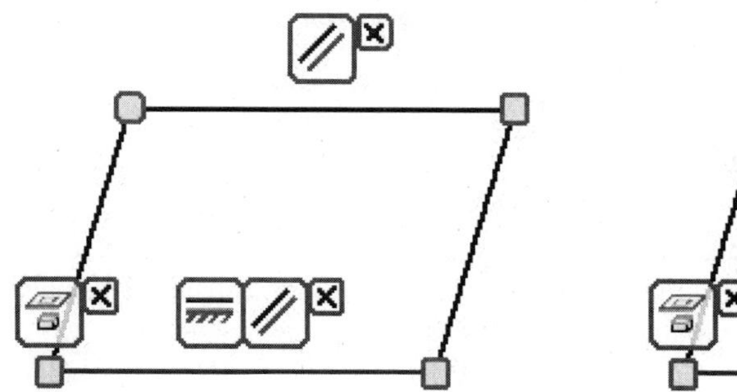

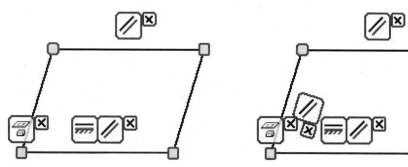

FIGURE 2.49

10. Select the top horizontal line in the sketch and drag the line. Notice how the sketch changes its size, but not its general shape. Try to drag the bottom horizontal line. The line cannot be dragged as it is constrained.
11. Select the endpoint on the bottom-right horizontal line, and drag the endpoint. The lines remain parallel due to the parallel constraints.
12. Place the cursor over the icon for the parallel constraint on the right angled line right-click, and click Delete from the marking menu as shown in Figure 2.50 on the left. The parallel constraint that applies to both angled lines is deleted.
13. On the Sketch Tab > Constrain panel, click the Perpendicular constraint icon.
14. Select the bottom horizontal line and the angled line on the right side. Even though it may appear that the rectangle is fully constrained, the left vertical line is still unconstrained and can move. Notice on the bottom-right of the Status Bar that 3 dimensions are needed to fully constrain the sketch.

15. While still in the Perpendicular Constraint command, select the bottom horizontal line and the left vertical line and then right-click and click OK on the marking menu. Notice on the bottom-right the Status Bar is down to 2 dimensions to fully constrain the sketch. Dimensions would be added to fully constrain the sketch.
16. Press the F8 key to refresh the visible constraints. Your screen should resemble Figure 2.50 on the right.
17. Click Hide All Constraints on the Status Bar, or press the F9 key.

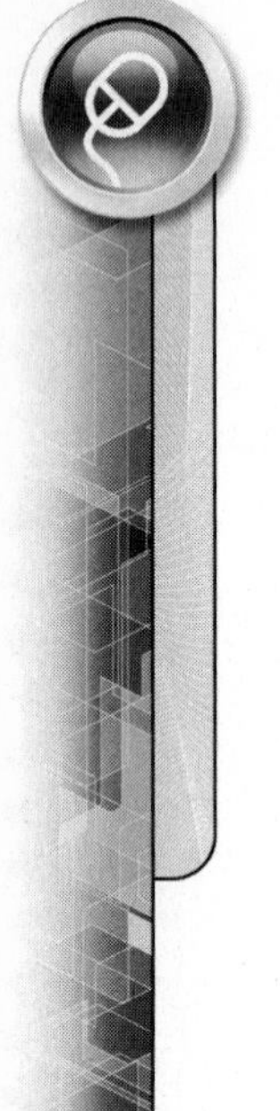

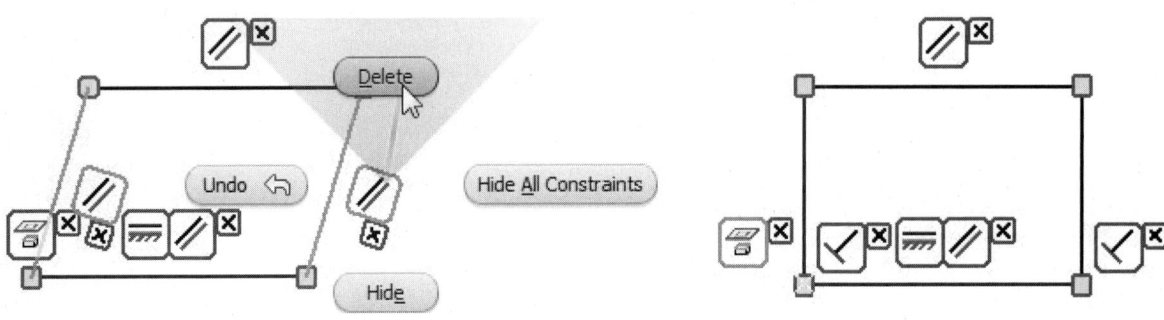

FIGURE 2.50

18. Drag the point at the upper-right corner of the sketch to verify that the rectangle can change size in both the horizontal and vertical directions, but its shape is maintained.
19. Press down the CTRL key and select the four lines and right-click and click Delete from the marking menu.

NOTE

If you would use the window or crossing selection method to select the rectangle, the origin point will also be deleted.

20. Use the Line command to sketch the geometry as shown with an approximate size of **2 inches** in the X and **1.375 inches** in the Y. Place the lower-left point of the sketch on the projected center point. Do not apply dimensions dynamically. Right-click in the graphics window, and then click Done.

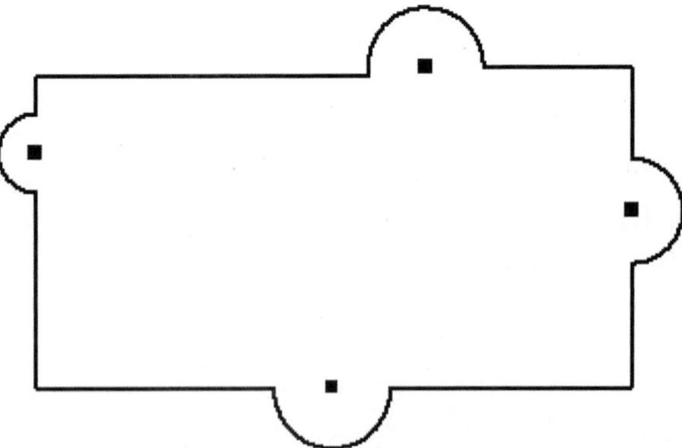

FIGURE 2.51

21. Inspect the constraints by dragging different points and edges.
22. Next you make the arcs equal in size. On the Constrain panel, click the Equal constraint command or press the = key on the keyboard.
23. Select the arc on the left and the bottom arc.
24. Select the arc on the left and the arc on the right side.
25. Select the arc on the left and the arc on the top.
26. Next you align the line segments. On the Constrain panel, click the Collinear constraint command.
27. Select the two bottom horizontal lines.
28. Select the two top horizontal lines.
29. Select the two left vertical lines.
30. Select the two right vertical lines.
31. To stop applying the collinear constraint, either right-click and click Cancel (ESC) from the marking menu or press the ESC key.
32. Next you will align the arcs. On the Constrain panel, click the Vertical constraint command.
33. Select the center point of the bottom arc, and then click the center point of the top arc.
34. On the Constrain panel, click the Horizontal constraint command.
35. Click the center point of the left arc, and then click the center point of the right arc.
36. To stop applying the constraints, right-click and click Cancel (ESC) from the marking menu, or press the ESC key.
37. Display all of the constraints by pressing the F8 key. Your screen should resemble Figure 2.52.

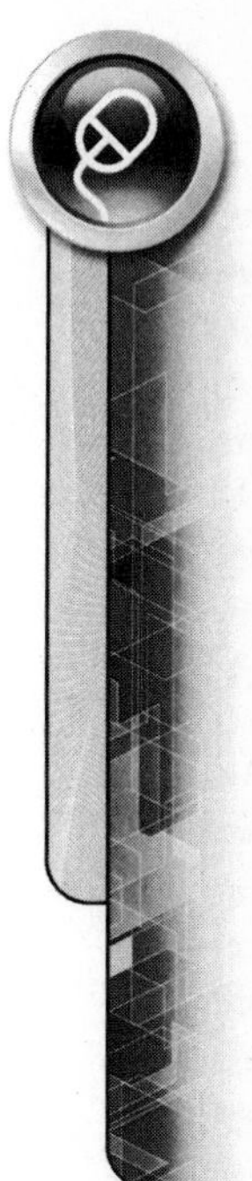

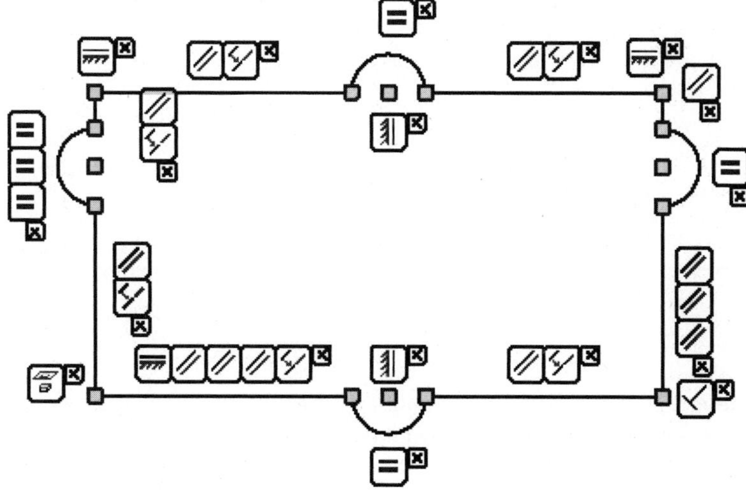

FIGURE 2.52

38. Hide all of the constraints by pressing the F9 key.
39. Click on an endpoint in the sketch and drag the endpoint. Try dragging different points, and notice how the sketch changes.

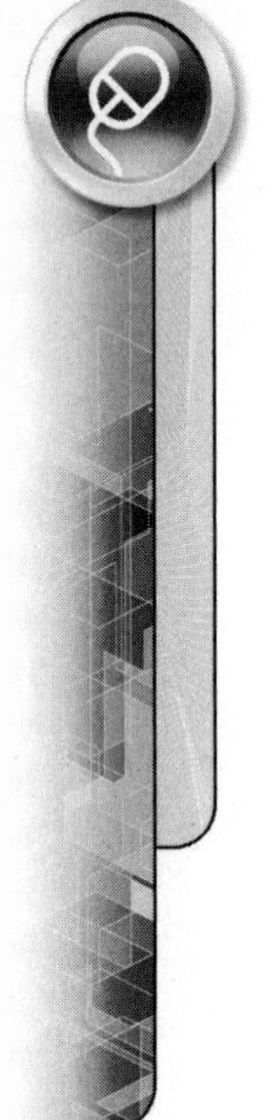

40. In this step you delete the geometry as shown in Figure 2.53 on the left. Press the Esc key twice to cancel any command, click a point above and to the left of the top arc, drag a window so it encompasses the arc on the right, release the mouse button, and press the Delete key on the keyboard.
41. Close the open line segments. Drag the open endpoints onto each other until your sketch resembles Figure 2.53 on the right. A green circle will appear when two endpoints are near each other; this applies a coincident constraint.

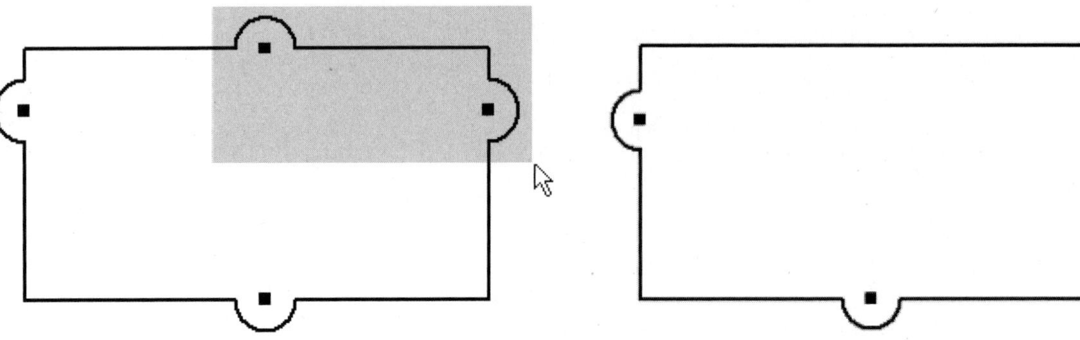

FIGURE 2.53

42. Next you center the arcs. On the Constrain panel, click the Vertical constraint command.
43. Click the center point on the bottom arc and the midpoint of the top horizontal line as shown in 2.54 on the left.
44. On the Constrain panel, click the Horizontal constraint command.
45. Click the center point on the left arc and the midpoint of the right vertical line as shown in Figure 2.54 on the right.

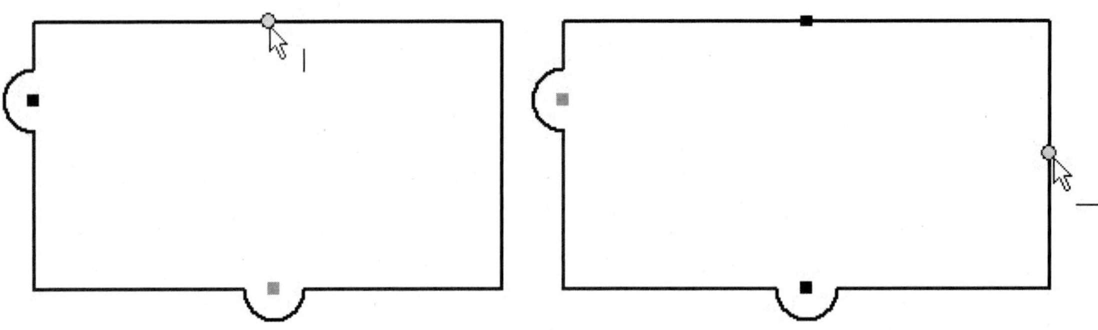

FIGURE 2.54

46. Click on different points and drag them, notice how the sketch changes shape, but the arcs are always centered as shown in Figure 2.55.

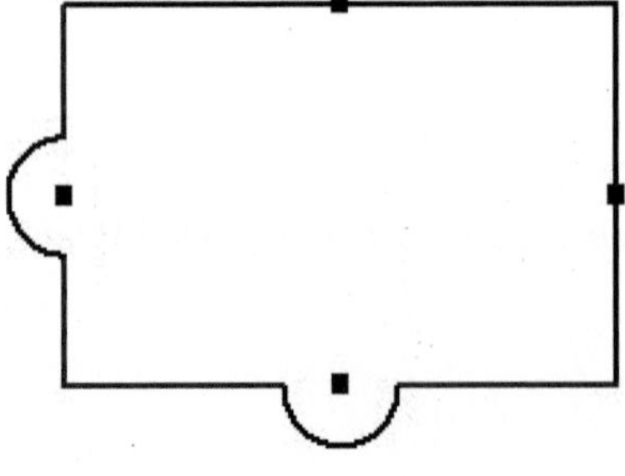

FIGURE 2.55

47. Close the file. Do not save changes. End of exercise. Note that dimensions would be added to fully constrain the sketch. Dimensions are covered in the next section.

STEP 3—ADDING DIMENSIONS MANUALLY

The last step to constraining a sketch is to add dimensions that were not added dynamically. The dimensions you place will control the size of the sketch and can also appear in the part drawing views when they are generated. When placing dimensions, try to avoid having extension lines go through the sketch, as this will require more cleanup when drawing views are generated. Click near the side from which you anticipate the dimensions will originate in the drawing views.

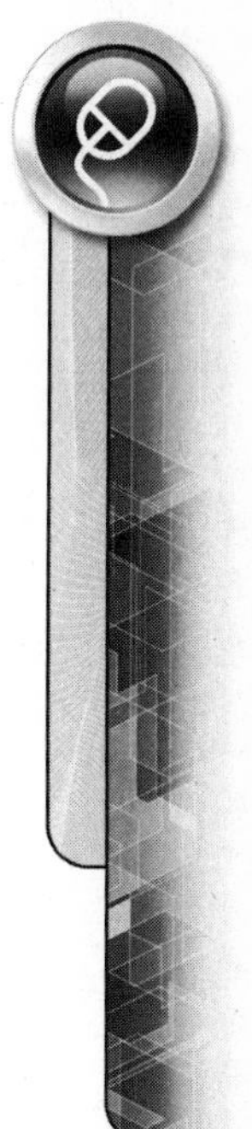

All dimensions that you create are parametric as well as the dynamic dimensions. Parametric means that they will change the size of the geometry.

Scale Sketch

If no dimension was dynamically added to the sketch when it was created, then the entire sketch will be uniformly scaled the same factor that the first dimension parametrically changes its geometry.

General Dimensioning

The General Dimension command can create linear, angle, radial, or diameter dimensions one at a time. Figure 2.56 on the left shows an example of a dimensioned sketch. To start the General Dimension command, follow one of these techniques:

- Click the Create Dimension command from the Sketch tab > Constrain panel as shown in Figure 2.56 on the right.
- Right-click in the graphics area and click General Dimension from the marking menu.
- Press the shortcut key D.

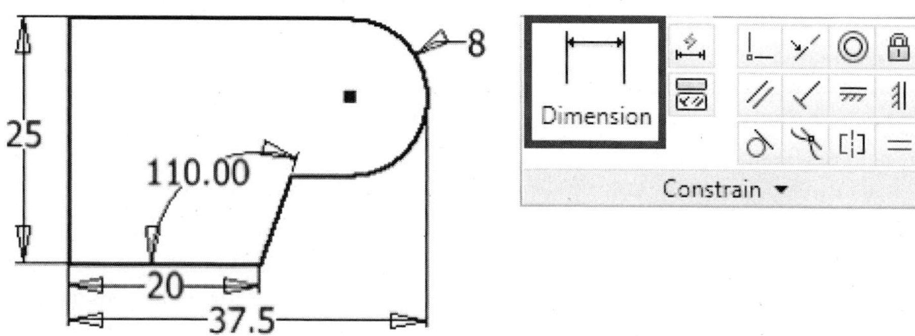

FIGURE 2.56

When you place a linear dimension, the extension line of the dimension will snap automatically to the nearest endpoint of a selected line; when an arc or circle is selected, it will snap to its center point. To dimension to a tangent point of an arc or circle, see "Dimensioning to a Tangent of an Arc or Circle," later in this chapter.

After you select the General Dimension command, follow these steps to place a dimension:

1. Click a point or points to locate where the dimension is to start and end.
2. After selecting the point(s) to dimension, a preview image will appear attached to your cursor showing the type of dimension. If the dimension type is not what you want, right-click, and then select the correct style from the menu. After changing the dimension type, the dimension preview will change to reflect the new style.
3. Click a point on the screen to place the dimension.

The next sections cover how to dimension specific objects and how to create specific types of dimensioning with the Dimension command.

Dimensioning Lines

There are two techniques for dimensioning a line. Issue the Dimension command and do one of the following:

- Click near two endpoints, move the cursor until the dimension is in the correct location, and click.
- To dimension the length of a line, click anywhere on the line; the two endpoints will be selected automatically. Move the cursor until the dimension is in the correct location and click.
- To dimension between two parallel lines, click one line and then the next, and then click a point to locate the dimension.
- To create a dimension whose extension lines are perpendicular to the line being dimensioned, click the line and then right-click. Click Aligned from the menu, and then click a point to place the dimension.

Dimensioning Angles

To create an angular dimension, issue the General Dimension command, click on two lines whose angle you want to define, move the cursor until the dimension is in the correct location, and place the dimension by clicking on a point.

Dimensioning Arcs and Circles

To dimension an arc or circle, issue the General Dimension command, click on the circle's circumference, move the cursor until the dimension is in the correct location, and click. By default, when you dimension a circle, the default is a diameter dimension; when you dimension an arc, the result is a radius dimension. To change the radial dimension to diameter or a diameter to radial, right-click before you place the dimension and select the other style from the Dimension Type found on the menu.

You can dimension the angle of the arc. Start the Dimension command, click on the arc's circumference, click the center point of the arc, and then place the dimension or click the center point and then click the circumference of the arc.

For arcs you can also add an arc length dimension by starting the dimension command: click on the arc, right-click and click Arc Length from the Dimension Type menu, and then click a point to locate the dimension.

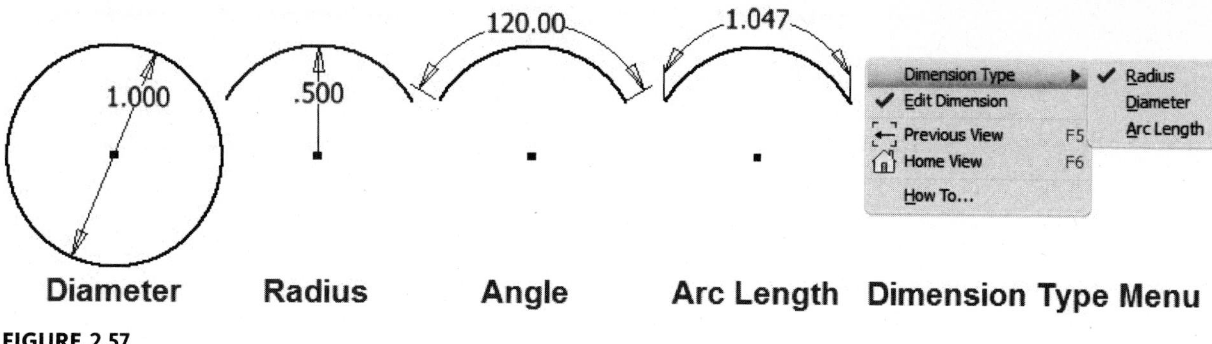

FIGURE 2.57

Dimensioning to a Tangent of an Arc or Circle

To dimension to a tangent of an arc or circle, follow these steps:

1. Issue the General Dimension command.
2. Click a line that is parallel to the tangent arc or circle that will be dimensioned.
3. Place the cursor over the tangent arc or circle that should be dimensioned.
4. Move the cursor over the tangent until the constraint symbol changes to reflect a tangent, as shown on the left side of Figure 2.58.
5. Click to accept the dimension type, and then move the cursor until the dimension is in the correct location. Click as shown on the right side of Figure 2.58.

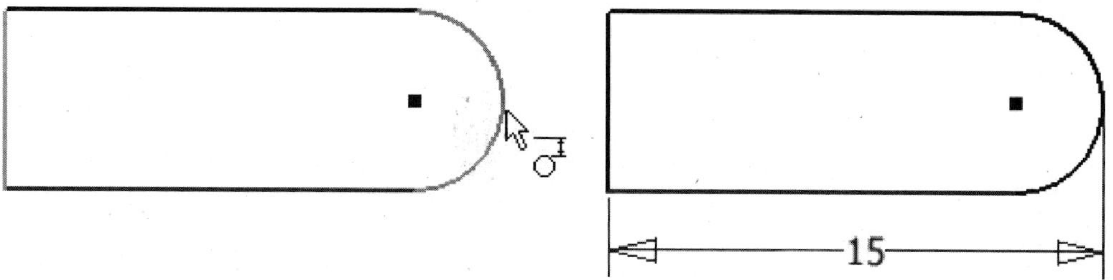

FIGURE 2.58

To dimension to two tangents, follow these steps:

1. Issue the General Dimension command.
2. Click an arc or circle that includes one of the tangents to which it will be dimensioned.
3. Place the cursor over the tangent edge of the second arc or circle to which it will be dimensioned.
4. Move the cursor over the area that is tangent until the constraint symbol appears, as shown on the left in Figure 2.59.
5. Click and then move the cursor until the dimension is in the correct location, and then click as shown in Figure 2.59 on the right.

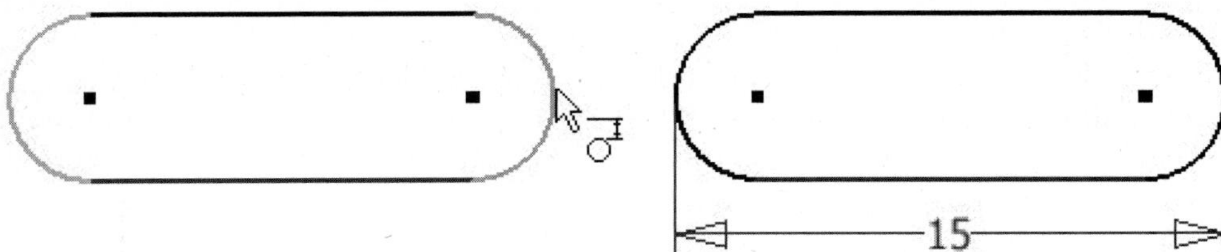

FIGURE 2.59

Entering and Editing a Dimension Value

After placing the dimension, you can change its value. Depending on your setting for editing dimensions when you created them, the Edit Dimension dialog box may or may not appear automatically after you place the dimension. To set the Edit Dimension option, do one of the following:

1. Click the Tools tab > Options panel > Application Options. On the Sketch tab of the Options dialog box, select Edit dimension when created as shown on the left side of Figure 2.60.
2. Or set this option by right-clicking in the graphics area while placing a dimension and clicking Edit Dimension from the menu as shown in Figure 2.60 on the right. This method will change the Application Option sketch tab Edit Dimension.

If the Edit dimension when created option is checked, the Edit Dimension dialog box will appear automatically after you place the dimension. Otherwise, the dimension will be placed with the default value, and you will not be prompted to enter a another value.

FIGURE 2.60

To edit a dimension that has already been created, double-click on the value of the dimension, and the Edit Dimension dialog box will appear, as shown in Figure 2.61. Enter the new value and unit for the dimension; then either press ENTER or click the checkmark in the Edit Dimension dialog box. If no unit is entered, the units that the file was created with will be used. When inputting values, enter the exact value—do not round up or down. The accuracy of the dimension that is displayed in a sketch is set in the Document Setting. For example, if you want to enter 4 1/16 decimally enter 4.0625 not 4.06.

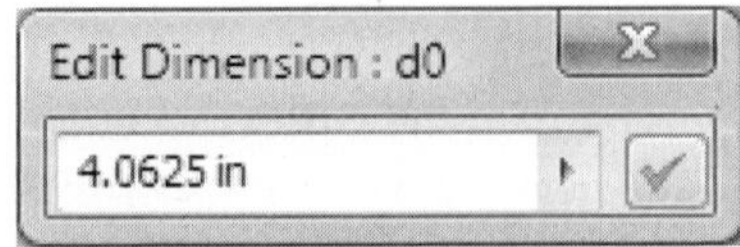

FIGURE 2.61

Fractions

Inventor also allows you to type in a fraction anywhere a value is required. When the length unit in the Tools tab > Options > Document Settings is set to a metric unit the decimal equivalent will be displayed in the graphics window but the fraction will be maintained in the edit dialog box. If the unit is set to any non-metric unit as shown in Figure 2.62 on the left and a fraction is entered, a fraction will display in the graphics window. After inputting a fraction you can click on the right-faced arrow and set the type of dimension to display Decimal, Fractional, or Architectural as shown in the middle of Figure 2.62. For example, enter 4 1/16, not 4-1/16. Inventor would interpret the - as part of an equation and would return the value 3.9375. Figure 2.62 on the right shows the fraction displayed in the graphics window.

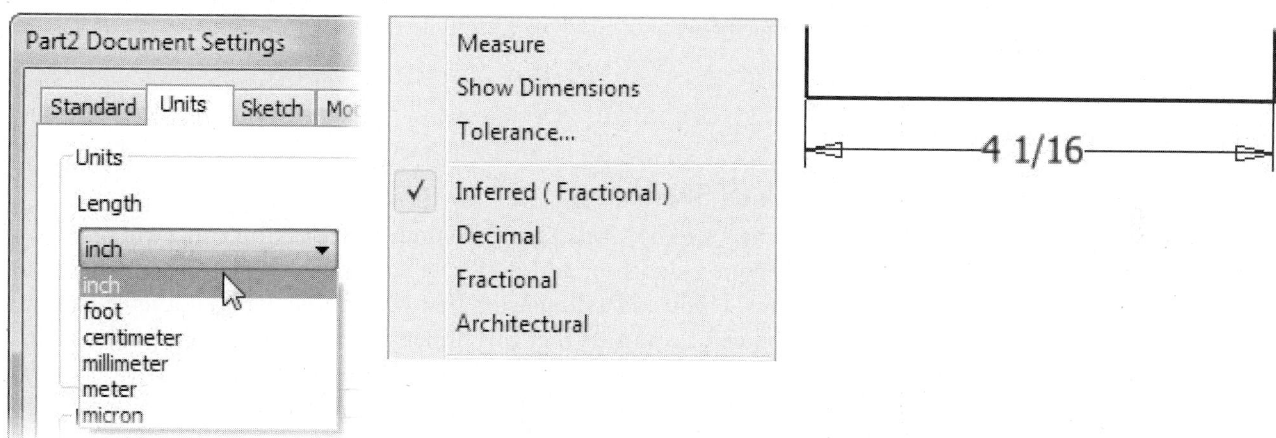

FIGURE 2.62

When placing dimensions, it is recommended that you place the smallest dimensions first. This will help prevent the geometry from flipping in the opposite direction.	**NOTE**

Repositioning a Dimension

Once you place a dimension, you can reposition it, but the origin points cannot be moved. Follow these steps to reposition a dimension:

1. Exit the current operation either by pressing ESC twice or right-clicking and then selecting Done from the menu.
2. Move the cursor over the dimension until the move symbol appears as shown in Figure 2.63.
3. With the left mouse button depressed, move the dimension or value to a new location and release the button.

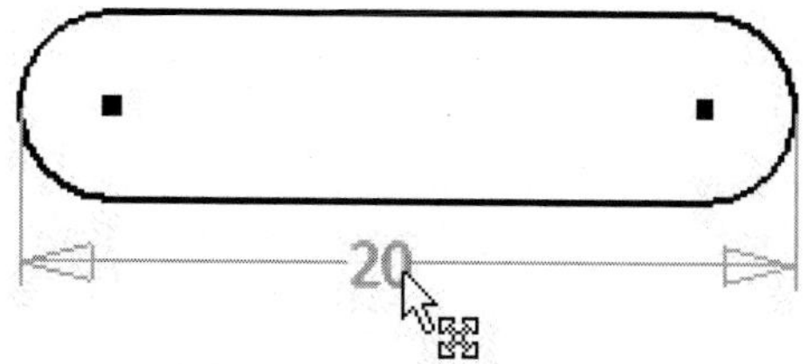

FIGURE 2.63

Fully Constrained Sketch

As you add constraints and dimensions the number of required dimensions is decreased. When no more dimensions are needed to constrain the sketch, the number of dimensions needed section on the bottom-right of the status bar will say Fully Constrained as shown in Figure 2.64 on the left. The icon to the left of the sketch in the browser will include a pushpin when the sketch is fully constrained as shown in Figure 2.64 on the right.

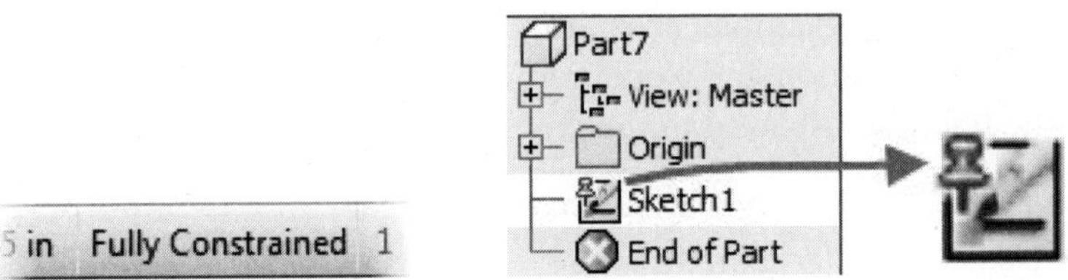

FIGURE 2.64

Over Constrained Sketch

As explained in the "Adding Constraints" section, Autodesk Inventor will not allow you to over-constrain a sketch or add duplicate constraints. The same is true when adding dimensions. If you add a dimension that will conflict with another constraint or dimension, you will be warned that this dimension will over-constrain the sketch or that it already exists. You can either cancel the operation and no dimension will be placed, or accept the warning and a driven dimension will be created.

A driven dimension is a reference dimension. It is not a parametric dimension—it reflects the size of the points to which it is dimensioned, if the part changes the driven dimension will be updated to reflect the new value. A driven dimension will appear with parentheses around the dimensions value—for example, (2.500). When you place a dimension that will over-constrain a sketch, a dialog box will appear similar to the one in Figure 2.65.

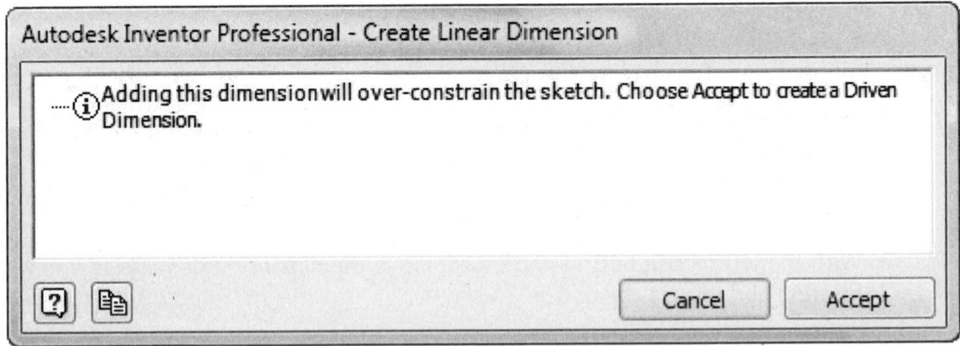

FIGURE 2.65

Another option for controlling the type of dimension that you create is to use the Driven Dimension command on the Format panel. If you issue the Driven Dimension command, any dimension you create will be a driven dimension. If you do not issue the command, you create a regular dimension. Figure 2.66 shows the Driven Dimension command on the Format panel in its normal condition.

The same Driven Dimension command can be used to change an existing dimension to either a normal or driven dimension by selecting the dimension and clicking the Driven Dimension command. Driven dimensions are represented in the sketch with parenthesis around the text as shown in Figure 2.66 on the right, whereas parametric dimensions do not have parenthesis around the text.

TIP

Avoid over using driven dimensions as they do not parametrically control the size of the profile, and reference dimensions are used only in the manufacturing process to verify the size of the part.

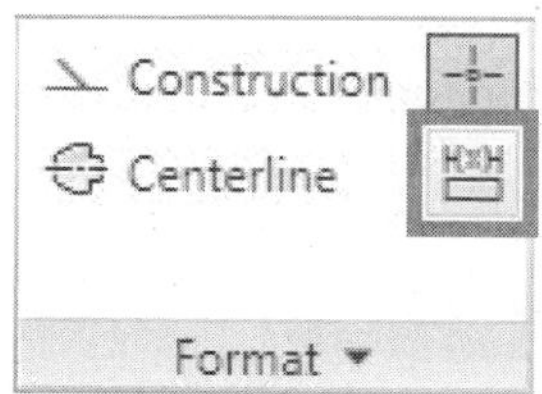

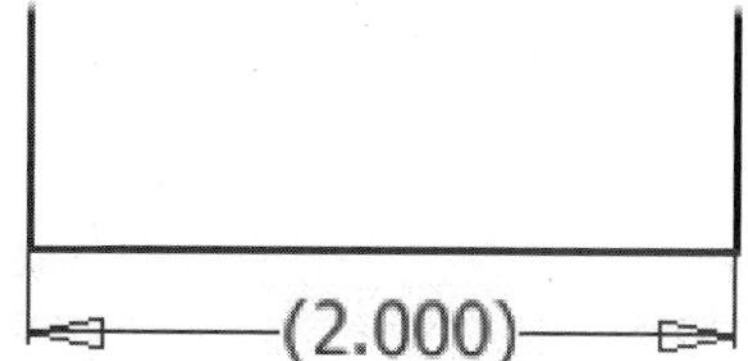

FIGURE 2.66

AUTO DIMENSION

At times when trying to fully constrain a part, it can be difficult to find the missing constraint or dimension. To aid this process, you can use the Auto Dimension command to create or remove dimensions or to add constraints to selected geometry automatically. Before using the Auto Dimension command, you should apply critical constraints and dimensions using the appropriate sketch constraint and the General Dimension command. The Auto Dimension command will not override or replace any existing constraints or dimensions. Click the Auto Dimension command on the Sketch tab > Constrain panel as shown in Figure 2.67 on the left. The Auto Dimension dialog box will appear as shown on the right.

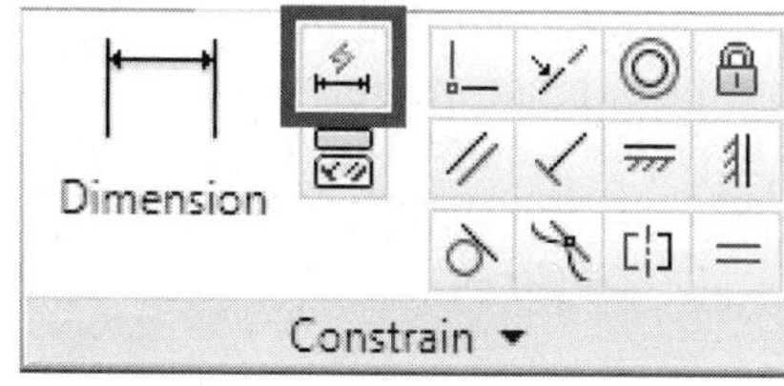

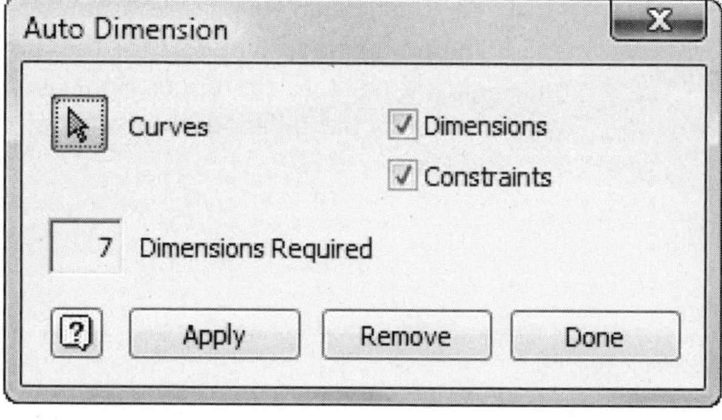

FIGURE 2.67

To use the Auto Dimension command, follow these steps:

1. Click the Auto Dimension command on the Sketch tab > Constrain panel.
2. The number of constraints and dimensions required to fully constrain the sketch appear in the lower-left corner of the dialog box.

3. Determine if you want to create dimensions or constraints or remove the dimensions or constraints that you previously added using the Auto Dimension command.
4. Click the Dimensions box and/or the Constraints box.
5. Click the Curves option, and then select the objects with which to work in the graphics window.
6. Click the Apply button to create the dimensions and/or apply constraints to the selected curves, or click the Remove button to delete the selected dimensions and/or constraints.
7. If you clicked Apply, you can change the values of the dimensions that you placed by double-clicking on the dimension text and entering in a new value in the Edit Dimension dialog box.
8. After the dimensions are placed, you can change their values by double-clicking on the numbers and entering in a new value.

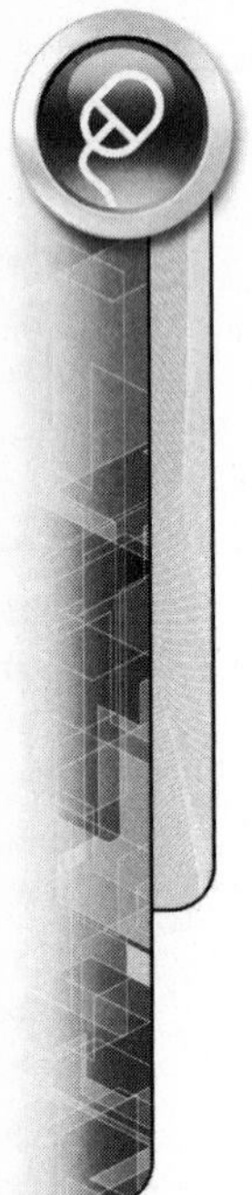

EXERCISE 2-4: CONSTRAINING AND DIMENSIONING A SKETCH

In this exercise, you add dimensional constraints to a sketch. Note: this exercise assumes that the "Edit dimension when created" and "Autoproject part origin on sketch create" options are checked in the Options dialog box under the Sketch tab. Experiment with Autodesk Inventor's color schemes to see how the sketch objects changes color to represent if they are constrained.

1. Click the New command and click the English folder, and then double-click *Standard (in).ipt*.
2. Click the Create 2D Sketch command on the 3D Model tab > Sketch panel and then select the XY origin plane.
3. Start sketching by starting the line command in the Sketch tab > Draw panel and click the origin point, move the cursor to the right, type **5** in the distance input field, press the Tab key, and then click a point when the horizontal constraint is previewed above the input field for degrees as shown in Figure 2.68 on the left.
4. Next you place an angle line and a dynamic dimension to define the angle. Press the tab key and type **150** for the angle input field, press the Tab key, and then click a point to the upper right as shown in Figure 2.68 on the right. The distance should be about 2 inches but the dimension is not needed to define this sketch.

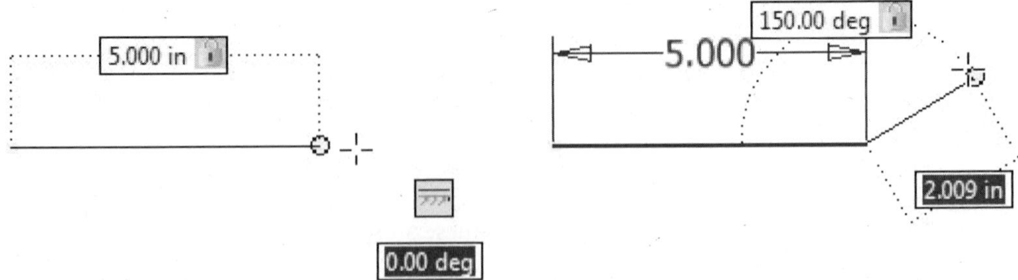

FIGURE 2.68

5. Sketch the geometry as shown in Figure 2.69. When sketching, ensure that a perpendicular constraint is not applied between the two angled lines. Hold down the CTRL key while sketching the top angle line to prevent sketch constraints from being applied. The arc should be tangent to both adjacent lines.

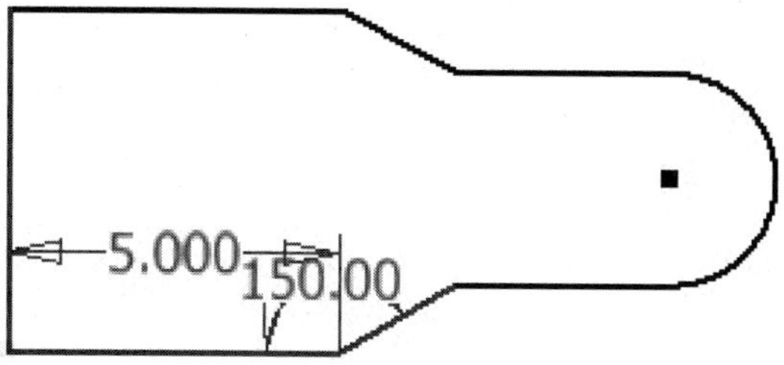

FIGURE 2.69

6. Add a horizontal constraint between the midpoint of the left vertical line and the center of the arc by clicking the Horizontal Constraint command from the Sketch tab > Constrain panel and select the midpoint of the line and the center point of the arc as shown in Figure 2.70 on the left.
7. Start the Vertical Constraint command from the Sketch tab > Constrain panel and add a vertical constraint between the endpoints of the angled lines nearest to the right side of the sketch as shown in Figure 2.70 on the right.

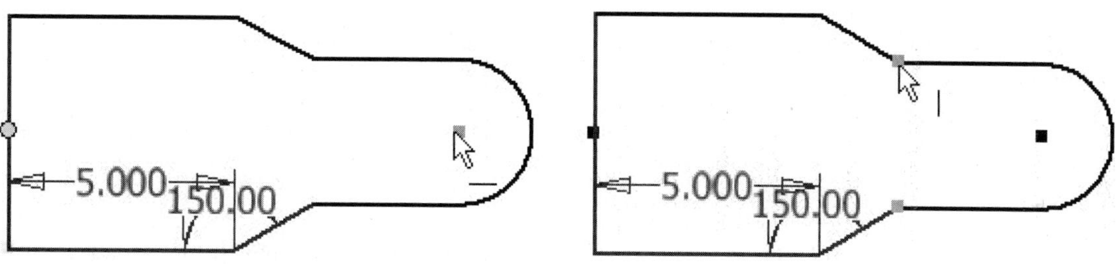

FIGURE 2.70

8. Add an equal constraint between the two angled lines by pressing the = key and then select the two angled lines.
9. Click the General Dimension command in the Sketch tab > Constrain panel.
10. Add a radial dimension by selecting the arc. Move the cursor until the dimension is positioned near the lower right corner of the sketch and then click a point and then enter **1.5** for its value (if the Edit Dimension dialog box did not appear, double-click on the dimension and change the dimension to 1.5), and click the checkmark in the Edit Dimension dialog box.

NOTE

To set the Edit Dimension option to appear when placing sketch dimensions, right-click in the graphics area while placing a dimension and select Edit Dimension from the menu.

11. Add a vertical dimension by clicking the vertical line. Position the dimension to the left, click a point to locate it, enter **5**, and click the checkmark. When complete, your sketch should resemble Figure 2.71 on the left. Notice on the bottom right of the status bar that the sketch requires 1 dimension as shown in Figure 2.71 on the right.

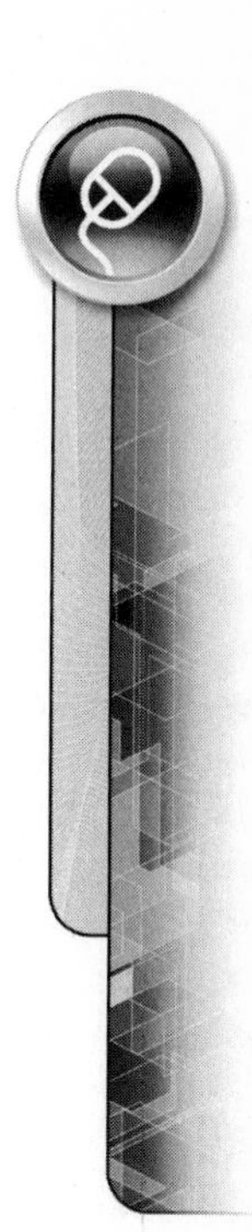

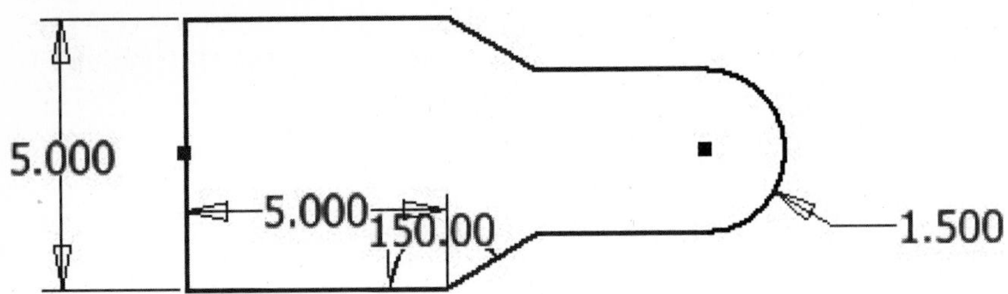

FIGURE 2.71

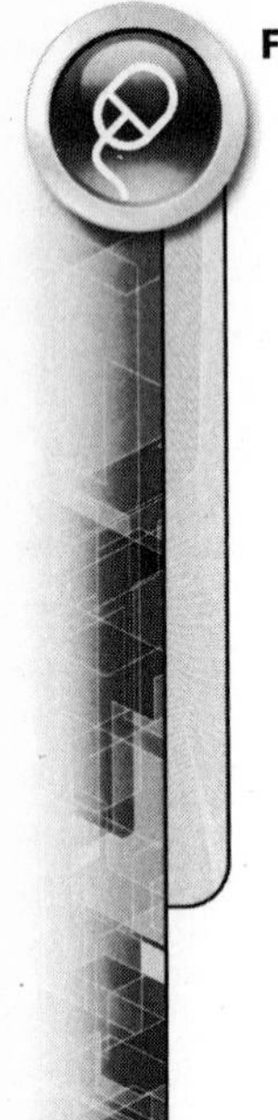

12. Add an overall horizontal dimension by clicking the vertical line (not an endpoint). Move the cursor near the right tangent point of the arc until the glyph of dimension to a circle appears, as shown in Figure 2.72 on the left. Click, drag the dimension down, click a point to locate it, enter **12**, and click the checkmark. When complete, your sketch should resemble Figure 2.72 on the right and in the status bar the, in the number of dimensions required section, it should state Fully Constrained, and the icon to the left of the sketch in the browser should have a push pin on it.

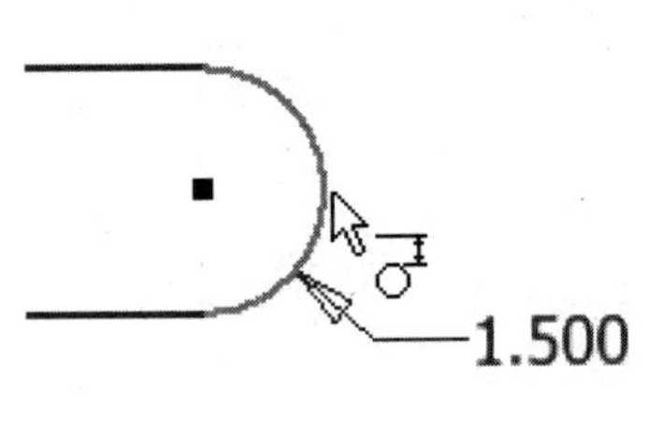

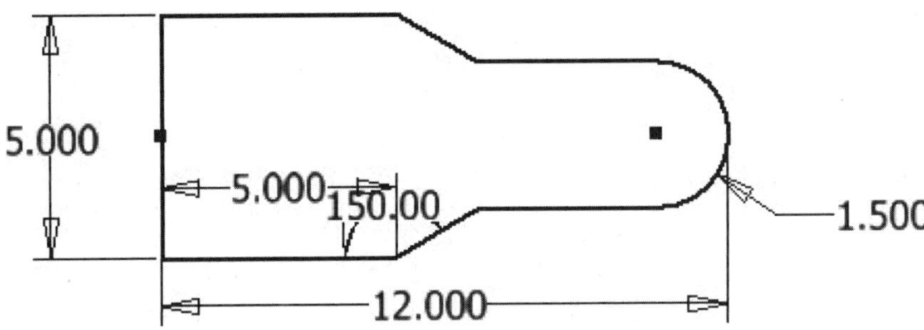

FIGURE 2.72

13. Edit the value of the dimensions by double-clicking on the dimension's value and type in a new value and then press ENTER on the keyboard or click the green check mark in the Edit Dimension dialog box, and examine how the sketch changes. The arc should always be in the middle of the vertical line. All of the dimensions are parametric, even the dimensions that were placed dynamically while sketching.
14. Delete the horizontal constraint between the center of the arc and the midpoint of the vertical line.
15. Delete the vertical constraint between the two angled lines.
16. Practice adding other types of constraints and dimensions.
17. Close the file. Do not save changes. End of exercise.

INSERTING AUTOCAD FILES

You may have legacy AutoCAD files or receive AutoCAD files that you need to convert into Inventor parts. In this section you learn how to insert AutoCAD data into a sketch. When importing a 2D DWG file into Autodesk Inventor, you can either copy the contents from the DWG file to the clipboard via Autodesk Inventor or AutoCAD and paste into Autodesk Inventor, or use an import wizard that guides you through the process. In this section you learn how to insert AutoCAD 2D data into a sketch in an Inventor part file.

AutoCAD does not need to be installed to import AutoCAD geometry into Inventor.

NOTE

INSERTING 2D AUTOCAD DATA INTO A SKETCH

In this section you learn how to insert AutoCAD 2D data into the active sketch in a part or drawing. To insert AutoCAD data into the active sketch, follow these steps:

1. Start a new part file or open an existing part file.
2. Create a new sketch or make an existing sketch active in the part file.
3. Click the Insert AutoCAD command on the Application Menu > Insert panel as shown in Figure 2.73.

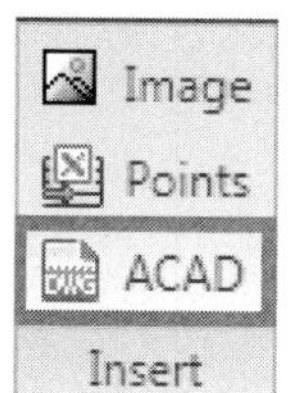

FIGURE 2.73

4. The Open dialog box will appear, browse to, and double-click the desired DWG file or click on the DWG file and then click Open. If this is a new file that has not been saved, click OK and Yes for the messages about the file not yet saved and not in the active project.

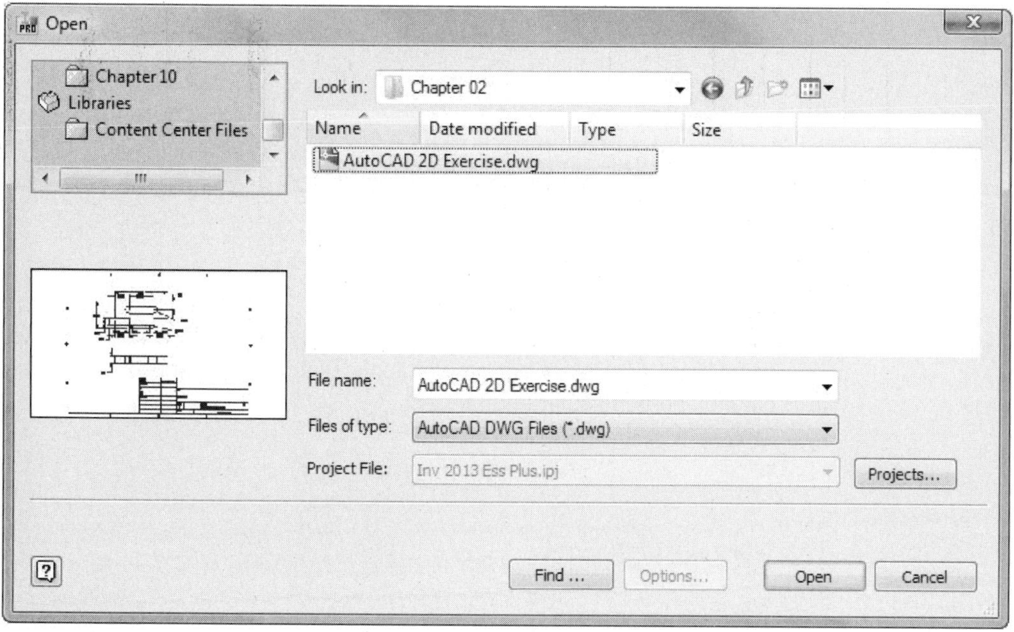

FIGURE 2.74

5. In the Selective import section, uncheck the layer names you do not want data from.
6. Click the Open button on the bottom of the dialog box. To select specific objects to insert, uncheck the All option, and then select the desired data in the preview window. In the preview window you can zoom and pan as needed.

7. You can change the background color of the preview image by clicking the black or white icon at the top of the dialog box. Import objects from Model Space or from a layout within the DWG file by clicking the tabs at the bottom of the preview window. The names of the tabs are identical to the tab names in the AutoCAD file. Click the Next button to go to the next step.

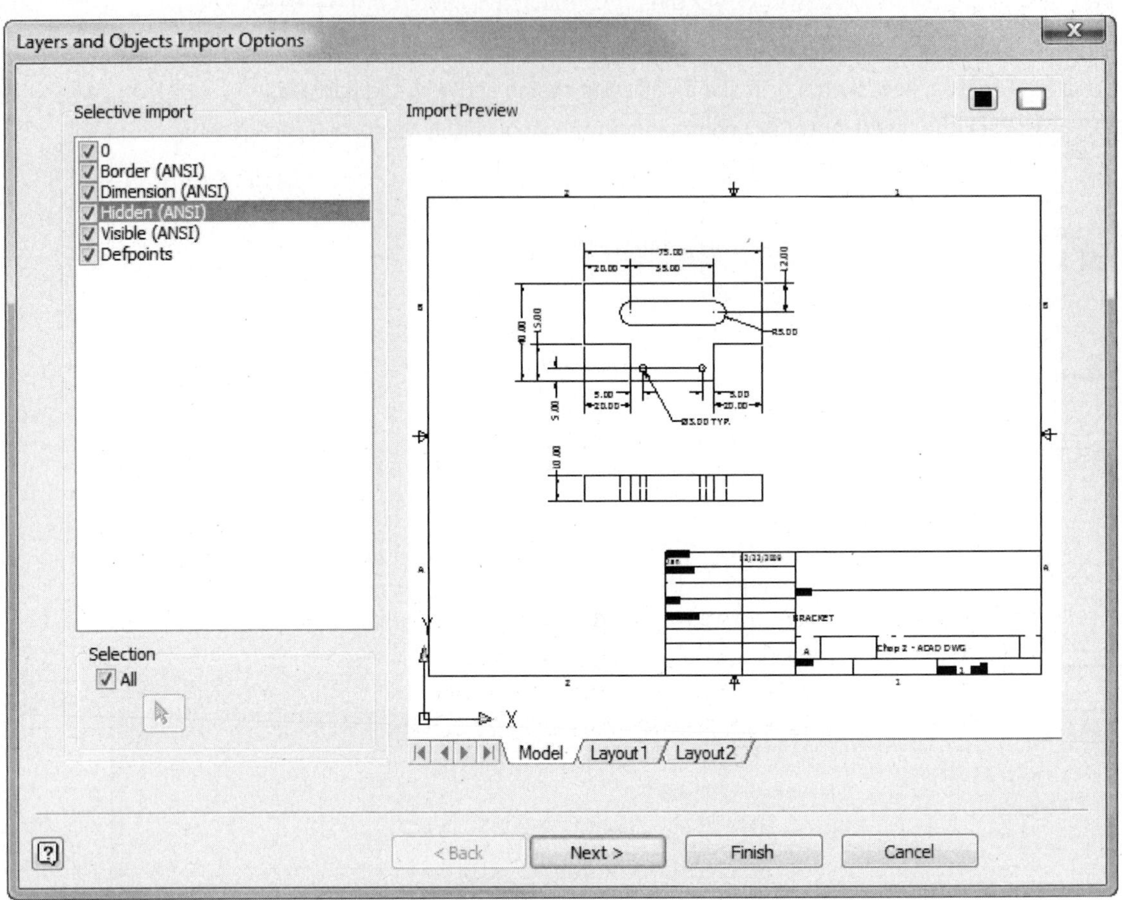

FIGURE 2.75

8. Specify the units in which the data was created.
9. Check the options to Constrain End Points and Apply geometric constraints. Applying geometric constraints option will apply sketch constraints to geometry that is parallel, perpendicular and tangent.

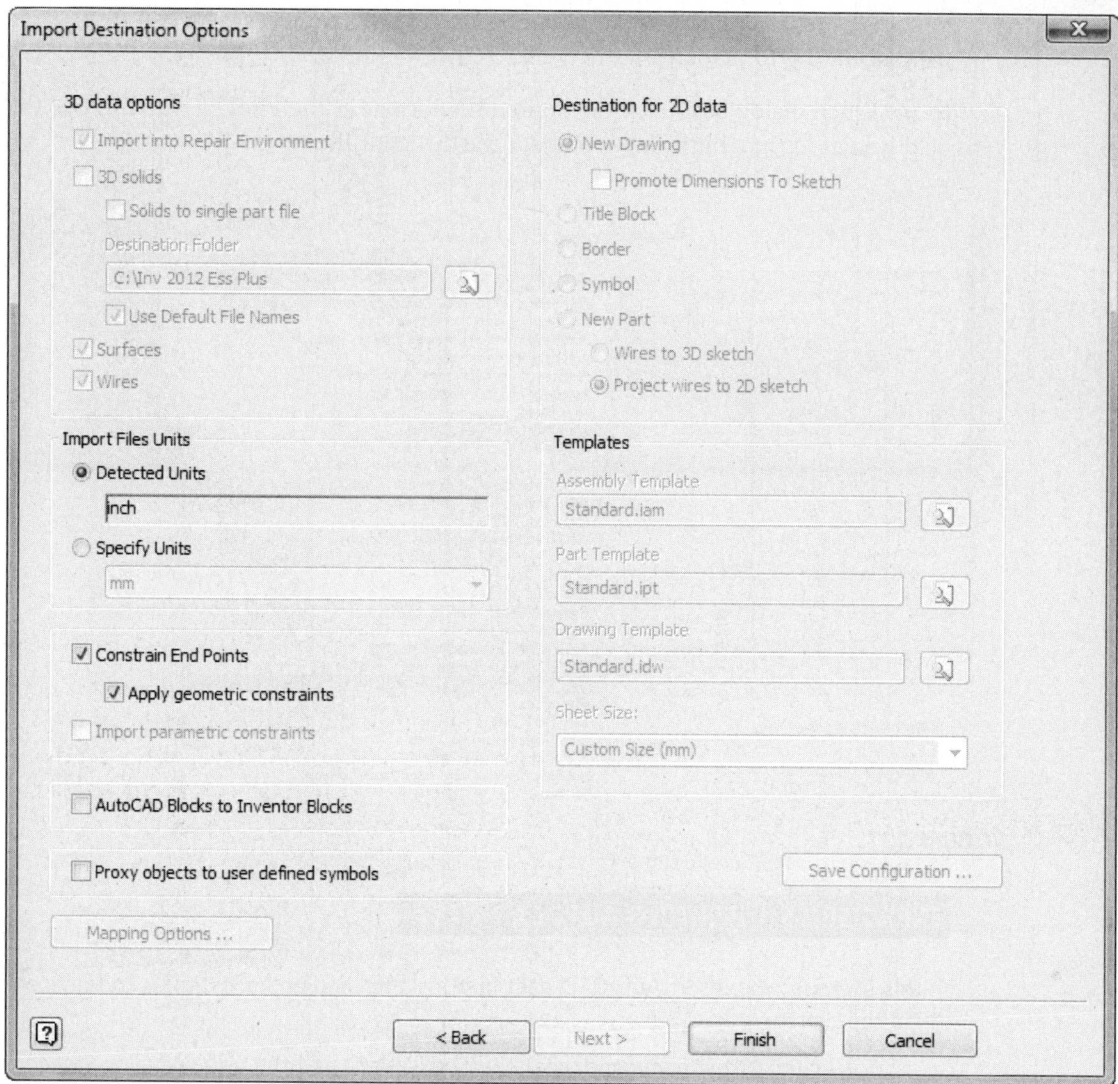

FIGURE 2.76

10. To complete the operation, click Finish.
11. Delete any unnecessary geometry, constraints and dimensions.
12. Add geometry (if needed), constraints and dimensions to fully constrain the sketch.

OPEN OTHER FILE TYPES

Autodesk Inventor can also open parts and assemblies exported from other CAD systems. Autodesk Inventor models created from these formats are base solids or surface models, and no feature histories or assembly constraints are generated when you import a file in any of these formats. You can add features to imported parts, edit the base solids using Autodesk Inventor's solids editing commands or Inventor Fusion, and add assembly constraints to the imported components. To open file types such as DWF (markup files), DXF, Alias, Catia, IDF Board File IGES, JT, Parasolids, PRO/E, SAT, STEP, SolidWorks, and Unigraphics NX, click Inventor Application Menu > Open or click Open on the Quick Access toolbar. You can also use the Import DWG command from the Application Menu > Open; this command will import AutoCAD data into a new drawing, title block, border, symbol or part file without having to first create a new file.

In the Open dialog box, click the desired file format in the Files of type list. See the help system for more information about the different file types.

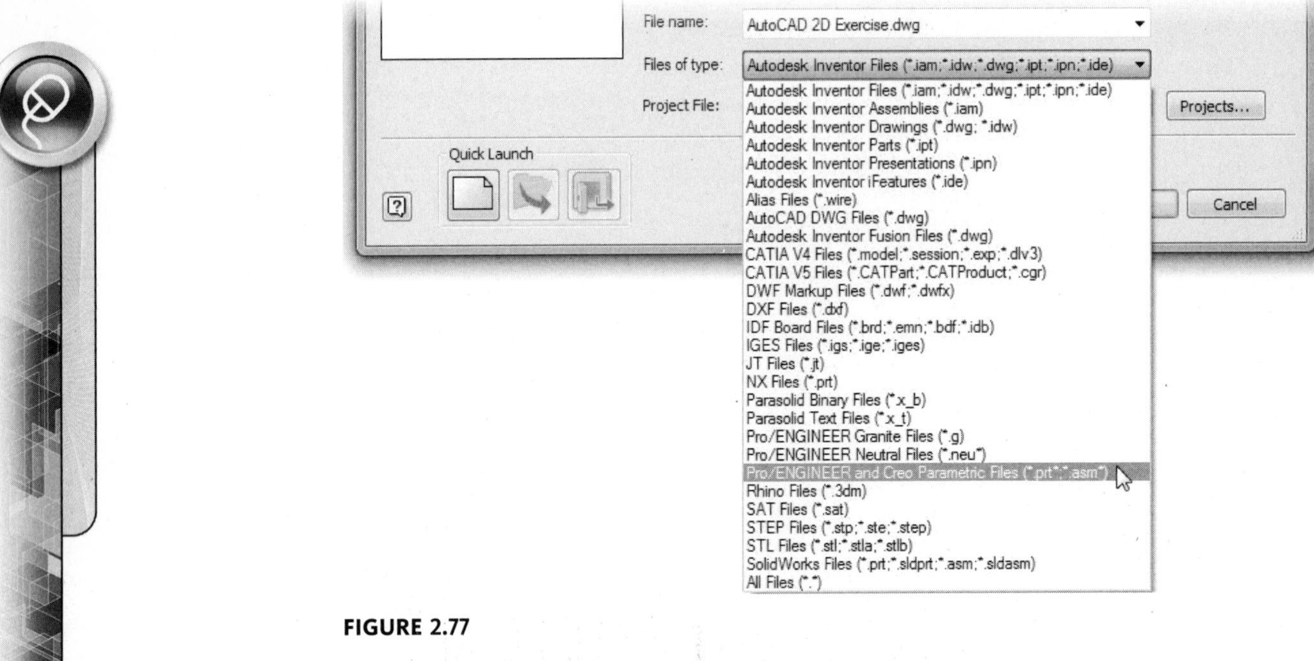

FIGURE 2.77

EXERCISE 2-5: INSERTING AUTOCAD DATA

In this exercise, you insert AutoCAD data into a sketch and add constraints to fully constrain the sketch.

1. Click the New command, click the English tab, and then double-click *Standard (in).ipt*, or if inch is the default unit; from the left side of the Quick Access toolbar you can click the down arrow of the New icon, and select Part.
2. Click the Create 2D Sketch command on the 3D Model tab > Sketch panel and then select the XY origin plane.

TIP

Click the Chapter 02 subfolder from the Frequently Used Subfolder area and then click the file in the file area.

3. Click the Insert AutoCAD command for the Sketch tab > Insert panel.

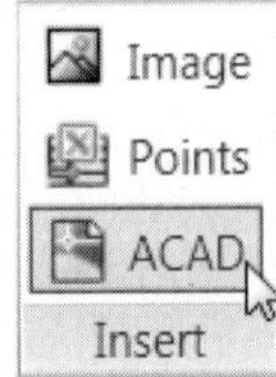

FIGURE 2.78

4. From the Frequently Used Subfolder area (upper left corner of the dialog box) click the Chapter 02 subfolder and then in the file area double-click on the file *AutoCAD 2D Bracket.dwg*

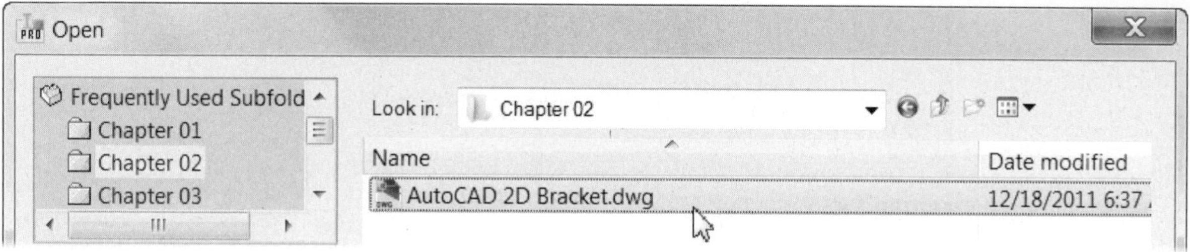

FIGURE 2.79

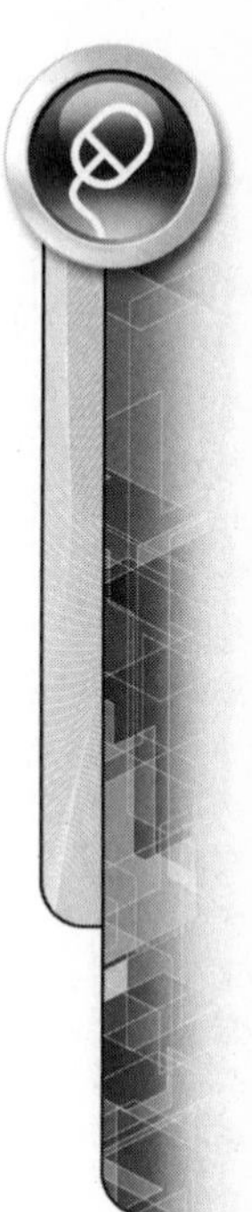

5. In the Layers and Objects dialog box you uncheck the layers that are not needed. In the upper left corner corner of the dialog box uncheck layers; 0, Border (ANSI) and Hidden (ANSI) as shown in Figure 2.80 labeled (1).
6. In the Selection area near the bottom left corner of the dialog box uncheck All, labeled (2) in Figure 2.80.
7. Select the geometry and dimensions to insert, use the window selection (click and drag from the left to right) technique to select the geometry and dimensions in the top view labeled (3) in Figure 2.80. There should should be 27 total objects selected as labeled (2) in Figure 2.80.

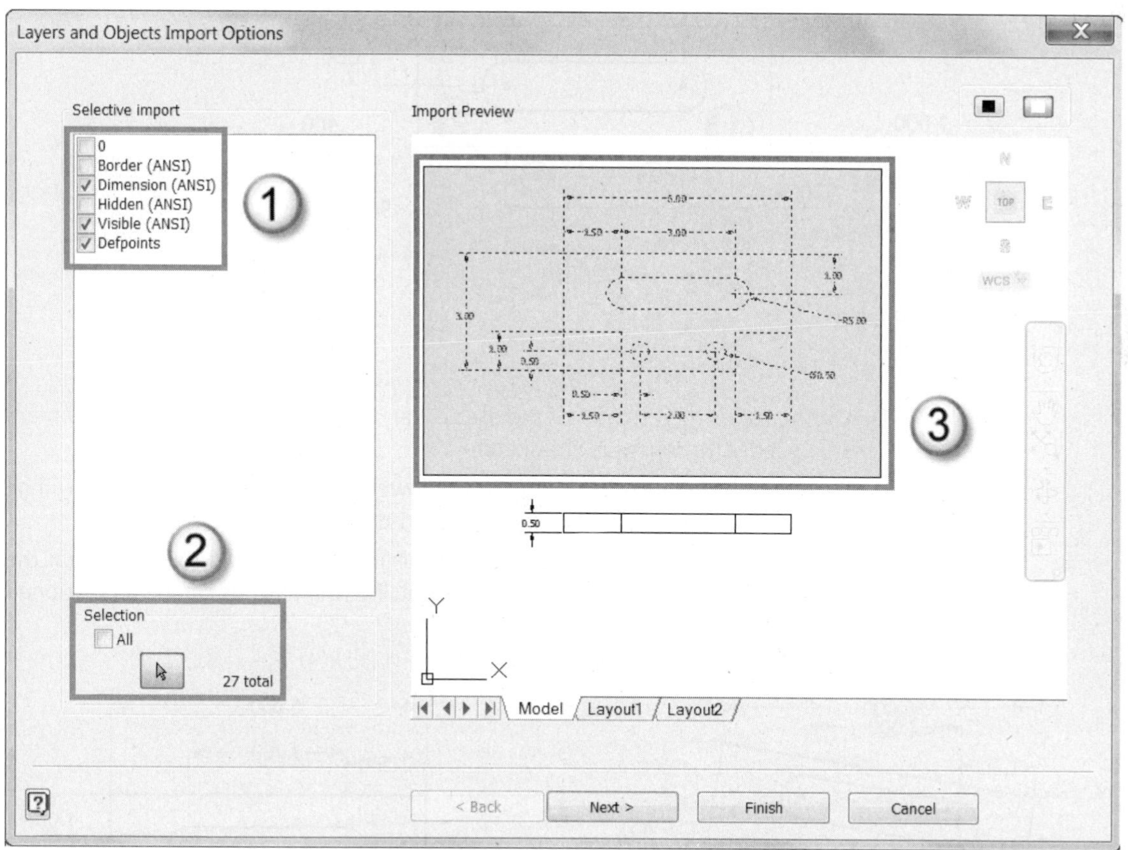

FIGURE 2.80

8. Click the Next button on the bottom of the dialog box.
9. In the Import Designation Dialog box check the Constrain End Points and the Apply geometric constraints on the lower left corner of the dialog box as shown in Figure 2.81.

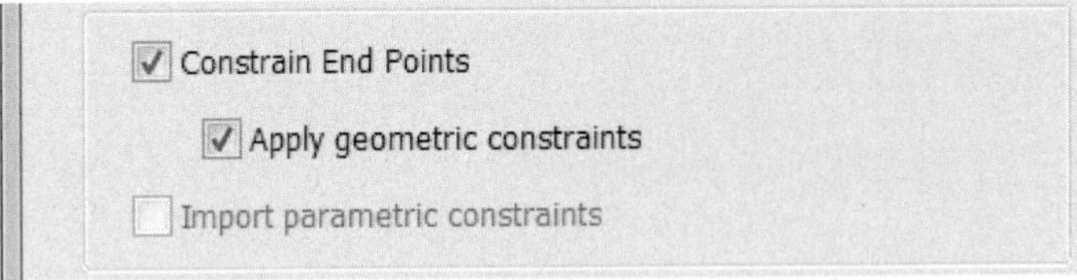

FIGURE 2.81

10. Click the Finish button on the bottom of the dialog box.
11. Apply a horizontal constraint between the center points of the two circles.
12. Apply a collinear constraint between the two middle horizontal lines labeled (1) and (2) as shown in Figure 2.82 on the left.
13. Press the Esc key twice to cancel the command.
14. The sketch is free to move. To constrain the sketch, drag the lower-left corner of the sketch labeled (3) in Figure 2.82 to the origin point of the sketch (0,0). Or you could add a coincident constraint between the endpoint and the origin point.

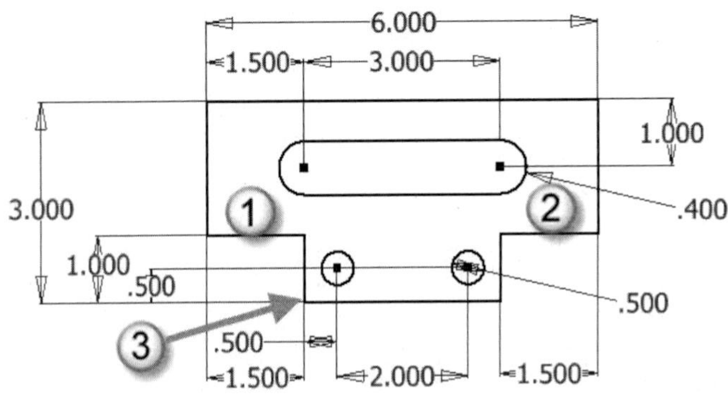

FIGURE 2.82

15. On the lower-right corner of the Status Bar, the text should state that 1 Dimension needed to constrain the sketch.
16. Drag up the right-side endpoint on the lower horizontal line; the sketch will be rotated slightly as shown in Figure 2.83 on the left.
17. Apply a horizontal constraint to the lower line, and this will fully constrain the sketch as shown in Figure 2.83 on the right. The dimensions can be repositioned as needed.

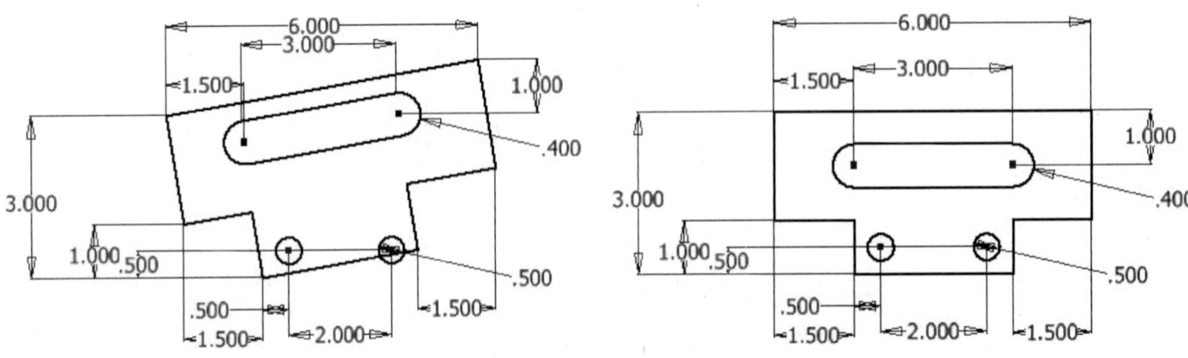

FIGURE 2.83

18. Click the F8 key to see all the constraints.
19. Click the F9 key to hide all the constraints.
20. The AutoCAD dimensions on the sketch are now parametric and can be edited. Practice editing the values of the dimensions by double-clicking on a dimension's value and enter a new value.
21. Close the file. Do not save changes. End of exercise.

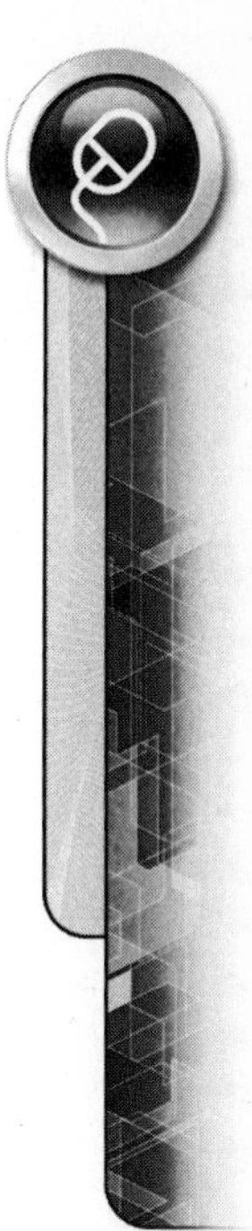

APPLYING YOUR SKILLS

Skill Exercise 2-1

In this exercise, you create a sketch and then add geometric and dimensional constraints to control the size and shape of the sketch. Start a new part file based on the *Standard (in).ipt*, create a sketch on the XY plane, and create the fully constrained sketch as shown in Figure 2.84. Assume that the top and bottom horizontal lines are collinear, the center points of the arcs are aligned vertically, and the sketch is symmetric about the left and right sides. One of the bottom angled lines should be coincident with the center point of the lower arc (if the arc is drawn via the line command, the center point of the arc will automatically be coincident with the line it was drawn from). When done, close the file and do not save the changes.

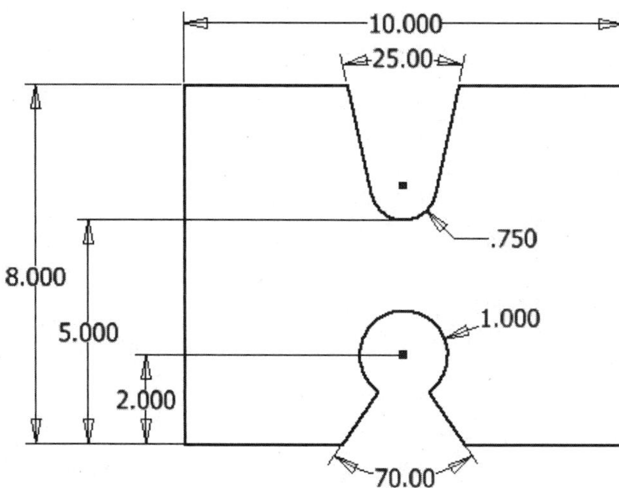

FIGURE 2.84

Skill Exercise 2-2

In this exercise, you create a sketch with linear and arc shapes, and then add geometric and dimensional constraints to fully constrain the sketch. Start a new part file based on the ***Standard (in).ipt***, create a sketch on the XY plane, and create the fully constrained sketch as shown in Figure 2.85. First create the two outer circles and align their center points horizontally. Then create the two lines, and place a vertical constraint between the line endpoints on both ends. When done, close the file and do not save the changes.

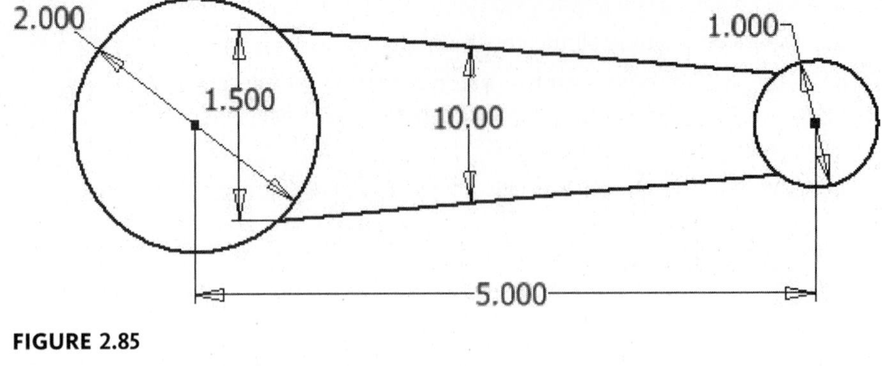

FIGURE 2.85

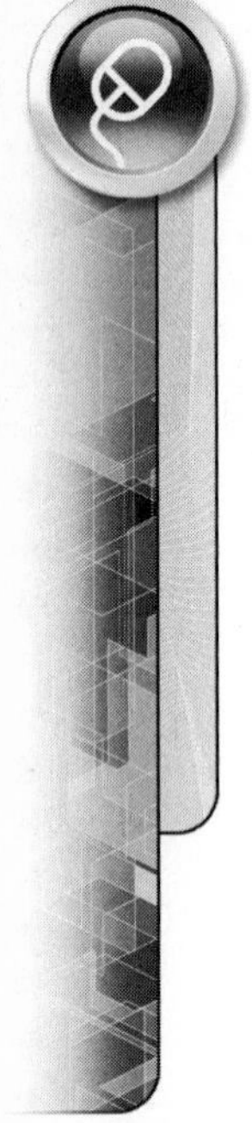

CHECKING YOUR SKILLS

Use these questions to test your knowledge of the material in this chapter.

1. True __ False__ When you sketch, constraints are not applied to the sketch by default.
2. True __ False __ When you sketch and a point is inferred, a constraint is applied to represent that relationship.
3. True __ False __ A sketch does not need to be fully constrained.
4. True __ False __ When working on an mm part, you cannot use English (inch) units.
5. True __ False __ After a sketch is constrained fully, you cannot change a dimension's value.
6. True __ False __ A driven dimension is another name for a parametric dimension.
7. True __ False __ Dimensions placed dynamically are not parametric.
8. True __ False __ You can import only 2D AutoCAD data into Autodesk Inventor.
9. Explain how to draw an arc while in the Line command.
10. Explain how to remove a geometric constraint from a sketch.
11. Explain how to change a vertical dimension to an aligned dimension while you place the dimension.
12. Explain how to create a dimension between two quadrants of two arcs.
13. True __ False __ AutoCAD needs to be installed to insert AutoCAD geometry.
14. True __ False __ When a sketch is extruded that contains construction geometry, the construction geometry is deleted.
15. Explain how to change the unit type in a part file.
16. True __ False __ To scale the entire sketch when placing the first dimension, you must right-click and check Scale from the menu.
17. True __ False __ When a Pushpin appears in the Sketch icon in the browser, the sketch is fully constrained.
18. True __ False __ An arc length dimension can only be a driven dimension.
19. Explain how to draw a rectangle that is centered on the origin point.
20. True __ False __ When creating the first 2D sketch, you must select an origin plane from the browser's Origin folder.

CHAPTER

3

Creating and Editing Sketched Features

INTRODUCTION

After you have drawn, constrained, and dimensioned a sketch, your next step is to turn the sketch into a 3D part. This chapter takes you through the process to create and edit sketched features and to create features using primitive shapes.

OBJECTIVES

After completing this chapter, you will be able to perform the following:

- Describe what a feature is used for in the modeling process
- Describe the functions of Autodesk Inventor's browser
- Use direct manipulation techniques to create and edit a part
- Extrude a sketch into a part
- Revolve a sketch into a part
- Edit features of a part
- Edit the sketch of a feature
- Create a sketch on a plane
- Create sketched features using one of three operations: cut, join, or intersect
- Project edges of a part
- Create primitive features

FEATURES

After creating, constraining, and dimensioning a sketch, the next step in creating a model is to turn the sketch into a 3D feature. The first sketch of a part that is used to create a 3D feature is referred to as the base feature. In addition to the base feature, you can create sketched features in which you draw a sketch on a planar face or work plane, and you can either add or subtract material to or from existing features in a part. Use the Extrude, Revolve, Sweep, or Loft commands to create sketched features in a part. You can also create placed features such as fillets, chamfers, and holes by applying them to features that have been created. Placed features will be covered in Chapter 4. Features are the building blocks in creating parts.

A plate with a hole in it, for example, would have a base feature representing the plate and a hole feature representing the hole. As features are added to the part, they appear in the browser, and the history of the part or assembly, that is, the order in which the features are created or the parts are assembled, is shown. Features can be edited, deleted from, or reordered in the part as required.

Consumed and Unconsumed Sketches

You can use any sketch as a profile in feature creation. A sketch that has not yet been used in a feature is called an unconsumed sketch. When you turn a 2D sketched profile into a 3D feature, the feature consumes the sketch. Figure 3.1 shows an unconsumed sketch in the browser on the left and a consumed sketch in the browser on the right. If desired, you can display more information in the browser about the features by clicking on the filter icon in the browser and click Show extended Names.

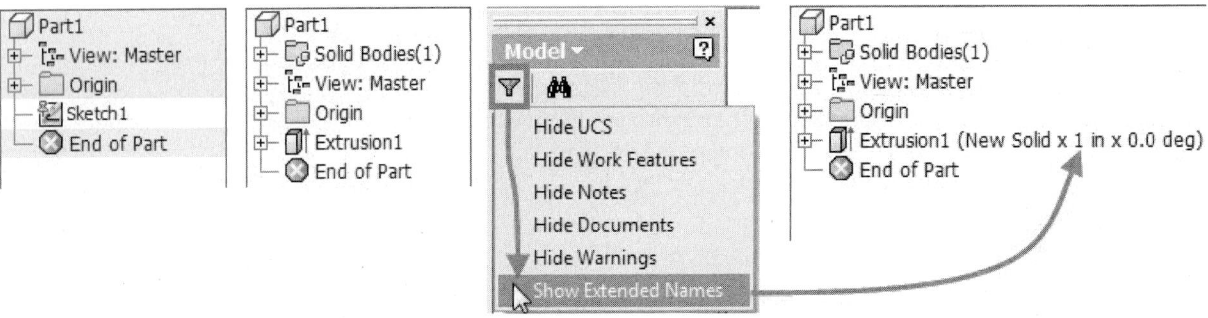

FIGURE 3.1

Although a consumed sketch is not visible as you view the 3D feature, you may need to access sketches and change their geometric or dimensional constraints in order to modify their associated features. A consumed sketch can be accessed from the browser by right-clicking and selecting Edit Sketch from the menu or by using the direct-manipulation mini-toolbar. The editing process will be covered later in this chapter. You may also access the sketch by starting the Sketch command and selecting the sketch from the browser. Figure 3.2 on the left shows the unconsumed sketch, and the image on the right shows the extruded solid that consumes the sketch.

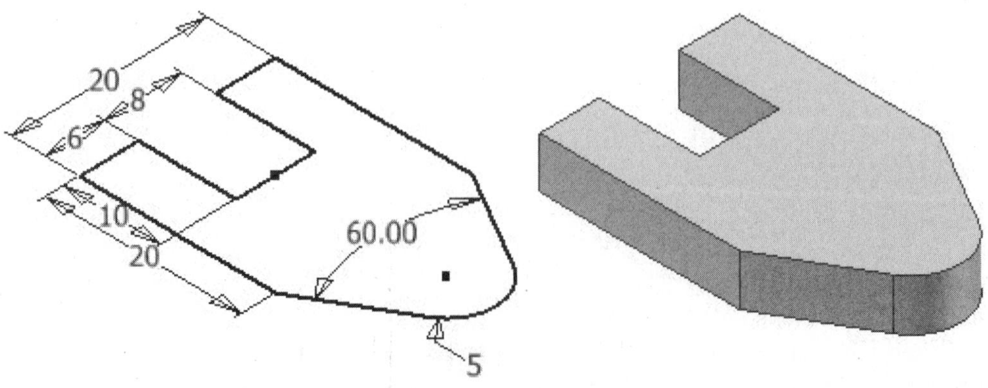

FIGURE 3.2

UNDERSTANDING THE BROWSER

The Autodesk Inventor browser, by default, is docked along the left side of the screen and displays the history of the file. In the browser you can create, edit, rename, copy, delete, and reorder features or parts. You can expand or collapse the browser to display the history of the features (the order in which the features were created) by clicking the + and – on the left side of the part or feature name in the browser. An alternate method of expanding the browser is to place your cursor on top of a feature icon but not click. The item in the browser automatically expands. To expand all the features, move the cursor into a blank area in the browser, right-click, and click Expand All from the menu.

Figure 3.3 shows a browser with features of the part expanded.

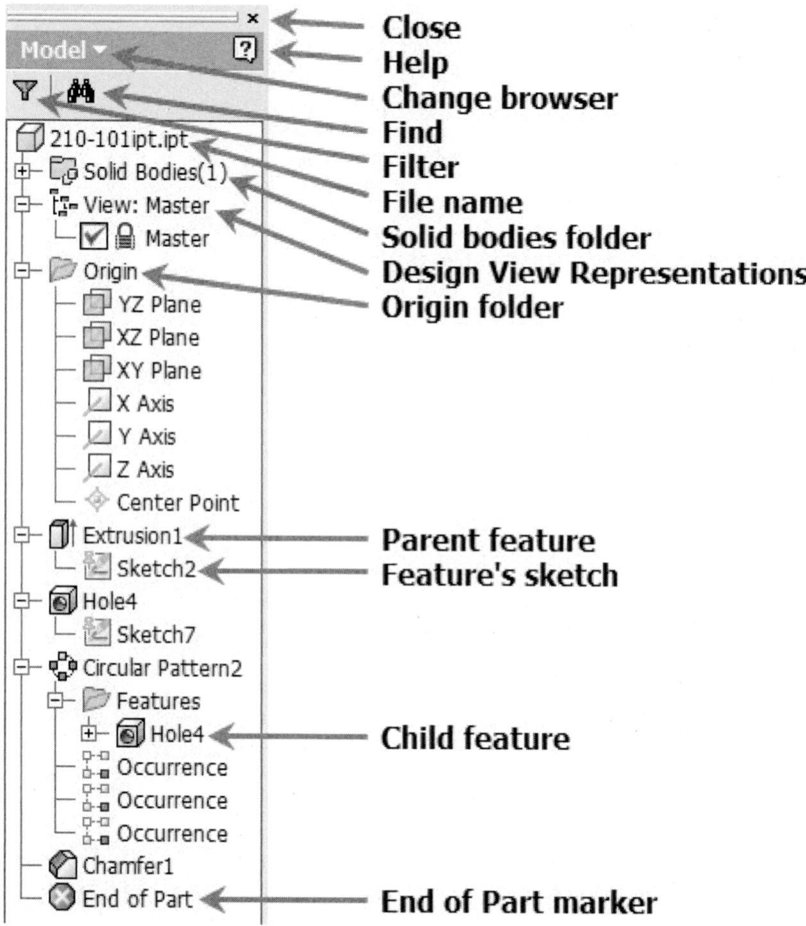

FIGURE 3.3

As parts grow in complexity, so will the information found in the browser. Dependent features are indented to show that they relate to the item listed above it. This is referred to as a *parent-child relationship*. If a hole is created in an extruded rectangle, for example, and the extrusion is deleted, the hole will also be deleted.

Each feature in the browser is given a default name. The first extrusion, for example, will be named Extrusion1, and the number in the name will sequence as you add similar features. The browser can also help you to locate features in the graphics area. To highlight a feature in the graphics window, simply move your cursor over the feature name in the browser.

To zoom in on a selected feature, right-click on the feature's name in the browser and select Find in Window on the menu or press the END key on your keyboard. The browser itself functions similarly to a toolbar except that you can resize it while it is docked. To close the browser, click the X in its upper-right corner. If the browser is not visible on the screen, you can display it by clicking the View tab > Windows panel > User Interface > Browser as shown in Figure 3.4.

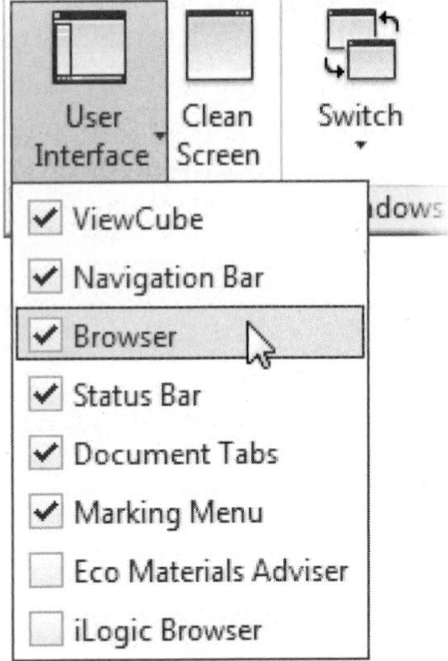

FIGURE 3.4

Specific functionality of the browser will be covered throughout this book in the pertinent sections. A basic rule is to either right-click or double-click the feature's name or icon to edit or perform a function on the feature.

SWITCHING ENVIRONMENTS

Up to this point, you have been working in the sketch environment where the work is done in 2D. The next step is to turn the sketch into a feature. To do so, you need to exit the *sketch environment* and enter the *part environment*. A number of methods can be used to accomplish this transition:

- Click the Finish Sketch command on the right side of the Sketch tab > Exit panel as shown in Figure 3.5 on the left.
- In a blank area in the graphics window right-click and click Finish Sketch from the menu as shown in the middle of Figure 3.5.
- Press the S key on the keyboard.
- Click the 3D Model tab as shown in Figure 3.5 on the right and click on a command.
- Enter a shortcut key to initiate one of the feature commands.

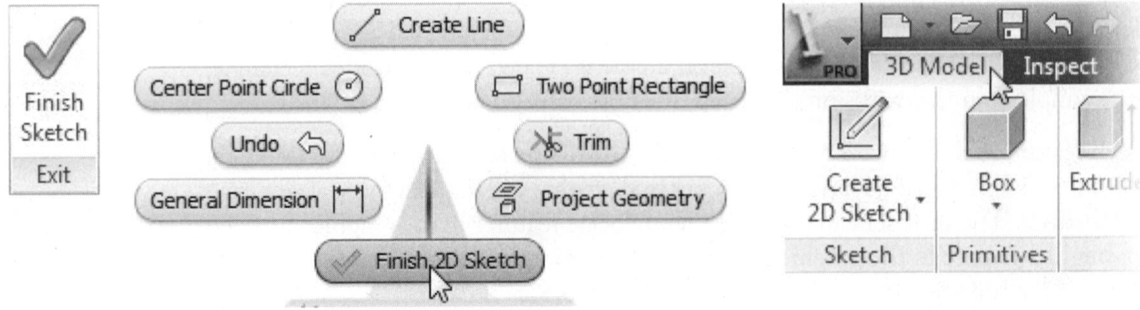

FIGURE 3.5

MODEL COMMANDS

When you exit the sketch environment, the Model Ribbon is active. Figure 3.6 shows the 3D Model tab. Many of these commands will be covered throughout this book.

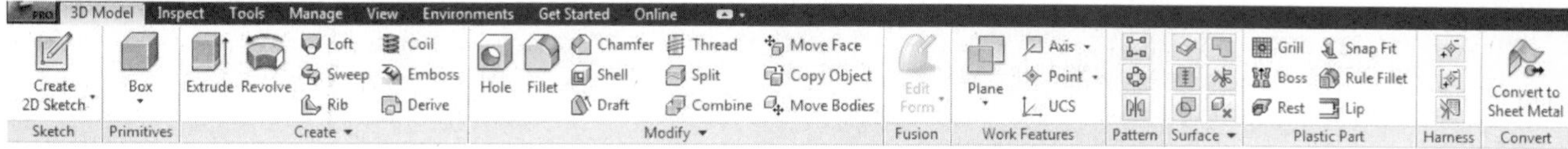

FIGURE 3.6

DIRECT MANIPULATION

Direct manipulation allows you to start common commands and operations by clicking directly on the geometry in the graphics window. The direct manipulation tools appear as mini-toolbars or in-canvas buttons. Each direct manipulation option is explained next.

Mini-Toolbars

While not in a command you have different options depending upon what geometry is selected. After selecting the geometry click on the desired option. The different options will be explained throughout the book as they pertain to the topic.

Sketch

While not in a command, not in the sketch environment, and you select a sketch, the mini-toolbar appears and provides you the option of performing Extrude, Revolve, Hole, or Edit Sketch operations.

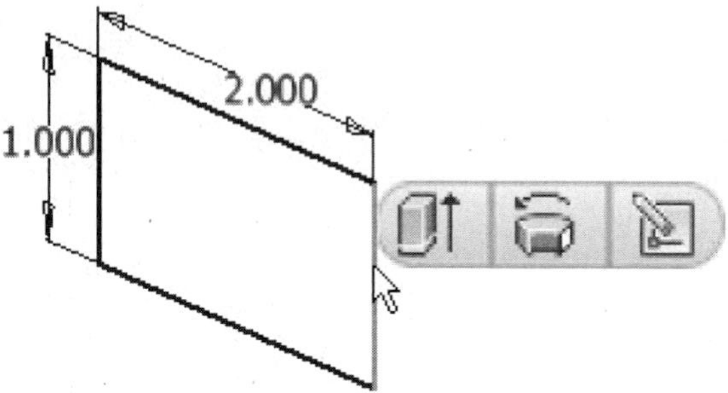

FIGURE 3.7

Face

While not in a command and you select a face, a mini-toolbar appears that allows you to quickly create a new sketch, edit an existing sketch, or edit a feature.

FIGURE 3.8

Edge

While not in a command and you select an edge, the mini-toolbar appears and provides you the option of performing a fillet or chamfer operation on the selected edge.

FIGURE 3.9

Mini-Toolbar–Command Options

After starting selected commands, the mini-toolbar will display command options in the graphics screen next to a selected object. The min-toolbar options are presented via buttons and selection tags. The in-canvas buttons and selection tags display command options and prompts such as profile, face, and axis. Min-toolbar display is enabled for Extrude, Revolve, Fillet, Chamfer, Hole, Face Draft and Shell commands. Figure 3.10 shows the min-toolbar buttons and the selection tag (profile) when extruding the first profile. You can move the mini-toolbar by clicking and dragging the vertical obround button in the upper left corner of the mini-toolbar. There are options to pin and fade the mini-toolbar by clicking on the options button on the bottom-right of the mini-toolbar as shown in Figure 3.10. When the Pin Mini-Toolbar Position option is unchecked the mini-toolbar's position will move as the arrow on the face is dragged. When checked the mini-toolbar will stay in its current position even when the arrow on the face is dragged. The Fade option will fade the mini-toolbar when the cursor moves away from the mini-toolbar. The min-toolbar display options are covered throughout the book where the command is covered.

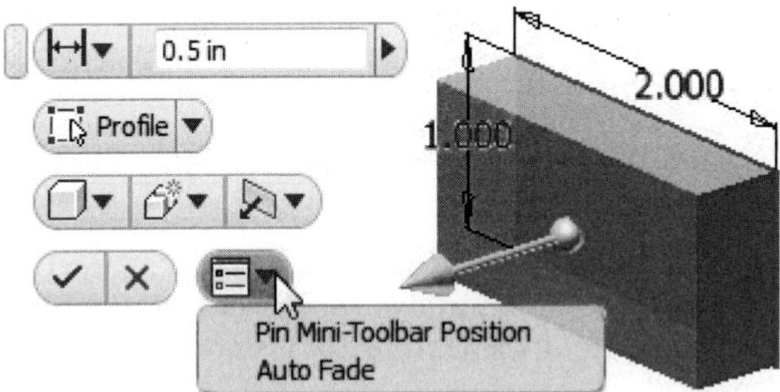

FIGURE 3.10

MINIMIZE DIALOG BOXES

For commands that have a mini-toolbar you have an option to minimize the command's dialog box. To minimize a dialog box, click the up arrow near the bottom of the dialog box as shown in Figure 3.11 on the left. The dialog box then displays only the horizontal title bar as shown in Figure 3.11 on the right. To maximize the dialog box, click the down arrow in the minimized dialog box as shown in Figure 3.11 on the right. This option is set for each dialog box.

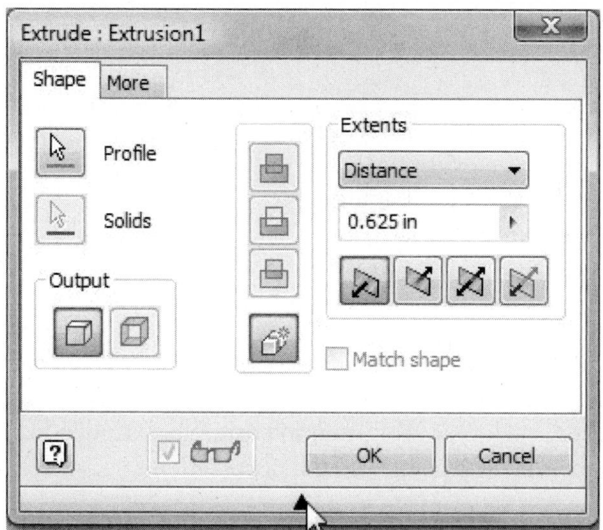

FIGURE 3.11

EXTRUDE A SKETCH

The most common method for creating a feature is to extrude a sketch and give it depth along the Z-axis. Before extruding, it is helpful to view the part in an isometric view. When you exit the first sketch the viewpoint will automatically change to the home view. Autodesk Inventor previews the extrusion depth and direction in the graphics window.

To extrude a sketch, finish the sketch, click the Extrude command on the 3D Model tab > Create panel bar as shown in Figure 3.12 on the left, while not in a sketch click on any geometry on the sketch and the mini-toolbar will appear, click the Create Extrude button as shown in Figure 3.12 in the middle, or press the shortcut key E (to use the shortcut key you can be in or out of the sketch). Alternately, you can right-click in the graphics window and select Extrude on the marking menu as shown in Figure 3.12 on the right.

FIGURE 3.12

After starting the Extrude command, the in-canvas display options will appear in the graphics window and the Extrude dialog box will appear. Figure 3.13 on the left shows the in-canvas extrude options that are available when extruding the first sketch and the image on the right shows the operation button that is available when creating additional extrude features. You can either enter data in the dialog box or via the in-canvas display. For the mini-toolbar buttons, click the down arrow next to each button to see more options and click and drag the arrow to change the distance or taper. The min-toolbar buttons will change depending upon what options are selected. If selection tags have a red arrow that command option needs to be satisfied before the command can be completed.

FIGURE 3.13

After you start the command, the Extrude dialog box also appears as shown in Figure 3.14. When you make changes in the dialog box, the shape of the sketch will change in the graphics window to represent these values and options. When you have entered the values and the options you need, click the OK button to create the extruded feature. The common extrude options will be explained in the next sections.

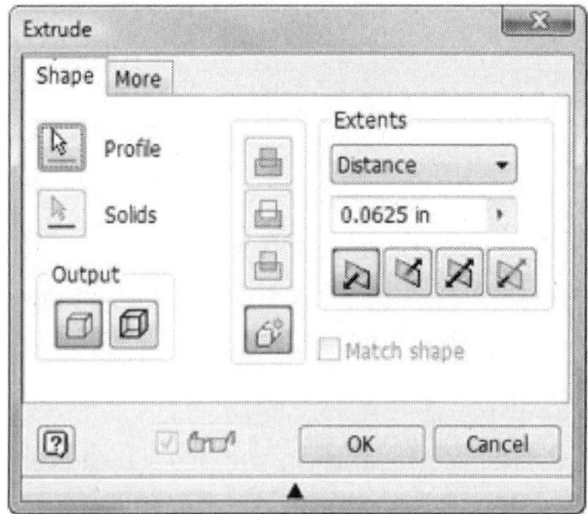

FIGURE 3.14

NOTE When an arrow is red in a dialog box or in a selection tag in the mini-toolbar area that prompts for profiles, faces, and axes, the color indicates that Autodesk Inventor needs input from you to complete that function.

Shape

The Shape tab gives you options to specify the profile to use, operation, extents, and output type. The options are described below.

	Profile	Click this button to choose the sketch that you wish to extrude. If there are multiple closed profiles, you will need to select which sketch area you want to extrude. If there is only one possible profile, Autodesk Inventor will select it for you, and you can skip this step. If you select the wrong profile or sketch area, click the Profile button again and choose the desired profile or sketch area. To remove a selected profile, hold the CTRL key and click the area you wish to remove.

Operation

This is the unlabeled middle column of buttons. If this is the first sketch that you create a solid from, it is referred to as a base feature, and only the Join button is available. The operation defaults to Join, which is the top button. Once the base feature has been established, you can extrude a sketch, adding or removing material from the part by using the Join or Cut options, or you can retain the common volume between the existing part and the newly defined extrude operation using the Intersect option.

	Join	Adds material to the part.
	Cut	Removes material from the part.
	Intersect	Removes material, keeping what is common to the existing part feature(s) and the new feature.
	New Solid	Creates a new solid body. The first solid feature created uses this option by default. Select to create a new body in a part file with an existing solid body.

Extents

The Extents option determines how the extruded sketch will be terminated. There are five options from which to choose as shown in Figure 3.15, but like the operation section, some options are not available until a base feature exists.

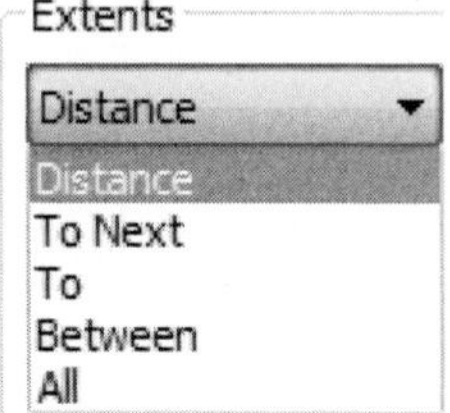

FIGURE 3.15

Distance

This option determines that the sketch will be extruded a specified distance.

To Next

This option determines that the sketch will be extruded until it reaches a plane or face. The sketch must be fully enclosed in the area to which it is projecting; if it is not fully enclosed, use the To termination with the Extend to Surface option. Click the Direction button to determine the extrusion direction.

To

This option determines that the sketch will be extruded until it reaches a selected face or plane. To select a point (midpoint or endpoint), plane, or face to end the extrusion, click the Select Surface to end the feature creation button, as shown in Figure 3.16, and then click a face or plane at which the extrusion should terminate.

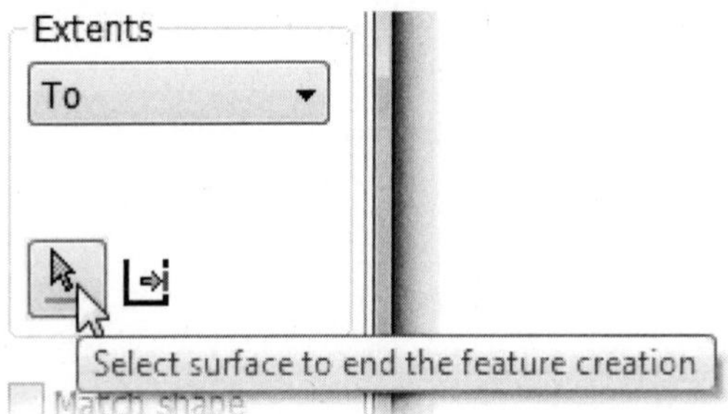

FIGURE 3.16

Between

This option determines that the extrusion will start at a selected plane or face and stop at another plane or face. Click the Select surface to start feature creation button as shown in Figure 3.17 on the left, and then click the face or plane where the extrusion will start. Then click the Select surface to end the feature creation button, as shown in Figure 3.17 on the right and then click the face or plane where the extrusion will terminate.

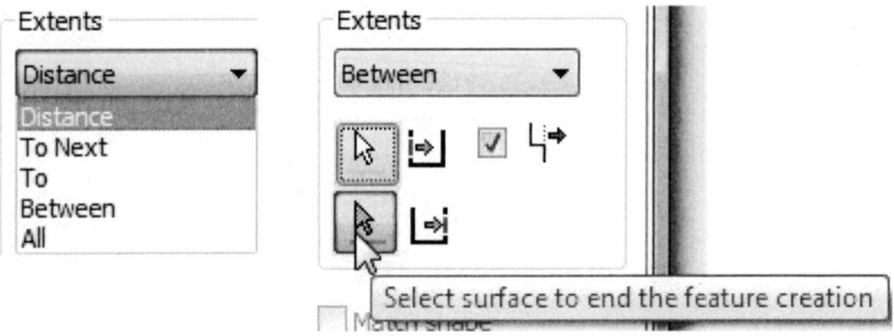

FIGURE 3.17

All

This option determines that the sketch will be extruded all the way through the part in one or both directions.

Distance Value

If the extent is set to Distance, you can either enter a value in the dialog box or in the in-canvas display where the sketch will be extruded or click and drag the arrow in the in-canvas area. Another option is to click the arrow to the right where the distance value is displayed, from here you have an option to click two points to determine a value, display the dimensions of previously created features to select from, or select from the list of the recent values used. After a value is determined, a preview image appears in the graphics area to show how the extrusion will look.

A preview image appears in the graphics area, and the corresponding value appears in the distance area in the dialog box and in the in-canvas area.

If values and units appear in red when you enter them, the defined distance is incorrect and should be corrected. For example, if you entered too many decimal places (e.g., 2.12.5) or an incorrect unit for the dimension value, the value will appear in red. You will need to correct the error before the extrusion can be created.

Direction

There are four buttons from which to choose for determining the direction. Choose from the first two to flip the extrusion direction, click the third button (symmetric) to have the extrusion go equal distances in the negative and positive directions (for example, if the extrusion distance is 2 inches, the extrusion will go 1 inch in both the negative and positive Z directions), and click the asymmetric option to define different distances for the negative and positive direction.

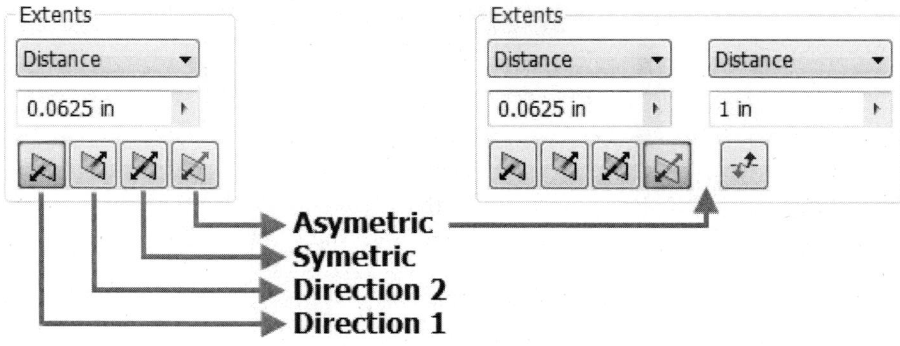

FIGURE 3.18

Output

Two options are available to define the type of output that the Extrude command will generate:

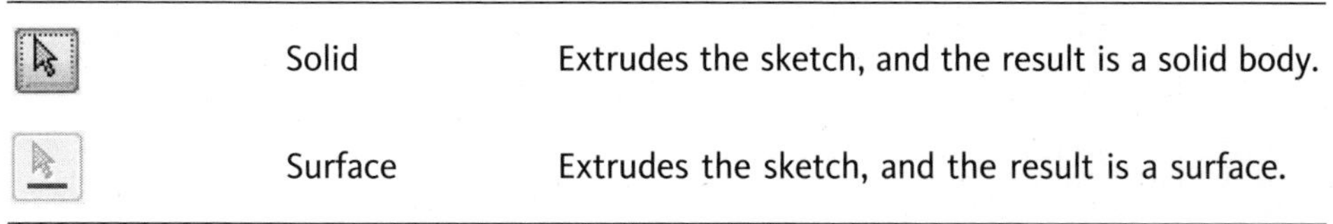

	Solid	Extrudes the sketch, and the result is a solid body.
	Surface	Extrudes the sketch, and the result is a surface.

More

The More tab, as shown in Figure 3.19, contains additional options to refine the feature being created:

Alternate Solution

Alternate Solution terminates the feature on the most distant solution for the selected surface. An example is shown in Figure 3.19.

Minimum Solution

Minimum Solution option when checked terminates the feature on the first possible solution for the selected surface. An example is shown in Figure 3.19.

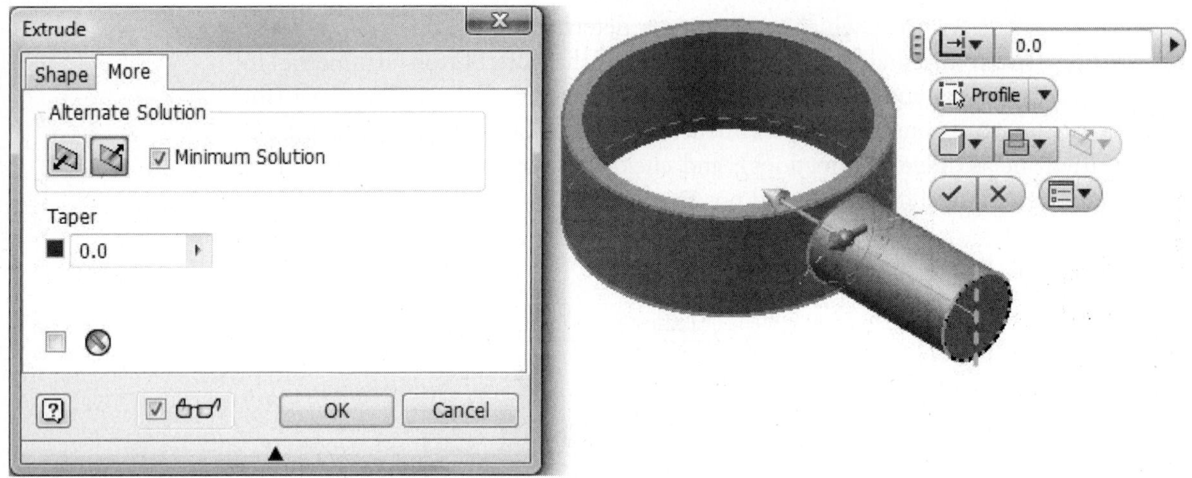

FIGURE 3.19

Taper

Taper extrudes the sketch and applies a taper angle to the feature. To extend the taper angle out from the part, give the taper angle a positive or negative number. This increases or decreases the volume of the resulting extruded feature. You can also click on the sphere in the mini-toolbar area and an arrow will appear as shown in Figure 3.20. Drag the arrow or enter a value.

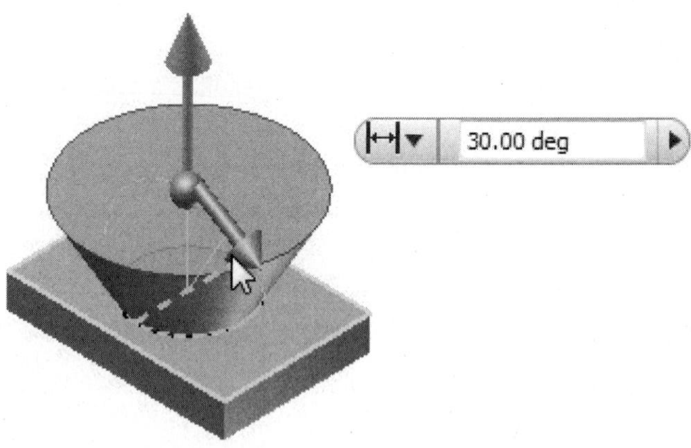

FIGURE 3.20

EXERCISE 3-1: EXTRUDING A SKETCH

In this exercise, you will create a base feature by extruding an existing profile. You will examine the direction options available in the Extrude dialog box.

1. Open *ESS_E03_01.ipt* from the Chapter 03 subfolder.
2. From the 3D Model tab > Create panel click the Extrude command or click on the sketch and in the mini-toolbar click the Create Extrude button. Since there is only one closed profile, the profile is automatically selected.
3. In the Extrude dialog box, set the Distance to **.5 in** (if needed expand the Extrude dialog box by clicking the down arrow at the bottom of the minimized Extrude dialog box).
4. In the graphics window select the arrow on the extrusion and drag the arrow until a distance of **1.000** is displayed in the value field of the Extrude dialog box and in mini-toolbar as shown in Figure 3.21, and then release the mouse button.

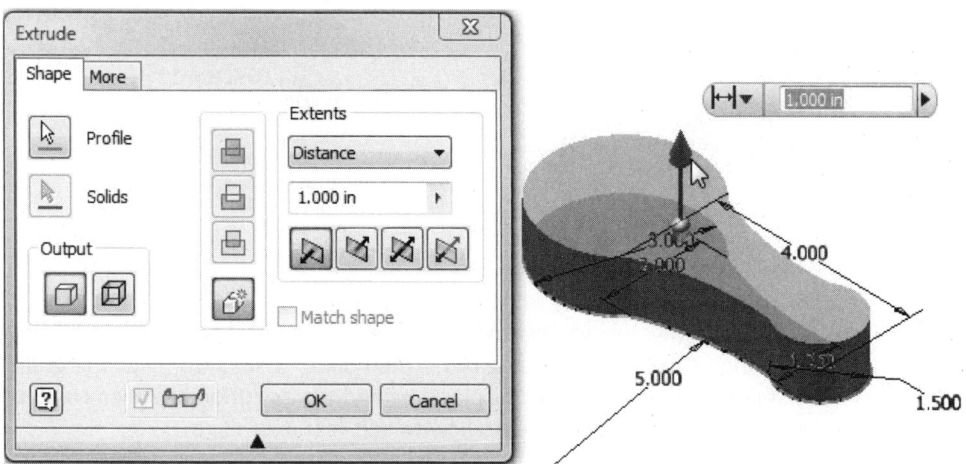

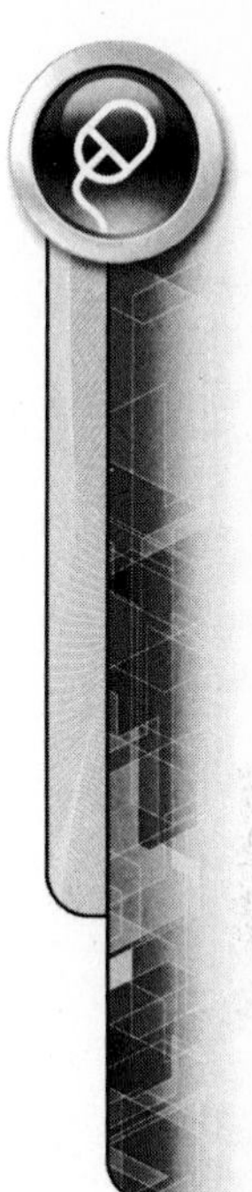

FIGURE 3.21

5. If needed click the down arrow on the bottom of the dialog box and click the More tab in the Extrude dialog box.
6. Change the taper by typing **20.00 deg** in the Taper area.

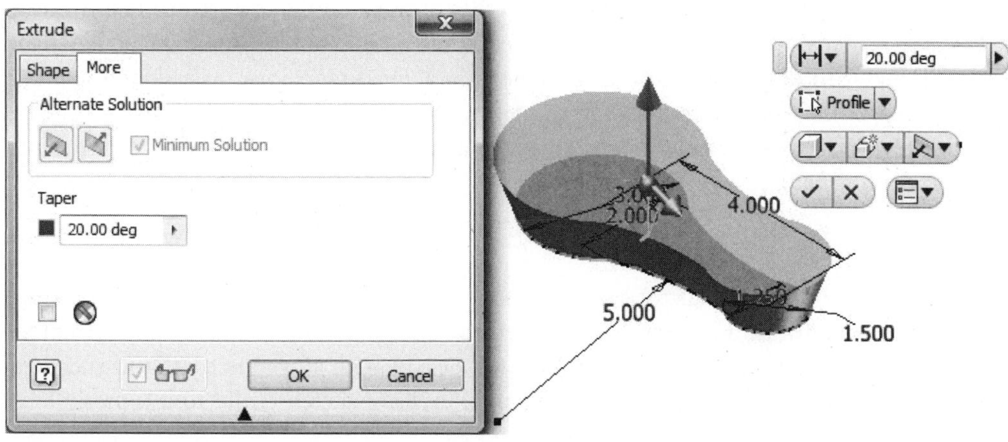

FIGURE 3.22

7. In the graphics window click and drag the sphere (for the taper) to adjust the taper to 10° and then to –10°.
8. Click the Shape tab in the Extrude dialog box.
9. Change the direction of the extrusion to go in the negative Y direction by selecting the Direction 2 button (second from the left button) on the direction area in the dialog box. This will flip the direction of the extrusion.
10. In the Extrude dialog box change the direction to asymmetric (extrude different values in each direction) and enter different values for each direction.
11. In the min-toolbar display area change the extent to symmetric as shown in Figure 3.23.
12. In the graphics window click and drag the arrows to different values.

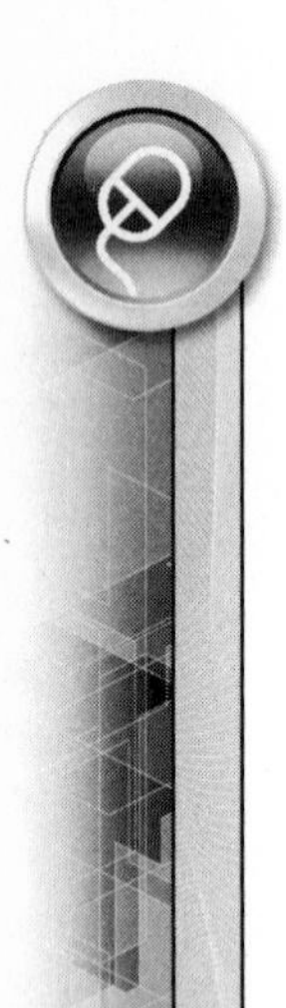

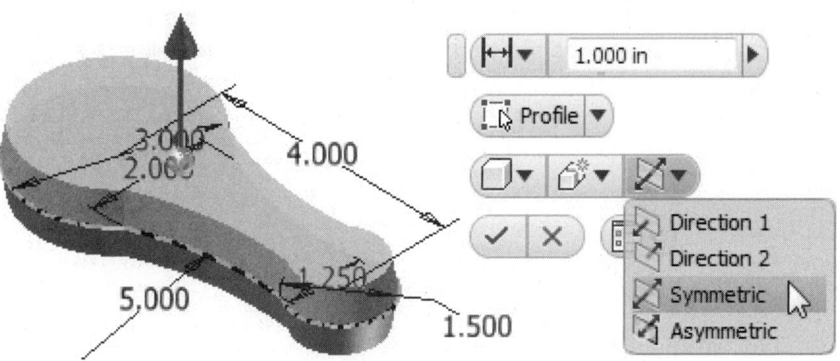

FIGURE 3.23

13. Practice changing the values and directions to see the results in the mini-toolbar and in the Extrude dialog box. When done, click OK or click the green check mark in the mini-toolbar display area and the extrusion has been created. Later in this chapter you will learn how to edit features.
14. Close the file. Do not save changes. End of exercise.

LINEAR DIAMETER DIMENSIONS

Another method for creating a part is to revolve a sketch around a straight edge or axis (centerline). You can use revolved sketches to create cylindrical features. To revolve a sketch, you follow similar steps that you did to extrude a sketch. You create the sketch but the sketch is a quarter section of the completed part. Add constraints and dimensions and then revolve the sketch. The dimensions can be radial or diameter to define the sketch. To place diameter dimensions on a sketch you use a centerline or a normal line to define the linear diameter dimension. The following sections explain how to place a diameter dimension with a centerline and a normal line.

Centerline

To create a centerline, activate the Centerline command on the Sketch tab > Format panel, as shown in Figure 3.24 on the left and then sketch a line. Click on the centerline to deactivate it. If the command is not deactivated, all geometry that is sketched will be a centerline. Or you can change an existing sketch entity to a centerline by selecting an existing sketched entity and then select the Centerline command. To create a linear diameter dimension by using a centerline, follow these steps:

1. Draw a sketch that represents a quarter section of the finished part.
2. Draw a centerline or change an existing line into a centerline that the sketch will be revolved around. The centerline will be treated as a regular line and can be used to define a closed profile.
3. Start the Dimension command.
4. Click the centerline, or a point or edge to be dimensioned.
5. Click the centerline, or a point or edge to be dimensioned. It does not matter the order that the entities are selected; however, one of the selections needs to be the centerline not just an endpoint of the centerline.
6. Move the cursor until the diameter dimension is in the correct location and click.
7. Figure 3.24 on the right shows the linear diameter dimension created using the centerline to define the axis of revolution.

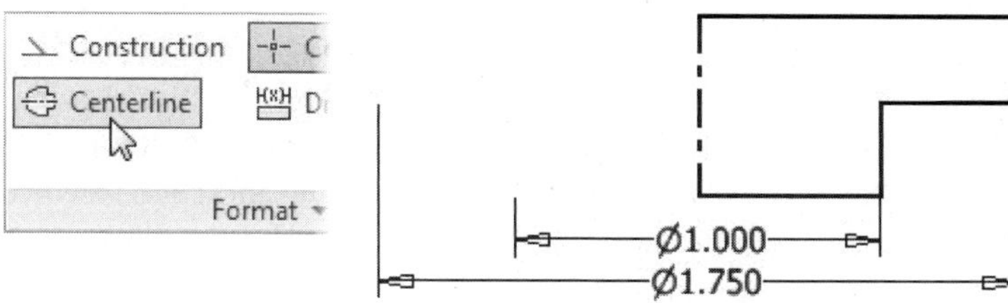

FIGURE 3.24

Normal Line

When a normal line is used to revolve the sketch around, you can create linear diameter (diametric) dimensions for sketches that represent a quarter outline of a revolved part. To create a linear diameter dimension, follow these steps:

1. Draw a sketch that represents a quarter section of the finished part.
2. Draw a line, if needed, or use a line in the sketch that the sketch will be revolved around.
3. Start the Dimension command.
4. Click the line (not an endpoint) that will be the axis of rotation.
5. Click the other point or line to be dimensioned.
6. Right-click and select Linear Diameter from the menu as shown in Figure 3.25 on the left.
7. Move the cursor until the diameter dimension is in the correct location and click. Figure 3.25 on the right shows a sketch that represents a quarter section of a part with a linear diameter dimension placed.

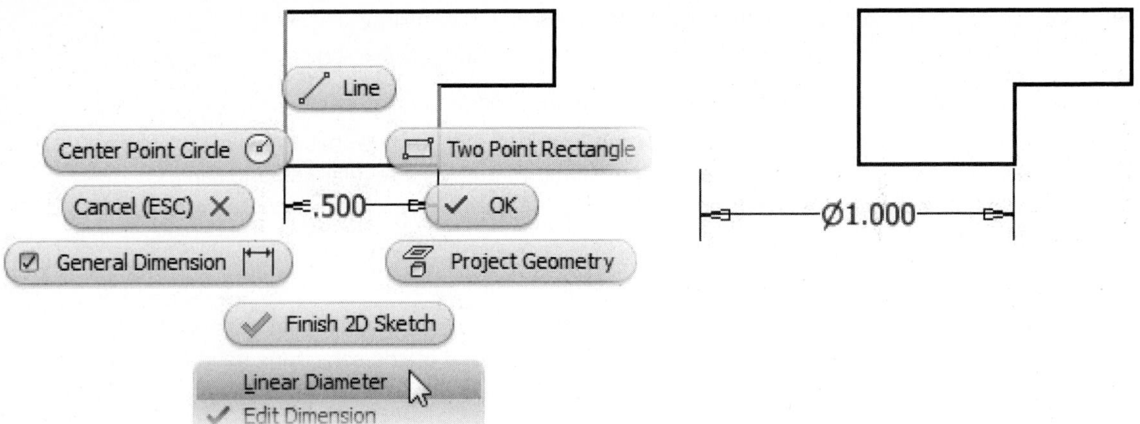

FIGURE 3.25

REVOLVE A SKETCH

After defining the sketch that will be revolved, click the Revolve command on the 3D Model tab > Create panel as shown in Figure 3.26 on the left; or while not in a sketch, in the graphics window click on the sketch geometry and the mini-toolbar will appear, click the Create Revolve button as shown in Figure 3.26 in the middle or press R. Alternately, you can right-click in the graphics window and click Revolve on the marking menu as shown in Figure 3.26 on the right. The Revolve dialog box appears, as shown in Figure 3.26 on the right.

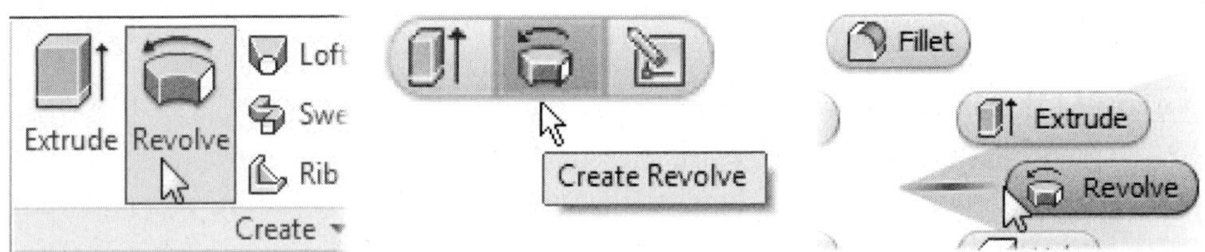

FIGURE 3.26

After starting the Revolve command, the mini-toolbar will appear in the graphics window and the Revolve dialog box will appear. Figure 3.27 shows the mini-toolbar revolve options that are available when revolving the first sketch and the image on the right shows the operation button that is available when creating additional revolve features. You can either enter data in the dialog box or via the in-canvas display. For the in-canvas buttons click the down arrow next to each button to see more options and click and drag the arrow to change the distance or taper. The in-canvas buttons will change depending upon what options are selected. If selection tags have a red arrow, that command option needs to be satisfied before the command can be completed.

FIGURE 3.27

The Revolve dialog box as shown in Figure 3.28 has five sections: Shape, Operation, Extents, Output, and Match Shape. When you make changes in the dialog box, the preview for the revolved feature changes in the graphics area to represent the values and options selected. When you have entered the values and the options that you need, click the OK button.

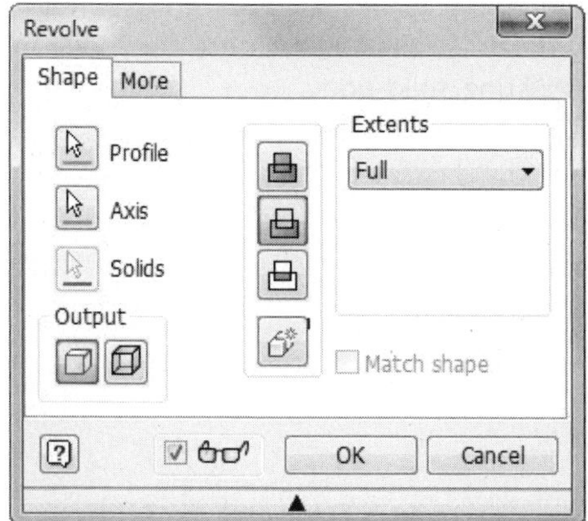

FIGURE 3.28

Shape

This section has two options: Profile and Axis.

	Profile	Click this button to choose the profile to revolve. If the Profile button is shown depressed, this is telling you that a profile or sketch needs to be selected. If there are multiple closed profiles, you will need to select the profile that you want to revolve. If there is only one possible profile, Autodesk Inventor will select it for you, and you can skip this step. If the wrong profile or sketch area is selected, click the Profile button, and choose the new profile or sketch area. To remove a selected profile, hold the CTRL key and click the profile to remove.
	Axis	Click a straight edge, centerline, work axis, or origin axis about which the profile(s) should be revolved. See the section below on how to create a centerline and create diametric dimensions.
	Solid	If there are multiple solid bodies, click this button to choose the solid body(ies) to participate in the operation.

Operation

This is the unlabeled middle column of buttons. If this is the first sketch that you create a solid from, it is referred to as a base feature, and only the top button is available. The operation defaults to Join (the top button). Once the base feature has been established, you can then revolve a sketch, adding or removing material from the part using the Join or Cut options, or you can keep what is common between the existing part and the completed revolve operation using the Intersect option.

	Join	Adds material to the part.
	Cut	Removes material from the part.
	Intersect	Removes material, keeping what is common to the existing part feature(s) and the new feature.
	New Solid	Creates a new solid body. The first solid feature created uses this option by default. Select to create a new body in a part file with an existing solid body.

Extents

The Extents determines if the sketch will be revolved 360°, a specified angle, stop at a specific plane, or start and stop at specific planes.

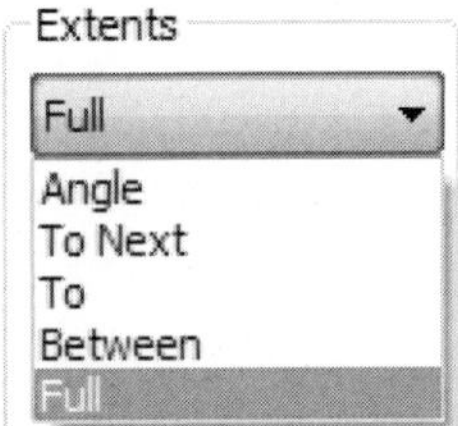

FIGURE 3.29

Full

Full is the default option; it will revolve the sketch 360° about a specified edge or axis.

Angle

Click this option from the drop-down list, and the Revolve dialog box displays additional options, as shown in Figure 3.30. Enter an angle for the sketch to be revolved. The three buttons below the degree area will determine the direction of the revolution. Choose the first two to flip the revolution direction, or click the right-side button to have the revolution go an equal distance in the negative and positive directions. If the angle was set to 90°, for example, the revolution will go 45° in both the negative and positive directions.

To Next

This option determines that the sketch will be revolved until it reaches a plane or face. The sketch must be fully enclosed in the area to which it is projecting; if it is not fully enclosed, use the To termination with the Extend to Surface option. Click the Direction button to determine the revolve direction.

To

This option determines that the sketch will be revolved until it reaches a selected face, plane, or point. To select a plane, face, or point to end the revolve, click the Select Surface button, as shown in Figure 3.30, and then click a face, plane, or point at which the revolve should terminate.

Between

This option determines that the revolve will start at a selected plane or face and stop at another plane or face. Click the Select surface to start feature creation button, and then click the face or plane where the revolve will start. Click the Select surface to end the feature creation button, and then click the face or plane where the extrusion or revolve will terminate.

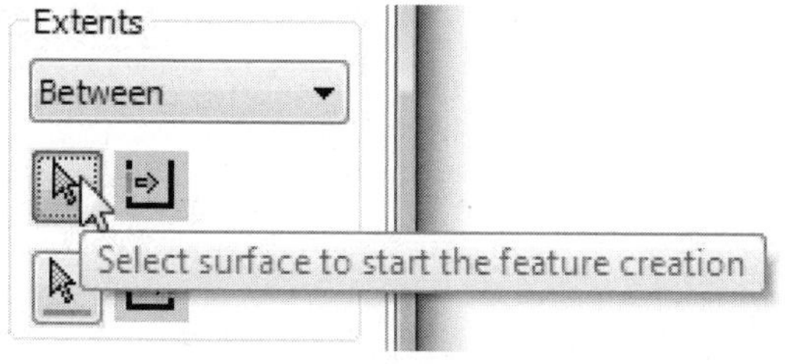

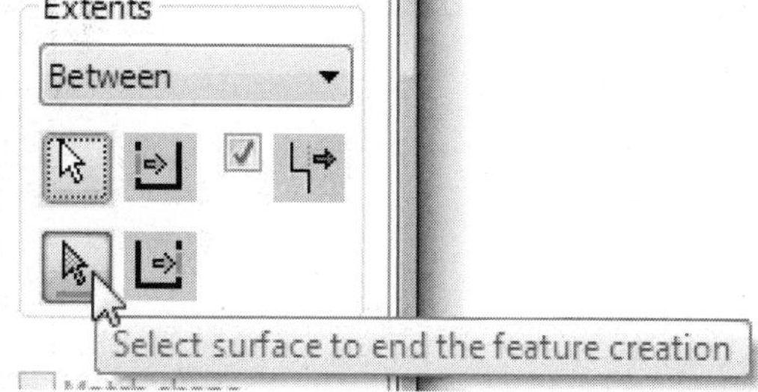

FIGURE 3.30

Direction

When the Extents is set to Angle there are four buttons from which to choose for determining the direction. Choose from the first two to flip the angle direction, click the third button (symmetric) to have the revolution go equal distances in the negative and positive directions, and click the asymmetric option to define different angles for the negative and positive direction. With the symmetric option, if the angle is 90 degrees, for example, the angle will go 45 degrees in both the negative and positive Z directions.

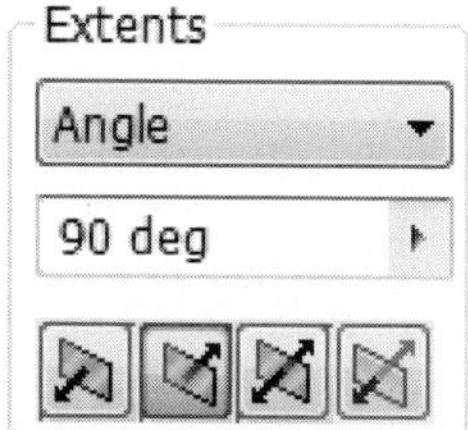

FIGURE 3.31

Output

Two options are available to select the type of output that the Revolve command will generate.

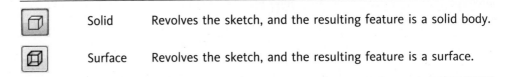

[icon]	Solid	Revolves the sketch, and the resulting feature is a solid body.
[icon]	Surface	Revolves the sketch, and the resulting feature is a surface.

More

The More tab, as shown in Figure 3.32, contains additional options to refine the feature being created.

Alternate Solution

Alternate Solution terminates the feature on the most distant solution for the selected surface. An example is shown in Figure 3.32.

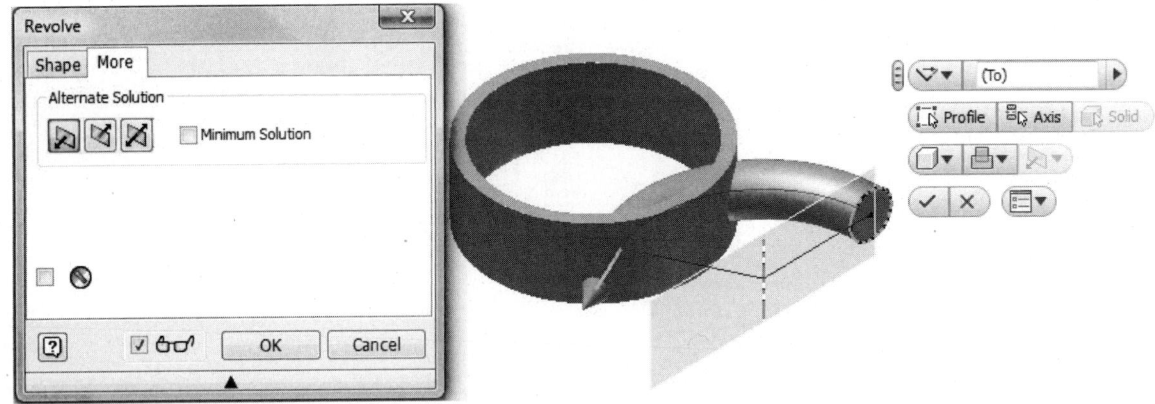

FIGURE 3.32

Minimum Solution

Minimum Solution terminates the feature on the first possible solution for the selected surface. An example is shown in Figure 3.33.

FIGURE 3.33

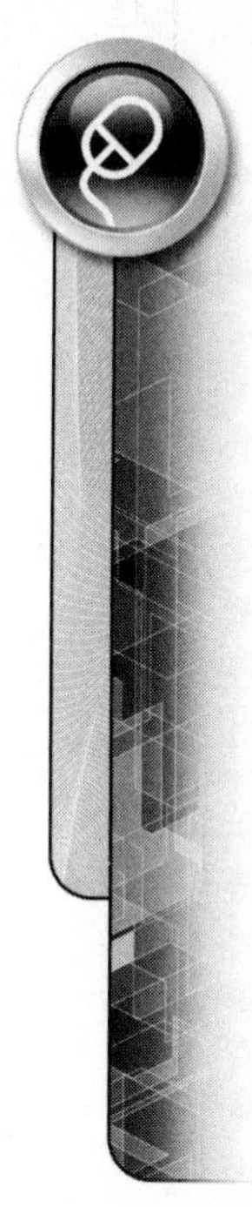

EXERCISE 3-2: REVOLVING A SKETCH

In this exercise, you will create a sketch and then create a revolved feature to complete a part. This exercise demonstrates how to revolve sketched geometry about an axis to create a revolved feature.

1. Click the New command on the Quick Access toolbar, click the English tab, and then double-click *Standard (in).ipt*, or if inch is the default unit; from the left side of the Quick Access toolbar you can click the down arrow on the New icon, and select Part.
2. Click the Create 2D Sketch command on the 3D Model tab > Sketch panel and then select the XY origin plane in the graphics window.
3. Create the sketch geometry as shown. Locate the lower endpoint of the geometry on the origin point. Select the left vertical line and then click the Centerline command on the Format panel to change the line to a centerline as shown in Figure 3.34 on the left.
4. Click the Dimension command on the Sketch tab > Constrain panel.
5. Add the linear diameter dimensions shown in Figure 3.34 on the right by selecting the centerline and selecting an endpoint on the sketch. Then click to place the dimension. Note, for clarity the remaining dimensions to fully constrain the sketch were not added.

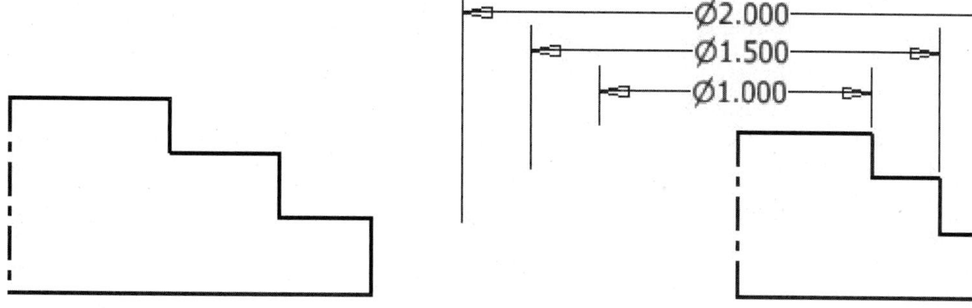

FIGURE 3.34

6. Right-click in the graphics window, and click Finish 2D Sketch from the marking menu.
7. From the 3D Model tab > Create panel bar, click the Revolve command or click on the sketch mini-toolbar and click the Create Revolve button. Since there is only one closed profile, the profile is selected for you. Since there is only one centerline the centerline is also automatically selected as the axis.
8. If needed expand the Revolve dialog box by clicking the down arrow at the bottom the minimized Revolve dialog box and change the Extents to Angle as shown in Figure 3.35.
9. Enter **45 deg** or drag the arrow on the model until **45.00 degrees** is displayed. The preview of the model updates to reflect the change as shown in Figure 3.35.

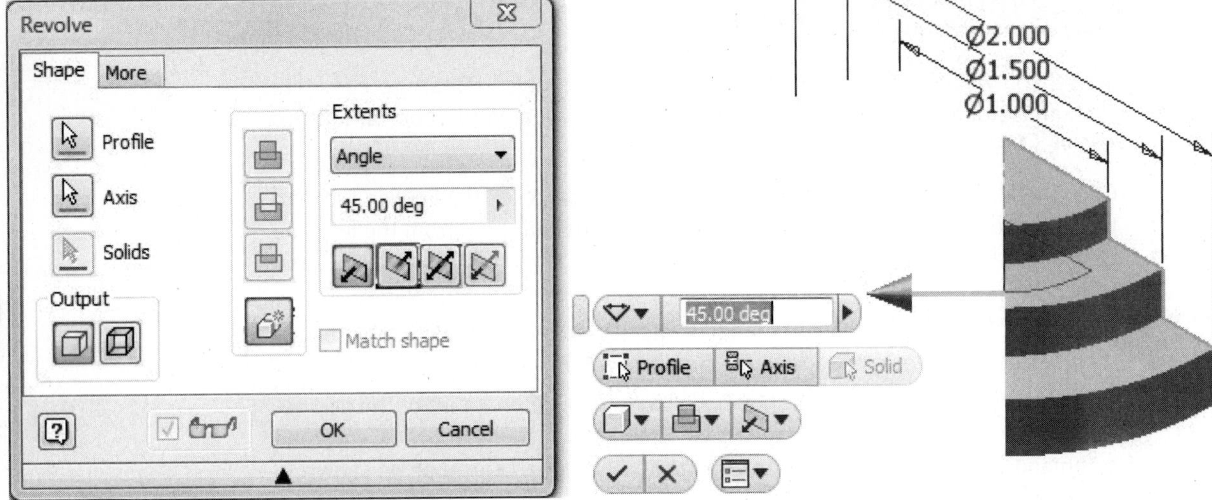

FIGURE 3.35

10. Change the direction of the revolve to go counter clockwise by selecting the left Direction 2 button in the dialog box or click and drag the arrow in the graphics screen to go **45.00 degrees** in the opposite direction. The preview image will reverse the direction counter clockwise.
11. Set the revolve direction to symmetric.
12. Enter **90 deg** for the angle. The preview should resemble Figure 3.36.

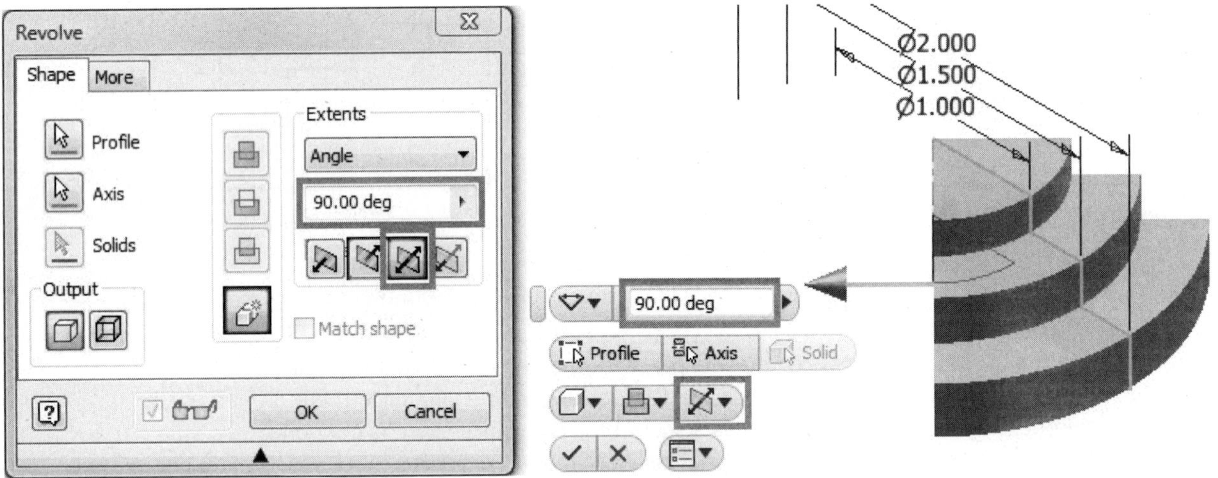

FIGURE 3.36

13. In the mini-toolbar change the extent to asymmetric.
14. In the graphics window click and drag the arrows to different values as shown in Figure 3.37.

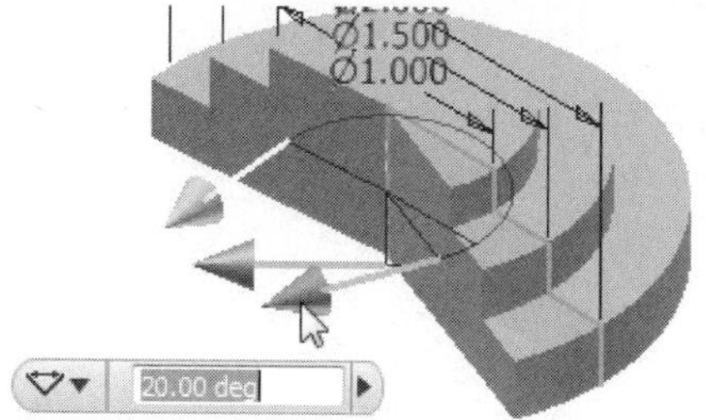

FIGURE 3.37

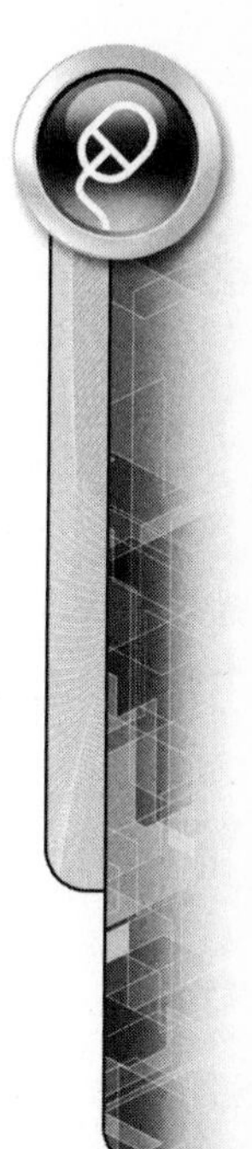

15. Change the Extents to Full on the first Extents column as shown in Figure 3.38 in the left.
16. Click the OK button or click the green check mark in the mini-toolbar to create the feature.
17. Your part should resemble Figure 3.38 on the right.

FIGURE 3.38

18. Close the file. Do not save changes. End of exercise.

PRIMITIVE FEATURES

Another method to quickly create basic shapes is to use one of the primitive commands; box, cylinder, sphere or torus as shown in Figure 3.39. A sketch and feature (extrusion for box and cylinder and revolution for sphere or torus) will be created for each primitive feature.

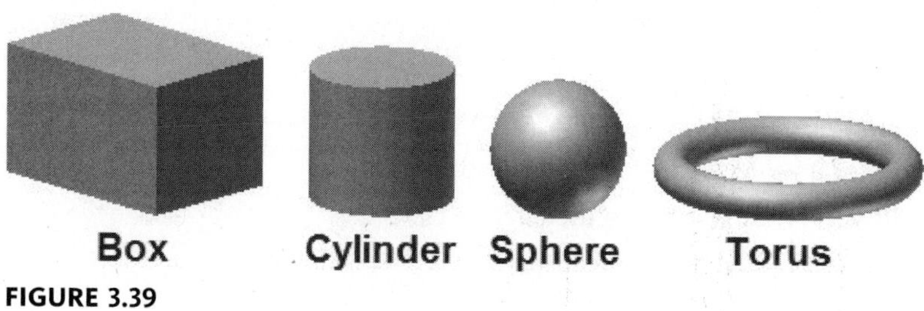

FIGURE 3.39

To create a primitive feature, follow these steps.

1. Click one of the primitive commands from the 3D Model tab > Primitives panel as shown in Figure 3.40.

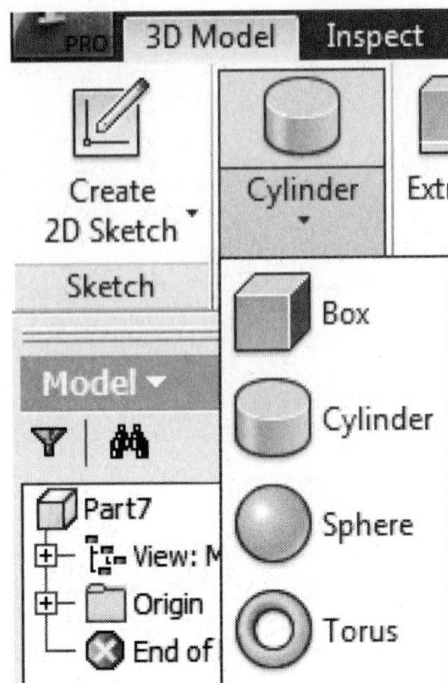

FIGURE 3.40

2. Select a plane to locate the primitive, the plane can be a origin plane, planar face on the part or a work plane.
3. Depending on the primitive that you are creating you will do the following.
   - **Box**
     (i) Select a point to locate center of the rectangle
     (ii) Enter a value for the horizontal size
     (iii) Press the Tab key
     (iv) Enter a value for the vertical size
     (v) Press ENTER to extrude the rectangle
     (vi) Enter a value for the extrusion distance
     (vii) Click OK to create the feature
   - **Cylinder**
     (i) Select a point to locate the center of the circle
     (ii) Enter a value for the diameter of the circle
     (iii) Press ENTER to extrude the circle
     (iv) Enter a value for the extrusion distance
     (v) Click OK to create the feature
   - **Sphere**
     (i) Select a point to locate the center of the circle
     (ii) Enter a value for the diameter of the circle
     (iii) Press ENTER to revolve the circle
     (iv) Click OK to create the feature
   - **Torus**
     (i) Select a point to locate the center of the torus

(ii) Enter a value and then either press ENTER or click a point to define the radius of the torus

(iii) Enter a value for the diameter of the circle

(iv) Press ENTER to revolve the circle

(v) Click OK to create the feature

NOTE

Primitive features can be edited like any other feature. If a primitive is a secondary feature you will need to edit the primitive's sketch to locate it. Editing a feature and a feature's sketch will be covered later in this chapter.

SECONDARY SKETCHED FEATURES

A secondary sketched feature is created from a sketch that you draw on a planar face or work plane and then add or remove material from the part. The basic steps to create a sketched feature are as follows:

1. Create a new 2D sketch or make an existing sketch active.
2. Draw the geometry that defines the sketch.
3. Add constraints and dimensions.
4. Perform a boolean operation that will add material to the part, remove material from the part, or keep what is common between the part and the feature that is being created. Figure 3.41 shows a sketch on the left and the affect the three operations would have on the part.

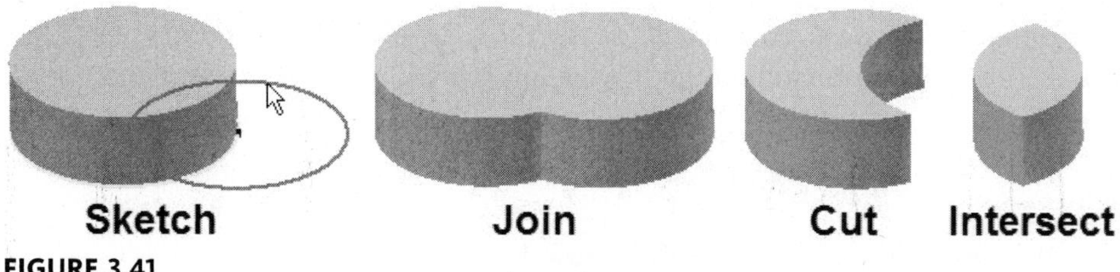

FIGURE 3.41

5. Use the Extents options in the extrude and revolve commands to control how the feature will terminate. Refer back to the Extents section in the Extrude a Sketch and the Revolve a Sketch section that were covered earlier in this chapter.

There are no limits to the number of sketched features that can be added to a part. Each sketched feature is created on its own plane, and multiple features can reference the same plane. You can sketch geometry, apply constraints, and apply dimensions exactly as you did with the first sketch. In addition to constraining and dimensioning the new sketch, you can also constrain the new sketch to the existing part. You can place dimensions to geometry that does not lie on the active plane; the dimensions, however, will be placed on the current plane. After the sketch has been constrained and dimensioned, it can be extruded, revolved, swept, or lofted. Exit the sketch environment by right-clicking in the graphics area and selecting Finish Sketch from the marking menu or by clicking the Finish Sketch command on the Sketch tab > Exit panel. Only one sketch can be active at a time.

In the following section, you learn how to create a sketch on a plane and create a feature from the sketch.

DEFINE THE ACTIVE SKETCH PLANE

As stated previously, each sketch must exist on its own plane. The active sketch has a plane on which the sketch is drawn. To assign a plane to the active sketch, the plane on which the sketch will be created must be a planar face, a work plane, or an origin plane. The planar face does not need to have a straight edge. A cylinder has two planar faces, one on the top and the other on the bottom of the part. Neither has a straight edge, but a sketch can be placed on either face.

To make a sketch active, use one of the following methods:

- Click the 2D Sketch command from the 3D Model tab > Sketch panel as shown in Figure 3.42 on the left, and then click the plane where you want to place the sketch.
- While not in the middle of an operation, right-click in the graphics window, and select New Sketch from the marking menu as shown in Figure 3.42 in the middle. Then click a planar face, a work plane, or an existing sketch in the browser.
- While not in a sketch, click on the planar face or workplane that you want to create the active sketch on. Then click the Create Sketch button from the mini-toolbar buttons as shown in Figure 3.42 on the right.
- Press the key S, and then click the plane where you want to place the sketch.

Once you have created a sketch, it appears in the browser with the name Sketch#, and sketch commands appear in the 2D Sketch panel bar. The number will sequence for each new sketch that is created. When you create a new sketch or make a sketch active, by default the view will automatically change so you are looking straight at the sketch. This is controlled by the Application Option > Sketch tab > Look at sketch plane on sketch creation. To manually adjust the viewpoint to look straight at a plane, click the Look At command on the Navigation Bar and then select the plane.

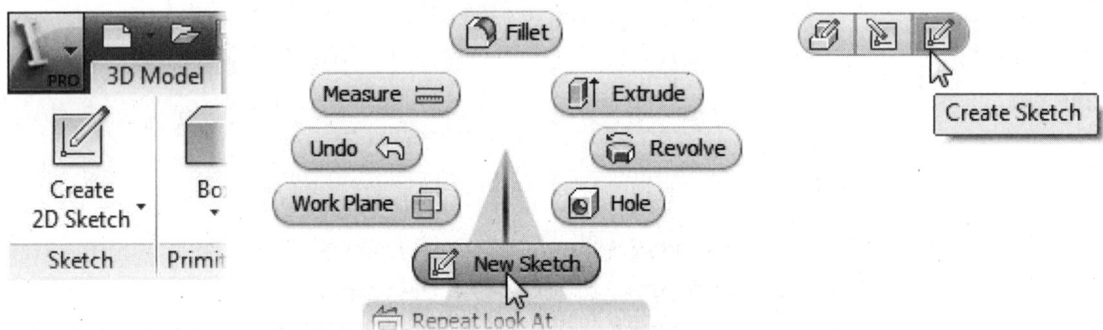

FIGURE 3.42

Select Other-Face Cycling

Autodesk Inventor has dynamic face highlighting that helps you to select the correct face, edge, part etc. to activate and to select objects. As you move the cursor over a given face, the edges of the face are highlighted. If you continue to move the cursor, different faces are highlighted as the cursor passes over them.

To cycle to a face that is behind another one, start a command and then move the cursor over the face that is in front of one that you want to select and hold the cursor still for two seconds. The Select Other tool appears, as shown in Figure 3.43 on the left. Select the drop down arrow and move the cursor over the available objects in the list until the correct face is highlighted and then click. The number of objects that appear in the list will depend upon the geometry and the location of the cursor. You can also access the Select Other command by right-clicking the desired location in the graphics window and click Select Other from the marking menu.

You can specify the amount of time before the Select Other command will appear automatically. Click Application Options on the Tools tab > Options panel and click the General tab. In the Selection area as shown in Figure 3.43 on the right, the "Select Other" delay (sec) feature can be specified in tenths of a second. If you do not want

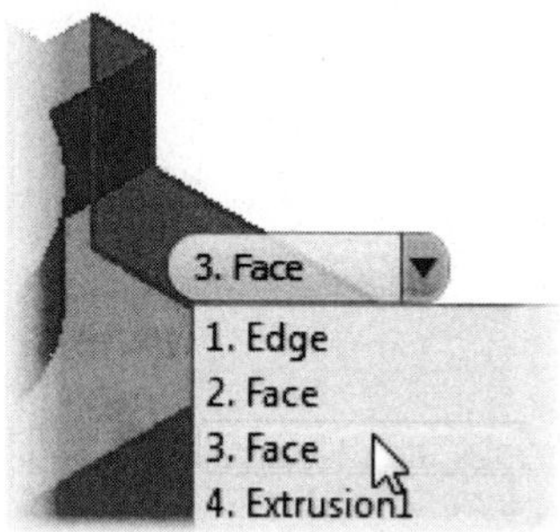

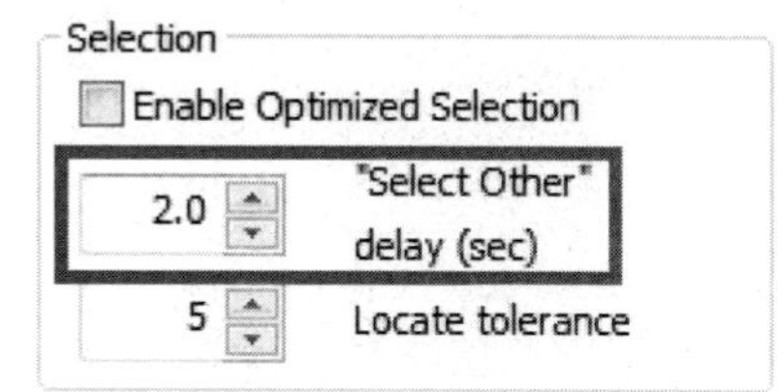

FIGURE 3.43

the Select Other command to open automatically, type OFF in the field. The default value is 2.0 second.

SLICE GRAPHICS

While creating parts, you may need to sketch on a plane that is difficult to see because features are obscuring the view. The Slice Graphics option will temporarily slice away the portion of the model that obscures the active sketch plane on which you want to sketch. Figure 3.44 on the left shows a revolved part with an origin plane visible and the Slice Graphics menu. The image on the right shows the graphic sliced and the origin plane's visibility turned off. To temporarily slice the graphics screen, follow these steps:

1. Make a planar face or work plane that the graphics of the active sketch will be sliced through.
2. Rotate the model so the correct side will be sliced, that is, the side of the model that faces up will be sliced away.
3. While editing the sketch, right-click and select Slice Graphics from the marking menu as shown in Figure 3.44 on the left or click Slice Graphics from the bottom of the Status Bar as shown in Figure 3.44 in the middle press the F7 key or click Slice Graphics on the View tab > Appearance panel. The part will be sliced on the active sketch plane as shown in Figure 3.44 on the right.
4. Use sketch commands from the Sketch tab to create geometry on the active sketch.
5. To exit the sliced graphics environment, right-click and select Slice Graphics from the marking menu, click Slice Graphics from the bottom of the Status Bar, press the F7 key or click the Finish Sketch button on the Sketch tab > Exit panel.

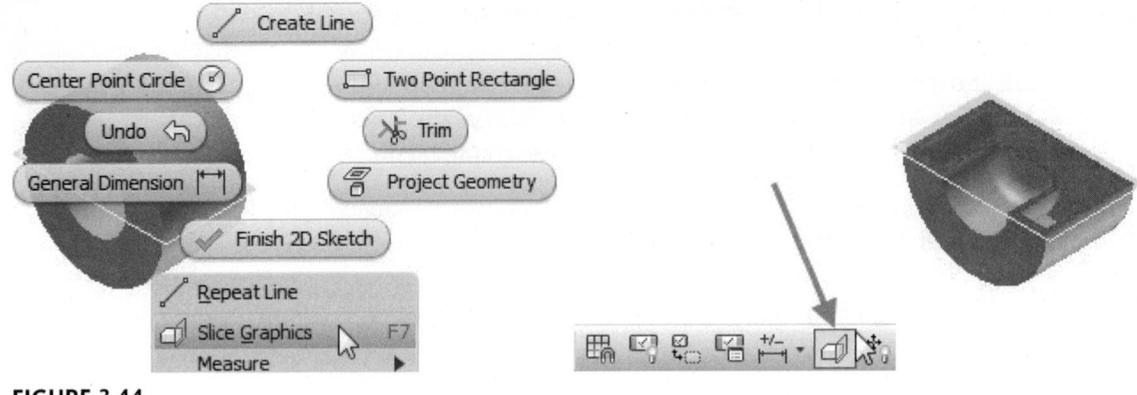

FIGURE 3.44

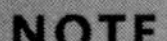

NOTE

You have additional Section commands available from the View tab > Appearance panel. These section commands are covered in chapter 7.

EXERCISE 3-3: SKETCH FEATURES

In this exercise, you will create a sketch on the angled face of a part and then create an oblong on the new sketch plane.

1. Open *ESS_E03_03.ipt* from the Chapter 03 folder.
2. Click the Create 2D Sketch command on the 3D Model tab > Sketch panel, and then click the top-inside angled face as shown in Figure 3.45 on the left. You could have clicked on the face and click the Create Sketch mini-toolbar button.
3. The view should change so you are looking straight at the sketch, if not click the View Face command from the Navigation bar, and then select Sketch6 in the browser.
4. Next you create a sketch for an obround in the part and center the slot vertically. Sketch a slot using dynamic input to create a **1.5 inch** long obround as shown in Figure 3.45 on the right. Both arcs should be tangent to the adjacent lines.

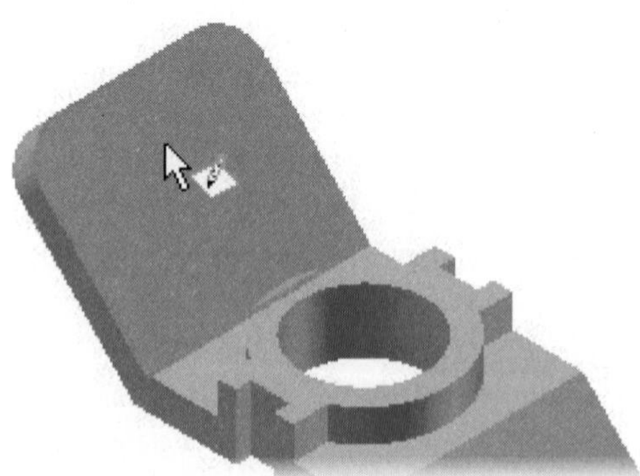

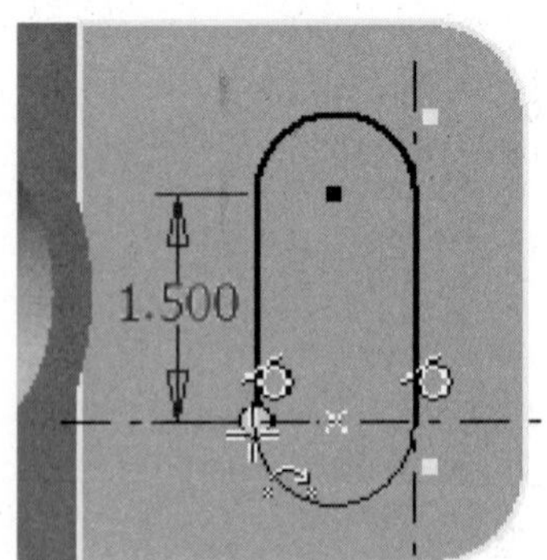

FIGURE 3.45

5. Click the Horizontal constraint command on the Constrain panel as shown in Figure 3.46 on the left.
6. Select the midpoint of the right vertical edge of the oblong and the midpoint of the right vertical edge of the part as shown in Figure 3.46 on the right. The obround is centered vertically using the horizontal constraint.

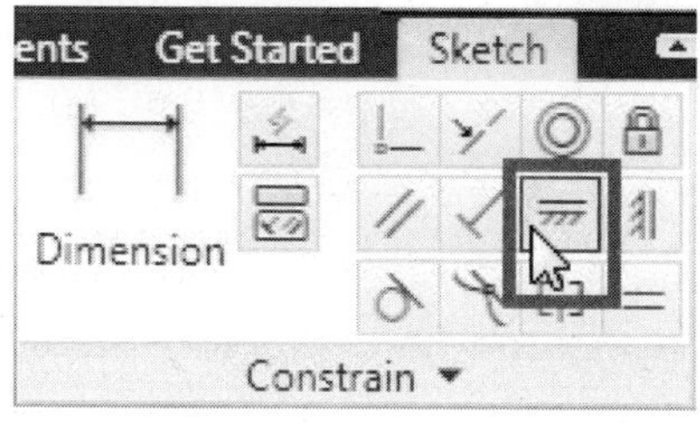

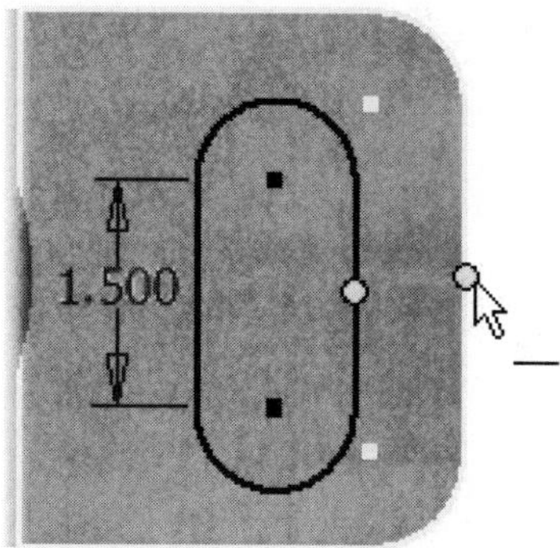

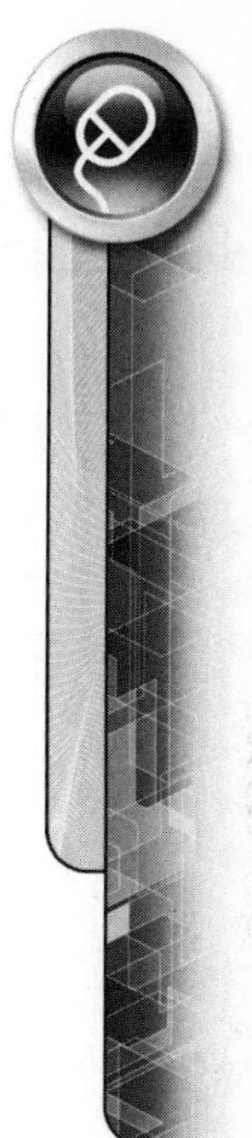

FIGURE 3.46

7. Click the Dimension command on the Constrain panel as shown in Figure 3.47 on the left. To fully constrain the sketch place the two dimensions as shown in Figure 3.47 on the right.

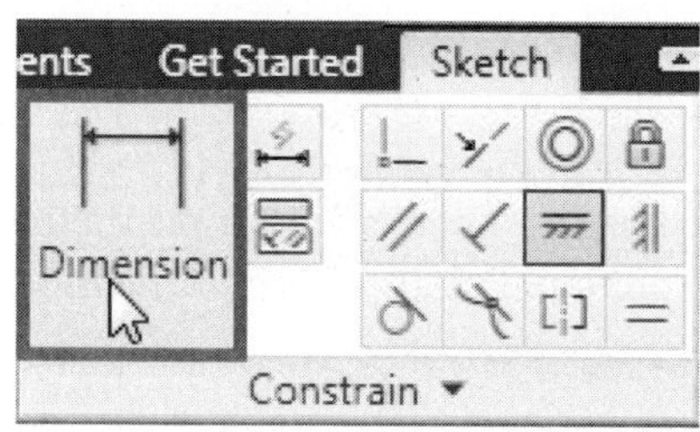

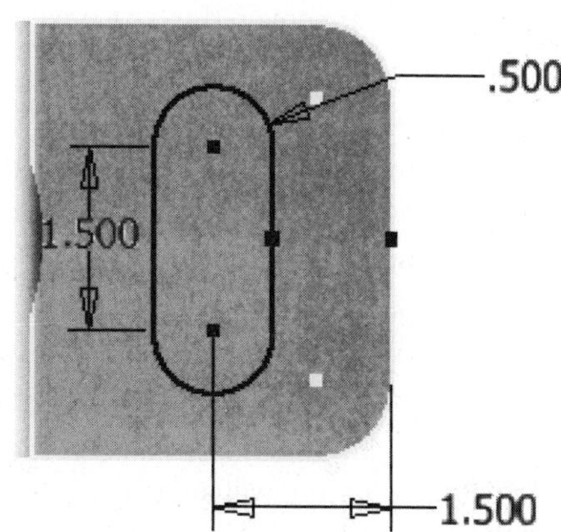

FIGURE 3.47

8. Click the Finish Sketch command on the Exit panel.
9. If the view did not automatically change to the home view press the F6 key to change to the Home View.
10. In the graphics window click on an entity on the obround and click Create Extrude from the mini-toolbar.
11. For the profile, click inside the oblong.
12. In the mini-tool click the Cut operation, and change the Extents to Through All. Ensure that the direction is pointing into the part as shown in Figure 3.48 on the left.
13. Click OK. The completed part is shown in Figure 3.48 on the right.

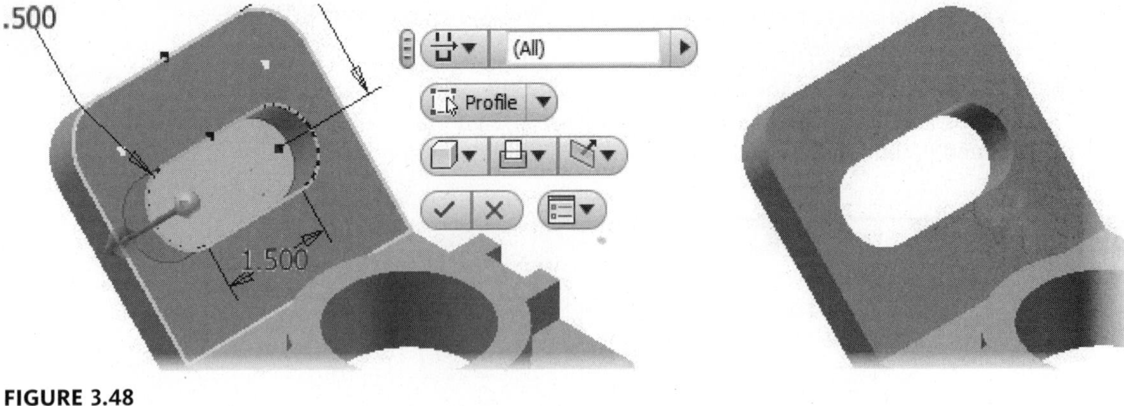

FIGURE 3.48

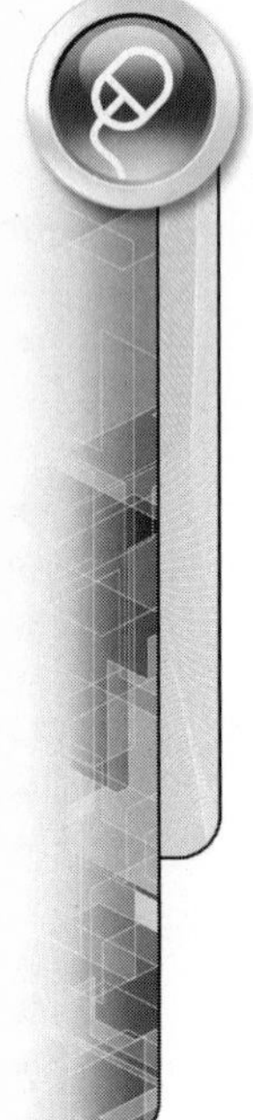

14. Practice creating sketched features trying different options.
15. Close the file. Do not save changes. End of exercise.

EDITING A FEATURE

After you create a feature, the feature consumes all of the dimensions that were visible in the sketch. If you need to change a feature's options and values such as operation, extents, distance, taper etc. you edit the feature. You can also edit the feature's sketch and change the values of the dimensions or add or remove geometry. Editing a feature and feature's sketch are covered in the next sections.

Editing Feature Options

To change a feature's options, you need to edit the feature. There are multiple methods that you can use to edit the feature. There is no preferred method; use the method that works best for your workflow.

- While not in a sketch, move the cursor over the feature in the graphics window and click. Click the Edit "Feature Name" button from the mini-toolbar. Figure 3.49 on the left shows the mini-toolbar button for editing an extrusion.
- Double-click the features name in the browser; the dialog box that was used to create the feature will appear.
- In the browser, right-click the feature's name and select Edit Feature from the menu as shown in Figure 3.49 on the right.

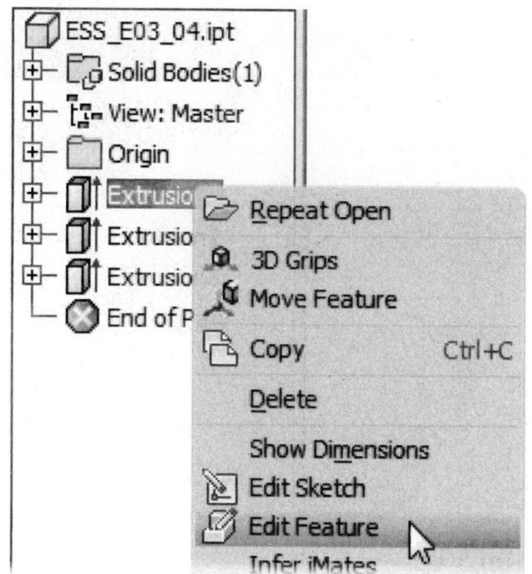

FIGURE 3.49

When editing a feature, the dialog box will appear that was used to create the feature. While editing you can change the feature's values and its options except the join operation on a base feature (first feature in the browser) and the output (solid or surface). Figure 3.50 on the left shows the menu that appears after right-clicking on Extrusion1.

Editing a Feature's Sketch

In the last section, you learned how to edit the dimensions and the settings in which the feature was created. In this section, you will learn how to add and delete constraints, dimensions, and geometry in the original 2D sketch. To edit the 2D sketch of a feature, do the following:

- While not in a sketch, move the cursor over the feature in the graphics window whose sketch you want to edit and click. Then click the Edit Sketch button from the mini-toolbar as shown in Figure 3.50 on the left.
- In the browser, expand the feature, right-click the name of the feature or sketch, and click Edit Sketch from the menu as shown in Figure 3.50 on the right.
- In the browser, expand the feature and double-click on the sketch icon.
- Another method to change the value of dimensions on a sketch is to right-click the feature's name in the browser, and select Show Dimensions from the menu.

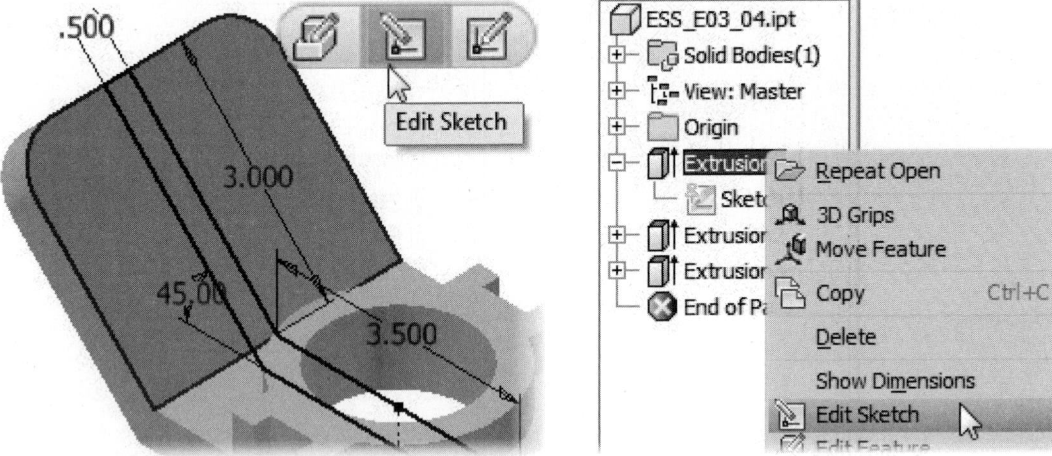

FIGURE 3.50

While editing the sketch, you can both add and remove objects. You can add geometry lines, arcs, circles, and splines to the sketch. To delete an object, right-click it and select Delete from the menu, or click it and press the DELETE key. If you delete an object from the sketch that has dimensions associated with it, the dimensions are no longer valid for the sketch, and they will also be deleted. You can also delete the entire sketch and replace it with an entirely new sketch. When replacing entire sketches, you should first delete other features that would be consumed by the new objects and re-create them.

You can also add or delete constraints and dimensions. When the dimensions are visible on the screen, double-click the dimension text that you want to edit. The Edit Dimension dialog box appears. Enter a new value, and then click the checkmark in the dialog box or press the ENTER key. Continue to edit the dimensions and, when finished click Finish Sketch from the ribbon or click the Local Update button on the Quick Access toolbar as shown in Figure 3.51. The sketch will be consumed by the feature and the new values will be used to regenerate the part.

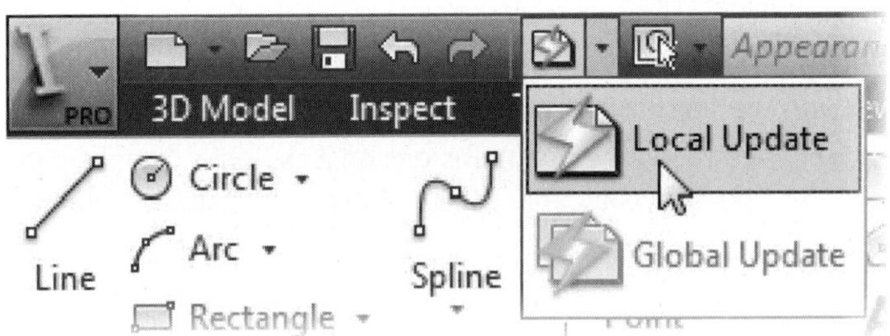

FIGURE 3.51

NOTE

If you receive an error after updating the part, make sure that the sketch forms a closed profile. If the appended or edited sketch forms multiple closed profiles, you will need to reselect the profile area.

Renaming Features and Sketches

By default, each feature is given a name. These feature names may not help you when trying to locate a specific feature of a complex part, as they will not be descriptive to your design intent. The first extrusion, for example, is given the name Extrusion1 by default, whereas the design intent may be that the extrusion is the thickness of a plate. To rename a feature, slowly double-click the feature name and enter a new name. Spaces are allowed.

Deleting a Feature

You may choose to delete a feature after it has been placed. To delete a feature, right-click the feature name in the browser, and select Delete from the menu, as shown in Figure 3.52. The Delete Features dialog box will then appear, and you should choose what you want to delete from the list. You can delete multiple features by holding down the CTRL or SHIFT key, clicking their names in the browser, right-clicking one of the names, and then selecting Delete from the menu. The Delete Features dialog box appears; it allows you to delete consumed or dependent sketches and features.

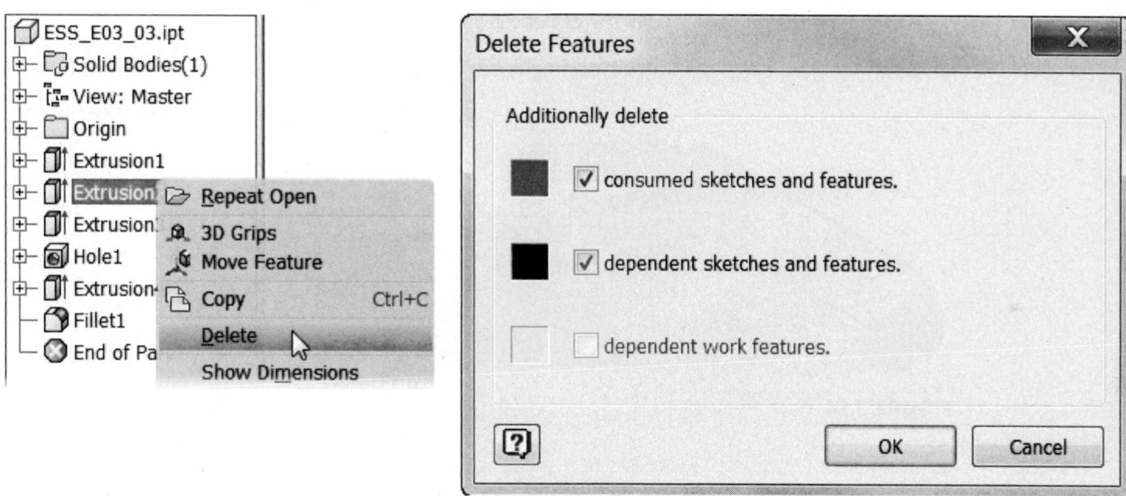

FIGURE 3.52

Failed Features

If the feature in the browser turns red and have a yellow triangle with an explanation point before the features icon as shown in Figure 3.53. After updating the part, this is an alert that the new values or settings were not regenerated successfully. To see what the feature looked like in its last successful state, move the cursor over the features name in the browser and it will highlight in the graphics window like what is shown in Figure 3.53 on the left of the last good fillet feature. You can then edit the values, enter new values, or select different settings to define a valid solution. Once you define a valid solution, the feature should regenerate without error. When you try to create a fillet, chamfer, shell, or a thicken/offset feature that cannot be calculated, an error glyph will appear on the right side of the mini-toolbar as shown in Figure 3.53 on the right.

FIGURE 3.53

EXERCISE 3-4: EDITING FEATURES AND SKETCHES

In this exercise, you will edit a consumed sketch in an extrusion and update the part. In the next section, you learn how to create sketched features that were created in this model.

1. Open *ESS_E03_04.ipt* from the Chapter 03 folder.
2. Edit the sketch of Extrusion1 by selecting the top planar face and then click Edit Sketch from the mini-toolbar as shown in Figure 3.54. You could have also right-clicked on Extrusion1 in the browser and clicked Edit Feature from the menu.

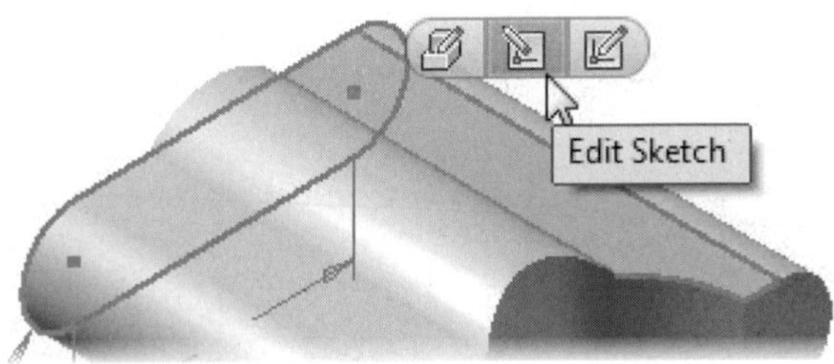

FIGURE 3.54

3. Double-click the .375 radius dimension. In the Edit Dimension dialog box delete the existing value and enter **.25**, and then press Enter or click the checkmark in the dialog box.
4. Double-click the 2.000 dimension, and change the value to **3** and then press Enter or click the checkmark in the dialog box. The sketch should resemble Figure 3.54 on the left.
5. Click the Finish Sketch command on the Sketch tab > Exit panel. The feature is updated with the new values as shown in Figure 3.54 on the right.

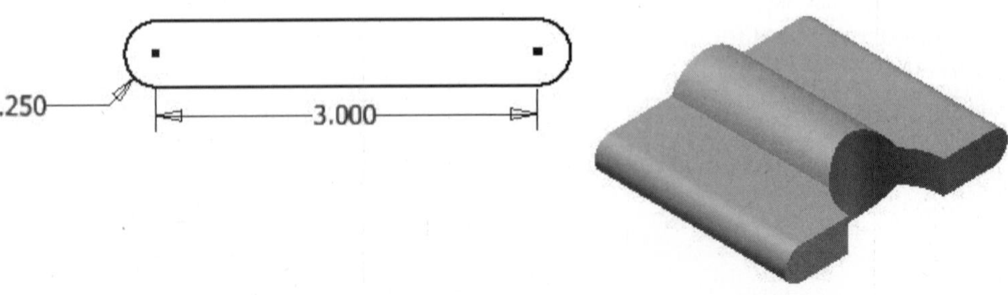

FIGURE 3.54A

6. You now edit the extents that Extrusion2 was created with. In the graphics window click on the inside circular face and then click on the Edit Extrude mini-toolbar button as shown in Figure 3.55 on the left labeled (1). The Extrude dialog box and mini-toolbar buttons are displayed.
7. Change the Operation to Join in the mini-toolbar as shown in Figure 3.55 on the right labeled (2).
8. Change the Extents option to To selected face/point labeled (3) in the Figure 3.55 on the right, then select the bottom face as shown in Figure 3.55 on the right. The extrusion will then stop at this face no matter what dimension changes in the part.

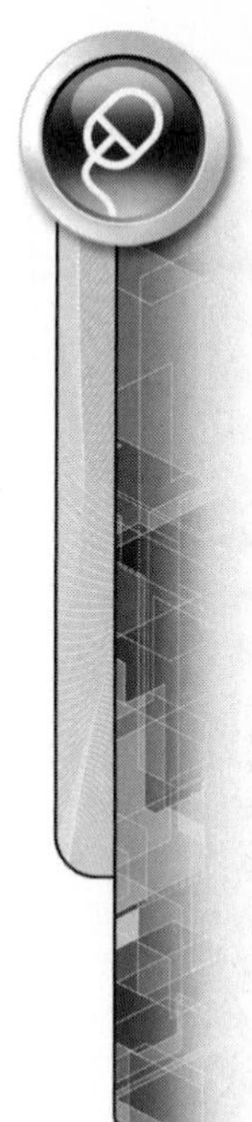

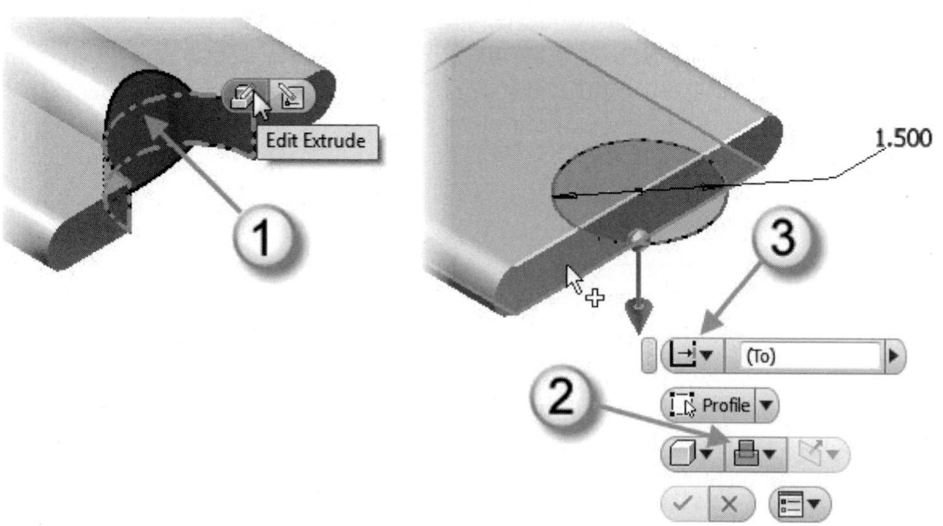

FIGURE 3.55

9. Click the green check mark in the mini-toolbar display. The feature will update as shown in Figure 3.56 on the left.
10. In the browser, right-click Extrusion3. Click Delete and then click OK in the dialog box to delete consumed sketches and features. When done, your part should resemble Figure 3.56 on the right.

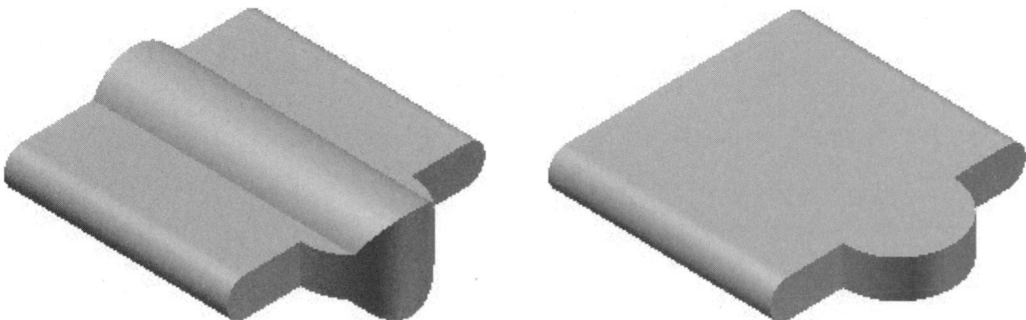

FIGURE 3.56

11. Practice editing the sketch dimensions and features.
12. Close the file. Do not save changes. End of exercise.

PROJECTING PART EDGES

Building parts based partially on existing geometry is done often, and you will frequently need to reference faces, edges, or loops from features that have been created.

While in a part file, you can project an edge, face, point, or loop onto a sketch. Projected geometry maintains an associative link to the original geometry that is projected. If you project the face of a feature onto another sketch, for example, and the parent sketch is modified, the projected geometry will update to reflect the changes.

PROJECT EDGES

In this section, you learn how to use the Project Geometry command that can project selected edges, vertices, work features, curves, the silhouette edges of another part in an assembly, or other features in the same part to the active sketch. There are four project commands available on the Sketch tab > Draw panel >: Project Geometry, Project Cut Edges, Project Flat Pattern and Project to 3D Sketch, as shown in Figure 3.57

Project Geometry

Use to project geometry from a sketch or feature onto the active sketch.

Project Cut Edges

Use to project part edges that touch the active sketch. The geometry is only projected if the uncut part would intersect the sketch plane. For example, if a sphere has a sketch plane in the center of the part and the Project Cut Edges command is initialized, a circle will be projected onto the active sketch.

Project Flat Pattern

Use to project a selected face or faces of a sheet metal part flat pattern onto the active sheet metal part sketch plane. Creation of sheet metal parts is covered in Chapter 10.

Project to 3D Sketch

Use to project geometry from the active 2D sketch onto selected faces to create a 3D sketch. 3D Sketches are covered in Chapter 7.

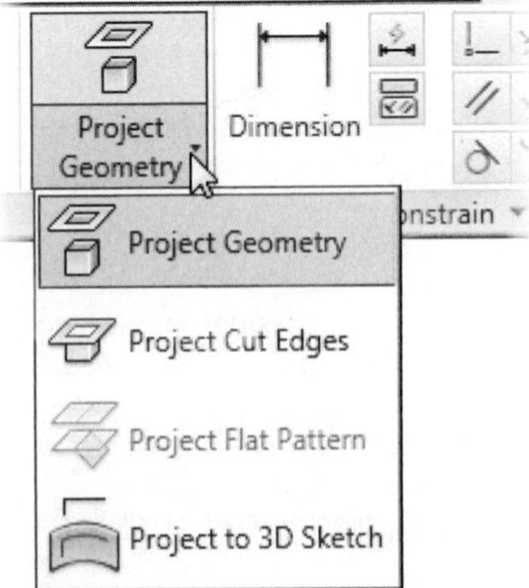

FIGURE 3.57

To project geometry, follow these steps:

1. Create a 2D sketch or make an existing 2D sketch active which the geometry will be projected onto.
2. Click the Project Geometry command on the Sketch tab > Draw panel.
3. Select the geometry to be projected onto the active sketch or click a point in the middle of a face, and all edges of the face will be projected onto the active sketch plane. If you want to project all edges that define the perimeter (a loop), use the Select Other / Face Cycling command to cycle through until they all appear highlighted, as shown in Figure 3.59 on the left. Figure 3.59 on the right shows the projected geometry.
4. To exit the operation, right-click and click OK from the marking menu or press the ESC key or click another command.

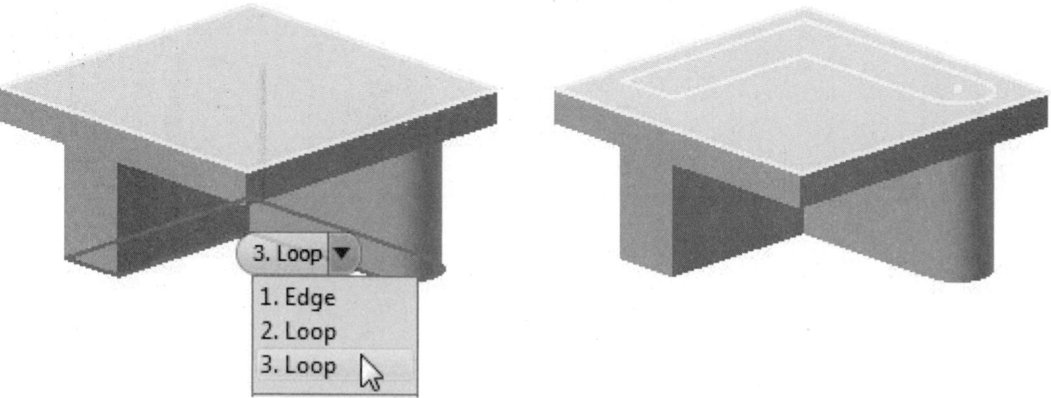

FIGURE 3.58

If you selected a loop for the projection, the sketch is updated to reflect the modification when any geometry of the profile changes. If a face is projected, the internal islands that are defined on the face are also projected and will update accordingly. For example, if the loop of edges of Figure 3.59 on the left is projected, the four lines that define the cutout will be projected and updated if the original feature is modified. To disassociate projected geometry from the original geometry, press the F8 key or click the Show Constraints option on the bottom of the Status Bar to show the sketch constraints and then delete the reference constraint that is created for each projected edge/curve.

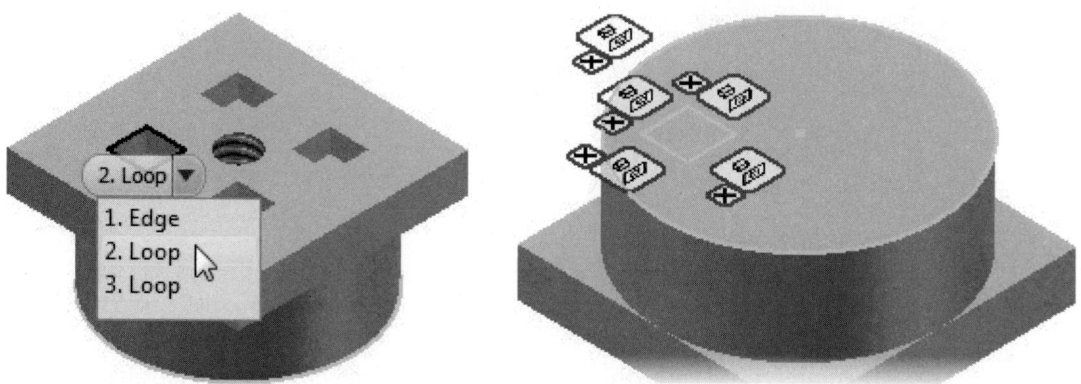

FIGURE 3.59

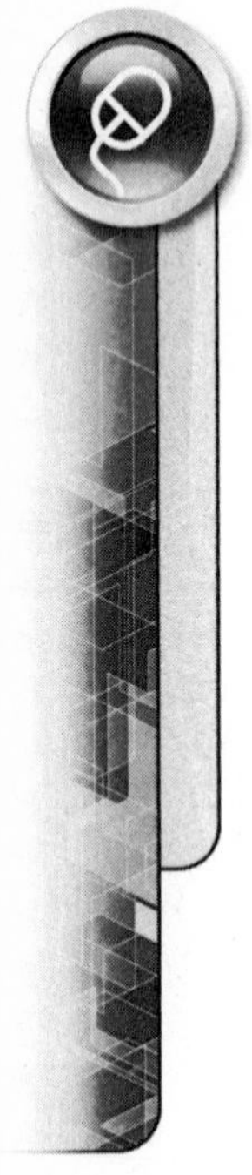

EXERCISE 3-5: PROJECT GEOMETRY

In this exercise, you will project geometry to create a feature, edit the original sketch and update the linked feature and then delete the sketch constraints that were created from the projected geometry.

1. Open ESS_E03_05.ipt from the Chapter 03 folder.
2. Create a sketch on the bottom face of the cylinder (opposite side of Extrusion2) as shown in Figure 3.60 on the left.
3. Rotate your viewpoint so you can see the front of the part.
4. Click the Project Geometry command Sketch tab > Draw panel and select the front face of Extrusion2 as shown in Figure 3.60 on the right.

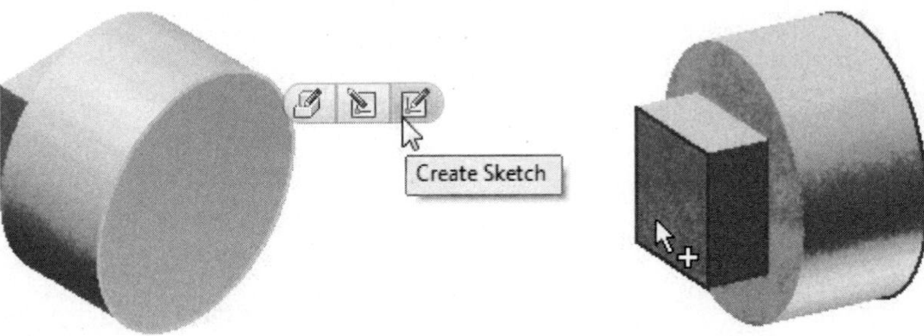

FIGURE 3.60

5. Again rotate your viewpoint so you can see the bottom of the part. Notice that the square has been projected onto the active sketch and is fully constrained.
6. Right-click and click Finish 2D Sketch from the marking-menu.
7. Click on one of the projected lines and click Create Extrude from the mini-toolbar and extrude the inside area of the projected geometry **.5** inches, cutting material into the part as shown in Figure 3.61 on the left.
8. In the browser right-click on Extrusion2 and click Edit Sketch from the menu (the original extruded square).
9. Double-click on the 1.000 dimension and type **.5** and enter and the sketch should resemble the following image on the right **.5**

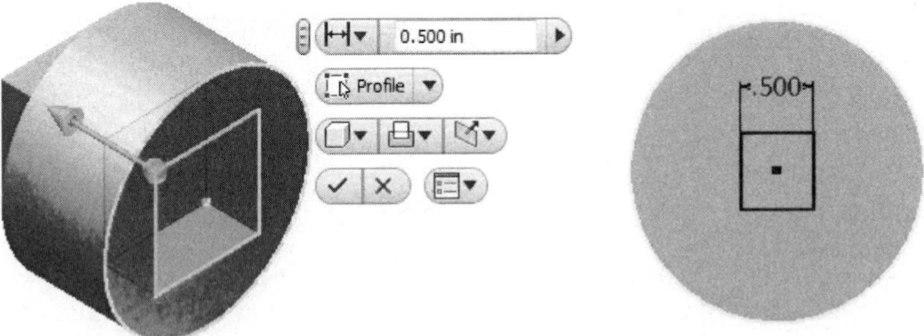

FIGURE 3.61

10. Click Finish Sketch from the Sketch tab > Exit panel and if needed rotate the viewpoint so you can see the bottom of the part. Notice that Extrusion3's sketch and feature have been updated to reflect the change to the geometry that was projected.
11. Edit the sketch of Extrusion 3 by clicking on one of the internal planar faces of the extrusion that removed material and click Edit Sketch from the mini-toolbar as shown in Figure 3.62.

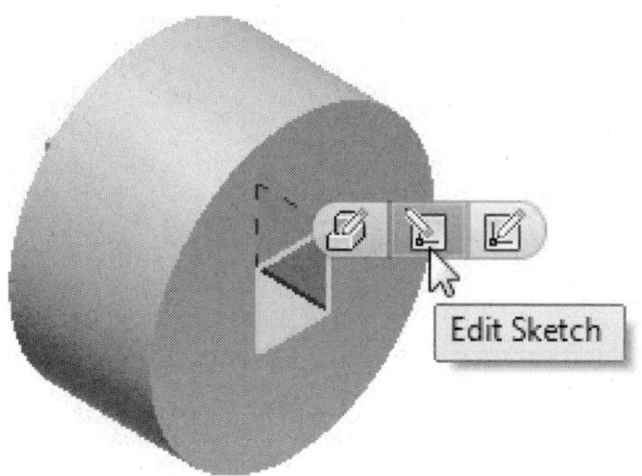

FIGURE 3.62

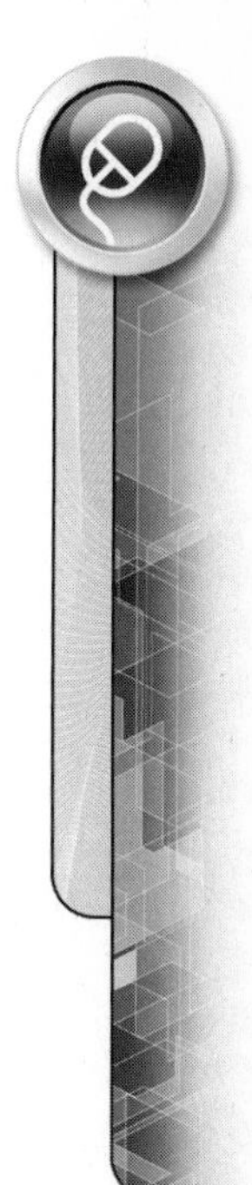

12. Next you will delete the reference constraints that were created when the geometry was projected. Press the F8 key or click Show All Constraints from the Status Bar to display the constraints as shown in Figure 3.63 on the left.
13. Delete the four reference constraints that surround the square (these constraints were automatically created when the geometry was projected onto this sketch). The sketch should now have 8 dimensions/constraints required to fully constrain the sketch.
14. Press the F8 key to refresh the constraints that are displayed, only the four coincident constraints that remain on the sketch are displayed.
15. Click the Automatic Dimension command from the Sketch tab > Constrain panel, select the four inside lines for the curves to Auto Dimension, uncheck the Dimension option (only constraints will be applied) as shown in Figure 3.63 on the right and then click Apply in the dialog box.

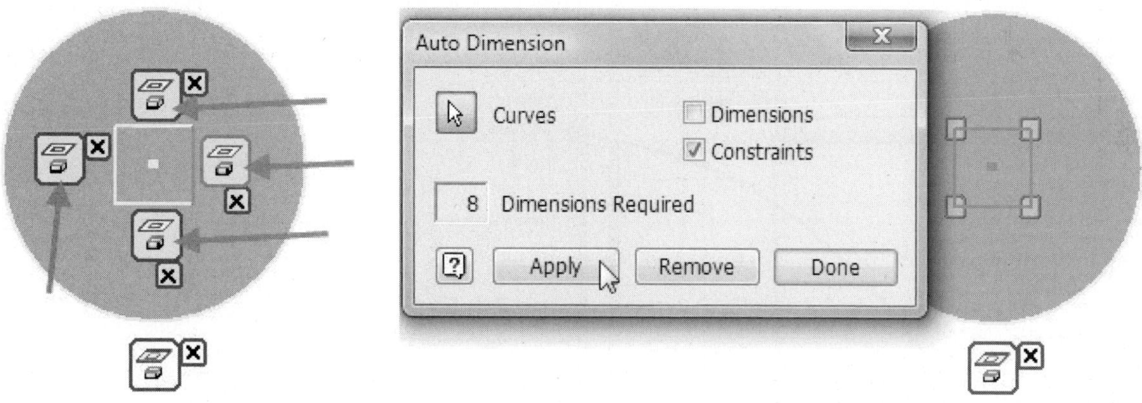

FIGURE 3.63

16. Click the Done button in the dialog box.
17. To display the constraints that were added in the last step press the F8 key.
18. Add a horizontal constraint to the bottom line; this will prevent the bottom line from rotating.
19. Press the F9 key to hide the constraints.
20. To fully constrain the sketch, add the four dimensions as shown in Figure 3.64 in the middle.
21. Finish the sketch and rotate the part so you see that Extrusion3 is a different size than Extrusion2 as shown in Figure 3.64 on the right.

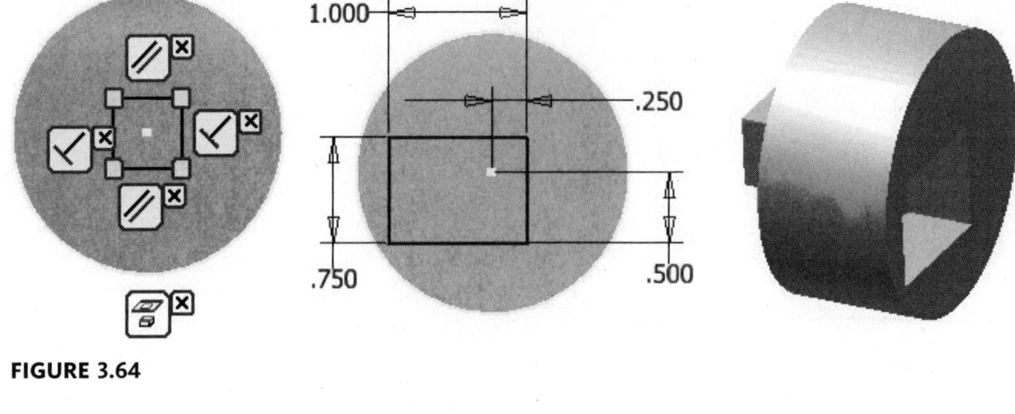

FIGURE 3.64

22. Close the file. Do not save changes. End of exercise.

PART MATERIAL, PROPERTIES AND APPEARANCE

To change a part's physical properties and appearance to a specific material, click down arrow in the Material area in the Quick Access toolbar as shown in Figure 3.66 on the left labeled (1). Another method is to click the Inventor Application Menu and click iProperties from the menu. You can also right-click on the part's name in the browser and click iProperties from the menu. Then in the iProperties dialog box click the Physical tab and select a material from the material drop-down list as shown in Figure 3.65 on the left. Click the Update button in the dialog box to update the properties of the file as shown in Figure 3.65 on the right.

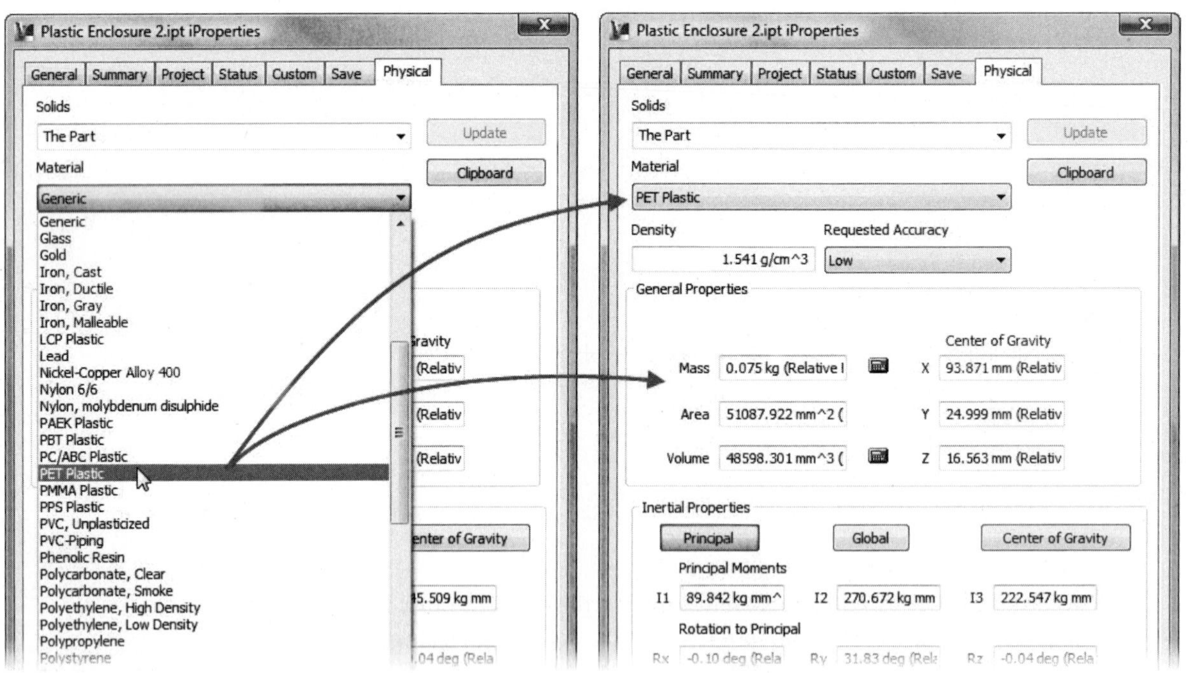

FIGURE 3.65

Click the OK button in the dialog box to complete the operation, and the appearance of the part change to reflect the selected material. You can also override the appearance (color) of a part by clicking on the down arrow in the Appearance area of the Quick Access toolbar as shown in Figure 3.66 on the left labeled (2). Then select an appearance from the list labeled (3) in Figure 3.66 on the left. The appearance override only changes how the part appears and has no affect on the physical material or the mass of the part.

If needed you can modify or create a new material by using the Material Editor or the Appearance Editor command found either in the Quick Access toolbar or in the Tools tab > Material and Appearance panel as shown in Figure 3.66 on the right.

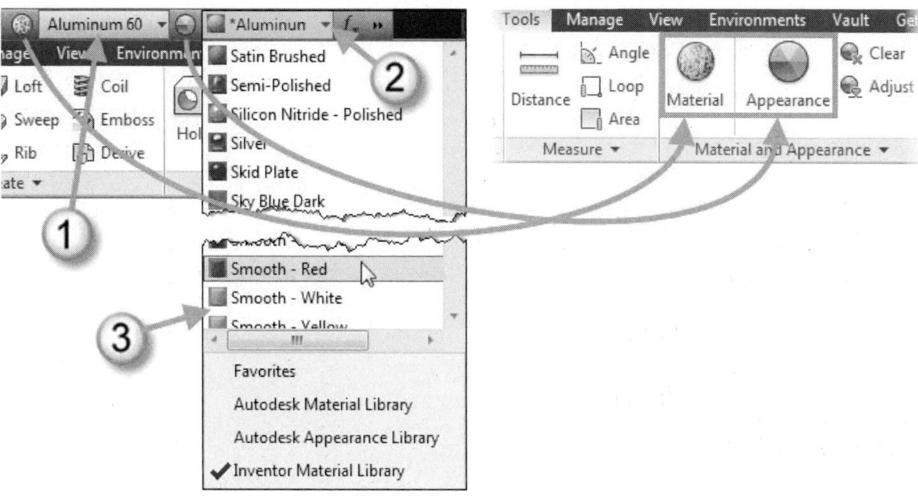

FIGURE 3.66

ADDITIONAL APPEARANCE COMMANDS

While working with a 3D part, you can also adjust the perspective, shading option, control the shadows. and easily change the visually style by using the commands found on the View tab > Appearance panel as shown in Figure 3.67. Following is brief description of each command.

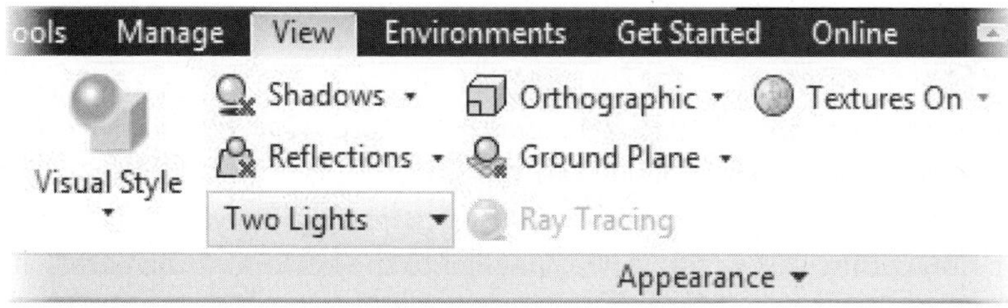

FIGURE 3.67

Visual Style

Set the appearance of model faces and edges in the graphics window. There are ten different styles to choose from as shown in Figure 3.68. The styles do not change the model properties; they just change how the model is displayed. Combine visual styles with the other appearance options to display your model(s) differently.

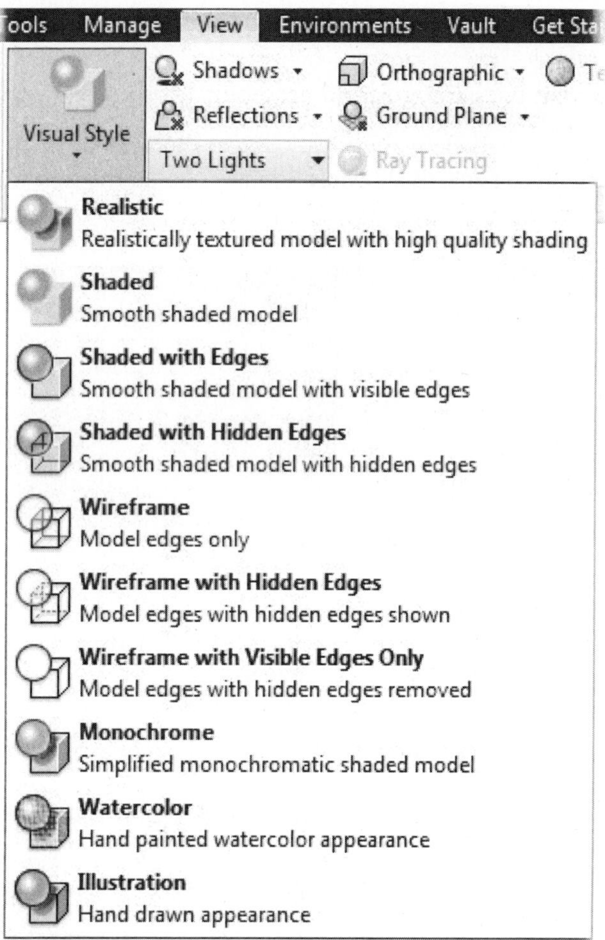

FIGURE 3.68

SHADOWS

To give your model a realistic look, you can choose different options for displaying shadows. By default shadows are turned off.

REFLECTIONS

Click the Reflection command to toggle reflections on / off. The reflection is displayed on the ground plane which is parallel to the Bottom plane of the view cube. You can change which plane is the top and bottom by adjusting the viewpoint with any of the navigation commands so the view is parallel to the bottom of the part that you want. Then right-click on the Viewcube and click Set Current View as > Top.

LIGHTS

Select or set different lighting options. From under the Lights command click the Settings option. From within the dialog box you can control the lights direction, color, add lights and control brightness and ambience. Also from this dialog box you can turn on Image Lighting, which allows you to set a background image and set the scale of the image. With a scene image set you can rotate your viewpoint 360 degrees. To move the image, select the Ground Plane Settings option, change the Position & Size option to Manual adjustment and a triad will appear. Select the arrows on the triad to move the image.

ORTHOGRAPHIC / PERSPECTIVE

Set the viewpoint to be orthographic or perspective. When set to orthographic the viewpoint is parallel, meaning that lines are projected perpendicular to the plane of projection. When set to perspective the geometry on the screen converges to a vanishing point similar to the way the human eye sees. You can adjust the perspective by pressing down the CTRL and SHIFT key and spin the wheel on the mouse.

GROUND PLANE

Click the Ground Plane command to toggle the ground plane on and off. You can change which plane is the ground plane by adjusting the viewpoint with any of the navigation commands so the view is parallel to the bottom of the part that you want. Then right-click on the Viewcube and click Set Current View as > Top. You can adjust the ground plane's appearance by clicking the down arrow next to Ground Plane and click Settings.

TEXTURES ON / TEXTURES OFF

If you have a material or an appearance that applied to a part that has a texture you can toggle the texture on and off. This is a helpful option to increase graphics performance.

APPLYING YOUR SKILLS

Skill Exercise 3-1

In this exercise, you will create a bracket from a number of extruded features. Assume that the part is symmetrical about the center of the horizontal slot.

1. Start a new part based on the English Standard (in).ipt template and create a sketch on the XY origin plane.
2. Sketch the outline for the base feature.
3. Add geometric constraints and dimensions. Make sure that the sketch is fully constrained.
4. Extrude the base feature.
5. Create the three remaining features to complete the part.

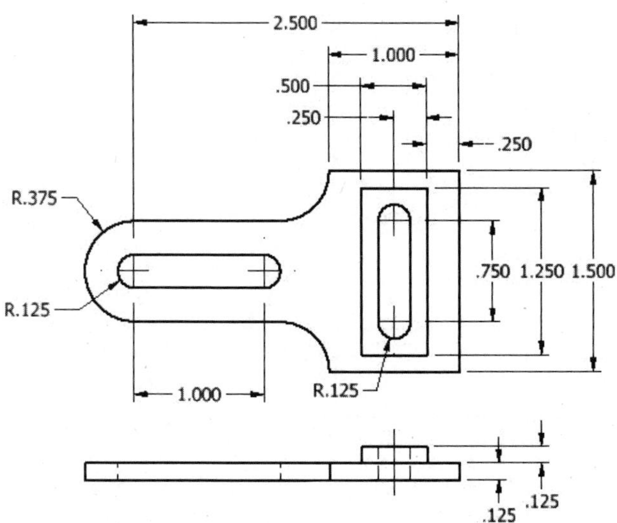

FIGURE 3.69

The completed part should resemble Figure 3.70. When done close the file. Do not save changes. End of exercise.

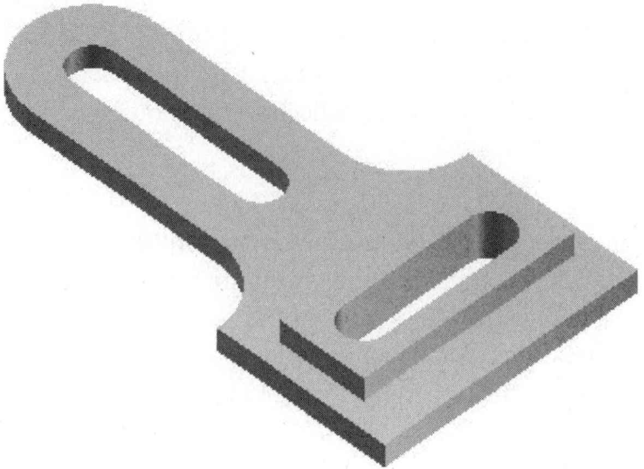

FIGURE 3.70

Skill Exercise 3-2

In this exercise, you will create a connecting rod and add draft to extrusions during feature creation.

1. Start a new part based on the English Standard (in).ipt template and create a sketch on the XY origin plane.
2. Sketch the outside shape of the connecting rod.
3. Add geometric constraints and dimensions to fully constrain the sketch.
4. Extrude the base feature using the Symmetric option, adding a −10° taper.
5. Create a separate feature for each pocket by projecting geometry (don't mirror the feature, the mirror command will be covered in Chapter 7). The sides of the pocket are parallel to the sides of the connecting rod.

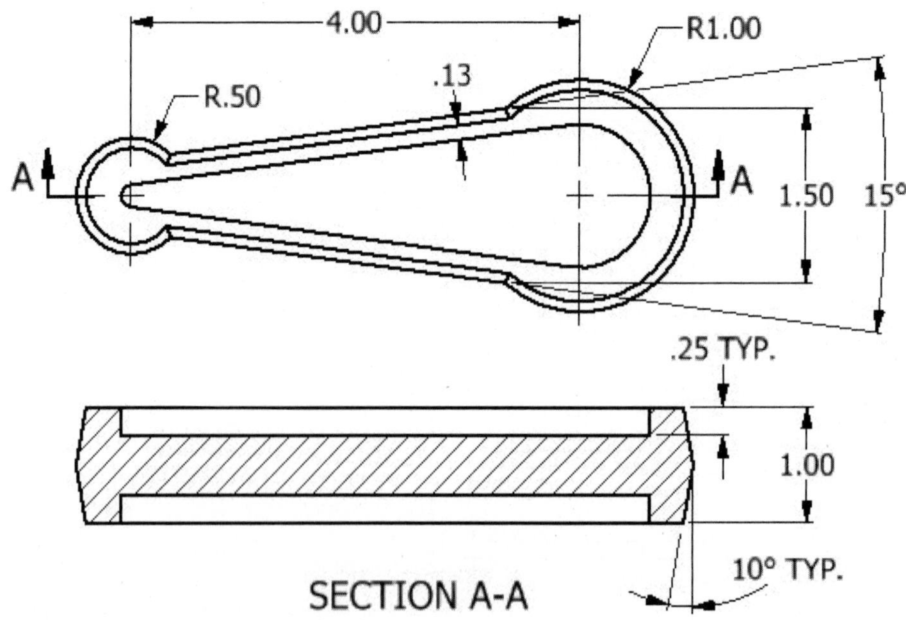

FIGURE 3.71

The completed part should resemble Figure 3.72. When done close the file. Do not save changes. End of exercise.

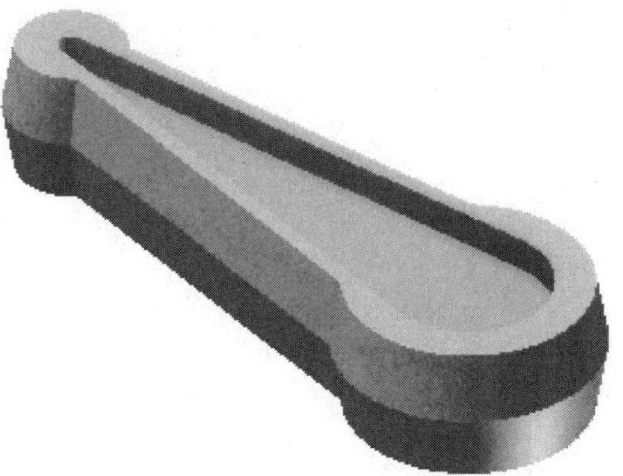

FIGURE 3.72

Skill Exercise 3-3

In this exercise, you will create a pulley using a revolved feature. Assume that the part is symmetric about the middle of the part vertically.

1. Start a new part based on the English Standard (in).ipt template and create a sketch on the XY origin plane.
2. Sketch the cross-section of the pulley.

> **TIP**
>
> 
>
> **To create a centerline, draw a line, select it, and then click the Centerline command on the Sketch tab > Format panel.**

3. Apply appropriate sketch constraints.
4. Add dimensions.
5. Revolve the sketch using the Full Extents.

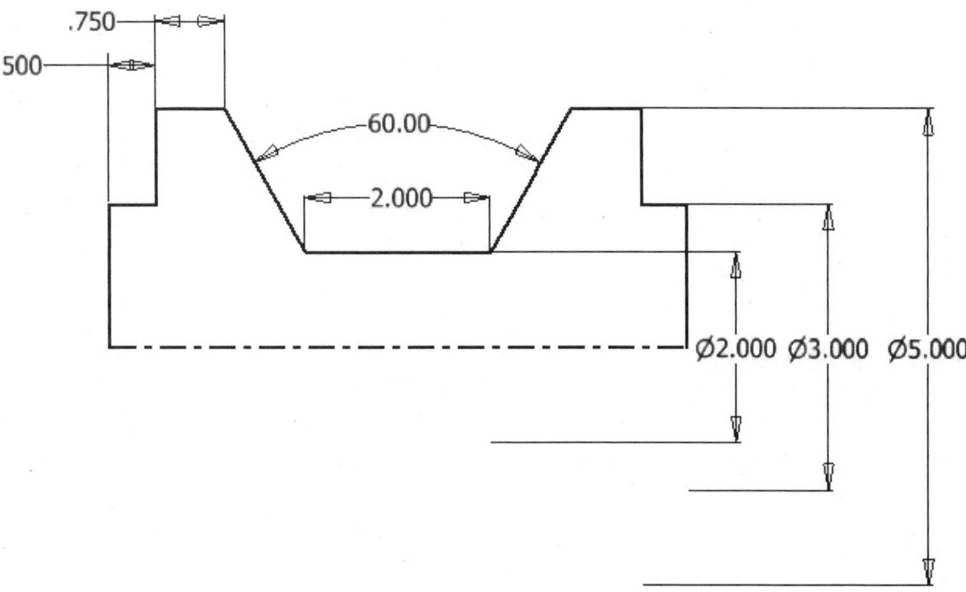

FIGURE 3.73

The completed part should resemble Figure 3.74. When done, close the file. Do not save changes. End of exercise.

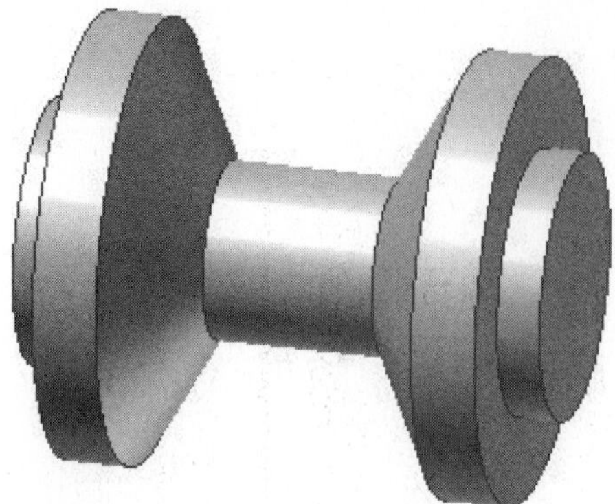

FIGURE 3.74

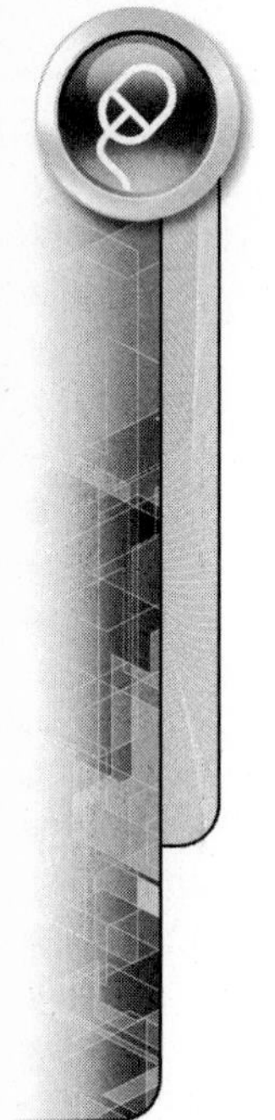

CHECKING YOUR SKILLS

Use these questions to test your knowledge of the material covered in this chapter.

1. What is a base feature?
2. True __ False __ When creating a feature with the Extrude or Revolve command, the mini-toolbar buttons only allows you to define the distance or angle.
3. When creating a revolve feature, which objects can be used as an axis of revolution?
4. Explain how to create a linear diameter (diametric) dimension on a sketch.
5. Name two ways to edit an existing feature.
6. True __ False __ Once a sketch becomes a base feature, you cannot delete or add constraints, dimensions, or objects to the sketch.
7. Name three operation types used to create sketched features.
8. True __ False __ A direct manipulation technique can only be started by clicking on a face of a part.
9. True __ False __ Once a sketched feature exists, its extents type cannot be changed.
10. True __ False __ By default geometry that is projected from one feature to a sketch will update automatically based on changes to the original projected geometry.
11. Explain what the asymmetric option does for the extrude and revolve commands.
12. Where do you set the physical material property of a part?
13. True __ False __ After setting a part's material properties, the Appearance setting in the Quick Access toolbar must be set to Color Matching for the appearance to match the material of the part.
14. True __ False __ The Project Geometry command is used to copy objects onto any selected face.
15. True __ False __ By default, when a feature is deleted, the feature's sketch will be maintained.

CHAPTER 4

Creating Placed Features

INTRODUCTION

In Chapter 3, you learned how to create and edit base and sketched features. In this chapter, you will learn how to create *placed* features. Many of the placed features are predefined except for specific values and only need to be located. You can edit placed features in the browser like sketched features. When you edit a placed feature, either the dialog box and mini-toolbar that you used to create it will open or feature values will appear in the graphics window.

When creating a part, it is usually better to use placed features instead of sketched features wherever possible. To make a through hole as a sketched feature, for example, you can draw a circle profile, dimension it, and then extrude it with the cut operation. By creating a hole as a placed feature, you can select the type of hole, size it, and then place it using a dialog box. When drawing views are generated, the type and size of the hole are easy to annotate, and they automatically update if the hole type or values change if a circle was extruded the only information that you can retrieve in a drawing is the hole's diameter.

OBJECTIVES

After completing this chapter, you will be able to perform the following:

- Create fillets
- Create chamfers
- Create holes
- Shell a part
- Create work axes
- Create work points
- Create work planes
- Create a UCS
- Pattern features

FILLETS

Fillet features consist of fillets and rounds. Fillets add material to interior edges to create a smooth transition from one face to another. Rounds remove material from exterior edges. Figure 4.1 shows a part without fillets on the left and the part with fillets on the right.

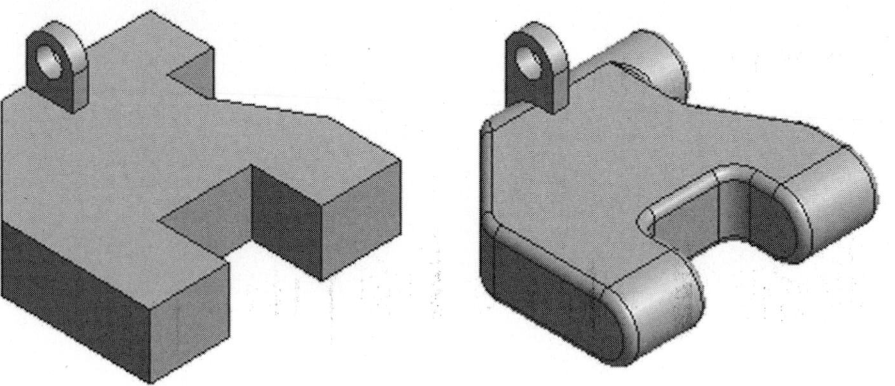

FIGURE 4.1

To create fillets, you select the edge that needs to be filleted; the fillet is created between the two faces that share the edge, or you can select two or three faces that a fillet will go between. When placing a fillet between two faces, the faces do not need to share a common edge. When creating a part, it is good practice to create fillets and chamfers as some of the last features in the part. Fillets add complexity to the part, which in turn adds to the size of the file. They also remove edges that you may need to place other features.

To create a fillet feature, click the Fillet command from the 3D Model tab > Modify panel, as shown in Figure 4.2 on the left, while not in a sketch click on an edge of a part and the mini-toolbar will appear click the Create Fillet button as shown in Figure 4.2 on the right or press the shortcut key F.

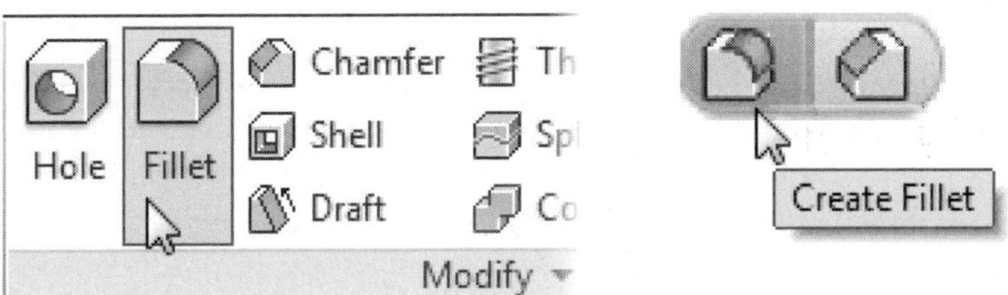

FIGURE 4.2

After you click the command, the Fillet dialog box appears as shown in Figure 4.3 on the left and the mini-toolbar will appear as shown in Figure 4.3 on the right. Along the left side of the Fillet dialog box, there are three types of fillets: Edge, Face, and Full Round. When the Edge option is selected, you see three tabs: Constant, Variable, and Setbacks. Each tab creates a fillet along an edge(s) with different options. The options for each of the tabs and fillet types are described in the following sections. A preview of a fillet will appear on the part when the preview option on the bottom of the dialog box is checked and a valid fillet can be created from the data you input into the Fillet dialog box or the mini-toolbar.

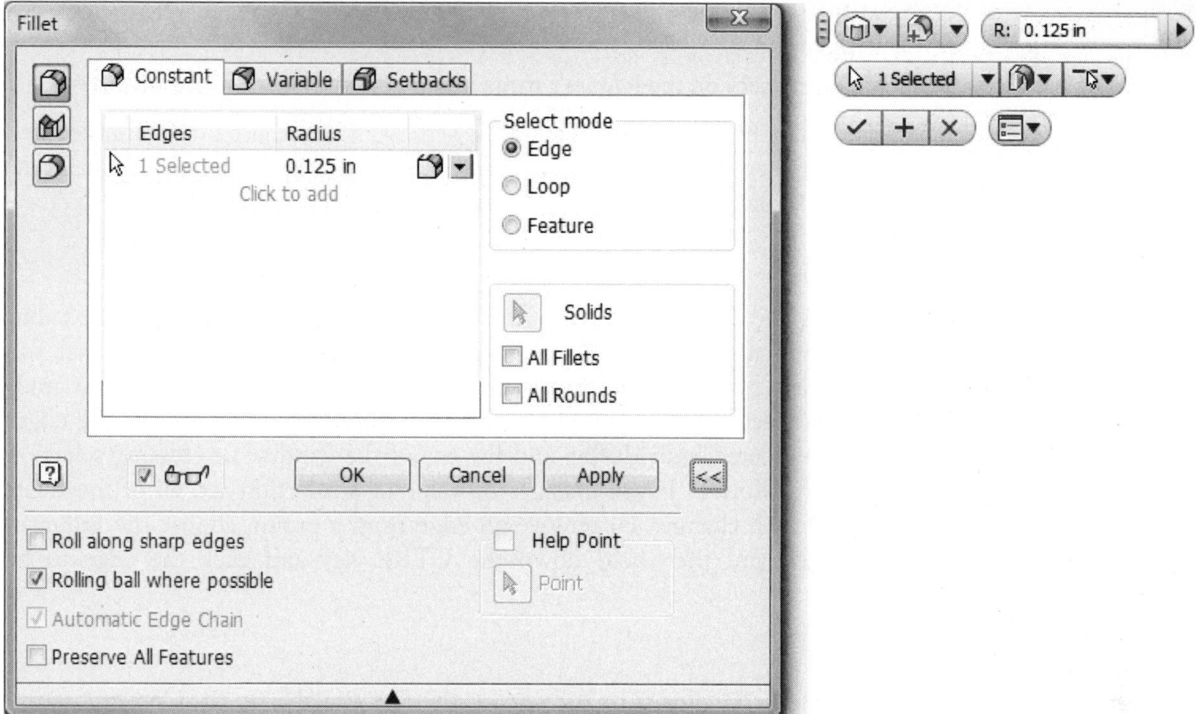

FIGURE 4.3

To create a fillet feature can click an edge or edges, or faces to fillet, or select the type of fillet to create. If you click an edge or face before you issue the Fillet command, it is placed in the first selection set. Selection sets contain the edges or faces that will be filleted when the OK button is clicked.

Each fillet feature can contain multiple selection sets, each having its own unique fillet value. There is no limit to the number of selection sets that can exist in a single instance of the feature. An edge, however, can only exist in one selection set. All of the selection sets included in an individual fillet command appear as a single fillet feature in the browser. Click the type of fillet that you want to create. To add edges to the first selection set, select the edges that you want to have the same radius. To create another selection set, select Click to add, and then click the edges that will be part of the next selection set. To remove an edge or face that you have selected, click the selection set that includes the edge or face, and the edges or faces will be highlighted. Hold down the CTRL key, and click the edge(s) or face(s) to be removed from the selection set. Enter the desired values for the fillet. As changes are made in the dialog box, a representation of the fillet is previewed in the graphics area. When the fillet type and value are correct, click the OK button to create the fillet.

To edit a fillet's type and radius, use one of the following methods to start the editing process:

- On the part click on the fillet to edit and click the Edit Fillet button in the mini-toolbar.

- Right-click on the fillet's name in the browser and selecting Edit Feature from the menu.
- Double-click on the feature's name in the browser.

The Fillet dialog box appears with all of the settings that you used to create the fillet. Change the fillet settings as needed. Then if needed click the Local Update command on the Quick Access toolbar to update the part.

Edge Fillet

Edge fillet is the default fillet type and creates fillets that have the same radius from beginning to end. There is no limit to the number of part edges that you can fillet with a constant fillet. You can select the edges as a single set or as multiple sets, and each set can have its own radius value. You need to select the edges that are to be filleted individually, and the use of the window or crossing selection method is not allowed. If you change the value of a selection set, all of the fillets in that group will change. To remove an edge from a group, choose the group in the Edges cell and then hold down the CTRL key and click the edge to be removed.

Constant Tab

The following section describes the options that are available on the Constant tab.

Select Edge, Radius, and Continuity

Edge

By default, after issuing the Fillet command, you can click edges, and they appear in the first selection set. You can continue to select multiple edges. To remove an edge from the set, hold down the CTRL key and select the edge.

Radius

Enter a size for the fillet. The size of the fillet will be previewed on the selected edges.

Continuity

To adjust the continuity of the fillet, select either a tangent or smooth (G2) condition (continuous curvature), as shown in Figure 4.4 in both the dialog box and in the mini-toolbar.

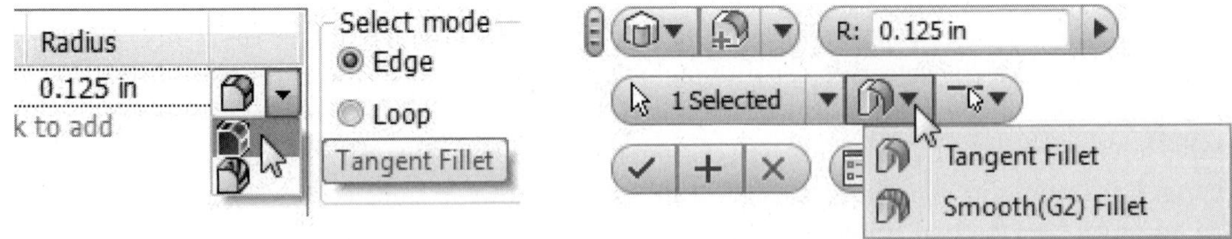

FIGURE 4.4

To create another selection set, select Click to add, and then click the edges that will be part of the next selection set. After clicking an edge, a preview image of the fillet appears on the edge that reflects the current values that are set in the dialog box or mini-toolbar, as shown in Figure 4.5.

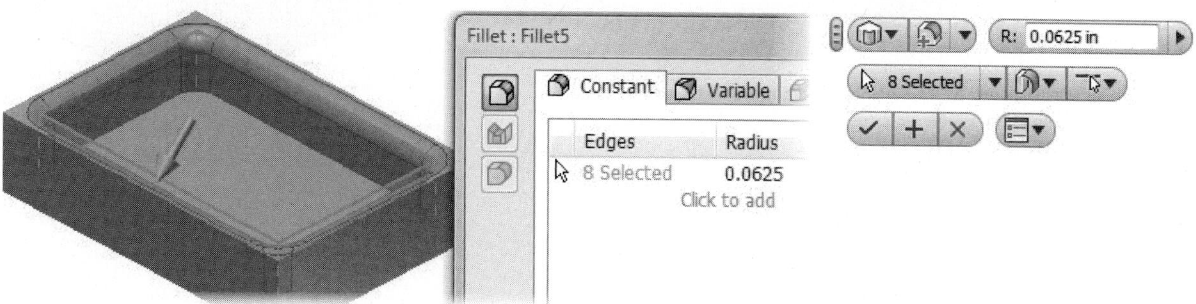

FIGURE 4.5

Select Mode - Additional Modes

Loop

Click the Loop mode to have all of the edges that form a closed loop with the selected edge filleted.

Feature

Click the Feature mode to select all of the edges of a selected feature.

Solids

If multiple solid bodies exist, select the solid body to apply All Fillets or All Rounds.

All Fillets

Click the All Fillets option to select all concave edges of a part that you have not filleted already—see Figure 4.6 on the left. The All Fillets option adds material to the part, and it requires a separate edge selection set to remove material from the remaining edges using the All Rounds option.

All Rounds

Click the All Rounds option to select all convex edges of a part that you have not filleted already, as seen in Figure 4.6 on the right. The All Rounds option removes material from the part, and it requires a separate edge selection set from All Fillets.

FIGURE 4.6

More (>>) Options. Click the More (>>) button to access other options, as shown in Figure 4.7.

Roll along Sharp Edges

Click this option to adjust the specified radius when necessary to preserve the edges of adjacent faces.

Rolling Ball Where Possible

Click this option to create a fillet around a corner that looks like a ball has been rolled along the edges that define a corner, as shown in Figure 4.7 (middle). When the Rolling ball where possible solution is possible, but it is not selected, a blended solution is used, as shown in Figure 4.7 on the right.

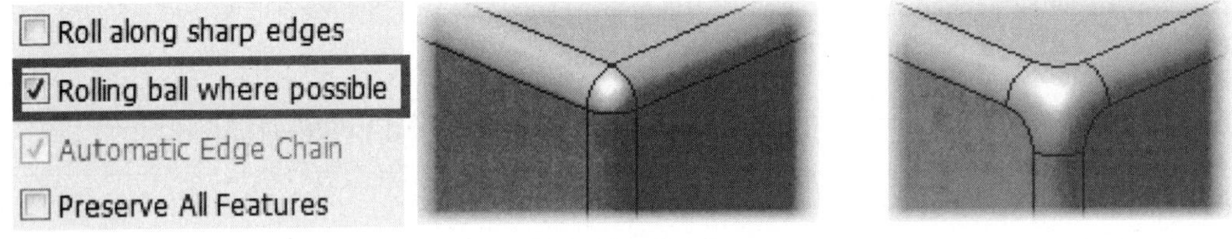

FIGURE 4.7

Automatic Edge Chain

Click this option to select tangent edges automatically when you click an edge.

Variable Tab

You can also create a variable radius fillet that has a different starting and ending radius and/or a different radius between the starting and ending radius. To create a variable radius fillet, click the Edge Fillet option in the upper-left corner of the dialog box as shown in the following image labeled (1), and click the Variable tab, as shown in Figure 4.8 labeled (2) on the left or click the Add Variable Set option from the mini-toolbar on the right labeled (2).

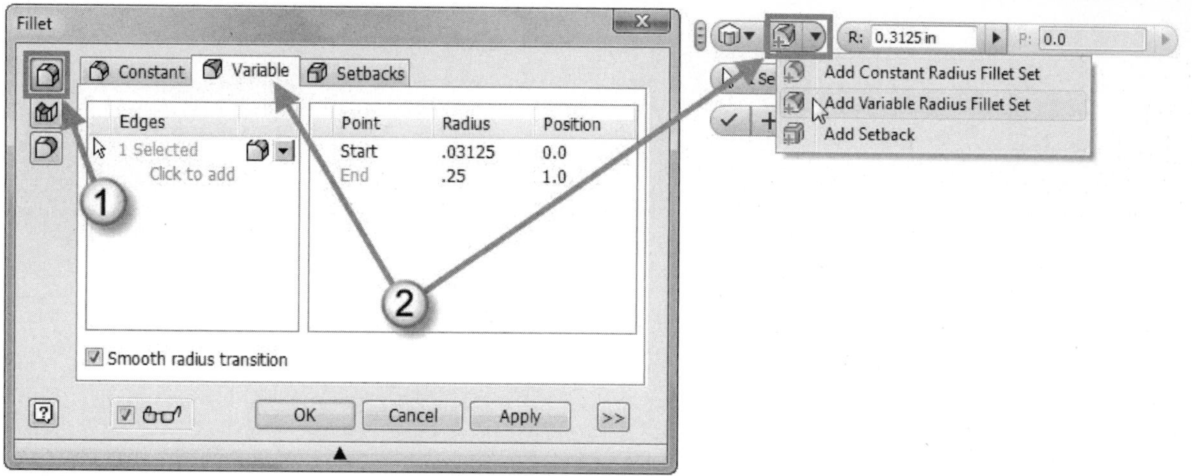

FIGURE 4.8

Figure 4.9 on the left shows a variable fillet with the smooth option, and the image on the right shows a variable fillet blending in a straight line.

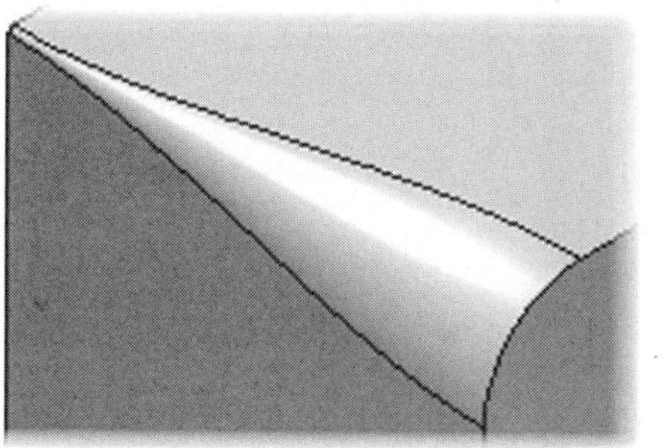

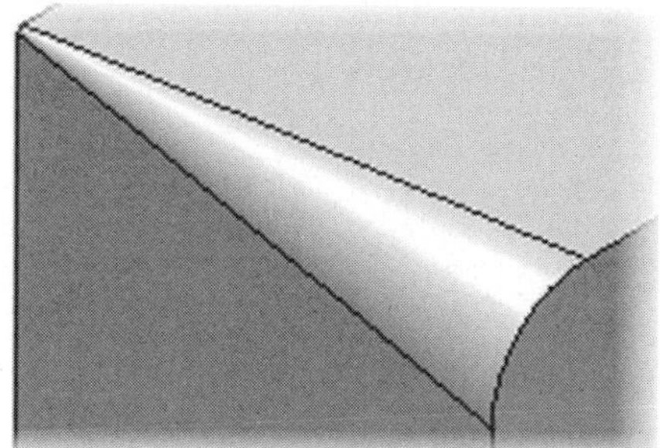

FIGURE 4.9

Setbacks Tab

You can specify the distance at which a fillet starts its transition from a vertex with the options on the Setbacks tab. Using these options, you can model special fillet applications where three or more edges converge, as shown in Figure 4.10. You can choose a different radius for each converging edge if needed. Click the minimal option to create a setback with the smallest possible fillet. You can only use setbacks where three or more filleted edges form a vertex. To setback fillets, follow these steps:

1. In the graphics window, select three or more edges to fillet. The fillets must converge at a point.
2. Click the Setbacks tab.
3. In the graphics window, click the vertex point.
4. In the Fillet dialog box, set the setback distance for each edge.

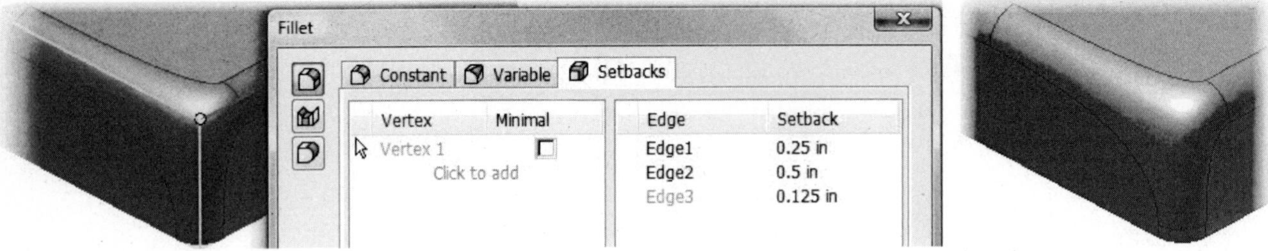

FIGURE 4.10

Face Fillet

With the Face fillet type selected, the dialog box will change, as shown in Figure 4.11 on the left labeled (1) and the mini-toolbar on the right labeled (2). Create a face fillet by selecting two or more faces; the faces do not need to be adjacent. If a feature exists that will be consumed by the fillet, the internal volume of the feature will be filled in by the fillet.

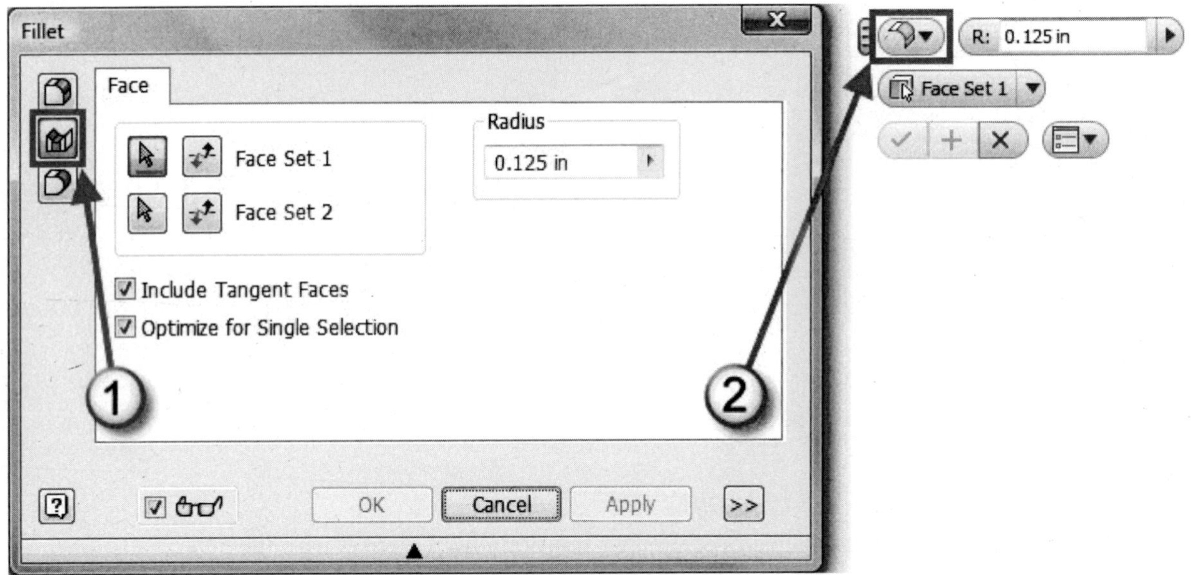

FIGURE 4.11

To create a face fillet, follow these steps:

1. With the Face Set 1 button active, select one or more tangent contiguous faces on the part to which the fillet will be tangent.
2. With the Face Set 2 button active, select one or more tangent contiguous faces on the part to which the fillet will be tangent.

Check the Include Tangent Faces option to automatically chain all faces that are tangent to faces in the selection set.

Figure 4.12 on the left shows a part that has a gap between the bottom and top extrusion, the middle image shows the preview of the face fillet, and the image on the right shows the completed face fillet. Notice the rectangular extrusion on the top face is consumed by the fillet.

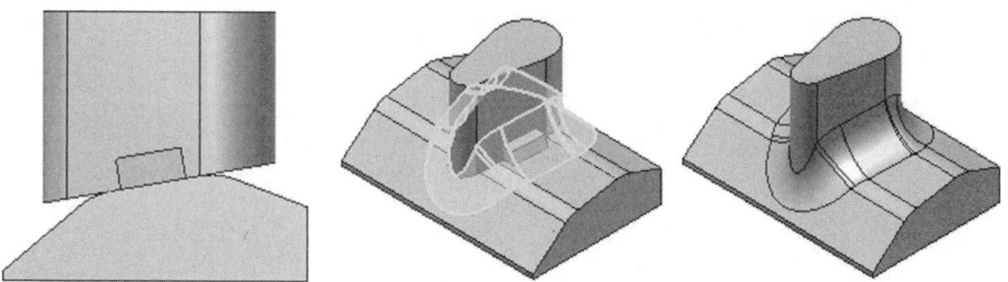

FIGURE 4.12

FullRound Fillet

Click the FullRound fillet type in the dialog box labeled (1) or in the mini-toolbar labeled (2) as shown in Figure 4.13. The full round fillet option creates a fillet that is tangent to three faces; the faces do not need to be adjacent.

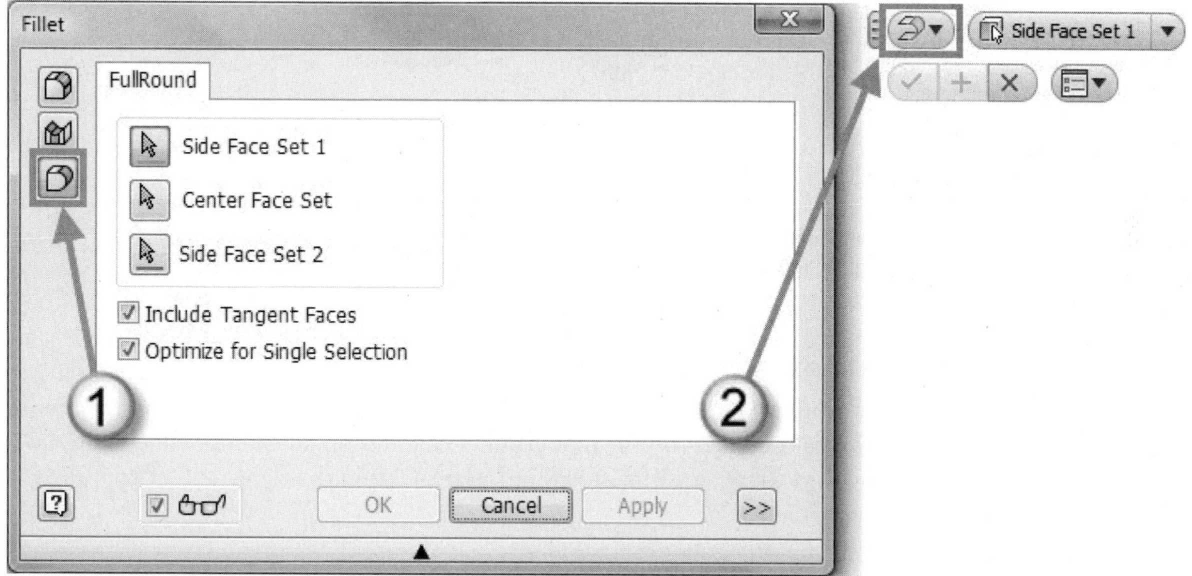

FIGURE 4.13

To create a FullRound fillet, follow these steps:

1. With the Side Face Set 1 button active, select a face on the part that the fillet will start at and to which it will be tangent.
2. With the Center Face Set button active, select a face on the part that the middle of the fillet will be tangent.
3. With the Side Face Set 2 button active, select a face on the part that the fillet will end at and to which it will be tangent.

Check the Include Tangent Faces option to automatically chain all faces that are tangent to faces in the selection set.

Check the Optimize for Single Selection to automatically make the next selection set button active after selecting a face.

Figure 4.14 shows three faces selected on the left and the resulting fullround fillet on the right.

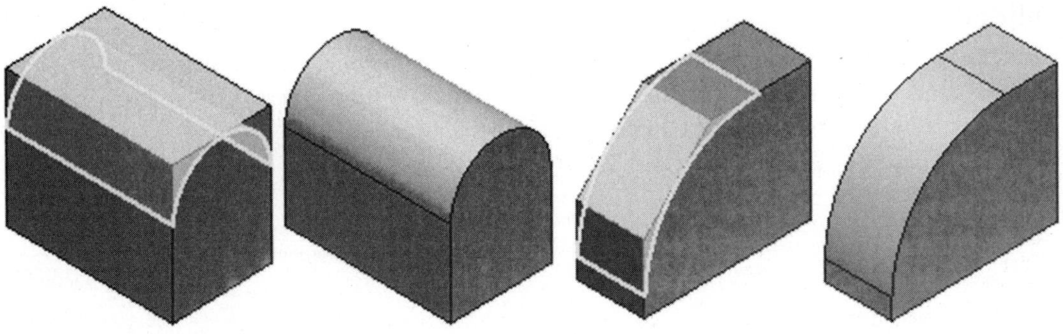

FIGURE 4.14

TIP If you get an error when creating or editing a fillet (an error glypg will appear on the right-side of the mini-toolbar), try to create it with a smaller radius. If you still get an error, try to create the fillet in a different sequence or create multiple fillets in one feature.

CHAMFERS

Chamfer features are used to bevel edges. When you create a chamfer on an interior edge, material is added to your model and a chamfer on an exterior edge, material is cut away from your model, as shown in Figure 4.15 on the right.

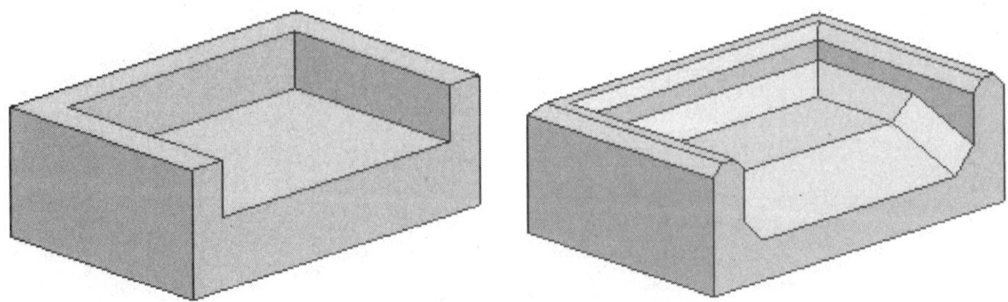

FIGURE 4.15

To create a chamfer feature, follow the same steps that you used to create fillet features. Click the common edge, and the chamfer is created between the two faces sharing the edge. To create a chamfer feature, click on the Chamfer command on the 3D Model tab > Modify panel, as shown in Figure 4.16 on the left, or while not in a sketch, click on an edge and the mini–toolbar will appear, then click the Create Chamfer button in the mini-toolbar, as shown in Figure 4.16 on the right.

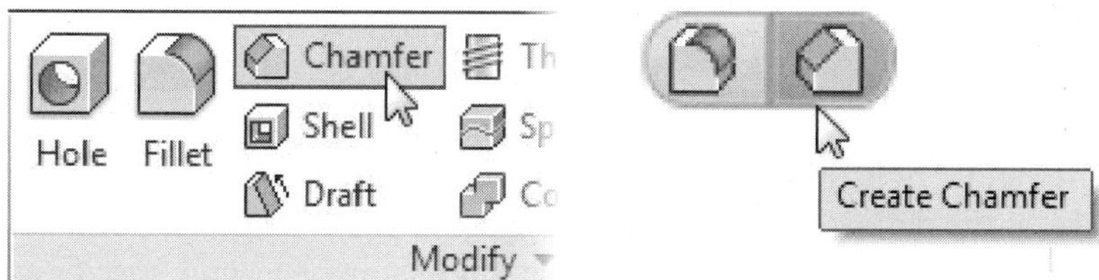

FIGURE 4.16

After you start the chamfer command, the Chamfer dialog box appears, as shown in Figure 4.17 on the left and the mini-toolbar appears as shown in Figure 4.17 on the right. As with fillet features, you can select multiple edges to be included in a single chamfer feature. From the dialog box or mini-toolbar, click a method, click the edge or edges to chamfer, enter a distance and/or angle, and then click OK.

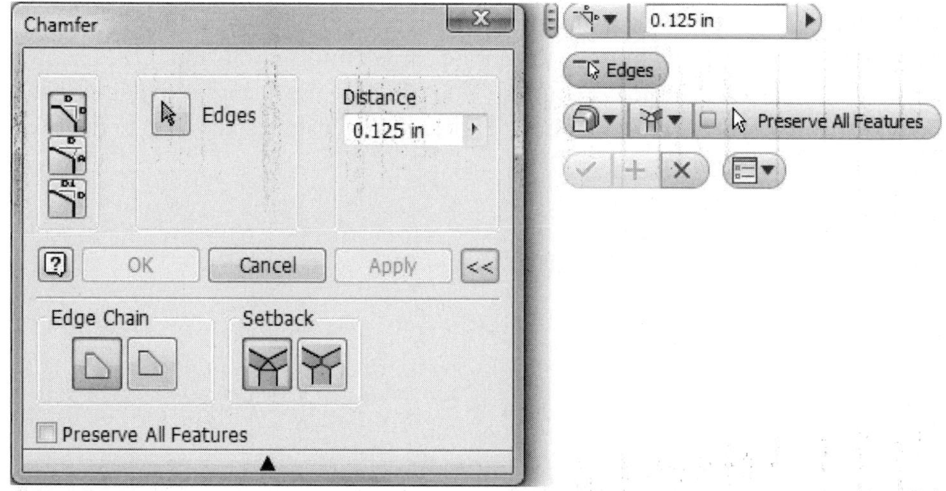

FIGURE 4.17

To edit the type of chamfer feature use one of the following methods to start the editing process and the Chamfer dialog box appears with all the settings that you used to create the feature. Adjust the settings as desired.

- In the graphics window click on the face chamfer to edit and click the Edit Chamfer button that appears in the mini-toolbar.
- Double-click the chamfer's name or icon in the browser.
- Right-click the chamfer's name in the browser, and select Edit Feature from the menu.
- Change the select priority to Feature Priority, and then double-click the chamfer on the part.

Method

Distance

Click the Distance option to create a 45° chamfer on the selected edge. You determine the size of the chamfer by typing a distance in the dialog box. The value is the offset from the common edge of the two adjacent faces. You can select a single edge, multiple edges, or a chain of edges. A preview image of the chamfer appears on the part. If you select the wrong edge, hold down the CTRL key and select the edge to remove. Figure 4.18 illustrates the use of the Distance option on the left.

Distance and Angle

Click the Distance and Angle option to create a chamfer offset from a selected edge on a specified face, at the defined angle. In the dialog box, enter an angle and distance for the chamfer, then click the face to which the angle is applied and specify an edge to be chamfered. You can select one edge or multiple edges. The edges must lie on the selected face. A preview image of the chamfer appears on the part. If you have selected the wrong face or edge, click on the Edge or Face button, and choose a new face or edge. Figure 4.18 in the middle illustrates the use of the Distance and Angle option.

Two Distances

Click the Two Distances option to create a chamfer offset from two faces, each being the amount that you specify. Click an edge first, and then enter values for Distance 1 and Distance 2. A preview image of the chamfer appears. To reverse the direction of the distances, click the Flip button. When the correct information about the chamfer is in the dialog box, click the OK button. You can only use a single edge or chained edges with the Two Distances option. Figure 4.18 illustrates the use of the Two Distances option on the right.

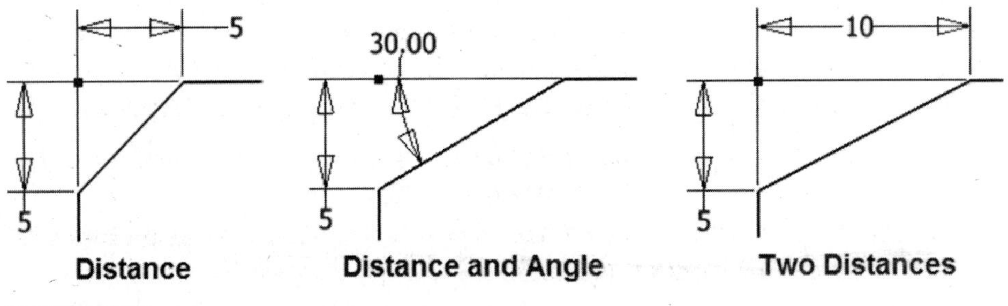

FIGURE 4.18

Edge and Face

Edges

Click an edge or edges to be chamfered.

Face

Click a face on which the chamfer will be based.

Flip

Click the button to reverse the direction of the distances for a Two Distances chamfer.

Distance and Angle

Distance

Enter a distance to be used for the offset.

Angle

Enter a value that will be used for the angle if creating the Distance and Angle chamfer type.

Edge Chain and Setback

Edge Chain

Click this option to include tangent edges in the selection set automatically, as shown in Figure 4.19.

Setback

When the Distance method is used and three chamfers meet at a vertex, click this option to have the intersection of the three chamfers form a flat edge (left button) or to have the intersection meet at a point as though the edges were milled (right button), as shown in Figure 4.19.

Preserve All Features

Click this option to check all features that intersect with the chamfer and to calculate their intersections during the chamfer operation, as shown in Figure 4.19. If the option's checkbox is clear, only the edges that are part of the fillet operation are calculated during the operation.

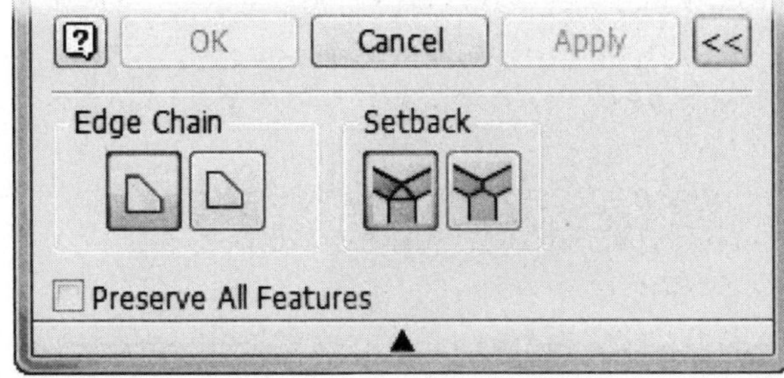

FIGURE 4.19

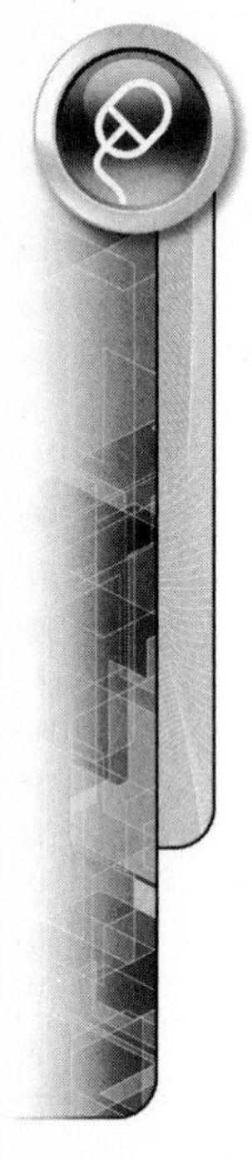

EXERCISE 4-1: CREATING FILLETS AND CHAMFERS

In this exercise, you will create constant radius fillets, full round fillets, face fillet, and chamfers.

1. Open *ESS_E04_01.ipt* in the Chapter 04 folder.
2. Click the Fillet command on the 3D Model tab > Modify panel or click the inside edge of the cutout and click Create Fillet from the mini-toolbar.
3. If needed expand the Fillet dialog box by clicking the down arrow at the bottom the minimized Fillet dialog box, click on the first entry in the Radius column, and type **.25** or in the mini-toolbar enter **.25** as shown in Figure 4.20.

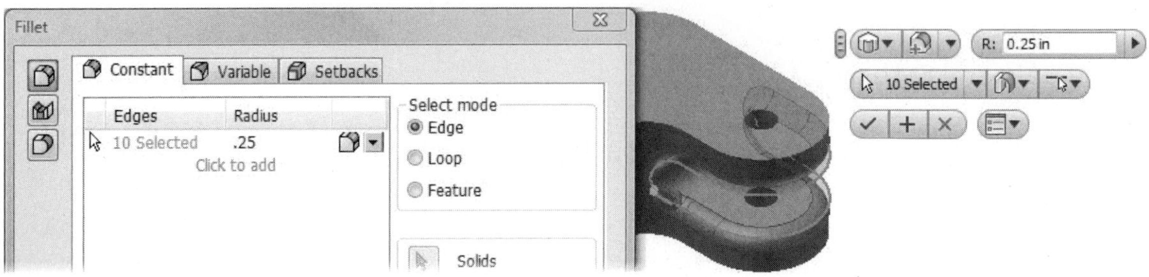

FIGURE 4.20

4. Click Apply in the dialog box or click the + in the mini-toolbar to create the fillet.
5. Next, create a Full Round Fillet that will be tangent to the front three faces. In the Fillet dialog box, click the Full Round Fillet option in the left column labeled (A) in **Figure 4.21**.
6. Click the front-inside face, then the left-front face, and then the back-vertical face of the part, as shown in Figure 4.21 labeled (1), (2), and (3).

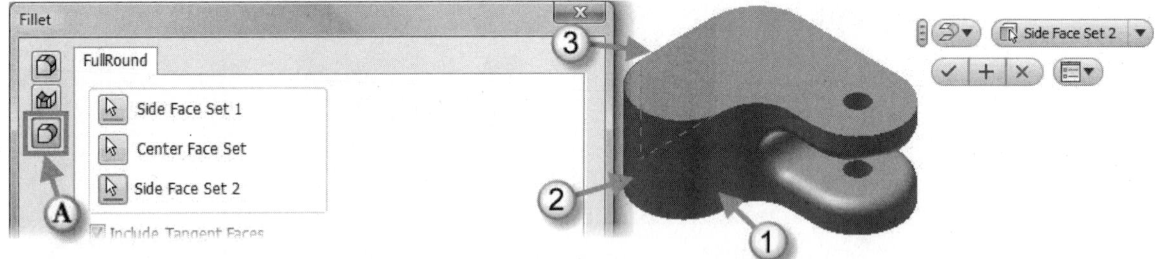

FIGURE 4.21

7. Click OK in the dialog box or click the check mark in the mini-toolbar to create the fillet.
8. In the browser, right-click on Extrusion4, and click Unsuppress Features.
9. Next, create a Face Fillet. Click the Fillet command in the Modify panel. In the Fillet dialog box, click the Face Fillet option in the left column as shown in **Figure 4.22** labeled (A).
10. Click the cylindrical face of Extrusion4 and then the full round fillet in the model you created in the last step, as shown in Figure 4.22 labeled (1) and (2).

11. In the Fillet dialog box, type **.375 in** in the Radius field, as shown in Figure 4.22.

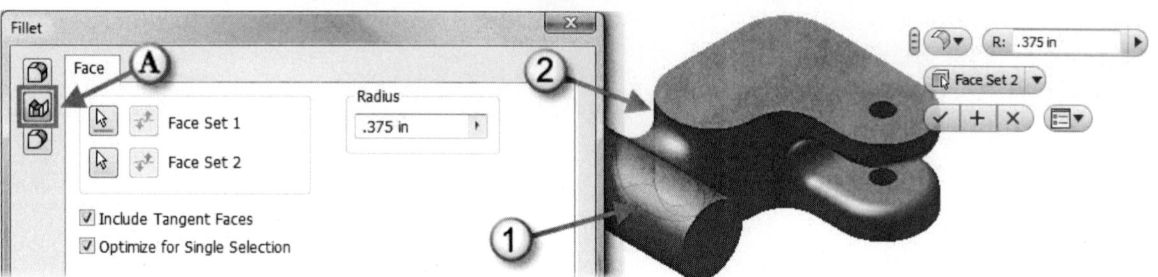

FIGURE 4.22

12. Click OK to create the fillet and rotate the viewpoint to examine the fillet features.
13. Move the cursor into a blank area in the graphics window, right-click, and click Repeat Fillet from the top of the context menu.
14. Next you fillet two edges. Click the top and bottom edges of the model, as shown in Figure 4.23.
15. In the mini-toolbar input radius field, type **.125**, as shown in Figure 4.23.

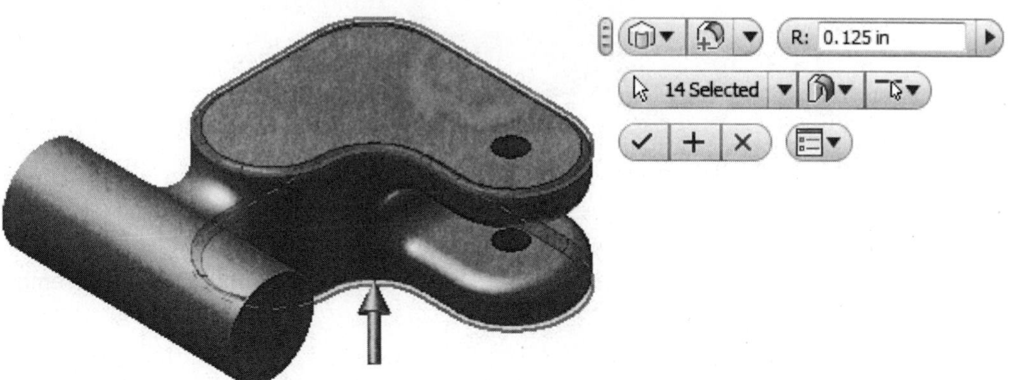

FIGURE 4.23

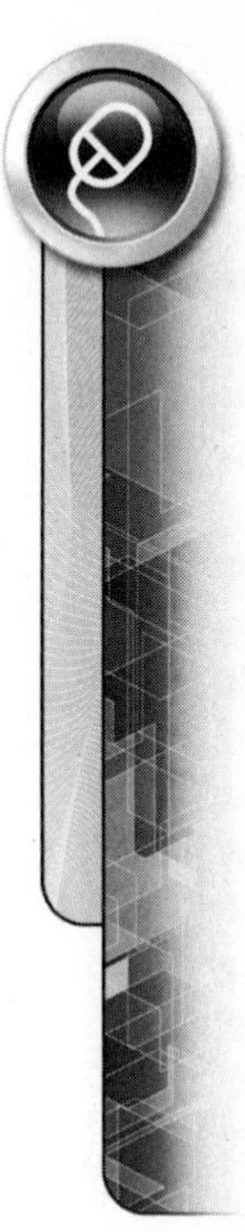

16. Click the green check mark (OK button) in the mini-toolbar to create the fillets.
17. Click the back cylindrical edge on Extrusion4 and click Create Chamfer from the mini-toolbar as shown in Figure 4.24 on the left.
18. In the mini-toolbar's distance field, type **.25**, as shown in Figure 4.24 on the right.
19. To create the chamfer and keep the dialog box open, click green plus button in the mini-toolbar or click the Apply button in the Chamfer dialog box (if needed expand the Chamfer dialog box by clicking the down arrow at the bottom of the minimized Chamfer dialog box).

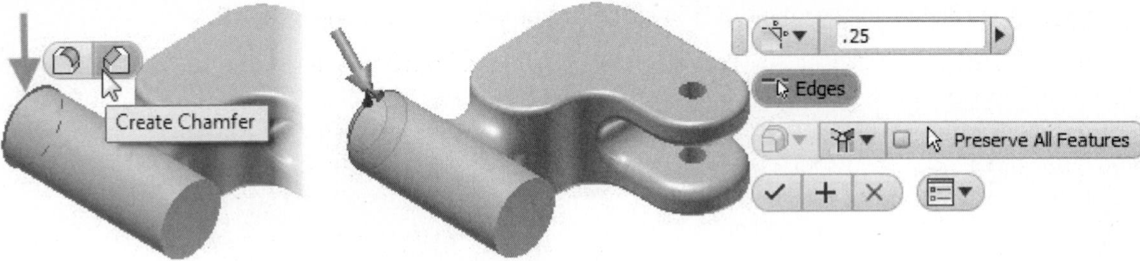

FIGURE 4.24

20. In the Chamfer dialog box or in the mini-toolbar, click the Distance and Angle option labeled (A) in **Figure 4.25**.
21. Click the front-circular face on Extrusion4 labeled (1) in **Figure 4.25**.
22. Click the front-circular edge of the same face labeled (2) in the following image.
23. In the Chamfer dialog box or in the mini-toolbar, enter **.25** in the Distance field and **60 deg** in the Angle field, as shown Figure 4.25.

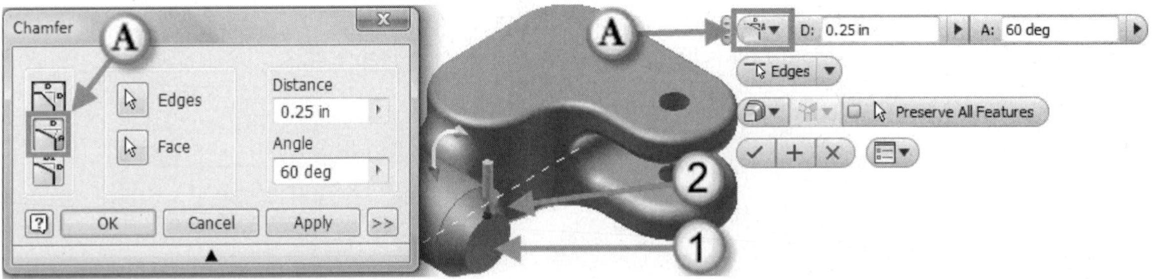

FIGURE 4.25

24. Click OK to create the chamfer. When done, your model should resemble Figure 4.26.

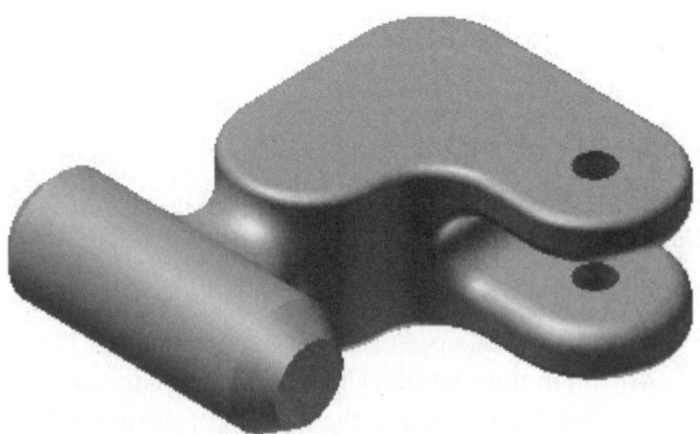

FIGURE 4.26

25. Practice editing and placing fillets and chamfers.
26. Close the file. Do not save changes. End of exercise.

HOLES

The Hole command lets you create drilled, counterbored, spotface, countersunk, clearance, tapped, and taper tapped holes, as shown in Figure 4.27. You can place holes using sketch geometry or existing planes, points, or edges of a part. You can also specify the type of drill point and thread parameters.

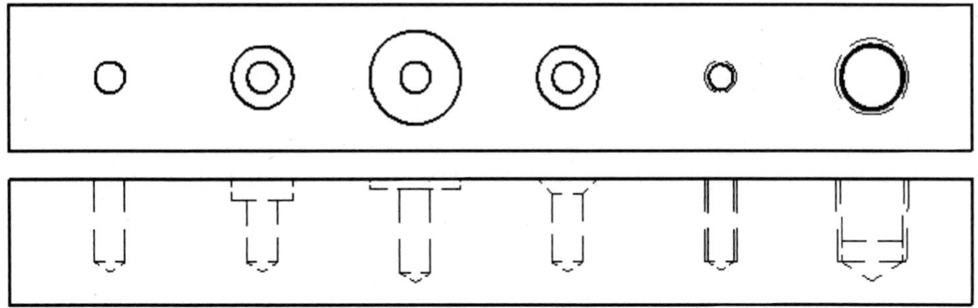

FIGURE 4.27

To create a hole feature, follow these steps:

1. Create a part that you want to place a hole on.
2. Click the Hole command from the 3D Model tab > Modify panel, as shown in Figure 4.28 on the left, or press the shortcut key H. You could also right-click and click Hole from the marking menu.
3. The Holes dialog box appears, as shown in Figure 4.28 on the right. Four placement options are available: From Sketch, Linear, Concentric, and On Point. The Placement options are covered in the next section. After you have chosen the hole placement options, select the desired hole style options in the Holes dialog box. As you change the options, the preview image of the hole(s) updates. When you are done making changes, click the Apply or the OK button to create the hole(s).

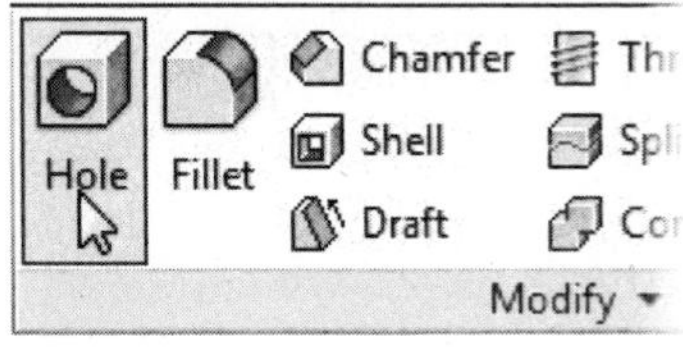

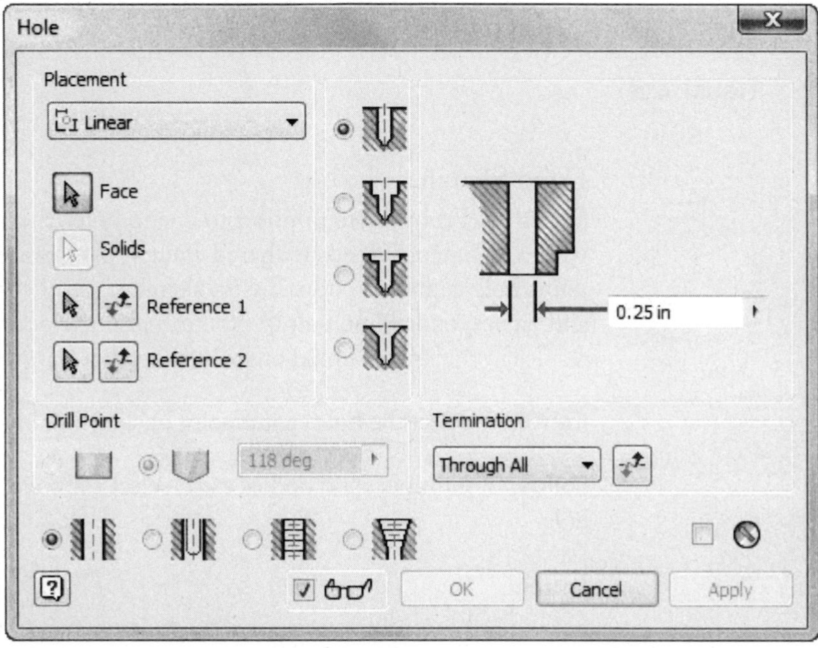

FIGURE 4.28

Editing Hole Features

To edit the type of hole feature use one of the following methods:

- Click on the circular face of the hole that you want to edit and click Edit Hole from the mini-toolbar.
- Double-click the feature's name or icon in the browser.
- Right-click the hole's name or icon in the browser, and select Edit Feature from the menu.

Holes Dialog Box

In the Holes dialog box, you establish the placement method, type of hole, its termination, and additional options such as type of drill point, angle, and tapped properties.

Placement

Select the appropriate placement method. If you select From Sketch, a sketch containing hole centers or any point, such as an endpoint of a line, must exist on the part. Hole centers are described in the next section. The Linear, Concentric, and On Point options do not require an unconsumed or shared sketch to exist in the model and are based on previously created features. Depending on the placement option you select, the input parameters will change, as shown in Figure 4.29.

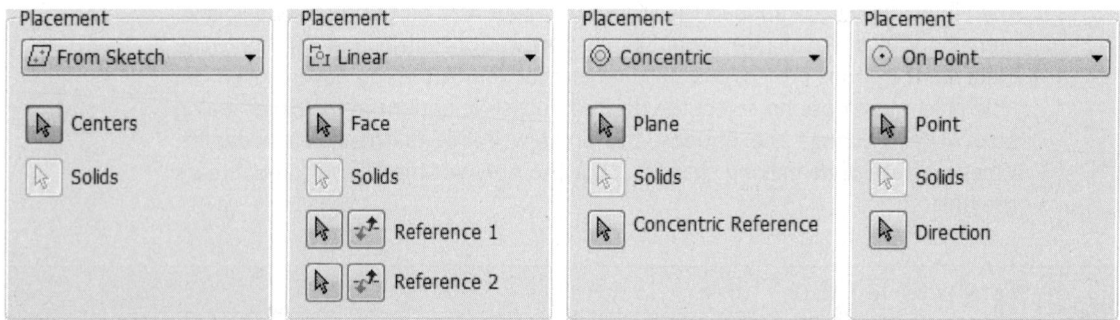

FIGURE 4.29

From Sketch

Select the From Sketch option to create holes that are based on a location defined within an unconsumed or shared sketch. You can base the center of the hole on a point/hole center or endpoints of sketched geometry like endpoints, centers of arcs and circles, and spline points. You can also use points from projected geometry that resides in the unconsumed or shared sketch.

Centers. Select the hole center point or sketch points where you want to create a hole.

Solids. If multiple solid bodies exist, select the solid body where you want to create a hole.

Linear

Select the Linear option to place the hole relative to two selected face edges.

Face. Select the face on the part where the hole will be created.

Solids. If multiple solid bodies exist, select the solid body where you want to create a hole.

Reference 1. Select a face edge as a positional reference for the center of the hole. When you select the edge, a dimension appears that can be edited to constrain the center of the hole dimensionally.

Reference 2. Select a face edge as a positional reference for the center of the hole. When you select the edge, a dimension appears that can be edited to constrain the center of the hole dimensionally.

Flip Side. Click this button to position the hole on the opposite side of the selected edge.

Concentric

Select the Concentric option to place the hole on a planar face and concentric to a circular or arc edge or a cylindrical face.

Plane. Select a planar face or plane where you want to create the hole.

Solids. If multiple solid bodies exist, select the solid body where you want to create the hole.

Concentric Reference. Select a circular or arc model edge or cylindrical face to constrain the center of the hole to be concentric with the selected entity.

On Point

Select the On Point option to place the center of the hole on a work point. The work point must exist on the model prior to selecting this option.

Point. Select a work point to position the center of the hole.

Solids. If multiple solid bodies exist, select the solid body where you want to create the hole.

Direction. Select a plane, face, work axis, or model edge to specify the direction of the hole. When selecting a plane or face, the hole direction will be normal to the face or plane.

Hole Options

Click the type of hole that you want to create: drilled, counterbore, spotface, or countersink, and enter the appropriate dimensions. Figure 4.30 shows the counterbore option.

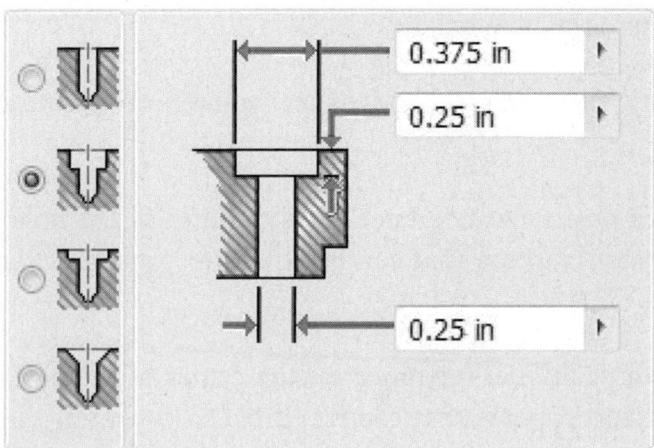

FIGURE 4.30

Termination

Select how the hole will terminate.

- *Distance:* Specify a distance for the depth of the hole.
- *Through All:* Choose to extend the hole through the entire part in one direction.
- *To:* Select a plane at which the hole will terminate.
- *Flip:* Reverse the direction in which the hole is drilled.

Drill Point

Select either a flat or angle drill point. If you select an angle drill point, you can specify the angle of the drill point.

Dimensions

To change the diameter, depth, countersink, counterbore diameter, countersink angle, or counterbore depth of the hole, click the dimension in the dialog box, and enter a desired value.

Hole Type

Click the type of hole you want to create. There are four options: Simple Hole, Clearance Hole, Tapped Hole, and Taper Tapped Hole. Figure 4.31 shows the tapped hole selected. After selecting the hole type, fill in the dialog box with the specific data for the hole you need to create.

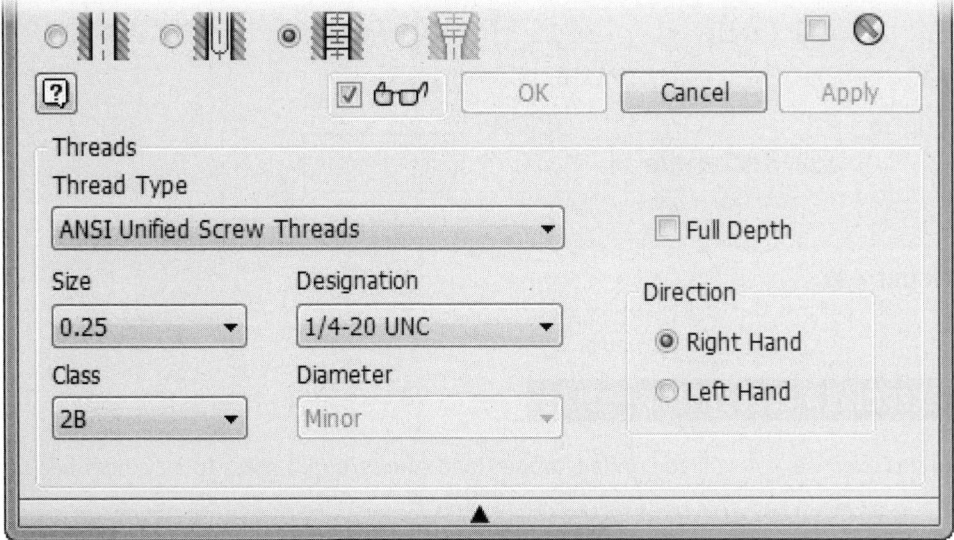

FIGURE 4.31

Simple Hole. Click the Simple Hole option to create a (drilled) hole feature with no thread features or properties.

Clearance Hole. Click the Clearance Hole option to create a simple hole feature that does not have thread.

Tapped Hole. Click the Tapped Hole option if the hole is threaded. Thread information appears in the dialog box area so that you can specify the thread properties, as shown in Figure 4.31.

Taper Tapped Hole. Click the Taper Tapped Hole option if the hole is tapered thread. The taper tapped hole information appears in the dialog box area so you can specify the thread properties.

Center Points

Center points are sketched entities that can be used to locate hole features. To create a hole center, follow these steps:

1. Make a sketch active.
2. Click the Point, Center Point command in the Sketch tab > Draw panel as shown in Figure 4.32.
3. Click a point to locate the hole center where you want to place the hole, and constrain and dimension the point as required.

When creating a hole(s) feature with the From Sketch option all center points that reside in the sketch are automatically selected as centers for the hole feature automatically. This can expedite the process of creating multiple holes in a single hole feature. You can also select endpoints of lines, arcs, splines, center points of arcs and circles, or spline control points as hole centers. Points can be deselected/selected by holding down the CTRL or SHIFT key and then selecting the points individually or use a window or crossing section method.

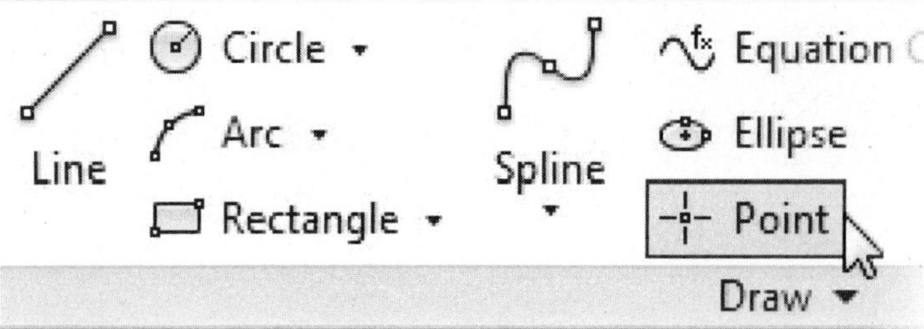

FIGURE 4.32

EXERCISE 4-2: CREATING HOLES

In this exercise, you will add drilled, tapped, and counterbored holes to a cylinder head.

1. Open *ESS_E04_02.ipt* in the Chapter 04 folder.
2. The first hole you place is a linear hole. Click the Hole command in the 3D Model tab > Modify panel.
   - For the Face, click the top planar face of the extruded rectangle (Extrusion2).
   - For Reference 1 and 2, select the far left top vertical and horizontal edges and place the point **.375 in** from each edge.
   - Verify that the Drilled hole option is selected.
   - Verify that the Termination to Through All is selected as the Termination.
   - Change the hole's diameter to **.25 in**, as shown in Figure 4.33.
   - Click OK to create the hole.

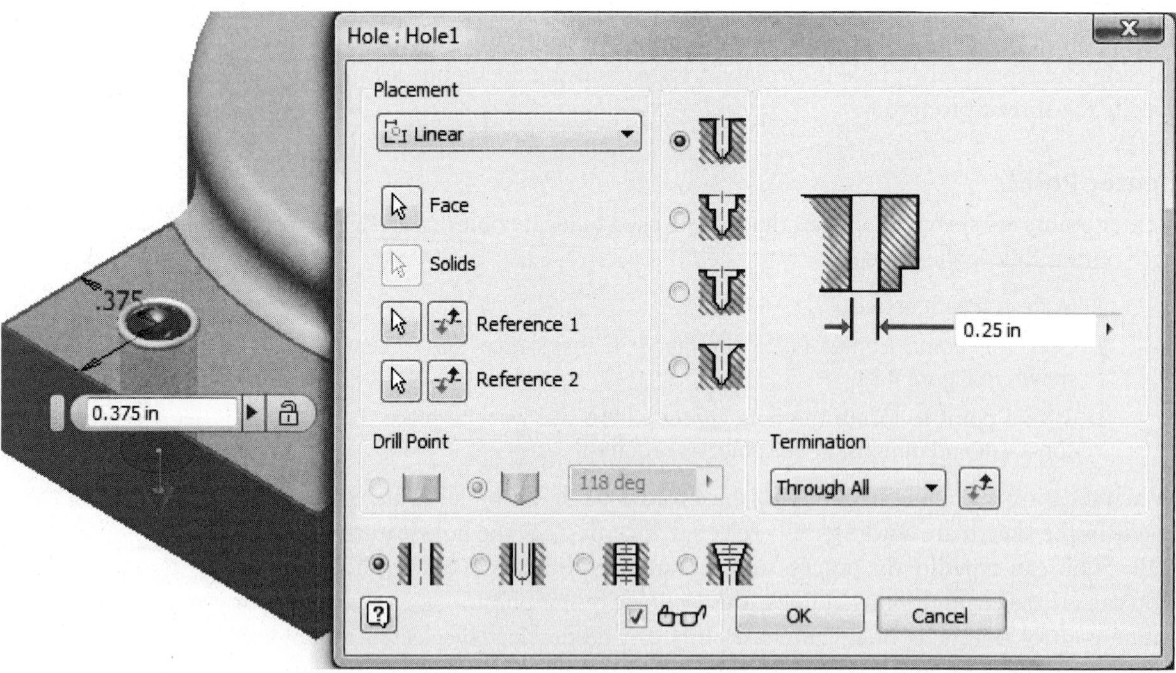

FIGURE 4.33

3. Next, you create multiple holes based on the Sketch option. Click the top planar face of Extrusion1 and click Create Sketch from the mini-toolbar as shown in the following image on the left.
4. Click the Point, Center Point command in the Sketch tab > 2D Sketch Panel, and place two points in the middle of the sketch labeled (1) and (2) in the following image on the right.
5. Add a horizontal constraint between the left point labeled (1) in **Figure 4.34** on the right and the center point that was automatically projected from the center of the right circular edges labeled (3) in **Figure 4.34** on the right. Add another horizontal constraint between the points labeled (2) and (3) in **Figure 4.34** on the right.
6. Add a **1.250** and a **1.000** dimension between the points as shown in **Figure 4.34** on the right.

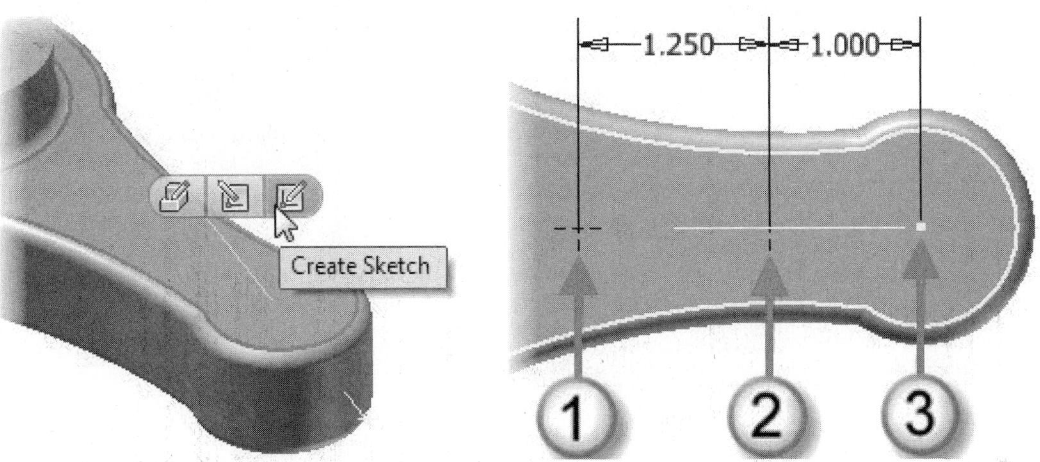

FIGURE 4.34

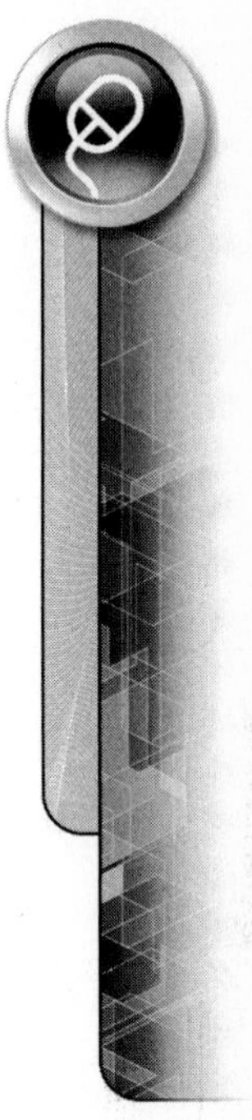

7. Right-click in the graphics window, click Finish 2D Sketch from the marking menu, the view should change to the home view. If not press the F6 key to change to the Home View.
8. Click the Hole command in the 3D Model tab > Modify panel bar or press the H key.
   - The Center Points are automatically selected as the Centers.
   - Add another point to add a hole to by clicking on the point labeled (3) in **Figure 4.34**.
   - Change the hole option to Countersink.

- Ensure the Termination is set to Through All.
- Set the size of the hole as shown in Figure 4.35.
- Click OK in the dialog box or the green check mark in the mini-toolbar to create the hole.

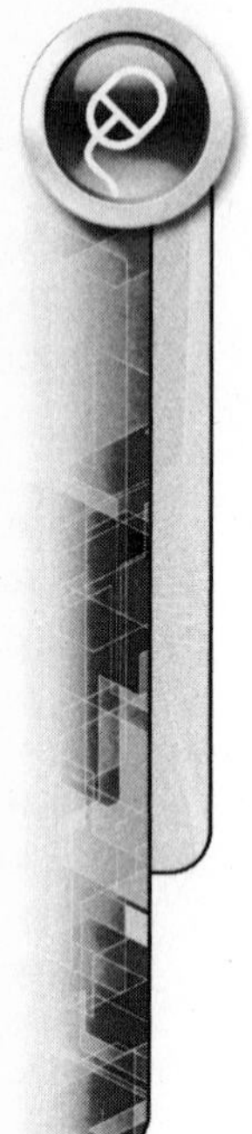

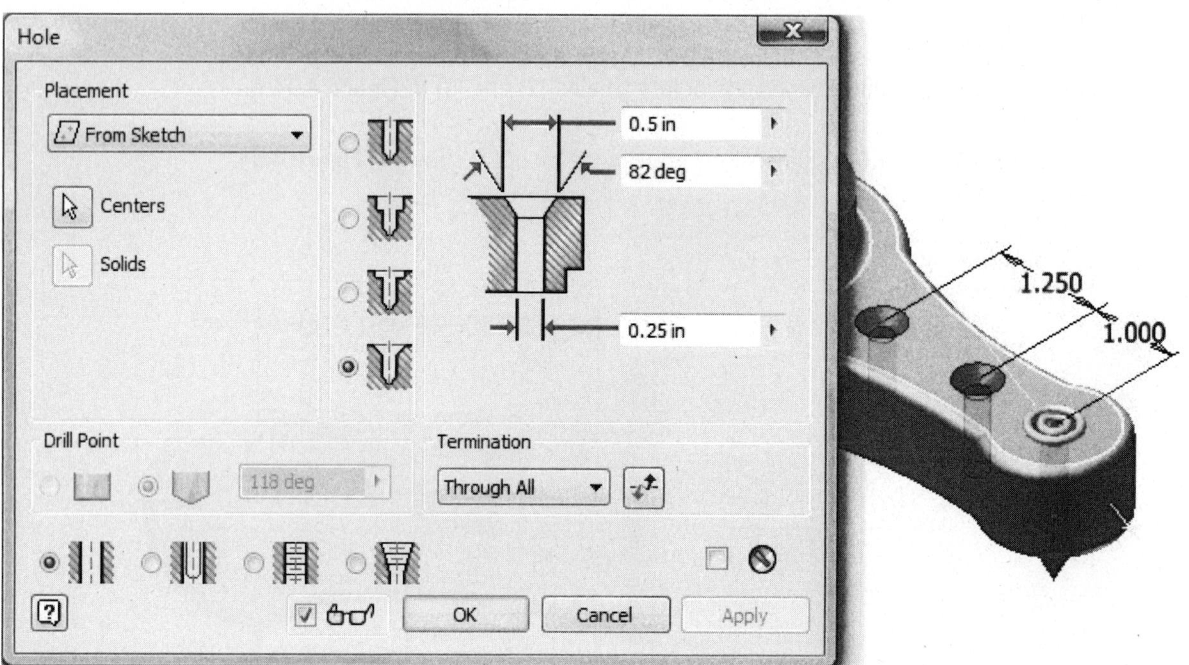

FIGURE 4.35

9. Next, you create a Taper Tapped hole that is concentric to the top of the left cylinder.

   - Press the ENTER key to restart the Hole command. For clarity use the labels in **Figure 4.36** that are denoted in these steps.
   - In the Hole dialog box change the Placement option to Concentric labeled (1).
   - For the Plane, select the top face of the cylinder Extrusion3 labeled (2).
   - For the Concentric Reference, select the top circular edge of the cylinder labeled (3).
   - Ensure the hole option is set to Drilled labeled (4).
   - Click the Taper Tapped Hole type labeled (5).
   - Change the Thread Type to NPT labeled (6).
   - Change the size to **1/2** labeled (7).
   - Ensure that the Termination is set to Through All labeled (8)
   - Click Apply in the Hole dialog box to create the hole.

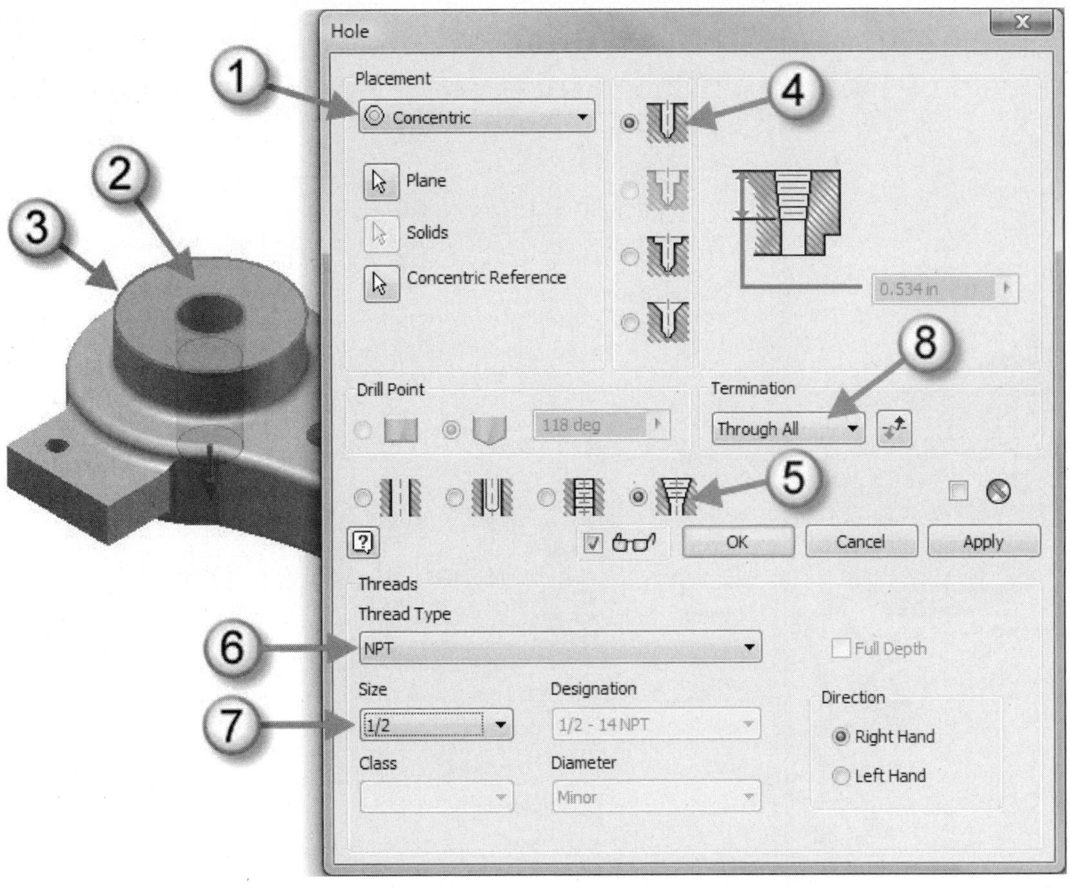

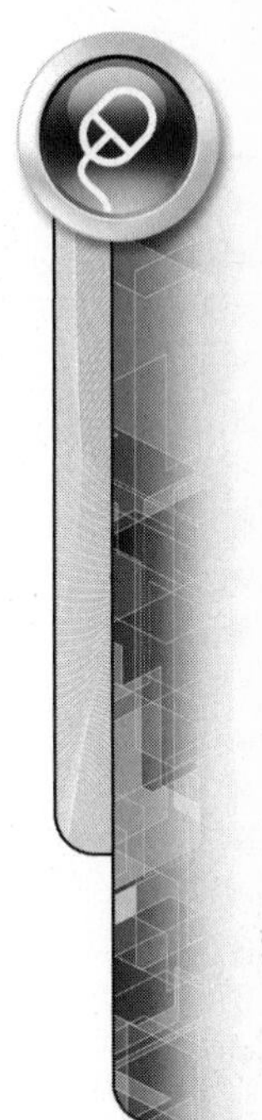

FIGURE 4.36

10. Lastly, you create a tapped hole that uses the work point and work axis that goes through the end of the part at an angle. Work point and work axis will be covered later in this chapter.
    - Change the Placement option to On Point labeled (1) in **Figure 4.37**.
    - For the Point, select the work point on the cylinder labeled (2).
    - For the Direction, select the work axis that goes through the cylinder labeled (3).
    - If needed click the Drilled Hole option labeled (4).
    - Change the Thread Type to Tapped Hole labeled (5).
    - Verify that the Thread Type is set to to **ANSI Unified Screw Threads** labled (6) if not make it current.
    - Change the size to **.25** labeled (7).
    - Change the Termination to Distance labeled (8).
    - Check the Full Depth option labeled (9).
    - Change the the distance value to **0.75** in labeled (10).
    - Click OK to create the hole.

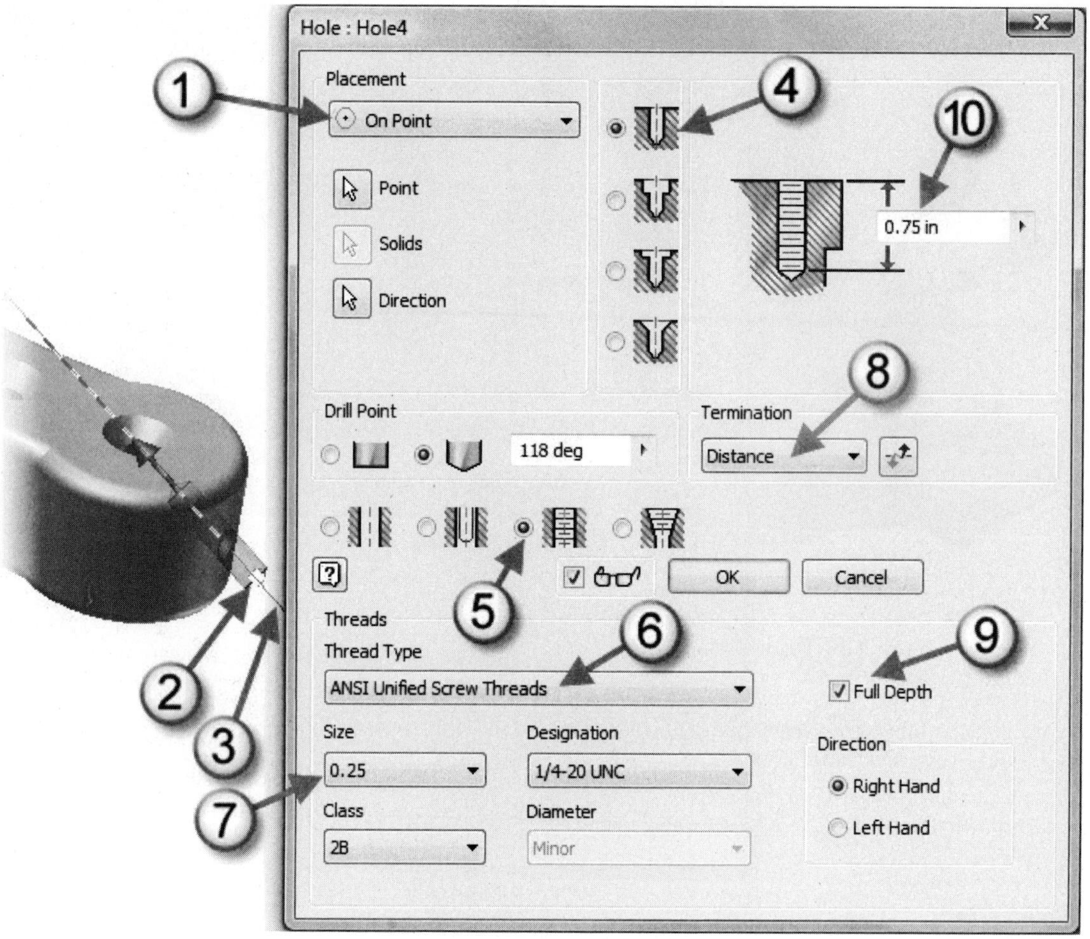

FIGURE 4.37

11. The completed part is shown in Figure 4.38 (the visibility of the work axis and work point have been turned off for clarity). Rotate the viewpoint part and examine the holes.

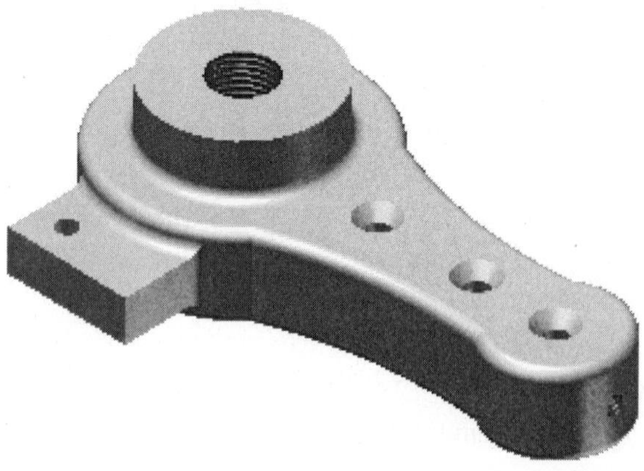

FIGURE 4.38

12. Practice editing the holes and placing new holes.
13. Close the file. Do not save changes. End of exercise.

SHELLING

As you design parts, you may need to create a model that is made with thin walls but is not a sheet metal part. The easiest way to create a thin-walled part is to create the main shape and then use the Shell command to remove material.

The term *shell* refers to giving a wall thickness to the outside shape of a part and removing the remaining material. Essentially, you are scooping out the inside of a part and leaving the walls a specified thickness, as shown in Figure 4.39. You can offset the wall thickness in, out, or evenly in both directions. If the part you shell contains a void, such as a hole, the feature will have the thickness built around it.

A part may contain more than one shell feature, and individual faces of the part can have different thicknesses. If a wall has a different thickness than the shell thickness, it is referred to as a unique face thickness. If a face that you select for a unique face thickness has faces that are tangent to it, those faces will also have the same thickness. You can remove faces from being shelled, and these faces remain open. If no face is removed, the part is hollow on the inside.

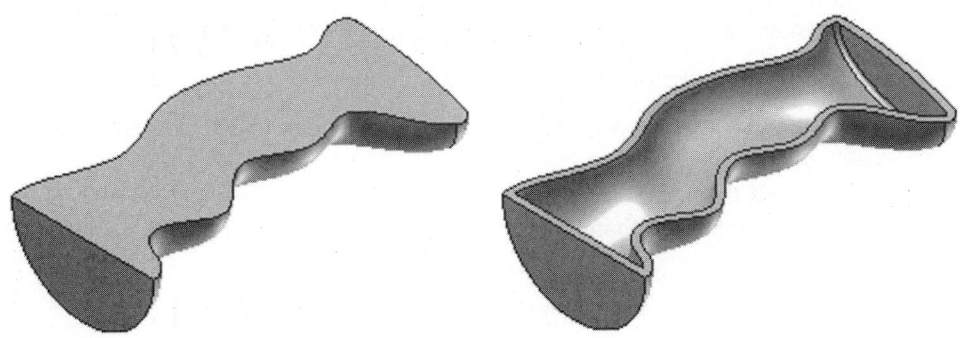

FIGURE 4.39

To create a shell feature, follow these steps:

1. Create a part that will be shelled.
2. Start the Shell command from the 3D Model tab > Modify panel, as shown Figure 4.40.
3. The Shell dialog box and the mini-toolbar appear, as shown in Figure 4.40 (the dialog box is shown expanded).

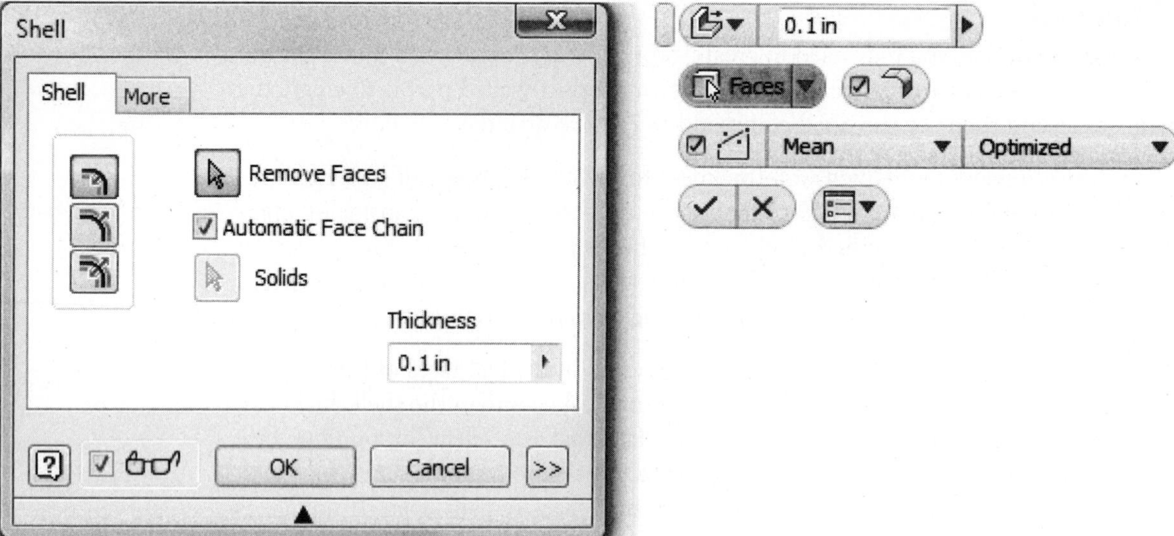

FIGURE 4.40

4. In the dialog box, select the direction for the shell, remove faces, and enter a thickness value and then click OK, and the part is shelled.

To edit a shell feature, use one of the following methods:

- In the graphics window select a face on the shell feature and click Edit Shell on the mini-toolbar.
- Right-click the name of the shell feature in the browser, and select Edit Feature from the menu. Alternately, you can double-click the feature's name or icon in the browser.

The following sections explain the options for the Shell command.

Direction

Inside

Click this button to offset the wall thickness into the part by the given value.

Outside

Click this button to offset the wall thickness out of the part by the given value.

Both

Click this button to offset the wall thickness evenly into and out of the part by the given value.

Remove Faces

Click the Remove Faces button, and then click the face or faces to be left open. To deselect a face, click the Remove Faces button, and hold down the CTRL key while you click the face.

Automatic Face Chain

When you are removing faces and this option is checked, faces that are tangent to the selected face are automatically selected. Uncheck this option to select only the selected face.

Solids

If multiple solid bodies exist, select the solid body to shell.

Thickness

Enter a value or select a previously used value from the drop-down list to be used for the shell thickness.

Unique Face Thickness

Unique face thickness is available by clicking the More (>>) button that is located on the lower-right corner of the dialog box, as shown in Figure 4.41 on the left.

To give a specific face a thickness, click on Click to add, select the face, and enter a value. A part may contain multiple faces that have a unique thickness, as shown in Figure 4.41 on the right.

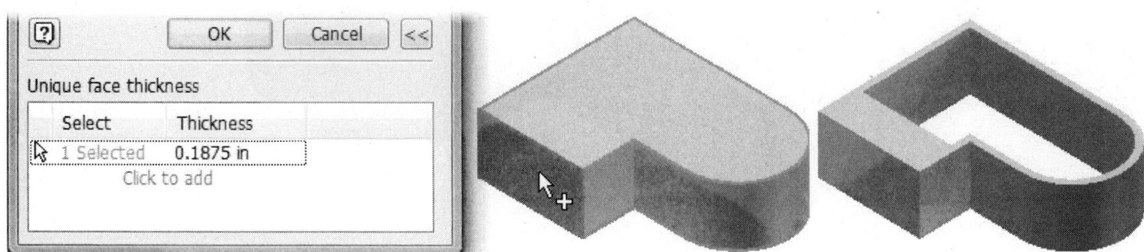

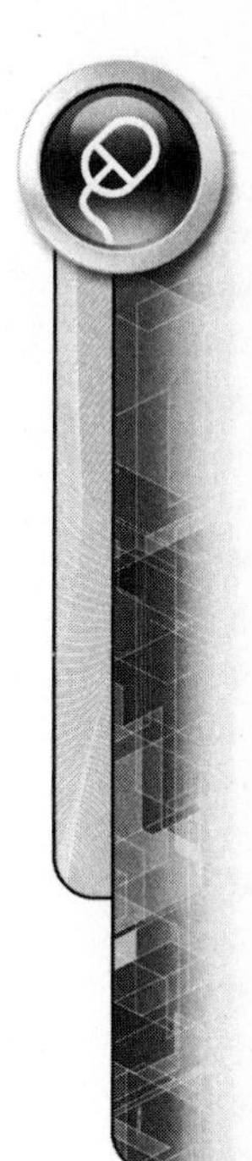

FIGURE 4.41

EXERCISE 4-3: SHELLING A PART

In this exercise, you will use the Shell command to create a shell on a part.

1. Open *ESS_E04_03.ipt* in the Chapter 04 folder.
2. Click the Shell command on the 3D Model tab > Modify panel.
3. Remove the top face by selecting the top face of the part.
4. Type a thickness of **.0625 in**, as shown in Figure 4.42.

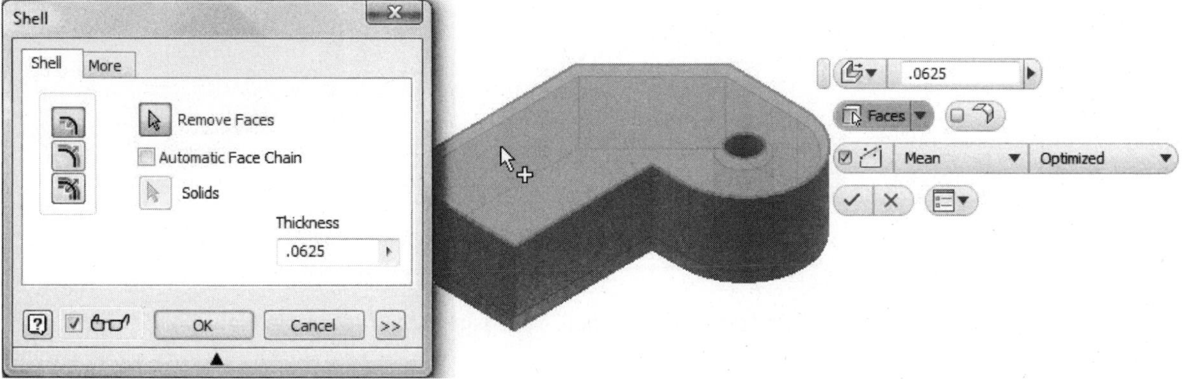

FIGURE 4.42

5. Click OK to create the shell feature.
6. Rotate the part and examine the shell. Only the top face should be removed.
7. Click the Home View command above the ViewCube.
8. Edit the Shell feature that you just created. Click on one of the inside faces of the shell and from the mini-toolbar click Edit Shell as shown in Figure 4.43.

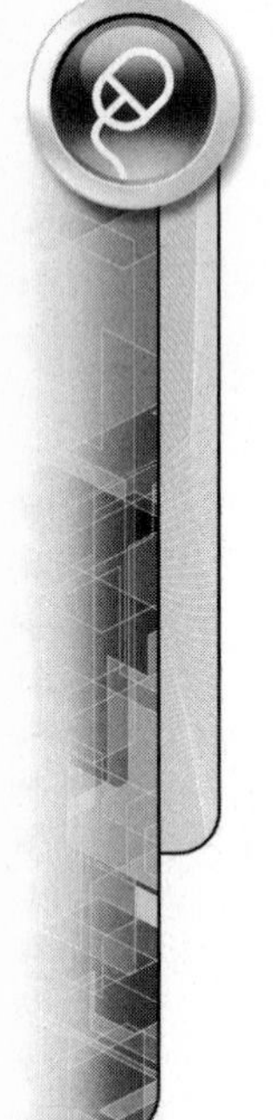

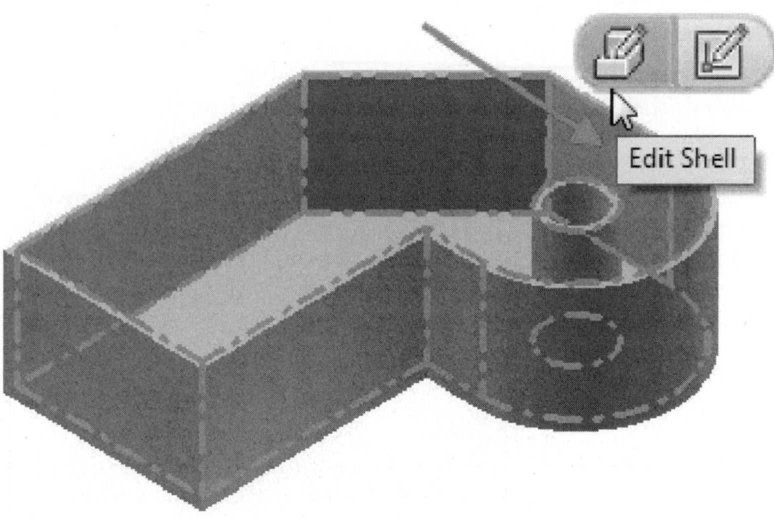

FIGURE 4.43

9. Click the More (>>) button on the bottom-right corner of the Shell dialog box, click in the "Click to add" area, and select the left outside-vertical face.
10. Enter a value of **.5**, as shown in Figure 4.44.

FIGURE 4.44

11. Click OK to create the shell. Your part should resemble Figure 4.45.

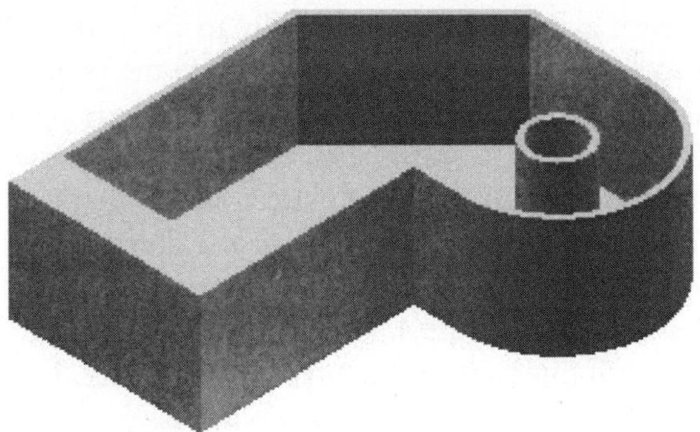

FIGURE 4.45

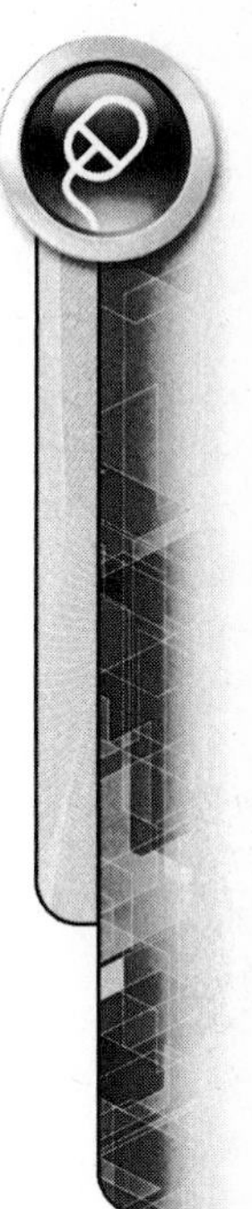

12. Practice editing the shell feature; change the thickness direction, thickness and delete the unique thickness by selecting the unique thickness entry in the dialog box and then press the Delete key.
13. Close the file without saving changes. End of exercise.

WORK FEATURES

When you create a parametric part, you define how the features of the part relate to one another; a change in one feature results in appropriate changes in all related features. Work features are special construction features that are attached parametrically to part geometry or other work features. You typically use work features to help you position and define new features in your model. There are three types of work features: work planes, work axes, and work points.

Use work features in the following situations:

- To position a sketch for new features when a planar part face is not available.
- To establish an intermediate position that is required to define other work features. You can create a work plane at an angle to an existing face, for example, and then create another work plane at an offset value from that plane.
- To establish a plane or edge from which you can place parametric dimensions and constraints.
- To provide an axis or point of rotation for revolved features and patterns.
- To provide an external feature termination plane off the part, such as a beveled extrusion edge, or an internal feature termination plane in cases where there are no existing surfaces.

Creating a Work Axis

A work axis is a feature that acts like a construction line. In the database, it is infinite in length but displayed a little larger than the model, and you can use it to help create work planes, work points, and subsequent part features. You can also use work axes as axes of rotation for polar arrays or to constrain parts in an assembly using assembly constraints. Their length always extends beyond the part—as the part changes size, the work axis also changes size. A work axis is tied parametrically to the part. As changes occur to the part, the work axis will maintain its relationship to the points, edge, or cylindrical face from which you created it. To create a work axis, use the Work Axis command on the 3D Model tab > Work Features panel or press the key / (forward slash). The Axis command will allow you to select all geometry in the graphics window to create an axis. Another option is to click the down arrow in the lower right corner of the Axis button and select an option for creating an axis using only certain types of geometry as shown in Figure 4.46. These options will only select the predetermined geometry and will filter out all other types of geometry. For example, if the Through Two Points option is selected, only points can be selected in the graphics window.

You can also create a work axis when using the Work Plane or Work Point commands; right-click and select Create Axis when one of these work feature commands is active. Then, after the work axis is created, the command you were running will be active. The created axis will be indented as a child of the work axis or the work plane in the browser.

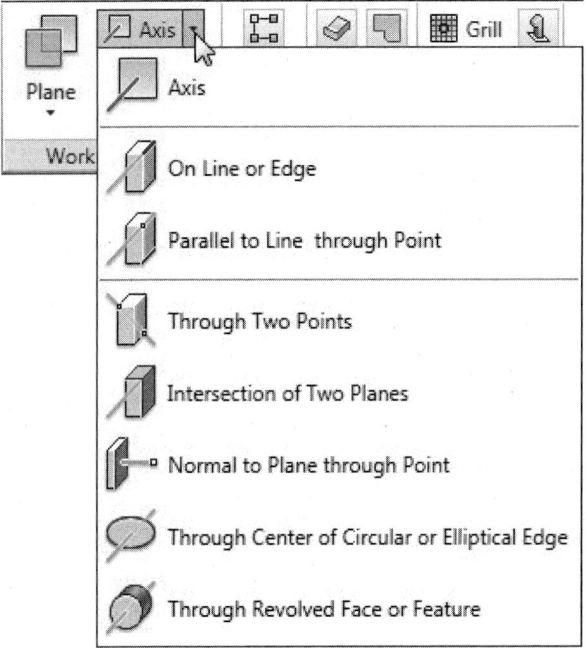

FIGURE 4.46

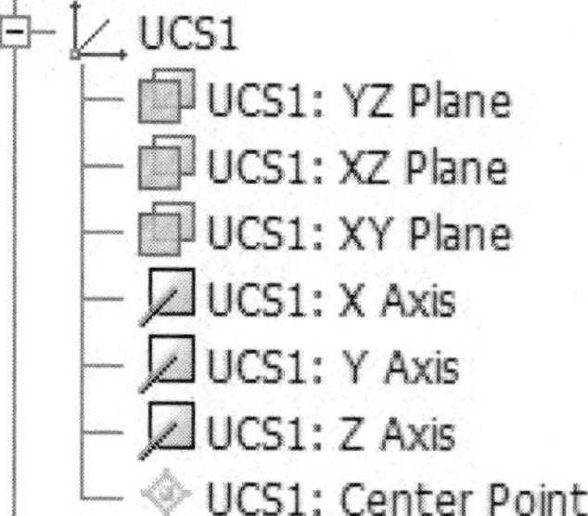

FIGURE 4.58

To edit a UCS, move the cursor over the UCS in the graphics window or in the browser and double-click or right-click, and then click Redefine Feature as shown in Figure 4.59 on the left. Select on the desired UCS segment: arrow, leg, or origin, and enter a new value or drag to a new location. When done relocating the UCS, right-click and click Finish from the menu. Do NOT click Done as this will cancel the operation. Figure 4.59 on the right shows the prompt to select a segment to edit.

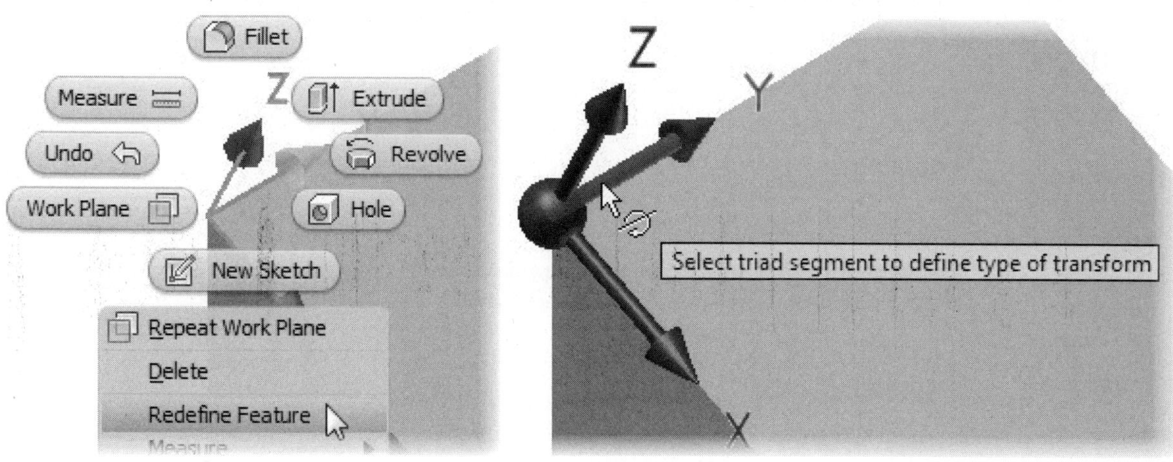

FIGURE 4.59

Feature Visibility

You can control the visibility of the origin planes, origin axes, origin point, user work planes, user work axes, user work points, and sketches by either right-clicking on them in the graphics area or on their name in the browser and selecting Visibility from the menu. You can also control the visibility for all origin planes, origin axes, origin point, user work planes, user work axes, user work points, sketches, solids, and UCS triad, planes, axis, and points from the View tab > Visibility panel > Object Visibility. Visibility can be checked to turn visibility on or cleared to turn it off, as shown in Figure 4.60.

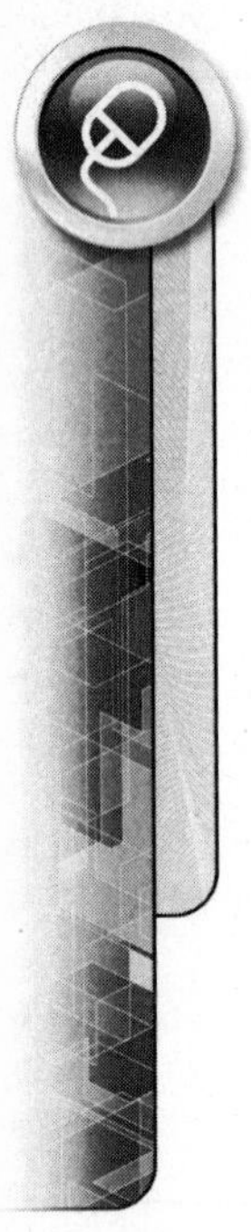

FIGURE 4.60

EXERCISE 4-5: CREATING WORK PLANES AND A UCS

In this exercise, you will create work planes in order to create a boss on a cylinder head and a slot in a shaft.

1. Open *ESS_E04_05-1.ipt* in the Chapter 04 folder.
2. From the 3D Model tab > Work Features panel click the down arrow on the lower right corner of the Plane command and from the list click Angle to Plane around Edge.
3. Click the top rectangular face and the top-back left edge, and then enter a value of −**30** as shown in Figure 4.61 on the left. Click the green checkmark in the mini-toolbar to create the work plane.
4. Create a sketch on the top face of the extrusion, do not select the workplane. Click on the top face of Extrusion1 and click Create Sketch command on the mini-toolbar.
5. Create and dimension a circle and apply a horizontal constraint between the center of the circle and the midpoint of the right vertical edge as shown in Figure 4.61 on the right. Finish the sketch.

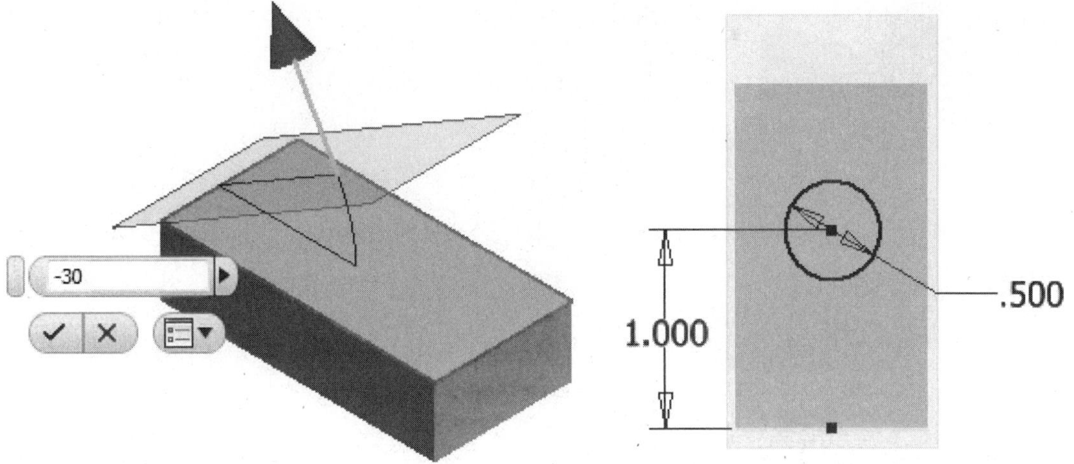

FIGURE 4.61

6. Extrude the circle with the Extents set to To, and select the work plane, as shown in Figure 4.62.

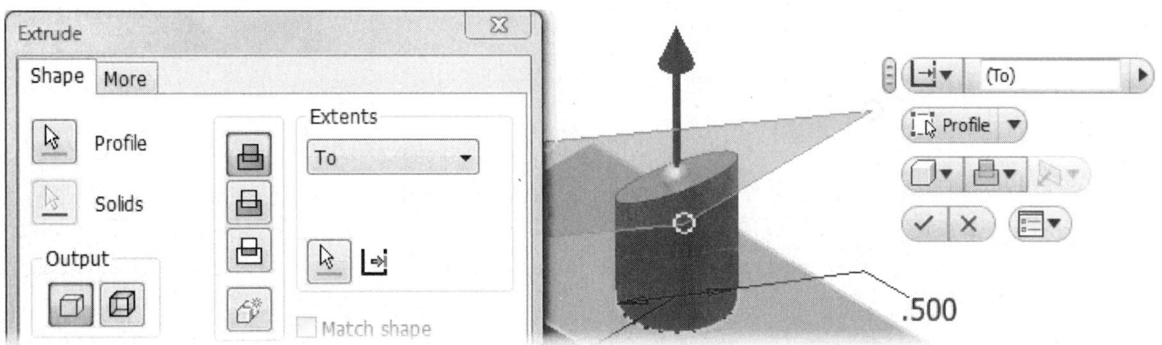

FIGURE 4.62

7. In the graphics window click on the work plane you created in step 3 click Edit Dimension from the mini-toolbar and enter different values for the angle and then click the green check mark in the mini-toolbar, if the part does not update, click the Update command on the Quick Access toolbar. Then change the angle of the work plane back to −**30 degrees**.
8. Turn off the visibility of the work plane by moving the cursor over an edge of the work plane in the graphics window, right-click, and click Visibility from the marking menu.
9. Another option to define a plane is to create a UCS. Click the UCS command on the 3D Model tab > Work Features panel.
10. Align the UCS on the top face of Extrusion1 by selecting the three vertices in the order as shown in Figure 4.63 on the left.
11. Edit the UCS by double-clicking on one the arrowheads on the UCS in the graphics window.
12. Click the X leg of the UCS and enter a value of −**40** as shown in Figure 4.63 on the right and press ENTER on the keyboard to see the change. To complete the edit, right-click and click Finish from the marking menu.

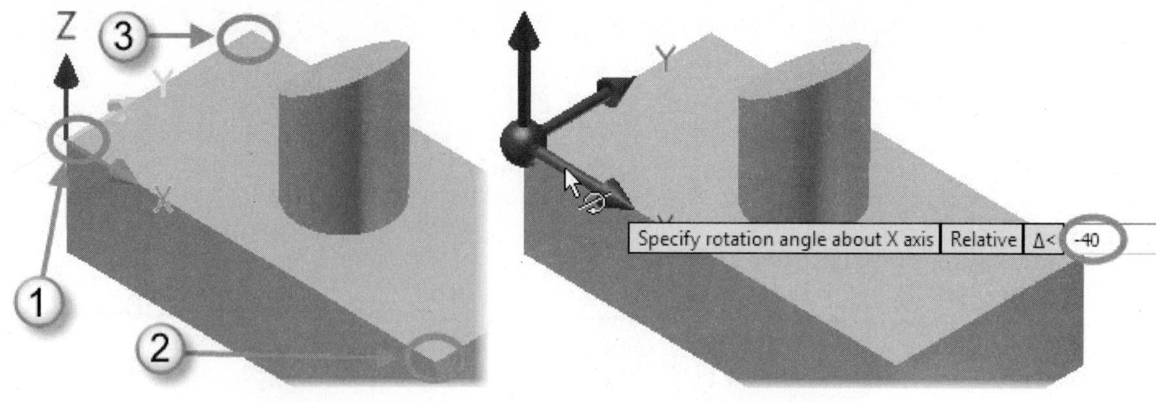

FIGURE 4.63

13. Reorder the UCS so it exists in the browser before Extrusion2. In the browser, click and drag the UCS entry so it is above Extrusion2 in the browser as shown in Figure 4.64. Reordering features is covered in Chapter 7.

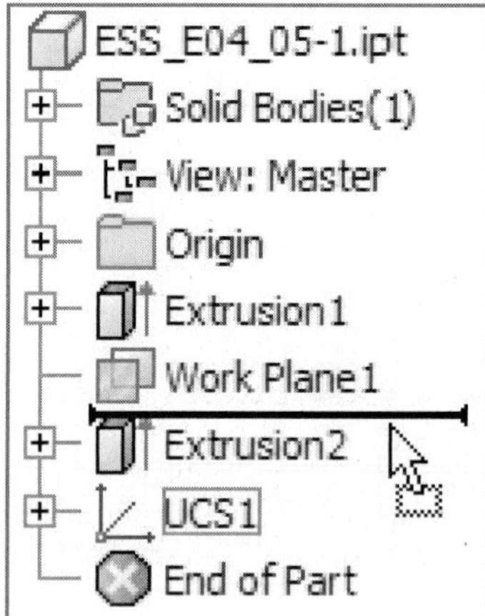

FIGURE 4.64

14. Edit Extrusion2 by clicking on a face of the circle that was extruded and click Edit Extrude from the mini-toolbar. In the Extrude dialog box click the arrow button for the option Select surface to end the feature creation and in the browser expand the UCS1 entry and then click UCS1: XY Plane as shown in Figure 4.65.

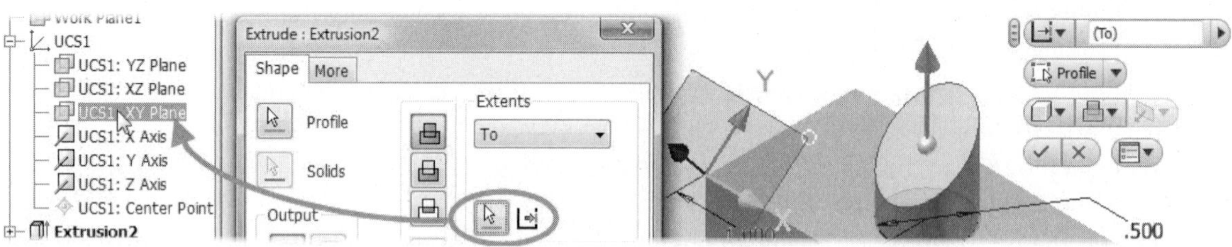

FIGURE 4.65

15. Click OK to complete the edit.
16. Create a work plane that is centered between two parallel planes. From the 3D Model tab > Work Features panel click the down arrow on the right side of the Plane command and from the list click Midplane between Two Parallel Planes.
17. Click the front-left face and the back-right face, as shown in Figure 4.66 on the left.

18. Click one of the edges of the work plane you just created and click Create Sketch from the menu.
19. Slice the graphics by pressing the F7 key click Slice Graphics from the Status Bar.
20. To project the edges of the part, click the Project Cut Edges command from the Draw panel (the Project Cut Edges command may be under the Project Geometry command).
21. Create and dimension a circle on the top edge of the projected geometry, as shown in Figure 4.66 on the right.

NOTE

If the circle is placed at the midpoint of the projected edge, the horizontal dimension cannot be placed.

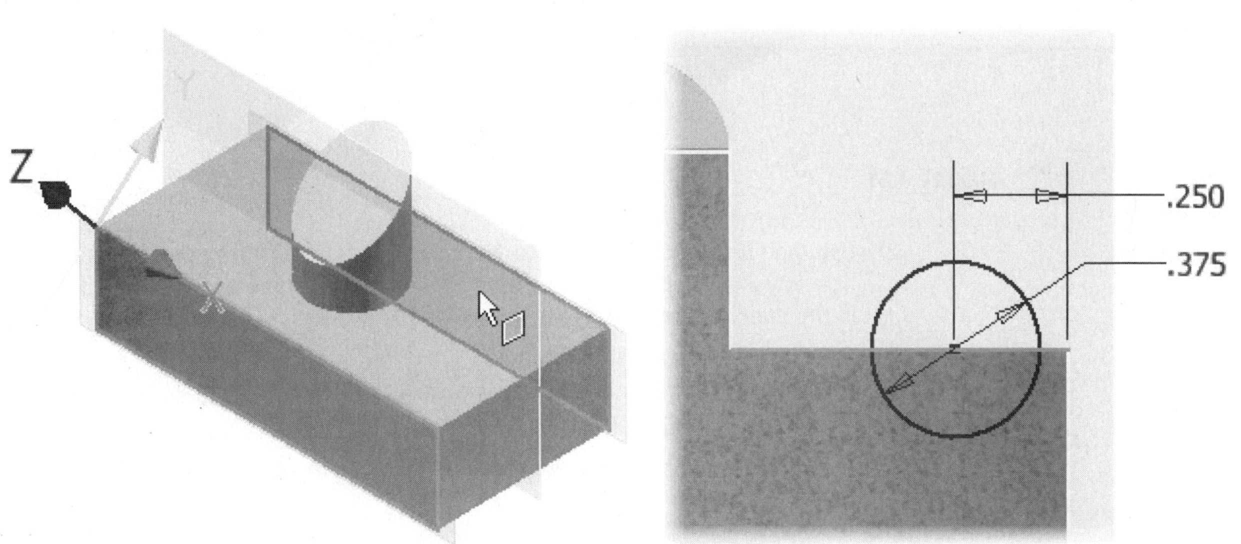

FIGURE 4.66

22. Finish the sketch, click Finish Sketch from the Sketch tab > Exit panel.
23. Extrude the circle **.5** with the symmetric option, as shown in Figure 4.67.

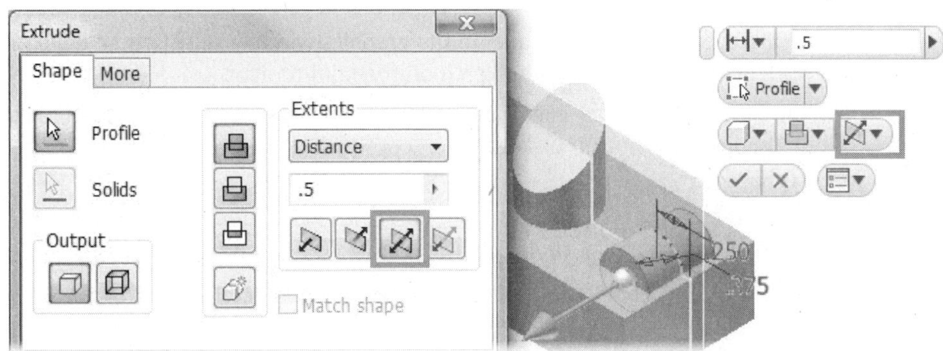

FIGURE 4.67

24. Turn off the visibility of the work plane and the UCS by moving the cursor over the work plane in the graphics window or in the browser and right-click and click Visibility from the marking menu and move the cursor over the UCS and right-click and click Visibility from the marking menu. Your screen should resemble Figure 4.68 on the left.

25. To verify that the symmetric extrusion will be maintained in the center of the part if the width changes, edit Sketch1 of Extrusion1, and change the 1 inch dimension to **1.5 inches**.
26. If needed click the Local Update command on the Quick Access toolbar. Your screen should resemble Figure 4.68 on the right.

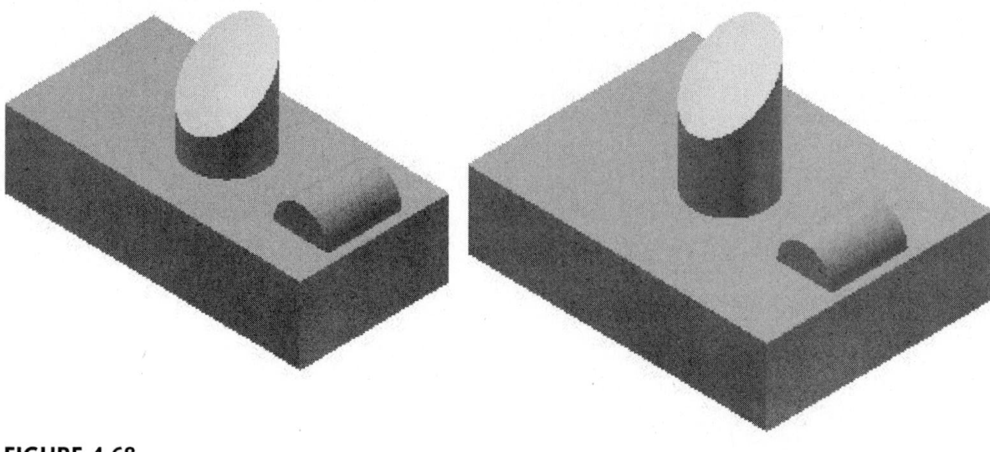

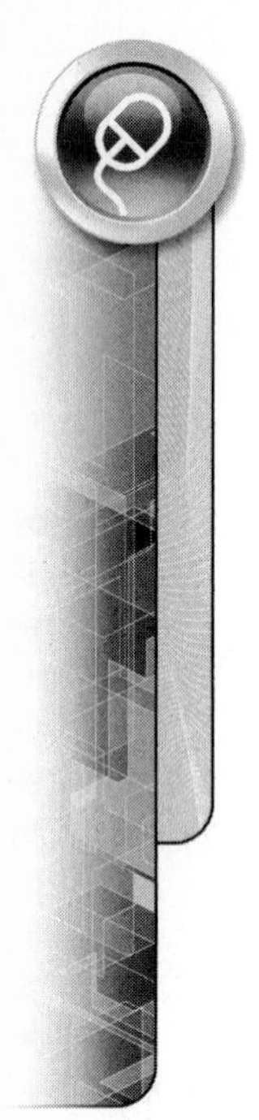

FIGURE 4.68

27. Use the Free Orbit command to view the part from different perspectives. Notice how Extrusion2 still terminates at the UCS XY plane and Extrusion3 is still in the middle of the part.
28. Close the file. Do not save changes.
29. In this portion of the exercise, you will place two holes on a cylinder. Open *ESS_E04_05-2.ipt* from the Chapter 04 folder.
30. Create a work plane that is parallel to an origin plane and tangent to the cylinder.
    - In the browser, expand the Origin folder.
    - Create a work plane on the outside of the cylinder. From the 3D Model tab > Work Features panel click the down arrow on the right side of the Work Plane command and from the list click Tangent to Surface and Parallel to Plane.
    - In the browser click the YZ plane under the Origin folder.
    - Click a point to the front right of the cylinder, as shown in Figure 4.69 on the left.
31. Create a sketch on the work plane. In the graphics window select on an edge of the work plane and click Create Sketch on the mini-toolbar.
32. Use the Project Geometry command from the Draw panel to project the Z axis of the Origin folder onto the sketch.
33. Place a Point, Center Point on the projected axis—this will center the point in the center of the cylinder horizontally—and add a dimension to locate it vertically, as shown in Figure 4.69 in the middle. Then finish the sketch.
34. Press H on the keyboard to start the Hole command and place a **.5 inch** Through All hole at the center point.
35. Turn off the visibility of the work plane and your screen should resemble Figure 4.69 on the right.

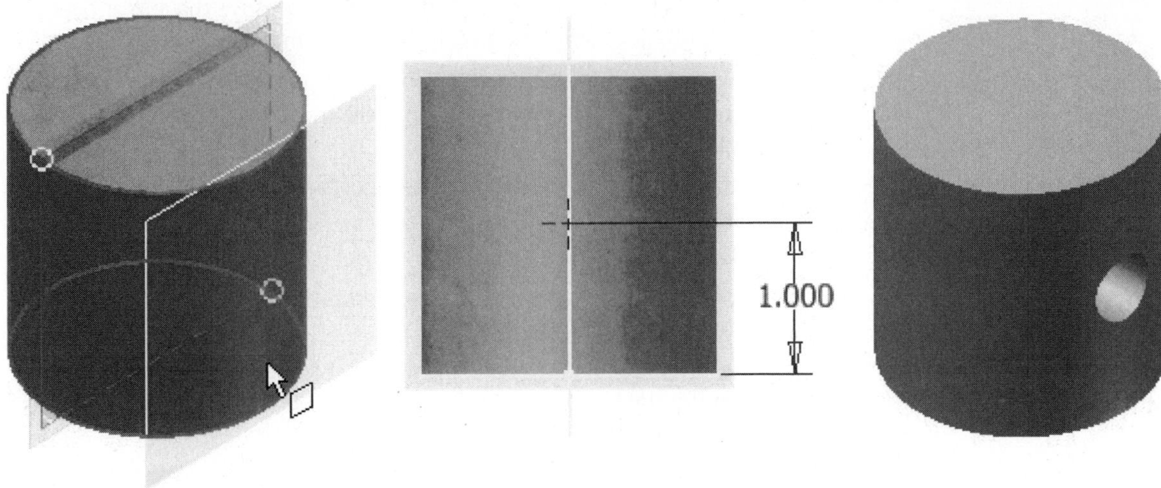

FIGURE 4.69

36. Next, you place a hole on the cylinder at an angle. First you create an angled work plane.
    - Create a work plane in the middle of the cylinder. From the 3D Model tab > Work Features panel click the down arrow on the right side of the Plane command and from the list click Angle to Plane around Edge.
    - In the browser, click the YZ plane under the Origin folder.
    - In the browser click, the Z Axis under the Origin folder.
    - In the Angle dialog box, enter **315** as shown in Figure 4.70 on the left.
    - Create the work plane, click the green check mark in the mini-toolbar.
37. Next, create a work plane that is parallel to an angled plane and tangent to the cylinder.
    - Create a work plane on the outside of the cylinder. From the 3D Model tab > Work Features panel click the arrow on the right side of the Plane command and from the list click Tangent to Surface and Parallel to Plane.
    - Click the angle work plane that you just created.
    - Click a point near the front face of the cylinder, as shown in Figure 4.70 on the right.

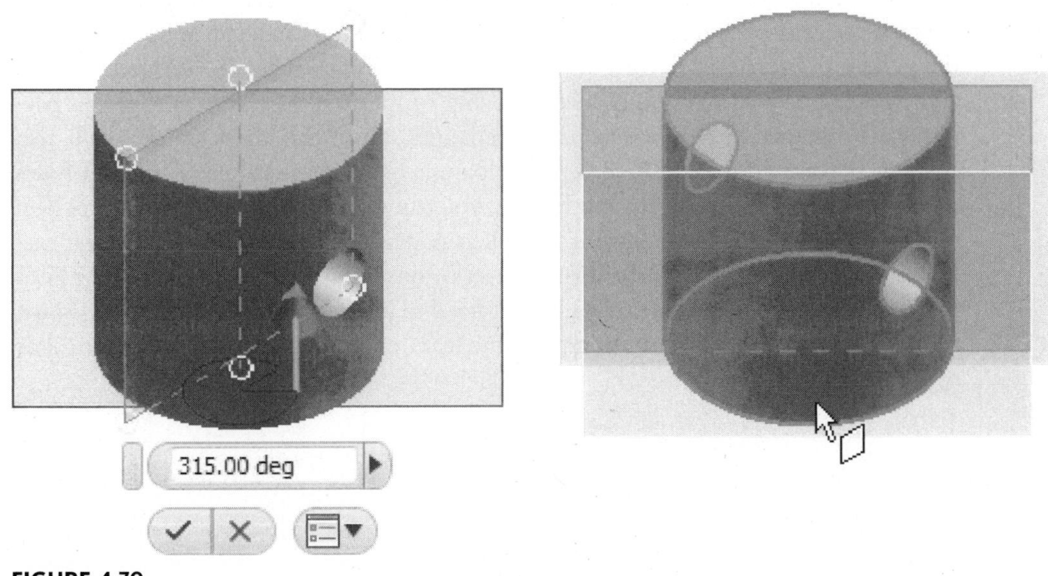

FIGURE 4.70

38. Turn off the visibility of the inside angled work planes. Move the cursor over the work plane. Right-click, and click Visibility from the marking menu.
39. Create a new sketch on the new work plane.
40. Use the Project Geometry command to project the Z axis of the origin folder onto the sketch.
41. Place a Point, Center Point on the projected axis, and dimension it, as shown in Figure 4.71 on the left.
42. Finish the sketch.
43. Start the Hole command, and place a counterbore hole with the size of your choice at the center point.
44. Turn off the visibility of the work plane and your screen should resemble Figure 4.71 on the right.

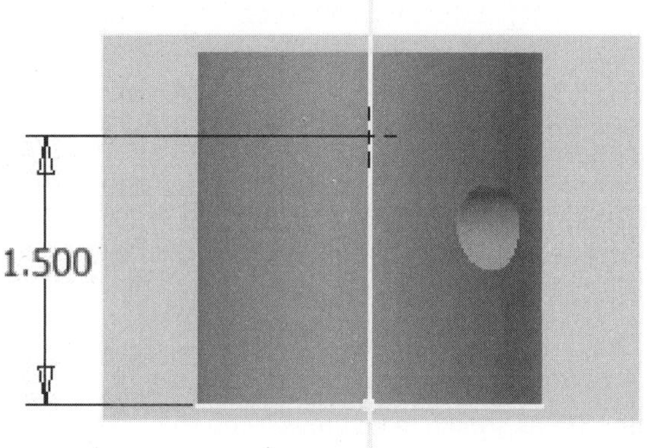

FIGURE 4.71

45. Expand Work Plane3 in the browser, double-click on Work Plane2, and enter a new value. Update the model to see the change.
46. Close the file. Do not save changes. End of exercise.

PATTERNS

There are two types of patterns that you can do; rectangular and circular. The pattern is represented as a single feature in the browser, and the original feature and individual feature occurrences are listed under the pattern feature. You can suppress the entire pattern or individual occurrences except for the first occurrence. Both rectangular and circular patterns have a child relationship to the parent feature(s) that you patterned. If the size of the parent feature changes, all of the child features will also change. If you patterned a hole, and the parent hole type changes, the child holes also change. Because a pattern is a feature, you can edit it like any other feature. You can pattern the base part or feature, as well as patterns. A rectangular pattern repeats the selected feature(s) along the direction set by one or two edges on the part or lines that reside in a sketch. These edges do not need to be horizontal or vertical, as shown in Figure 4.72. A circular pattern repeats the feature(s) around an axis, a cylindrical or conical face, or an edge.

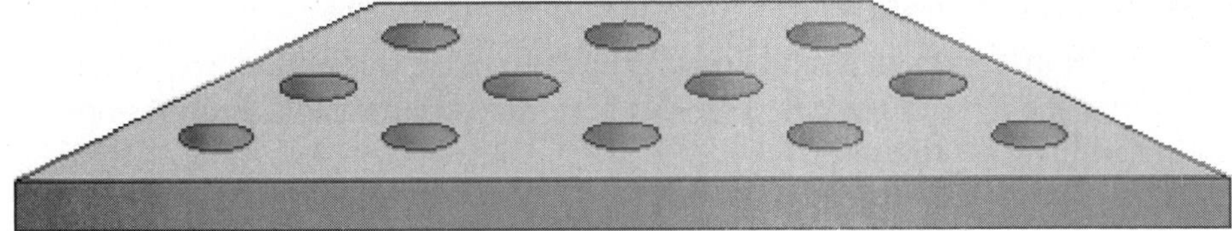

FIGURE 4.72

Rectangular Patterns

To create a rectangular pattern, follow these steps.

1. Click the Rectangular Pattern command on the 3D Model tab > Pattern panel, as shown in **Figure 4.73** on the left.
2. The Rectangular Pattern dialog box appears, as shown in **Figure 4.73** on the right.
3. Click one or more features to pattern.
4. Click the Direction 1 button in the dialog box to define the direction that the feature will travel.
5. 5. Click an edge or line on a sketch to define the direction.
6. Define the count and the spacing or distance by entering values in the Column Count and Column Spacing cells in the dialog box.
7. If needed click the Direction 2 button in the dialog box to define the direction that the feature will travel in a direction that is not parallel to Direction1 and enter values for the count and spacing.
8. As you enter values the pattern is previewed in the graphics window.
9. Click OK to create the pattern.

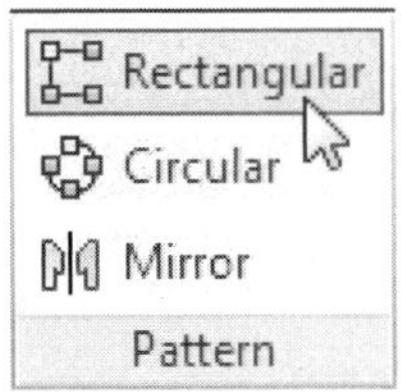

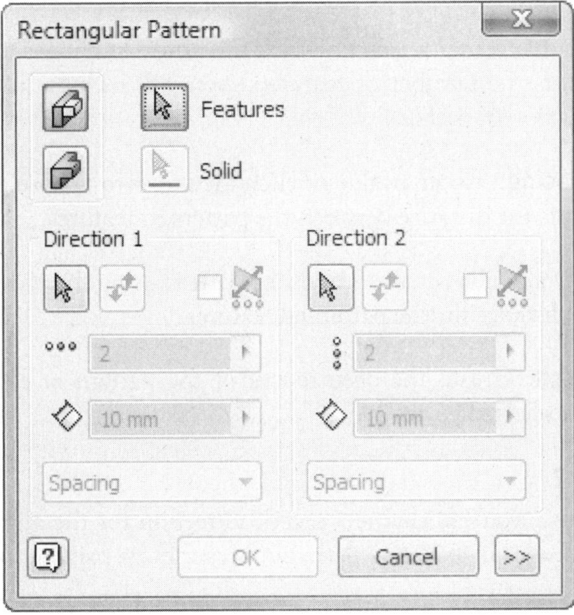

FIGURE 4.73

The following options are available for rectangular patterns.

Pattern Individual Features

Click this button to pattern a feature or features. When you select this option, you activate a features button, as described below.

Pattern the Entire Solid

Click this button to pattern a solid body. When you select this option, you select the entire part as the item to pattern. You also have the Include Work Features option when patterning an entire solid.

Features

Click this button, and then click a feature or features to be patterned from either the graphics window or the browser. You can add or remove features to or from the selection set by holding down the CTRL key and clicking them.

Solid. If multiple solid bodies exist, select the solid body that you want the feature(s) patterned on.

Direction 1

In Direction 1, you define the first direction for the alignment of the pattern. It can be an edge, an axis, or a path.

Path. Click this arrow button, and then click an edge or sketch that defines the alignment along which you will pattern the feature.

Flip. If the preview image shows the pattern going in the wrong direction, click this button to reverse its direction.

Midplane. Check this option to have the occurrences patterned on both sides of the selected feature. The midplane option is independent for both Direction 1 and Direction 2.

Column Count. Enter a value or click the arrow to choose a previously used value that represents the number of feature(s) you will include in the pattern along the selected direction or path.

Column Spacing. Enter a value or click the arrow to choose a previously used value that represents the distance between the patterned features.

Distance. Define the occurrences of the pattern using the provided dimension as the total overall distance for the patterned features.

Curve Length. Create the occurrences of the pattern at equal spacing along the length of the selected curve.

Direction 2

In Direction 2, you can define a second direction for the alignment of the pattern. It can be an edge, an axis, or a path, but it cannot be parallel to Direction 1.

Path. Click this button, and then click an edge or sketch that defines the alignment along which you will pattern the feature.

Flip. If the preview image shows the pattern going in the wrong direction, click this button to reverse its direction.

Midplane. Check this option to have the occurrences patterned on both sides of the selected feature. The midplane option is independent for both Direction 1 and Direction 2.

Row Count. Enter a value or click the arrow to choose a previously used value that represents the number of feature(s) that you will include in the pattern in the second direction.

Row Spacing. Enter a value or click the arrow to choose a previously used value that represents the distance between the patterned features.

Distance. Define the occurrences of the pattern using the provided dimension as the total overall distance for the patterned features.

Curve Length. Create the occurrences of the pattern at equal spacing along the length of the selected curve.

Options for the start point of the direction, compute type, and orientation method, as shown in Figure 4.74, are available by clicking the More (>>) button located in the bottom-right corner of the Rectangular Pattern dialog box.

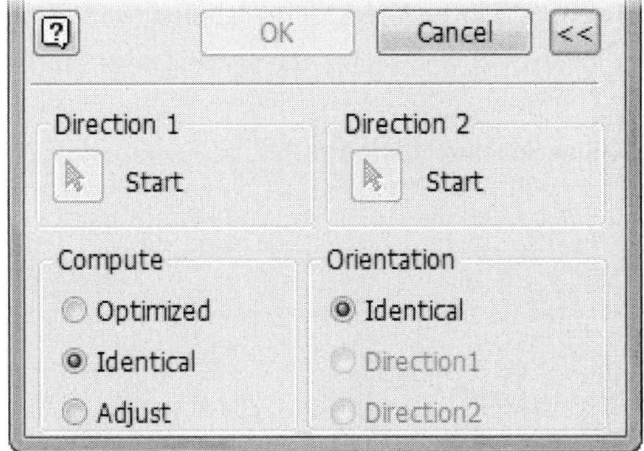

FIGURE 4.74

Start

Click the Start button to specify where the start point for the first occurrence of the pattern will be placed. The pattern can begin at any selectable point on the part. You can select the start points for both Direction 1 and Direction 2.

Compute

Optimized. Click this option to pattern the feature's faces instead of the feature(s) to calculate all of the occurrences in the pattern. This option is ideal when the occurrences you are creating do not intersect and are all identical. It can improve the performance of pattern creation.

Identical. Click this option to use the same termination as that of the parent feature(s) for all of the occurrences in the pattern. This is the default option.

Adjust. Click this option to calculate the termination of each occurrence individually. Since each occurrence is calculated separately, the processing time can increase. You must use this option if a parent feature terminates to a face or plane.

Orientation

Identical. Click this option to orient all of the occurrences in the pattern the same as the parent feature(s). This is the default option.

Direction 1. Click this option to control the position of the patterned features by the selected direction. Each occurrence of the pattern is rotated to maintain proper orientation with the 2D tangent vector of the path.

Direction 2. Click this option to control the position of the patterned features by the selected direction. Each occurrence of the pattern is rotated to maintain proper orientation with the 2D tangent vector of the path.

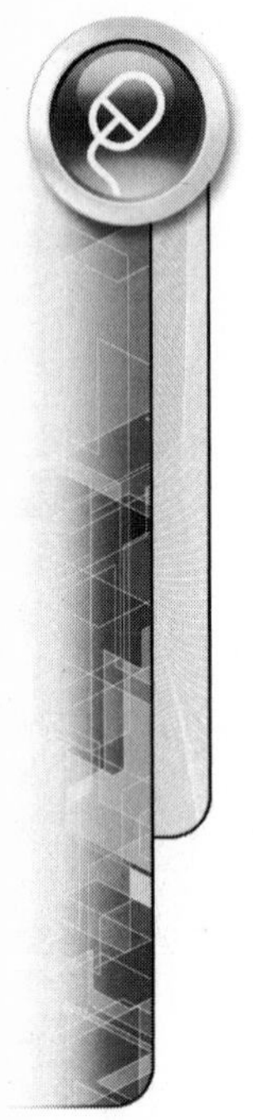

EXERCISE 4-6: CREATING A RECTANGULAR PATTERN

In this exercise, you will create a rectangular pattern of a hole in a cover plate.

1. Open *ESS_E04_06.ipt* in the Chapter 04 folder.
2. Click the Rectangular Pattern command in the 3D Model tab > Pattern panel.
3. Click the small hole feature as the feature to be patterned labeled (1) in Figure 4.75.
4. In the Direction 1 area of the dialog box, click the Direction Path 1 Path button labeled (2), and then select the bottom horizontal edge of the part labeled (3). A preview of the pattern is displayed.
5. Click the Flip button labeled (4).
6. Enter **5** for the Count labeled (5) and **.625 in** for the Spacing labeled (6).
7. Click the Direction 2 Path button labeled (7), and select the vertical edge on the left side of the part labeled (8).
8. Enter **4** Count labeled (9) and **.625 in** Spacing labeled (10).

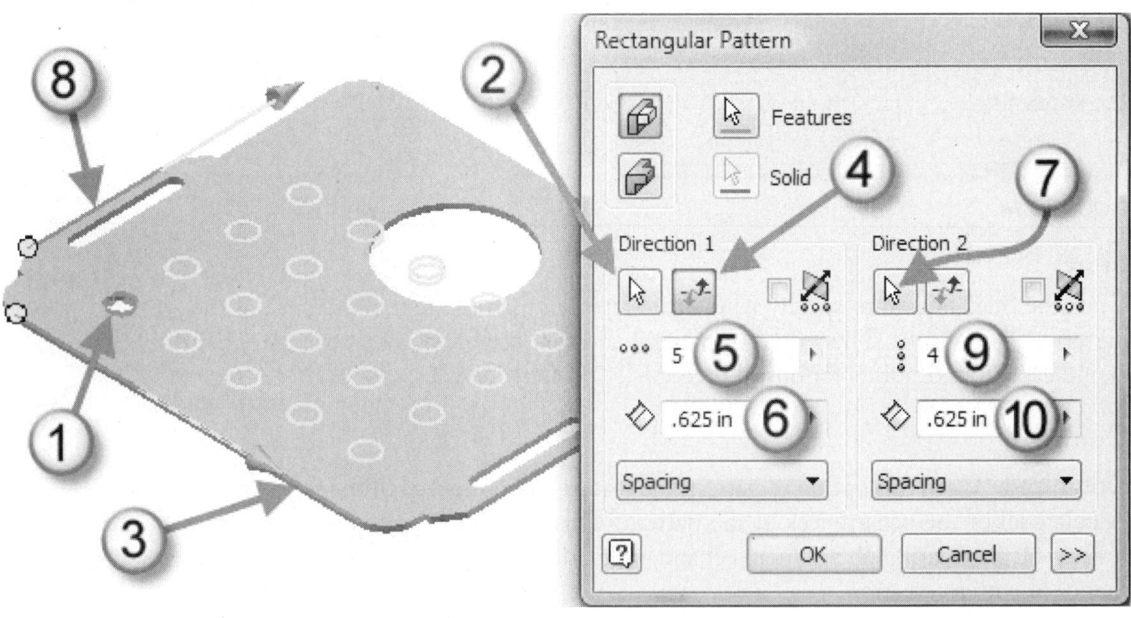

FIGURE 4.75

9. Click OK to create the pattern.
10. You now suppress three of the holes that are not required in the design. Expand the rectangular pattern feature entry in the browser to display the occurrences.
11. In the browser, move the cursor over the occurrences. Each occurrence highlights in the graphics window as you point to it in the browser.
12. Hold the CTRL key down, and click the three occurrences the holes will highlight on the model, as shown in Figure 4.76.
13. Right-click on any one of the highlighted occurrences in the browser, and click Suppress, as shown in Figure 4.76.

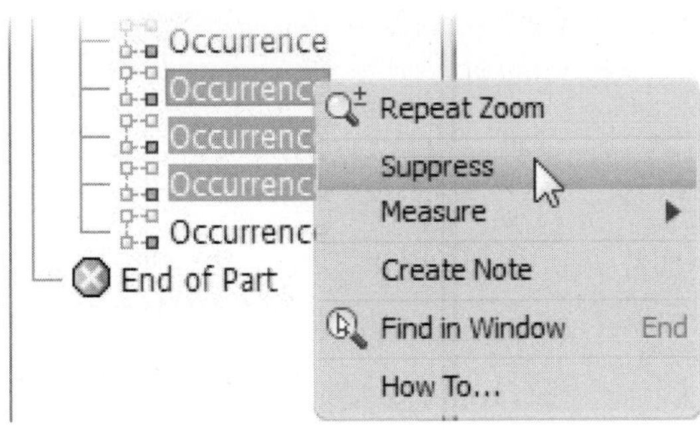

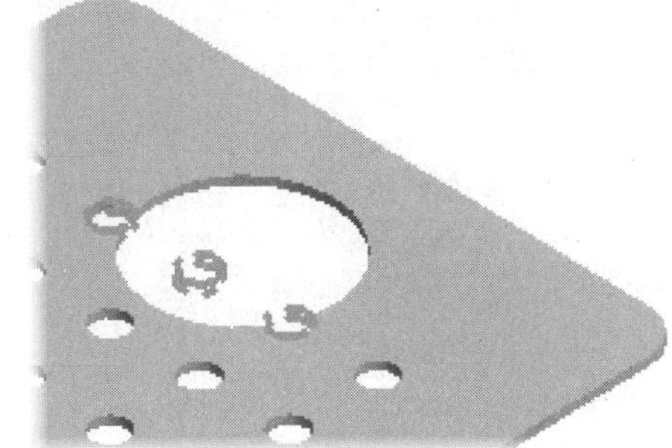

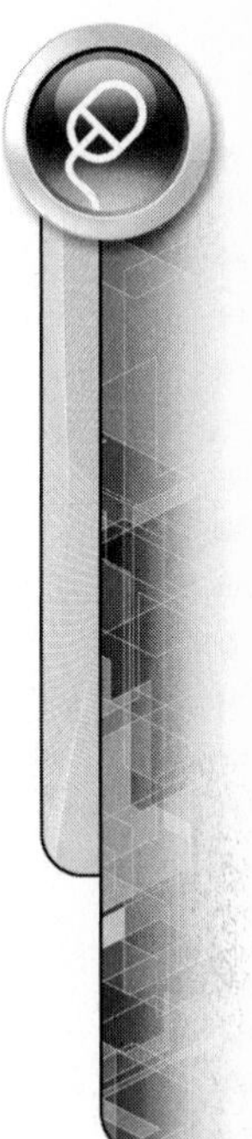

FIGURE 4.76

14. The holes are suppressed in the model, as shown in Figure 4.77.

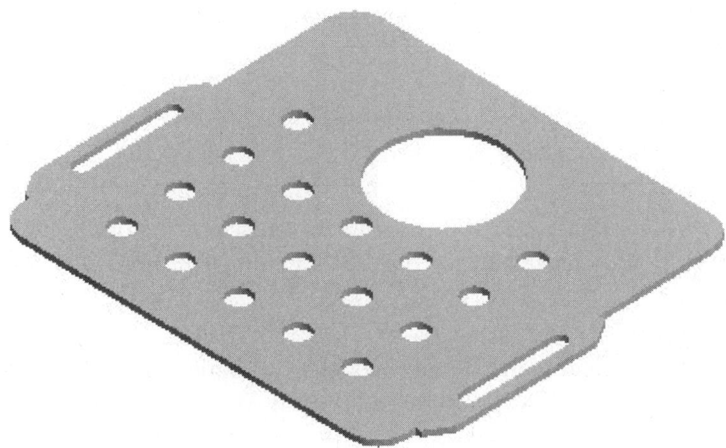

FIGURE 4.77

15. Practice editing the feature pattern, and change the count and spacing for each direction. The suppressed occurrences are still suppressed.
16. Close the file. Do not save changes. End of exercise.

Circular Patterns

When creating a circular pattern, you must have a work axis, a part edge, or a cylindrical face about which the features will be patterned around. To create a circular pattern follow these steps.

1. Click the Circular Pattern command on the 3D Model tab > Pattern panel, as shown in Figure 4.78 on the left.
2. The Circular Pattern dialog box appears, as shown in Figure 4.78 on the right.
3. Click one or more features to pattern.
4. Click the Rotation Axis button in the dialog box to define the axis of rotation that the feature will be patterned around.
5. Click a work axis, an edge, or a cylindrical face (the centerline of the cylinder will be used) to define the axis.
6. Define the Occurrence Count and Occurrence Angle by entering values in the Occurrence Count and Occurrence Angle cells in the dialog box.
7. As you enter values the pattern is previewed in the graphics window.
8. Click OK to create the pattern.

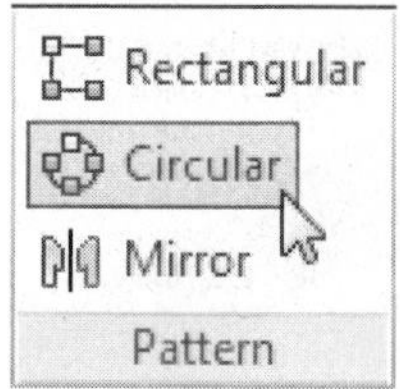

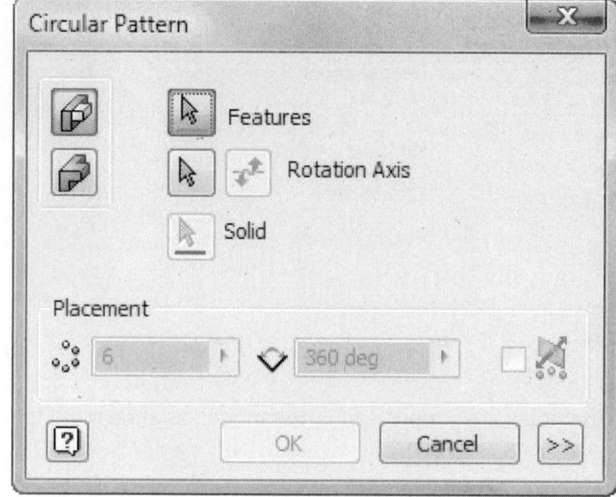

FIGURE 4.78

The following options are available for circular patterns.

Pattern Individual Features

Click this button to pattern a feature or features. When you select this option, the features button is available and operates as described below.

Pattern the Entire Solid

Click this button to pattern a solid body. When you select this option, you select the entire part as the item to pattern. You also have the Include Work Features option when patterning an entire solid.

Features

Click this button, and then click a feature or features to be patterned. You can add or remove features to or from the selection set by holding down the CTRL key and clicking them.

Rotation Axis

Click the button and then click an edge, axis, or cylindrical face (center) that defines the axis about which the feature(s) will rotate.

Flip. If the preview image shows the pattern going in the wrong direction, click this button to reverse its direction.

Solid. If multiple solid bodies exist, select the solid body you want the feature(s) patterned on.

Placement

Occurrence Count. Enter a value or click the arrow to choose a previously used value that represents the number of feature(s) that you will include in the pattern. A positive number will pattern the feature(s) in the clockwise direction; a negative number will pattern the feature in the counterclockwise direction.

Occurrence Angle. Enter a value or click the arrow to choose a previously used value that represents the angle that you will use to calculate the spacing of the patterned features.

Midplane. Check this option to have the occurrences patterned evenly on both sides of the selected feature.

By clicking the More (>>) button, located in the bottom-right corner of the Circular Pattern dialog box, you can access options for the creation method and positioning method of the feature, as shown in Figure 4.79.

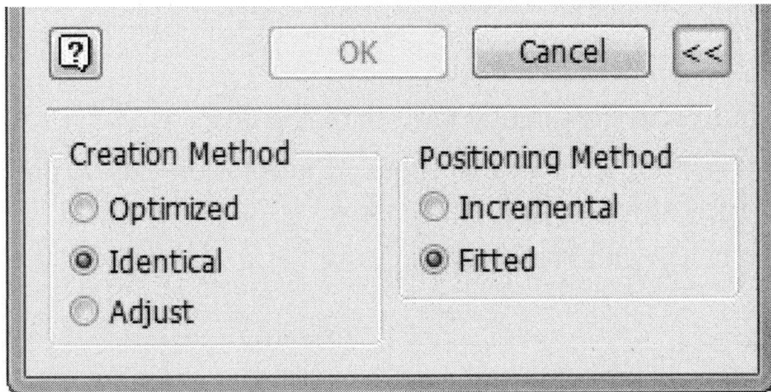

FIGURE 4.79

Creation Method

Optimized. Click this option to pattern the feature's faces instead of the feature(s) to calculate all of the occurrences in the pattern. This option is ideal when the occurrences you are creating do not intersect and are all identical. It can improve the performance of pattern creation. This method is recommended when there are 50 or more occurrences.

Identical. Click this option to use the same termination as that of the parent feature(s) for all of the occurrences in the pattern. This is the default option.

Adjust. Click this option to calculate each occurrence termination individually. Because each occurrence is calculated separately, the processing time can increase. You must use this option if a parent feature terminates to a face or plane.

Positioning Method

Incremental. Click this option to separate each occurrence by the number of degrees specified in Angle in the dialog box.

Fitted. Click this option to space each occurrence evenly within the angle specified in Angle in the dialog box.

TIP

A work axis, an edge, or a cylindrical face (the centerline of the circular face will be used) about which the feature will rotate must exist before you create a circular pattern.

Linear Patterns—Pattern along a Path

You can also use the Rectangular Pattern command to pattern a feature about a path. You can define a path by a complete or partial ellipse, an open or closed spline, or a series of curves (lines, arcs, splines, etc.).

To pattern along a path, click the Path button, and use the options described above for rectangular patterns. The path you use can be either 2D or 3D.

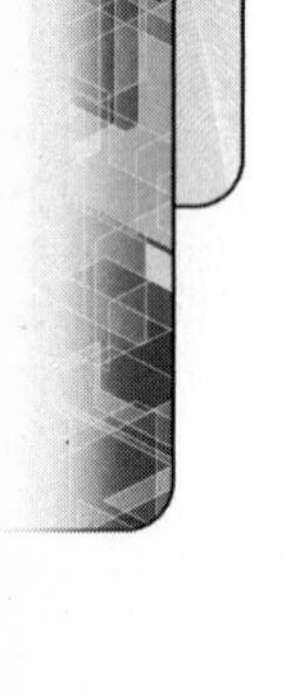

EXERCISE 4-7: CREATING CIRCULAR PATTERN

In this exercise, you will create a circular pattern of a counterbore hole.

1. Open *ESS_E04_07.ipt* in the Chapter 04 folder.
2. Edit the Hole1 feature to verify that the termination for the hole is Through All.
3. Click the Cancel button to close the Hole dialog box.
4. Click the Circular Pattern command in the 3D Model tab > Pattern panel.
5. Click the hole feature as the feature to pattern.
6. Click the Rotation Axis button in the Circular Pattern dialog box.
7. In the graphics window click the work axis to specify the rotation axis. A preview of the pattern is displayed in the graphics window.
8. Type **8** in the Count field as shown in Figure 4.80.

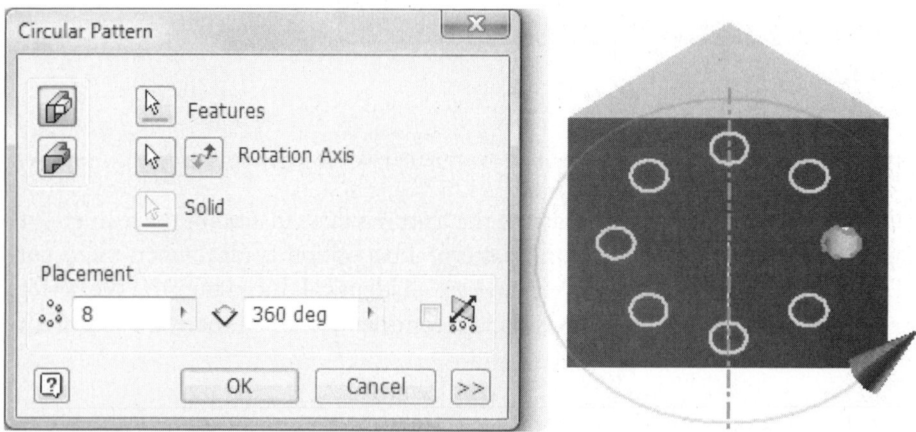

FIGURE 4.80

9. Click the OK button to create the pattern.
10. Rotate your viewpoint so you can see the back of the part, as shown in Figure 4.81. The other six holes do not go through because the Identical Creation Method was selected; that is, the hole that is patterned is identical to the original hole. If desired, change the visual style to Wireframe to verify that all the holes are identical.

FIGURE 4.81

11. In the browser right-click on Circular Pattern1 and click Edit Feature from the menu or click on one of the circular faces of a patterned hole (not the original hole) and click Edit Circular Pattern from the mini-toolbar.
12. Click the More (>>) button.
13. Under Creation Method, select the Adjust option, as shown in Figure 4.82 on the left.
14. Click OK to create the circular pattern. When done, your model should resemble Figure 4.82 on the right.

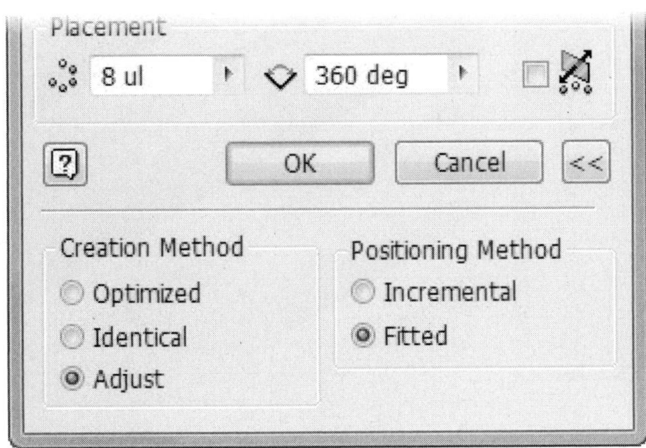

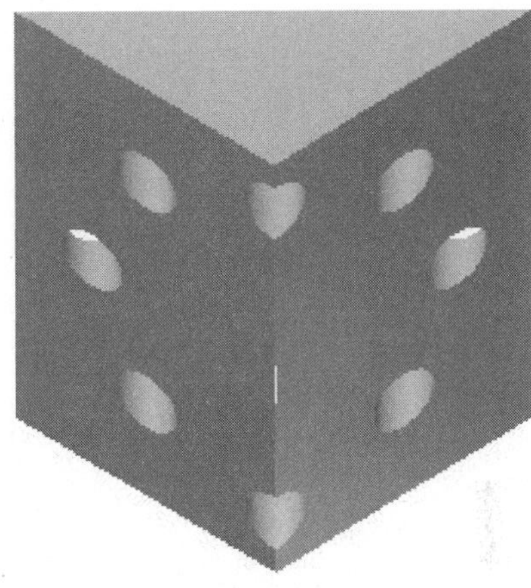
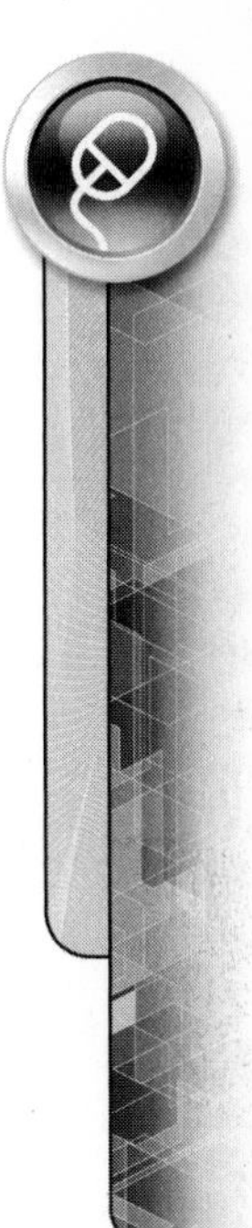

FIGURE 4.82

15. Edit the circular pattern, and try different combinations of count, angle, creation method, and positioning method.
16. Close the file. Do not save changes. End of exercise.

EXERCISE 4-8: CREATING A PATTERN ALONG A NONLINEAR PATH

In this exercise, you will pattern a boss and hole along a nonlinear path.

1. Open *ESS_E04_08.ipt* in the Chapter 04 folder. Note that the visibility of the sketch that the pattern will follow is on.
2. Click the Rectangular Pattern command in the 3D Model tab > Pattern panel.
3. For the features to pattern, in the browser or in the graphics window, click both Extrusion2 and Hole2 feature as shown in the following image on the left.
4. In the Direction 1 area in the Rectangular Pattern dialog box, click the Direction Path arrow button. To select the path in the graphics window click a line or an arc in the visible sketch. A preview of the pattern is displayed, as shown in Figure 4.83 on the right.

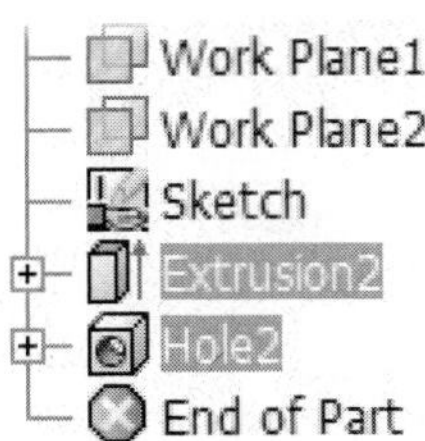

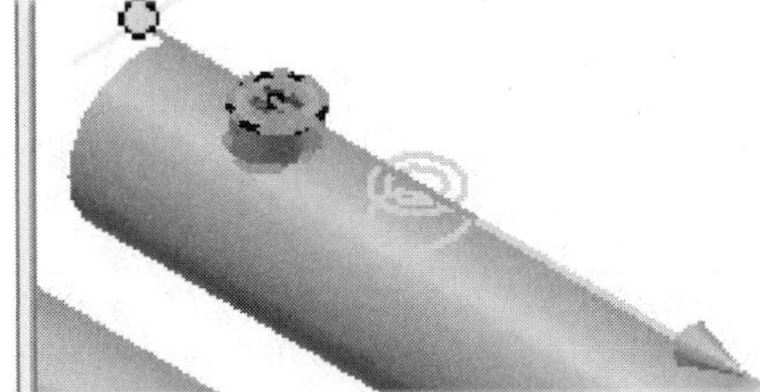

FIGURE 4.83

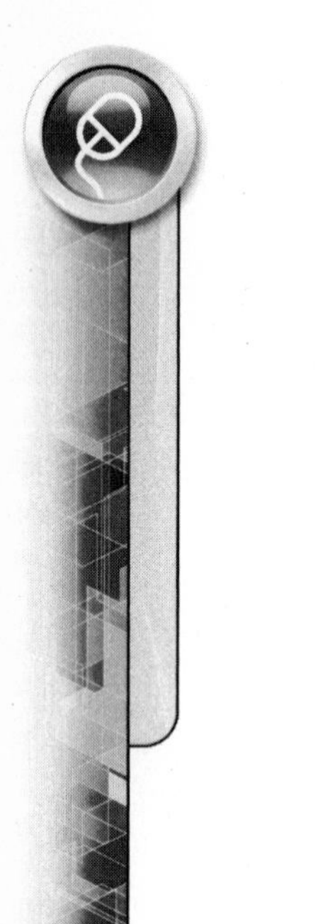

5. In Column Count cell, enter **40**, and in Column Spacing cell , enter **1.25**. The preview shows 1.25 inches between each occurrence as shown in **Figure 4.84** on the right.

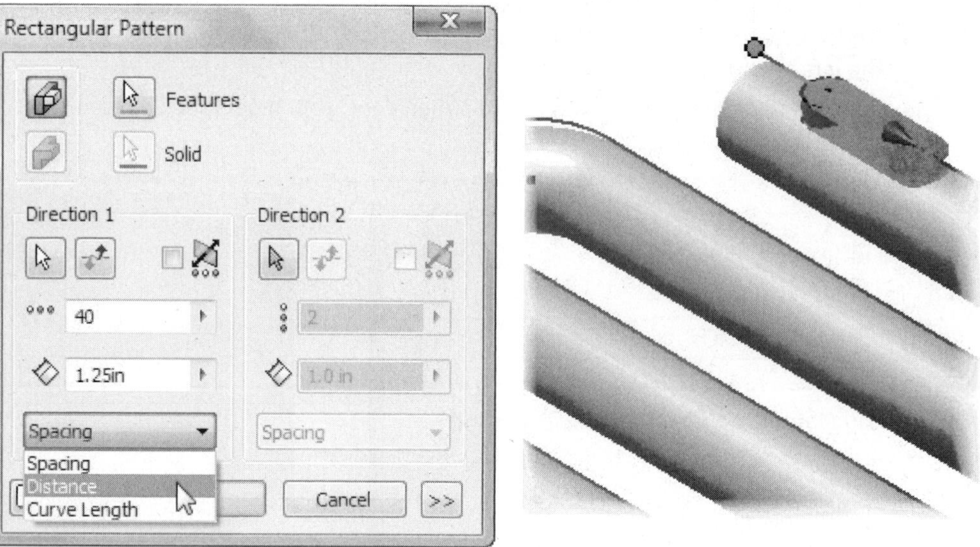

FIGURE 4.84

6. From the Spacing drop-down list, select Distance as shown in **Figure 4.85** on the left. The preview shows 40 occurrences fit within 1.25 inches.
7. From the Distance drop-down list, select Curve Length. The preview shows the 40 occurrences fitting within the entire length of the path.
8. Rotate the model to verify that the occurrences on the right side are hanging off the model; this is because the first occurrence is spaced 1 inch away from the start of the path.
9. To solve the starting point issue, click the More (>>) button.
10. In the Direction 1 area in the bottom area of the dialog box, click the Start button, and then click the center point of the first hole, as shown in Figure 4.85 on the left. The preview updates to show that all occurrences are now located on the part, as shown in Figure 4.85 on the right. However, the last occurrence is on the edge of the part.

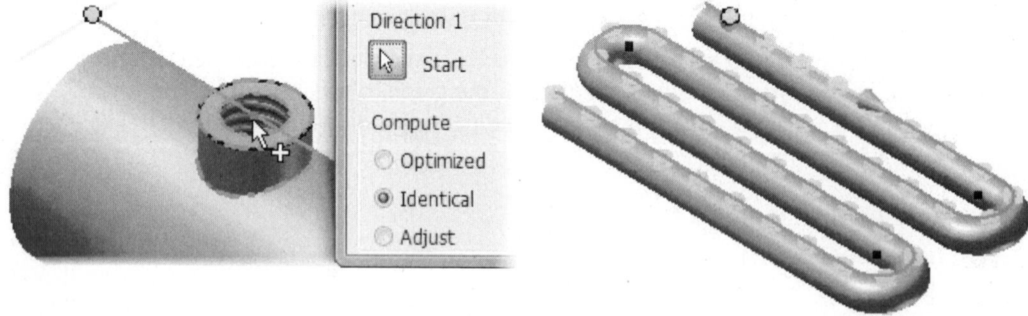

FIGURE 4.85

11. From the Curve Length drop-down list, select Distance. The curve length remains in the distance area but the value can be edited.
12. Click in the distance area and arrow to the right of the value of the four lines and three arcs. Subtract **–1 in** from the distance, as shown in Figure 4.86 on the left or enter a value of **55.425**.
13. Click OK to create the pattern.

14. In the browser, right-click on the entry Sketch, and uncheck Visibility from the menu. When done, your model should resemble Figure 4.86 on the right.

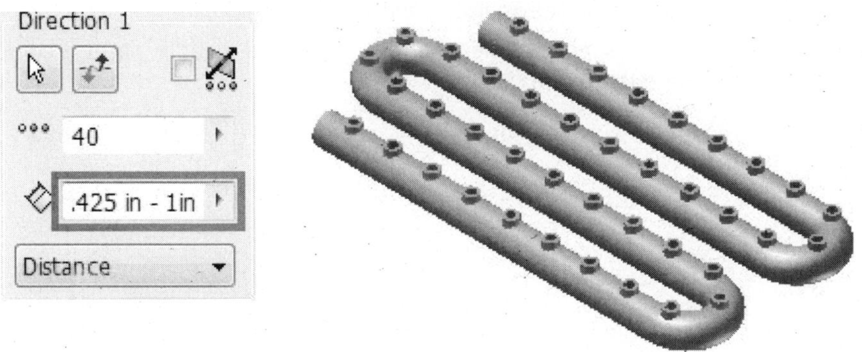

FIGURE 4.86

15. Edit the pattern trying different combinations.
16. Close the file. Do not save changes.

APPLYING YOUR SKILLS

Skill Exercise 4-1

In this exercise, you will create a drain plate cover.

1. Start a new part based on the English *Standard (in).ipt* template.
2. Use the extrude, shell, hole, rectangular pattern, and the Fillet commands to create the part. Only fillet the inside vertical edges.

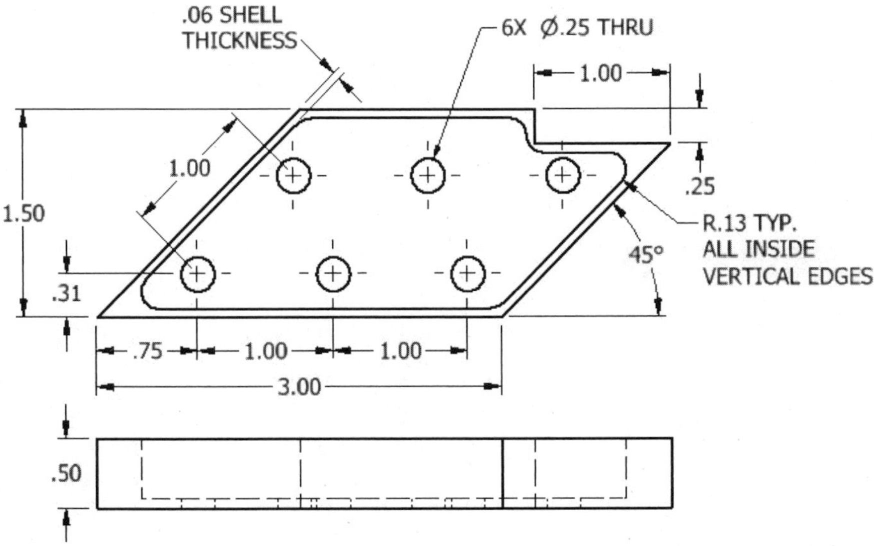

FIGURE 4.87

Skill Exercise 4-2

In this exercise, you will create a connector.

1. Start a new part based on the English *Standard (in).ipt* template.
2. Use the revolve, work plane, hole, chamfer, fillet, and circular pattern commands to complete the part.

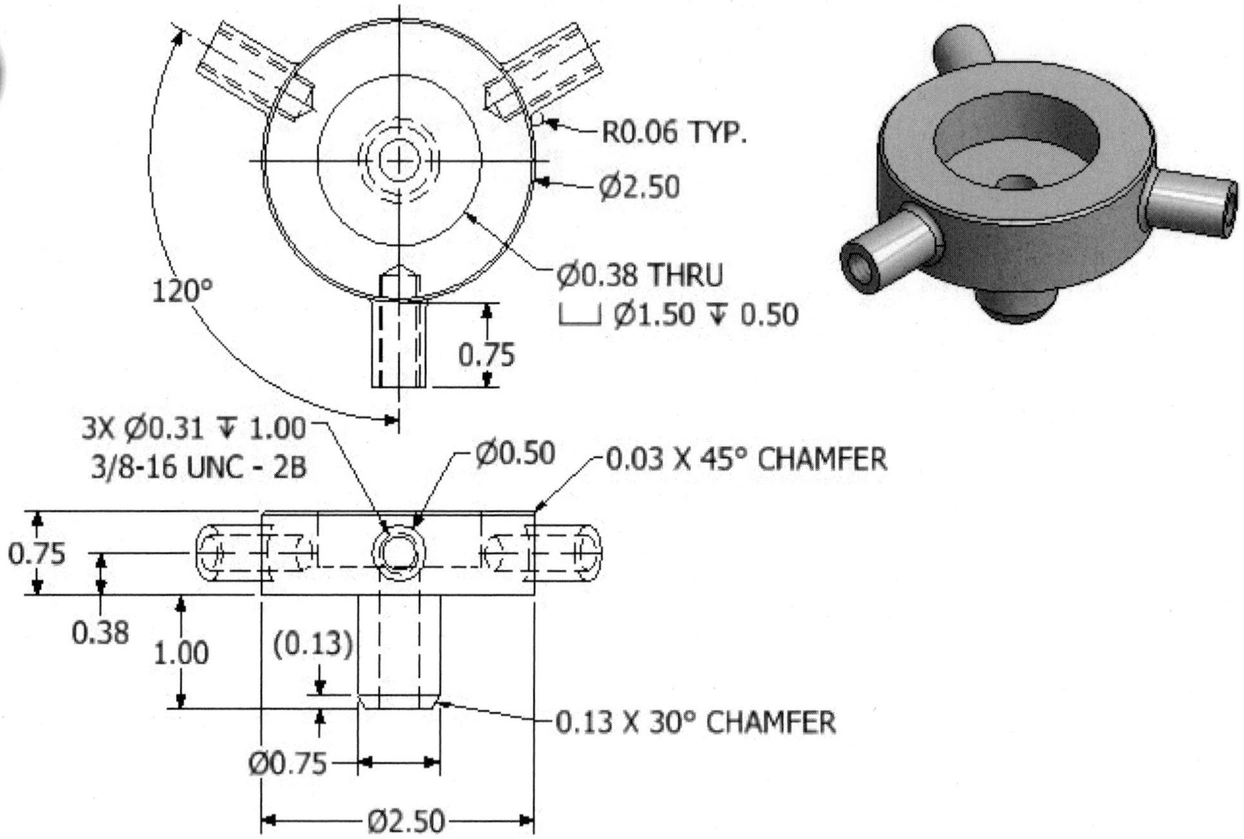

FIGURE 4.88

CHECKING YOUR SKILLS

Use these questions to test your knowledge of the material covered in this chapter.

1. True __ False __ When creating a fillet feature that has more than one selection set, each selection set appears as an individual feature in the browser.
2. In regard to creating a fillet feature, what is a smooth radius transition?
3. True __ False __ When you are creating a fillet feature with the All Fillets option, material is removed from all concave edges.
4. True __ False __ When you are creating a chamfer feature with the Distance and Angle option, you can only chamfer one edge at a time.
5. True __ False __ When you are creating a hole feature, you do not need to have an active sketch.
6. What is a Point, Center Point used for?
7. True __ False __ When shelling a part you can only have one unique face thickness in the part.

8. True __ False __ The only method to create a work axis is by clicking a cylindrical face.
9. True __ False __ You need to create every new sketch on a work plane.
10. Explain the steps to create an offset work plane.
11. True __ False __ A work plane is only used to create a sketch on.
12. True __ False __ You cannot create work planes from the origin planes.
13. True __ False __ A UCS can only be placed on existing geometry.
14. True __ False __ When you are creating a rectangular pattern, the directions along which the features are duplicated must be horizontal or vertical.

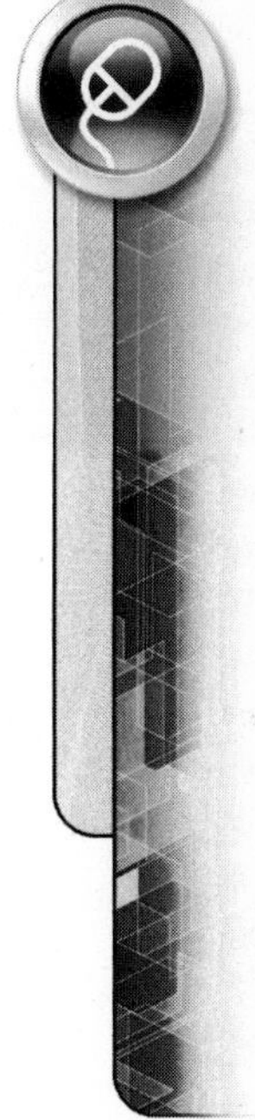

15. True __ False __ When you are creating a circular pattern, you can only use a work axis as the axis of rotation.
16. True __ False __ To start the Fillet command from the mini-toolbar, select the two planes that define the location of the fillet.
17. True __ False __ Use the Filet command with the Face option to create a fillet that is tangent to three faces.
18. While in the shell command, explain how to create a unique face thickness within the shell command.

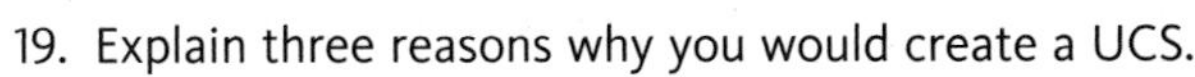

19. Explain three reasons why you would create a UCS.

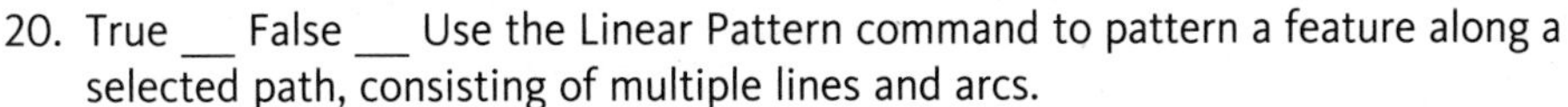

20. True __ False __ Use the Linear Pattern command to pattern a feature along a selected path, consisting of multiple lines and arcs.

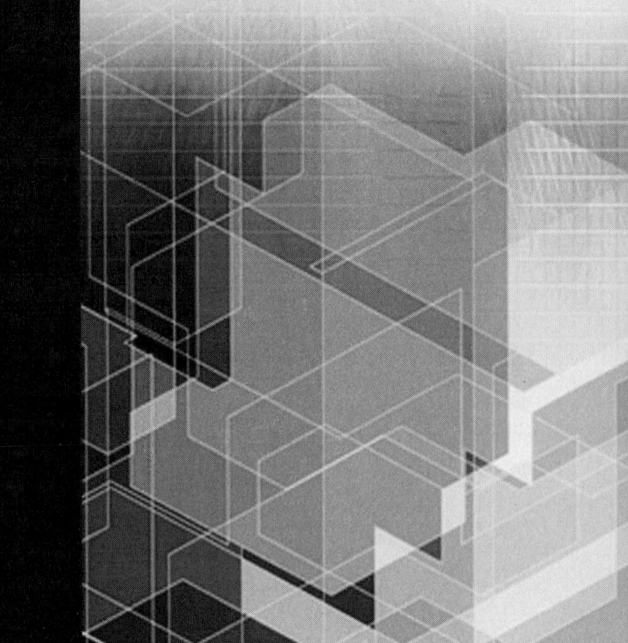

CHAPTER 5

Creating and Editing Drawing Views

INTRODUCTION

After creating a part or assembly, the next step is to create 2D drawing views that represent that part or assembly. To create drawing views, start a new drawing file, select a 3D part or assembly on which to base the drawing views, project orthographic and isometric views from the part or assembly, and then add annotations to the views. You can create drawing views at any point after a part or assembly exists. The part or assembly does not need to be complete because the part and drawing views are associative in both directions (bidirectional). This means that if the part or assembly changes, the drawing views will automatically be updated. If a parametric dimension changes in a drawing view, the part will be updated before the drawing views get updated. This chapter will guide you through the steps for setting up styles, creating drawing views of a single part, editing dimensions, and adding annotations.

When creating drawings you adhere to specific drawing standards that communicate information about designs in a consistent manner. There are many different drawing standards, such as ANSI and ISO. Each company usually adopts a drawing standard and modifies it to meet their requirements. This book follows the ANSI drawing standard. More information about drawing standards and good practices are covered throughout this chapter.

OBJECTIVES

After completing this chapter, you will be able to perform the following:

- Create base and projected drawing views from a part
- Create auxiliary, section, detail, broken, break out, and cropped views
- Edit the properties and location of drawing views
- Retrieve and arrange model dimensions for use in drawing views
- Edit, move, and hide dimensions
- Add automated centerlines
- Add general dimensions, baseline, chain, and ordinate dimensions

- Add annotations such as text, leaders, Geometric Dimensioning & Tolerancing (GD&T), surface finish symbols, weld symbols, and datum identifiers
- Create hole notes
- Open a model from a drawing
- Open a drawing from a model
- Create a hole table and a general table

Drawing Options

Before drawing views are created, the drawing options should be set to your preferences. To set the drawing options, click Application Options on the Tools panel. The Options dialog box will appear. Click the Drawing tab, your screen should resemble Figure 5.1. Make changes to the options before creating the drawing views, otherwise the changes may not affect drawing views that you have already created. A description of the common options follows. For more information about the Drawing Application Options consult the help system.

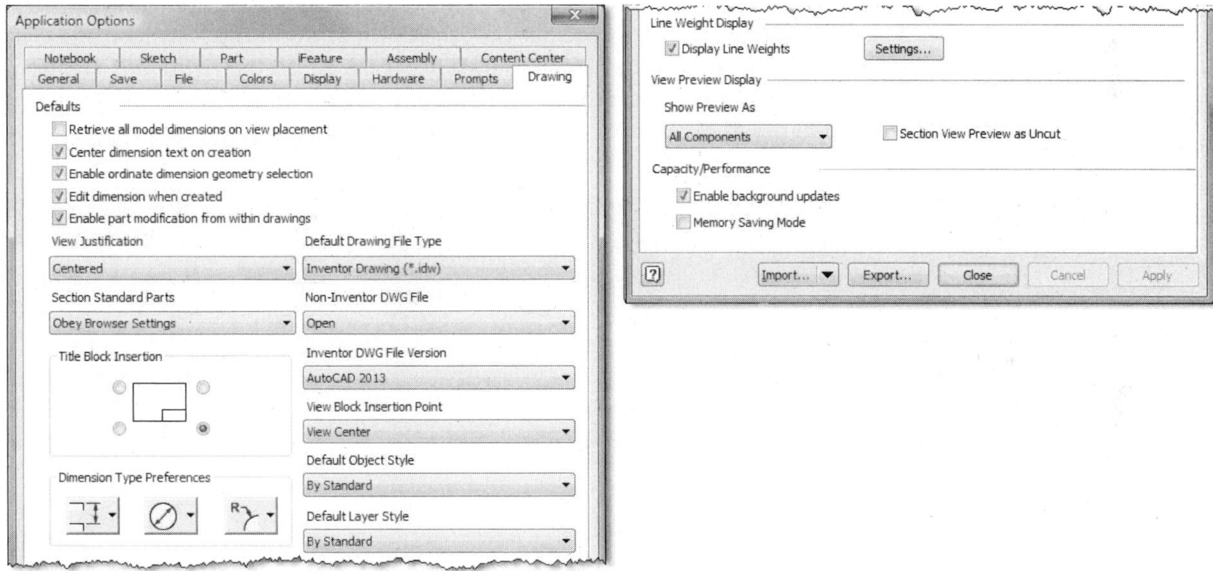

FIGURE 5.1

Retrieve All Model Dimensions on View Placement

Click this box to add applicable model dimensions to drawing views when they are placed. If the box is clear, no model dimensions will be placed automatically. You can override this setting by manually selecting the All Model Dimensions in the Drawing View dialog box when creating base views.

Center Dimension Text on Creation

Click this option to have dimension text centered as you create the dimension.

Edit Dimension When Created

Click this option to see the Edit Dimension dialog box everytime a dimension is placed in a drawing.

Default Drawing File Type

In this section select either Inventor drawing IDW or DWG as the default file format. Also set how Inventor should handle DWG and drawing data.

Inventor DWG File Version

Set the Inventor DWG file version to save.

Creating a Drawing

The first step in creating a drawing from an existing part or assembly is to create a new drawing IDW or DWG file by one of the following methods. You can use the following methods to create a new part file:

- Click New on the Inventor Application menu, and then click the desired folder on the left side of the Create New File dialog box, and in the Drawing create an annotated document area click the DWG or IDW template file.
- Click the New icon in the Quick Launch area of the Open dialog box, and then click the desired folder on the left side of the Create New File dialog box, and in the Drawing, create an annotated document area click the DWG or IDW template file.
- From the Quick Access toolbar click the down arrow of the New icon and click Drawing from the drop list as shown in Figure 5.2 on the right. This creates a new drawing file based on the default drawing file type (DWG or IDW that is set in the Application Options > Drawing tab.

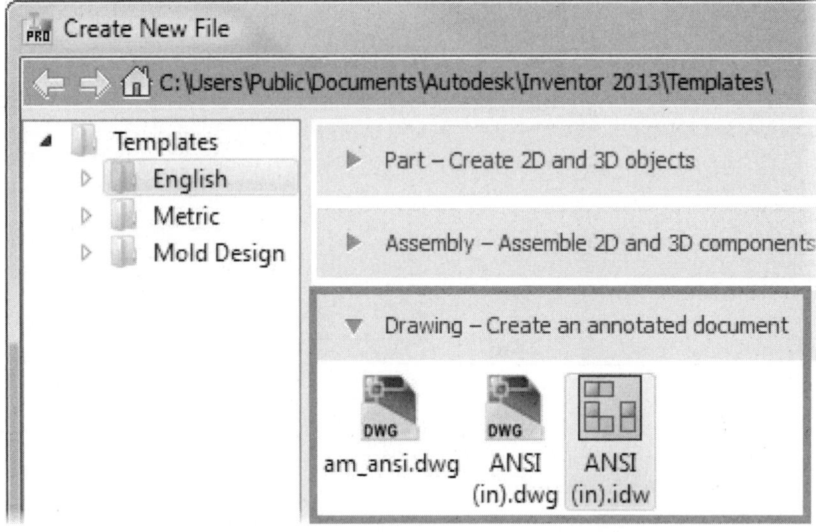

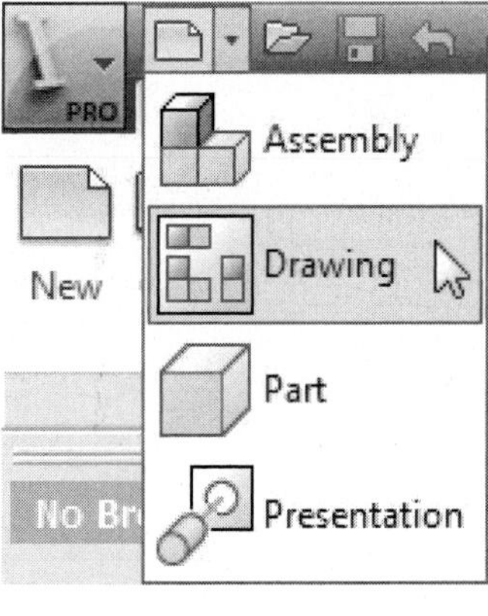

FIGURE 5.2

When the Metric template folder is selected there are seven drafting standards to select from. The drafting standards are:

- ANSI (American National Standards Institute)
- BSI (British Standards Institute)
- DIN (The German Institute for Standardization)
- GB (The Chinese National Standard)
- GOST (The Russian Standard)
- ISO (International Organization for Standardization)
- JIS (Japan Industrial Standard)

DWG TrueConnect

In Inventor, when a drawing file is created from a DWG template, an Inventor DWG is created and AutoCAD will be able to open this file without translation. In AutoCAD you can view, plot, and add annotations to the Inventor DWG file. You cannot edit the drawing views as these are controlled by Inventor. Also from the Design Center in AutoCAD you can insert Inventor drawing from an Inventor DWG file and as each view is represented a block. If the drawing views change in Inventor the blocks in AutoCAD can be updated via the Design Center. In Inventor you can open an AutoCAD 2D drawing and you can view, plot, measure and copy data to the clipboard that can be pasted into a part file.

DRAWING SHEET PREPARATION

When you create a new drawing file using one of the provided template files or a template file that you created. Inventor displays a default drawing sheet with a default title block and border. The template DWG or IDW file that is selected determines the default drawing sheet, title block, and border. The drawing sheet represents a blank piece of paper on which you can alter the border, title block, and drawing views. There is no limit to the number of sheets that can exist in the same drawing, but you must have at least one drawing sheet. To create a new sheet, click the New Sheet command from the Place Views tab > Sheets panel as shown in Figure 5.3 on the left. Alternately, you can right-click in the browser and click New Sheet from the menu as shown in Figure 5.3 in the middle, or on the current sheet in the graphics window right-click and click New Sheet from the menu as shown in Figure 5.3 on the right.

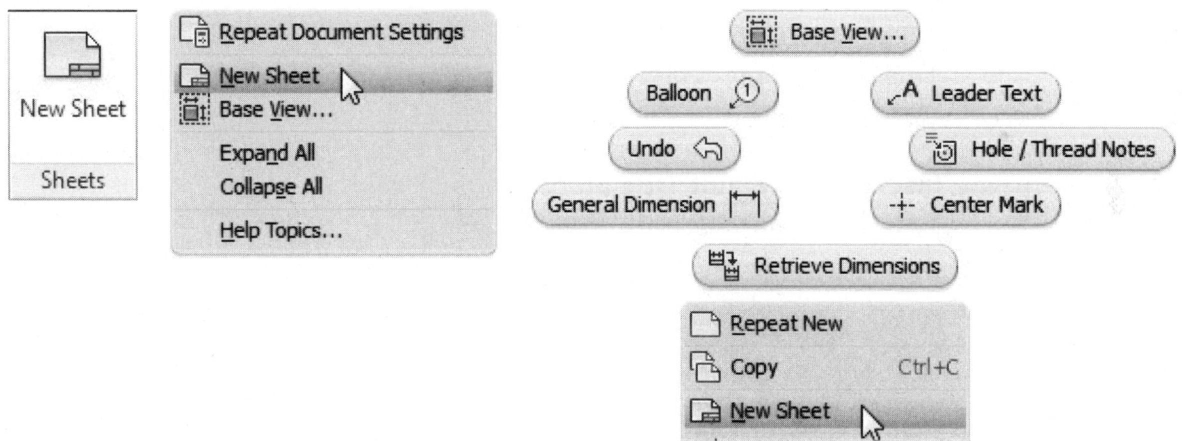

FIGURE 5.3

A new sheet will appear in the browser, and the new sheet will appear in the graphics window with the same size, border and title block of the active sheet. To edit the sheet's size right-click on the sheet name in the browser, and select Edit Sheet from the menu, as shown in Figure 5.4 on the left. Then select a size from the list, as shown in Figure 5.4 on the right. To use your own values, select Custom Size from the list, and enter values for height and width. From the dialog box you can also change the sheet's name or slowly double-click on its name in the browser and then enter a new name.

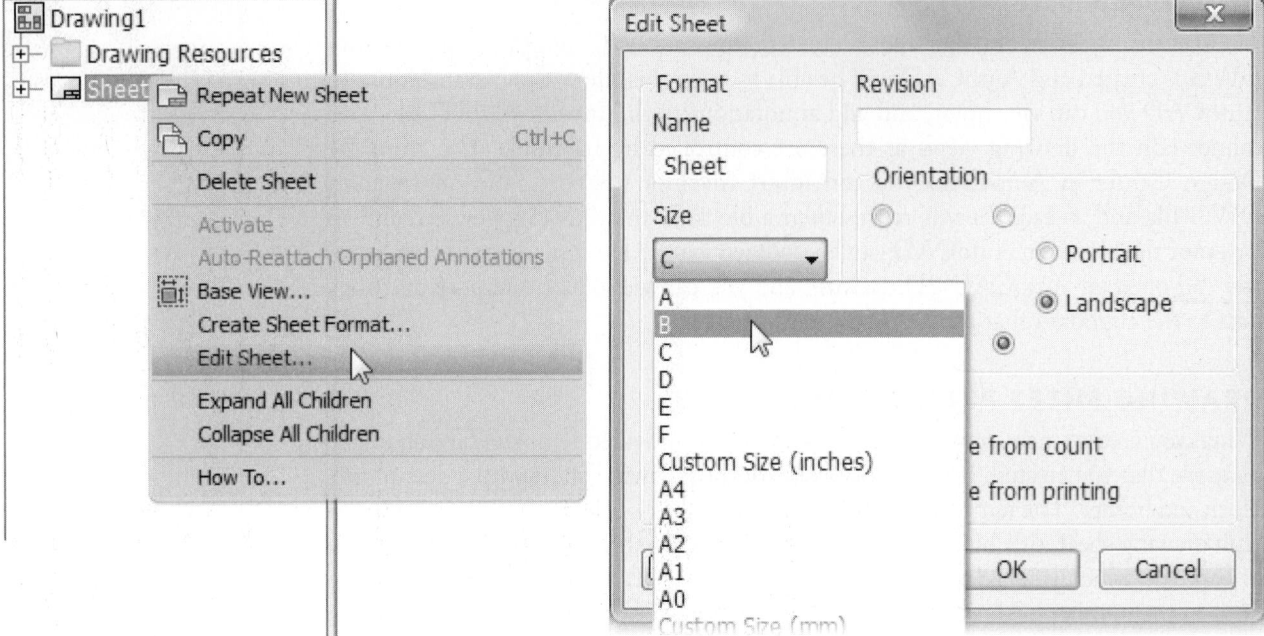

FIGURE 5.4

NOTE	The sheet size is inserted full scale (1:1) and should be plotted at 1:1. The drawing views will be scaled to fit the sheet size.

TITLE BLOCKS

To change a title block on a drawing sheet, you can either insert a default title block or construct a customized title block and insert it into a drawing sheet.

Inserting a Default Title Block

To insert a default title block, follow these steps:

1. If a title block exists in the sheet, it must be deleted before a new title block can be inserted. Make the sheet active, from the browser right-click on the title block entry and click Delete from the menu as shown in Figure 5.5 on the left.
2. Insert a title block by expanding Drawing Resources > Title Blocks in the browser, as shown in Figure 5.5. Either double-click on the title block's name, as shown in Figure 5.5 on the right, or right-click on the title block's name and select Insert from the menu.

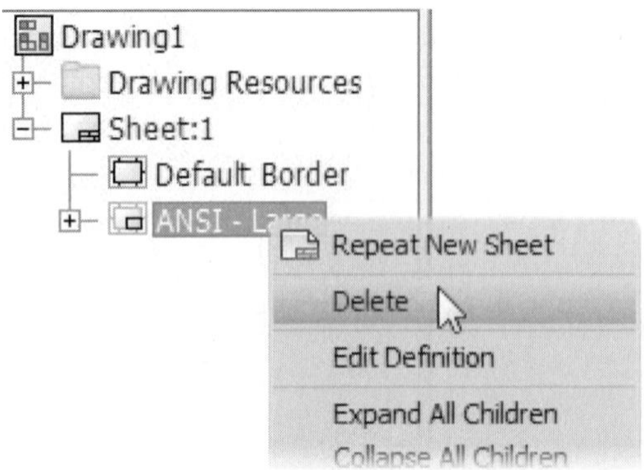

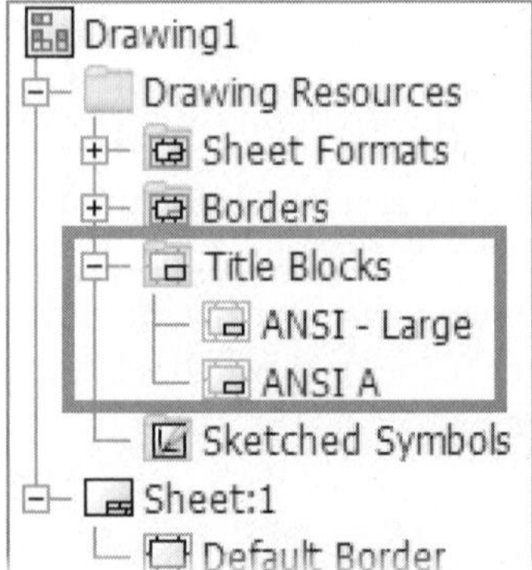

FIGURE 5.5

BORDER

To change a border on a drawing sheet, you can either insert a default border or construct a customized border and insert it into a drawing sheet.

Inserting a Default Border

To insert a default border, follow these steps:

1. If a border exists in the sheet, it must be deleted before a new border can be inserted. Make the sheet active, from the browser right-click on the Default Border entry and click Delete from the menu as shown in Figure 5.6 on the left.
2. Insert a border by expanding Drawing Resources > Borders in the browser.
   - To insert a generic border double-click on the border's name and a generic border will be inserted.
   - To control the border's appearance right-click on Borders > Default Border and click Insert Drawing Border, as shown in Figure 5.6 in the middle. The Default Drawing Border Parameters dialog box will appear, modify the options as needed, click the More (>>) button to control text and layer data as well as the sheet's margin as shown in Figure 5.6 on the right.
   - Consult the help system to learn how to create a new border.

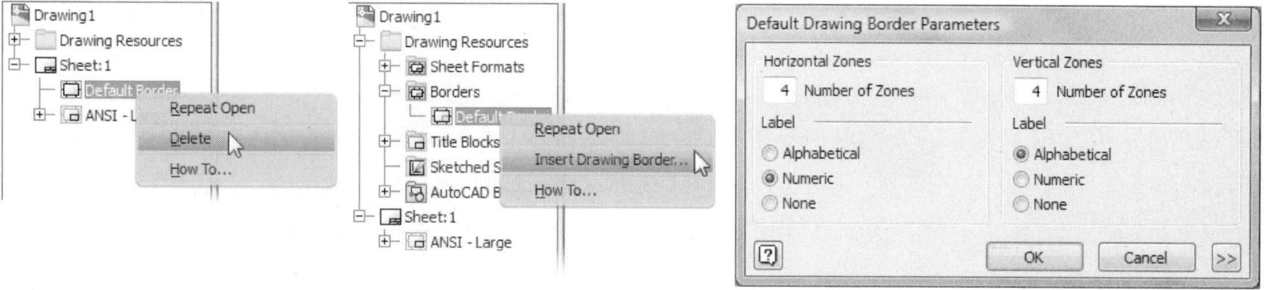

FIGURE 5.6

Edit Property Fields Dialog Box

The time will come when you will need to fill in title block information. Expanding the default title block in the browser and double-click or right-click on the Field Text entry and click Edit Field Text from the menu as shown in Figure 5.7 on the left will display the Edit Property Fields dialog box. By default, the following information will already be filled in: Sheet Number, Number of Sheets, Author, Creation Date, and Sheet Size. To fill in other title block information such as Part Number, Company Name, Checked By, and so on, select the iProperties command button as shown in Figure 5.7 in the middle. The Drawing Properties dialog box will appear, as shown in Figure 5.7 on the right, and you can fill in the information as needed. You can find most title block information under the Summary, Project, and Status tabs.

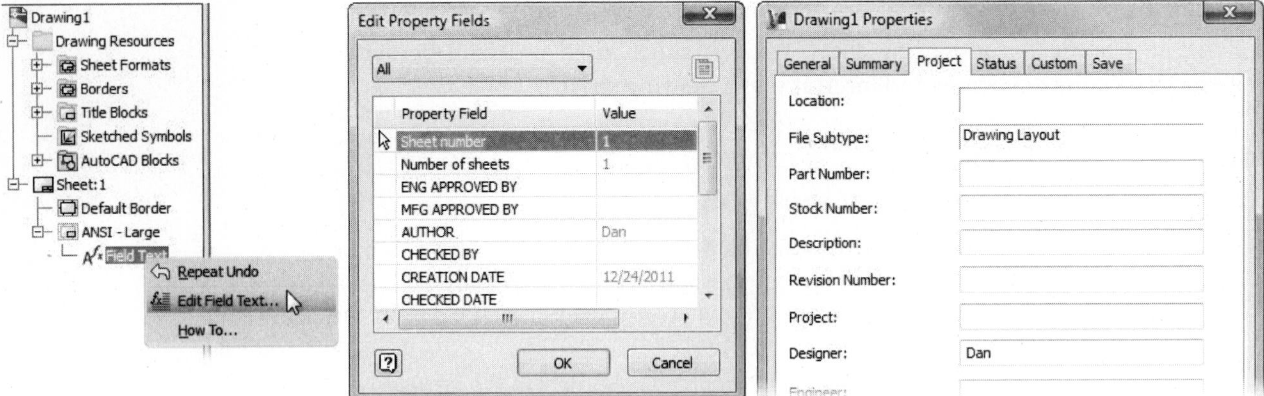

FIGURE 5.7

SAVE DRAWING DATA TO A TEMPLATE

After getting a drawing setup, you can save the drawing to the template folder so you can create a new drawing that contains the changes. From the Application Menu click Save As > Save Copy As Template as shown in Figure 5.8. Save the file to the correct folder; Templates (the default location), English and Metric or create a new folder. The location of the template files is set in the Application Options > File tab > Default templates or overridden in the active project's Folder Options > Templates.

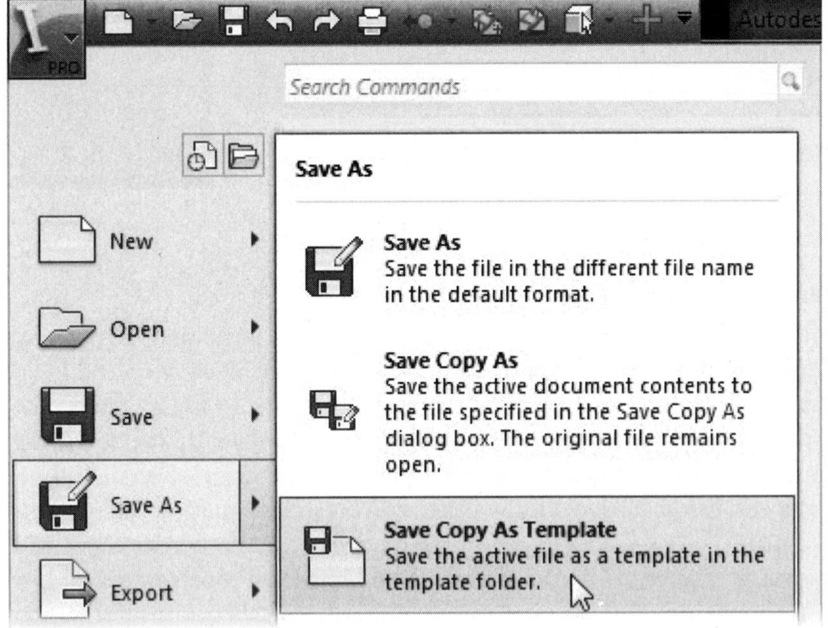

FIGURE 5.8

CREATING DRAWING VIEWS

After you have set the drawing sheet format, border, and title block you can create drawing views from an existing part, assembly, or presentation file. The file from which you will create the views does not need to be open when a drawing view is created. It is suggested, however, that both the file and the associated drawing file be stored in the same directory and that the directory be referenced in the project file. When creating drawing views, you will find that there are many different types of views you can create. The following sections describe these view types.

Base View. This is the first drawing view of an existing part, assembly, or presentation file. It is typically used as a basis for generating the following dependent view types. You can create many base views in a given drawing file.

Projected View. This is a dependent orthographic or isometric view that is generated from an existing drawing view.

- Orthographic (Ortho View): A drawing view that is projected horizontally or vertically from another view.
- Iso View: A drawing view that is projected at a 30° angle from a given view. An isometric view can be projected to any of four quadrants.

Auxiliary View. This is a dependent drawing view that is perpendicular to a selected edge of another view.

Section View. This is a dependent drawing view that represents the area defined by a slicing plane or planes through a part or assembly.

Detail View. This is a dependent drawing view in which a selected area of an existing view will be generated at a specified scale.

Broken View. This is a dependent drawing view that shows a section of the part removed while the ends remain. Any dimension that spans over the break will reflect the actual object length.

Break Out View. This is a drawing view that has a defined area of material removed in order to expose internal parts or features.

Crop View. This drawing view allows a view to be clipped based on a defined boundary.

Overlay View. This drawing view uses positional representations to show an assembly in multiple positions in a single view.

NOTE

When you are creating drawing views Inventor will place a temporary raster image for each view while precise views are being calculated. While a precise view is being calculated a small horizontal and a vertical line will appear in the corners of the view and two circular arrows are displayed in the browser in front of the views that are being calculated. While the view(s) are being calculated you can continue to work on the drawing.

CREATING A BASE VIEW

A base view is the first view that you create from the selected part, assembly, or presentation file. When you create a base view, the scale is set in the dialog box, and from this view, you can project another drawing view. There is no limit to the number of base views you can create in a drawing based on different parts, assemblies, or presentation files. As you create a base view, you can select the orientation of that view from the Orientation list on the Component tab of the Drawing View dialog box. By default there is an option to create projected views immediately after placing a base view.

To create a base view, follow these steps:

1. Click the Base View command on the Place Views tab Create panel as shown in Figure 5.9 on the left, or right-click in the graphics window and select Base View from the marking menu.

2. The Drawing View dialog box will appear, also shown in Figure 5.9. On the Component tab, select an open document from the Select Document list labeled (1) click the Open an existing file icon labeled (2) to navigate to and select the part, assembly, or presentation file from which to create the base drawing view. After making the selection, a preview image will appear attached to your cursor in the graphics window. Do not place the view until the desired view options have been set.
3. If needed you can change the orientation of the base **view** by selecting a view from the Orientation list labeled (3).
4. Select the scale for the view labeled (5).
5. If needed you can change the view orientation to a custom view by clicking on the Change view orientation button labeled (4).
6. Select the style for the view labeled (6).
7. By default the option is on that allows you to create projected views immediately after placing a base view labeled (7).
8. If needed you can display the view's identifier (name) and the scale, labeled (8). You can also enter a new View Identifier.
9. Locate the base view by selecting a point in the graphics window. Move the cursor in the direction to place an orthographic or isometric view and click. Continue locating views and when done right-click and click Create from the menu.

As shown in Figure 5.9 on the right, the Drawing View dialog box has three tabs: Component, Model State, and Display Options. The following sections describe these tabs.

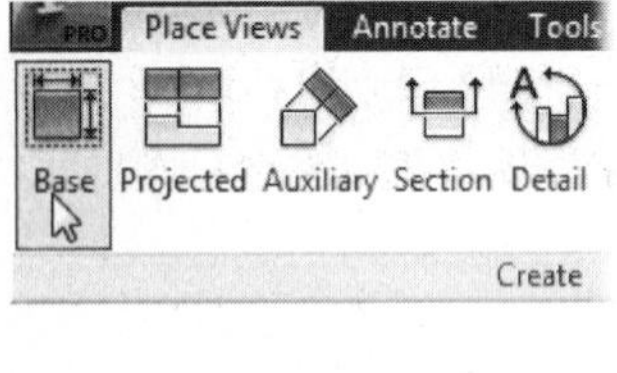

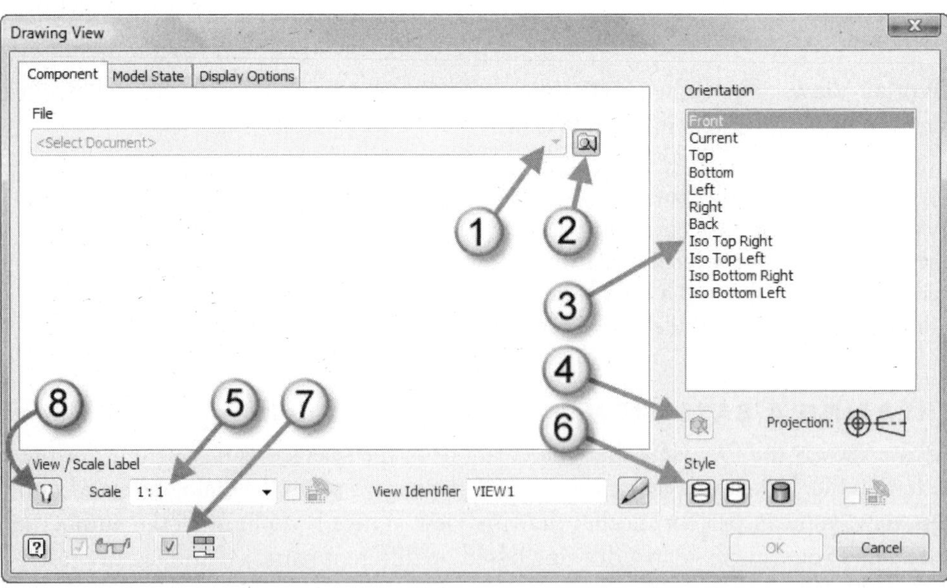

FIGURE 5.9

The Component Tab

The following categories are explained in detail under the Drawing View Component tab.

File. Any open part, assembly, or presentation files will appear in the drop-down list. If multiple files are open, the last open part, assembly, or presentation file will be the default file. You can also click the Explore directories icon and navigate to and select a part, assembly, or presentation file.

Orientation. After selecting the file, choose the orientation in which to create the view. After selecting an orientation, a preview image will appear in the graphics window attached to your cursor. If the preview image does not show the view orientation that you want, select a different orientation view by selecting another option. A symbol that identifies the type of drawing projection is also displayed under the Orientation area and is a symbol that identifies the type of drawing projection, such as First Angle or Third Angle Projection.

The following view orientations are available:

- Front, Current (current orientation of the part, assembly, or presentation in its file), Top, Bottom, Left, Right, Back, Iso Top Right, Iso Top Left, Iso Bottom Right, and Iso Bottom Left. A Change view orientation button is also present that allows you to create a custom view.

View/Scale Label (Light Bulb). Click to display or hide the view scale label. Clicking the checkbox displays the scale label, and leaving the box unchecked hides the scale label.

Scale. Enter a number for the scale in which to create the view. Note that the drawing sheet will be plotted at full scale (1:1), and the drawing views are scaled as needed to fit the sheet. You can edit the scale of the views after the views have been generated.

Scale from Base. Click to set the scale of a dependent view to match that of its base view, as shown in Figure 5.10 on the left. To change the scale of a dependent view, clear the option's box, and modify the scale value for that view. This feature activates when you edit existing views.

View Identifier. Use to include and/or change the label for the selected view. When you create a view, a default label is determined by the active drafting standard. To change the label, select the label in the box and enter the new label.

Style. Choose how the view will appear. There are three choices: Hidden Line, Hidden Line Removed, and Shaded. The preview image will not update to reflect the style choice. When the view is created, the chosen style will be applied. The style can be edited after the view has been generated.

Style from Base. Click to set the display style of a dependent view to be the same as that of its base view, as shown in Figure 5.10 on the right. To change the display style of a dependent view, clear the checkbox and modify the style for that view. This feature activates when you edit existing views.

Feature Preview. Click to preview the drawing view before it is created. This option is unavailable if Show Preview As option is set to All Components on the Drawing tab of the Application Options dialog box.

Create Projected Views Immediately after Base View Creation. By default the option is on that allows you to create projected views immediately after placing a base view. Uncheck this option if you want to only place a base view.

The Model State Tab

Use this tab to specify the weldment state and member states of an iAssembly or iPart to use in a drawing view, as shown in Figure 5.10. Other items are used to control line style and hidden line calculations.

Weldment. Enable this area when you select a document that is a weldment. Weldments have four states: Assembly, Preparations, Welds, and Machining.

Member. Allows you to select the member of an iAssembly or iPart to represent in a drawing view.

Line Style. Three options are presented: As Reference Parts, As Parts, and Off. When you select As Reference Parts for shaded view types, the reference data will appear transparently shaded with tangent edges turned off and all part edges as phantom lines. The reference data will appear on top of the product data that is shaded. When you select As Parts, reference data will have no special display characteristics. The reference data will appear on top of product data, and the reference data will appear with tangent edges turned off and all other edges as the phantom line type. When you select Off, reference data will not appear.

Hidden Line Calculation. Specifies if hidden lines are calculated for Reference Data Separately or for All Bodies.

Margin. To see more reference data, set the value to expand the view boundaries by a specified value on all sides.

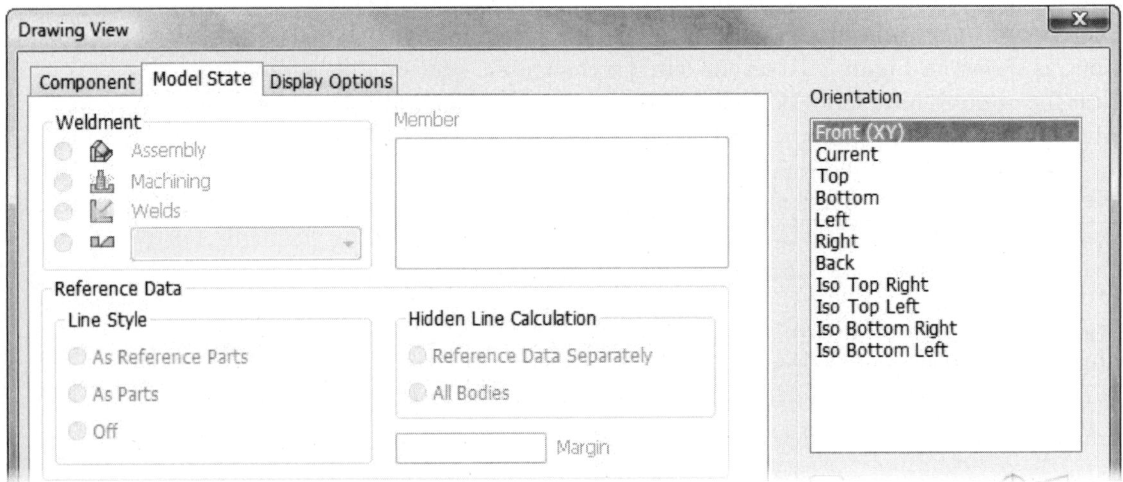

FIGURE 5.10

The Display Options Tab

Figure 5.11 shows the Display Options tab of the Drawing View dialog box. The following sections describe the tab's options.

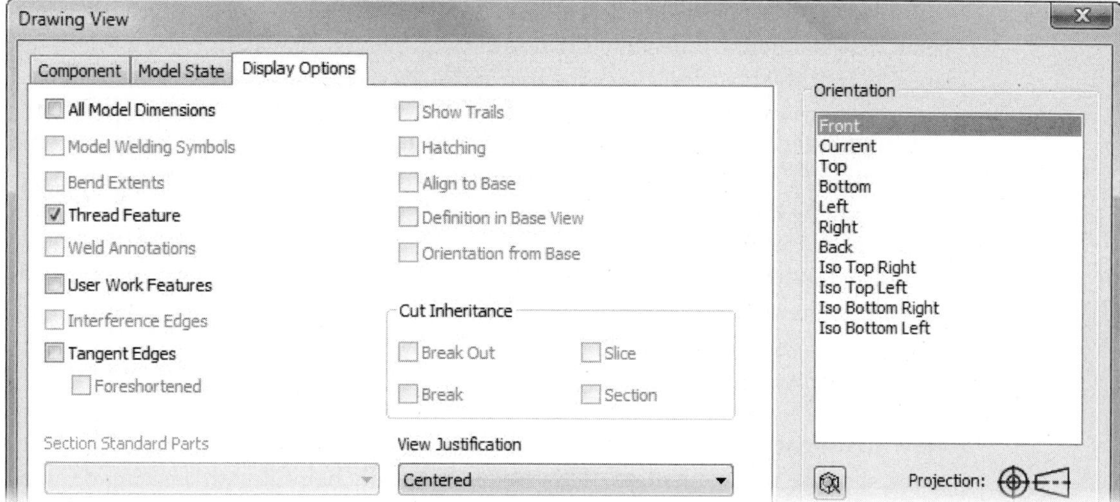

FIGURE 5.11

All Model Dimensions. Click to see model dimensions in the view, which is only active upon base view creation. If the option's box is clear, model dimensions will not be placed automatically upon view creation. When checked, only the dimensions that are parallel to the view and have not been retrieved in existing views on the sheet will appear.

Model Welding Symbols. This box is active only if you are creating a drawing view of a weldment. Click to retrieve welding symbols placed in the model in the drawing.

Bend Extents. This box is active only if you are creating a drawing view of a sheet metal part flat pattern. Click to control the visibility of bend extent lines or edges.

Thread Feature. This box is active only if you are creating a view of an assembly model. Click to set the visibility of thread features in the view.

Weld Annotations. This box is active only if creating a view of a weldment. Click to control the display of weld annotations.

User Work Features. Click to display work features in the drawing view.

Interference Edges. When selected, a drawing view will display both hidden and visible edges due to an interference condition such as a press or interference fit.

Tangent Edges. Click to set the visibility of tangent edges in a selected view. Checking the box displays tangent edges, and leaving the box clear hides them. If enabled, tangent edges can also be adjusted to display Foreshortened.

Show Trails. Click to control the display of trails in drawing views based on a presentation file.

Hatching. Click to set the visibility of the hatch lines in the selected section view. Checking the box displays hatch lines, and leaving the box clear hides them.

Align to Base. Click to remove the alignment constraint of a selected view to its base view. When the box is checked, alignment of views exists. Leaving the box clear breaks the alignment and labels the selected view and the base view.

Definition in Base View. Click to display or hide the section view projection line or detail view boundary. Checking the box displays the line, circle, or rectangle and text label; leaving the box clear hides the line, circle, or rectangle and text label.

Cut Inheritance. Use this option to turn on and off the inheritance of a Break Out, Break, Section, and Slice cut for the edited view. Selecting the appropriate checkbox will inherit the corresponding cut from the parent view.

Section Standard Parts. Use this option to control whether or not standard parts placed into an assembly from the supplied Inventor parts library, such as nuts, bolts, and washers, are sectioned. Three options are available in this area: Always, Never, and Obey Browser Settings.

View Justification. Use to control the position of drawing views when the size or position of the model changes. This area is especially helpful with creating drawing views from assembly models. This area contains two modes: Center and Fixed. The Center mode keeps the model image centered in the drawing view. If, however, the model's view increases or decreases in size, the drawing view will shift on the drawing sheet. In some cases, this shift could overlap the drawing border or title block. The Fixed mode keeps the drawing view anchored on the drawing sheet. In the event that the model image increases, an edge of the drawing view will remain fixed.

Creating Projected Views

A projected view can be an orthographic or isometric view that you project from a base view or any other existing view. When you create a projected view, a preview image will appear, showing the orientation of the view you will create as the cursor moves to a location on the drawing. There is no limit to the number of projected views you can create. To create a projected drawing view, follow these steps:

1. Click the Projected command on the Place Views tab > Create panel as shown in Figure 5.11A on the left. You could also right-click inside the bounding area of an existing view box, displayed as dashed lines when the cursor moves into the view and click Projected View from the marking menu as shown in the following image on the right.

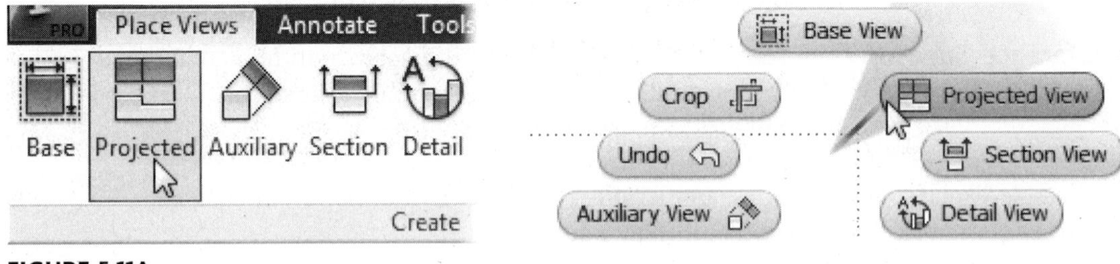

FIGURE 5.11A

2. If you selected the Projected command from the Create panel, click inside the desired view to project a view from.
3. Move the cursor horizontally, vertically, or at an angle (to create an isometric view) to get a preview image of the view you will generate. Keep moving the cursor until the preview matches the view that you want to create, and then press the left mouse button. Continue placing projected views.
4. When finished, right-click, and select Create from the marking menu.

Creating Auxiliary Views

An auxiliary view is a view that is projected perpendicular to a selected edge or line in a base view. It is designed primarily to view the true size and shape of a surface that appears foreshortened in other views.

To create an auxiliary drawing view, follow these steps:

1. Click the Auxiliary command on the Place Views tab, as shown in Figure 5.19 on the left. You can also right-click inside the bounding area of an existing view box, shown as a dotted box when the cursor moves into the view, and then select Auxiliary View from the marking menu as shown in the following image on the right.

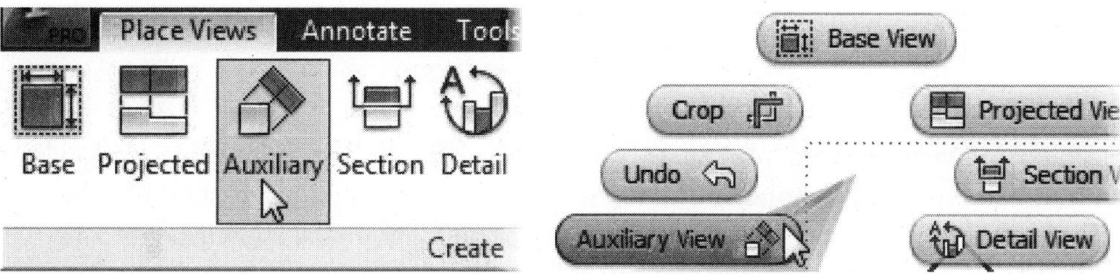

FIGURE 5.19

2. If you selected the Auxiliary command from the Create panel, click inside the view from which the auxiliary view will be projected. The Auxiliary View dialog box will appear, as shown in Figure 5.20 on the left. Type in a name for Label and a value for Scale, and then select one of the Style options: Hidden, Hidden Line Removed, or Shaded.
3. In the selected drawing view, select a linear edge or line from which the auxiliary view will be perpendicularly projected, as shown in Figure 5.20 on the right.

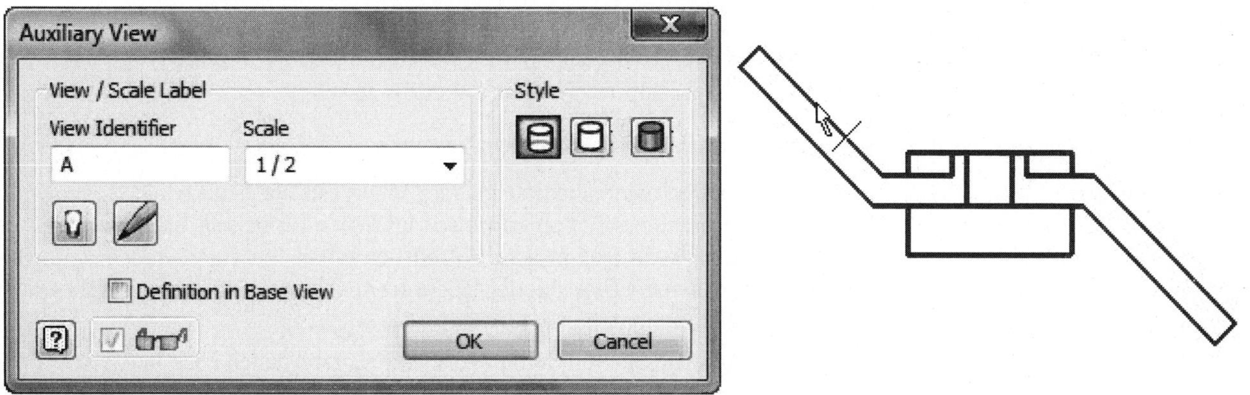

FIGURE 5.20

4. Move the cursor to position the auxiliary view, as shown in Figure 5.21 on the left.
5. Click a point on the drawing sheet to create the auxiliary view. The completed auxiliary view layout is shown in Figure 5.21 on the right.

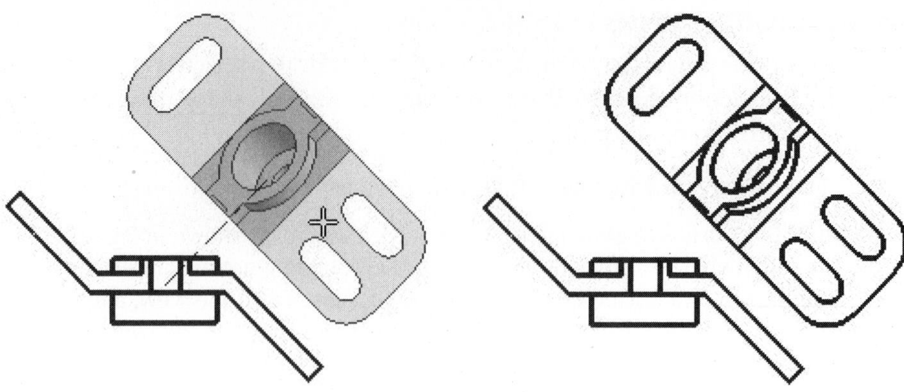

FIGURE 5.21

Creating Section Views

A section view is a view you create by defining a section line or multiple section lines that will represent the plane(s) that will be cut through a part or assembly. The view will represent the surface of the cut area and any geometry shown behind the cut face from the direction being viewed. When defining a section, you sketch line segments that are horizontal, vertical, or at an angle. You cannot use arcs, splines, or circles to define section lines. When you sketch the section line(s), geometric constraints will automatically be applied between the line being sketched and the geometry in the drawing view. You can also infer points by moving the cursor over (or scrubbing) certain geometry locations, such as centers of arcs, endpoints of lines, and so on, and then moving the cursor away to display a dotted line showing that you are inferring or tracking that point. To place a geometric constraint between the drawing view geometry and the section line, click in the drawing when a green circle appears; the glyph for the constraint will appear. If you do not want the section lines to have constraint(s) applied to them automatically when they are created, hold down the CTRL key when sketching the line(s). Because the area in the section view that is solid material appears with a hatch pattern by default, you may want to set the hatching style before creating a section view.

To create a section drawing view, follow these steps:

1. Click the Section command on the Place Views tab > Create panel, as shown in Figure 5.22 on the left. You can also right-click inside the bounding area of an existing view, displayed as a dotted box when the cursor moves into the view. You can then click Section View from the marking menu as shown in Figure 5.22 in the middle.
2. If you clicked the Section command in the Create panel, click inside the view from which to create the section view.
3. Sketch a line or lines that define where and how you want the view to be cut. In Figure 5.22 on the right, a vertical line is sketched through the center of the object. When sketching and a green dot appears a coincident constraint will be applied.

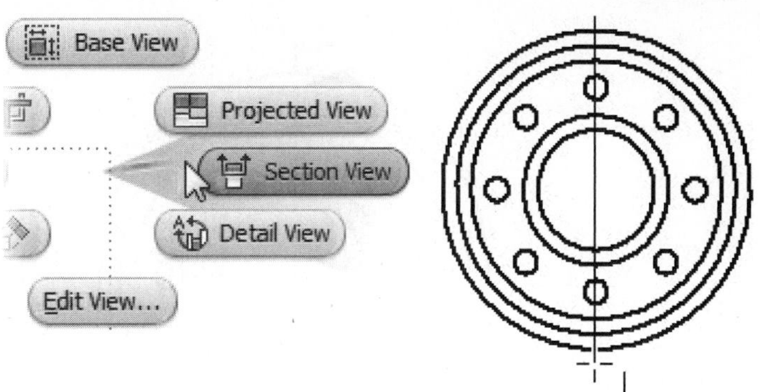

FIGURE 5.22

4. When you finish sketching the section line(s), right-click and select Continue from the marking menu as shown in Figure 5.23 on the left.
5. The Section View dialog box will appear, as shown in Figure 5.23 on the right. Fill in the information for how you want the label, scale, and style to appear in the drawing view. When the Include Slice option is checked, a section view will be created with some components sliced and some components sectioned, depending on their browser attribute settings. Placing a check in the box next to Slice All parts will override any browser component settings and will slice all parts in the view according to the Section line geometry. Components that are not crossed by the Section Line will not be included in the section operation.

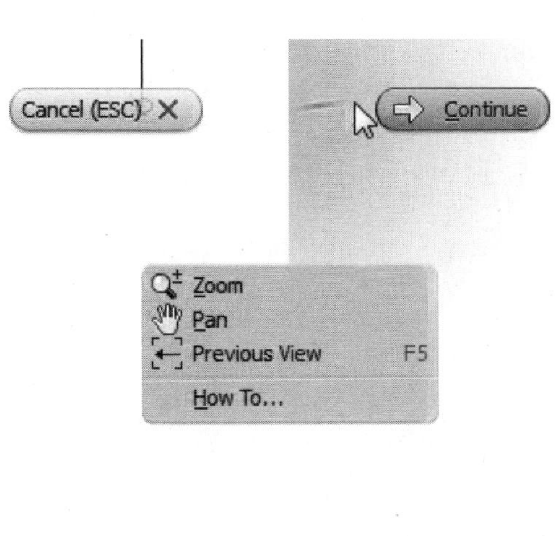

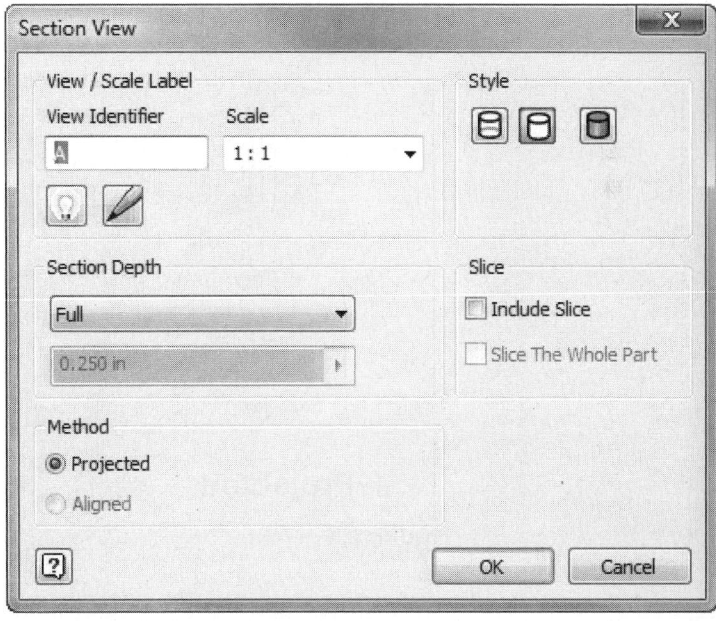

FIGURE 5.23

6. Move the cursor to position the section view (the section arrows will flip direction depending upon the location of the cursor), as shown in Figure 5.24, and select a point to place the view.

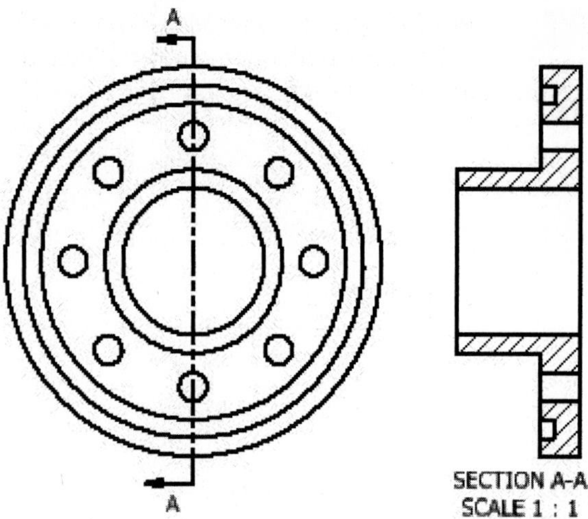

FIGURE 5.24

Creating Aligned Sections

Aligned sections take into consideration the angular position of details or feature of a drawing instead of projecting the section view perpendicular to the view, as shown in Figure 5.25 on the left. As illustrated on the left in Figure 5.25, it is difficult to obtain the true size of the angled elements of the bottom area in the section view. In the Side view, they appear foreshortened or not to scale. Hidden lines were added as an attempt to better clarify the view.

An aligned section view creates a section view that is perpendicular to its section lines. This prevents objects in the section view from being distorted.

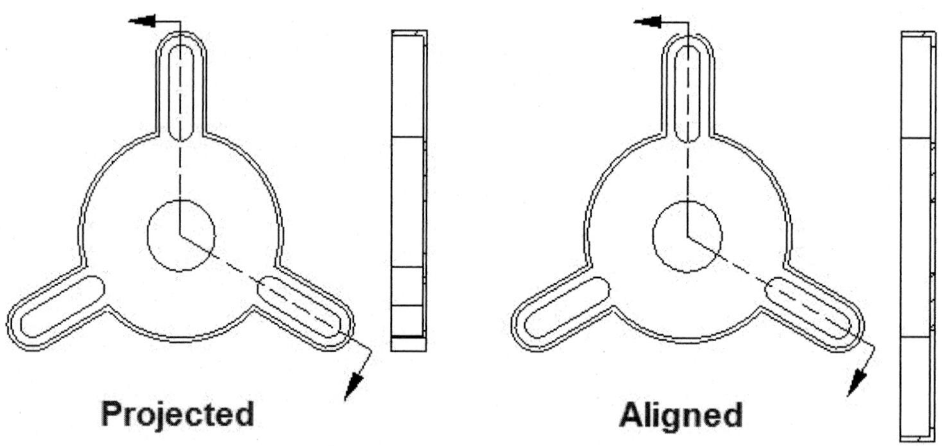

FIGURE 5.25

To create an aligned section view, follow these steps:

1. Click the Section command on the Place Views tab. You can also right-click inside the bounding area of an existing view box, shown as a dotted box when the cursor moves into the view. You can then select Create View > Section from the menu.
2. If you started the Section command from the Create panel, click inside the view from which to create the aligned section view.

3. Sketch the line or lines that define where and how you want the view to be cut. In Figure 5.26, a vertical line is sketched up to the center of the object. Then, the line changes direction to catch the counterbore feature as shown in Figure 5.26 on the left. When sketching and a green dot appears a coincident constraint will be applied.
4. When you finish sketching the section line(s), right-click and click Continue.
5. The Section View dialog box will appear, as shown in Figure 5.26 on the right. Fill in the information for how you want the label, scale, and style to appear in the drawing view. Verify that the type of section being created is Aligned. Components that are not crossed by the Section Line will not be included in the slicing operation.

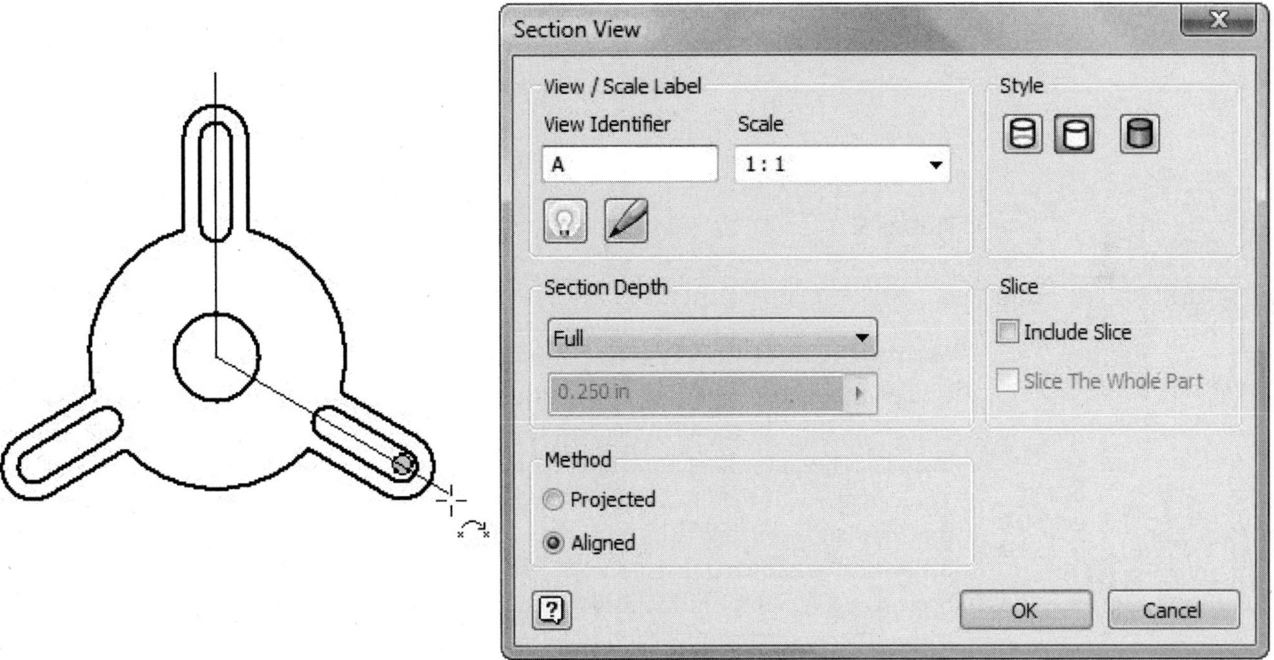

FIGURE 5.26

6. The completed aligned section is displayed in Figure 5.27. The bottom-right slot was rotated so it is a full section.
7. To edit the location of the section line(s) right-click on the section line and click Edit from the menu. The sketch environment will be current, add or delete constraints as needed. Dimensions can also be added between a section line and existing geometry. Before adding dimensions use the Project Geometry command to project the existing geometry onto the active sketch.

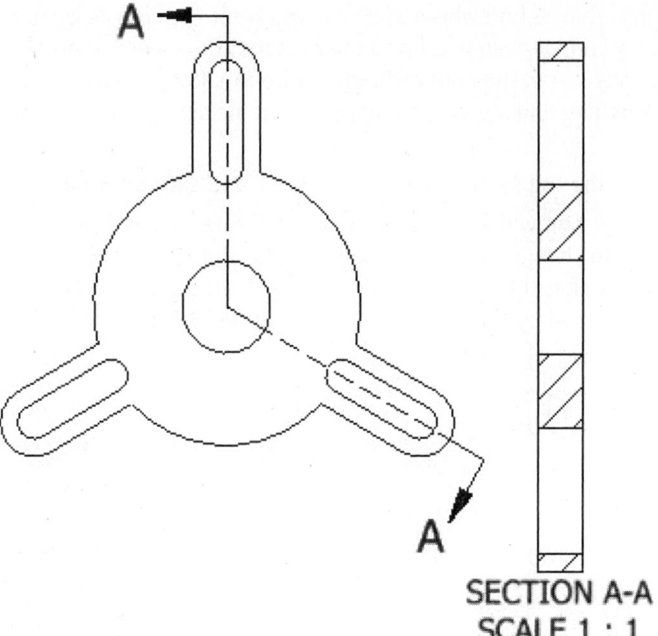

FIGURE 5.27

Modifying Section View and a Hatch Pattern

You edit the section view by right-clicking in the section view and clicking Edit Section Properties. In the Edit Section dialog box you can change the section depth, to include a slice and change the method to Projected or Aligned if the view supports it. You can also edit the hatch pattern by right-clicking on the hatch pattern in the section view and selecting Edit from the menu. This will launch the Edit Hatch Pattern dialog box, as shown in Figure 5.28. Make the desired changes in the Pattern, Angle, Scale, Line Weight, Shift (shifts the hatch pattern a specified distance, but stills stays within the section boundary), Double, and Color areas. Click the OK button to reflect the changes in the drawing.

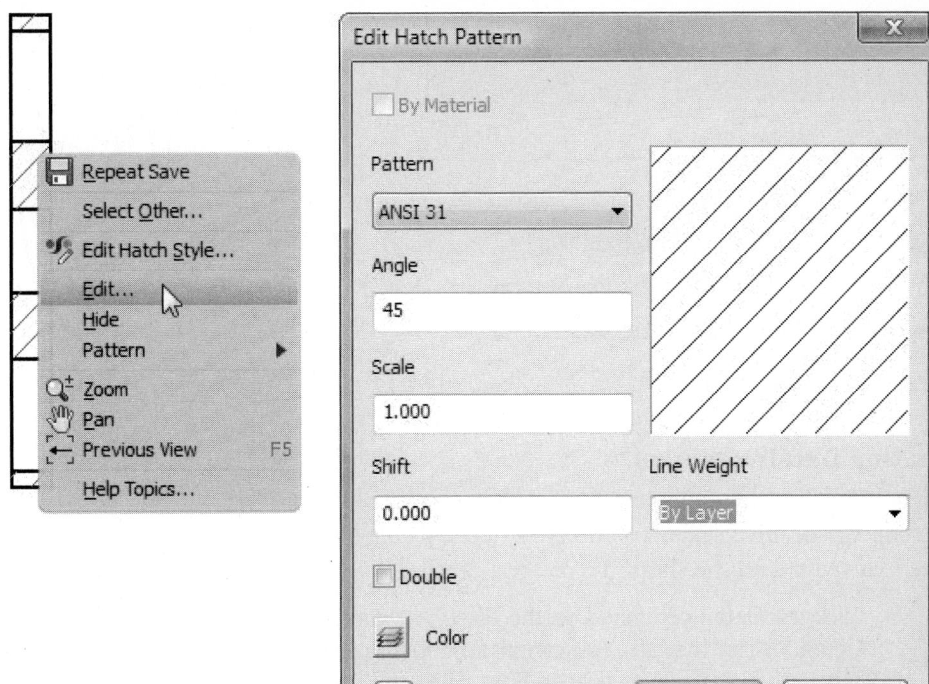

FIGURE 5.28

Hatching Isometric Views

In addition to applying crosshatching to orthographic view, as shown in Figure 5.29 on the left, you can also apply crosshatching to an isometric view. To perform this operation, double-click or right-click inside of the bounding box of the isometric view, and click Edit View from the menu, as shown in Figure 5.29.

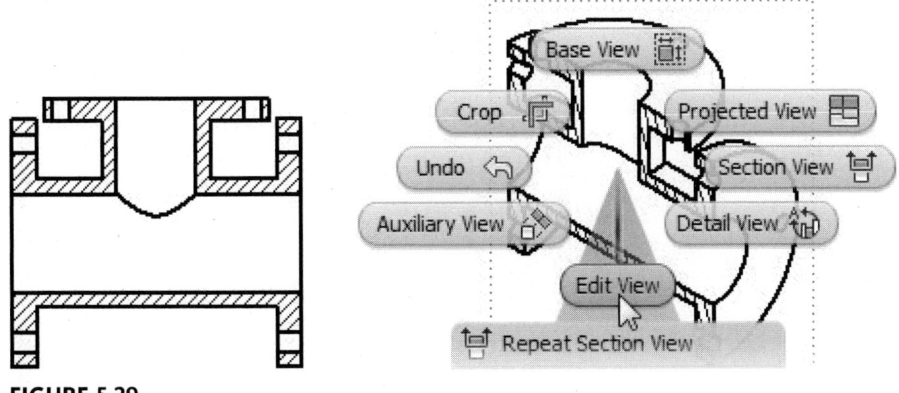

FIGURE 5.29

When the Drawing View dialog box appears, click on the Display Options tab, and place a check in the box next to Hatching, as shown in Figure 5.30 on the left. The results are illustrated in Figure 5.30 on the right with the hatching applied to the isometric view.

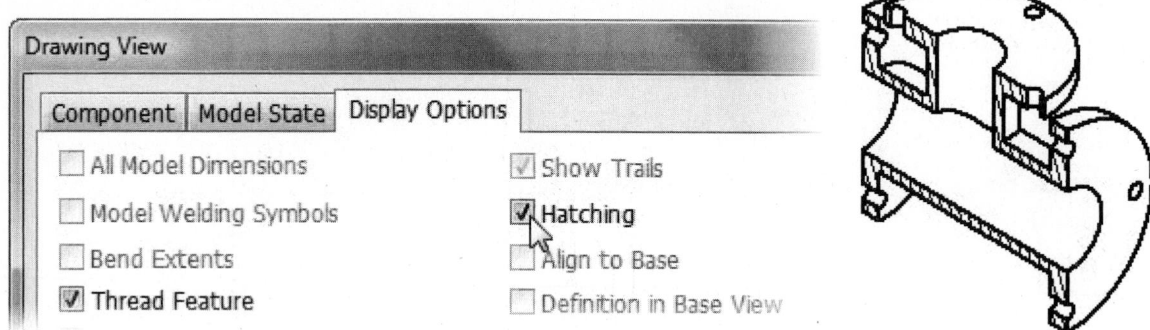

FIGURE 5.30

Creating Detail Views

A detail view is a drawing view that isolates an area of an existing drawing view and can reflect a specified scale. You define a detailed area by a circle or rectangle and can place it anywhere on the sheet. To create a detail drawing view, follow these steps:

1. Click the Detail command on the Place Views tab > Create panel as shown in Figure 5.31 on the left. You can also right-click inside the bounding area of an existing view shown as dashed lines when the cursor moves into the view, and right-click and click Detail View from the marking menu.
2. If you started the Detail command from the Create panel, click inside the view from which you will create the detail view.
3. The Detail View dialog box will appear, as shown in Figure 5.31 on the right. Fill in information according to how you want the label, scale, and style to appear in the drawing view, and pick the desired view fence shape to define the view boundary. Do not click the OK button at this time, as doing so will generate an error message.

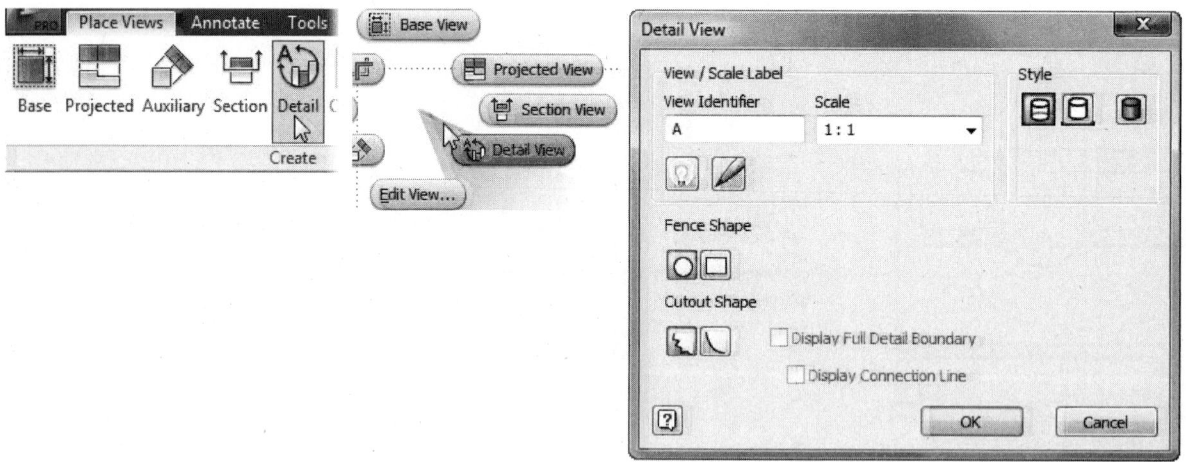

FIGURE 5.31

4. In the selected view, select a point to use as the center of the fence shape that will describe the detail area. Figure 5.32 on the left shows the icon that appears when creating a circular fence. The image on the right shows the icon used for creating a rectangular fence.

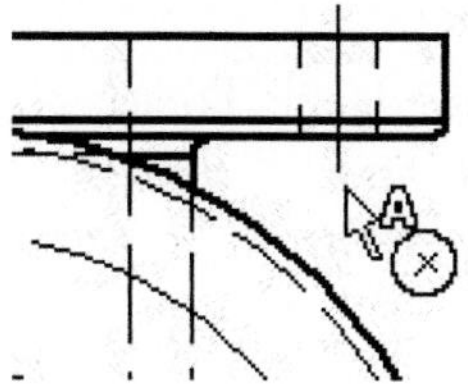

FIGURE 5.32

NOTE

You can right-click at this point and select Rectangular Fence to change the bounding area of the detail area definition.

5. Click another point that will define the radius of the detail circle, as shown in Figure 5.33 on the left, or corner of the rectangle, as shown in Figure 5.33 on the right. As you move the cursor, a preview of the boundary will appear.

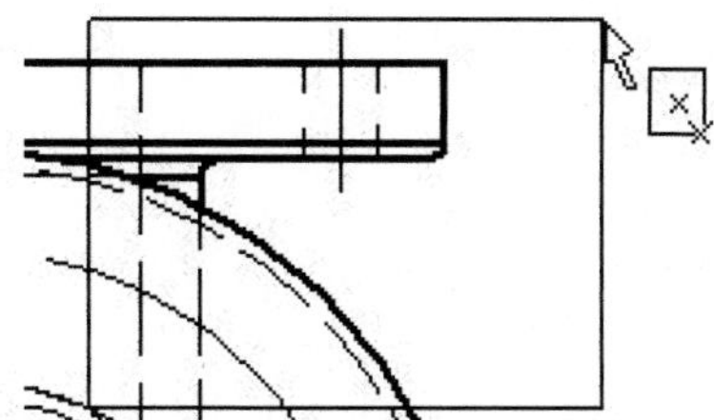

FIGURE 5.33

6. Select a point on the sheet where you want to place the view. There are no restrictions on where you can place the view. Figure 5.34 on the left shows the completed detail view based on a circular fence; a similar detail based on a rectangular fence is shown on the right.

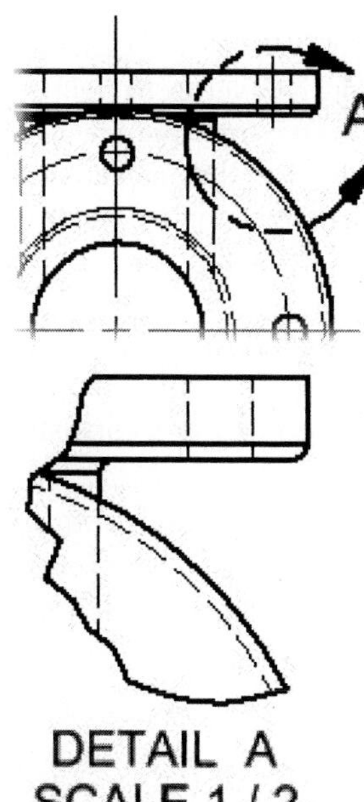

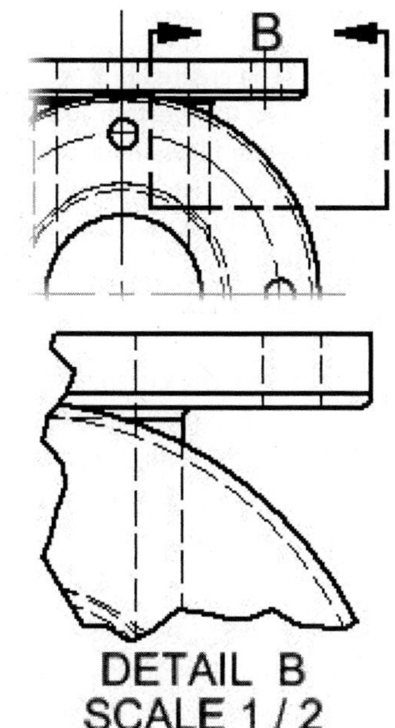

FIGURE 5.34

Modifying a Detail View

After a detail view is created, options are available to fine-tune the detail. To access these controls, right-click on the edge of the detail circle and choose options, as shown in Figure 5.35 on the right. The three detail view options are explained below.

Smooth Cutout Shape: Placing a check in this box will affect actual detail view. In Figure 5.35, instead of the cutout being ragged, a smooth cutout shape is present.

Full Detail Boundary: This option displays a full detail boundary in the detail view when checked, as shown in Figure 5.35.

Connection Line: This option will create a connection line between the main drawing and the detail view, as shown in Figure 5.35.

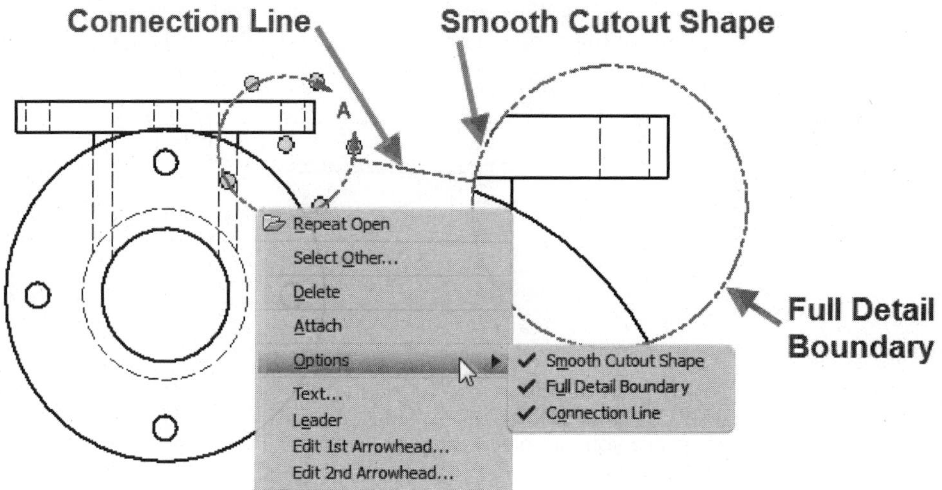

FIGURE 5.35

> **NOTE**
>
> Once a connection line has been created in a detail view, you can add a new vertex by right-clicking the detail boundary or connection line and clicking Add Vertex from the menu. Picking a point on the connection line will allow you to add the vertex and then move the new vertex to the new position. You can even remove a vertex by right-clicking a vertex and choosing Delete Vertex from the menu.

Moving Drawing Views

To move a drawing view, move the cursor over the view until a bounding box consisting of dotted lines appears, as shown in Figure 5.36. Press and hold down the left mouse button, and move the view to its new location. Release the mouse button when you are finished. As you move the view, a rectangle will appear that represents the bounding box of the drawing view. If you move a base view, any projected children or dependent views will also move with it as required to maintain view alignment. If you move an orthographic or auxiliary view, you will only be able move it along the axis in which it was projected from the part edge or face. You can move detail and isometric views anywhere in the drawing sheet.

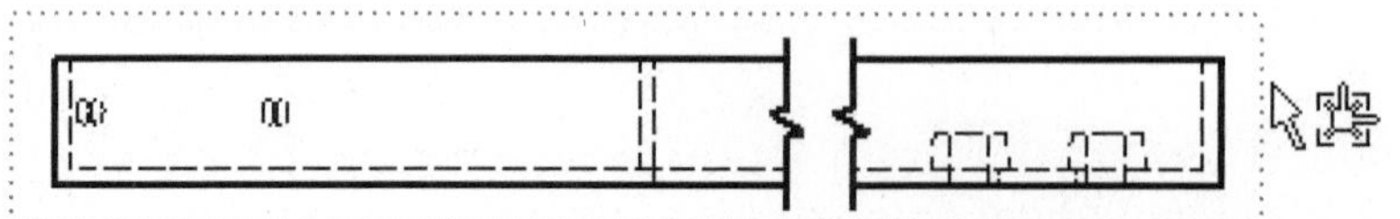

FIGURE 5.36

EXERCISE 5-2: CREATING AUXILIARY, SECTION, AND DETAIL VIEWS

In this exercise, you create a variety of drawing views from a model of a cover.

1. Open the file *ESS_E05_02.idw* from the Chapter 05 folder. Use the Zoom command to zoom in on the top view.
2. First you create an auxiliary view. Click the Auxiliary command on the Place Views tab > Create panel.
3. In the graphics window, click in the top view.
4. Select the left-outside angled line, as shown in Figure 5.37 on the left, to define the projection direction.
5. Place the view to the upper-left of the front view as shown in Figure 5.37 on the right.

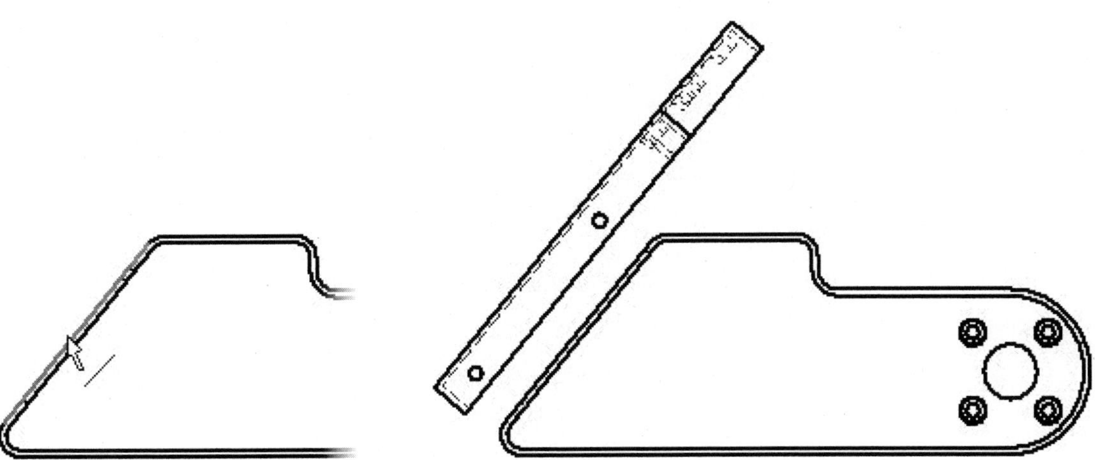

FIGURE 5.37

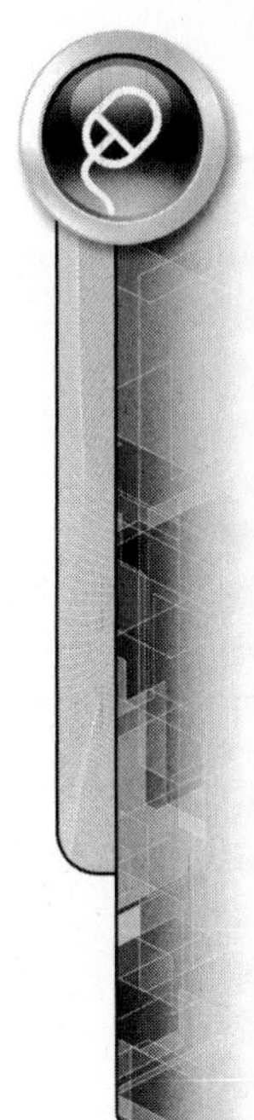

6. In the graphics window, select and drag the border of the views to fit them in the drawing.
7. You now create a section view. Click the Section command on the Place Views tab > Create panel.
8. In the graphics window, click in the top view.
9. To define the first point of the section line, hover the cursor over the center hole, and click a point directly above the hole (the dotted lines represent the inferred point), as shown in Figure 5.38.
10. Click a point below the geometry to create a vertical section line.
11. Right-click and click Continue from the marking menu.
12. Place the section view with the default settings to the right of the top view, as shown in Figure 5.38 on the right.
13. If the section view does not show the top portion of the part, drag the endpoints of the section line in the top view so it is above the view.

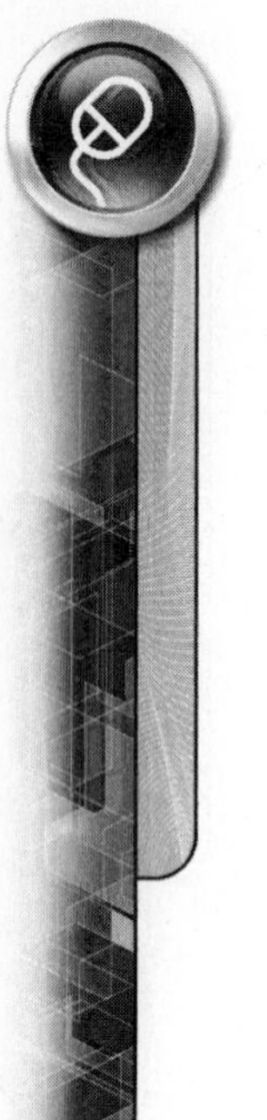

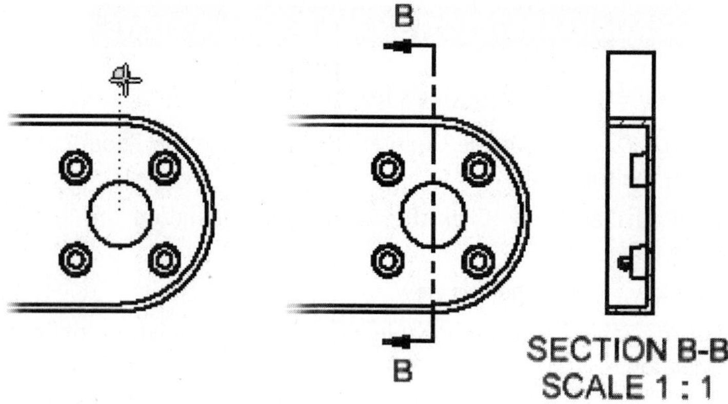

FIGURE 5.38

14. Try to drag the section line in the top view. If the section line can move follow these steps.
    - Right-click on the section line and click Edit from the menu.
    - Use the project geometry command and project the circle that is in the center of the part.
    - Apply a coincident constraint between the section line and the center point of the projected circle.
    - Click the Finish Sketch command from the Exit panel.

NOTE **If the sketch line cannot move and you need to reposition it, edit the section line as described in step 14, but delete a constraint that is holding the geometry in place.**

15. If desired, project an isometric view from the sectioned view.
16. Delete the section view by right-clicking in the section view and click Delete from the menu and then click OK confirm that the view will be deleted.
17. Next you create an aligned section view. Click the Section command on the Place Views tab > Create panel.
18. In the graphics window, click in the top view.
19. To define the first point of the section line, hover the cursor over the center hole, and then click a point directly above the hole (the dotted lines represent the inferred point), as you did in step 9.
20. Click the center point of the center circle.
21. Move the cursor over the center point of the lower-right circle (a green dot will appear) and then click a point outside of the geometry as shown in Figure 5.39 on the left.
22. Right-click and click Continue from the menu.
23. Place the aligned section view with the default settings to the right of the top view, as shown in Figure 5.39 on the right.

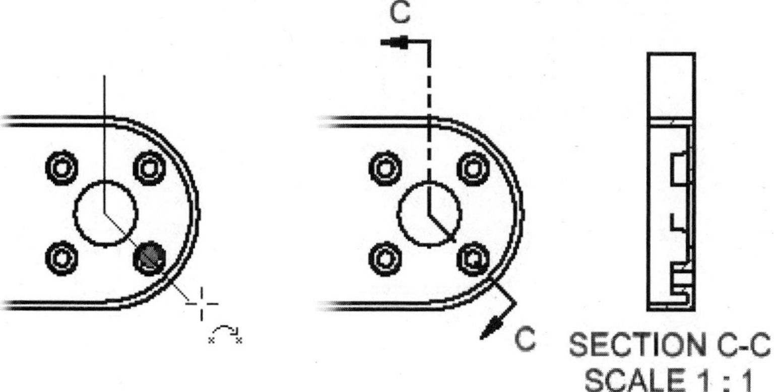

FIGURE 5.39

24. If the section view does not show the entire part, drag the endpoints of the section line in the top view so the entire part is displayed.
25. Finally, create a detail view. Click the Detail command from the Place Views tab > Create panel.
26. In the graphics window, click in the top view.
27. Select a point near the left corner of the view to define the center of the circular boundary of the detail view.
28. Drag the circular boundary to include the entire lower-left corner of the view geometry, as shown in Figure 5.40 on the left.
29. Click to position the detail view below the break view. Figure 5.40 in the middle shows the completed detail in the top view and the image on the right shows the completed detail view.

NOTE Your view identifier (letter) may be different than what is shown.

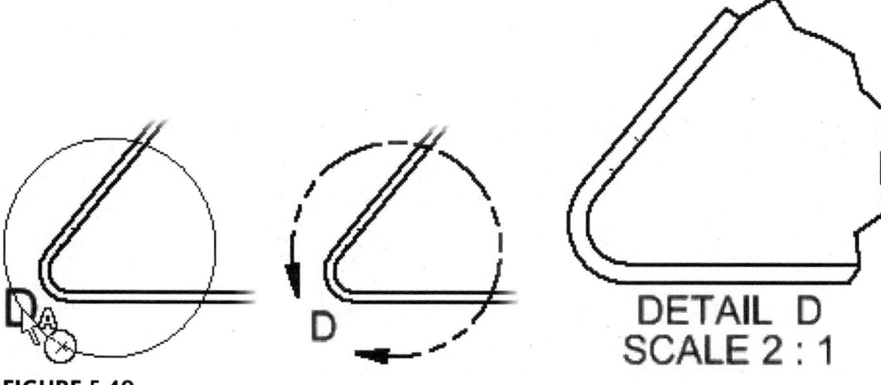

FIGURE 5.40

30. Reposition the drawing views as shown in Figure 5.41.
31. Close all open files. Do not save changes. End of exercise.

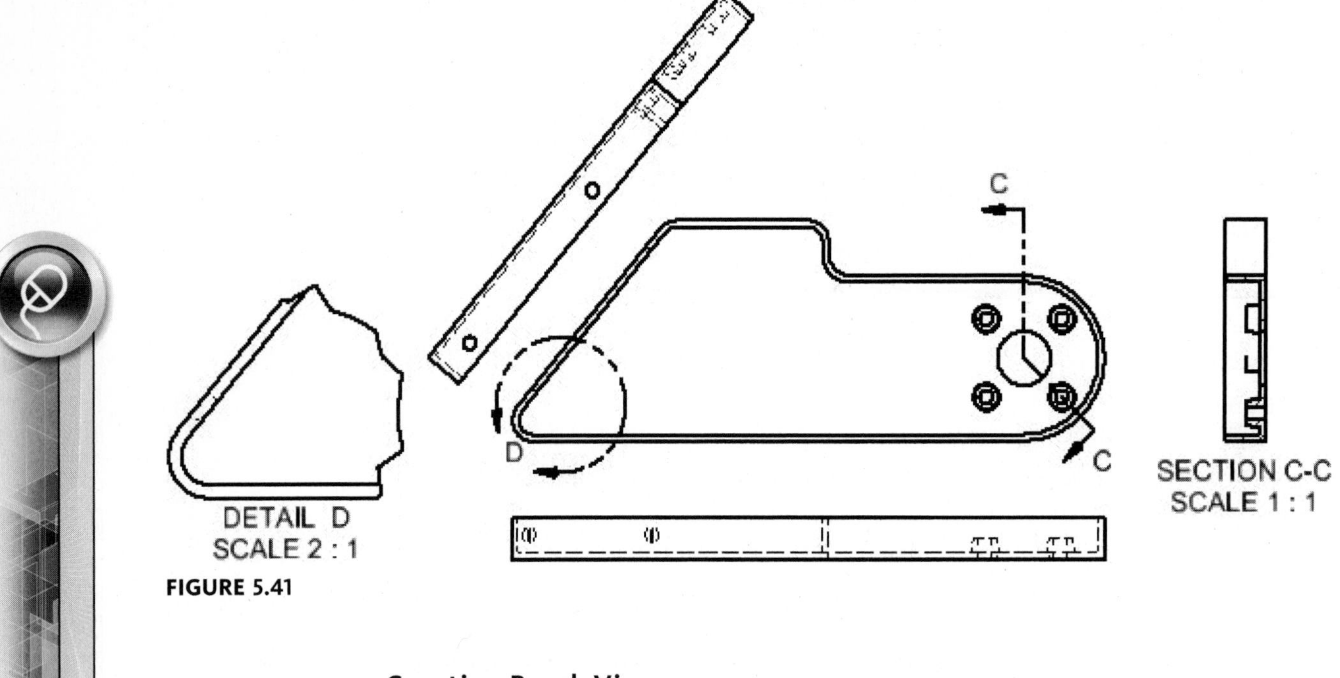

FIGURE 5.41

Creating Break Views

When creating drawing views of long parts, you may want to remove a section or multiple sections from the middle of the part in a drawing view and show just the ends. This type of view is referred to as a "break view." You may, for example, want to create a drawing view of a 2 × 2 × 1/4 angle, as shown in Figure 5.42, which is 48 inches long and has only the ends chamfered. When you create a drawing view, the detail of the ends is small and difficult to see because the part is so long.

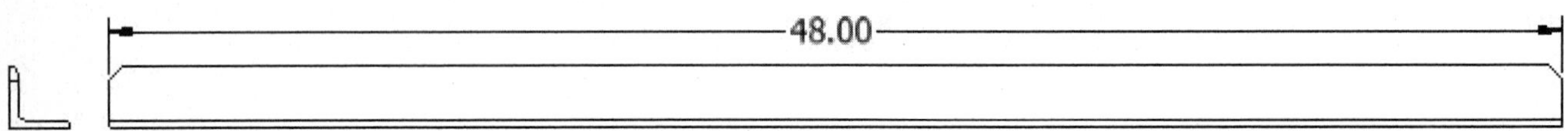

FIGURE 5.42

In this case, you can create a break view that removes the middle of the angle and leaves a small section on each end. When you place an overall length dimension that spans the break, it appears as 48 inches, and the dimension line shows a break symbol to note that it was based from a break view, as shown in Figure 5.43.

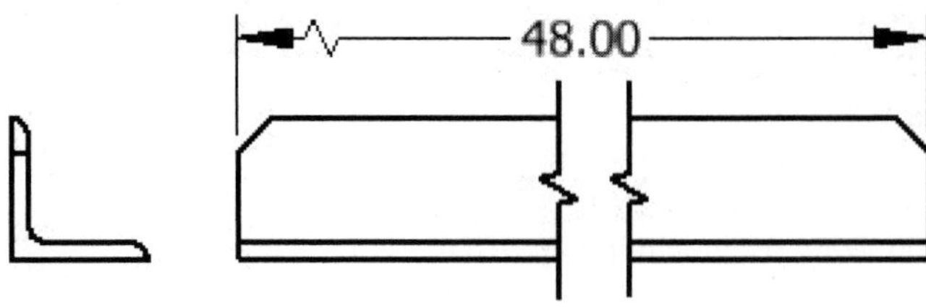

FIGURE 5.43

You create a break view by adding as many breaks to an existing drawing view as needed. The view types that can be changed into break views are as follows: part views, assembly views, projected views, isometric views, auxiliary views, section views, break out views, and detail views. After creating a break view, you can move the breaks dynamically to change what you see in the broken view.

To create a break view, follow these steps:

1. Create a base or projected view or one that will eventually be shown as broken.
2. Click the Break command on the Place Views tab > Modify panel, as shown in Figure 5.44 on the left.
3. If you selected the Break command on the Place Views tab > Modify panel, click inside the view from which to create the break view.
4. The Break dialog box will appear, as shown in Figure 5.44 on the right. Do not click the OK button at this time, as this will end the command.

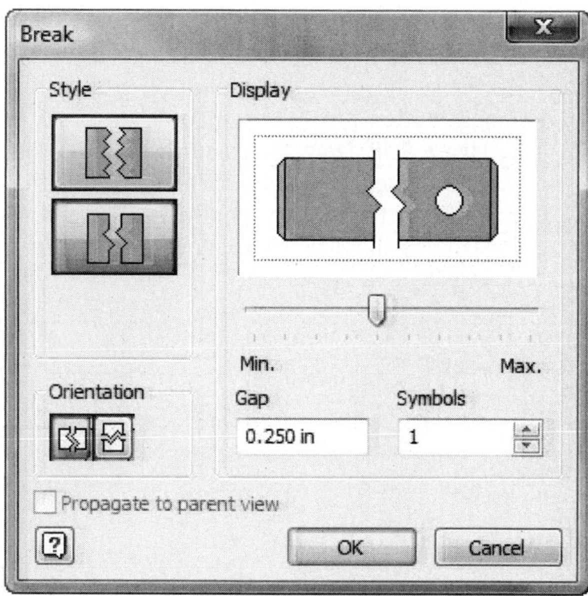

FIGURE 5.44

5. In the drawing view that will be broken, select a point where the break will begin, as shown in Figure 5.45.

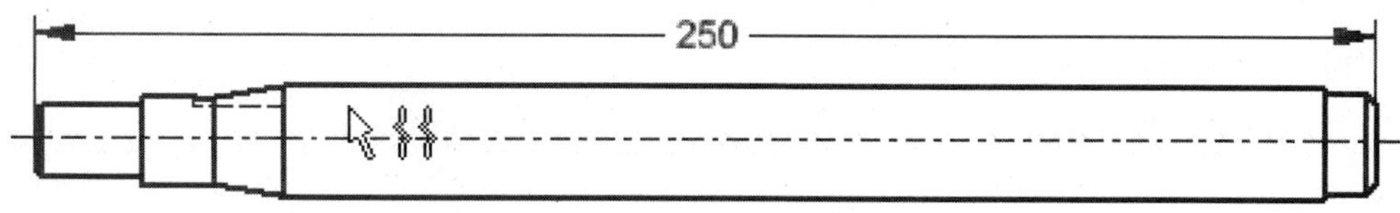

FIGURE 5.45

6. Select a second point to locate the second break, as shown in Figure 5.46. As you move the cursor, a preview image will appear to show the placement of the second break line.

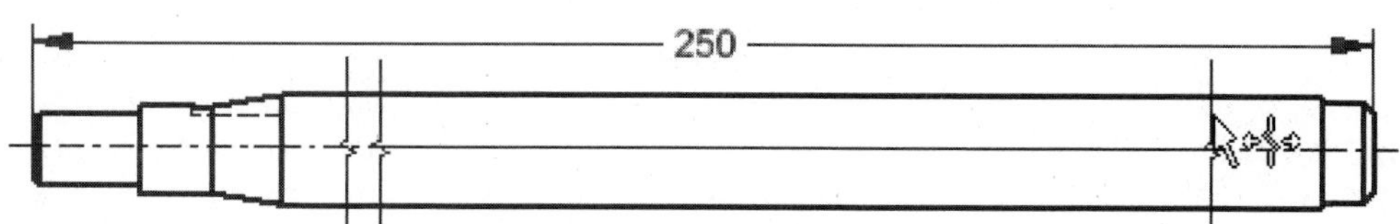

FIGURE 5.46

7. Figure 5.47 illustrates the results of creating a broken view with a dimension added. Notice that the dimension line appears with a break symbol to signify that the view is broken.

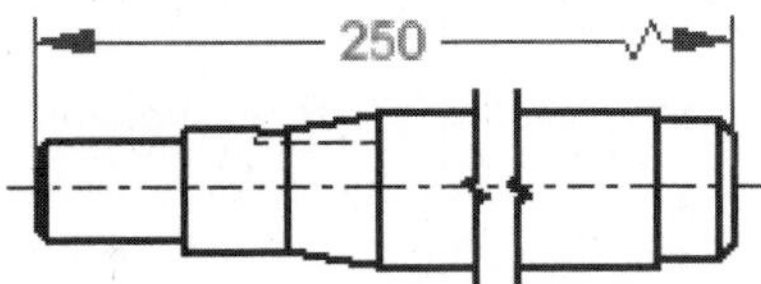

FIGURE 5.47

8. To edit the properties of the break view, move the cursor over the break lines, and a green circle will appear in the middle of the break. Right-click and select Edit Break from the menu. The same Break dialog box will appear. Edit the data as needed, and click the OK button to complete the edit.
9. To move the break lines, click on one of the break lines and, with the left mouse button pressed down, drag the break line to a new location, as shown in Figure 5.48. Drag the cursor away from the other break line to reduce the amount of geometry that is displayed, drag the cursor into the other break line to increase the amount geometry that is displayed. In either case the other break line will follow to maintain the gap size.

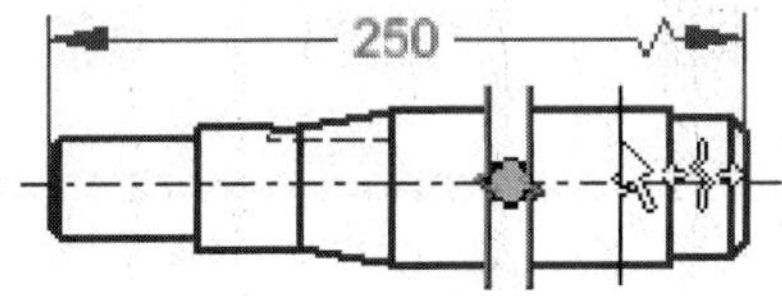

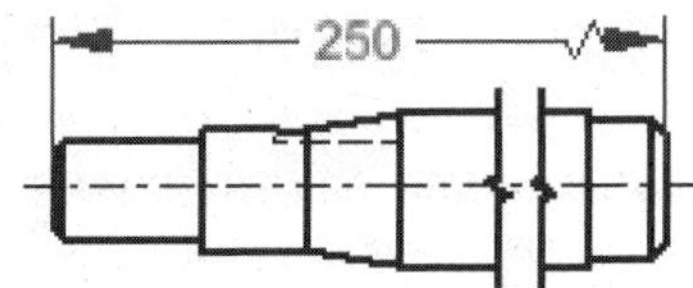

FIGURE 5.48

Creating Break Out Views

When you want to expose internal components or features, Autodesk Inventor allows you to cut or peel away a body and expose those internal parts; this is called a "break out view." It is not unusual for assemblies to have housings or covers that hide internal components. Break out views make it possible to expose these components. You can create break out views on assemblies as well as on part files.

To create a break out view, follow these steps.

1. Create a sketch based on a view in which you will break out a section.
2. Draw a closed profile that defines the area to break out.
3. Click the Break Out command on the Place Views tab > Modify panel as shown in the following image on the left.
4. Click a point in the view to create a break out.
5. The Break Out dialog box will appear, as shown in the following image on the right.
6. In the Break Out dialog box click a termination options that determines where the break out view will terminate; From Point, To Sketch, To Hole, or Through Part.

The process to create a break out view using each termination option is covered in the next section.

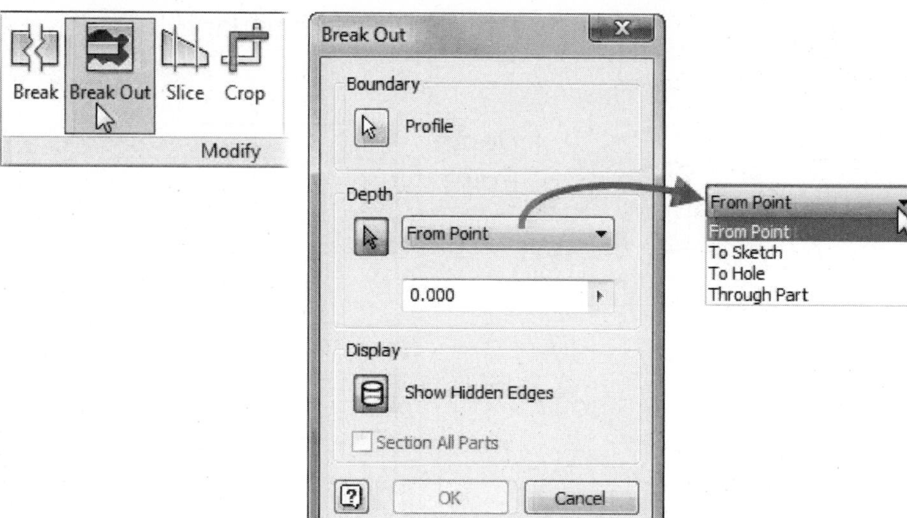

FIGURE 5.49

From Point

Select this option to terminate a break out at a specified distance from a point located in a view that is orthographic from the view that the sketch is created in. The point can be selected in a base view or in a projected view.

To create a break out view using the From Point option, follow these steps:

1. Click on the view in which to create the break out section.
2. Click on the Create Sketch command on the Place Views tab > Sketch panel and sketch a closed profile over the area you want broken out, as shown in Figure 5.50. When done, right-click, and click Finish Sketch from the marking menu.

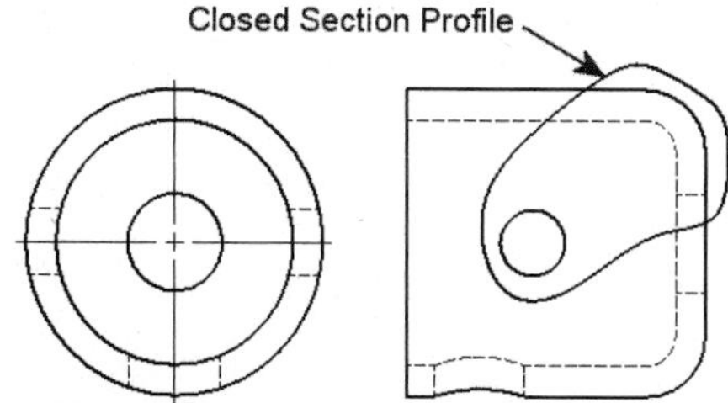

FIGURE 5.50

3. Click the Break Out command from the Place Views tab > Modify panel; this will activate the Break Out dialog box. Select the view in which the break out will occur, and select the defined boundary. Select the sketch you just created as the boundary in the view that will contain the break out.
4. Select the From Point option in the dialog box, click on the Depth arrow, and select a point in the adjacent orthographic view. This point will be used to calculate the depth of cut based on the distance, as shown in Figure 5.51.

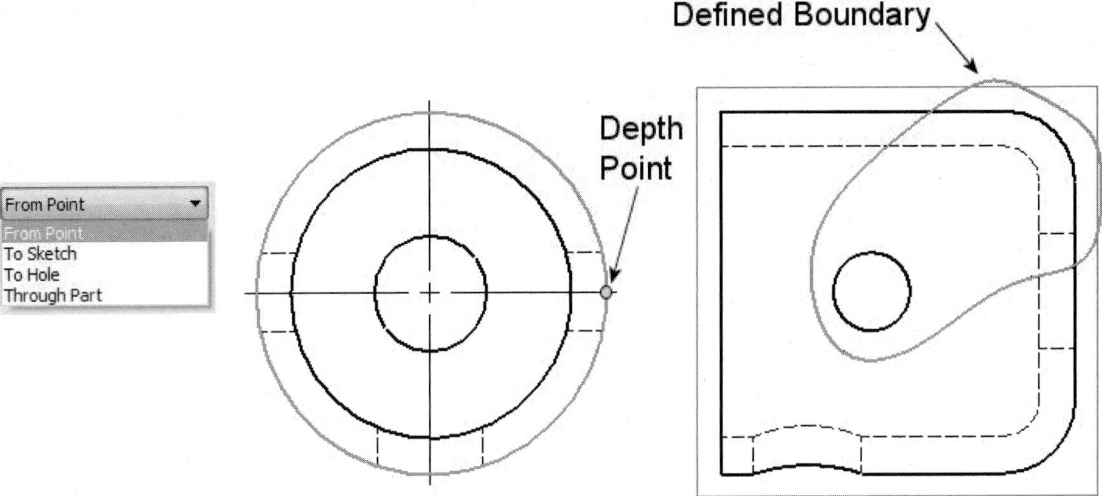

FIGURE 5.51

5. Figure 5.52 shows the results of using the From Point option for creating a break out. The depth of 1 inch from the selected point cuts the view at the middle of the circular features, thus displaying the wall thickness of the part.

NOTE

You can also select a point at the center of the circle for the depth of the cut. In this case, the distance value would change to 0, because the depth of the cut was specified by a point.

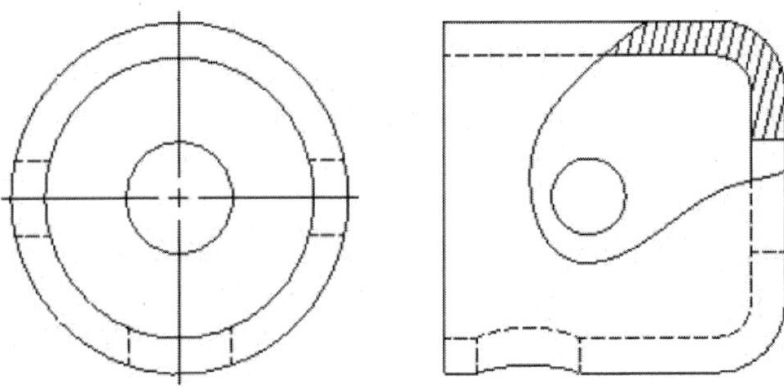

FIGURE 5.52

To Sketch

Select this option to define the depth of the break out view based on sketch geometry associated in another view.

To create a break out view using the To Sketch option, follow these steps:

1. Click in or on the bounding box of the view in which to create the break out view and click on the Create Sketch command on the Place Views tab > Sketch panel.
2. Sketch a closed profile around the area you want broken out, as shown in the following image. Then finish the sketch.
3. Click in an adjacent orthographic view and create a new sketch. This sketch will be used to determine the depth of the cut for the break out, as shown in Figure 5.53. When finished with this operation, right-click, and select Finish Sketch.

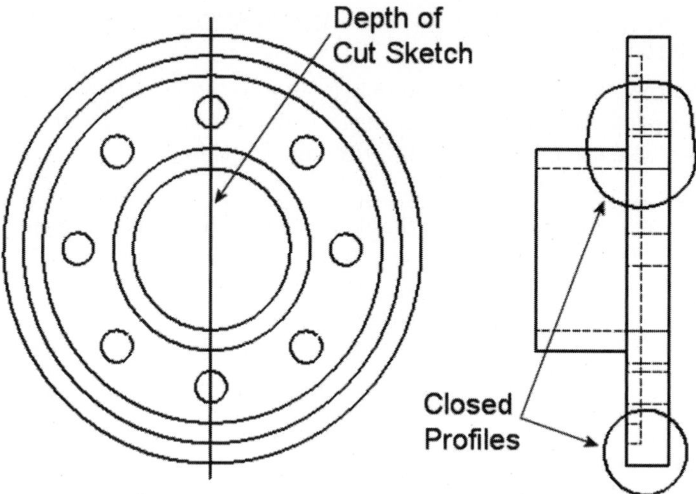

FIGURE 5.53

4. Click the Break Out command from the Place Views tab > Modify panel, and select the view that contains the closed sketch.
5. When the Break Out dialog box appears in the Depth area of the dialog box, change the option to To Sketch. In the adjacent orthographic projected view, select the sketch that will determine the depth of the cut for creating the break out view, as shown in Figure 5.54 on the left.
6. Figure 5.54 on the right shows the break out view created based on the first group of sketched boundaries as well as the second sketch, which determined the depth of the cut.

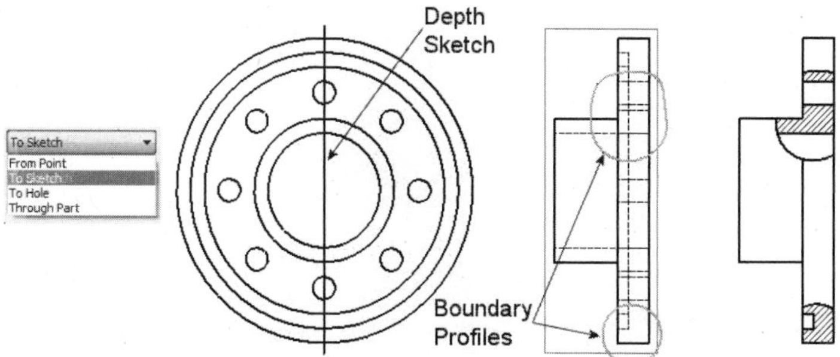

FIGURE 5.54

To Hole

Use the To Hole option for creating break out views is similar to the To Sketch option. Instead of basing the break out on a sketch, it will base it on a hole feature. The axis of the hole feature determines the termination depth.

To create a break out view using the To Hole option, follow these steps:

1. Click in or on the bounding box of the view in which to create the break out view and click on the Create Sketch command on the Place Views tab > Sketch panel > Modify panel.
2. Sketch a closed profile around the area you want broken out, as shown in the following image on the left. Then finish the sketch.
3. Click the Break Out command from the Place Views tab > Modify panel and click in the view that contains the closed profile. When the Break Out dialog box appears, click to select the defined boundary. The boundary profile must be on a sketch associated to the selected view. In the Break Out dialog

box, click the arrow next to the Depth type box, and select the To Hole option, as shown in Figure 5.55 on the left.

4. Click the select arrow, and then select the hole feature in the graphic window, as shown in the middle in Figure 5.55. The depth is defined by the axis of the hole. The selected hole can be displayed as a front or side view.

NOTE **If the hole feature is hidden, edit the view, and click the Show Hidden Edges button to temporarily display hidden lines. You can also select the hole in another view.**

5. When the view is defined fully, click OK to create the view, as shown in Figure 5.55 on the right.

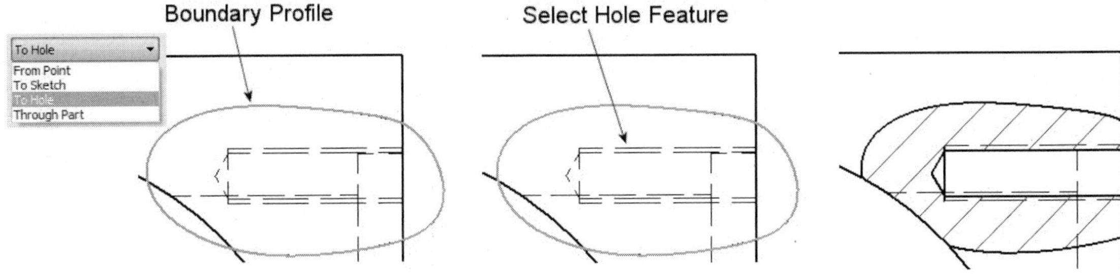

FIGURE 5.55

Through Part

In Figure 5.56, a housing hides internal details of an assembly. You can cut away a segment of the housing and select the parts to cut through to exposing obscured parts or features using the Through Part option of the Break Out command.

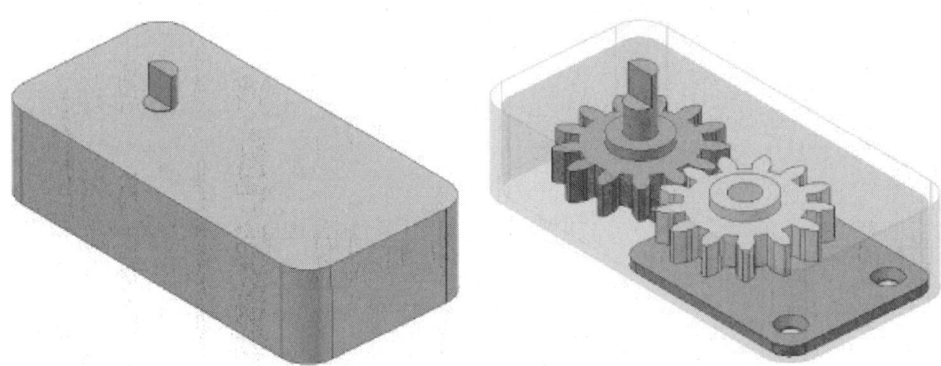

FIGURE 5.56

To create a break out view using the Through Part option, follow these steps:

1. Click in or on the bounding box of the view in which to create the break out view and click on the Create Sketch command from the Place Views tab > Sketch panel.
2. Next sketch a closed profile, as shown in Figure 5.57 on the left, as the area through which you wish to cut. When finished, right-click, and select Finish Sketch.
3. Click the Break Out command from the Place Views tab > Modify panel and click in the view that contains the closed profile. The Break Out dialog appears, the boundary profile consisting of the previous sketch will be selected automatically. In the Depth area of the dialog box, click on the Through Part option, as shown in Figure 5.57 in the middle.

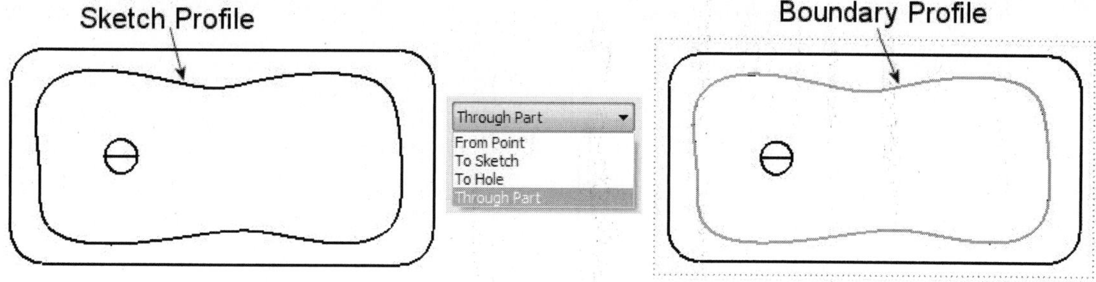

FIGURE 5.57

4. In the view that will contain the break out view select edge of the part or parts (you can also select the parts in the browser under the view that the sketch was based on) and select this edge, as shown in Figure 5.58 on the left.
5. When finished, click the OK button. Figure 5.58 on the right illustrates the results of using the Through Part option. The gear cover has been sliced away, based on the sketched boundary, to expose the internal workings of the gear mechanism.

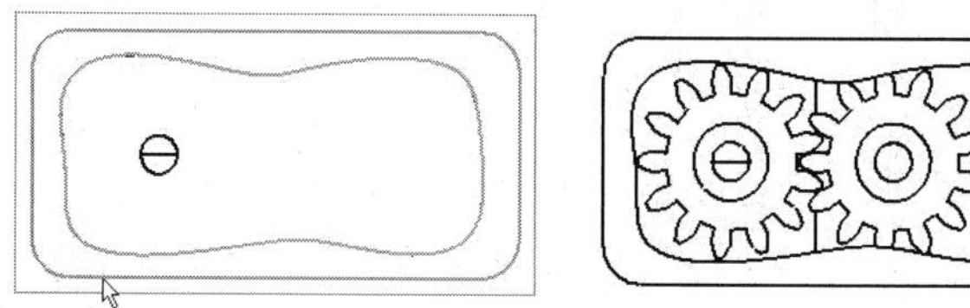

FIGURE 5.58

The Through Part option can be very effective when viewed in an isometric view. As shown in Figure 5.59, a spline sketch was created on an isometric view. Notice how the sketch covers the top and sides of the isometric. The image on the right in the figure illustrates the completed break out view. All sides surrounded by the spline sketch break out to display the internal components.

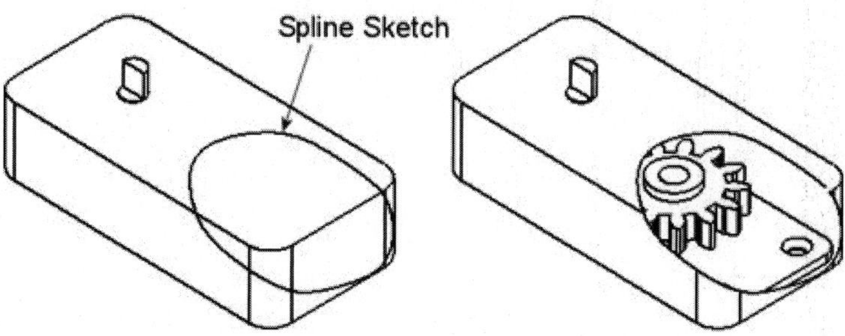

FIGURE 5.59

Creating Cropped Views

Cropped views are used to show a portion of an existing view. Figure 5.60 on the left illustrates a complete view. After cropping this view, only a portion of the view is shown in Figure 5.60 on the right. This is different from a detail view, which was explained previously. A detail view is generated by a parent view and references this parent, complete with callout information. The cropped view will not create a callout and is meant for stand-alone use.

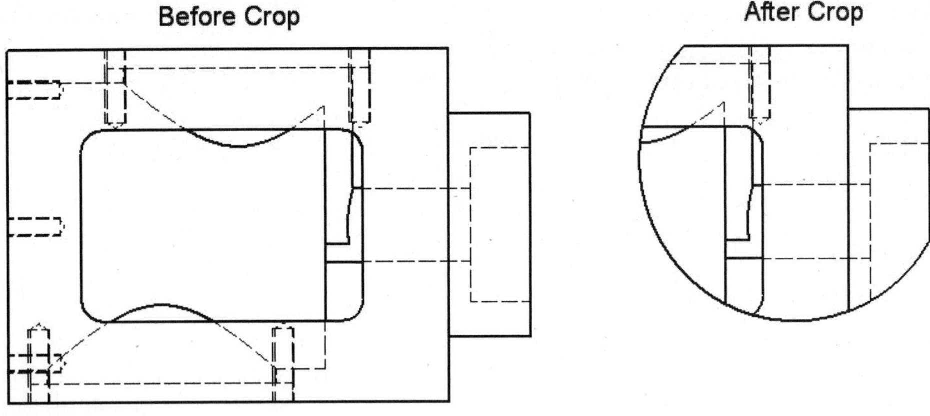

FIGURE 5.60

The steps used for creating a cropped view are as follows.

1. On the Place Views tab, click the Crop command from the Modify panel as shown in the following image on the left or right-click in a view to crop and click Crop from the marking menu as shown in Figure 5.61 on the right.

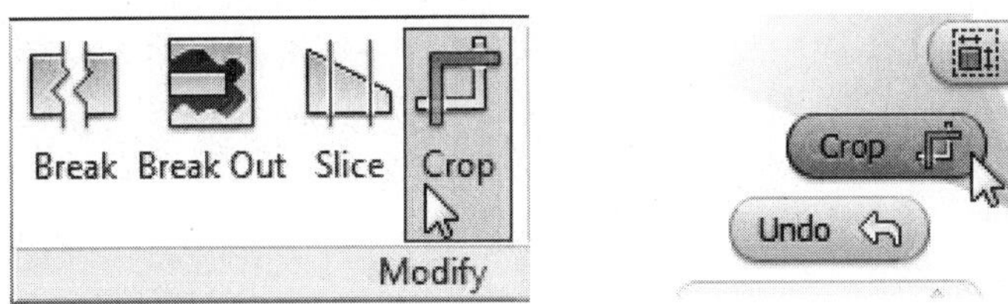

FIGURE 5.61

2. If the Crop command was started from the Modify panel select the view to crop by clicking in the view or on the edge of the view.
3. By default the view will be cropped to a rectangular shape. You can change this option by right-clicking and in the menu click either circular or rectangular. When specifying the circular boundary, a center point acts as the boundary center; for rectangular boundaries, two diagonal points are used to create the rectangle as shown in Figure 5.62.

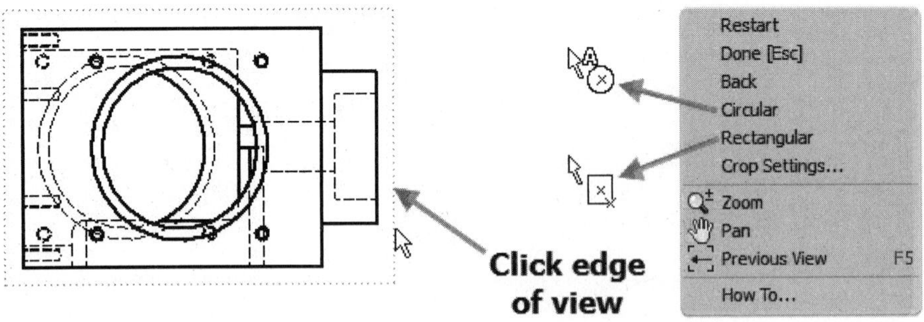

FIGURE 5.62

displayed, drag the cursor into the other break line to increase the amount of geometry that is displayed. In either case the other break line will follow to maintain the gap size.

8. Next you will create a dimension. Press the D key on the keyboard and place the dimension on the two outside points as shown in Figure 5.68 on the left. Notice the break symbol in the dimension. Dimensioning a drawing will be covered later in this chapter.
9. When done, your screen should resemble Figure 5.68 on the right.

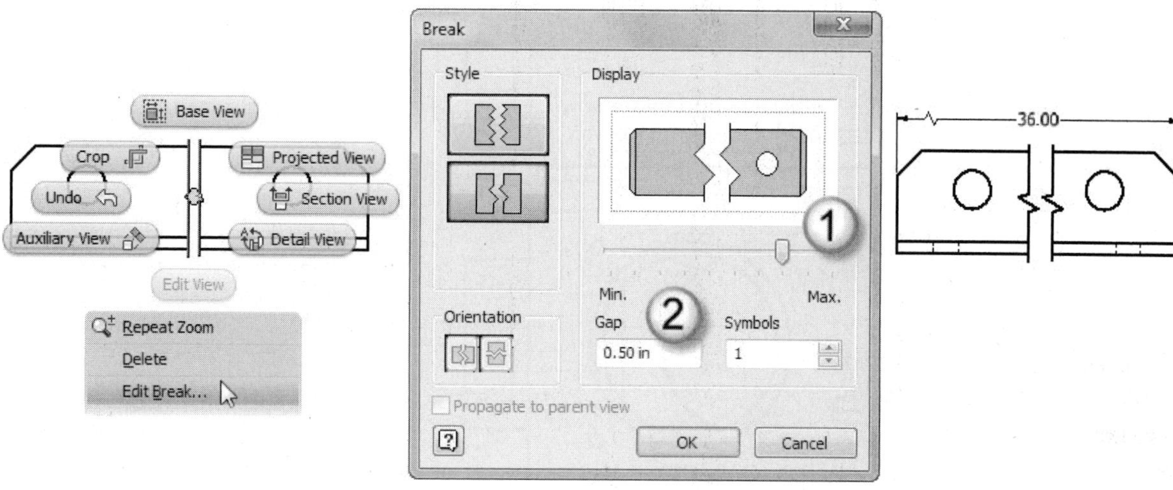

FIGURE 5.68

10. Close all open files. Do not save changes.
11. Open the file *ESS_E05_03-2.idw*. In this portion of the exercise you create a break out view of a part. This drawing contains a front and top view. Note that a break out view can also be created for assemblies.
12. First you create a sketch based on the top view. Select the bounding edge of the top view as shown in Figure 5.69 on the left and click Create Sketch from the Place Views tab > Sketch panel. If the edge of the top view is not selected, the sketch will not be associated to the view and the break out view will not be able to be created.
13. Create a closed profile of your choice. This closed profile will be used to cut away a section in the view. Figure 5.69 on the right shows a closed spline.

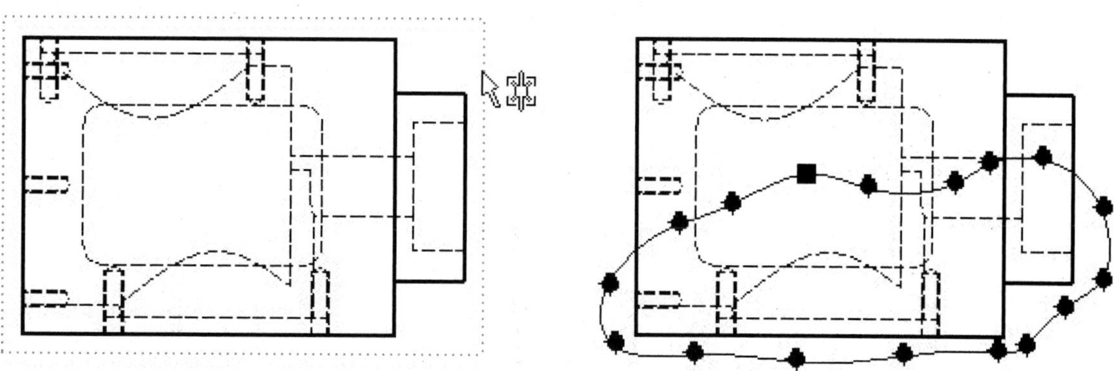

FIGURE 5.69

14. Exit the sketch by clicking Finish Sketch from the Sketch tab > Exit panel.
15. You now create a break out view. First click the Break Out command on the Place Views tab > Modify panel.

16. In the graphics window click in the top view, this is the view that has the closed profile for the break out view.
17. The Break Out dialog box will appear. The default Depth option is From Point, in the front view click on the point at the center of the hole displayed as a side view as shown in Figure 5.70.

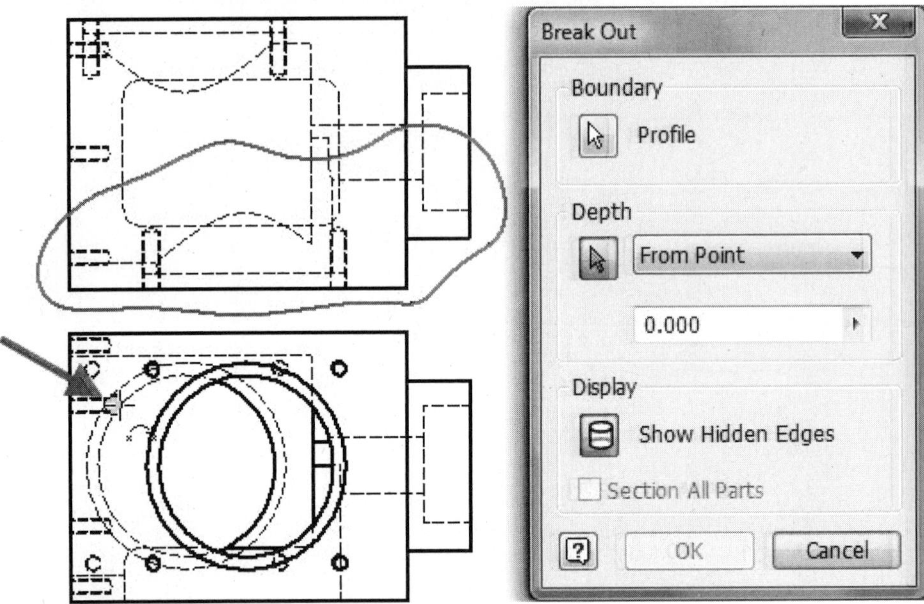

FIGURE 5.70

18. Click OK in the Break Out dialog box, the closed profile will break out a section in the top view as shown in Figure 5.71 on the left.
19. Next you project an isometric view of the top broken out view. Click the Projected View command on the Place Views tab > Create panel.
20. Click in the top view and then locate the isometric view by clicking a point to the lower-left.
21. Move the isometric view to the right side of the drawing and edit the isometric view and turn on the Shaded style (in the Style area in the lower-right corner of the dialog box) and from the Display Options tab check the Hatching option and click OK to complete the Change the scale of isometric view by unchecking the Scale from Base option to the right of the Scale in the dialog box and type **.75.** When done the views should resemble the Figure 5.71.

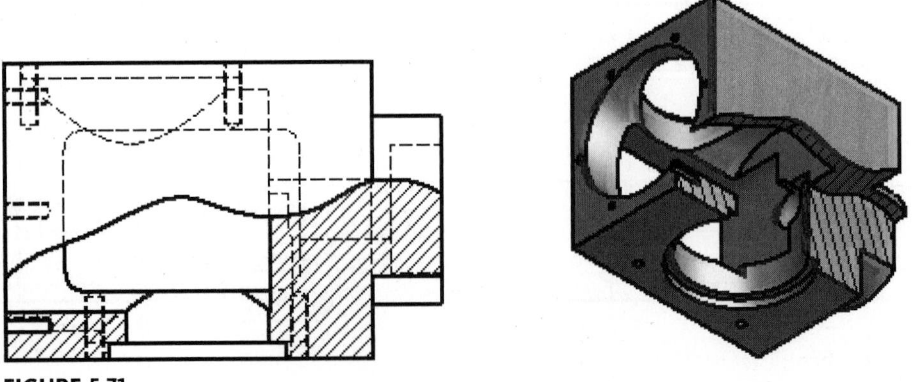

FIGURE 5.71

22. Edit the break out view by right-clicking on the Break Out view in the browser and click Edit Break Out from the menu as shown in Figure 5.72 on the left.

23. In the Break Out dialog box change the Depth type to To Hole and select the lower left hole in the front view as shown in Figure 5.72 in the middle.
24. Click OK in the Break Out dialog box and the top and isometric view will be updated. Figure 5.72 on the right shows the updated isometric view.

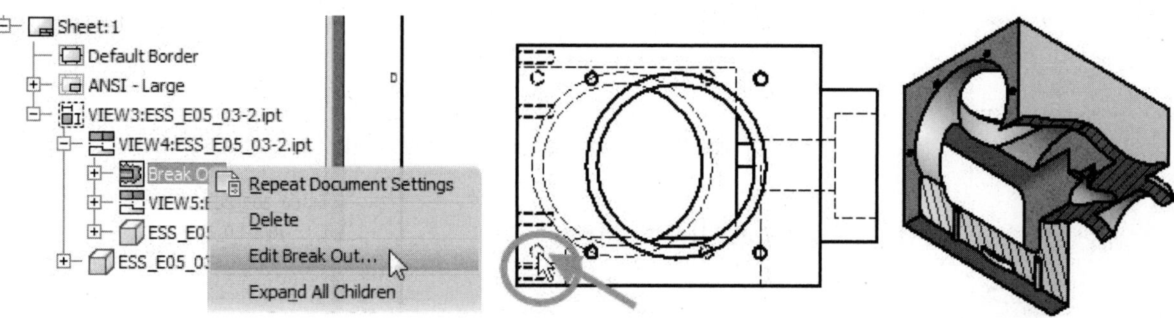

FIGURE 5.72

25. Edit the break out view by right-clicking on the Break Out View in the browser as you did in step 22.
26. In the Break Out dialog box change the Depth type to Through Part; select an edge of the part in the top view is shown in Figure 5.73 on the left.
27. Click OK in the Break Out dialog box and the top and isometric view will update as shown in Figure 5.73.

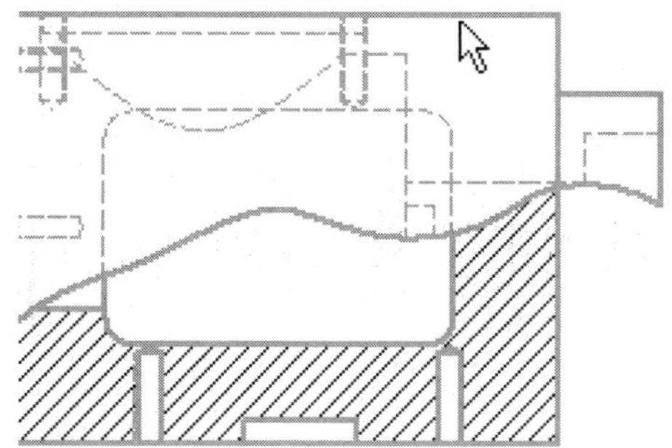

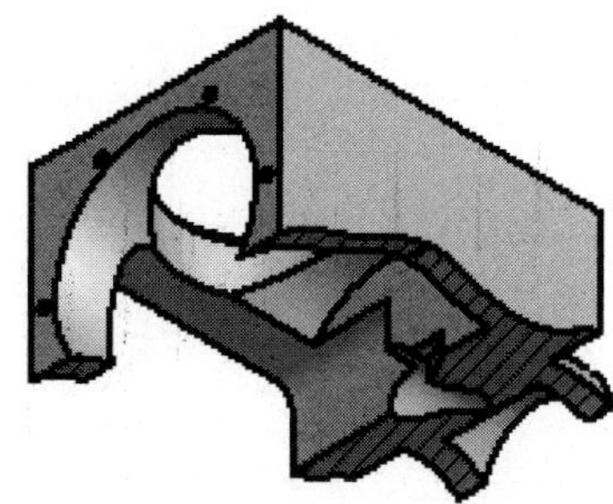

FIGURE 5.73

28. If desired, you can edit the sketch that the breakout view used by double-clicking on the Sketch entry under the Break Out view in the browser. Then reposition / resize the sketch as needed. When done, click the Finish Sketch in the Exit panel.
29. Crop the front view by clicking the Crop command on the Place Views tab > Modify panel.
30. Click in the front view and define the area to crop to by clicking near the two points labeled (1) and (2) as shown in Figure 5.74 on the left.
31. The view will be cropped similar to the view shown in Figure 5.74 in the middle.
32. Remove the crop lines by right-clicking on the Crop view entry in the browser and click Display Crop Cut Lines and the cut lines will be removed from the cropped view.
33. If desired edit the sketch under the Crop entry in the browser and change the size of the area that is cropped. If you cannot change the size of the sketch you need to delete a sketch constraint. When done finish the sketch.

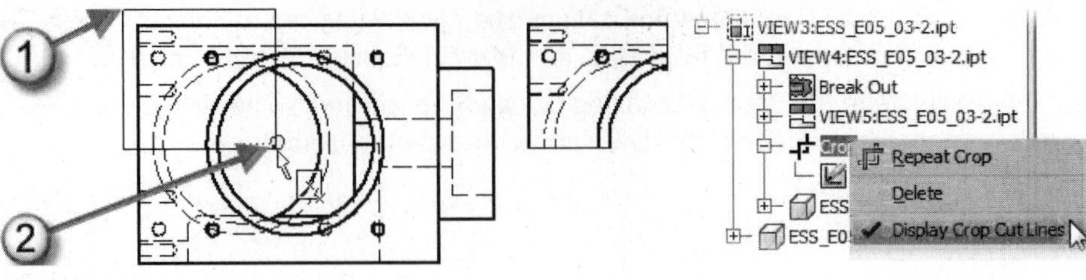

FIGURE 5.74

34. Close all open files. Do not save changes. End of exercise.

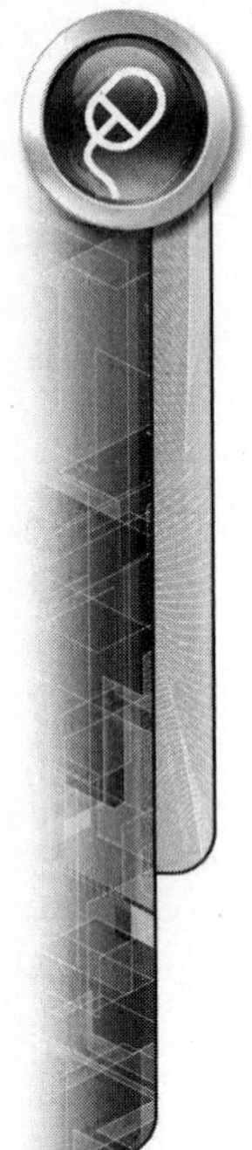

EDITING DRAWING VIEWS

After creating the drawing views, you may need to edit the properties of, delete a drawing view, or replacing the component that a drawing references. The following sections discuss these options.

Editing Drawing View Properties

After creating a drawing view, you may need to change the label, scale, style, or hatching visibility, break the alignment constraint to its base view, or control the visibility of the view projection lines for a section or auxiliary view. To edit a drawing view, follow one of these steps:

- Double-click in the bounding area of the view.
- Double-click on the icon of the view you want to edit in the browser.
- Right-click in the drawing view's bounding area or on its name, and select Edit View from the marking menu, as shown in Figure 5.75.

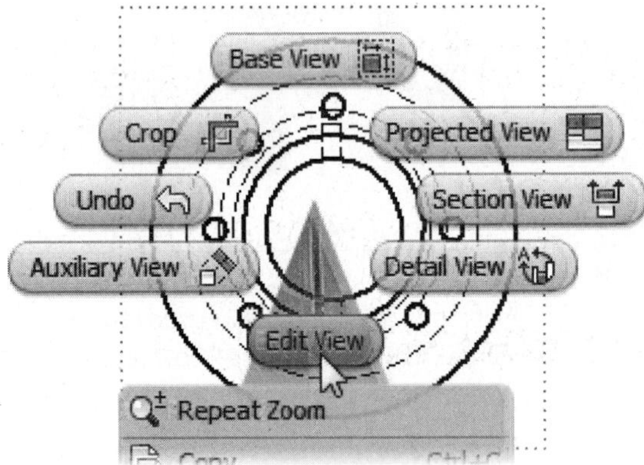

FIGURE 5.75

When you perform the operations listed above, the Drawing View dialog box will appear, as shown in Figure 5.76. This is the same dialog box used to create drawing views. Depending on the view that you selected, certain options may be grayed out from the dialog box. Make the necessary changes, and click the OK button to complete the edit. When you change the scale in the base view, all the dependent views will be scaled as well. You can also change the orientation of the base view; dependent views will then be updated to reflect the new orientation.

CREATING CHAIN DIMENSIONS

Another option for placing dimensions is to create chain dimensions, dimensions whose extensions lines are shared between two dimensions. Chain dimensions can be arranged horizontal or vertical. Like the baseline dimensions commands there are two Chain dimension commands, Chain and Chain Set.

The Chain command allows you to create a group of horizontal or vertical dimensions. However, the group of dimensions are not considered one group of objects; rather, each dimension that makes up the chain dimension are considered single dimensions.

To create and chain dimensions, follow these steps:

1. Select the Chain dimension command from the Annotate tab > Dimension panel as shown in Figure 5.125.

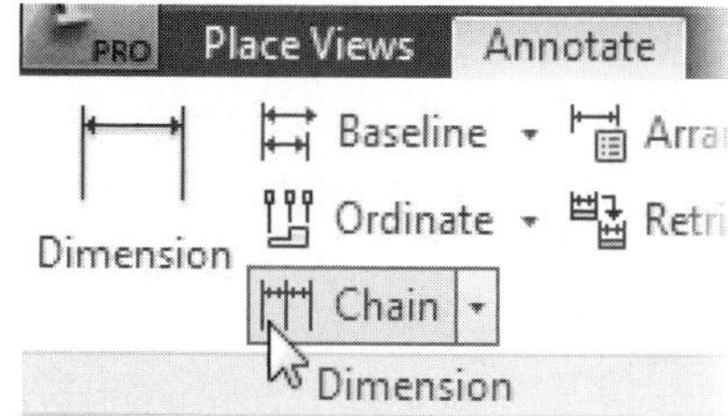

FIGURE 5.125

2. Individually select or drag a selection window around the geometry that you want to dimension. To window select, move the cursor into position where the first point of the box will be. Press and hold down the left mouse button, and move the cursor so the preview box encompasses the geometry that you want to dimension. In Figure 5.126 on the left, four individual objects were selected.
3. When you are finished selecting geometry, right-click, and click Continue from the marking menu.
4. Move the cursor to a position where the dimensions will be placed, as shown in Figure 5.126 on the right. When the dimensions are in the correct place, click the point with the left mouse button to anchor the dimensions.

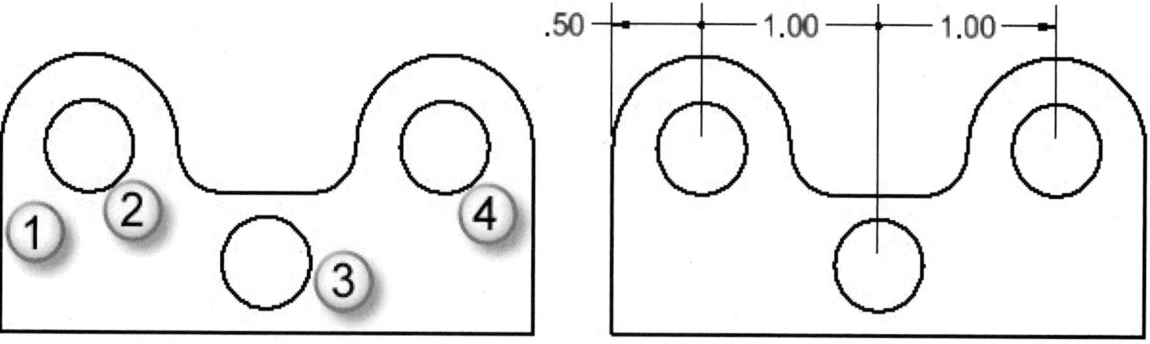

FIGURE 5.126

5. To complete the operation, right-click and click Create from the marking menu.
6. After creating dimensions using the Chain dimension command, the same editing operations can be performed on these dimensions as general dimensions.

Creating a Chain Dimension Set

Creating a Chain dimension Set is similar to Chain dimensions as explained in the previous section. With the Chain Set command, however, dimensions are grouped together into a set. Select the Chain Set command from the Annotate tab > Dimension panel > click the down arrow next to the Chain dimension command, as shown in Figure 5.127.

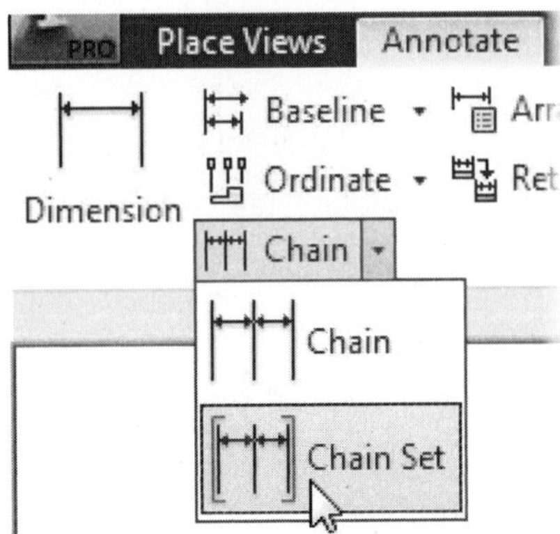

FIGURE 5.127

Members can be added and deleted from the set. Right-click on any dimension in the set will highlight all dimensions and display the menu, as shown in Figure 5.128.

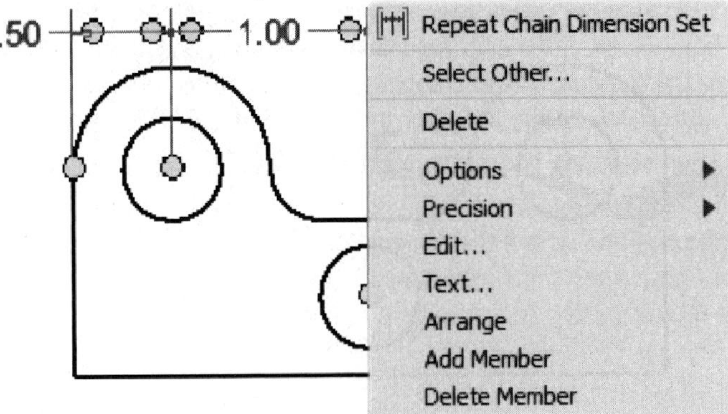

FIGURE 5.128

The main options for editing a chain dimension set are described as follows:

Delete. This option will delete all chain dimensions in the set. Move your cursor over any dimension in the set. When the green circles appear on each dimension, right-click, and select Delete from the menu.

Options. From the Options area you can lock the chain, control the arrowheads and add a leader to a dimension that the cursor was over when you right-click.

Add Member. This option will add a drawing dimension to the chain dimensions. Move the cursor over the chain dimension set, when the green circles appear on the dimensions, right-click and click Add Member from the menu and then click a point that you want to add a dimension to. The dimension will be added to the set of chain dimensions and rearranged in the proper order.

Delete Member. This option will erase a dimension from the set of chain dimensions. Move your cursor over the dimension in the set that you want to delete. When the green circles appear on each dimension, right-click, and then select Delete Member from the menu. The dimension will be deleted.

EXERCISE 5-7: CREATING BASELINE AND CHAIN DIMENSIONS

In this exercise, you will add annotations using baseline and chain dimensions.

1. Open *ESS_E05_07.dwg* in the Chapter 05 folder.
2. In this step you add a baseline dimension set. Click the Baseline Dimension Set command from the Annotate tab > Dimension panel > click the down arrow next to the Baseline Dimension command.
3. Select the three edges as shown in Figure 5.129 on the left.
4. Right-click, and click Continue from the marking menu.
5. Click a point above the geometry to position the dimension set as shown in Figure 5.129 middle image.
6. Right-click, and click Create from the marking menu. The baseline dimension set is created.
7. However, all the baseline dimensions need to reference the opposite edge of the object. To perform this task, move the cursor over the extension line on the right, right-click, and click Make Origin from the menu. Notice how the dimensions are regenerated from the new origin, as shown in Figure 5.129 on the right.

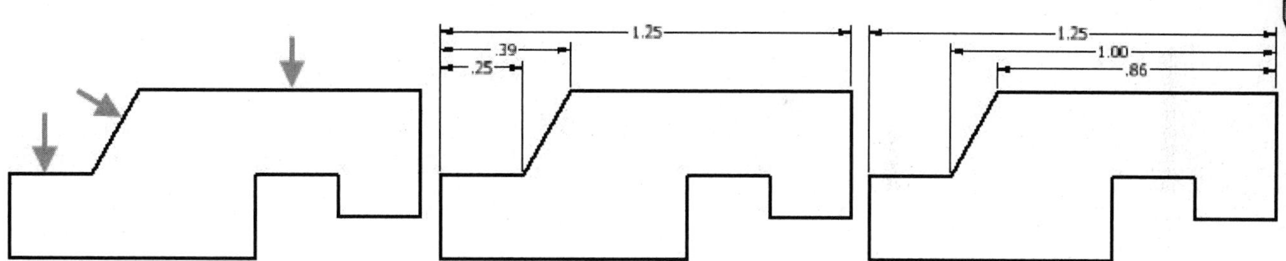

FIGURE 5.129

8. To add a dimension to the set move the cursor over the baseline dimension set, when the green circles appear on the dimensions, right-click and click Add Member from the menu and then click the point as shown in Figure 5.130 on the left. Right-click and click Done from the menu and the dimension will be added to the set and the dimensions will be rearranged as shown in Figure 5.130 on the right.

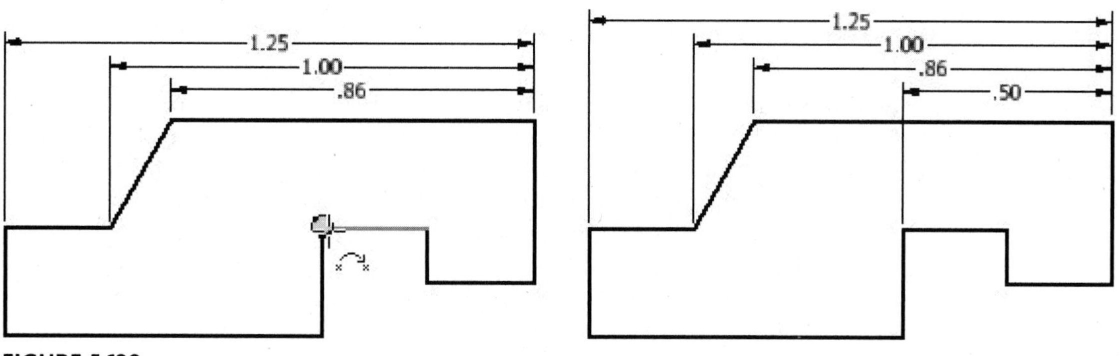

FIGURE 5.130

9. Delete a dimension, move the cursor over the **.86** dimension, right-click and click Delete Member from the menu. The dimension will be deleted as shown in Figure 5.131 on the left. Notice the dimensions were not automatically arranged.
10. Next you arrange the dimensions, move the cursor over the dimension set, when the green circles appear on the dimension right-click and click Arrange from the menu. The dimension will be rearranged as shown in Figure 5.131 on the right.

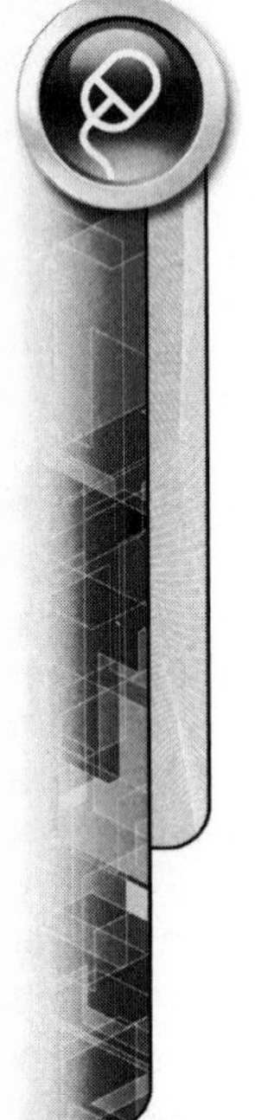

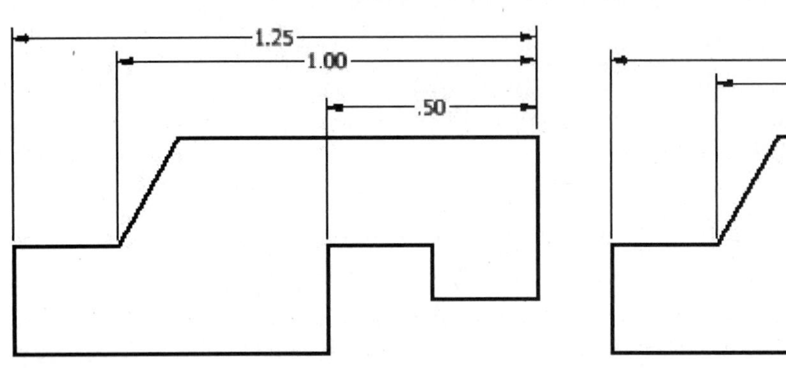

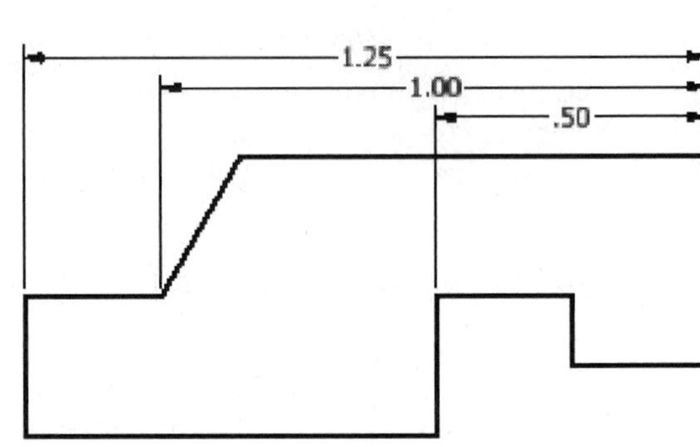

FIGURE 5.131

11. Next delete the baseline dimension set. Move the cursor over the dimension set, when the green circles appear on the dimension right-click and click Delete from the menu. The dimension set will be deleted.
12. Next you create a chain dimension set. Select the Chain Set command from the Annotate tab > Dimension panel > click the down arrow next to the Chain Set dimension command.
13. Window select the edges as shown in Figure 5.132 on the left.
14. Right-click, and click Continue from the marking menu.

    Click a point below the geometry to position the dimension set as shown in Figure 5.132 on the right.
15. Right-click, and click Create from the marking menu and the chain dimension set will be created.

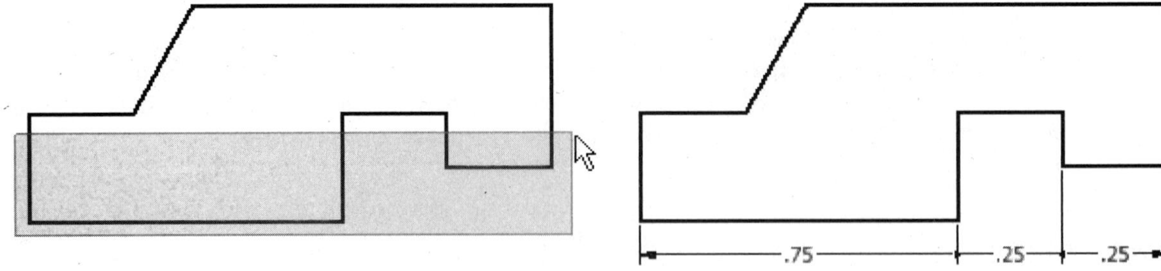

FIGURE 5.132

16. To add a dimension to the set move the cursor over the chain dimension set, when the green circles appear on the dimensions, right-click and click Add Member from the menu and then click the point as shown in Figure 5.133 on the left. Right-click and click Done from the menu. The dimension will be added to the set and the dimensions will be rearranged as shown in Figure 5.133 on the right.

NOTE

If a dimension does not fit within its extension lines you can right-click on the dimension and from the menu click Options > Leader and then drag the dimension and leader as needed.

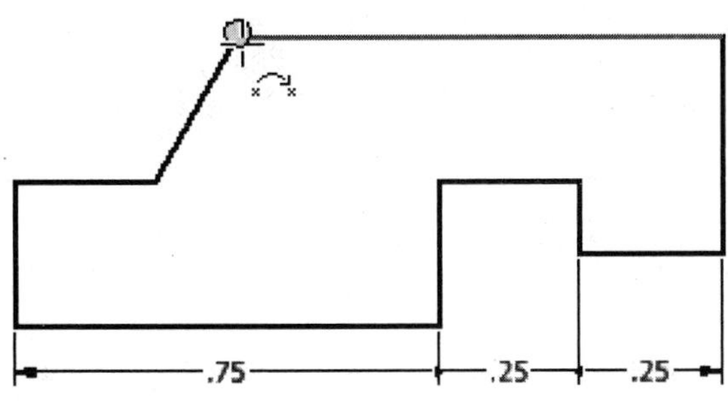

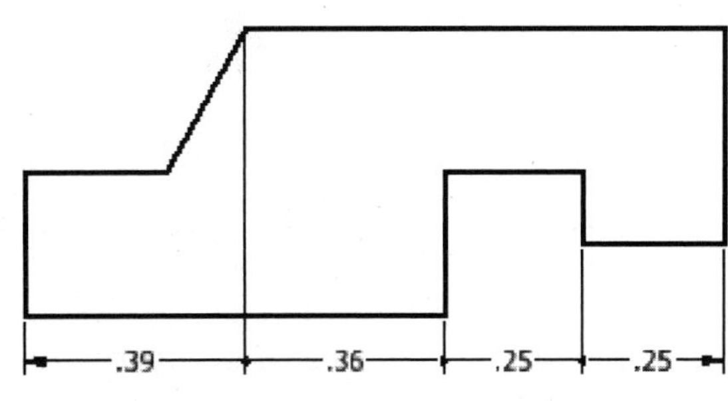

FIGURE 5.133

17. Delete a dimension, move the cursor over the **.39** dimension, right-click, and click Delete Member from the menu. To center the text, right-click on the dimension set and click Arrange from the menu. Notice the dimensions were automatically arranged as shown in **Figure 5.134**.

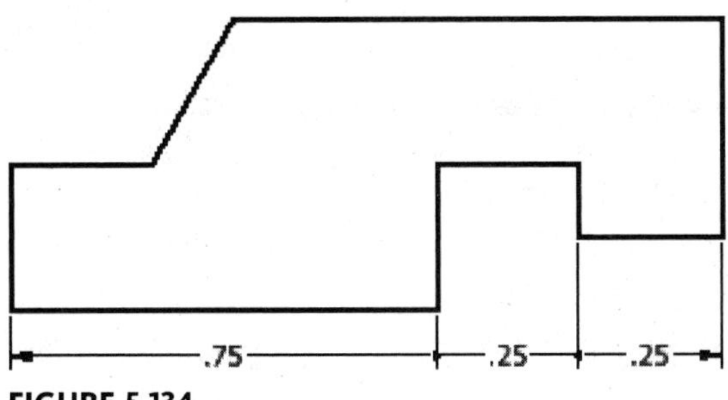

FIGURE 5.134

18. Close all open files. Do not save changes. End of exercise.

CREATING ORDINATE DIMENSIONS

Ordinate dimensions are used to indicate the location of a particular point along the X or Y axis from a common origin point. This type of dimensioning is especially suited for describing part geometry for numerical control tooling operations. Two commands are available for creating ordinate dimensions: Ordinate Dimension and Ordinate Dimension Set. Both commands will create drawing dimensions that reference the geometry and will be updated to reflect any changes in the geometry to which they are dimensioned. Ordinate dimensions can be placed on circular or straight edges.

Ordinate Dimensions

Ordinate dimensions created with the Ordinate Dimension command are recognized as individual objects, and an origin indicator will be created as part of the operation. If the origin indicator location is moved, the other ordinate dimensions will be updated to reflect the change.

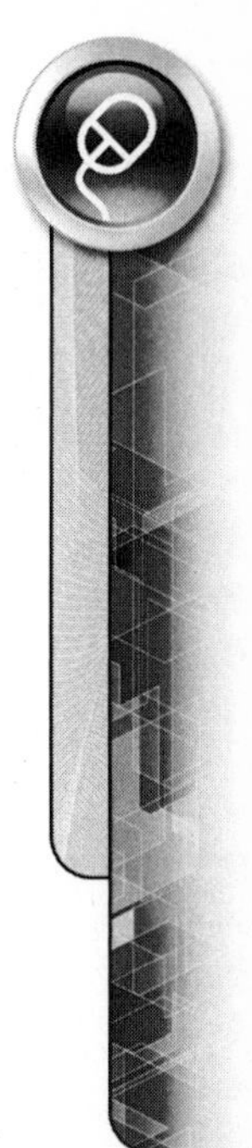

To create an ordinate dimension using the Ordinate Dimension command, follow these steps:

1. Create a drawing view.
2. Click the Ordinate Dimension command on the Annotate tab > Dimension panel, as shown in Figure 5.135 on the left.
3. In the graphics window, select in the view to dimension.
4. Select a point to set the origin indicator for the dimensions (the zero), as shown in Figure 5.135.

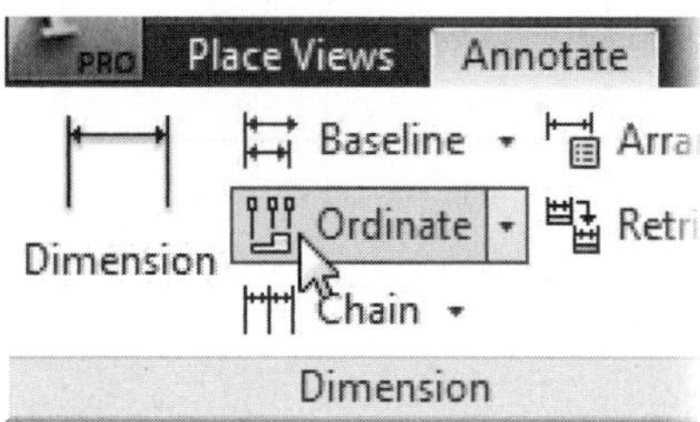

FIGURE 5.135

5. Select geometry that the ordinate dimensions will be applied to. When done selecting geometry, right-click and click Continue from the marking menu.
6. Locate the dimensions horizontally or vertically by moving the cursor and then click a point when the dimensions are previewed in the correct orientation, as shown in Figure 5.136 on the left.
7. To create the dimensions and end the operation, right-click, and click OK from the menu.
8. To edit a dimension, right-click on a dimension and click the desired option from the menu, as shown in Figure 5.136 on the right.

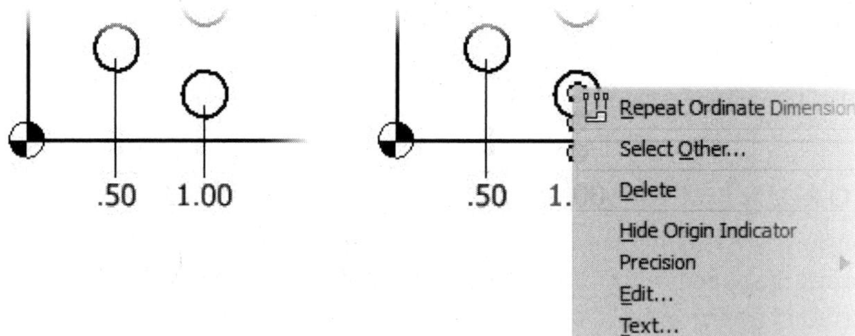

FIGURE 5.136

9. To move an ordinate dimension click on the dimension, and then click and drag on an anchor point (a green circle), and drag it to the desired location.
10. To edit the origin indicator, do one of the following:
    - Move the cursor over the origin indicator, and drag it to the desired location.
    - Double-click the origin indicator, and enter the precise location in the Origin Indicator dialog box.

Ordinate Dimension Set

In an ordinate dimension set, all the ordinate dimensions that are created in a single operation will be grouped together and can be edited individually or as a set. When creating an ordinate dimension set, the first dimension created will be used as the origin. The origin dimension needs to be a member of the set and can later be changed to a different dimension. If the location of the origin or the origin member changes, the other members will be updated to reflect the new location.

When you place ordinate dimensions, they will automatically be aligned to avoid interfering with other ordinate dimensions.

To create an ordinate dimension set, follow these steps:

1. Create a drawing view.
2. Click the Ordinate Dimension Set command on the Annotate tab > Dimension panel > click the down arrow next to the Ordinate Dimension command, as shown in Figure 5.137 on the left.
3. Select a point on the geometry to set the origin for the dimensions.
4. Select geometry that the ordinate dimensions will be applied to. When done selecting geometry, right-click and click Continue from the marking menu.
5. Locate the dimensions horizontally or vertically by moving the cursor and then click a point when the dimensions are previewed in the correct orientation, as shown in Figure 5.137 on the right.

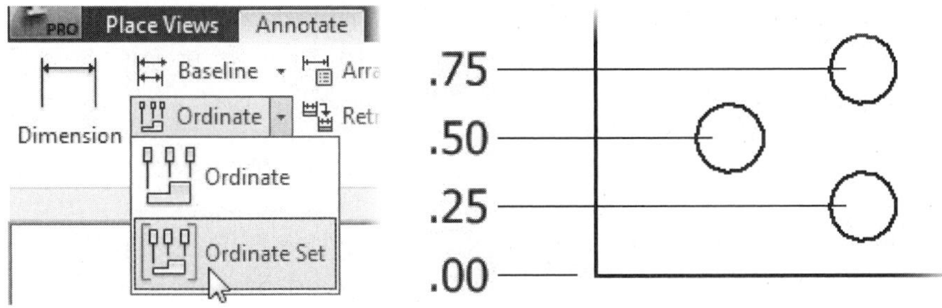

FIGURE 5.137

6. To change options for the dimension set, right-click, and then click Options, as shown in Figure 5.138 on the left. This can also be done after creating the dimensions by moving the cursor over the dimension set, right-click and click Options from the marking menu.
7. To create the dimensions set and end the command, right-click, and click Create from the menu.
8. To edit the origin, right-click on the dimension that will be the origin and click Make Origin from the menu as shown in Figure 5.138 on the right. The other dimension values will be updated to reflect the new origin.
9. To edit a dimension set, right-click on a dimension in the set and click the desired option from the menu as shown in Figure 5.138 on the right.

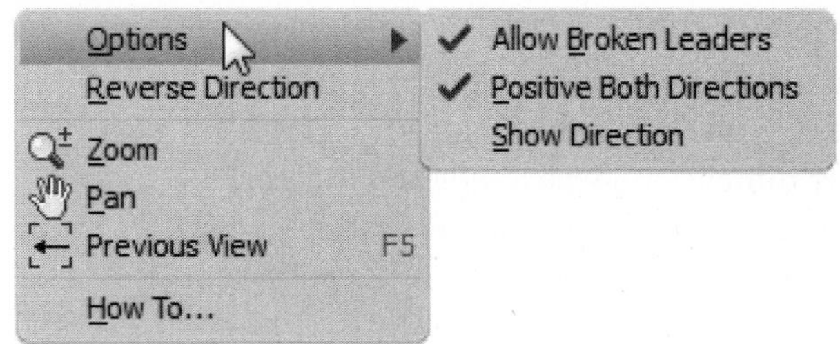

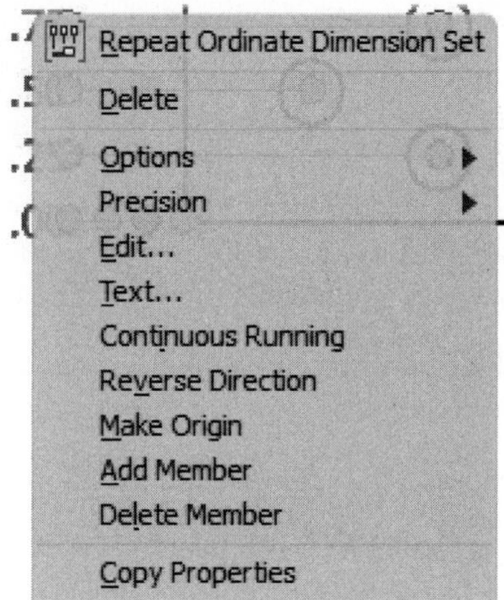

FIGURE 5.138

Text and Leader Text

To add text to the drawing, click either the Text or the Leader Text command on the Annotate tab > Text panel, as shown in Figure 5.139.

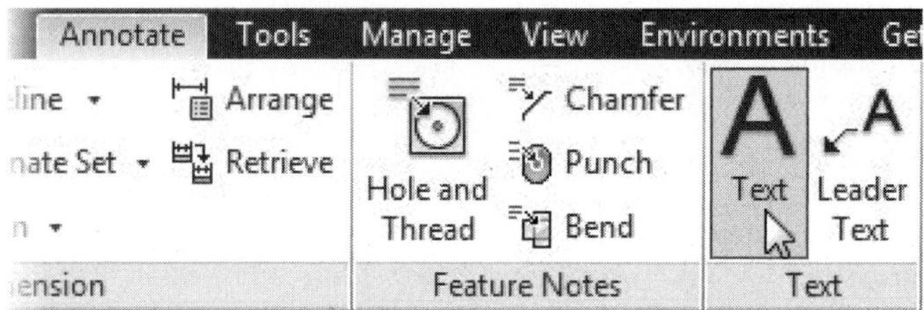

FIGURE 5.139

The Text command will add just text while the Leader Text command will add a leader with the text. Select the desired text command, and define the leader points and/or text location. Once you have chosen the location in the graphics window, the Format Text dialog box will appear. When placing text through the Format Text dialog box, select the orientation and text style as needed and type in the text, as shown in Figure 5.140. Click the OK button to place the text in the drawing. To edit the text or text leader position, move your cursor over it, click one of the green circles that appear, and drag to the desired location. To edit the text or text leader content, right-click on the text, and select Edit Leader Text or Edit Text from the menu.

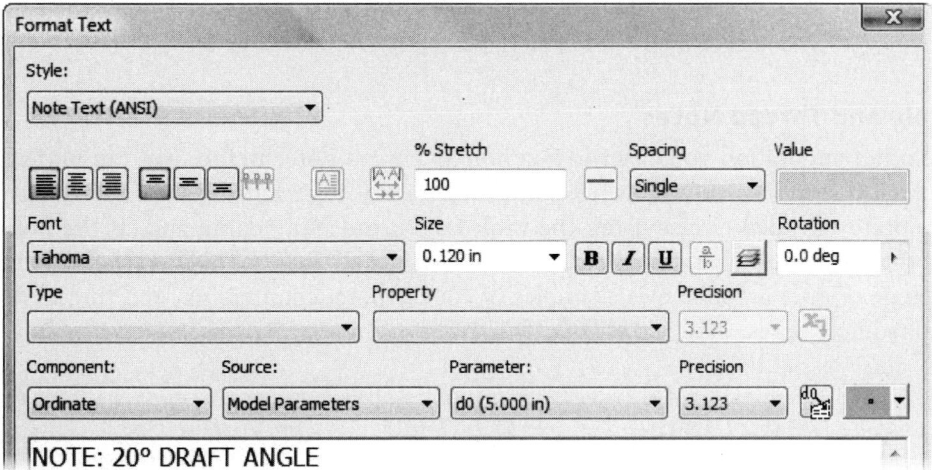

FIGURE 5.140

Additional Annotation Commands

To add more detail annotations to your drawing, you can add surface texture symbols, welding symbols, feature control frames, feature identifier symbols, datum identifier symbols, and datum targets by clicking the corresponding command on the Annotate tab > Symbols panel, as shown in Figure 5.141. The Import AutoCAD Block command is only available when you are in an Inventor DWG file.

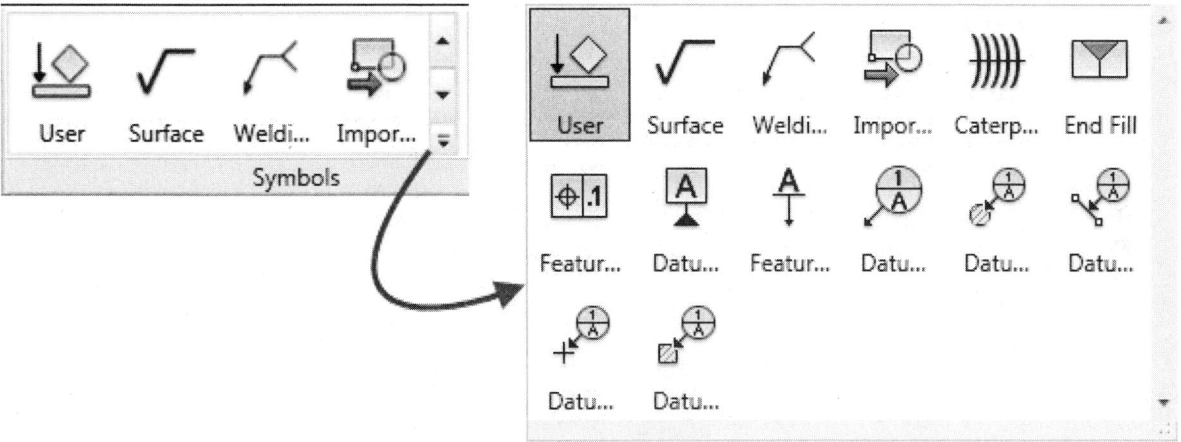

FIGURE 5.141

Follow these steps for placing symbols:

1. Click the appropriate symbol command on the Annotate tab > Symbols panel.
2. Select a point at which the leader will start. If you don't want a leader, click a point that the symbol will be placed and proceed to step 4.
3. Continue selecting points to position the leader lines.
4. Right-click, and select Continue from the marking menu.
5. Fill in the information as needed in the dialog box.
6. When done, click the OK button in the dialog box.
7. To complete the operation, right-click, and click Cancel [ESC] from the marking menu.

8. To edit a symbol, move the cursor over it. When the green circles appear, right-click, and click the corresponding Edit option from the menu.

Hole and Thread Notes

Another annotation you can add is a hole or thread note. Before you can place a hole or thread note in a drawing, a hole or thread feature must exist. You can also annotate extruded circles using the Hole or Thread Note command. If the hole or thread feature changes, the note will be updated automatically to reflect the change. Figure 5.142 shows examples of counterbore, countersink, and threaded/tapped hole notes.

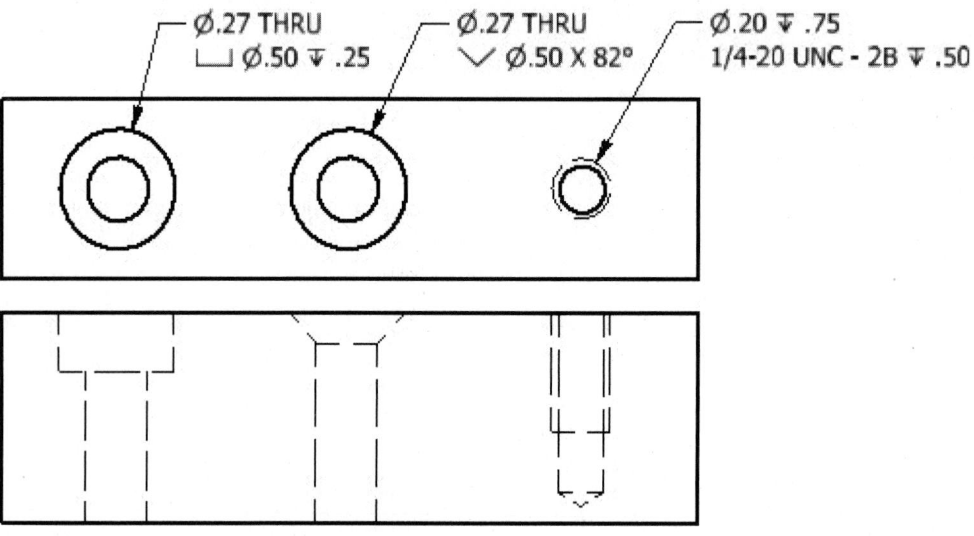

FIGURE 5.142

To create a hole or thread note, follow these steps:

1. Click the Hole/Thread Notes command on the Annotate tab > Feature Notes panel, as shown in Figure 5.143.

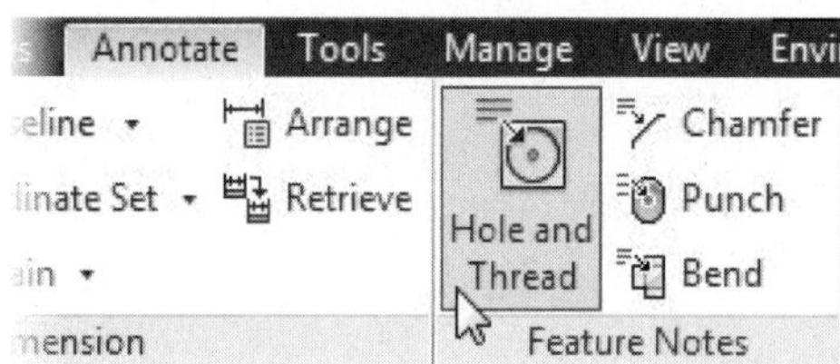

FIGURE 5.143

2. In a drawing view select a hole or thread feature to annotate.
3. Click a second point to locate the leader and the note.
4. To complete the operation, right-click, and select OK from the marking menu.

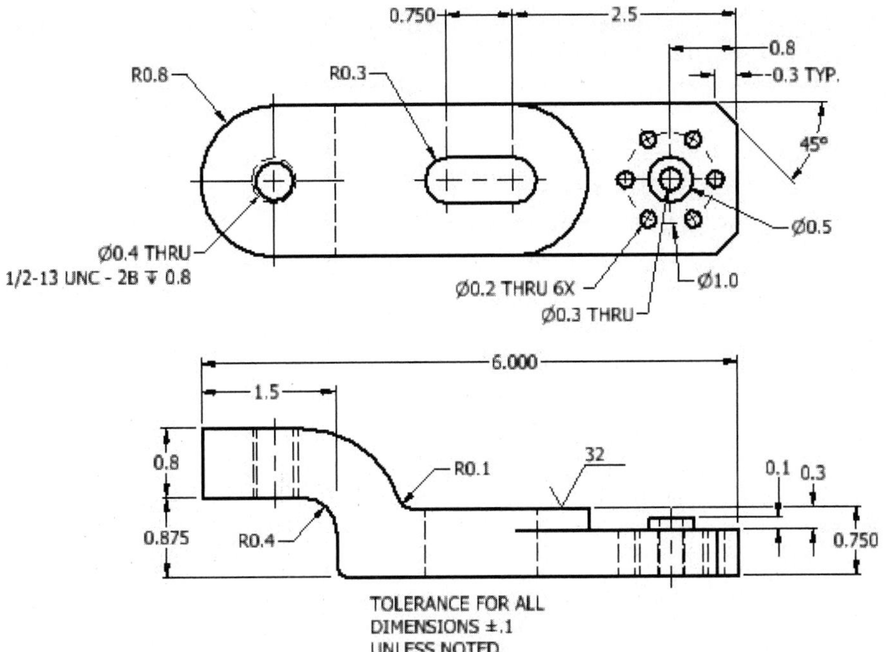

FIGURE 5.152

21. Close all open files. Do not save changes. End of exercise.

OPENING A MODEL FROM A DRAWING

While working inside of a complex drawing view, you may need to make changes to a part, assembly, or presentation file. Rather than close down the drawing and open the individual file, you can open a part, assembly, or presentation file directly from the drawing using a number of techniques.

One technique is to right-click in a drawing view to display the menu in Figure 5.153 on the left, and then click Open to open the model file. In this example, an assembly file was opened from the drawing view.

Another technique involves expanding the drawing view in the browser to expand the assembly and continue expanding the browser until the file you want to open is visible in the list. Right-click the desired file in the browser and select Open from the menu to display the part model as shown in Figure 5.153 on the right.

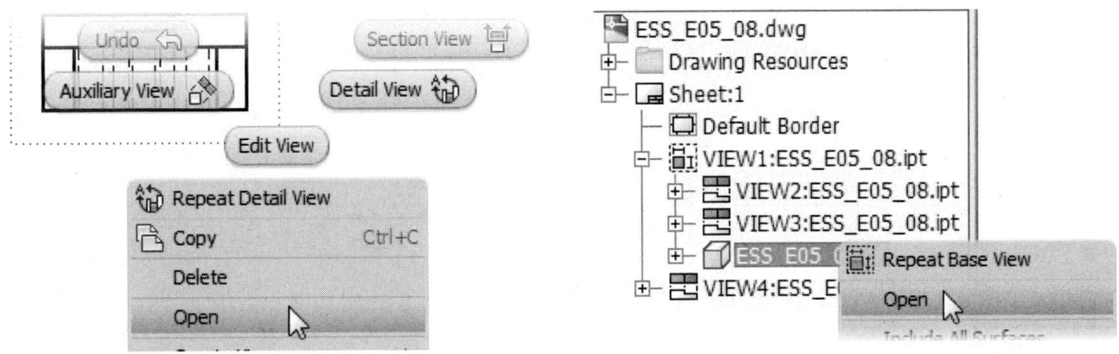

FIGURE 5.153

A third technique used to open a part file from a drawing is to first change the selection filter to Part Priority. After doing so, right-click on the specific part in the drawing view and select Open from the menu. This will open the source file.

OPENING A DRAWING FROM A MODEL

You can open a drawing from a part or assembly model by right-clicking on the part, assembly presentation file name and click Open Drawing from the menu as shown in Figure 5.154. This will open the drawing file associated with the part, assembly or presentation file. The drawing file must have the same name as the part, assembly or presentation file. Use this technique to prevent laborious searches for drawing files.

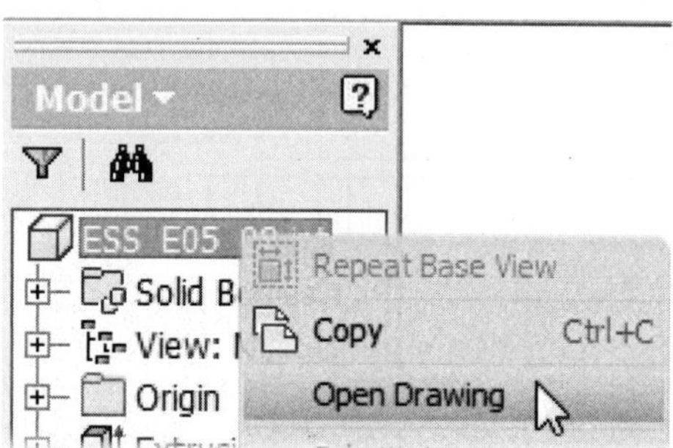

FIGURE 5.154

CREATING HOLE TABLES

If the drawing view that you are dimensioning contains holes, you can locate them by placing individual dimensions or by creating a hole table that will list the location and size of all the holes, or just the selected holes, in a view. The hole locations will be listed in both X- and Y-axis coordinates with respect to a hole datum that will be placed before creating the hole table.

After placing a hole table, if you add, delete, or move a hole, the hole table will automatically be updated to reflect the change after the part is saved and the drawing is the active document. Each hole in the table is automatically given an alphanumeric tag as its name. It can be edited by double-clicking on the tag in the drawing view or hole table. Alternately, you can right-click on the tag in the drawing view or hole table, and select Edit Tag from the menu. Type in a new tag name, and the change will appear in the drawing view and hole table. There are three commands for creating a hole table:

- Hole Table—Selection: Hole table is created from selected holes.
- Hole Table—View: Hole table is created from all holes in the selected view.
- Hole Table—Selected Feature: Hole table is created from selected hole(s) and the holes that are identical to the selected hole(s).

The commands are located on the Annotate tab > Table panel, as shown in Figure 5.155.

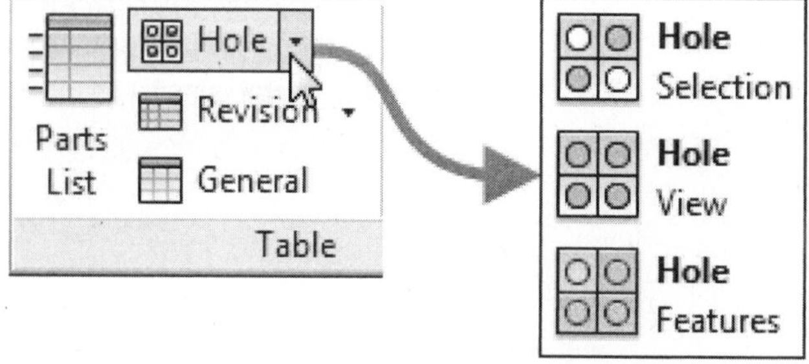

FIGURE 5.155

To create a hole table based on an existing drawing view that shows holes in a plan view, follow these steps:

1. Select the desired Hole Table command from the Hole drop list on the Annotate tab > Table panel.
2. Select the view on which the hole table will be based.
3. Select a point to locate the origin (0,0). Typical origins include the corners of rectangular objects and the centers of drill holes used for data.
4. For the Hole Selection and the Hole Features commands select the hole(s) to create the hole table on and then right-click and click Create from the marking menu.
5. Click a point in the drawing to locate the table.
6. The contents of the hole table can be edited by either double-clicking or right-clicking on the hole table and click Edit Hole Table from the menu.

When using the Hole Selection command, a hole table is created from only the selected holes, as shown in Figure 5.156.

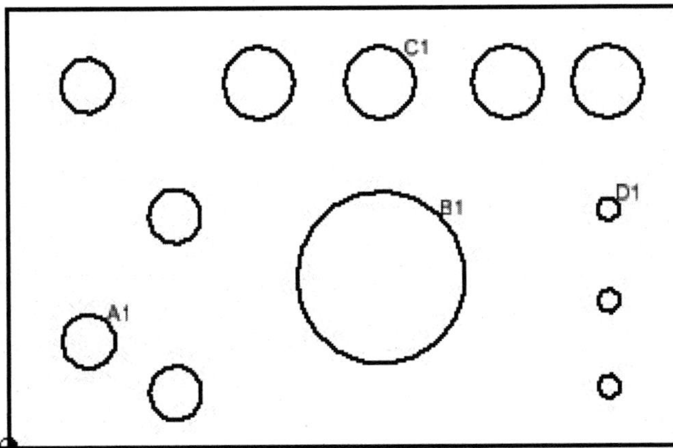

Hole Table

HOLE	XDIM	YDIM	DESCRIPTION
A1	14.28	18.40	Ø0.38 THRU
B1	66.53	29.23	Ø1.18 THRU
C1	66.53	63.09	Ø0.50 THRU
D1	106.89	40.89	Ø0.16 THRU

FIGURE 5.156

When using the Hole View command, a hole table is created based on all the holes in a selected view, as shown in Figure 5.157. Notice that all holes are given an alphanumeric identifier, which locates each hole by X- and Y-axis coordinates.

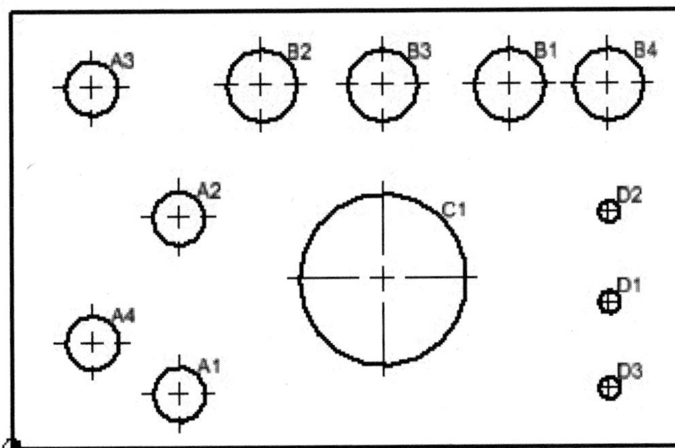

Hole Table

LOC	XDIM	YDIM	SIZE
A1	29.77	9.39	Ø0.38 THRU
A2	29.77	40.02	Ø0.38 THRU
A3	14.28	62.73	Ø0.38 THRU
A4	14.28	18.40	Ø0.38 THRU
B1	89.23	63.09	Ø0.50 THRU
B2	44.91	63.09	Ø0.50 THRU
B3	66.53	63.09	Ø0.50 THRU
B4	106.89	63.09	Ø0.50 THRU
C1	66.53	29.23	Ø1.18 THRU
D1	106.89	25.12	Ø0.16 THRU
D2	106.89	40.89	Ø0.16 THRU
D3	106.89	10.11	Ø0.16 THRU

FIGURE 5.157

When using the Hole Features command, a hole table is created based on holes that are identical to selected hole. In Figure 5.158, one of the larger holes located in the upper part of the object (A4) is selected using the Hole Features command.

When the hole table is created, all holes that share the same type and size will be added to the hole table, as shown in Figure 5.158.

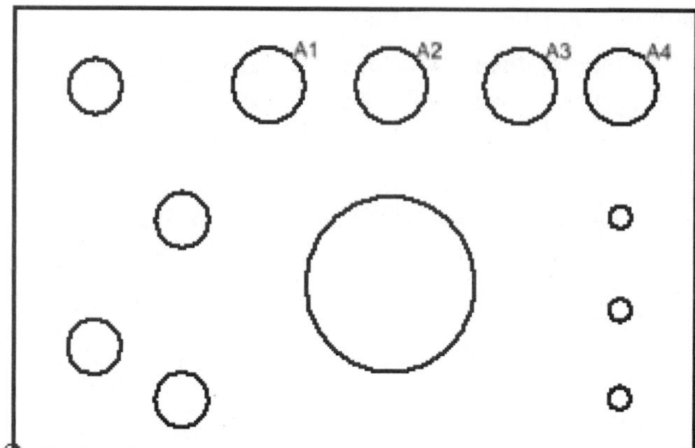

Hole Table			
HOLE	XDIM	YDIM	DESCRIPTION
A1	44.91	63.09	Ø0.50 THRU
A2	66.53	63.09	Ø0.50 THRU
A3	89.23	63.09	Ø0.50 THRU
A4	106.89	63.09	Ø0.50 THRU

FIGURE 5.158

EXERCISE 5-9: CREATING HOLE TABLES

In this exercise, you will create a hole chart for a shim plate. After creating the hole table, you will modify the plate model to add new holes and update the hole table accordingly.

1. In preparation for creating the hole table, first open the existing drawing, *ESS_E05_09.idw* in the Chapter 05 folder.
2. To create the hole table, Click the Annotate tab > Table panel > click the down arrow next to Hole and click the Hole Table View command.
3. Click in the front view of the plate as the view to which the hole table will be associated. To place the hole table datum, aquire the theoretical intersection of the lower left corner of the front view, as shown in Figure 5.159 on the left. Then place the hole table on the right side of the sheet, as shown in Figure 5.159 on the right.

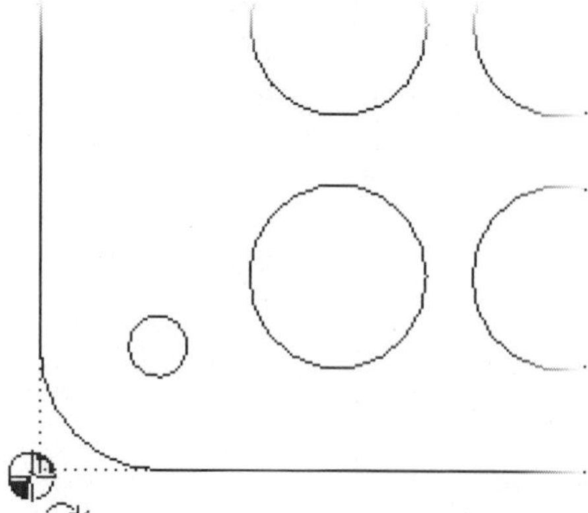

HOLE TABLE			
HOLE	XDIM	YDIM	DESCRIPTION
A1	.25	.25	Ø0.13 THRU
B1	.63	.39	Ø0.38 THRU
B2	1.10	.39	Ø0.38 THRU
B3	1.57	.39	Ø0.38 THRU
B4	2.05	.39	Ø0.38 THRU
B5	2.52	.39	Ø0.38 THRU
B6	.63	.91	Ø0.38 THRU
B7	1.10	.91	Ø0.38 THRU
B8	1.57	.91	Ø0.38 THRU
B9	2.05	.91	Ø0.38 THRU
B10	2.52	.91	Ø0.38 THRU
C1	.38	2.00	Ø0.25 THRU
C2	2.75	2.00	Ø0.25 THRU

FIGURE 5.159

4. The top two holes in the plate need to be changed to Spotface holes. Open the part file by right-clicking in the front view and click Open from the menu as shown in Figure 5.160.

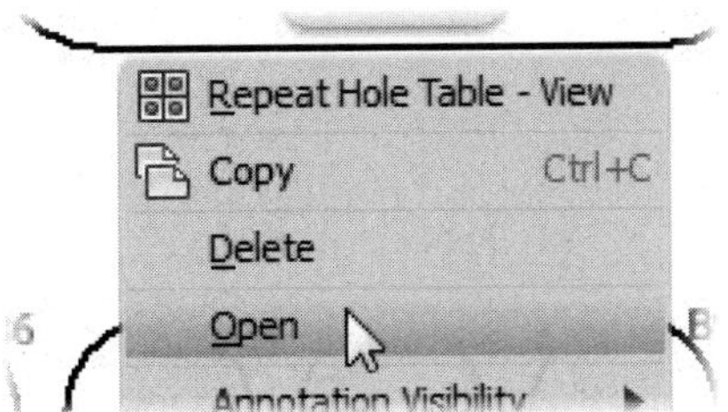

FIGURE 5.160

5. In the browser double-click the Hole1 feature.
6. In the Hole dialog box, change the hole type to Spotface and change the hole specifications, as shown in Figure 5.161 on the left. Click the OK button to update the holes.
7. Save the part file and then switch back to the drawing, click the tab near the bottom-left corner of the graphics window for the file *ESS_E05_09.idw*. Note the updated hole table; the holes labeled with the letter C reflect the change to the holes, as shown in Figure 5.161 on the right.

Another method to switch between windows in an application is to hold down the CTRL key and press the TAB key. **NOTE**

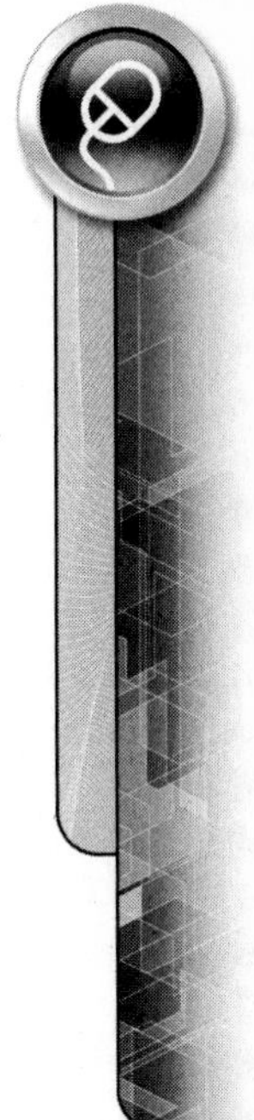

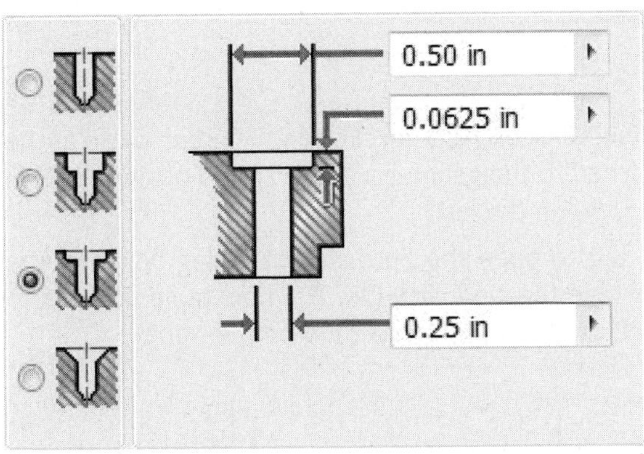

HOLE TABLE			
HOLE	XDIM	YDIM	DESCRIPTION
A1	.25	.25	Ø0.13 THRU
B1	.63	.39	Ø0.38 THRU
B2	1.10	.39	Ø0.38 THRU
B3	1.57	.39	Ø0.38 THRU
B4	2.05	.39	Ø0.38 THRU
B5	2.52	.39	Ø0.38 THRU
B6	.63	.91	Ø0.38 THRU
B7	1.10	.91	Ø0.38 THRU
B8	1.57	.91	Ø0.38 THRU
B9	2.05	.91	Ø0.38 THRU
B10	2.52	.91	Ø0.38 THRU
C1	.38	2.00	Ø0.25 THRU ⌴ Ø0.50
C2	2.75	2.00	Ø0.25 THRU ⌴ Ø0.50

FIGURE 5.161

8. A series of mounting holes needs to be created in the model consisting of countersink holes. Switch back to *ESS_E05_09.ipt* to view the part. Add two countersink through all holes using the following specifications, as shown in Figure 5.162.

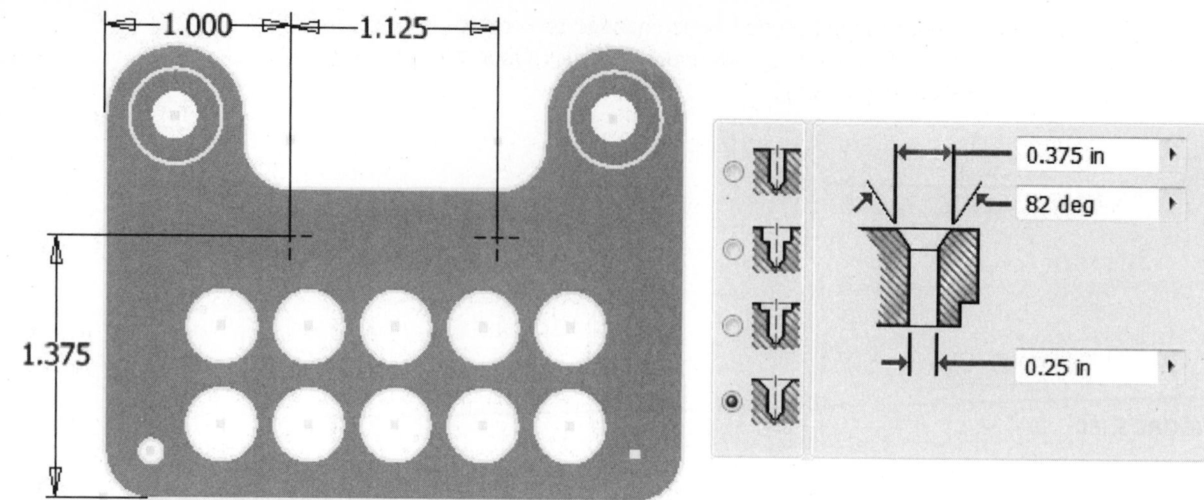

FIGURE 5.162

9. Save the part file and then make the file *ESS_E05_09.idw* current to view the drawing. Note in Figure 5.163 that the new holes have been automatically added to the hole table.

			⌴ Ø0.50
C2	2.75	2.00	Ø0.25 THRU ⌴ Ø0.50
D1	1.00	1.38	Ø0.25 THRU ∨ Ø0.38 X 82°
D2	2.13	1.38	Ø0.25 THRU ∨ Ø0.38 X 82°

FIGURE 5.163

10. A hole tag will now be modified. Click on A1, then click on the green circle at the location where the tag is attached to the geometry and drag and position the A1 hole tag, as shown in Figure 5.164 on the left.
11. Next double-click on the A1 tag to open the Format Text dialog box, add the text-**TOOLING HOLE** to the hole tag, and click OK. The hole table updates to include the Tooling Hole label, as shown in Figure 5.164 on the right.

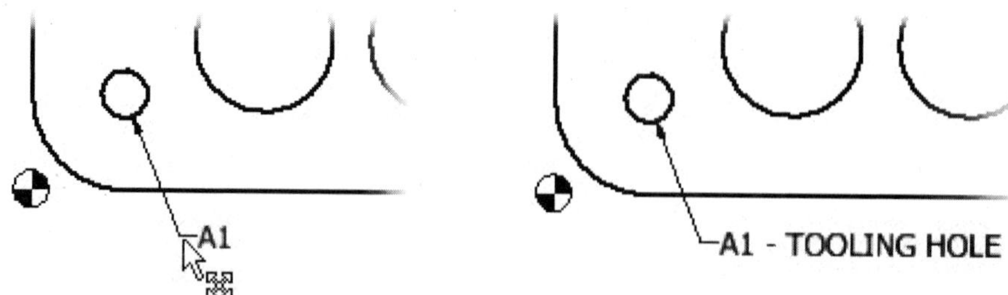

FIGURE 5.164

12. In the hole table notice that the information is updated to include the -Tooling Hole label. You can also resize the columns by clicking and dragging on a cell's border, as shown in Figure 5.165.

HOLE TABLE			
HOLE	XDIM	YDIM	DESCRIPTION
A1 - TOOLING HOLE	.25	.25	Ø0.13 THRU
B1	.63	.39	Ø0.38 THRU

FIGURE 5.165

13. If needed you can combine the descriptions that are identical in the hole table. Combine the descriptions by right-clicking on the hole table and click Edit Hole Table from the menu. In the Edit Hole Table dialog box, on the Options tab check the Combine Notes option.
14. Close all open files. Do not save changes. End of exercise.

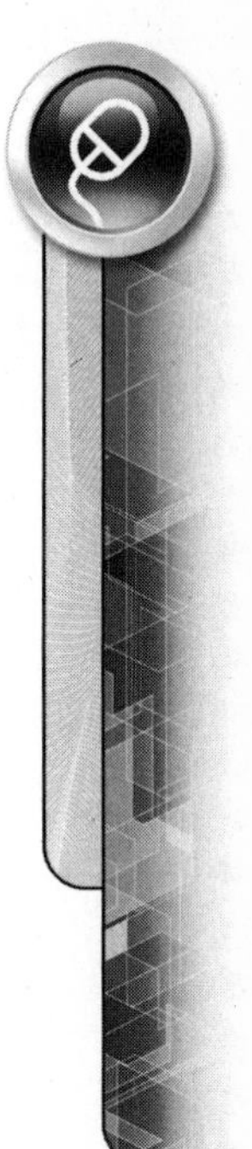

CREATING A TABLE

In addition to organizing hole patterns, tables can also contain data from sheet metal flat pattern, iPart or iAssembly. A table can also be generated manually or created from a Microsoft Excel spreadsheet.

Creating a General Table

To create a generic table that is blank, use the following steps:

To activate the Table command click the General (Table) command on the Annotate tab > Table panel, shown in Figure 5.166.

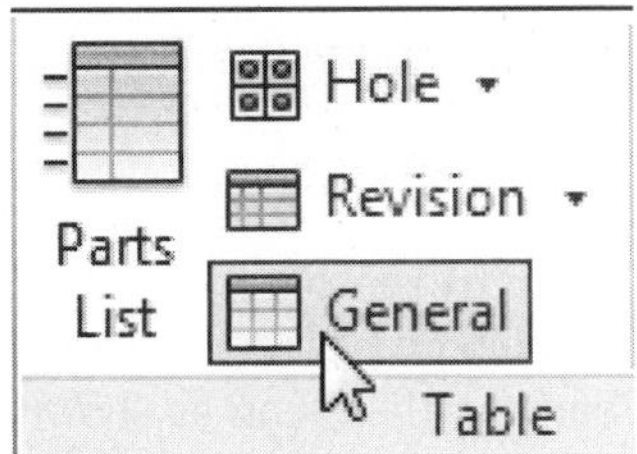

FIGURE 5.166

1. When the Table dialog box appears, verify that <Empty table> is selected from the source list. Then specify the number of rows and columns, as shown in Figure 5.167 on the left. On the Table dialog box, select <Empty table> from the source list. Clicking the OK button will create the table, as shown in Figure 5.167 on the right. Notice that the title of the table is blank. Notice also the column headers labeled Column 1 and so on. Both of these items are based on the current table style.

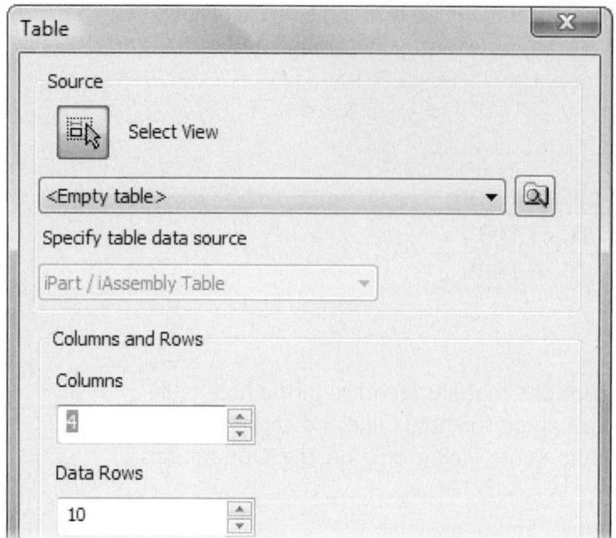

TABLE			
Column 1	Column 2	Column 3	Column 4

FIGURE 5.167

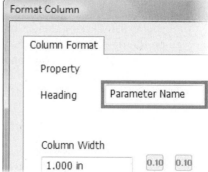

2. To edit the title to the table, double-click the table, and from the Table dialog box click the Table Layout button as shown in Figure 5.168 on the left and in the Table Layout dialog box type in a new title name, as shown in Figure 5.168 on the right. Click OK to complete the edit.

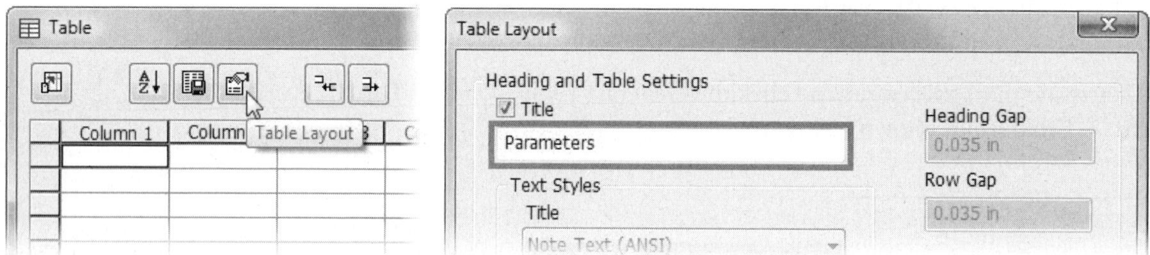

FIGURE 5.168

3. While still in the Table dialog box you can change the column headers by right-clicking on a column header and click Format Column from the menu as shown in Figure 5.169 on the left. Then in the Format Column dialog box type in a new column heading name as shown in Figure 5.169 on the right. You can also adjust column width from this dialog box. Click OK to complete the edit.

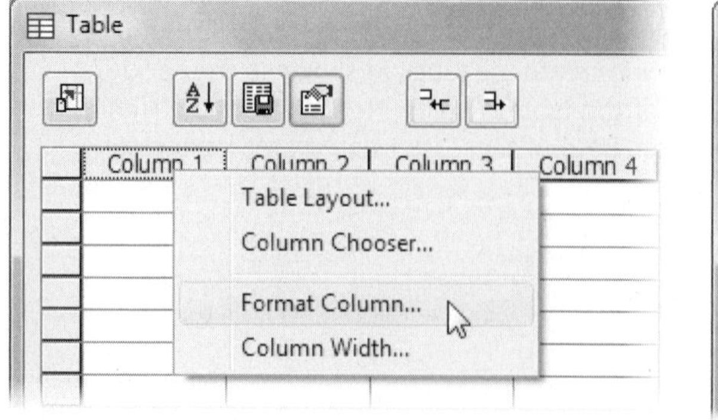

Format Column
Column Format
Property
Heading Parameter Name
Column Width
1.000 in

FIGURE 5.169

4. While still in the Table dialog box you can add information to the individual table cells by double-clicking on a cell and type in new information, like what is shown in Figure 5.170 on the left. Click OK to complete the edit. Figure 5.170 on the right shows the edited table. You can change the column width by dragging the edges of the table.

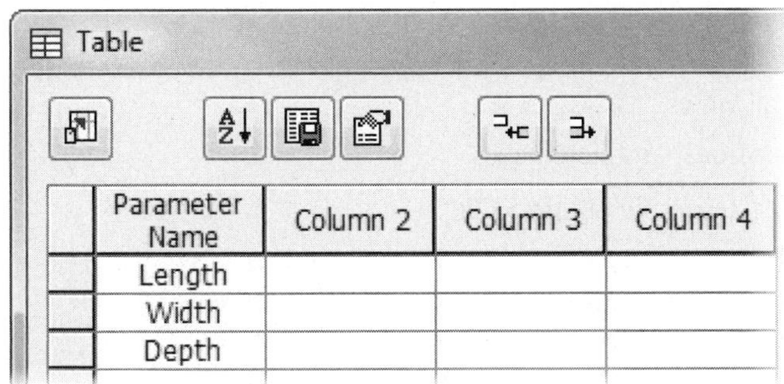

Parameters			
Parameter	Column 2	Column 3	Colu
Length			
Width			
Depth			

FIGURE 5.170

Creating a Table from an Excel Spreadsheet

Tables can also be created from an external Excel spreadsheet using the following steps:

1. To activate the Table command click the General button on the Annotate tab > Table panel.
2. When the Table dialog box appears, click on the Browse for file button to locate the Excel file as the source.
3. In the Table dialog box, specify the start cell and column header row, as shown in Figure 5.171 on the left.
4. Click OK to place the table on the drawing sheet, as shown in Figure 5.171 on the right.

Make sure that the Excel file is saved in the current project path(s). **NOTE**

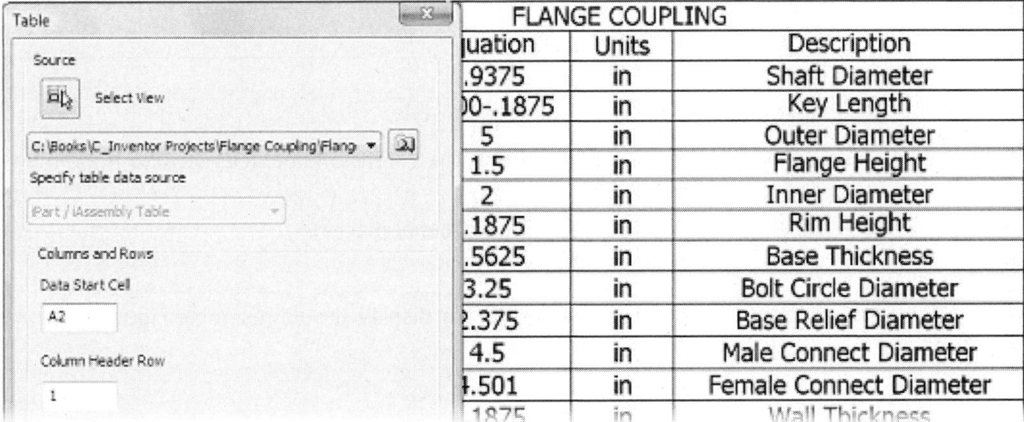

FLANGE COUPLING		
[illegible]uation	Units	Description
.9375	in	Shaft Diameter
[illegible]0-.1875	in	Key Length
5	in	Outer Diameter
1.5	in	Flange Height
2	in	Inner Diameter
.1875	in	Rim Height
.5625	in	Base Thickness
3.25	in	Bolt Circle Diameter
[illegible].375	in	Base Relief Diameter
4.5	in	Male Connect Diameter
[illegible].501	in	Female Connect Diameter

FIGURE 5.171

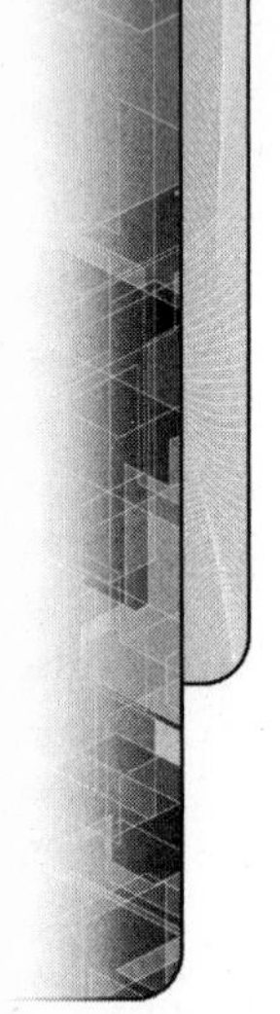

APPLYING YOUR SKILLS

Skill Exercise 5-1

In this exercise, you will create a drawing of a part.

1. Create a new drawing based on the *English ANSI (in).idw* template.
2. Change the sheet size to A.
3. Create drawing views from the file *ESS_Skills_5-1.ipt* from the Chapter 05 folder. Use a scale of 1:1 for all views.
4. Add centerlines, dimensions, and hole notes.

When done your drawing should resemble Figure 5.172.

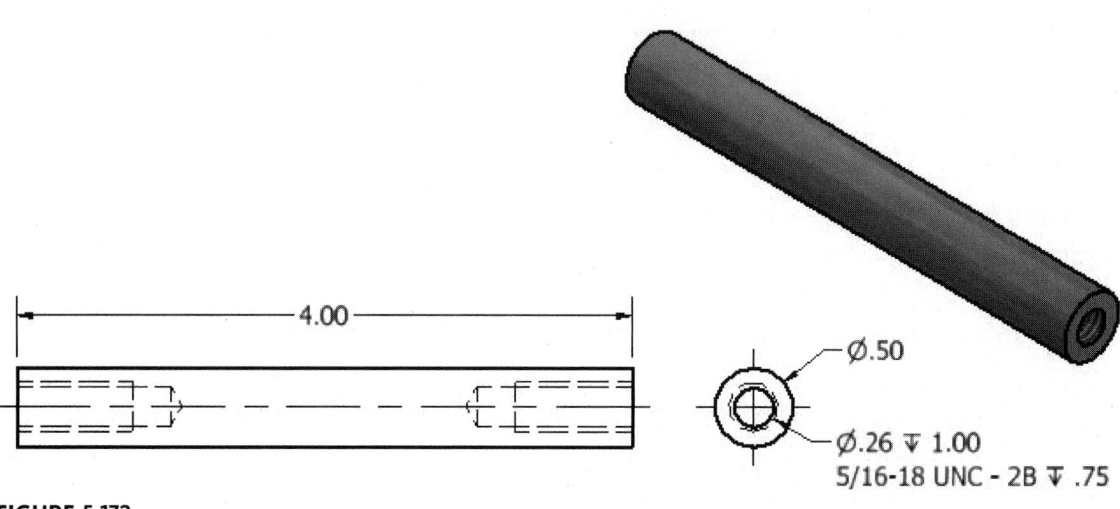

FIGURE 5.172

Skill Exercise 5-2

In this exercise, you will create a drawing for a drain plate cover.

1. Create a new drawing using a *English ANSI (in).dwg template.*
2. Change the sheet size to A.
3. Create three views of the part *ESS_Skills_5-2.ipt* in the Chapter 05 folder. Use a scale of 1:1 for all views.
4. Add center marks to the views.
5. Change the Default (ANSI) dimension style:
   - Change the Extension length that the extension line goes beyond the dimension leader line to **.0625 in**.
   - Set the precision to **3.123** (three decimal places).
   - Set text height to **.100 in**.
6. Add dimensions and annotations. Your display should resemble Figure 5.173.

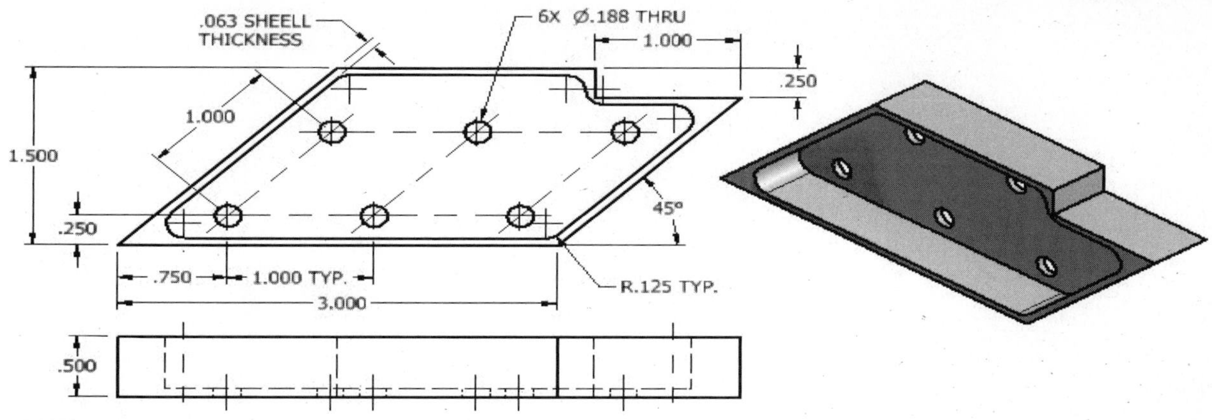

FIGURE 5.173

CHECKING YOUR SKILLS

Use these questions to test your knowledge of the material covered in this chapter.

1. True ___ False ___ A drawing can have an unlimited number of sheets.
2. Explain how to change a sheet's size.
3. True ___ False ___ There can only be one base view per sheet.
4. True ___ False ___ An inclined view is a view that is projected perpendicular to a selected edge or line in a base view.
5. True ___ False ___ An isometric view can only be projected from a base view.
6. True ___ False ___ A section view is a view created by sketching a line or multiple lines that will define the plane(s) that will cut through a part or assembly.
7. True ___ False ___ Drawing dimensions can drive dimensional changes parametrically back to the part.
8. Explain how to shade an isometric drawing view.
9. True ___ False ___ When creating a hole note using the Hole/Thread Notes command, circles that are extruded to create a hole can be annotated.
10. Explain how to open a part file from within a drawing.
11. True ___ False ___ When you create a title block in a drawing, the new title block will automatically be written back to the template file.
12. True ___ False ___ When creating a section view that goes through holes at different angles, you use the projected option to prevent the holes from being distorted.
13. True ___ False ___ A hatch pattern in a section view is changed via the Document Settings.
14. True ___ False ___ Create a Segment View to create a section view that has no depth.
15. True ___ False ___ When creating a Detail View you can use a circular or rectangular fence.
16. True ___ False ___ Use the Break Out command to remove a middle section from a drawing view of long part.
17. True ___ False ___ The Styles and Standards Editor only controls dimension styles.
18. True ___ False ___ The Centerline Bisector command can only be used to create a centerline between two parallel lines.
19. True ___ False ___ Use the General Table command to insert data from an Excel file into a drawing.
20. True ___ False ___ After creating dimensions with the Baseline Dimension command you cannot change the origin.

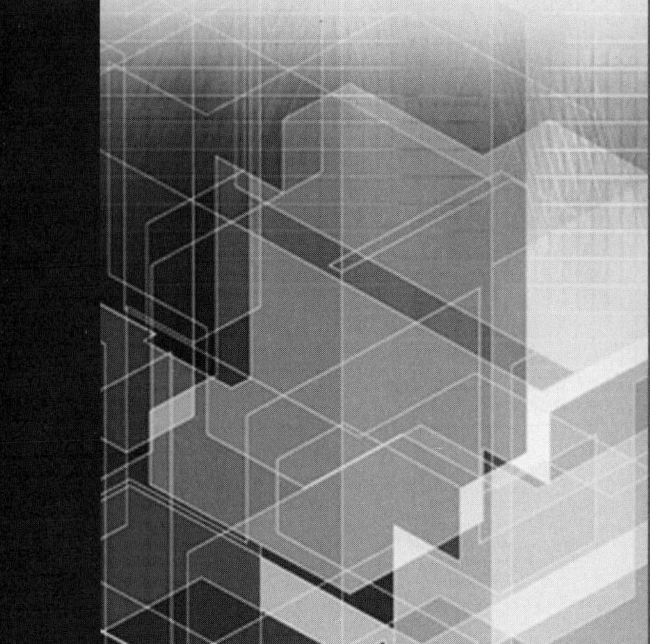

CHAPTER 6

Creating and Documenting Assemblies

INTRODUCTION

In the first four chapters, you learned how to create a component in its own file. In this chapter, you will learn how to place individual component files into an assembly file. You will also learn to create components in the context of the assembly file. After creating components, you will learn how to constrain the components to one another using assembly constraints, edit the assembly constraints, check for interference, and create presentation files that show how the components are assembled or disassembled. Manipulating and editing a Bill of Materials (BOM) is also discussed, including the placement of a Parts List and identifying balloons.

OBJECTIVES

After completing this chapter, you will be able to:

- Understand the assembly options
- Place components into an assembly
- Create components and assemblies
- Create subassemblies
- Constrain components together using assembly constraints
- Edit assembly constraints
- Pattern components in an assembly
- Check parts in an assembly for interference
- Drive constraints
- Create a presentation file
- Manipulate and edit the Bill of Materials (BOM)
- Create individual and automatic balloons
- Create and perform edits on a Parts List of an assembly

CREATING ASSEMBLIES

As you have already learned, component files have the *.ipt* extension, and they can only have one component each. In this chapter, you will learn how to create assembly files (*.iam* file extension). An assembly file holds the information needed to assemble the components together. All of the components in an assembly are *referenced in*, meaning that each component exists in its own component IPT file, and its definition is linked into the assembly. You can edit the components while in the assembly, or you can open the component file and edit it. When you have made changes to a component and saved the component, the changes will be reflected in the assembly after you open or update it. There are three methods for creating assemblies: bottom-up, in-place, and a combination of both. *Bottom-up* refers to an assembly in which all of the components were created in individual component files and are referenced into the assembly. The *in-place* approach refers to an assembly in which the components are created from within the context of the assembly. In other words, the user creates each component from within the top-level assembly. Each component in the assembly is saved to its own IPT file. The following sections describe the bottom-up and in-place assembly techniques.

Note that assemblies are not made up only of individual parts. Complex assemblies are typically made up of subassemblies, and therefore you can better manage the large amount of data that is created when building assemblies.

To create a new assembly, click New on the Get Started tab, and then click the Standard.iam icon, as shown in Figure 6.1. Alternately, you can click the down arrow of the Inventor icon located in the upper-left corner and click the New icon. After issuing the new assembly operation, Autodesk Inventor's commands will change to reflect the new assembly environment. The assembly commands appear on the ribbon, and these commands are covered throughout this chapter.

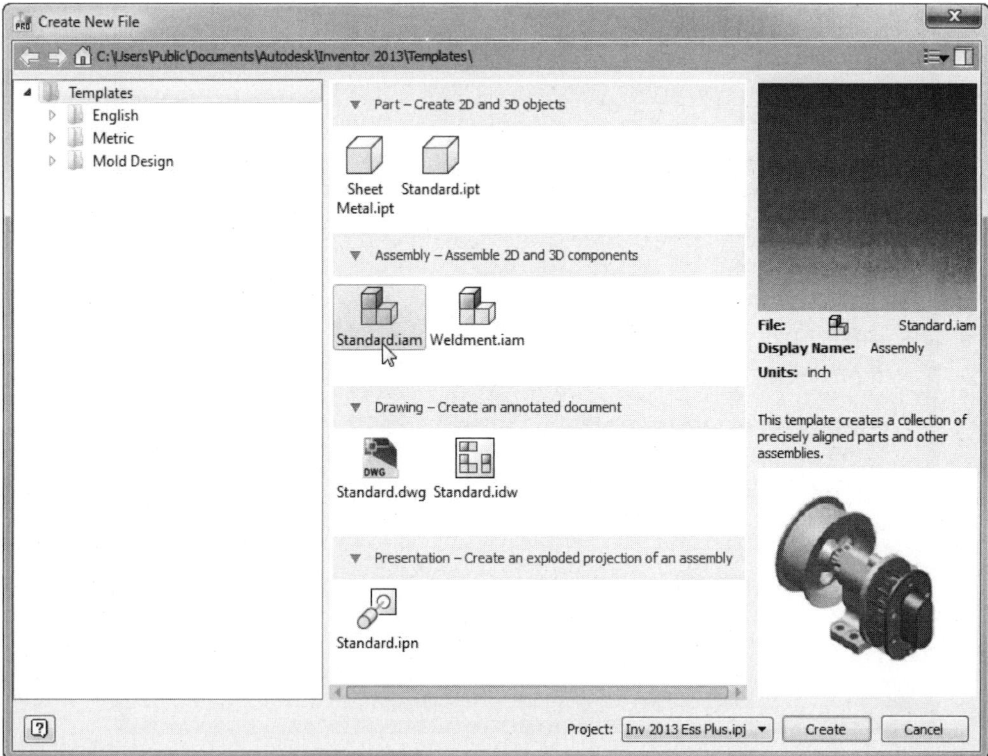

FIGURE 6.1

NOTE

There are a number of valid approaches to creating an assembly. You will determine which method works best for the assembly that you are creating based on experience. You can create an assembly using the bottom-up technique, the in-place technique, or a combination of both. Whether you place or create the components in the assembly, all of the components will be saved to their own individual IPT files, and the assembly will be saved as an IAM file.

Assembly Options

Before creating an assembly, review the assembly option settings. On the Tools tab, click Application Options, and the Application Options dialog box will appear. Click on the Assembly tab, as shown in Figure 6.2. Consult the help system for information about the Assembly Application Options. These settings are global and will affect how new components are created, referenced, analyzed, or placed in the assembly.

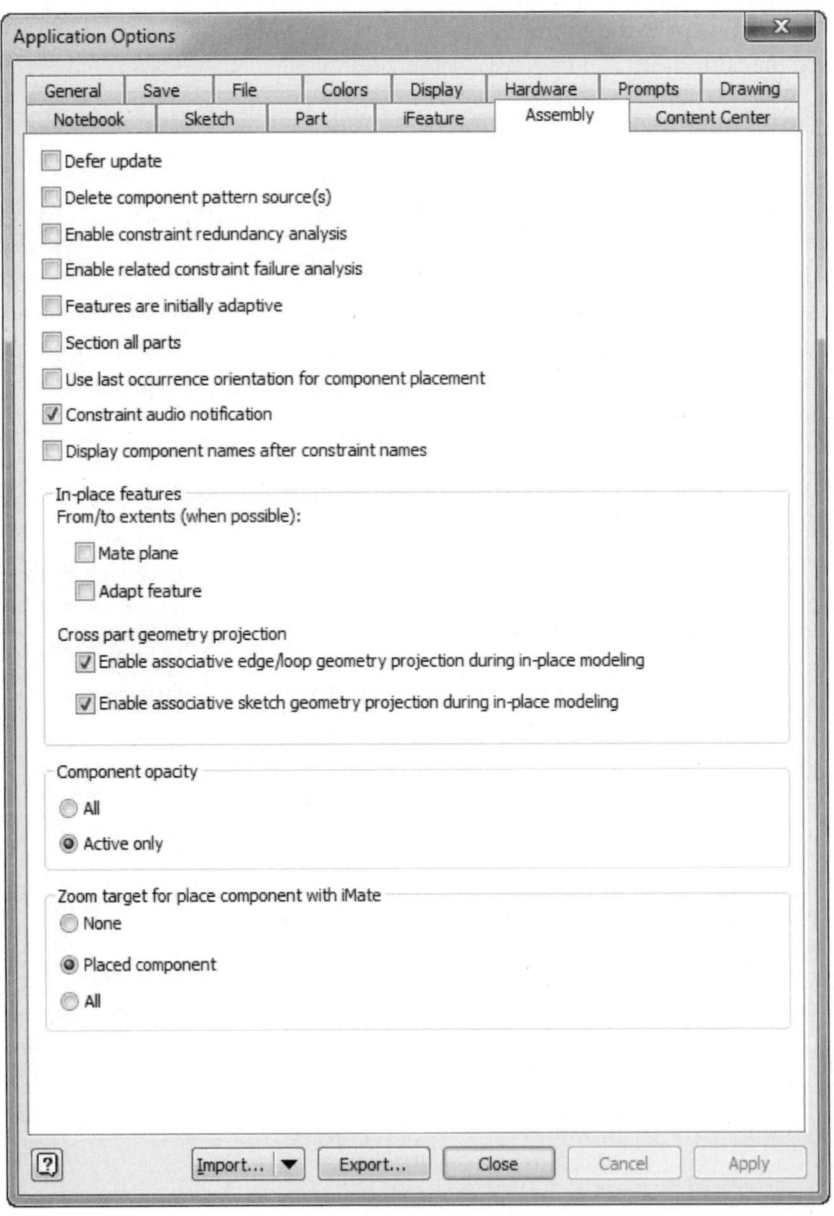

FIGURE 6.2

Placing Components

To place a component(s) you create the components in their individual files and then place them into an assembly. If you place an assembly file into another assembly, it will be brought in as a subassembly. Be sure to include the path(s) for the file location(s) of the placed components in the project file; otherwise, Autodesk Inventor may not be able to locate the referenced component when you reopen the assembly.

To insert a component into the current assembly, click the Place Component command on the Assemble tab, as shown in Figure 6.3 on the left, press the shortcut key P, or right-click in the graphics window, and select Place Component from the marking menu.

The Place Component dialog box will appear, as shown in Figure 6.3. Select the component to place or select multiple components by holding down the CTRL or SHIFT key and select the components from the list. If the component you are placing is the first in the assembly, an instance of it will be placed into the assembly automatically. If the component is not the first, you pick a point in the assembly where you want to place the component. If you need multiple occurrences of the component in the assembly, continue selecting placement points. When done, press the ESC key, or right-click and select Done from the menu.

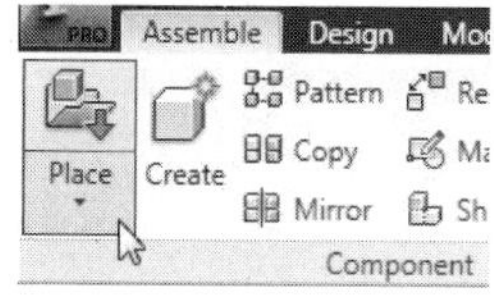

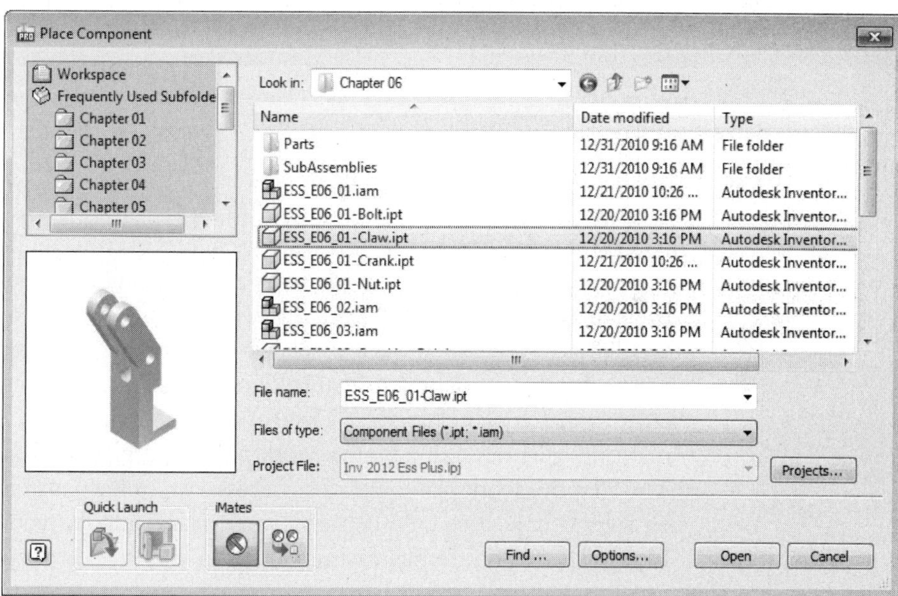

FIGURE 6.3

When placing or creating components in an assembly, it is recommended to list them in the order in which they are assembled. The order is important when placing assembly constraints. To reorder a component, in the browser click and drag on a component to move it up or down.

CREATING PARTS IN PLACE

Most components in the assembly environment are created in relation to existing components in the assembly. When creating an in-place component, you can sketch on the face of an existing assembly component or a work plane. Click Application Options on the Tools tab, and then click on the Part tab to set the default sketch plane.

When you create a new component, you can select an option in the Create In-Place Component dialog box to constrain the sketch plane to the selected face or work plane automatically, as shown in Figure 6.4. After you specify the location for the sketch, the new part immediately becomes active, and the browser and ribbon switch to the part environment.

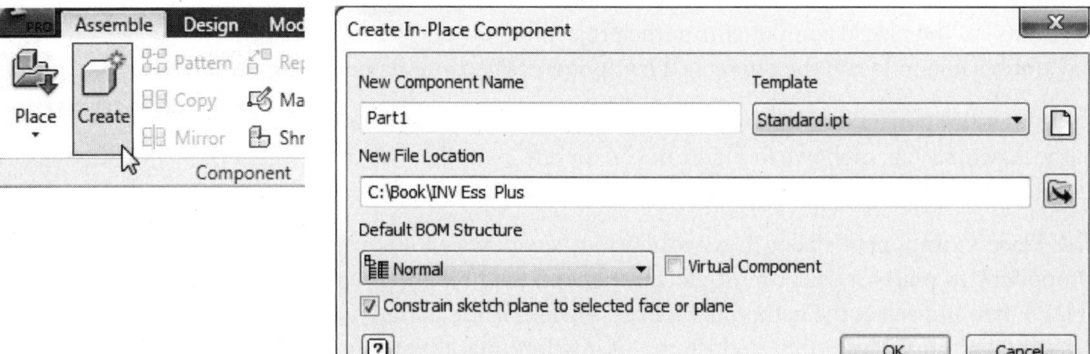

FIGURE 6.4

Notice also that the Sketch tab, as shown in Figure 6.5, is available to create sketch geometry in Sketch1 of your new part.

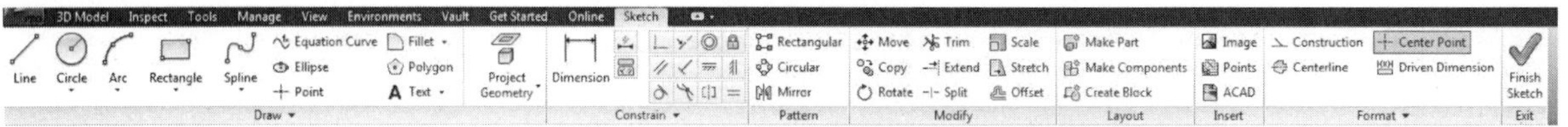

FIGURE 6.5

After you create the base feature of your new part, you can define additional sketches based on the active part or other parts in the assembly. When defining a new sketch, you can click a planar face of the active part or another part to define the sketch plane on that face. You can also click a planar face and drag the sketch away from the face to create the sketch plane automatically on the resulting offset work plane. When you create a sketch plane based on a face of another component, Autodesk Inventor automatically generates an adaptive work plane and places the active sketch plane on it. The adaptive work plane moves as necessary to reflect any changes in the component on which it is based. When the work plane adapts, your sketch moves with it. Features based on the sketch then adapt to match its new position.

After you finish creating a new part, you can return to assembly mode by double-clicking the assembly name in the browser. In assembly mode, assembly constraints become visible in the browser. If you selected the Constrain Sketch Plane to Selected Face option when you created your new part, a flush constraint will appear in the Assembly browser. As with all constraints, you can delete or edit this constraint at any time. No flush constraint is generated if you create a sketch by clicking in the graphics window or if the box is clear when selecting an existing part face.

The Assembly Browser

The Assembly browser displays the hierarchy of all parts and subassembly occurrences and assembly constraints in the assembly, as shown in Figure 6.6. Each occurrence of a component is represented by a unique name. In the browser, you can select a component for editing, move components between assembly levels, reorder assembly components, control component status, rename components, edit assembly constraints, and manage design views and representations.

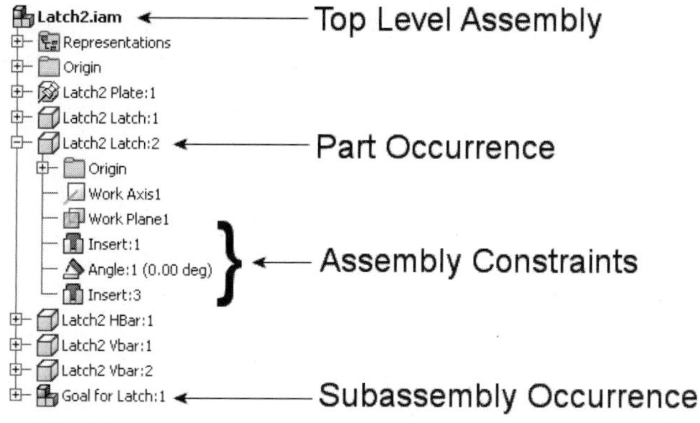

FIGURE 6.6

Occurrences

An occurrence, or instance, is a copy of an existing component and has the same name as the original component with a colon and a sequenced number. If the original component is named Bracket, for example, the first occurrence in the assembly will be Bracket:1 and a subsequent occurrence will be Bracket:2, as shown in Figure 6.7. If the original component changes, all of the component instances will reflect the change.

To create an occurrence, place the component. If the component already exists in the assembly, you can click the component's icon in the browser, and drag an additional occurrence into the assembly. You can also use the copy-and-paste method to place additional components. You right-click on the component name in the browser, or on the component itself in the graphics window, and then select Copy from the menu. Then right-click and select Paste from the menu. You can also use the Windows shortcuts CTRL-C and CTRL-V to copy and paste the selected component. If you want an occurrence of the original component to have no relationship with its source component, make the original component active, use the Save Copy As command. This command is found by clicking the Inventor Application button located in the upper left corner of the display screen. Then click the arrow next to Save As and click Save Copy As, where you will enter a new name. The new component will have no relationship to the original, and you can place it in the assembly using the Place Component command.

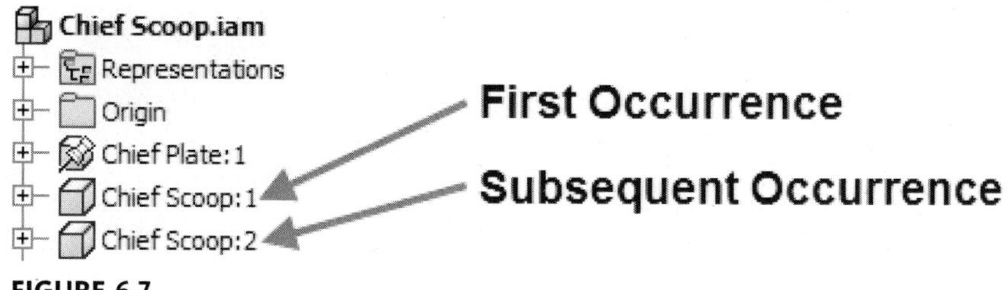

FIGURE 6.7

The Assembly Capacity Meter

To provide feedback on the resources used by an assembly model, use the Assembly Capacity Meter. This meter is located in the lower-right corner of the display screen and is present only in the assembly environment. Three different types of information are displayed in the status bar of this meter. The first numbered block deals with the number of occurrences or instances in the active assembly. The second numbered block displays the total number of files open in order to display the assembly.

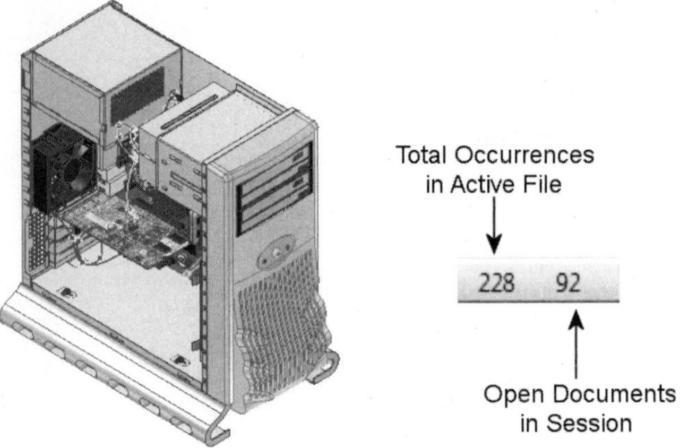

FIGURE 6.8

EDITING A COMPONENT IN PLACE

To edit a component while in an assembly, activate the component. Only one component in the assembly can be active at a time. To make a component active, double-click on the component in the graphics window, or double-click on the file name or icon in the browser. Alternately, right-click on the component name in the browser or graphics window, and select Edit from the menu. Once the component is active, the other component names in the browser will appear shaded, as shown in Figure 6.9. If Component Opacity, in Application Options on the Assembly tab, is set to Active Only, the other components in the assembly will take on a faded appearance in the graphics window.

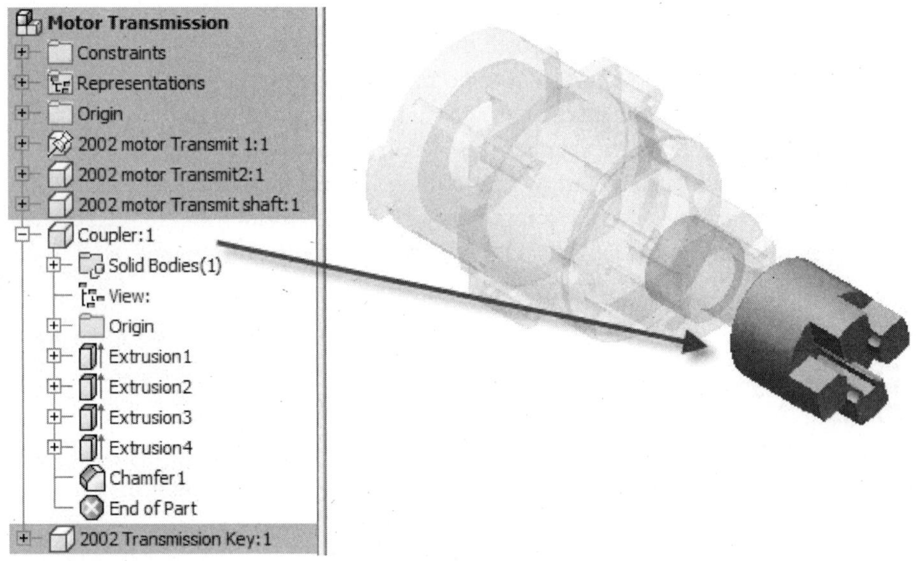

FIGURE 6.9

You can edit the component and save the changes by using the Save command. Only the active component will be saved. To make the assembly active, click the Return button on the Return panel, as shown in Figure 6.10 on the left which will take you up one level, double-click on the assembly name in the browser, or right-click in the graphics window and select Finish Edit from the marking menu, as shown in Figure 6.10 on the right. Under the Return button there are two other buttons, as shown in the middle of Figure 6.10; Return to Parent that will return you up one level (from the part level to the part subassembly) and Return to Top that will return you to the top level in the browser no matter how deep you are in the browser tree.

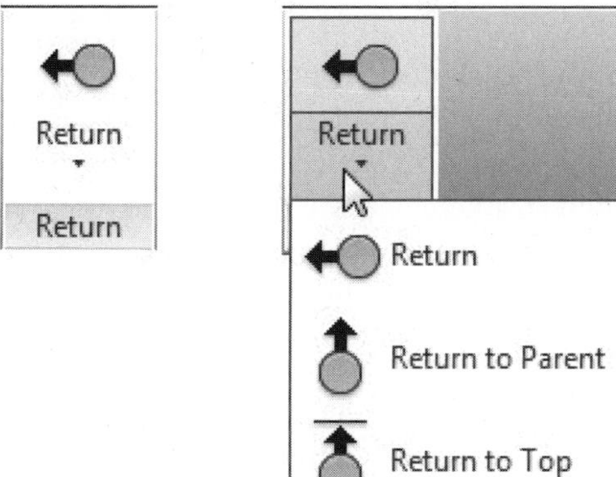

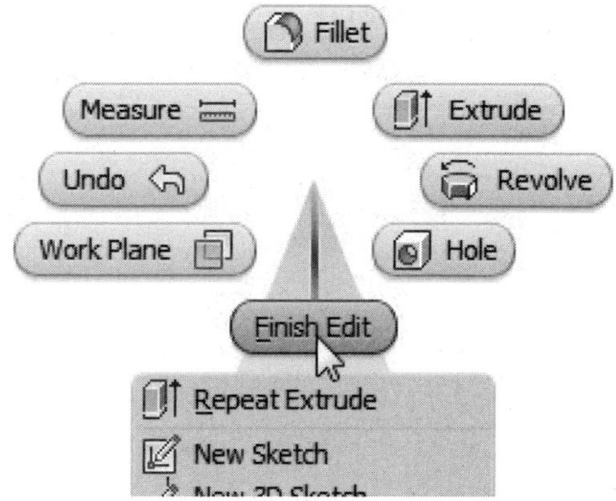

FIGURE 6.10

OPENING AND EDITING ASSEMBLY COMPONENTS

Another way to edit a component in the assembly is to open the component in another window. Click the Open command on the Inventor icon, or right-click on the component's name in the browser or on the component in the graphics window. Select Open from the menu, as shown in Figure 6.11. The component will appear in a new window. Edit the component as needed, save the changes, activate the assembly file, and the changes will appear in the assembly.

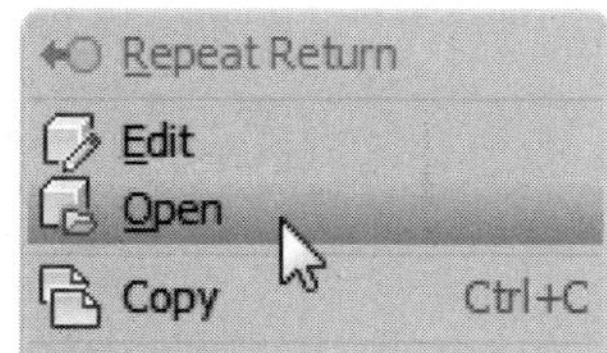

FIGURE 6.11

Grounded Components

When assembling components, you may want to make a component or multiple components grounded or stationary, meaning that they will not move. When applying assembly constraints, the unconstrained components will be moved to the grounded component(s). By default, the first component placed in an assembly is grounded. There is no limit to how many components can be grounded. It is strongly recommended that at least one component in the assembly be grounded; otherwise, the assembly can move. A grounded component is represented with a pushpin superimposed on its icon in the browser, as shown in Figure 6.12 on the left. To ground or unground a component, right-click on the component's name in the browser, and select or deselect Grounded from the menu, as shown in Figure 6.12 on the right.

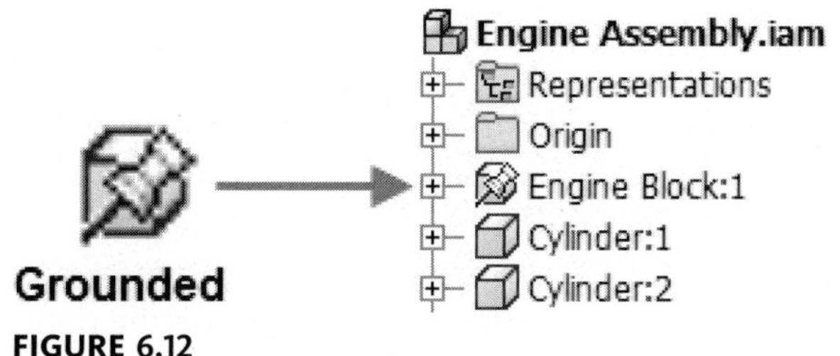

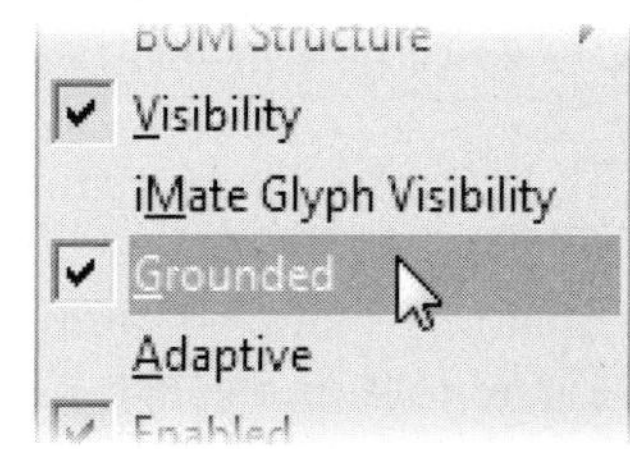

FIGURE 6.12

DEGREES OF FREEDOM (DOF)

You have now learned how to create assembly files, but the components had no relationship to one another except for the relationship that was defined when you created a component from the context of an assembly and in reference to a face on another component. For example, if you placed a bolt in a hole and the hole moved, the bolt would not move to the new hole position. Use assembly constraints to create relationships between components. With the correct constraint(s) applied, if a hole moves, the bolt will move to the new hole location.

In a previous chapter, you learned about geometric constraints. When you apply geometric constraints to sketches, they reduce the number of dimensions or constraints required to fully constrain a profile. When you apply assembly constraints, they reduce the degrees of freedom (DOF) that allow the components to move freely in space.

There are six degrees of freedom: three are translational and three are rotational. Translational means that a component can move along an axis X, Y, or Z. Rotational means that a component can rotate about an axis X, Y, or Z. As you apply assembly constraints, the number of the DOF decreases.

To see a graphical display of the DOF remaining on all of the components in an assembly, select Degrees of Freedom on the View tab, as shown in Figure 6.13.

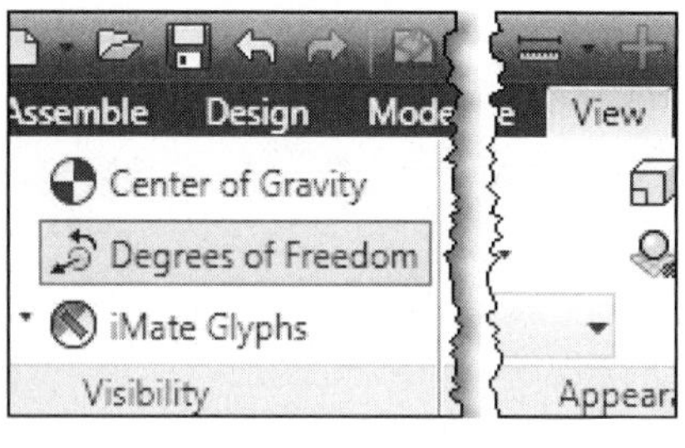

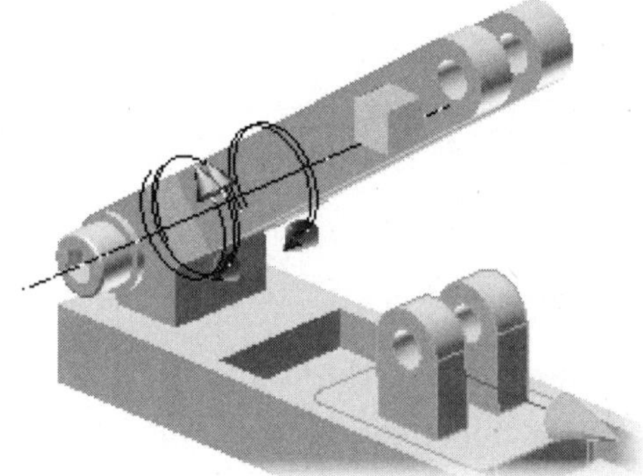

FIGURE 6.13

An icon will appear in the center of the component that shows the DOF remaining on the component. The line and arrows represent translational freedom, and the arc and arrows represent rotational freedom. To turn off the DOF icons, again click Degrees of Freedom on the View menu.

TIP

You can turn on the DOF symbols for single or multiple component(s) by right-clicking on the component's name in the canvas or in the browser, clicking iProperties from the menu, and clicking Degrees of Freedom on the Occurrence tab. If the symbols are turned on, following the same steps will toggle them off.

To better see the remaining degrees of freedom (DOF) you can animate them. To animate the remaining degrees of freedom (DOF) click the Degree of Freedom Analysis command from the Assemble tab > Productivity panel as shown in Figure 6.14 on the left. The Degree of freedom Analysis dialog box will appear that lists the remaining degrees of freedom (DOF) as shown in Figure 6.14 on the right. To animate components remaining degrees of freedom (DOF) check the Animate Freedom option and then select a component from the list in the dialog box.

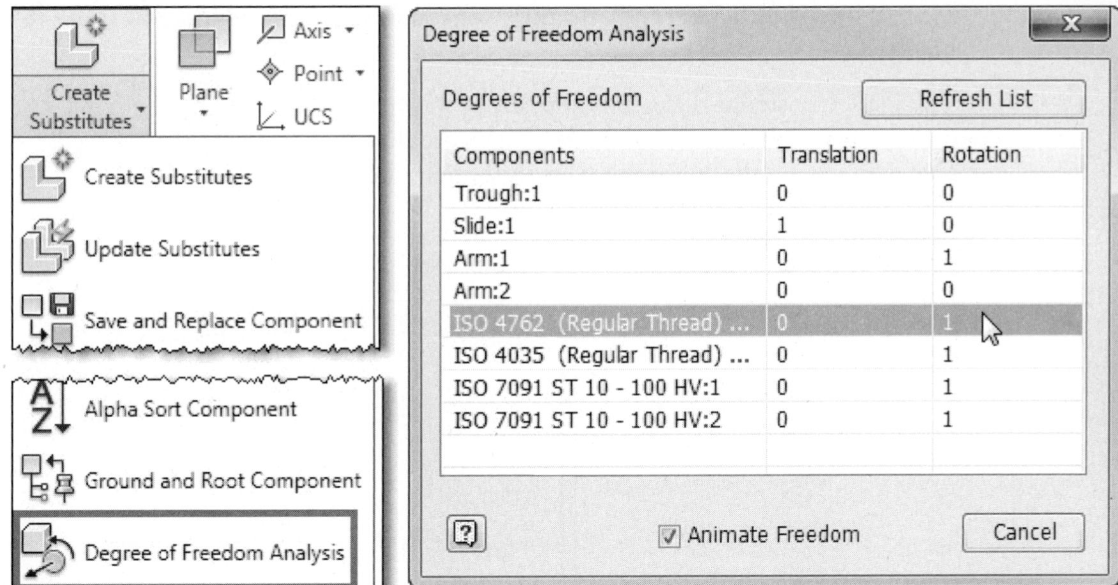

FIGURE 6.14

ASSEMBLY CONSTRAINTS

When constraining components to one another, you will need to understand the terminology. The following terminology is used with assembly constraints:

Line. This can be the centerline of an arc, a circular edge, a cylindrical surface, a selected edge, a work axis, or a sketched line.

Normal. This is a vector that is perpendicular to a planar face.

Plane. This can be defined by the selection of a plane or face to include the following: two noncollinear but coplanar lines or axes, three points, or one line or axis and a point that does not lie on the line or axis. When you use edges and points to select a plane, this creates a work plane, and it is referred to as a construction plane.

Point. This can be an endpoint or midpoint of a line, the center or end of an arc or circular edge, or a vertex created by the intersection of an axis and a plane or face.

Offset. This is the distance between two selected lines, planes, or points, or any combination of the three.

Autodesk Inventor does not require components to be fully constrained. By default, the first component created or added to the assembly will be grounded and will have zero DOF. As discussed earlier, more than one component can be grounded. Other components will move in relation to the grounded component(s).

Assembly Constraint Types

Autodesk Inventor uses four types of assembly constraints (mate, angle, tangent, and insert), two types of motion constraints (rotation and rotation-translation), a transitional constraint, and a constraint set. You can access the constraints through the Constraint command found on the Assemble tab, as shown in Figure 6.15 on the left, by right-clicking and selecting Constraint from the menu, or by using the hot key C.

The Place Constraint dialog box appears, as shown in Figure 6.15 on the right. The dialog box is divided into four areas, which are described in the following sections. Depending upon the constraint type, the option titles may change.

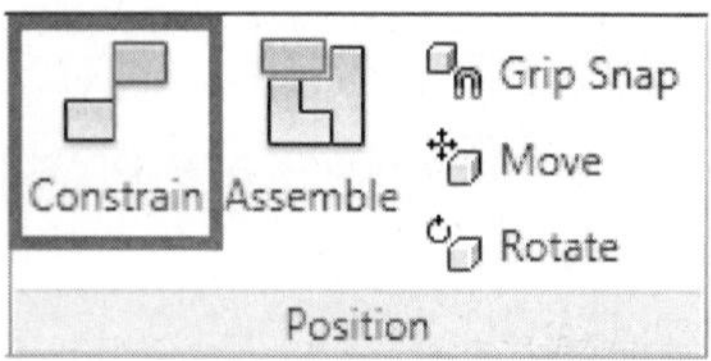

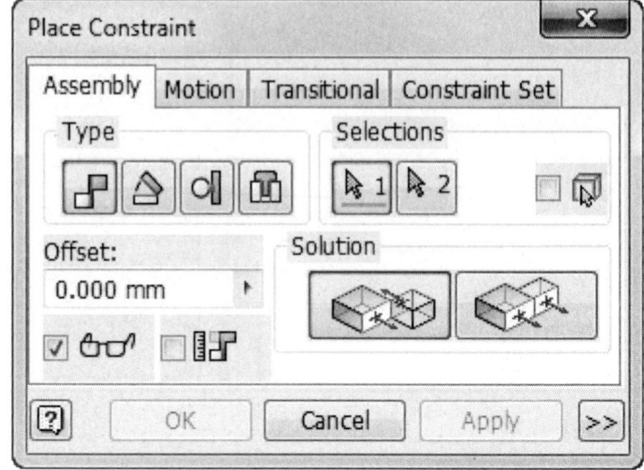

FIGURE 6.15

Assembly Tab

Type. Select the type of assembly constraint to apply: mate, angle, tangent, or insert.

Selections. Click the button with the number 1, and select a component's edge, face, point, and so on, on which to base the constraint type. Then click the button with the number 2, and select a component's edge, face, point, and so on, on which to base the constraint type. By default, the second arrow will become active after you have selected the first input.

Color coding is also available to assist with the assembly process. For example, when picking a face with the number 1 button, the color blue is associated with this selection. In the same way, the color green is associated with the number 2 button selection. This schema allows you to better recognize the selections, especially if they need to be edited.

You can edit an edge, face, point, and so on, of an assembly constraint that has already been applied by clicking the number button that corresponds to the constraint and then selecting a new edge, face, point, and so on. While working on complex assemblies, you can click the box on the right side of the Selections section called Pick part first. If the box has a check, select the component before selecting a component's edge, face, point, and so on.

Offset/Angle. Enter or select a value for the offset or angle. The Offset option changes to Angle when the Angle constraint is being applied.

Solution. Select how the constraint will be applied; the normals will be pointing in the same or opposite directions.

Show Preview. Click and when constraints are applied to two components, you will see the under-constrained components previewed in their constrained positions. If you leave the box clear, you will not see the components assembled until you click the Apply button.

Predict Offset and Orientation. Click to display the existing offset distance between two components. This allows you to accept this offset distance or enter a new offset distance in the edit box.

Motion Tab

Type. Select the type of assembly constraint to apply: rotation or rotation-translation.

Selections. Click the button with the number 1 and select a component's face or axis on which to base the constraint type. You will see a glyph in the graphics window previewing the direction of rotation motion. Click the button with the number 2, and select the component's axis or face on which to base the constraint type. A second glyph appears showing the direction of rotation.

Ratio. Enter or select a value for the ratio.

Solution. Select how the constraint type will be applied. The components will rotate in the same or opposite directions as previewed by the graphics window glyphs.

Transitional Tab

A transitional constraint will maintain contact between the two selected faces. You can use a transitional constraint between a cylindrical face and a set of tangent faces on another part.

Type. Select transitional as the type of assembly constraint to apply.

Selections. Click the First Selection button, and select the first face on the part that will be moving. Click the Second Selection button, and select a face around which the first part will be moving. If there are tangent faces, they will become chained automatically as part of the selected face.

Constraint Set Tab

This option will constrain two UCSs together.

Type. Select the type of assembly constraint to apply: UCS to UCS.

Selections. Select two UCSs on different parts.

Constraint Limits

While placing assembly constraints the geometry may not perfectly fit with the value of the constraint but may fit within a tolerance that is OK. By setting constraint limits you specify the distance or rotation that the geometry can deviate from an exact location without having the constraint fail. To set a limit for a constraint click the More button >> on the lower-right corner of the Place Constraint dialog box and the Limits section will appear as shown in Figure 6.16 on the left. An assembly constraint that has limits will appear in the browser with a +/− symbol appended to its name as shown in Figure 6.16 on the right.

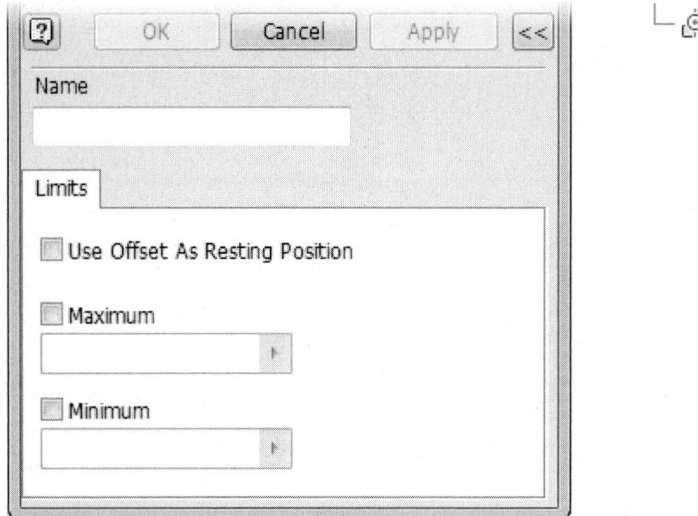

FIGURE 6.16

Name. Enter a unique name for the constraint in the browser. If left blank a default name is used.

Use Offset or Angle as Resting Position. This option sets a value as the default position of a constraint with limits. When checked, the value entered in either the Offset or Angle field is used as the resting position.

Maximum. This option sets the maximum value that a constraint can move or rotate.

Minimum. This option sets the minimum value that a constraint can move or rotate.

Assembly Constraint Types

This section explains each of the assembly constraint types.

Mate

There are three types of mate constraints: plane, line, and point.

Mate Plane. The mate plane constraint assembles two components so that the faces on the selected planes will be planar to and opposing one another. In Figure 6.17, the mate condition is being applied; it is selected in the Solution area of the Place Constraint dialog box.

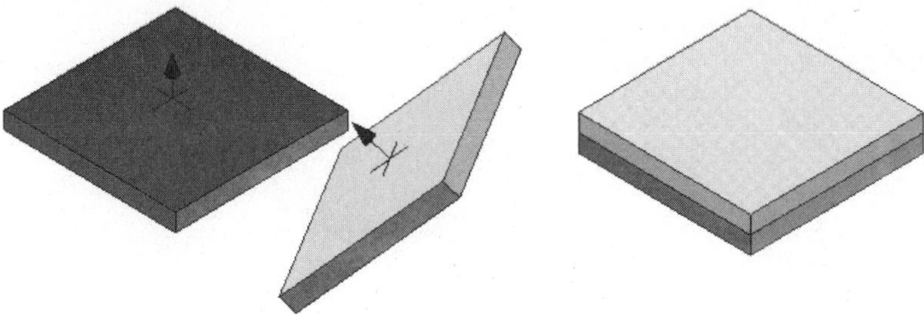

FIGURE 6.17

Mate Line. The mate line constraint assembles the edges of lines to be collinear, as shown in Figure 6.18.

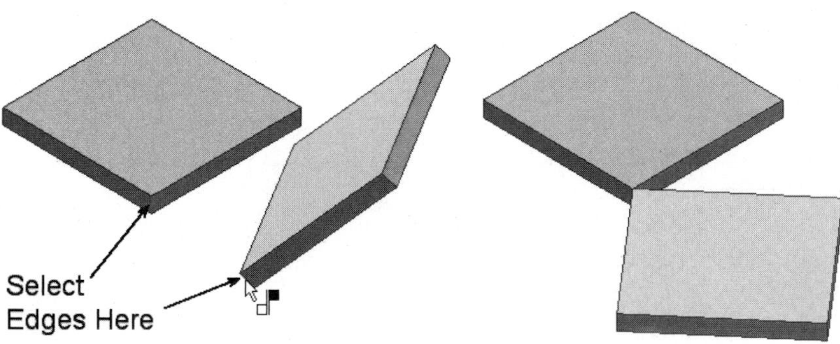

FIGURE 6.18

You can also use the mate line constraint to assemble the center axis of a cylinder with a matching hole feature or axis, as shown in Figure 6.19.

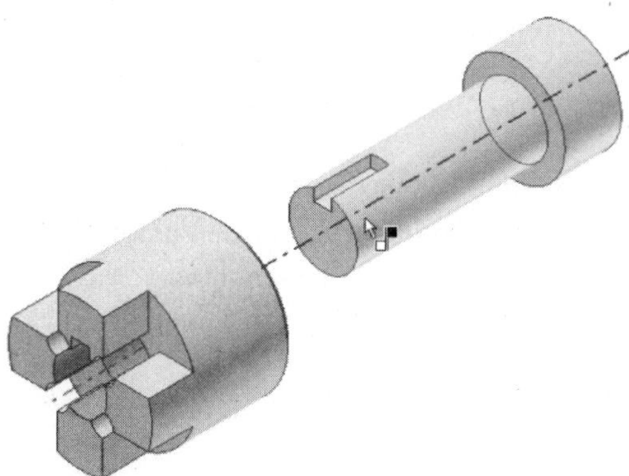

FIGURE 6.19

Mate Point. The mate point constraint assembles two points, such as centers of arcs and circular edges, endpoints, and midpoints, to be coincident, as shown in Figure 6.20.

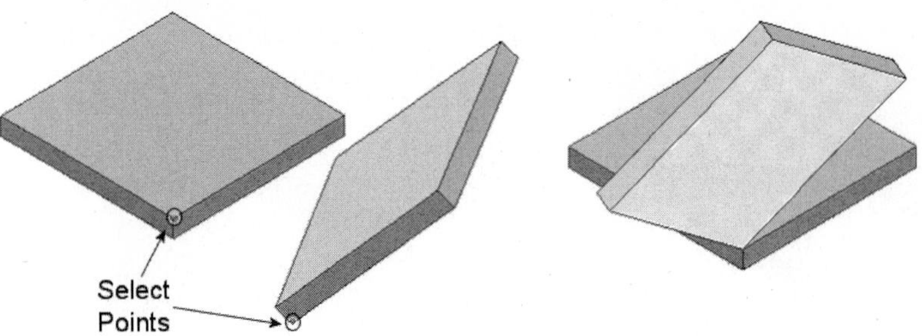

FIGURE 6.20

Mate Flush Solution

The mate flush solution constraint aligns two components so that the selected planar faces or work planes face the same direction or have their surface normals pointing in the same direction, as shown in Figure 6.21. Planar faces are the only geometry that can be selected for this constraint.

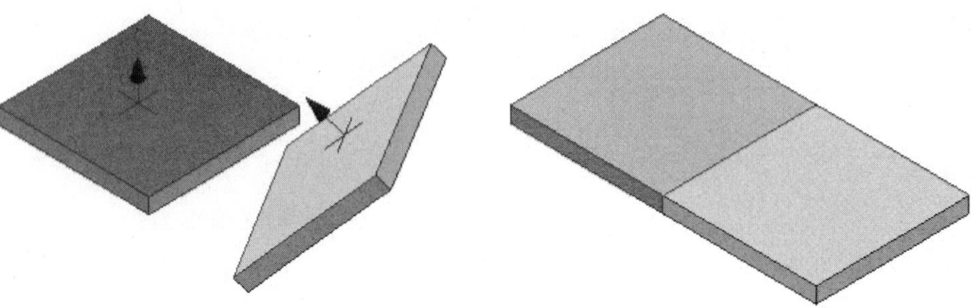

FIGURE 6.21

Angle

The angle constraint specifies the degrees between selected planes or faces or axes. Figure 6.22 shows the angle constraint with two planes selected and a 30° angle applied.

Three solutions are available when placing an angle constraint: directed angle, undirected angle, and Explicit Reference Vector. The Explicit Reference Vector option requires a third selection that defines the Z axis. You can experiment with the solutions, especially when driving the angle constraint and observing the behavior of the assembly.

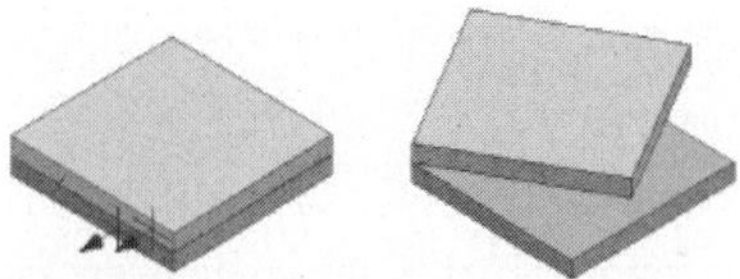

FIGURE 6.22

Tangent

The tangent constraint defines a tangent relationship between planes, cylinders, spheres, cones, and ruled splines. At least one of the faces selected needs to be a curve, and you can apply the tangency to the inside or outside of the curve.

Figure 6.23 shows the tangent constraint applied to one outside curved face and a selected planar face, as well as the piece with the outer and inner solutions applied.

FIGURE 6.23

Insert

The insert constraint takes away five degrees of freedom (DOF) with one constraint, but it only works with components that have circular edges. Select the circular edges of two different components. The centerlines of the selected circles or arcs will be aligned, and a mate constraint will be applied to the planes defined by the circular edges. Circular edges define a centerline/axis and a plane. Figure 6.24 shows the insert constraint with two circular edges selected and the opposed solution applied.

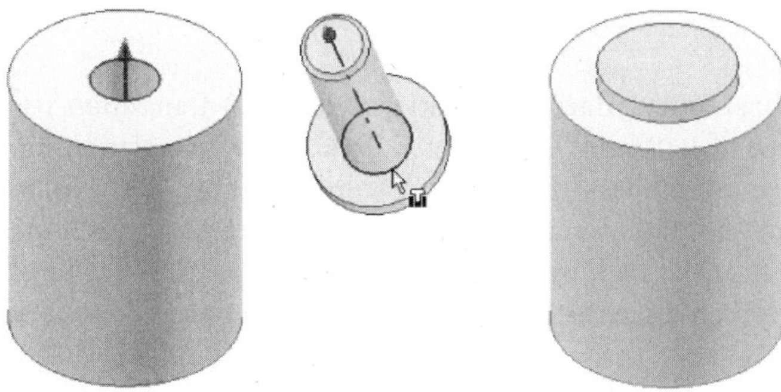

FIGURE 6.24

Motion Constraint Types

There are two types of motion constraints: rotation and rotation-translation, as shown in the Type section of Figure 6.25. Motion constraints allow you to simulate the motion relationships of gears, pulleys, rack and pinions, and other devices. By applying motion constraints between two or more components, you can drive one component and cause the others to move accordingly.

Both types of motion constraints are secondary constraints, which means that they define motion but do not maintain positional relationships between components. Constrain your components fully before you apply motion constraints. You can then suppress constraints that restrict the motion of the components you want to animate.

Rotation

The rotation constraint defines a component that will rotate in relation to another component by specifying a ratio for the rotation between the two components. Use this constraint for showing the relationship between gears and pulleys. Selecting the tops of the gear faces displays the rotation glyph, as shown in Figure 6.25. You may also have to change the solution type from Forward to Backward, depending on the desired results.

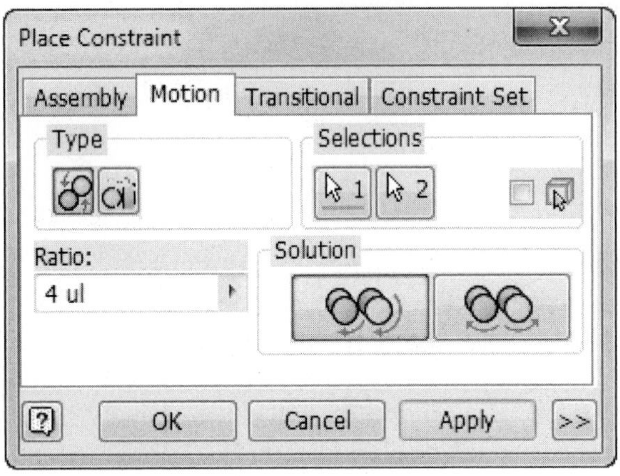

FIGURE 6.25

Rotation-Translation

The rotation-translation constraint defines the rotation relative to translation between components. This type of constraint is well suited for showing the relationship between rack and pinion gear assemblies. In a rack and pinion assembly, as shown in Figure 6.26, the top face of the pinion and one of the front faces of the rack are selected. You supply a distance the rack will travel based on the pitch diameter of the pinion gear, and then you can drive the constraints and test the travel distance of the mechanism.

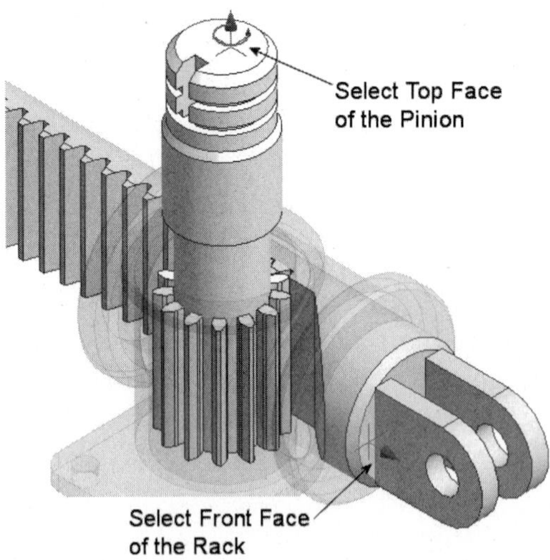

FIGURE 6.26

Creating a Transitional Constraint

The transitional constraint specifies the intended relationship between, typically, a cylindrical part face and a contiguous set of faces on another part, such as a cam follower in a cam slot. The transitional constraint maintains contact between the faces as you slide the component along open DOF. Access this constraint type through the Transitional tab of the Place Constraint dialog box, as shown in Figure 6.27.

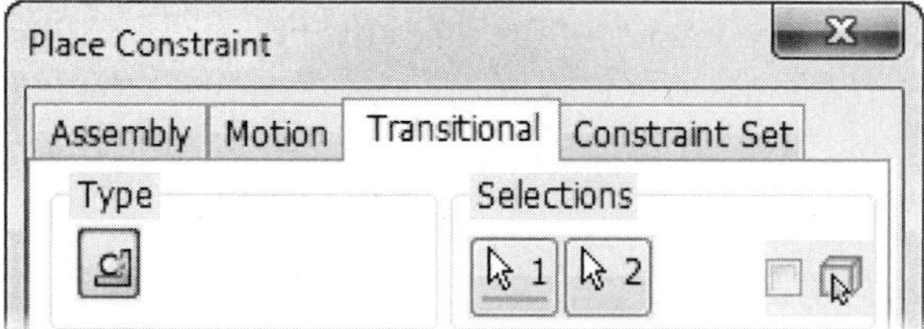

FIGURE 6.27

Select the moving face first on the cam as shown in Figure 6.28 on the left. Next, select the transition face, as shown in Figure 6.28 in the middle. The transitional face will now contact and follow the cam rotation, as shown on the right in Figure 6.28.

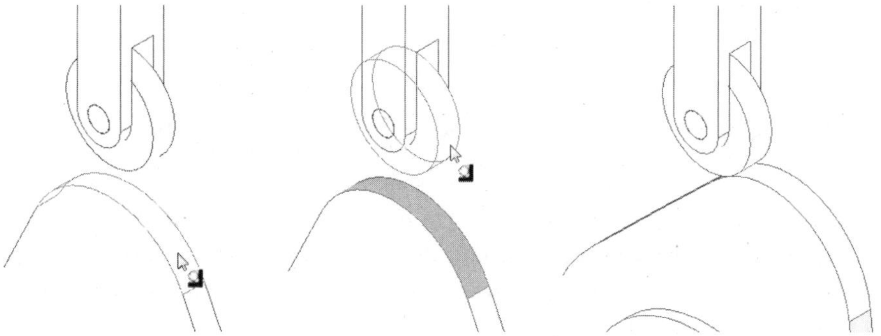

FIGURE 6.28

Creating a Constraint Set

If User Coordinate Systems were defined in individual part or assembly files, these UCSs can be constrained together. The buttons found under the Type and Selections areas of the dialog box allow for the selecting of individual UCSs and having the constraints be applied to the selections.

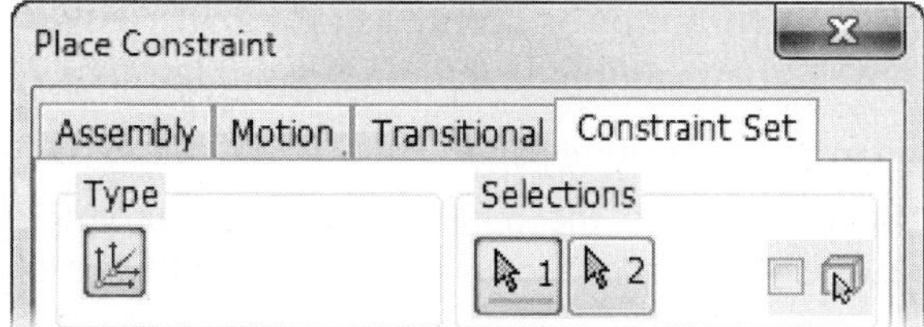

FIGURE 6.29

Selecting Geometry and the Select Other Command

After selecting the type of assembly constraint that you want to apply, the Selections button with the number 1 will become active; if it does not automatically become active, click the button. Position the cursor over the face, edge, point, and so on, to apply the first assembly constraint.

You may need to cycle through the selection set using the Select Other command, as shown in Figure 6.30, until the correct location is highlighted. Cycle by clicking on the down arrow of the command. If you select an option in the menu (list), then that geometry is selected. To preview the geometry in the list, hover over the option from the menu until you see the desired constraint condition, and then press the left mouse button.

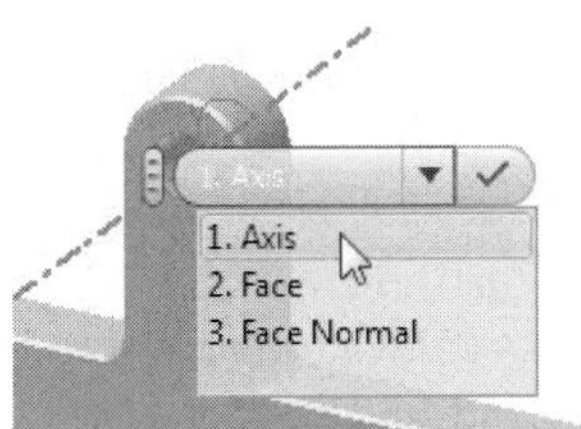

FIGURE 6.30

The next step is to position the cursor over the face, edge, point, and so on, and select the second geometry input for the assembly constraint. Again, you may need to cycle through the selection set until the correct location is highlighted. If the Show Preview option is selected in the dialog box, the components will move to show how the assembly constraint will affect the components, and you will hear a snapping sound when you preview the constraint. To change either selection, click on the button with the number 1 or 2, and select the new input. Enter a value as needed for the offset or angle, and select the correct Solution option until the desired outcome appears. Click the Apply button to complete the operation. Leave the Constraint dialog box active to define subsequent constraint relationships.

Direct Select - Assembly Constraints

Instead of placing assembly constraint via a dialog box you can use the Assemble command from the Assemble tab > Position panel as shown in Figure 6.31 on the left. The Assemble command can be used to place Mate, Flush, Directed Angle, Tangent, Insert, and UCS constraints by selecting geometry and the constraint types are inferred based on the selections. After starting the command a mini toolbar displays in the graphics window as shown in the center of Figure 6.31. It allows you to choose a specific constraint option and only geometry that matches that option is selectable. For example, the Insert Opposed option will only select circular edges. To create a constraint with the Assemble command, follow these steps

1. Click the Assemble command from the Position panel.
2. If needed select a constraint option from the mini toolbar. The Automatic option, which is also the top level command, allows you to select all types of geometry.

3. Select geometry for the first selection set (this part will move).
4. Move the cursor over your second selection. Keep moving the cursor over the geometry until the solution type is previewed then click.
5. The mini toolbar will change. The new state of the mini toolbar will allow you to change the solution type and enter a value for the offset or angle.
6. When the options are set either press the ENTER key, press the Apply, or the OK button on the mini toolbar as shown in Figure 6.31 on the right.

NOTE

The Assemble command always moves the first component to the second one. If the first component is grounded, it will move to the new location and maintain its grounded status.

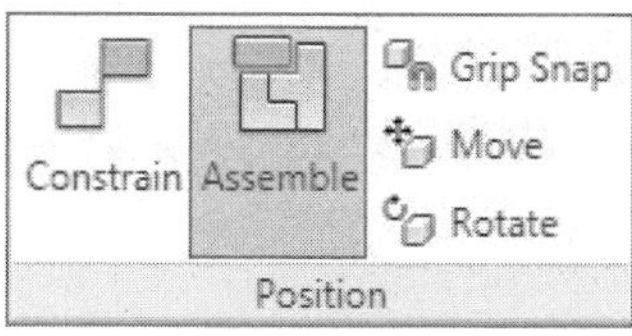

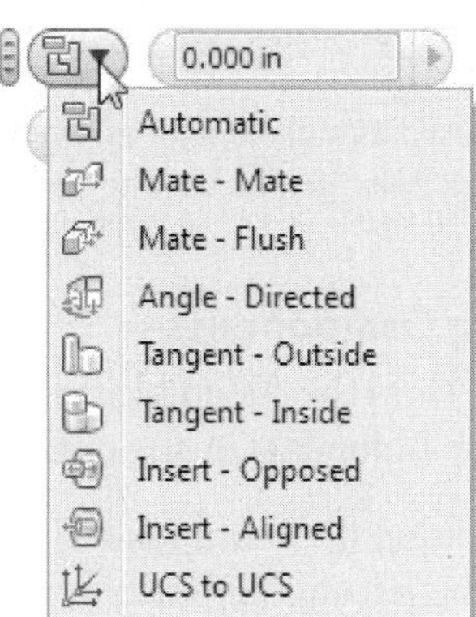

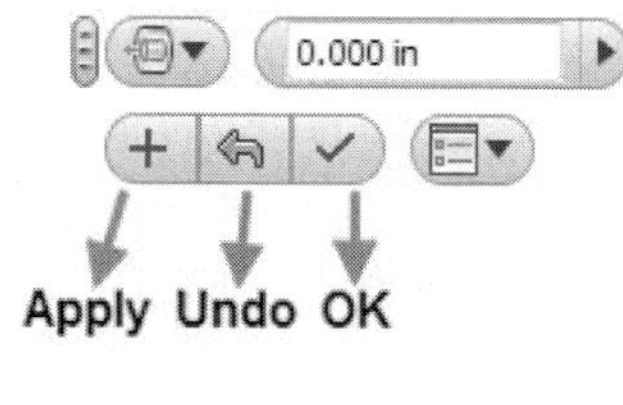

FIGURE 6.31

ALT + DRAG CONSTRAINING

Another way to apply an assembly constraint is to hold down the ALT key while dragging a part edge or face to another part edge or face; no dialog box will appear. The key to dragging and applying a constraint is to select the correct area on the part. Selecting an edge will create a different type of constraint than if a face is selected. If you select a circular edge, for example, an insert constraint will be applied. To apply a constraint while dragging a part, you cannot have another command active.

To apply an assembly constraint, follow these steps:

1. While holding down the ALT key, select the face, edge, or other part on the part that will be constrained.
2. Select a planar face, linear edge, or axis to place a mate or flush constraint. Select a cylindrical face to place a tangent constraint. Select a circular edge to place an insert constraint.
3. Drag the part into position. As you drag the part over features on other parts, you will preview the constraint type. If the face you need to constrain to is behind another face, pause until the Select Other command appears. Cycle through the possible selection options, and then click to accept the selection.

To change the constraint type previewed while you drag the part, release the ALT key and press the space bar to flip to the solution direction and if needed press one of the following shortcut keys:

Key	Constraint Type
M or 1	Mate
A or 2	Angle
T or 3	Tangent
I or 4	Insert
R or 5	Rotation-Motion
S or 6	Rotation-Translation
X or 8	Transitional

NOTE

A work plane can also be used as a plane with assembly constraints, a work axis can be used to define a line, and a work point can be used to define a point.

Moving and Rotating Components

Use the Move command from the Assemble tab > Position panel, as shown in Figure 6.32, to drag individual components in any direction in the viewing plane.

To perform a move operation on a component, activate the Move Component command. Click and hold the left mouse button on the component to drag it to a new location. Drop the component at its new location by releasing the button.

Moved components will follow these guidelines:

- An unconstrained component remains in the new location when moved until you constrain it to another component.
- A full or partially constrained component initially remains in the new location. When the assembly is updated, the component adjusts its location to comply with the constraints that you have already applied.

Use the Rotate command on the Assemble tab > Position panel, as shown in Figure 6.32, to rotate an individual component. This command is very useful when constraining faces that are hidden from your view.

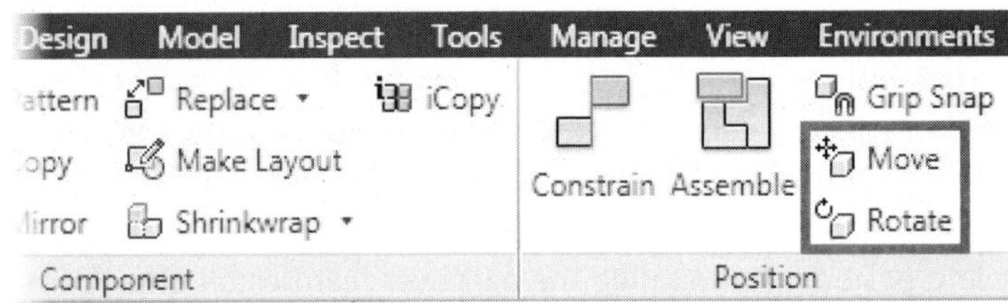

FIGURE 6.32

Follow these steps for rotating a component in an assembly:

1. Activate the Rotate command, and select the component to rotate. Notice the appearance of the 3D rotate symbol on the selected component in Figure 6.33.

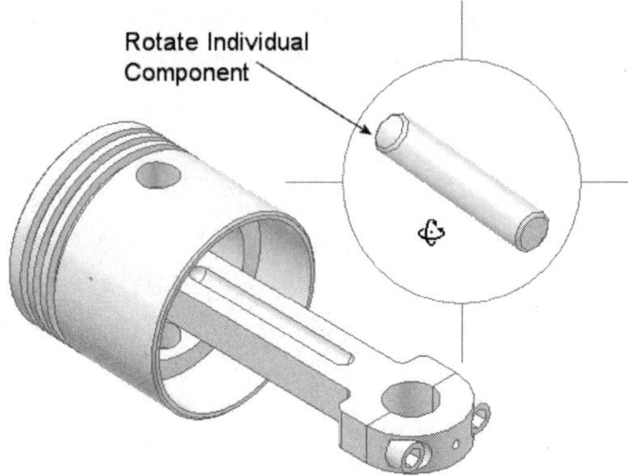

FIGURE 6.33

2. Drag your cursor until you see the desired view of the component.

- For free rotation, click inside the Dynamic Rotate command, and drag in the desired location.
- To rotate about the horizontal axis, click the top or bottom handle of the Dynamic Rotate command, and drag your cursor vertically.
- To rotate about the vertical axis, click the left or right handle of the Dynamic Rotate command, and drag your cursor horizontally.
- To rotate planar to the screen, hover over the rim until the symbol changes to a circle, click the rim, and drag in a circular direction.
- To change the center of rotation, click inside or outside the rim to set the new center.

Release the mouse button to drop the component into the rotated position.

NOTE

If you click the Update button after moving or rotating components in an assembly, any components constrained to a grounded component will snap to their constrained positions in the new location. A fully constrained component can be moved or rotated temporarily. Once the assembly constraints are updated, the component will resolve the constraints and return to a fully constrained location/orientation.

EDITING ASSEMBLY CONSTRAINTS

After you have placed an assembly constraint, you may want to edit, suppress, or delete it to reposition the components. There are two ways to edit assembly constraints.

Both methods are executed through the browser. In the browser, activate the assembly or subassembly that contains the component that you want to edit. Expand the component name, and you will see the assembly constraints, as shown in Figure 6.34 on the left. Double-click on the constraint name, and an Edit Dimension dialog box will appear, allowing you to edit the constraint offset value. Refer to the following Constraint Offset Value Modification section for more information. You can also right-click on the assembly constraint's name in the browser and select Delete, Edit, or Suppress from the menu, as shown in Figure 6.34.

If you select Edit, the Edit Constraint dialog box will appear. If you select Suppress, the assembly constraint will not be applied. Select Drive Constraint to drive a constraint through a sequence of steps, simulating mechanical motion. Select Delete, and the assembly constraint will be deleted from the component. If you try to place or edit an assembly constraint and it cannot be applied, a warning window that explains the problem will appear. You will have to either select new options for the operation or suppress or delete another assembly constraint that conflicts with it.

NOTE

When editing offset dimension values, click once with the left mouse button on the constraint with the offset value. Then change the offset value from the edit box that will appear at the bottom of the Assembly browser.

If an assembly constraint is conflicting with another, a triangular, yellow icon with an exclamation point will appear in the browser, as shown in Figure 6.34 on the right. To edit a conflicting constraint in the browser, either double-click on its name or right-click on its name and select Recover from the menu, as shown in Figure 6.34 on the right. The Design Doctor will appear and guide you through the steps to fix the problem.

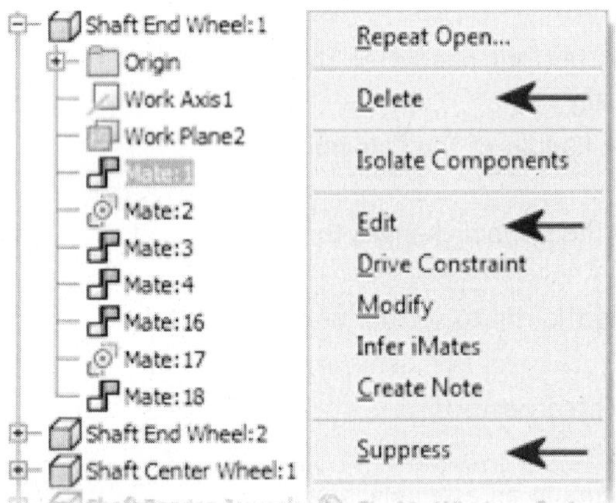

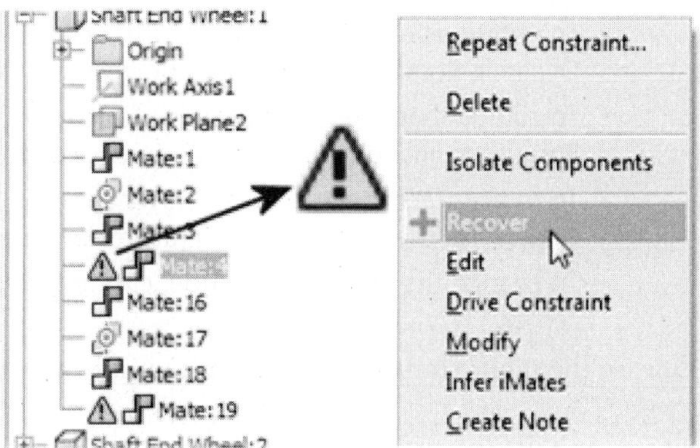

FIGURE 6.34

Constraint Offset Value Modification

When editing a constraint offset value, use the standard value edit control. This process is similar to editing work plane offsets and sketch dimensions, and it will allow you to measure while editing constraint offset values. Either double-click or right-click on the constraint to edit in the browser, and select Modify from the menu, as shown in Figure 6.35 on the left. The Edit Dimension dialog box will appear; edit the offset value.

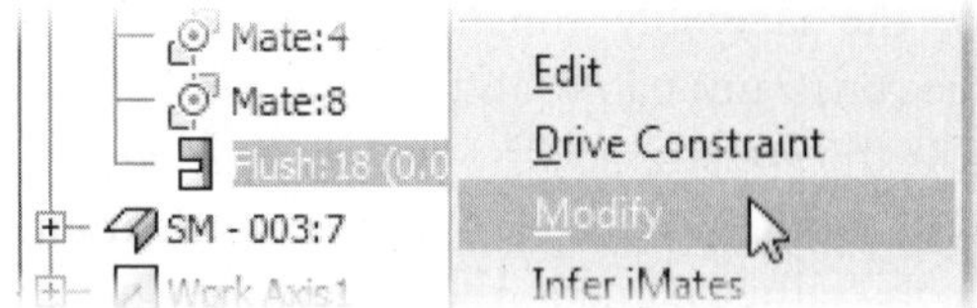

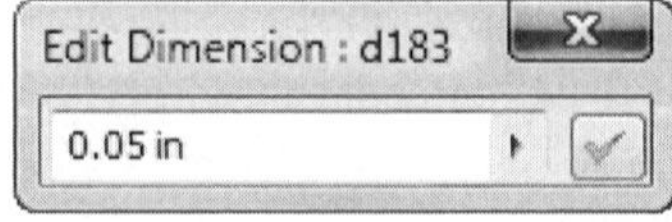

FIGURE 6.35

You could also click on the constraint in the browser and a dialog box appears at the bottom of the Assembly browser, enabling you to make changes to the constraint's value.

NOTE

EXERCISE 6-1: ASSEMBLING PARTS

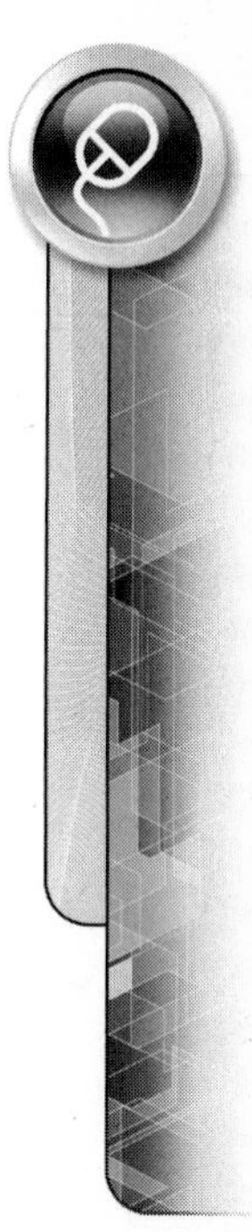

1. Open *ESS_E06_01.iam*.

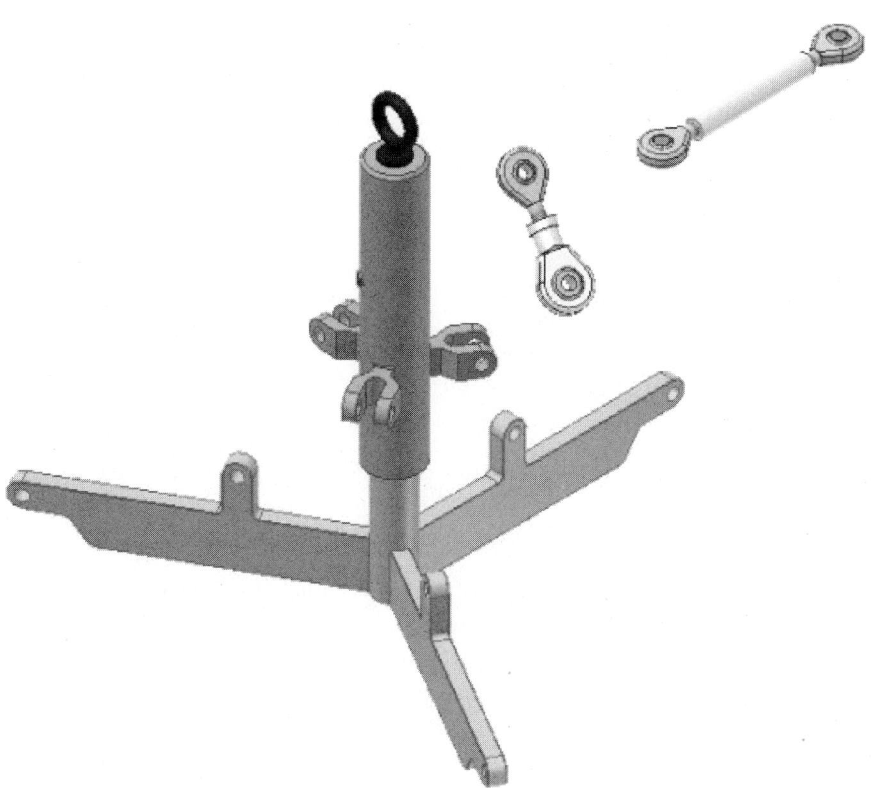

FIGURE 6.36

2. Begin assembling the connector and sleeve:
   a. Right-click in the graphics window, click Home View or press the F6 key.
   b. Zoom in on the small connector and sleeve.
   c. Drag the connector so that the small end is near the sleeve, as shown in Figure 6.37.

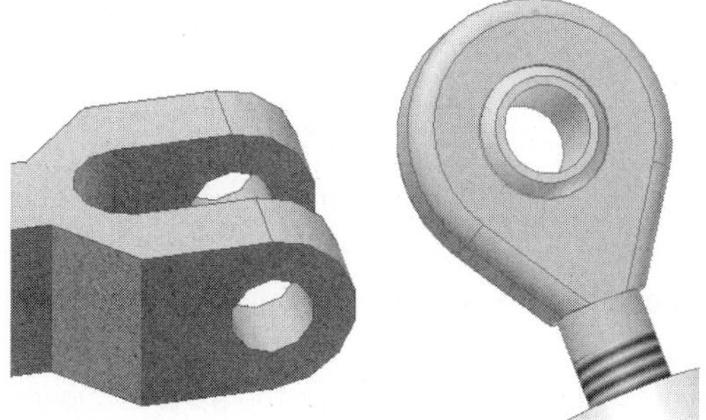

FIGURE 6.37

3. Next add a mate constraint between the centerlines of both components. Click the Constrain command from the Assemble tab > Position panel. Mate is the default constraint.
4. Move the cursor over the hole in the arm on the sleeve. Click when the centerline displays, as shown in Figure 6.38.
5. Move the cursor over the hole in the link. Click when the centerline displays, as shown in Figure 6.38.

NOTE

If the green dot displays, move the cursor until the centerline displays, or use Select Other to cycle through the available choices to select the axis.

6. Click Apply to accept this constraint.

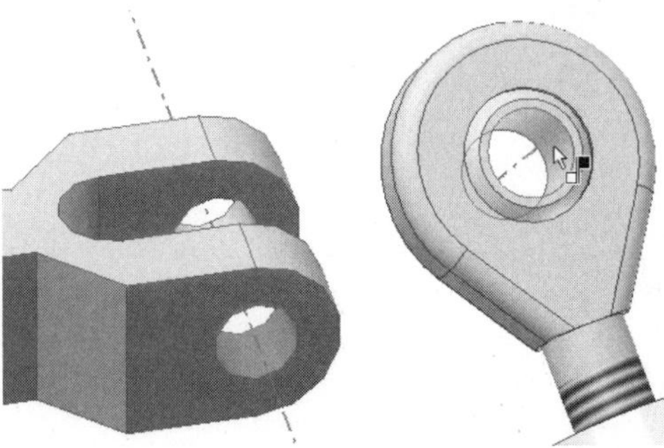

FIGURE 6.38

7. Now add a mate between the faces of both parts. Click the small flat face on the ball in the link end, as shown in Figure 6.39.
8. Click the inner face of the slot on the sleeve, as shown in Figure 6.39.
9. Click OK to create this constraint and close the Place Constraint dialog box.

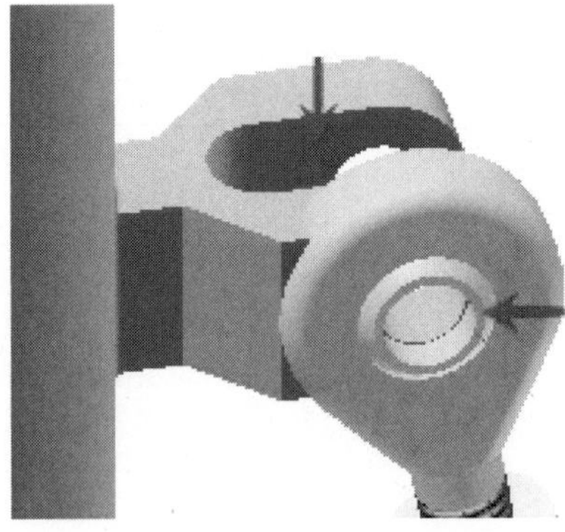

FIGURE 6.39

10. Click on the small link arm, and drag it to view the effect of the two constraints.
11. Place the crank in the assembly by clicking the Place command on the Assemble tab > Component panel and opening the file *ESS_E06_01-Crank.ipt.* from the Chapter 06 folder.
12. Move the component near the spyder arm, and click to place the part, as shown in Figure 6.40. Right-click, and click OK.

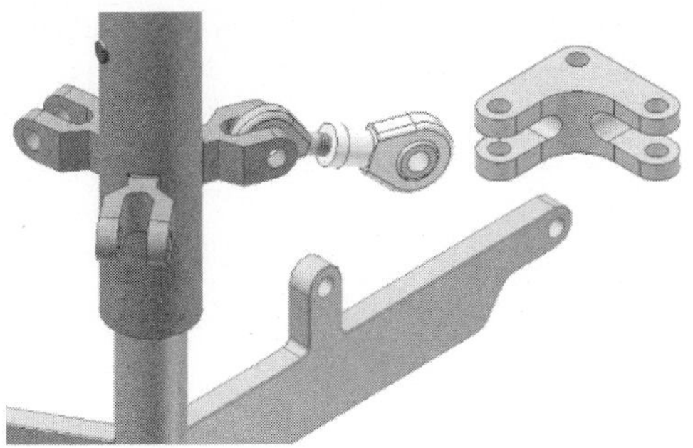

FIGURE 6.40

13. To assemble the crank to the spyder you will use the Assemble command with the Automatic option. Begin by rotating your viewpoint of the model so you can see the inner edge of a hole on the crank and the spyder as shown in Figure 6.41.

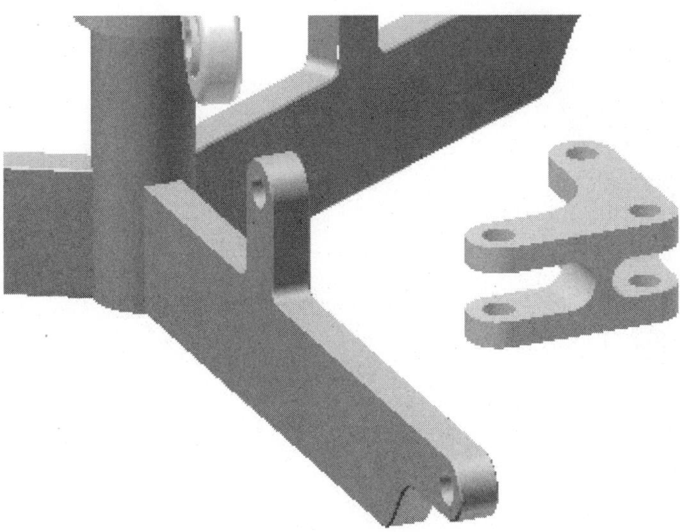

FIGURE 6.41

14. Click the Assemble command from the Assemble tab > Position panel. By default, the Automatic option is shown in the heads-up display.

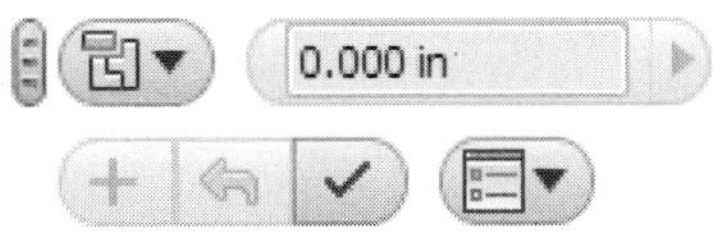

FIGURE 6.42

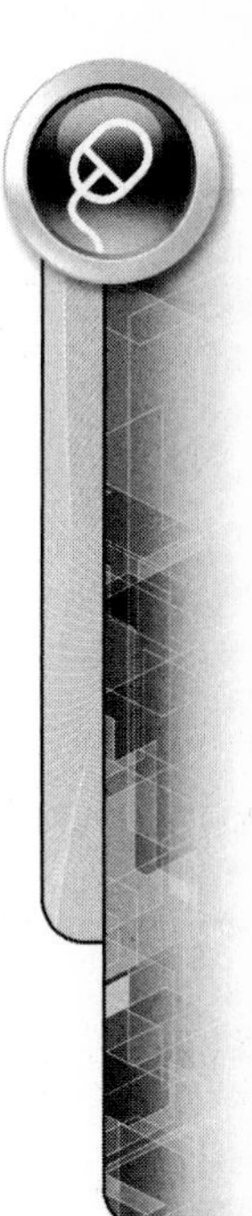

15. Move the cursor over the hole in the arm on the crank, and click when the centerline displays.

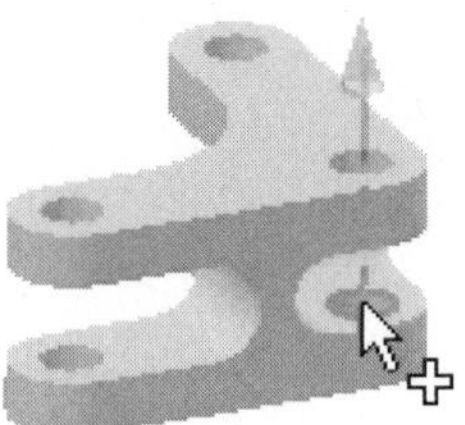

FIGURE 6.43

16. Move the cursor over the hole on the spyder and click when the centerline is displayed as shown in Figure 6.44.

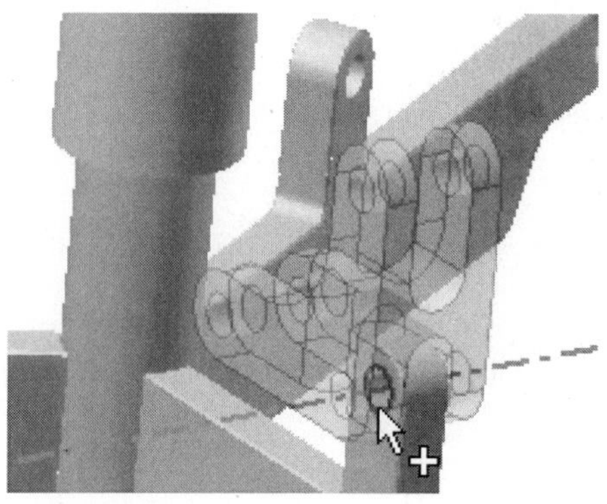

FIGURE 6.44

17. Click OK (green check mark) on the heads-up display or press ENTER to add the constraint.
18. Next, you will assemble the crank and link. Return to the Home View.
19. Drag the crank so you can see the affect of the constraint and to position it approximately as shown in Figure 6.45.

FIGURE 6.45

20. Drag the end of the small link arm close to the crank, as shown in Figure 6.46.
21. Zoom in to the crank and link.

22. Use the Constrain command to place a mate constraint between the centerlines of the two holes, as shown in Figure 6.46.

FIGURE 6.46

23. You will now place a claw in the assembly. Click the Place command from the Assemble tab > Component panel, and open the file *ESS_E06_01-Claw.ipt*.
24. Move the component near the end of the spyder arm, and click to place the part. Right-click, and select OK. Your display should appear similar to Figure 6.47.

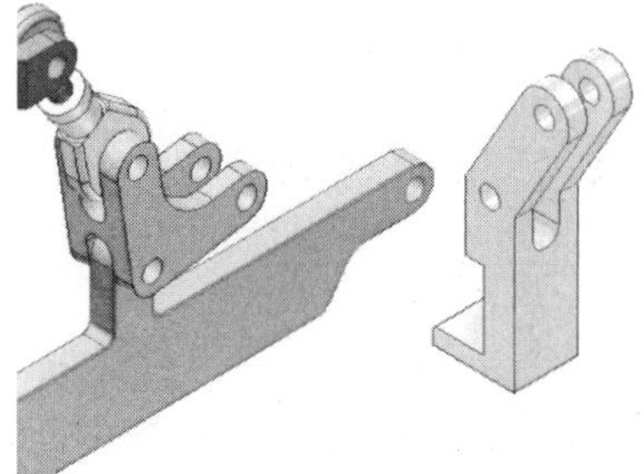

FIGURE 6.47

25. Constrain the claw to the spyder. Click the Constrain command on the Assembly tab > Position panel.
26. Click Insert from the type area of the Place Constraint dialog box.
27. Select the hole on the end of the spyder arm. Then select the inside edge of the hole on the claw as shown in Figure 6.48.

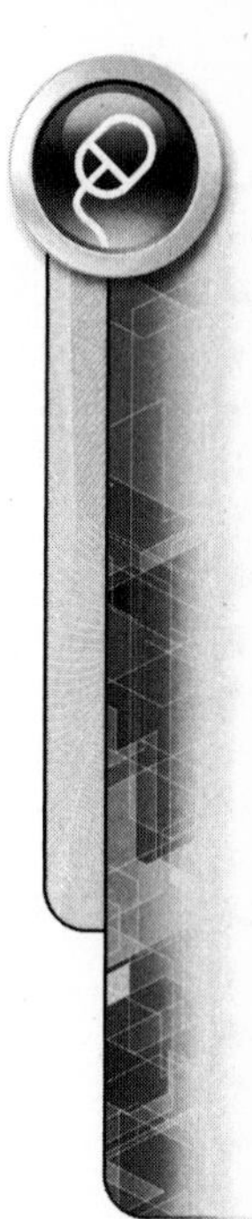

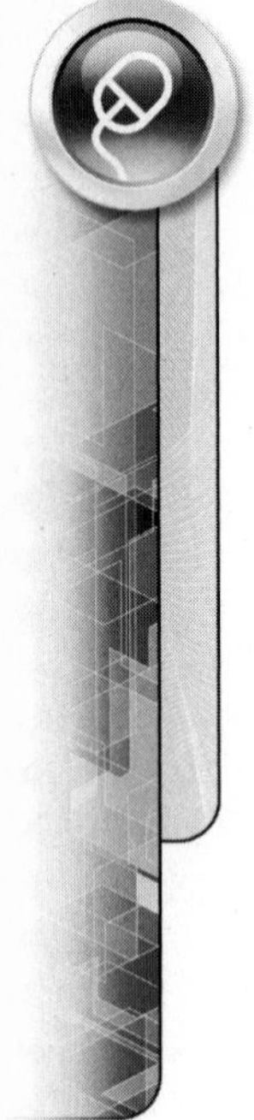

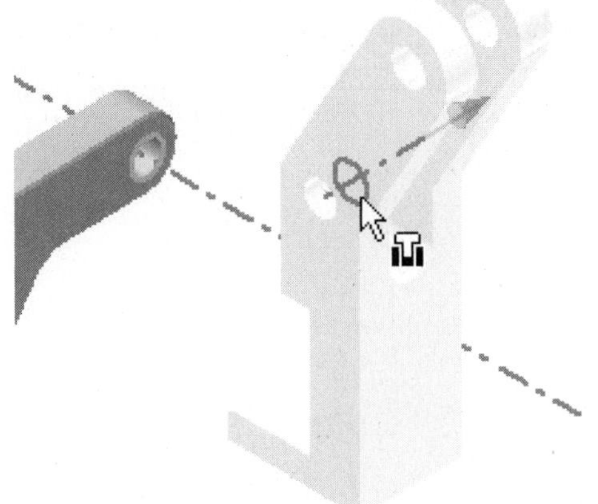

FIGURE 6.48

28. Click OK to place the constraint and close the Place Constraint dialog box.
29. Assemble the link rod to the crank and claw. First drag the claw and link rod into the position as shown in Figure 6.49.

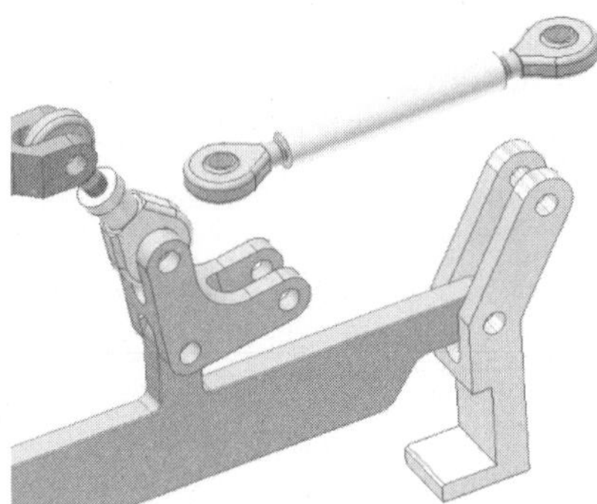

FIGURE 6.49

30. Next you use the Assemble command as an alternate method of placing an insert constraint between the link rod and the crank. Click the Assemble command from the Assemble tab > Position panel.
31. Select Insert – Opposed from the heads-up display drop-down list.

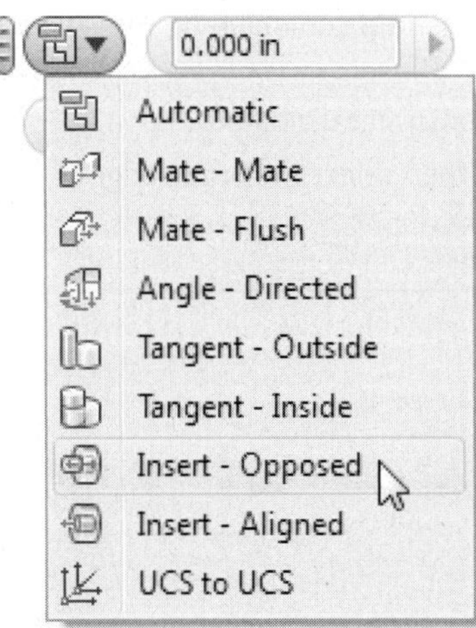

FIGURE 6.50

32. Select the top edge of the ball in the end of the link rod as shown in Figure 6.51 on the left. This component must be selected first. With the Assemble command the first selected component will move.
33. Select the inner edge of the hole in the crank as shown in Figure 6.51 on the right.

FIGURE 6.51

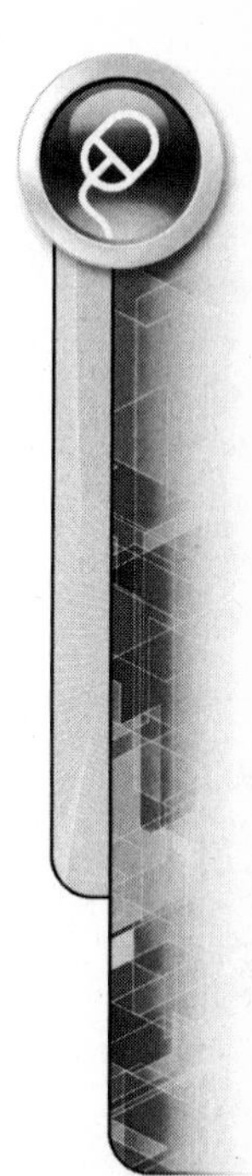

34. Press ENTER or click OK to place the constraint.
35. Click the Constrain command from the Assemble tab > Position panel.
36. Add a mate constraint between the centerlines of the claw and end of the link rod.
37. Click OK to apply the constraint and close the Place Constraint dialog box.
38. Drag the claw to view the effect of the assembly constraints, as shown in Figure 6.52.

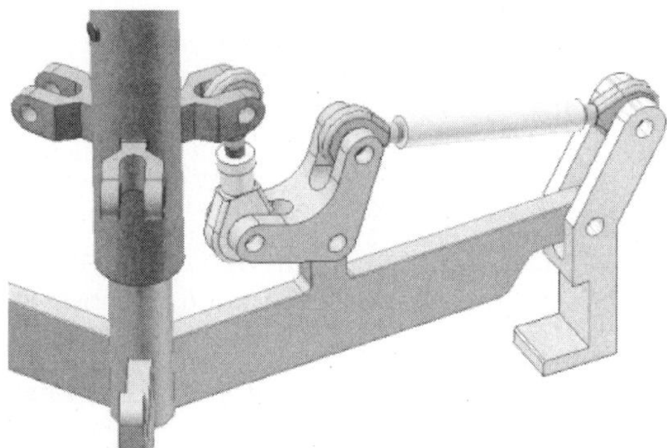

FIGURE 6.52

39. You will now add bolts and nuts to the spyder assembly. Begin by clicking the Place Component command on the Assemble tab > Component panel. Hold down the CTRL key and select ESS_E06_01-Bolt.ipt and ESS_E06_01-Nut.ipt. Click Open.
40. Place six nuts and bolts near their final position in the assembly, as shown in Figure 6.53. Right-click and click OK.

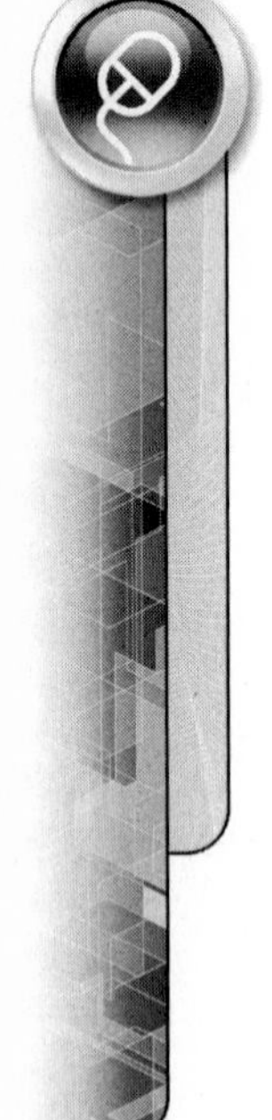

FIGURE 6.53

41. Constrain the bolts into the assembly by clicking the Constrain command from the Assemble Tab > Position panel and in the Type area, click Insert to select an insert constraint.
42. Insert this bolt by selecting the top of the bolt's shank and then the edge of the hole, as shown in Figure 6.54.
43. Click Apply to place the constraint and then continue to add the appropriate insert constraints to the remainder of the bolts.

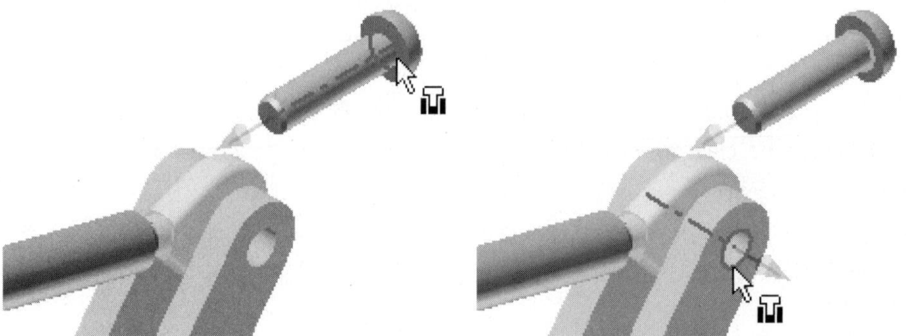

FIGURE 6.54

44. Now constrain the nuts to the end of the bolts. Rotate the viewpoint so that you can see the back of the claw. Drag two nuts near their final location, as shown in Figure 6.55.
45. Use either the Constrain or Assemble command to place an insert constraint between the circular edge of the hole in the nut and the edge of the hole on the claw as shown in Figure 6.55 on the right.

NOTE

You must select an edge on the face of the nut. Do not select the inner chamfered edge.

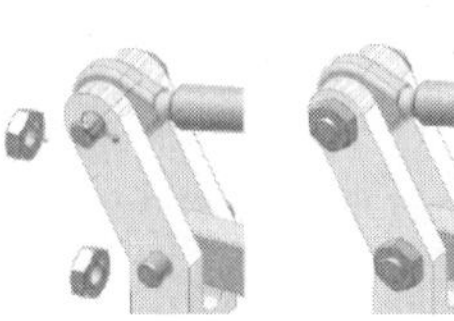

FIGURE 6.55

46. Follow the same process to add the appropriate Insert constraints to the remainder of the nuts.
47. The completed assembly is displayed in Figure 6.56.

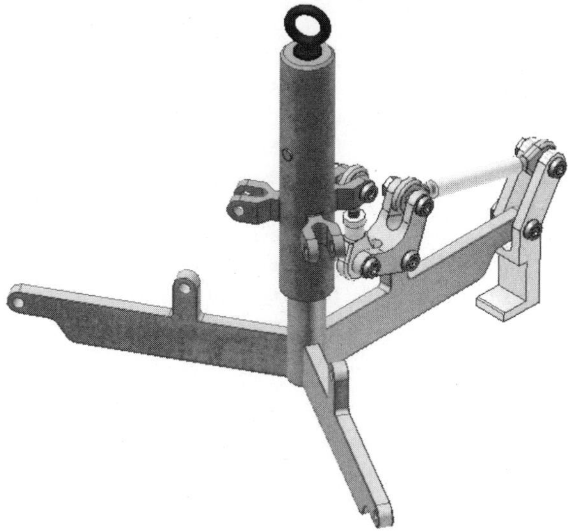

FIGURE 6.56

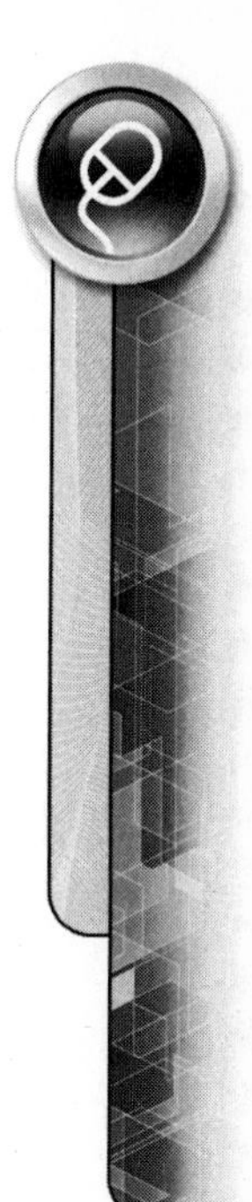

48. Close all open files. Do not save changes. End of exercise.

ADDITIONAL CONSTRAINT COMMANDS

Additional commands available to manipulate, navigate, and edit assembly constraints include browser views, Other Half, Constraint Offset Value Modification, Isolating Assembly Components, and Isolating Constraint Errors. The following sections describe these commands.

Browser Views

Two modes for viewing assembly information are located in the top of the browser toolbar: Assembly View and Modeling View. Assembly View, the default mode, displays assembly constraint symbols nested below both constrained components, as shown in Figure 6.57 on the left. In this mode, the features used to create the part are hidden.

When Modeling View is set, assembly constraints are located in a Constraints folder at the top of the assembly tree, as shown in Figure 6.57 on the right. In this mode, the features used to create the part are displayed just as they are in the original part file.

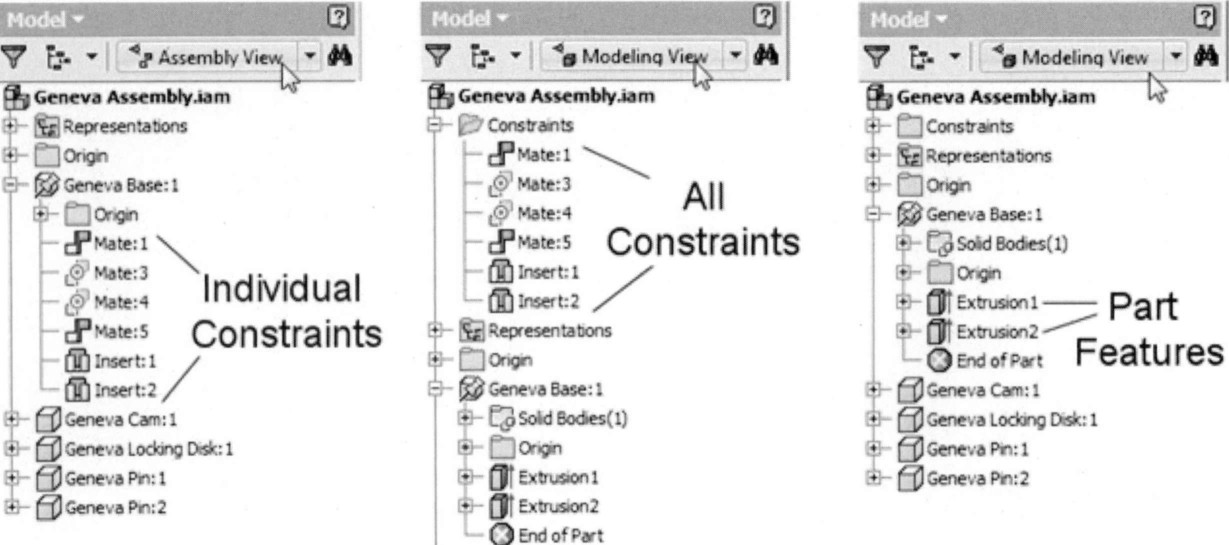

FIGURE 6.57

Other Half

You can use the Other Half command to find the matching part that participates in a constraint placed in an assembly. As you add parts over time, you may wish to highlight an assembly constraint and find the part(s) to which it is constrained. As shown in Figure 6.58 on the left, a mate constraint has been highlighted. Half of this constraint has been applied to a part called Engine Block:1. To view the part sharing a common constraint, right-click on the constraint in the browser, and select Other Half from the menu, as shown in Figure 6.58 on the left. The browser will expand and highlight the second half of the constraint, as shown in Figure 6.58 in the middle. In this example, the other half of the mate constraint is a part called Cylinder:1.

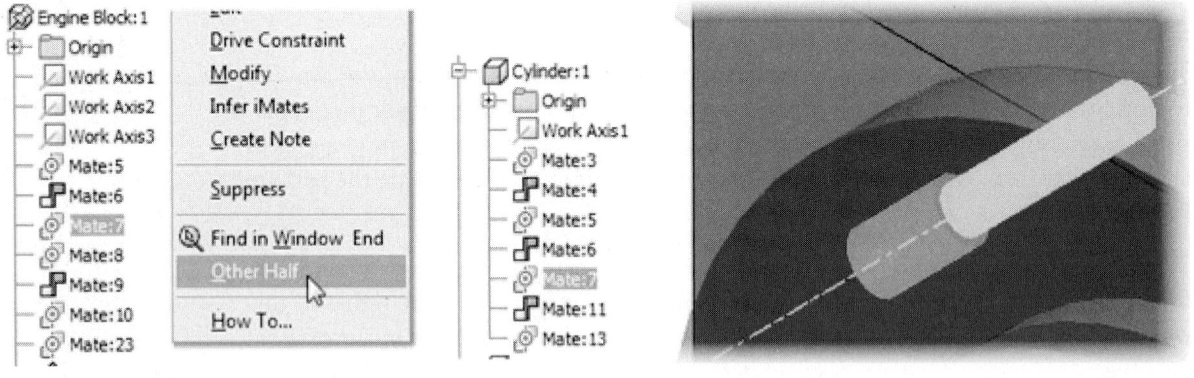

FIGURE 6.58

Constraint Tooltip

To display all property information for a specific constraint, move your cursor over the constraint icon, and a tooltip will appear, as shown in Figure 6.59. Although the constraint name is highlighted in Figure 6.59, you must hover your cursor over the constraint icon to view the tooltip information.

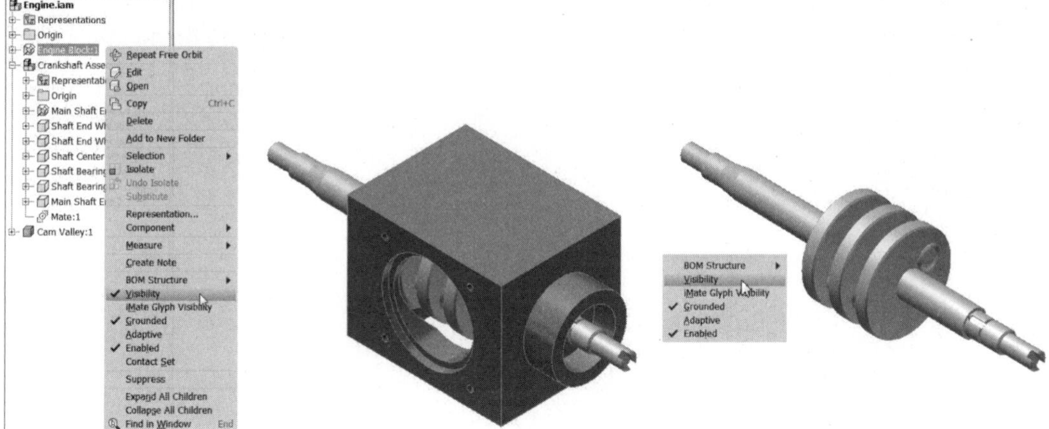

FIGURE 6.65

ADAPTIVITY

This section will introduce you to the concept of adaptivity. To learn more about adaptivity see the Help system. Adaptivity is the Autodesk Inventor function that allows the size of a part to be determined by setting up a relationship between two parts in an assembly. Adaptivity allows under-constrained sketches—features that have undefined angles or extents, hole features, and subassemblies, which contain parts that have adaptive sketches or features—to adapt to changes. The adaptivity relationship is defined by applying assembly constraints between an adaptive sketch or feature and another part. If a sketch is fully constrained, it cannot be made adaptive. However, the extruded length or revolved angle of the part can be. A part can only be adaptive in one assembly at a time. In an assembly that has multiple placements of the same part, only one occurrence can be adaptive. The other occurrences will reflect the size of the adaptive part.

An example of adaptivity would be determining the diameter of a pin from the size of a hole. You could determine the diameter of the hole from the size of the pin, and you can turn adaptivity on and off as needed. Once a part's size is determined through adaptivity and adaptivity is no longer useful you may want to turn its adaptivity off. If you want to create adaptive features, Autodesk Inventor includes settings that will speed up the process of creating them. These settings are not covered in detail in this book, but can be found on the Assembly tab of the Application Options dialog box.

Under-constrained Adaptive Features

The next series of figures show how parts adapt when you apply assembly constraints. In Figure 6.66, a rectangular sketch for a small plate is not dimensioned (unconstrained) along its length.

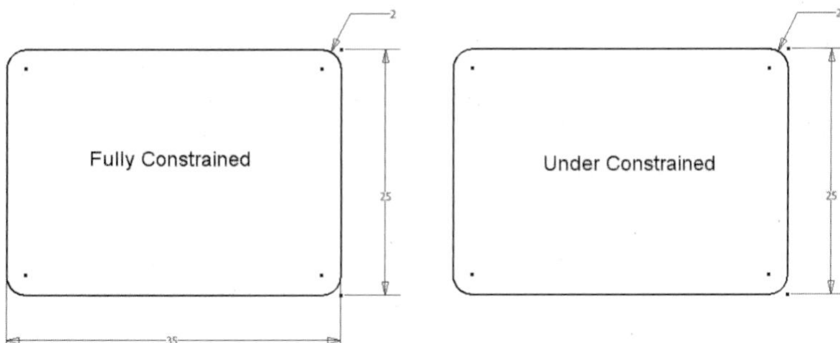

FIGURE 6.66

The extruded feature is then defined as adaptive by right-clicking on the extrusion in the browser as shown in Figure 6.67. Selecting Adaptive from the menu applies the adaptive property to the sketch and the extrusion.

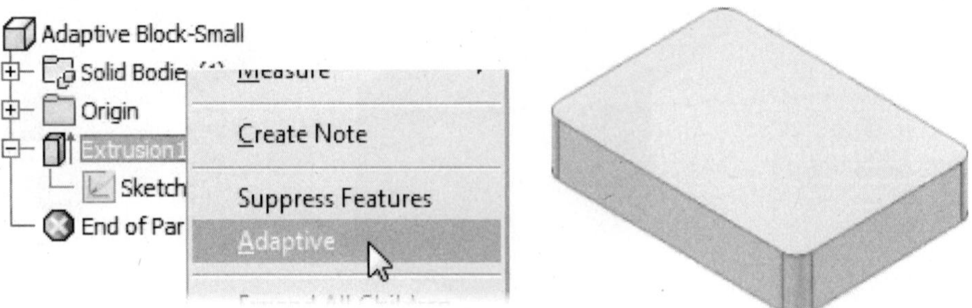

FIGURE 6.67

Once you have placed the parts in an assembly, right-clicking on the component in the browser activates a menu. Selecting Adaptive from the menu adds an icon consisting of two arcs with arrows next to the component, as shown in Figure 6.68 on the left. The presence of this icon in the browser means that part is now adaptive. As flush constraints are placed along the edge faces of the plates, the small plate will adapt its length to meet the length of the large plate.

The results are shown in Figure 6.68 on the right.

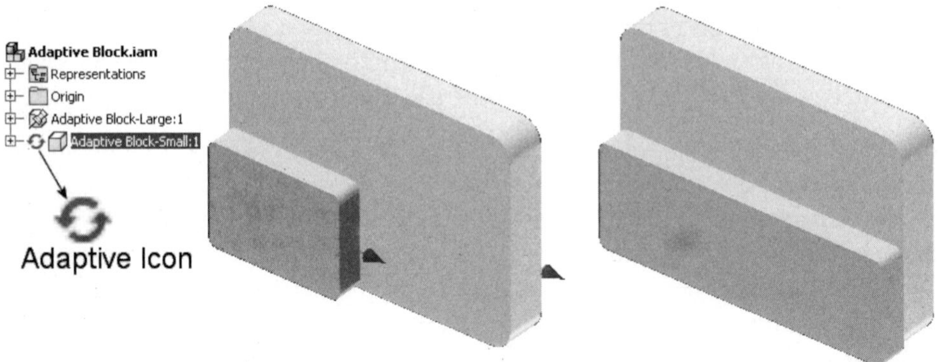

FIGURE 6.68

ASSEMBLY SECTIONS

Assembly section views can be used to visualize portions of an assembly, especially where components are hidden or obscured within chambers. While the assembly is sectioned, part and assembly tools can be used to create or modify parts within the assembly. You can use the exposed cut edges, along with the Project Cut Edges command, to assist in sketch construction for additional features that need to be created in a part. There are four assembly section tools available on the View tab > Appearance panel as shown in Figure 6.69 on the left. The other two images show a before and after view of an assembly that is using the Half Section View tool.

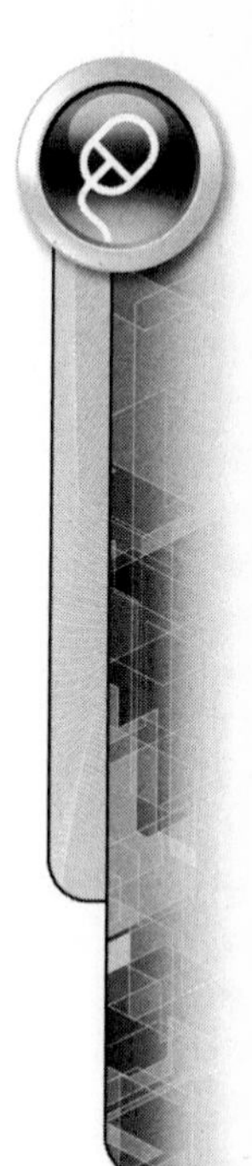

FIGURE 6.69

The general process for using the section tools are:

1. Select the desired assembly section tool.
2. Select any existing planar face(s) or work plane(s).
3. Specify which side of the selected section remains visible by right clicking and using the Flip Section option, shown in Figure 6.70.
4. Right-click and select Done.

You can change the position of section planes in a section view using the Virtual Movement option from the context menu, as shown in Figure 6.70. When defining the section plane, you can enter a distance in the Offset dialog box, drag the section plane in the canvas, or use the mouse wheel while your cursor is on top of the offset field in the Offset dialog box, to move the section plane. The Scroll Step Size option can be used to adjust the distance of the mouse wheel scroll.

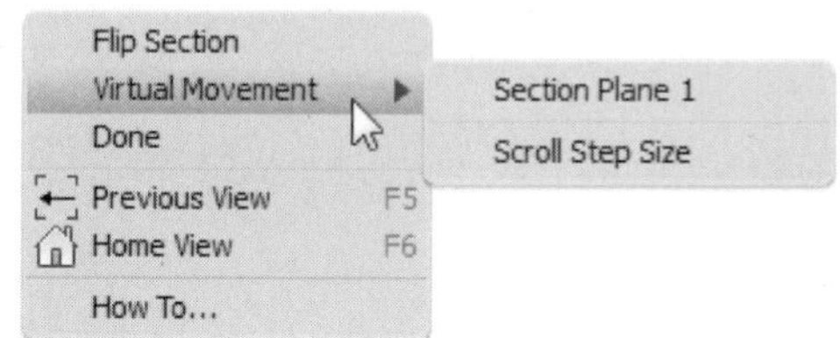

FIGURE 6.70

EXERCISE 6-2: DESIGNING PARTS IN THE ASSEMBLY CONTEXT

In this exercise, you create a lid for a container based on the geometry of the container. This cross part sketch geometry is adaptive, and it automatically updates to reflect design changes in the container.

1. In this exercise you will open an existing assembly and then create a new part based in the context of the assembly and use cross part sketch geometry to define the profile of the part. Begin by opening the file *ESS_E06_02.iam,* in the Chapter 06 folder as shown in Figure 6.71 on the left.
2. Begin the process of creating a new component in the context of an assembly by clicking the Create command from the Assemble tab > Component panel.
3. Under New Component Name, enter *ESS_E06_02-Lid.*
4. Click the Browse Templates button, and in the English tab, click *Standard (in).ipt.* Click OK. There should be a checkmark beside Constrain Sketch Plane to Selected Face or Plane at the bottom of the Create In-Place Component dialog box.

5. Click the OK button to exit the dialog box, and then select the top planar face of the container, as shown in Figure 6.71 on the right.

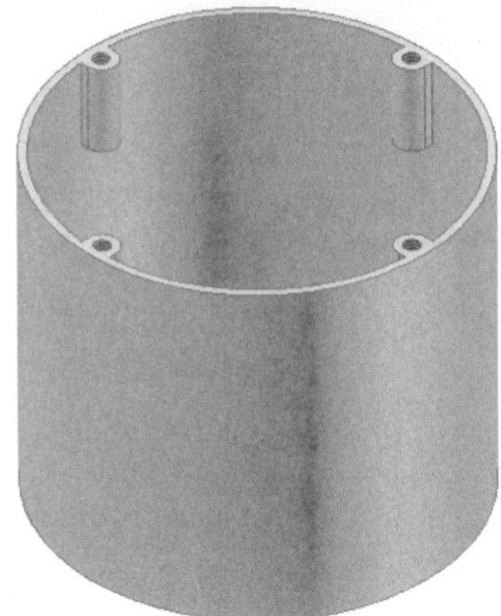

FIGURE 6.71

6. Create a new sketch by clicking the Create 2D Sketch command on the 3D Model tab > Sketch panel. The origin planes for the new part will be displayed in the graphics window. Select the XY plane as shown in **Figure 6.72.** The selected plane becomes the active sketch plane for the new part.

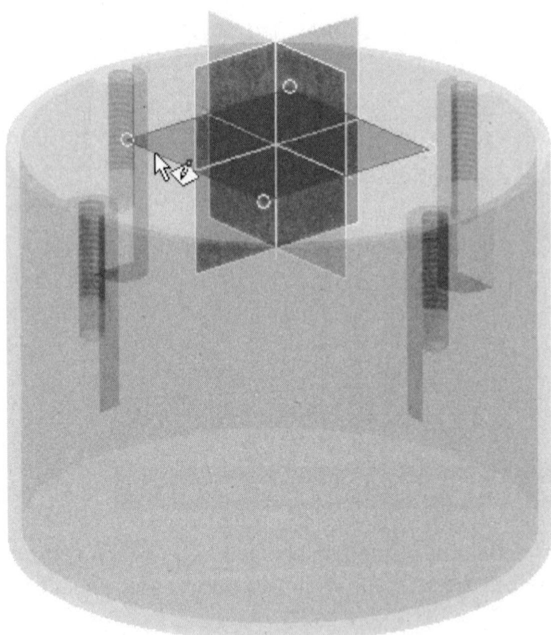

FIGURE 6.72

7. Press the F6 key to change the viewpoint to the Home View. You will now project all geometry contained in this face. First click the Project Geometry command from the Sketch tab > Draw panel.
8. Move the cursor onto the face of the base part until the profile of the entire planar face is highlighted, as shown in Figure 6.73 on the left.

9. Click to project the edges. Your display should appear similar to Figure 6.73 on the right.

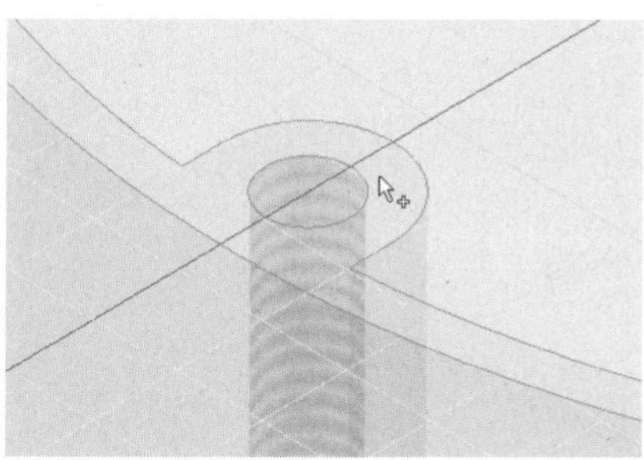

FIGURE 6.73

10. You will now create a series of clearance holes. First zoom into a tapped hole.
11. Create a circle that is concentric to and larger than the projected circle, as shown in Figure 6.74 on the left.
12. Create three additional circles centered on the remaining projected holes, as shown in Figure 6.74 on the right.

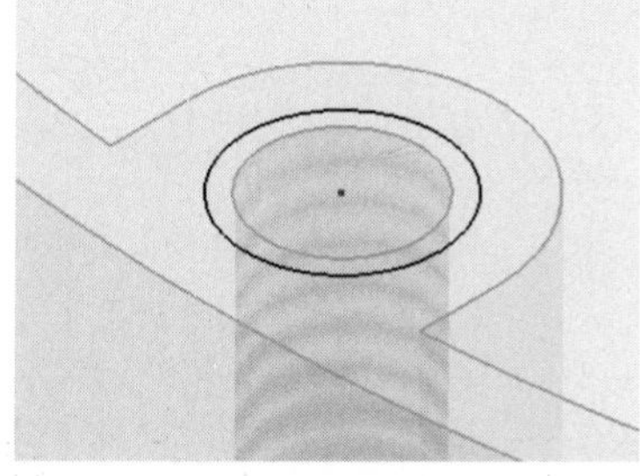
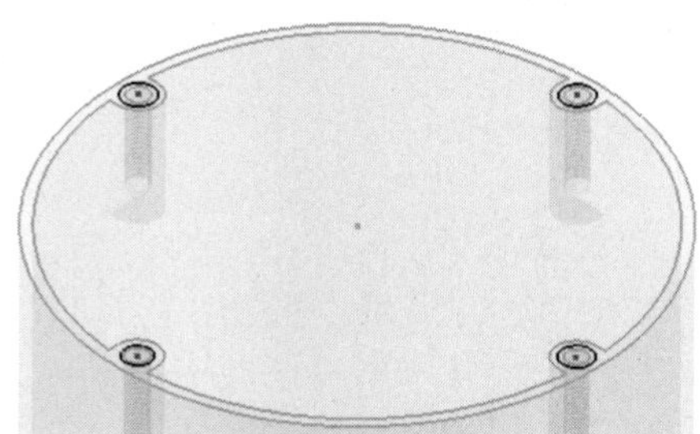

FIGURE 6.74

13. All circle diameters need to be made equal. Perform this operation by clicking the Equal constraint command from the Sketch tab > Constrain panel.
14. Click one circle, and then click one of the other circles. Click the first circle again, and click a third circle. Click the first circle again, and then click the fourth circle. These steps will make all circles equal to each other in diameter.

NOTE

The first entity selected is the source size that the next selection will adapt and become equal to.

15. Now add a dimension to one of the circles to fully constrain the sketch. First click the Dimension command from the Sketch tab > Constrain panel.
16. Click the edge of one of the circles, and click outside the circle to place the dimension.
17. Enter **.15 in** in the dimension edit box, and press ENTER to apply the dimension. Your display should appear similar to Figure 6.75 on the left.

18. Next click the Finish Sketch button to exit the Sketch environment. Click the Extrude command from the 3D Model tab > Create panel, and select the two profiles as shown in Figure 6.75 on the right.

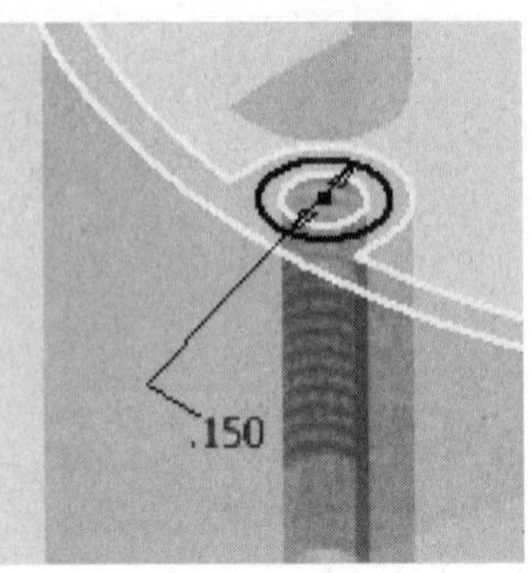

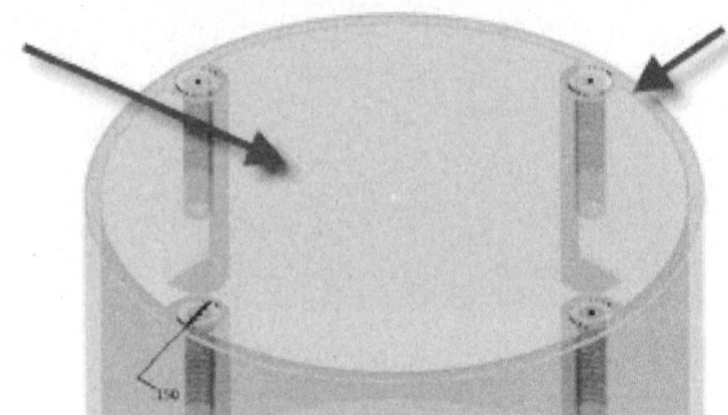

FIGURE 6.75

19. In either the Extrude dialog box or the heads-up display, enter a distance of **.25 in**, and then click OK.
20. Click the Return command to activate the assembly. Your display should appear similar to Figure 6.76 on the right.

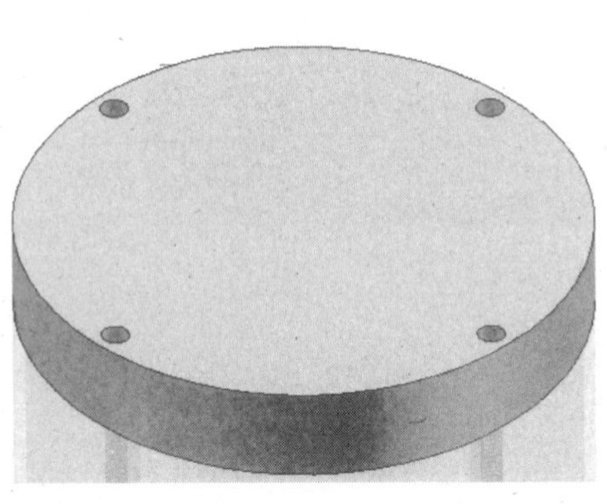

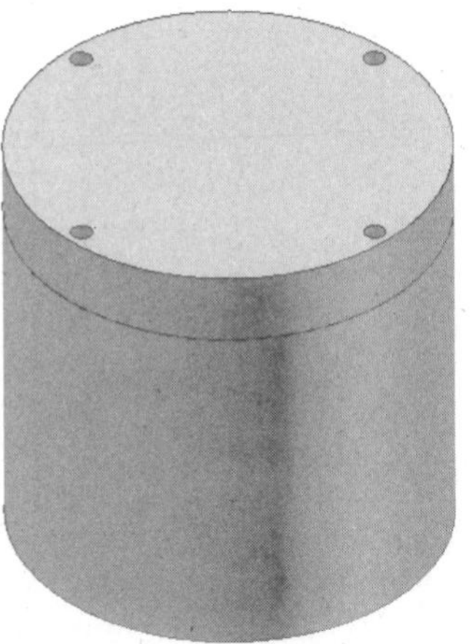

FIGURE 6.76

21. You will now modify the container base and observe how this affects the lid. In the browser, right-click ESS_E06_02-Container:1, and then select Edit. Your display should appear similar to Figure 6.76.

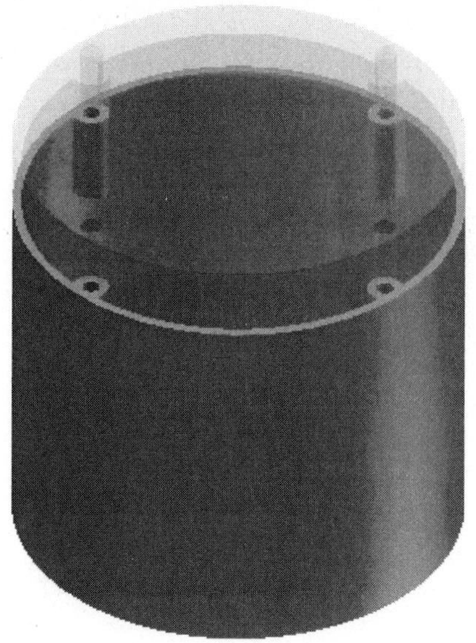

FIGURE 6.77

22. In the browser, right-click Extrusion1, and then select Edit Sketch.
23. Double-click the 2 in dimension, and change the value to **2.5 in**.
24. Click the Finish Sketch command. The model should appear as shown in Figure 6.78 on the left.
25. Click the Return command again to return to the assembly context. Notice that the lid adapts to the modified dimensions of the base, as shown in Figure 6.78 on the right.

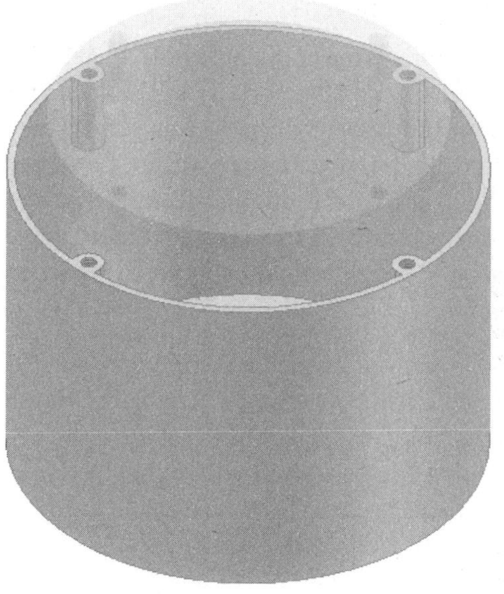

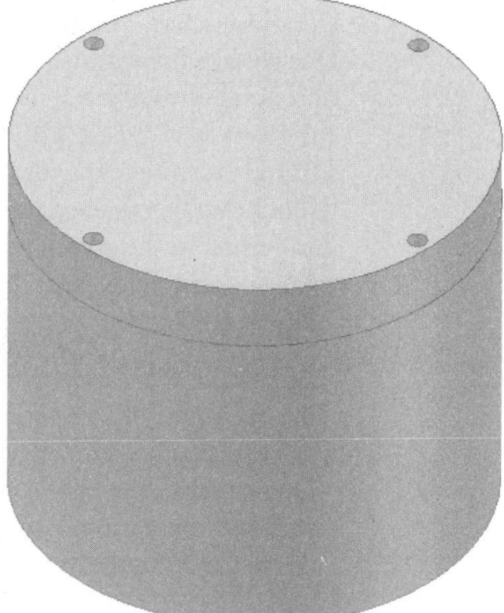

FIGURE 6.78

26. Close all open files. Do not save changes. End of exercise.

PATTERNING COMPONENTS

You can use the Pattern command on the Assemble tab > Component panel, as shown in Figure 6.79, when you place multiple occurrences of selected parts and subassemblies that match a feature pattern on another part (a component pattern) or that have a set of circular or rectangular part patterns in an assembly (an assembly pattern).

Three tabs are available in the Pattern Component dialog box: Associate, Rectangular, and Circular, as shown in Figure 6.79. The Associate tab is the default.

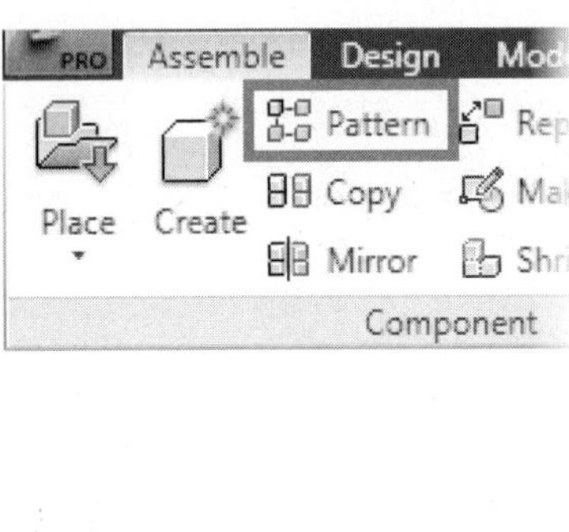

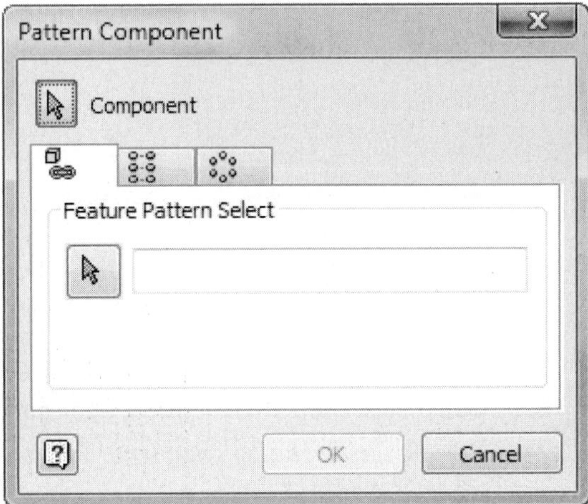

FIGURE 6.79

Associated Patterns

An associated component pattern will maintain a relationship to the feature pattern that you select. For example, a bolt is component-patterned to a part bolt-hole circular pattern that consists of four holes. If the feature pattern, the bolt-hole, changes to six holes, the bolts will move to the new locations, and two new bolts will be added for the two new holes. To create an associative component pattern, there must be a feature-based rectangular or circular pattern, and the part that will be patterned should be constrained to the parent feature, that is, the original feature that was patterned in the component. Issue the Pattern command from the Assemble tab > Component panel, and the Pattern Component dialog box will appear, as shown in Figure 6.80 on the left.

By default, the Component selection option is active. Select the component, such as the cap screw, or components to pattern. Next click the Feature Pattern Select button in the dialog box and select a feature, such as a hole, that is part of the feature pattern. Do not select the parent feature. After selecting the pattern, it will highlight on the part and the pattern name will appear in the dialog box. When done, click OK to create the component pattern, as shown in Figure 6.80 on the right.

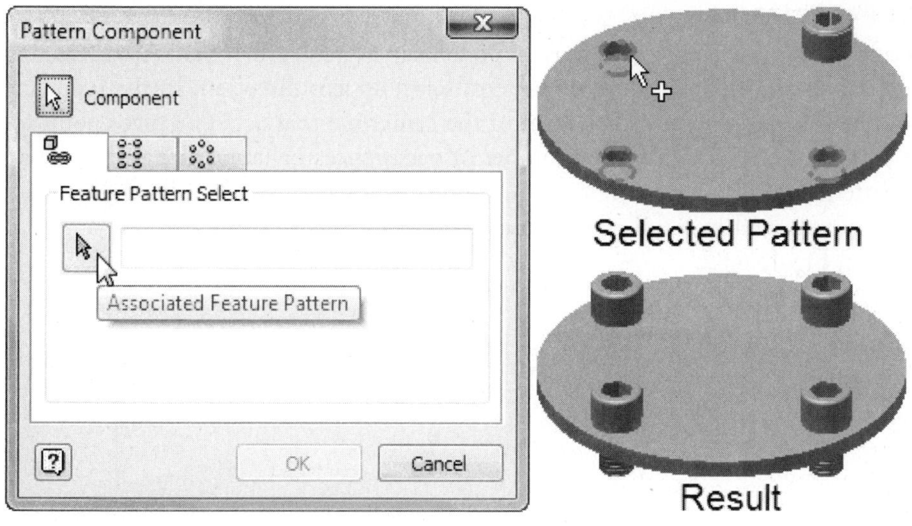

FIGURE 6.80

Rectangular Patterns

A rectangular pattern operation is illustrated in Figure 6.81. When performing this operation, two edges are chosen that define the direction of the columns and rows of the pattern. Enter the number of columns and rows and then the spacing or distance between these rows and columns. The resulting pattern acts like a feature pattern. After creating it, you can edit it to change its numbers, spacing, and so on.

In Figure 6.81, because one of the part edges is inclined, work axes are turned on and used for defining the X and Y directions of the pattern.

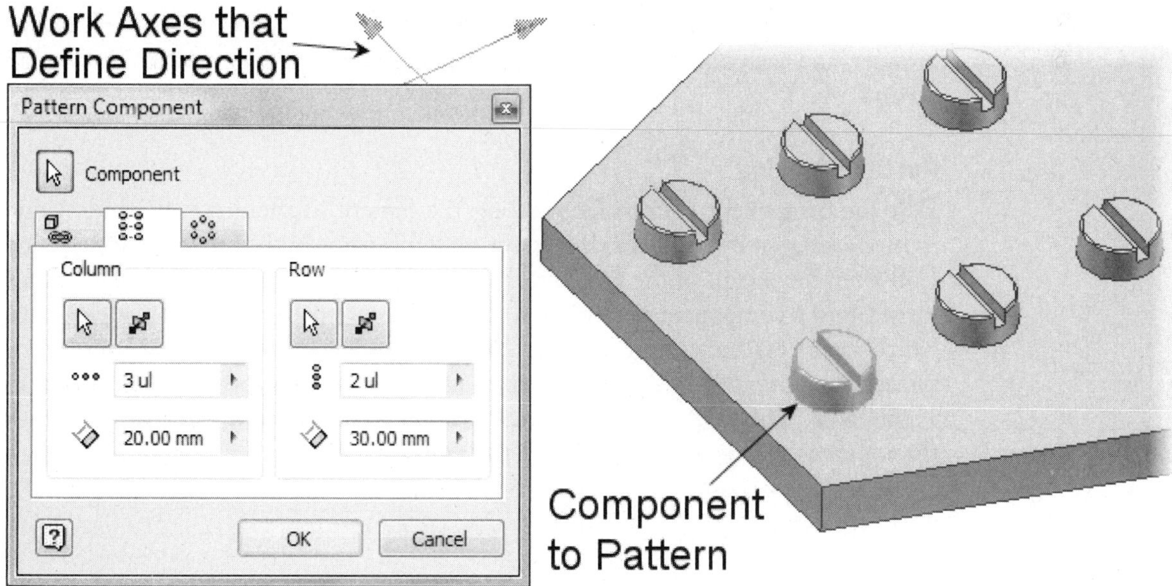

FIGURE 6.81

Circular Patterns

Circular patterns will copy selected components in a circular direction. After selecting the component or components to pattern, select an axis direction. In the following example, this element takes the form of the centerline that acts as a pivot point for the pattern. You then enter the number of occurrences or items that will make up the pattern and the circular angle. In the following example, since the bolt component is being patterned in a full circle, enter 360° as the value for the circular angle, as shown in Figure 6.82.

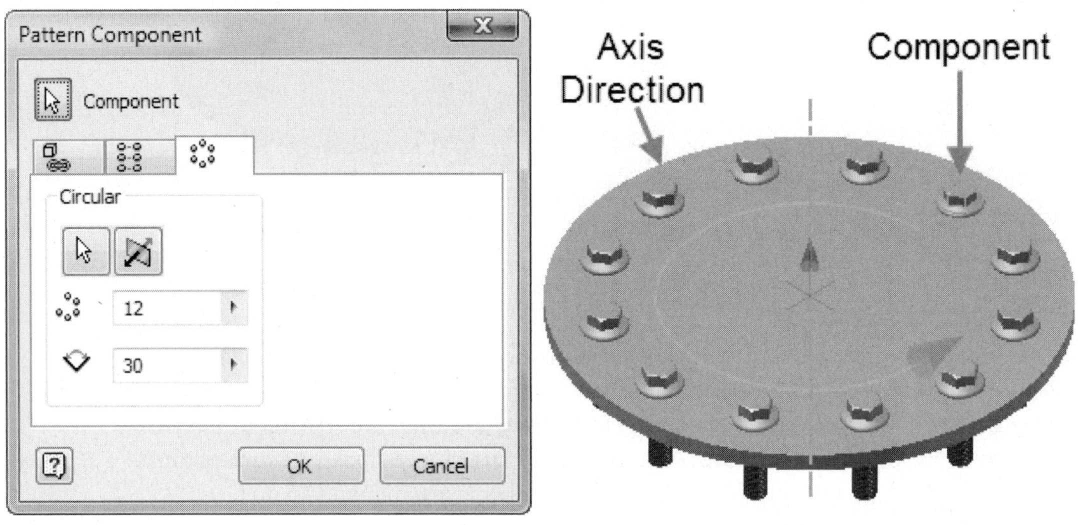

FIGURE 6.82

The completed pattern then acts as a single part. If one part moves, all parts move. The patterned component will be consumed into a component pattern in the browser, and each of the part occurrences will also appear as an element that you can expand.

Pattern Editing

Edit the component pattern by selecting the pattern in the graphics window and right-clicking, or by right-clicking on the pattern's name in the browser and selecting Edit from the menu. In the browser, the component that you patterned will be consumed into a component pattern, and the part occurrences will appear as elements, which can be expanded to see the part. You can suppress an individual pattern component by right-clicking on it and selecting Suppress from the menu, as shown in Figure 6.83. The suppressed component will be grayed out and a line will be struck through it as shown in Figure 6.83 on the right.

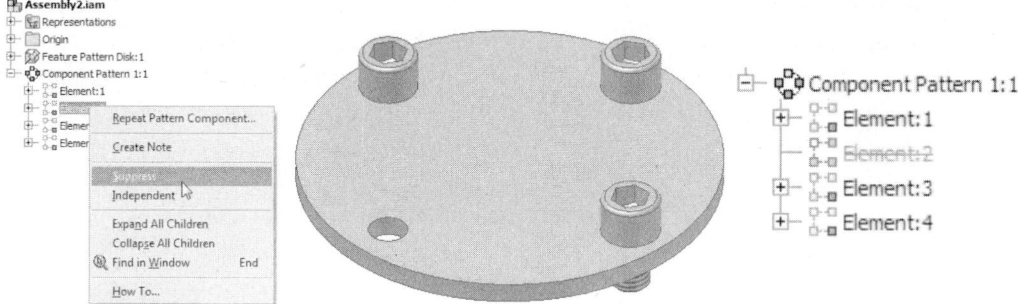

FIGURE 6.83

You can break an individual part out of the pattern by right-clicking on it and selecting Independent from the menu, as shown in Figure 6.84. Once a part is independent, it no longer has a relationship with the pattern.

FIGURE 6.84

Replacing a Component Pattern

When replacing a component in a component pattern, you replace all occurrences in the selected component pattern with a newly selected component. Expand one of the elements in the browser, and select the component to replace. With this item highlighted in the browser, right-click and select Replace from the menu, as shown in Figure 6.85 on the left. The Open dialog box will appear and will enable you to select the replacement component.

Figure 6.85 on the right shows the result of performing the operation to replace a pattern component. Since the use of component patterns allows for better capture of design intent, replacement of all occurrences maintains this design intent without manually replacing each component in the pattern. Overall ease of assembly use is improved, and component patterns can be completed more quickly.

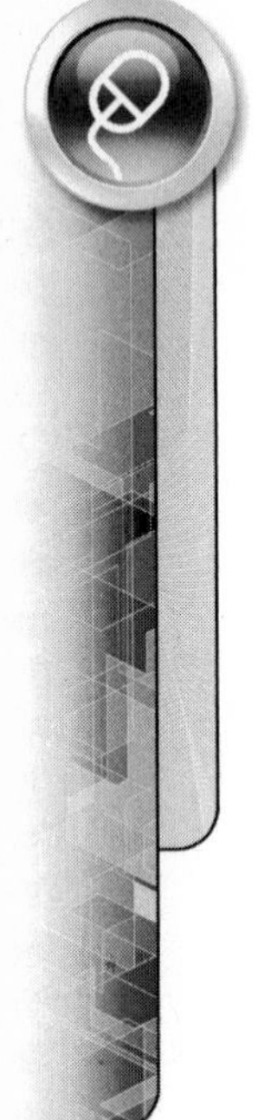

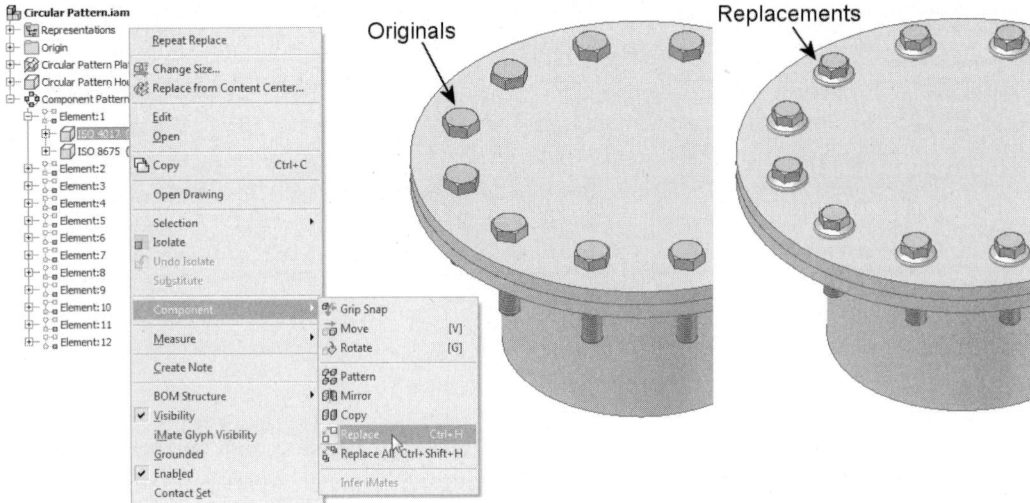

FIGURE 6.85

EXERCISE 6-3: PATTERNING COMPONENTS

In this exercise, you create a fastener component pattern to match an existing hole pattern and then replace the bolt in the pattern. You complete the exercise by demoting components to create a subassembly with the cap plate and patterned fasteners.

1. Open *ESS_E06_03.iam*, as shown in Figure 6.86 on the left. The file contains a T pipe assembly.
2. In the browser, expand the parts *ESS_E06_03-Cap_Bolt.ipt:1* and *ESS_E06_03-Cap_Nut.ipt:1*. Notice that the parts already have insert constraints applied to them, as shown in Figure 6.86 on the right.

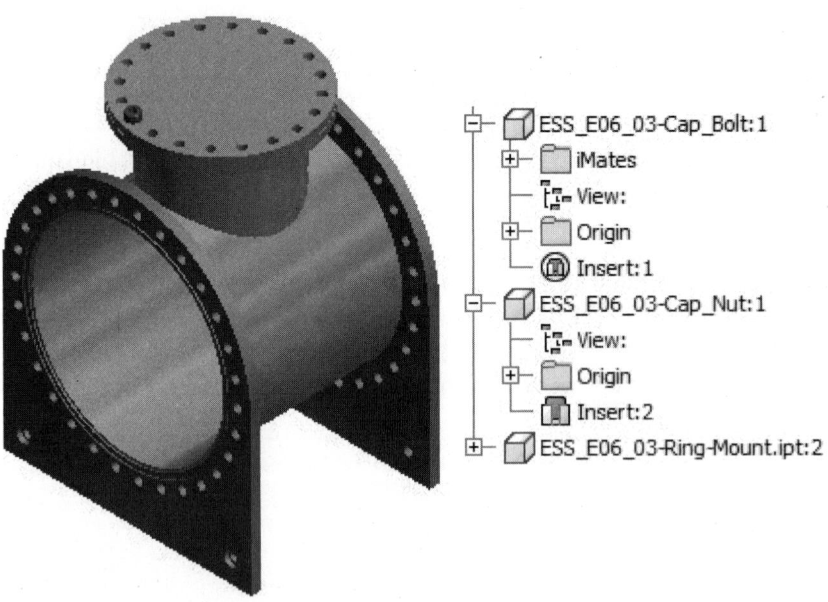

FIGURE 6.86

3. Now create a circular pattern consisting of a collection of nuts and bolts. Click the Pattern Component command from the Assemble tab > Component panel.
4. Press and hold the CTRL key and select the *ESS_E06_03-Cap_Bolt.ipt:1* and *ESS_E06_03-Cap_Nut.ipt:1* parts in the browser.
5. Click the Associated Features Pattern button found under the Feature Pattern Select area of the Pattern Component dialog box. In the graphics window, move the cursor over any hole in the Cap_Plate part. When the circular pattern of holes in the cap plate is highlighted, select the edge of one of the holes, as shown in Figure 6.87 on the left.
6. You should see a preview of the component pattern, as shown in Figure 6.87 on the right.

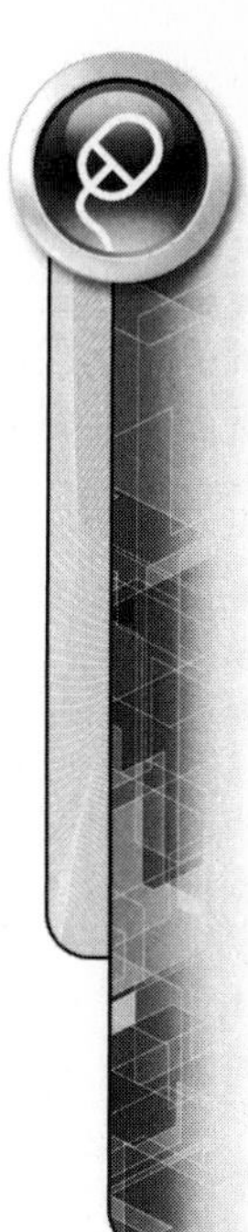

FIGURE 6.87

7. Click the OK button to create the component pattern. Expand the display of the Component Pattern 1 feature in the browser.
8. In the browser, expand Element:1 and the parts beneath it. Notice that the constraints on the original components were retained, as shown in Figure 6.88.

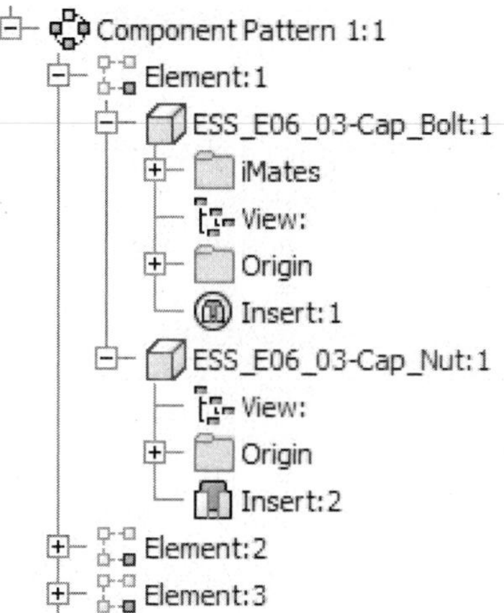

FIGURE 6.88

9. Now replace the bolt in the pattern with a similar but different fastener. Under Element:1 in the browser, right-click the *ESS_E06_03-Cap_Bolt,* click Component, and click Replace from the menu.

10. In the Place Component dialog box, select *ESS_E06_03-Cap_Hex_Bolt.ipt*, from the Chapter 06 folder and click Open. A warning message box will appear to notify you that some assembly constraints may be lost.
11. Click the OK button. The pattern is updated to reflect the new bolt, as shown in Figure 6.89.

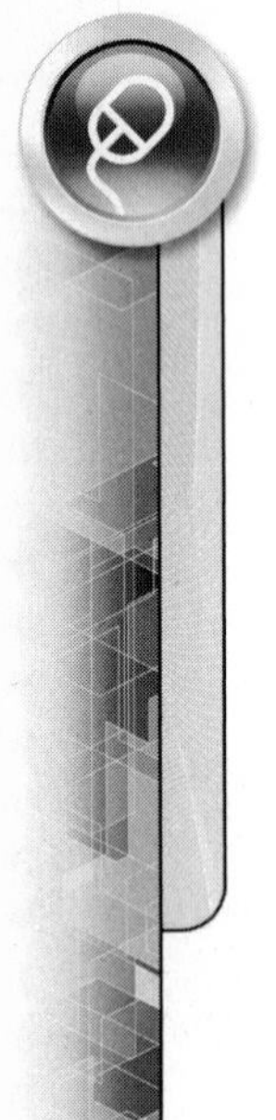

NOTE

The fastener in this exercise is constrained to the plate with an Insert iMate constraint. Since the replacement fastener has a similar Insert iMate, the constraint to the plate is retained.

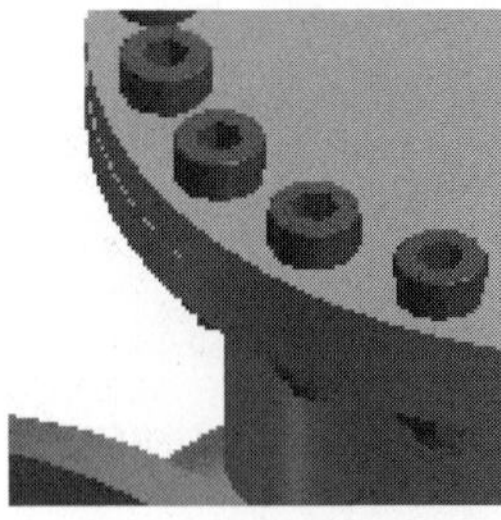

FIGURE 6.89

12. Next, you demote a number of components; this will create a subassembly with the cap plate and component pattern.
13. Hold down CTRL while you click *ESS_E06_03-Cap_Plate:1* and Component Pattern 1 in the browser.
14. Right-click, and click Component followed by Demote from the menu. In the Create In-Place Component dialog box, enter *Cap_Kit_Assembly* in the New Component Name field. Change the New File Location to C:\Inv 2013 Ess Plus\Chapter 06\ When finished, click the OK button. A warning message box appears to notify you that some assembly constraints may be lost. Click Yes.
15. In the browser, expand Cap_Kit_Assembly:1. Notice that the cap plate and component pattern are now part of the new subassembly. Notice also that the constraints on the original components were retained. Your display should appear similar to Figure 6.90.

FIGURE 6.90

16. Close all open files. Do not save changes. End of exercise.

ANALYSIS COMMANDS

Various commands are available that assist in analyzing sketch, part, and assembly models. You can calculate minimum distance between components, calculate the center of gravity of parts and assemblies, and perform interference detection.

The Minimum Distance Command

You can easily obtain the minimum distance between any components, parts, or faces in an assembly. While in the assembly, select the Measure Distance command from the Inspect tab > Measure panel as shown in Figure 6.91 on the left. Then change the selection priority, depending on what you are measuring. The three available selection modes are shown in Figure 6.91 on the right.

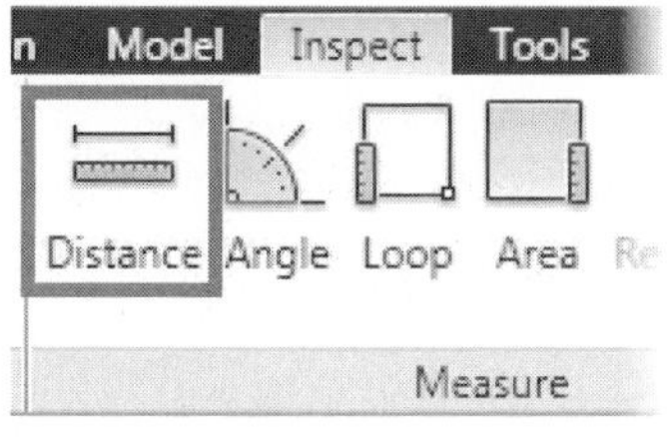

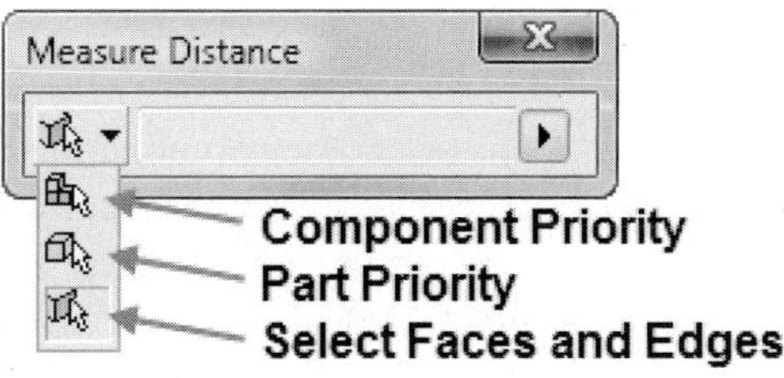

FIGURE 6.91

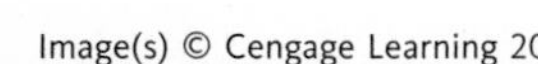

Identifying the Center of Gravity of an Assembly

A center of gravity placed in an assembly could be critical to the overall design process of that assembly, whether it is used in the next assembly or in the main assembly. On the View tab > Visibility panel, click the Center of Gravity command as shown in Figure 6.92 on the left. The image on the right shows the center of gravity icon applied to an assembly model. This icon actually consists of a triad displaying the X, Y, and Z directions. Three selectable work planes and a selectable work point area are also available for the purpose of measuring distances and angles.

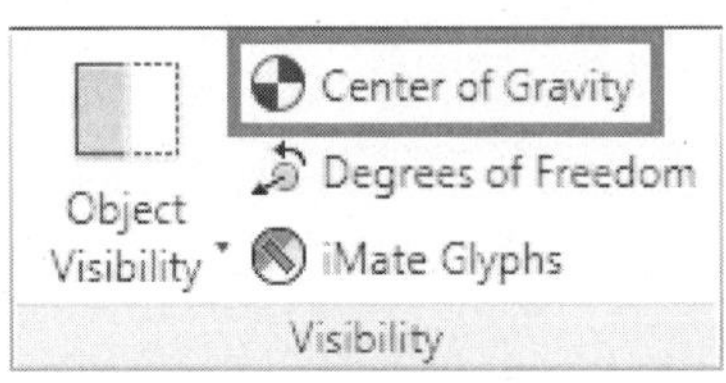

FIGURE 6.92

Interference Checking

You can check for interference in an assembly using one or two sets of objects. To check the interference among sets of stationary components, make the assembly or subassembly in question active. Then click the Analyze Interference command on the Inspect tab > Interference panel as shown in Figure 6.93 on the left. The Interference Analysis dialog box will appear, as shown in Figure 6.93 on the right.

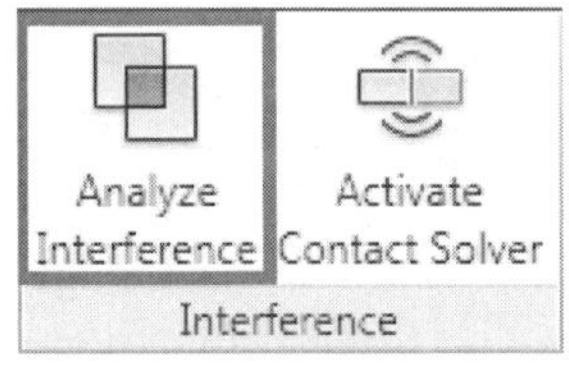

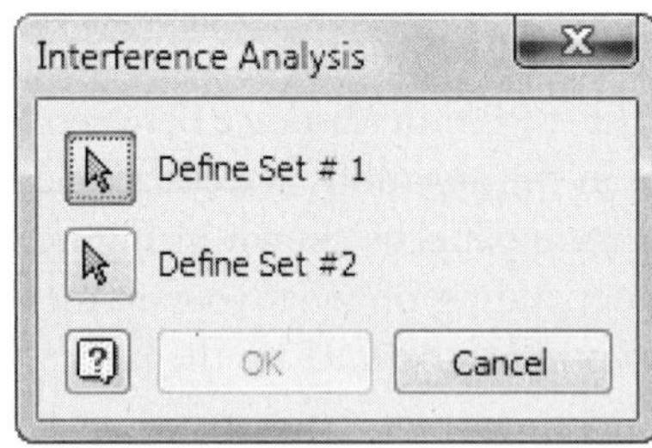

FIGURE 6.93

Click on Define Set #1, and select the components that will define the first set. Click on Define Set #2, and select the components that will define the second set. A component can exist in only one set. To add or delete components from either set, select the button that defines the set that you want to edit. Click components to add to the set, or press the CTRL key while selecting components to remove from the set.

NOTE

Use only Define Set #1 if you want to check for interference in a single group of objects.

Once you have defined the sets, click the OK button. The order in which you selected the components has no significance. If interference is found, the Interference Detected dialog box will appear, as shown in Figure 6.94.

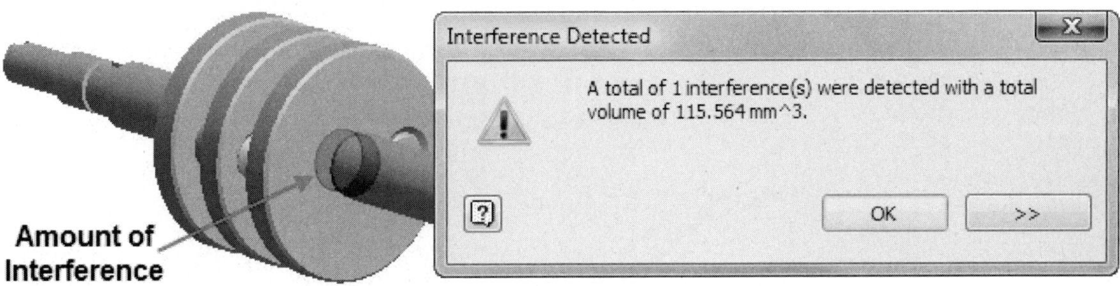

FIGURE 6.94

The information in the dialog box defines the X, Y, and Z coordinates of the centroid of the interfering volume. It also lists the volume of the interference and the components that interfere with one another. A temporary solid will also be created in the graphics window that represents the interference. You can copy the interference report to the clipboard or print it from the commands in the Interference Detected dialog box. When the operation is complete, click the OK button, and the interfering solid will be removed from the screen.

Performing an interference detection does not fix the interfering problem; it only presents a graphical representation of the problem. After analyzing and finding an interference, edit the assembly or components to remove the interference. You can also detect interference when driving constraints.

EXERCISE 6-4: ANALYZING AN ASSEMBLY

In this exercise, you analyze a partially completed assembly for interference between parts and then check the physical properties to verify design intent.

1. Open the assembly *ESS_E06_04.iam*. The linkage is displayed, as shown in Figure 6.95 on the left.
2. Begin the process of checking for interferences in the assembly. First zoom into the linkages, as shown in Figure 6.95 on the right.

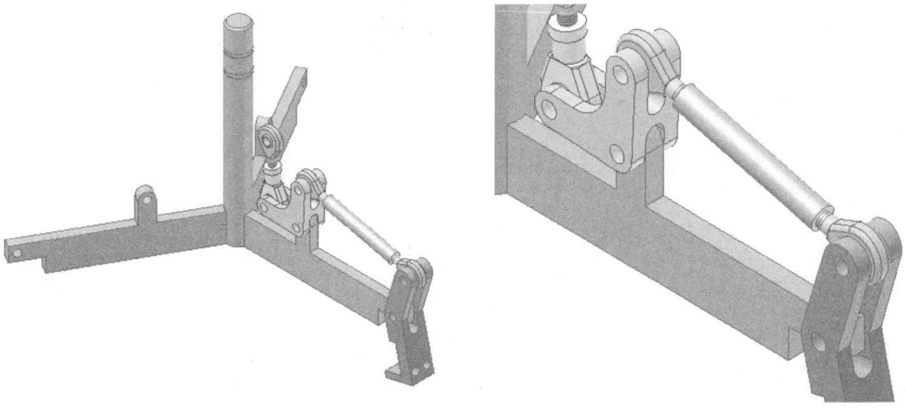

FIGURE 6.95

3. Click Analyze Interference from the Inspect tab > Interference panel.
4. When the Analyze Interference dialog box displays, select the two components for Define Set #1, as shown in Figure 6.96 on the left.
5. Click the Define Set #2 button, then select the spyder, as shown in Figure 6.96 on the right.

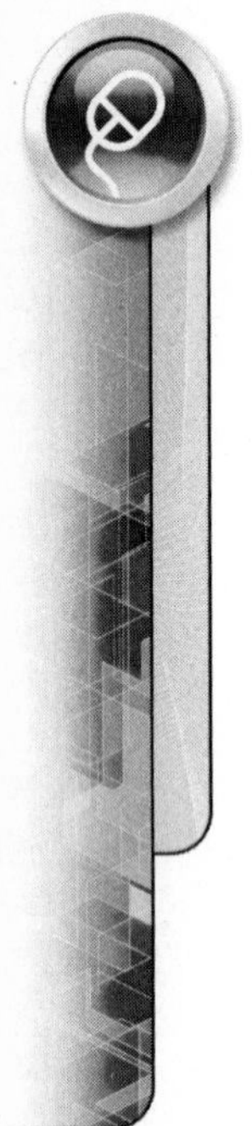

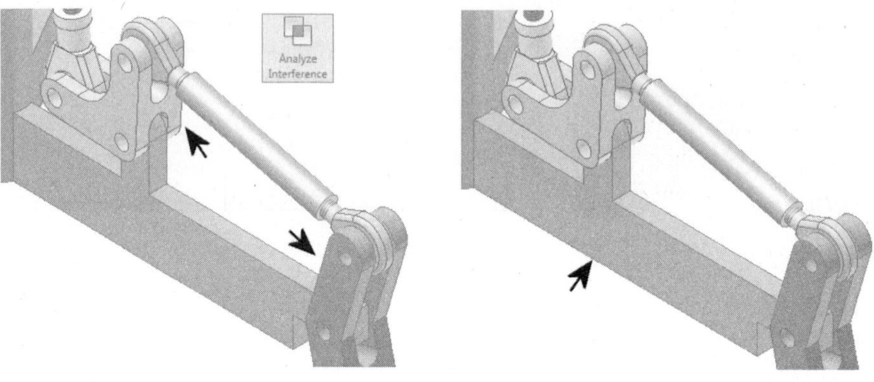

FIGURE 6.96

6. Click the OK button. Notice that the Interference Detected dialog box is displayed, and the amount of interference displays on the parts of the assembly, as shown in Figure 6.97.

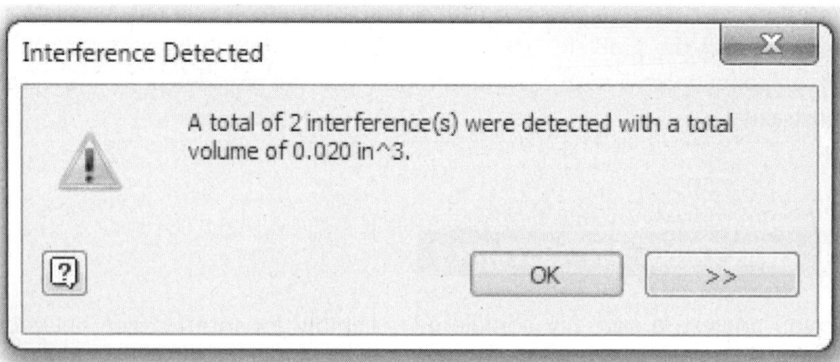

FIGURE 6.97

7. Click the More (>>) button to expand the dialog box and display additional information.
8. Click the OK button to close the Interference Detected dialog box.
9. The design intent of the lifting mechanism relies on the correct selection of materials. This impacts the strength and mass of the assembly. You will now display the physical properties of the entire assembly. Return to the Home View of the assembly.
10. In the browser, right-click *ESS_E06_04.iam*, and then select iProperties, as shown in Figure 6.98.

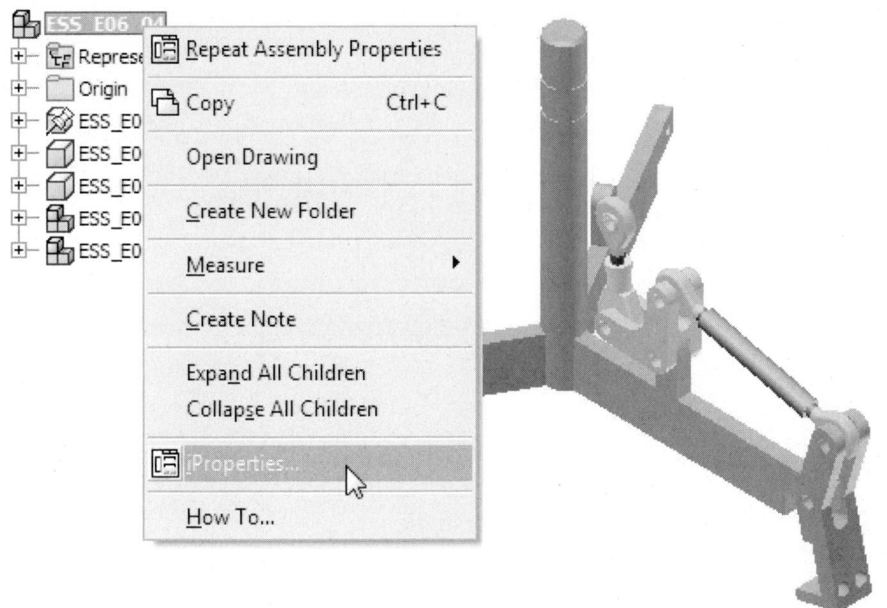

FIGURE 6.98

11. When the iProperties dialog box displays, click the Physical tab.
12. Click the Update button. Notice that the physical properties of the assembly are displayed.

To get a more accurate account of the assembly properties, you can change the Requested Accuracy setting to Very High. The updating action may take some time, depending on the size of the assembly being analyzed.

13. Click the Close button to dismiss this dialog box.
14. Close all open files. Do not save changes. End of exercise.

DRIVING CONSTRAINTS

You can animate or drive mechanical motion using the Drive Constraint command. To simulate motion, an angle, mate, tangent, or insert assembly constraint must exist. You can only drive one assembly constraint at a time, but you can use equations to create relationships to drive multiple assembly constraints simultaneously. To drive a constraint, right-click on the desired constraint in the browser, and select Drive Constraint from the menu, as shown in Figure 6.99 on the left.

The Drive Constraint dialog box will appear, as shown in Figure 6.99 on the right. Depending on the constraint that you are driving, the units may be different. Enter a Start value; the default value is the angle or offset for the constraint. Enter a value for End and a value for Pause Delay if you want a dwell time between the steps. In the dialog box, you can also choose to create an animation (AVI or WMV) file that will record the assembly motion. You can replay the recorded file without having Autodesk Inventor installed.

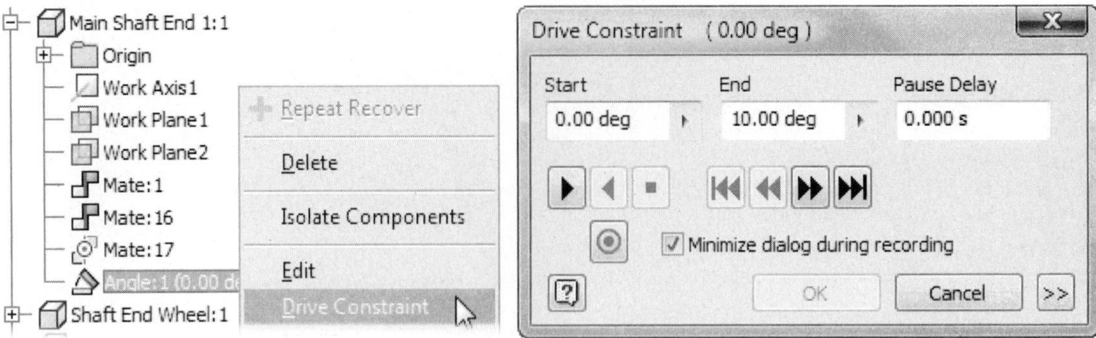

FIGURE 6.99

The following sections describe the Start, End, and Pause Delay controls.

Start. Sets the start position of the offset or angle.

End. Sets the end position of the offset or angle.

Pause Delay. Sets the delay between steps. The default delay units are seconds.

Use the following Motion and video buttons to control the motion and to create a (PE) recorded video file.

	Forward	Drives the constraint forward, from the start position to the end position.
	Reverse	Drives the constraint in reverse, from the end position to the start position.
	Pause	Temporarily stops the playback of the constraint drive sequence.
	Minimum	Returns the constraint to the starting value and resets the constraint driver. This button is not available unless the constraint driver has been run.
	Reverse Step	Reverses the constraint driver one step in the sequence. This button is not available unless the drive sequence has been paused.
	Forward Step	Advances the constraint driver one step in the sequence. This button is not available unless the drive sequence has been paused.
	Maximum	Advances the constraint sequence to the end value.
	Record	Begins capturing frames at the specified rate for inclusion in an animation file named and stored to a location that you define.

To set more conditions on how the motion will behave, click the More (>>) button. This action will display the expanded Drive Constraint dialog box, as shown in Figure 6.100. The following sections describe the options in the dialog box.

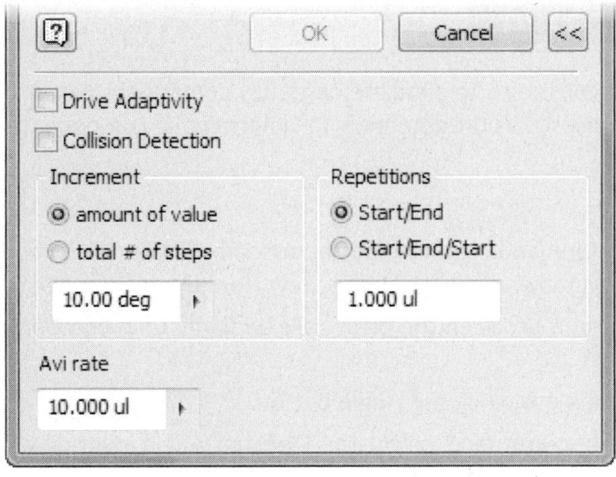

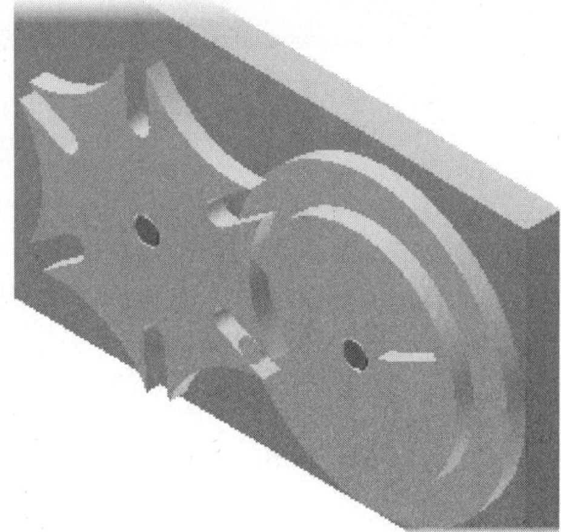

FIGURE 6.100

Drive Adaptivity. Click this option to adapt the component while the constraint is driven. It only applies to assembly components for which adaptivity has been defined and enabled.

Collision Detection. Click this option to drive the constraint until a collision is detected. When interference is detected, the drive constraint will stop and the components where the collision occurs will be highlighted. It also shows the constraint value at which the collision occurred.

Increment

The options and value in this area determine the method and value that the constraint will be incremented during the animation.

Amount of Value. Increments the offset or angle value by this value for each step.

Total # of Steps. The constraint offset or angle is incremented by the same value per step based on the number of steps entered and the difference between the start and stop values.

Repetitions

Sets how the driven constraint will act when it completes a cycle and how many cycles will occur.

Start/End. Drives the constraint from the start value to the end value and resets at the start value.

Start/End/Start. Drives the constraint from the start value to the end value and then drives it in reverse from the end value to the start value.

AVI Rate. Specifies how many frames are skipped before a screen capture is taken of the motion that will become a frame in the completed AVI file.

NOTE

If you try to drive a constraint and it fails, you may need to suppress or delete another assembly constraint to allow the required component degrees of freedom. To reduce the size of a WMV or AVI movie file, reduce the screen size before creating the file and use a solid background color in the graphics window.

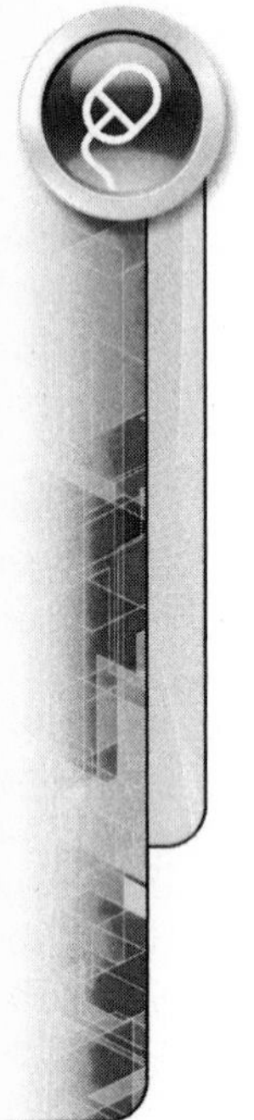

EXERCISE 6-5: DRIVING CONSTRAINTS

In this exercise, you drive an angle constraint to simulate assembly component motion. Then you drive the constraint and use collision detection to determine if components interfere.

1. Open *ESS_E06_05.iam*.
2. Use the Orbit and Zoom commands to examine the assembly. When finished, right-click in the graphics window, and click Home View from the menu.
3. Now create an angle constraint between the pivot base and arm. Click the Constrain command.
4. In the Place Constraint dialog box, click the Angle button.
5. Select the Base face and then arm face, select the Directed Angle solution as shown in Figure 6.101.
6. Enter an angle of **45°**. Click the OK button.

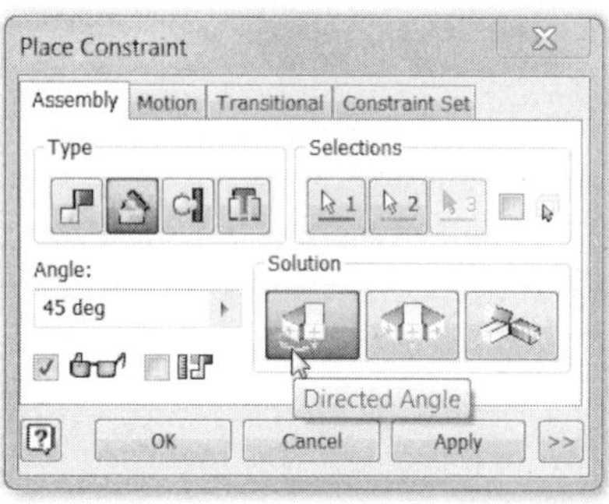

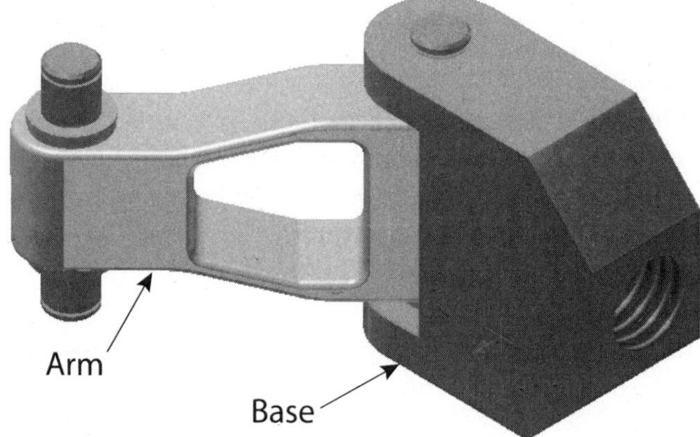

FIGURE 6.101

7. Now drive the angle constraint you just placed. In the browser, expand *ESS_E06_05-Pivot_Base:1.*
8. Right-click on the Angle constraint, and then click Drive Constraint from the menu.
9. In the Drive Constraint dialog box, verify that 45.00 deg is the Start value and enter **120.00 deg** as the End value.
10. Click the More (>>) button.
11. In the Increment section, enter **2.00 Deg**, as shown in Figure 6.102.
12. Click the Forward button. Notice that the arm assembly interferes with the pivot base between 45.00° and 120.00° but Inventor continues to drive the constraint.
13. Click the Reverse button.

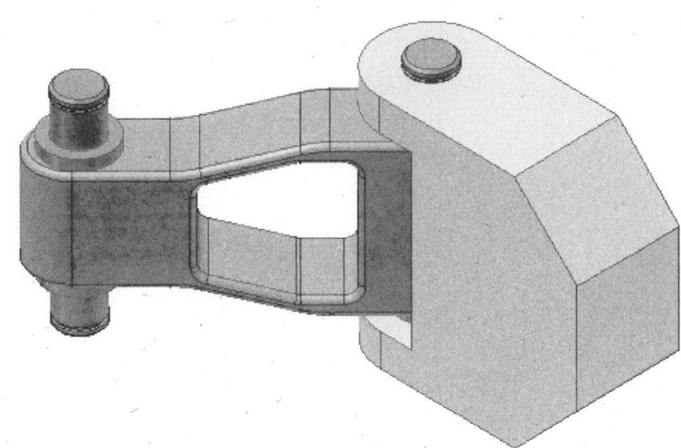

FIGURE 6.102

14. Now perform a collision detection operation. In the Drive Constraint dialog box click the >> more button and place a check in the Collision Detection box.
15. Change the Increment value to **.1 deg**.
16. Click the Forward button. Notice that a collision is detected at 80.7° and that the interfering parts are highlighted, as shown in Figure 6.103. Click OK to return to the Drive Constraint dialog box.

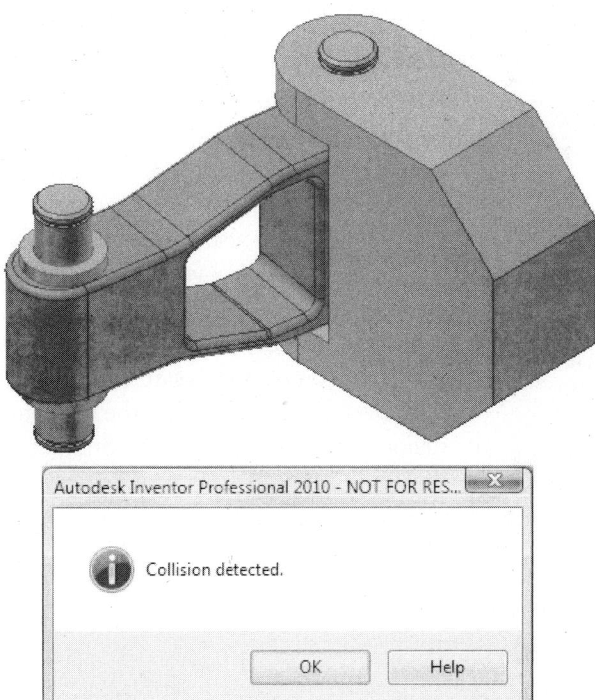

FIGURE 6.103

17. Now perform a check on the full range of motion for the arm. In the Drive Constraint dialog box, enter **80.00 deg** for the Start value and **–80.00 deg** for the End value.
18. Ensure that the Collision Detection option is still checked.
19. Change the Increment value to **2.00 deg**.
20. Select Start/End/Start and enter **4** in the Repetitions area of the Drive Constraint dialog box.
21. Click the Reverse button to test the full range of motion for interference. No collision is detected in this range of motion.
22. In the Drive Constraint dialog box, click the OK button. The 80.00°, which represents the maximum angle before collision, is applied to the Angle constraint, as shown in Figure 6.104.

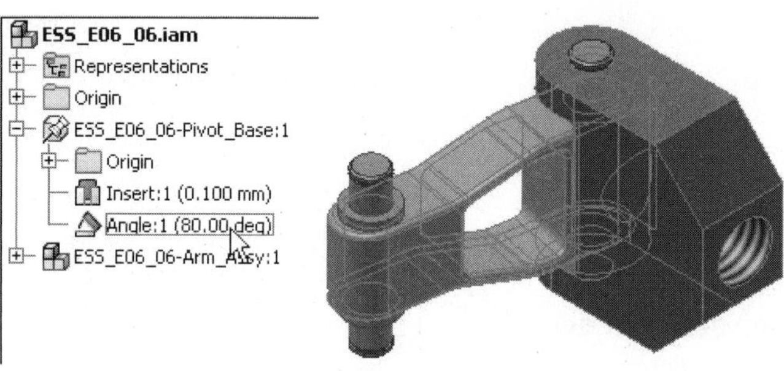

FIGURE 6.104

23. Close all open files. Do not save changes. End of exercise.

CREATING PRESENTATION FILES

After creating an assembly, you can create drawing views based on how the parts are assembled. If you need to show the components in different positions similar to an exploded view, hide specific components, or create an animation that shows how to assemble and disassemble the components, you need to create a presentation file. Components can also be hidden and later retrieved through design view representations. This topic is covered in a later chapter. A presentation file is separate from an assembly and has a file extension of *.ipn*. The presentation file is associated with the assembly file, and changes made to the assembly file will be reflected in the presentation file. You cannot create components in a presentation file. To create a new presentation file, click New on the Get Started tab > Launch panel. When the New File dialog box appears, select the Standard.ipn icon, as shown in Figure 6.105.

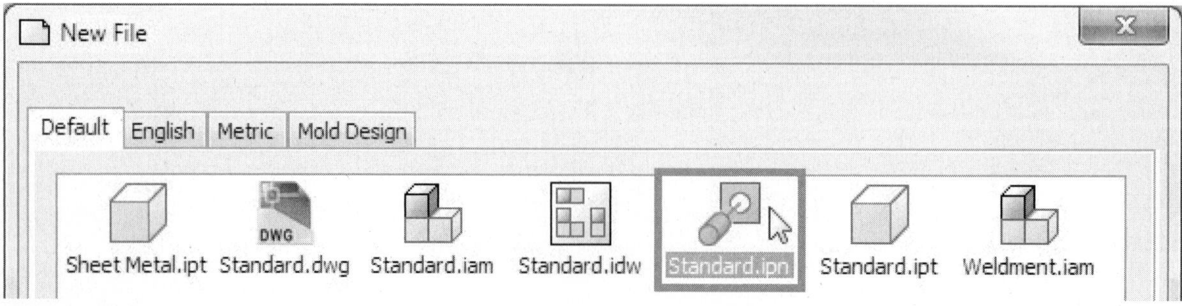

FIGURE 6.105

By default, the ribbon will show the presentation commands available to you, as shown in Figure 6.106. The following sections explain these commands.

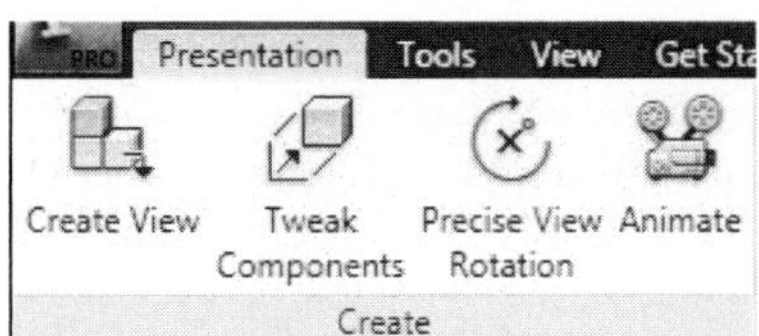

FIGURE 6.106

Create View. Click to create views of the assembly. Created views are based on the same assembly model. Once created, a view will be listed as an explosion in the browser.

Tweak Components. Click to move and/or rotate parts in the view.

Precise View Rotation. Click to rotate the view by a specified angle and direction using a dialog box.

Animate. Click to create an animation of the assembly tweaks and saved camera views; an AVI or WMV file can be output.

The basic steps for creating a presentation view are to reposition (tweak) the parts in specific directions and then create an animation, if needed. The following sections discuss these steps.

Creating Presentation Views

The first step in creating a presentation is to create a presentation view. Issue the Create View command on the Presentation tab > Create panel as shown in Figure 6.107 on the left, or right-click and select Create View from the menu. The Select Assembly dialog box appears, as shown in Figure 6.107 on the right. The dialog box is divided into two sections: Assembly and Explosion Method.

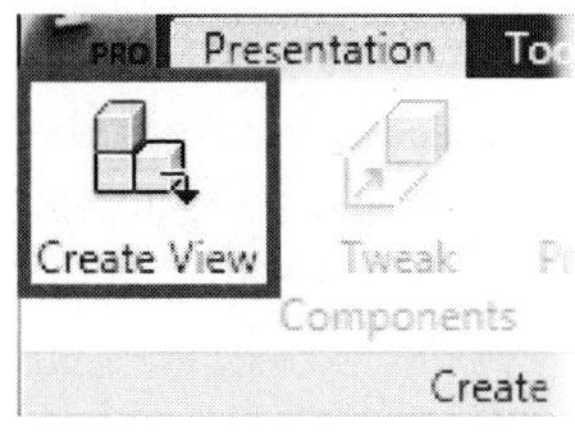

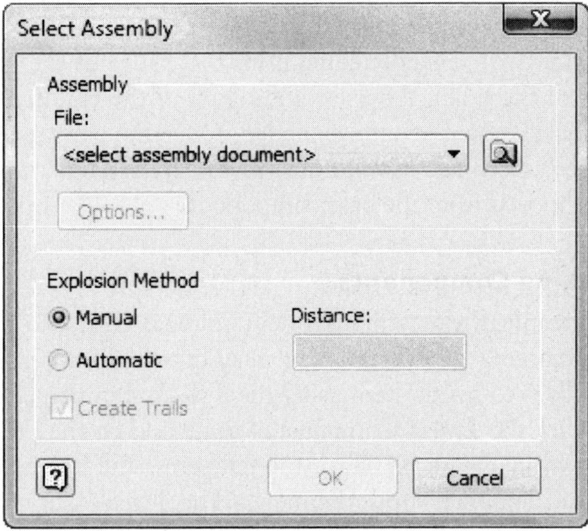

FIGURE 6.107

Assembly

In this section, you determine on which assembly and design view to base the presentation view.

File. If an assembly file or files are open they will appear in the drop-down list. Select an assembly file or navigate to and select an assembly file on which to base the presentation view.

Options. Click to display the File Open Options dialog box. Use this dialog box to select design view representations, positional representations, and level of detail representations, which are covered in Chapter 9.

Explosion Method

In this section, you can choose to manually or automatically explode the parts, meaning that you separate the parts a given distance.

Manual. Click so that the components will not be exploded automatically. After the presentation view is created, you can add tweaks that will move or rotate the parts.

Automatic. Click so that the parts will be exploded automatically with a given value.

Create Trails. If you clicked Automatic, click to have visible trails. Trails are lines that show how the parts are exploded.

Distance. If Automatic is selected, enter a value that the parts will be exploded.

After making your selection, click the OK button to create a presentation view. The components will appear in the graphics window and a presentation view will appear in the browser. If you clicked the Automatic option, the parts will be exploded automatically. To determine how the parts will be exploded, Autodesk Inventor analyzes the assembly constraints. If you constrained parts using the mate plane option and the arrows perpendicular to the plane (normals) for both parts were pointing outward, they will be exploded away from each other the defined distance. Think of the mating parts as two magnets that want to push themselves in opposite directions. The grounded component in the browser will be the component that stays stationary, and the other components will move away from it. If you are creating trails automatically, they will generally come from the center of the part, not necessarily from the center of the holes. After expanding all of the children in the browser, you can see numerically how the parts exploded. Once the parts are exploded, the distance is referred to as a tweak. The number by each tweak reflects the distance that the component is moved from the base component.

Tweaking Components

After creating the presentation view, you may choose to edit the tweak of a component or move or rotate the component an additional distance. To reposition the components manually is to tweak them using the Tweak command. To manually tweak a component, click the Tweak Components command on the Presentation tab > Create panel as shown in Figure 6.108 on the left, press the hot key T, or right-click and select Tweak Component from the menu. The Tweak Component dialog box will appear, as shown in Figure 6.108 on the right. The Tweak Component dialog box has two sections: Create Tweak and Transformations.

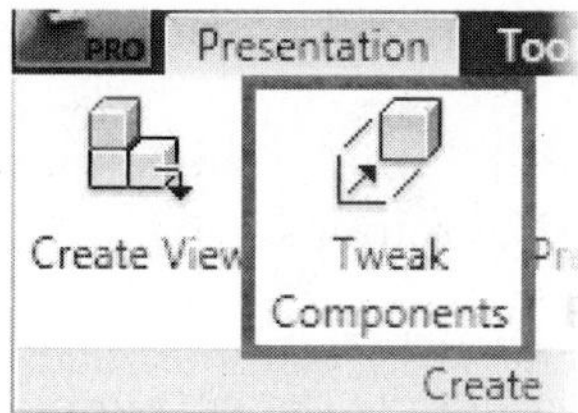

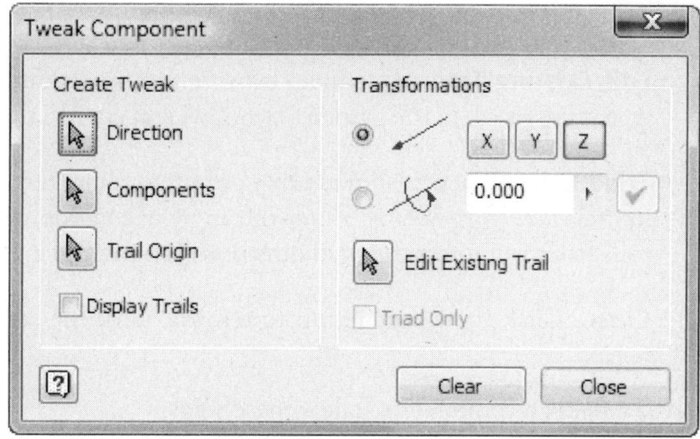

FIGURE 6.108

Create Tweak

In this section, you select the components to tweak, set the direction and origin of the tweak, and control trail visibility.

Direction. Determine the direction or axis of rotation for the tweak. After clicking the Direction button, select an edge, face, or feature of any component in the graphics window to set the direction triad (X, Y, and Z) for the tweak. The edge, face, or feature that you selected does not need to be on the components you are tweaking.

Components. Select the components to tweak. Click the Components button, and then click the components in the graphics window or browser to tweak. If you selected a component when you started the Create Tweak operation, it will be included in the components automatically. To remove a component from the group, press and hold the CTRL key and click the component.

Trail Origin. Set the origin for the trail. Click the Trail Origin button, and then click in the graphics window to set the origin point. If you do not specify the trail origin, it will be placed at the center of mass for the part.

Display Trails. Click if you want to see the tweak trails for the selected components. Clear the checkbox to hide the trails.

Transformations

In this section, you set the type and value of a tweak.

Linear. Click the button next to the arrow and line to move the selected components in a linear fashion.

Rotation. Click the button next to the arrow and arc to rotate the selected components.

X, Y, Z. Click the X, Y, or Z coordinate button to determine the direction for a linear tweak or the axis for a rotational tweak. Alternately, you can select the arrow on the triad that represents the X, Y, or Z direction.

Tweak Distance/Tweak Angle Field. Enter a positive or negative value for the tweak distance or rotation angle, or click a point in the graphics window and move the cursor with the mouse button depressed to set the distance.

Apply. After making all selections, click the Apply button to complete the tweak.

Edit Existing Trail. To edit an existing tweak, click the Edit Existing Trail button. Select the tweak in the graphics window, and change the desired settings.

Triad Only. Click the Triad Only option to rotate the direction triad without rotating selected components. Enter the angle of rotation, and then click the Apply button. After you rotate the triad direction, you can use it to define tweaks.

Clear. Click the Clear button to remove all of the settings and set up for another tweak.

To tweak a component, follow these steps:

1. Issue the Tweak Component command.
2. Determine the direction or axis of rotation for the tweak by clicking the Direction button and selecting an edge, face, or feature.
3. Select the components to tweak by clicking the Components button and clicking the components in the graphics window or browser that will be tweaked.
4. Select any additional trail origin points.
5. Determine whether or not you want trails to be visible.
6. Set the type of tweak to linear or rotation.
7. Click the X, Y, or Z coordinate button to determine the direction for a linear tweak or the axis for a rotational tweak.
8. Enter a value for the tweak in the text box, or select a point on the screen and drag the part into its new position.
9. Click the Apply button in the Tweak Component dialog box.

To edit an individual tweak, click on its value in the browser, as shown in Figure 6.109. Enter a new value in the text box in the lower-left corner of the browser, and press ENTER to use the new value.

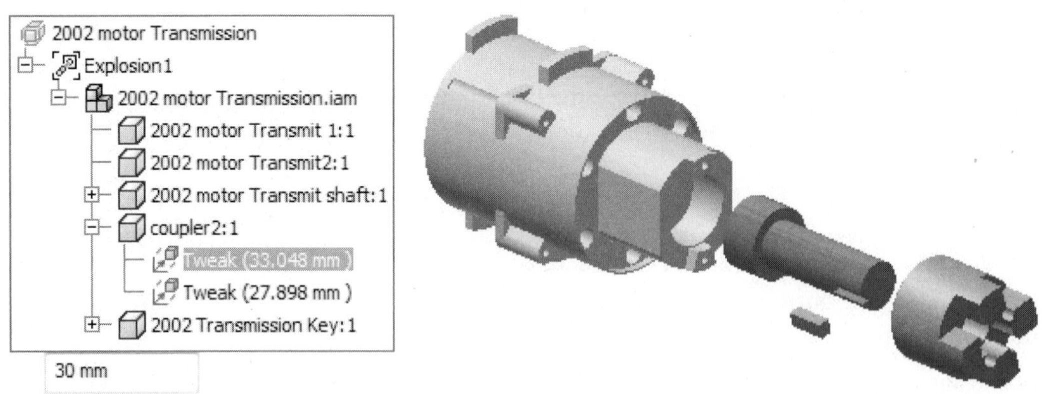

FIGURE 6.109

To extend all of the tweaks the same distance, right-click on the assembly's name in the browser, and select Auto Explode from the menu, as shown in Figure 6.110 on the left. Enter a value, and click the OK button, as shown in Figure 6.110 on the right, and all of the tweaks will be extended the same distance. You cannot use negative values with the Auto Explode method.

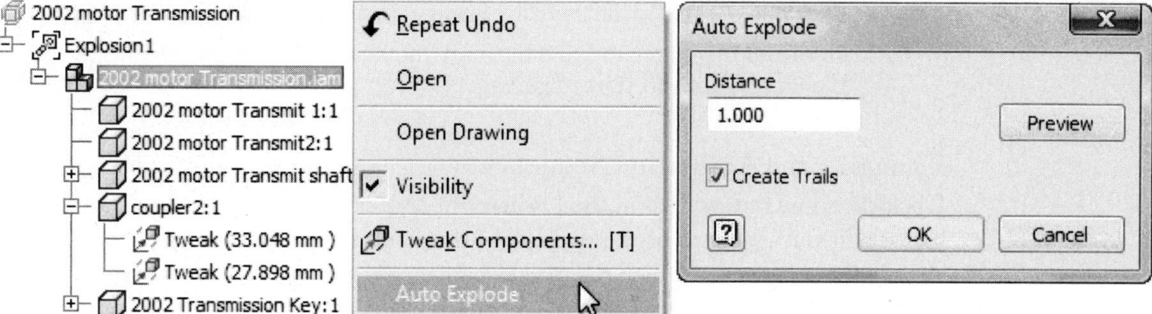

FIGURE 6.110

Animation

After you have repositioned the components, you can animate the components to show how they assemble or disassemble. To create an animation, click the Animate command on the Presentation tab > Create panel as shown in Figure 6.111 on the left. The Animation dialog box will appear as shown in Figure 6.111. The Animation dialog box has three sections: Parameters, Motion, and Animation Sequence, which appears under the More (>>) button.

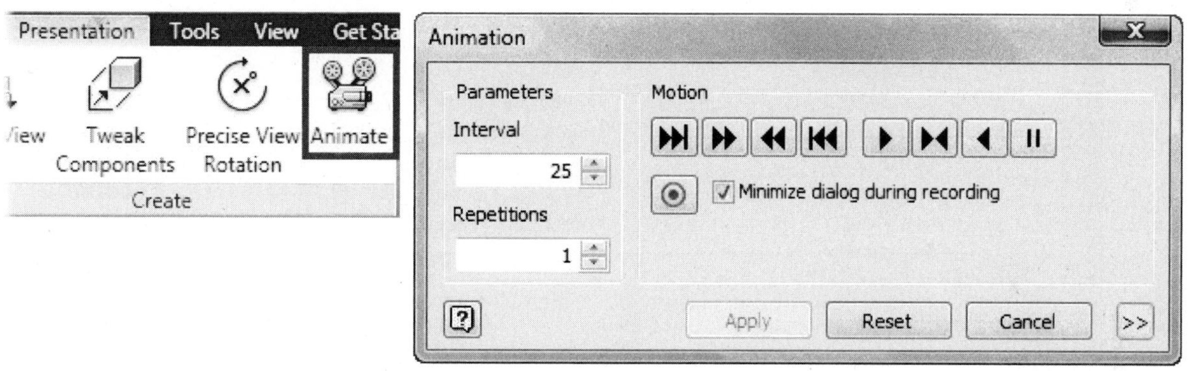

FIGURE 6.111

Parameters

In this section, you specify the playback speed and the number of repetitions for the animation.

Interval. Set this value for the playback speed of the animation in frames. The higher the number, the greater the number of steps in the tweak and the slower the animation. A smaller number will speed up the animation.

Repetitions. Set the number of times to repeat the playback. Enter the desired number of repetitions, or use the up or down arrow to select the number.

NOTE

To change the number of repetitions after playing the motion, click the Reset button on the dialog box, and then enter a new value.

To animate components, follow these steps:

1. Issue the Animation command.
2. Set the number of repetitions.
3. Adjust the tweaks as needed.

4. Click one of the play buttons to view the animation in the graphics window.
5. To record the animation to a file, click the Record button, and then click one of the Play buttons to start recording.

Changing the Animation Sequence

Click the more button >> in the lower-right corner of the Animation dialog box to see or edit the sequence of the tweaks. In the Animation Sequence section, you can change the sequence in which the tweaks happen, select the tweak, and then select the needed operation. Expanding the Animation dialog box displays Figure 6.112. For example, in the image on the left, each component will move until its complete cycle is over before another component starts to move. In the image on the right, notice all sequence numbers have changed to a value of 1. This means when the animation begins, all components will move at the same time at the same rate of speed. Experiment with the sequencing of an exploded assembly to gain more dramatic animation results.

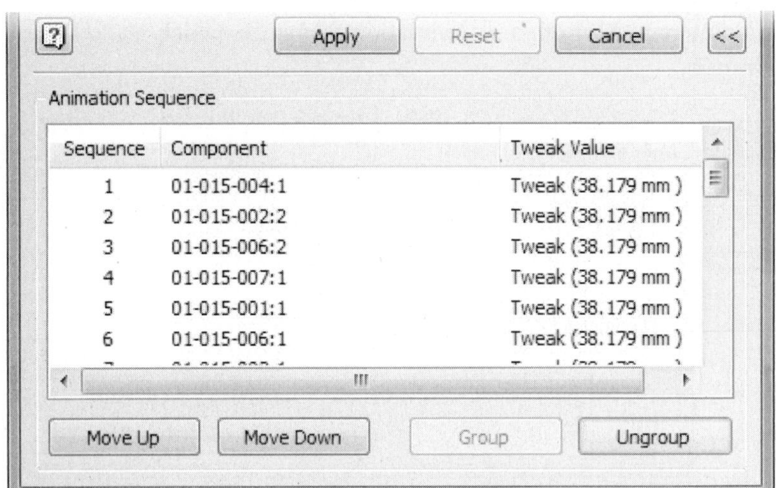

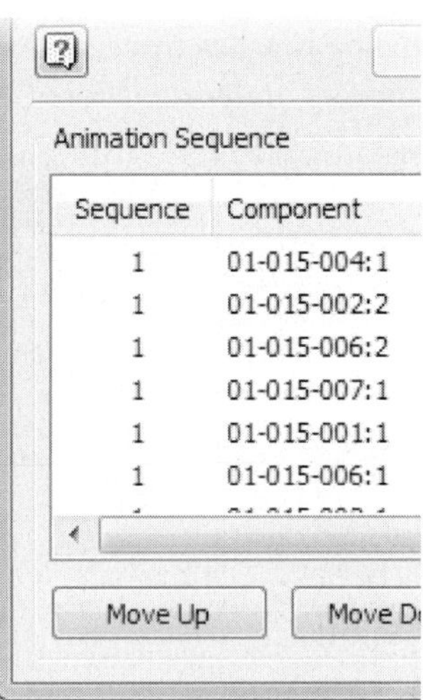

FIGURE 6.112

Move Up. Click to move the selected tweak up one place in the list.

Move Down. Click to move the selected tweak down one place in the list.

Group. Select a number of tweaks, and then click the Group button. When tweaks are grouped, all of the tweaks in the group will move together as you change the sequence. The group assumes the sequence order of the lowest tweak number.

Ungroup. After selecting a tweak that belongs to a group, you can click the Ungroup button, and the tweak can then be moved individually in the list. The first tweak in the group assumes a number that is one higher than the group. The remaining tweaks are numbered sequentially following the first.

TIP

You can search the Help system for *Sequence View* or *Edit Tasks and Sequences* to discover additional browser views and editing capabilities when working with Presentation files.

AUTODESK INVENTOR PUBLISHER

Although beyond the scope of this book, another product, Autodesk Inventor Publisher may be useful for you to know about as it relates to illustrating or documenting your assemblies. Publisher is an application that leverages CAD models to easily create product documentation—maintaining an associative link so that updates are easily incorporated as your designs change.

After importing your CAD data into Inventor Publisher, you will find tools allowing you to easily reposition and change the appearance of components in different process steps and views, automatically explode 3D assemblies, add annotations, and more, in order to quickly and easily communicate detailed process steps from any viewpoint.

Some of the authoring tools include:

- Component move/pull reposition tools make it easy to position individual parts and restore them to their home position.
- Storyboarding a sequence of instructions that can be animated or viewed at specified speed and transition.
- Automated and Manual tools will explode all parts or just sub-assemblies.
- Edit multiple snapshots in the storyboard at the same time, re-order, or one-step reverse the storyboard.
- Text annotations can be manually entered or inherited from the CAD model.
- Imported images and labels can be used to annotate and supplement the instructions.
- Pre-defined styles for the appearance of the assembly or individual parts can be used to obtain the required appearance.

Designs can be published for use with a wide range of applications, viewers, and media players so your documentation can be accessed by anyone. An entire document, particular storyboard or a snapshot can be selected to be published. Snapshots are published as views and storyboards are published as animations or a sequence of views. Figure 6.113 shows an example of a file opened and being edited in Inventor Publisher.

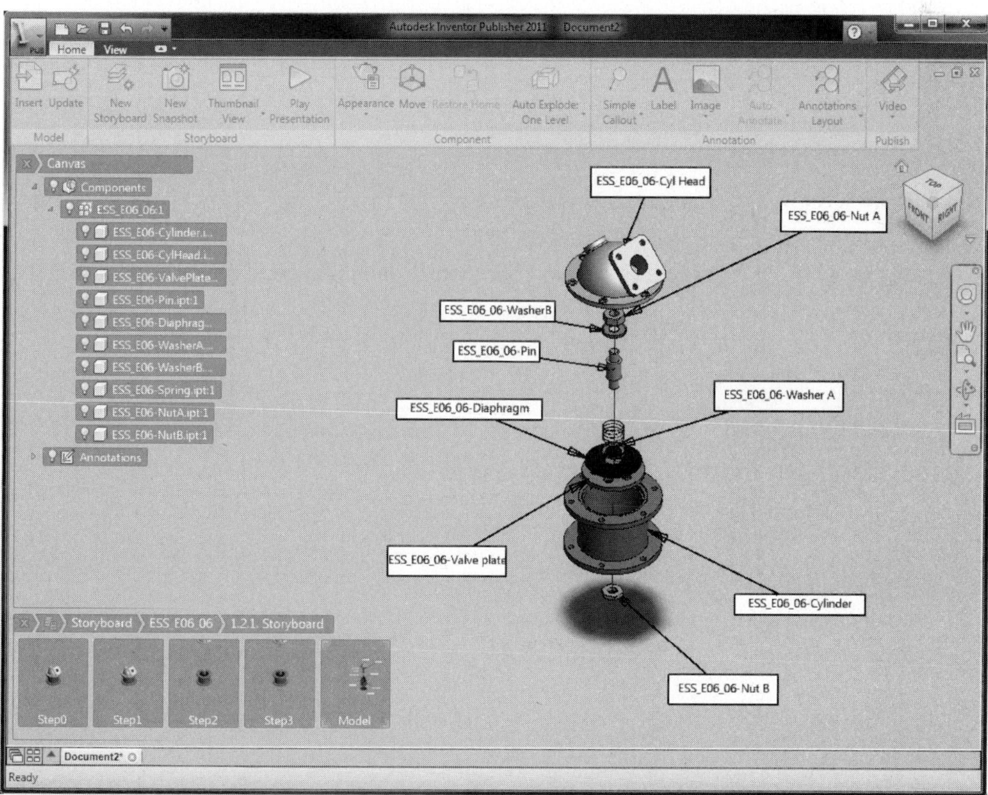

FIGURE 6.113

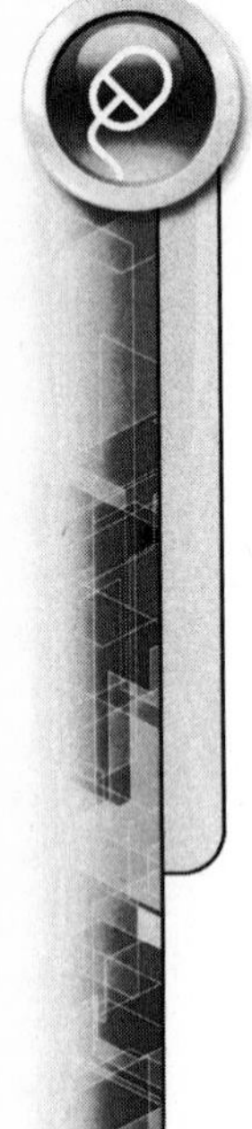

EXERCISE 6-6: CREATING PRESENTATION VIEWS

In this exercise, you create a presentation view of an existing assembly, and you add tweaks to the assembly components to create an exploded view.

1. Begin creating a presentation by clicking the New command.
2. Select the English template folder, and in the Presentation area double-click Standard (in).ipn.
3. Click the Create View command on the Presentation tab > Create panel.
4. In the Select Assembly dialog box, verify that the Explosion Method is set to Manual. Then click the Open an Existing File button next to the File field. Open *ESS_E06_06.iam*, in the Chapter 06 folder and click the Options button in the Select Assembly dialog box.
5. In the Design View Representation field, select Internal Components, as shown in Figure 6.114 and click OK.

FIGURE 6.114

6. Click OK to create the view. Your display should appear similar to Figure 6.115.

FIGURE 6.115

7. In the browser, click the plus sign in front of Explosion1 and the plus sign in front of *ESS_E06_06.iam* to expand their components. The assembly components are displayed so you can select components in the graphics screen or in the browser.
8. Click the Tweak Components command on the Presentation tab > Create panel.
9. In the Tweak Component dialog box, verify that Direction is selected and Display Trails is checked.
10. Move the cursor over a cylindrical face until the temporary Z axis aligns with the axis of the assembly, as shown in Figure 6.116.
11. Click to accept the axis orientation.

FIGURE 6.116

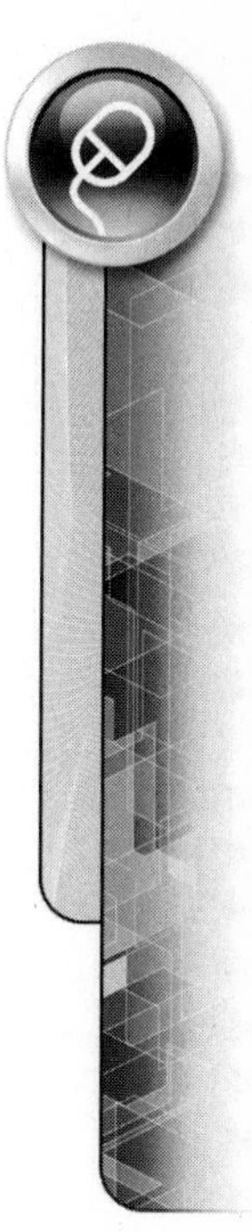

12. Click on the nut at the base of the assembly to select it. In the browser, observe that *ESS_E06-NutB* is highlighted.
13. In the Tweak Components dialog box, verify that the Z button is selected, and then enter **–1.75** in the Tweak Distance field.
14. Click Apply, the green checkmark, to create the tweak. Your display should appear similar to Figure 6.117.

FIGURE 6.117

TIP

After you define the tweak axes, you can select a component or group of components and then drag them to create a tweak. This way, you can quickly arrange your components visually. You can later edit the values of the tweaks from the browser to define their final positions.

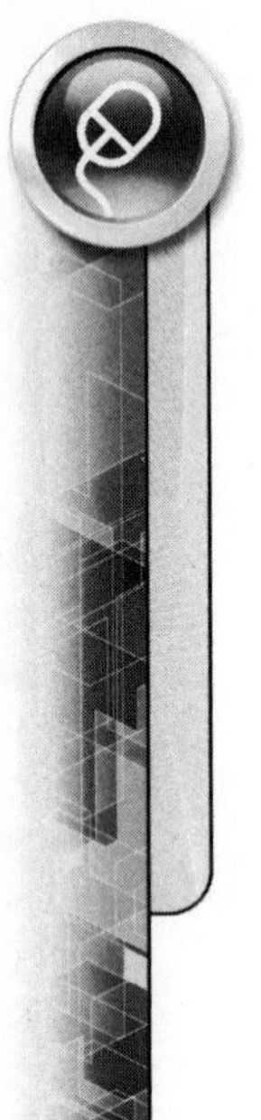

15. Next click the valve plate to select it. Hold down the CTRL key, and click the nut to deselect it. Observe that the valve plate is the only entry highlighted in the browser.
16. Enter **–0.875** in the Tweak Distance field of the dialog box.
17. Click Apply to create the tweak. Your display should appear similar to Figure 6.118.

FIGURE 6.118

TIP

You can also select components by clicking on them directly in the browser, which is useful when trying to select hidden components.

18. Hold down the CTRL key and click the valve plate to deselect it. The next component to tweak is the top nut. In the browser, click *ESS_E06-NutA* to select it.
19. Enter **4.5** in the Tweak Distance field, and click Apply to create the tweak. Your display should appear similar to Figure 6.119.

FIGURE 6.119

TIP

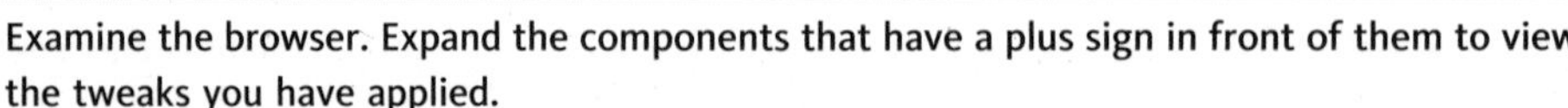

Examine the browser. Expand the components that have a plus sign in front of them to view the tweaks you have applied.

20. Complete this phase of the exercise by adding the following tweaks:

Part Name	Tweak
ESS_E06-WasherB.ipt:1	4
ESS_E06-Pin.ipt:1	3.125
ESS_E06-Spring.ipt:1	1.5
ESS_E06-WasherA.ipt:1	1
ESS_E06-Diaphragm.ipt:1	0.5

21. Close the Tweak Component dialog box when you are finished. Your display should appear similar to Figure 6.120.

FIGURE 6.120

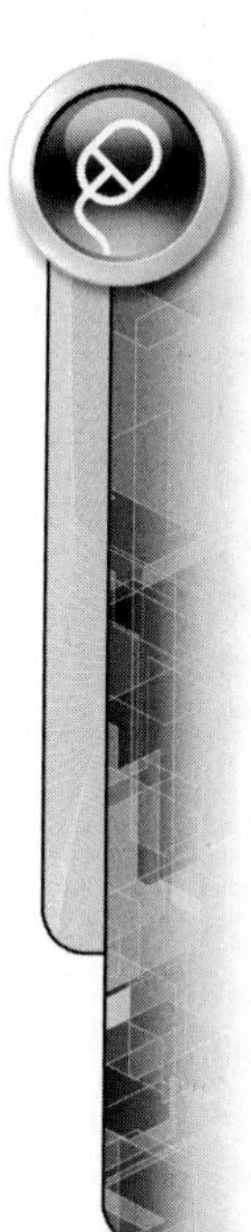

22. Now adjust an existing tweak. In the browser, expand *ESS_E06-Diaphragm.ipt:1*, *ESS_E06-Spring.ipt:1*, and *ESS_E06-WasherA.ipt:1*. The hierarchy displays tweaks nested below each component definition.
23. Right-click each of the tweaks, and select Visibility to turn off the trail display for the tweak.
24. Click Tweak (0.500 in) under *ESS_E06-Diaphragm.ipt:1*.
25. Below the browser, enter a new value of **0.125** in the Offset field, and then press ENTER. Observe that the name of the tweak reflects the change and the diaphragm moves to its new position.
26. Click the Animate command from the Presentation tab > Create panel.
27. Click the Play Forward button to view the animation. When finished, click Cancel in the Animation dialog box.
28. Close all open files. Do not save changes. End of exercise.

CREATING DRAWING VIEWS FROM ASSEMBLIES AND PRESENTATION FILES

After creating an assembly or presentation, you may want to create drawings that use the data. You can create drawing views from assemblies using information from an assembly or a presentation file. You can select from them a specific design view or presentation view. To create drawing views based on assembly data, start a new drawing file or open an existing drawing file. Use the Base View command and select the IAM or IPN file from which to create the drawing. If needed, specify the design view or presentation view in the dialog box. Figure 6.121 shows a presentation view being selected after a presentation file was selected.

FIGURE 6.121

THE BILL OF MATERIAL (BOM) EDITOR

A Bill of Materials, usually referred to as a BOM, is a table that contains information about the components inside an assembly. It can include item number, quantity, part number, description, vendor, and other information needed to describe the assembly. The BOM is also considered associative; when a component is added or removed from the assembly, the quantity field in the BOM will update to reflect this change in the assembly.

To facilitate managing item numbers, every item is automatically assigned a number, and the item number can be changed or reordered if necessary. Whenever these types of changes are made in the BOM, the same changes are updated automatically in the Parts List and balloons.

Values in the BOM can be changed in several ways. The first way is to add information while inside each individual part through the Properties dialog box. Another way is to add the part information directly into the BOM with the BOM Editor; then, when the BOM is saved, the information supplied inside of the BOM is also saved to each individual part. This is an efficient way to edit the properties of multiple parts at the same time.

In order to activate the BOM Editor, you must be inside an assembly model or editing a Parts List in a drawing. In an assembly click Bill of Materials on the Assemble Tab > Manage panel as shown in Figure 6.122 on the left and the Bill of Materials dialog box appears as shown in Figure 6.122 on the right.

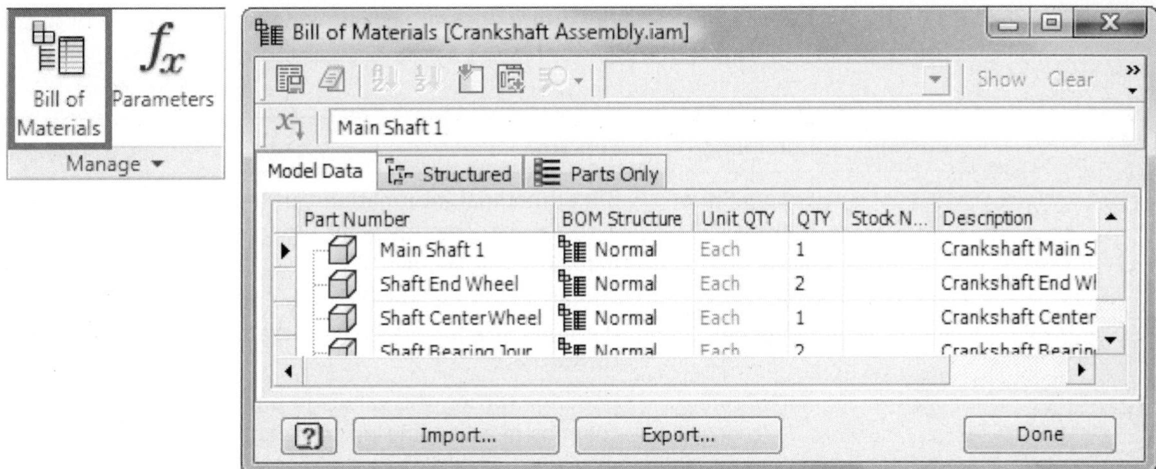

FIGURE 6.122

There are three tabs in the BOM Editor: Model Data, Structured, and Parts Only. The Model Data tab is shown in Figure 6.123 on the left. Here the components that make up the assembly are arranged in a format that resembles the assembly browser. Crankshaft Assembly can be expanded to display its individual parts. The information under this tab is not meant to be reported to a Parts List.

The Structured tab for this example is shown in Figure 6.123 in the middle. This is the actual BOM data that is reported to the Parts List. When arranging an assembly in Structured mode, Crankshaft Assembly is considered an individual item along with the other parts listed.

When the Parts Only tab is selected, Crankshaft Assembly is not listed. However, the individual parts that make up Crankshaft Assembly are listed along with the other parts, as shown in Figure 6.123 on the right. This arrangement of parts is referred to as a flat list.

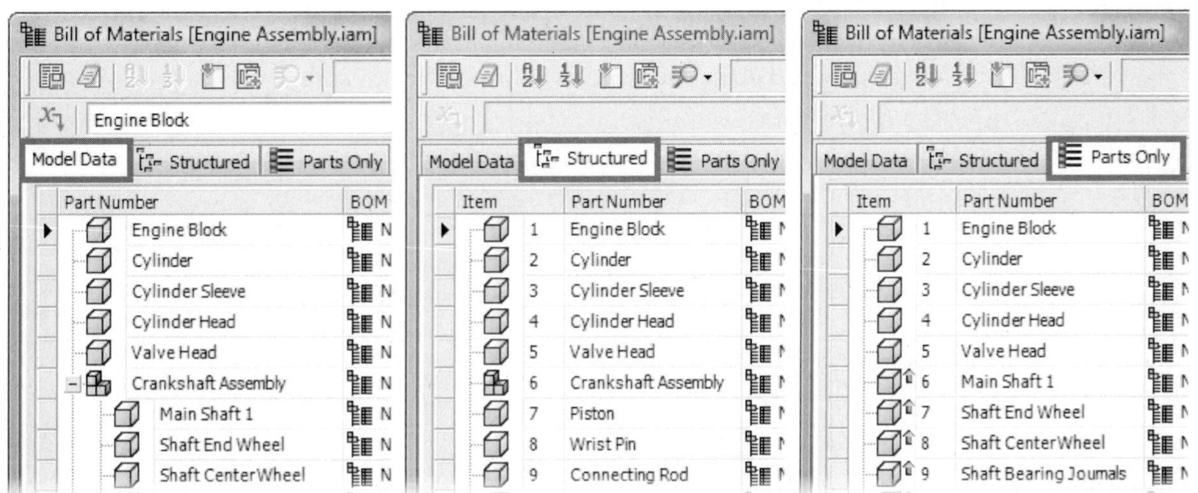

FIGURE 6.123

The BOM Editor also contains several buttons that provide further control over the information supplied in the BOM. These buttons are described as follows:

Button	Command	Function
	Export Bill of Materials	Bill of material information can be exported out to an external file. The following file formats are supported: mdb, xls, dbf, txt, and csv.
	Engineer's Notebook	The BOM can be added to the Engineer's Notebook in the form of a table as a note that is linked to an Excel spreadsheet.
	Sort Items	Used for sorting the values in one or more BOM columns in ascending or descending order.
	Renumber Items	Allows items to be renumbered based on the current sort order of the BOM.
	Choose Columns	Allows extra informational columns to be added or removed from the BOM.
	Add Custom iProperties Column	Adds the new values to the Custom tab in the corresponding iProperties dialog box.
	View Options	Enables or disables the BOM information from being displayed in the Structured and Parts Only tabs.
	Part Number Row Merge Settings	Allows different components with the same part number to be treated as a single component.
	Update Mass Properties	Calculates the mass and volume of all items in the BOM.

Sorting Items in the BOM

Clicking on the Sort button located in the BOM Editor will display the Sort dialog box, as shown in Figure 6.124. Sorting can be performed only when in the Structured and Parts Only tabs. In the Sort by area of the dialog box, select the first column to sort by followed by selecting ascending or descending order. Columns can also be arranged by a secondary sort and, if desired, a tertiary sort.

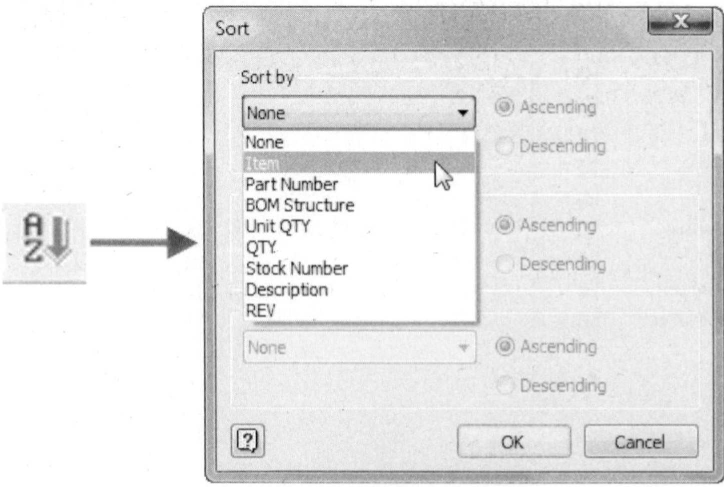

FIGURE 6.124

Renumbering Items in the BOM

Clicking on the Renumber Items button will display the Item Renumber dialog box, as shown in Figure 6.125. This operation is usually performed after sorting information in the BOM Editor.

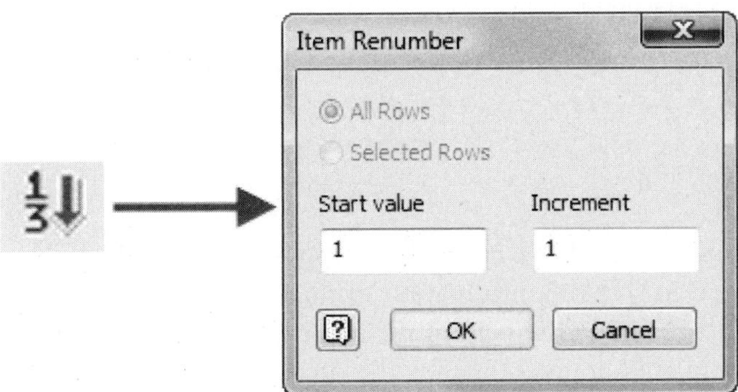

FIGURE 6.125

Choosing Extra Columns for the BOM

Clicking on the Choose Columns button will display a Customization dialog box in the lower-right corner of the BOM Editor, as shown in Figure 6.126. Choose the desired column from the list, and then drag and drop this listing onto an existing column in the BOM Editor. This action will add the custom column to the BOM. You can drag existing column headers to remove them from the BOM Editor; drop the column when a large X appears at your cursor position. Removed headers are added to the Customization dialog so that you can restore them to the BOM Editor.

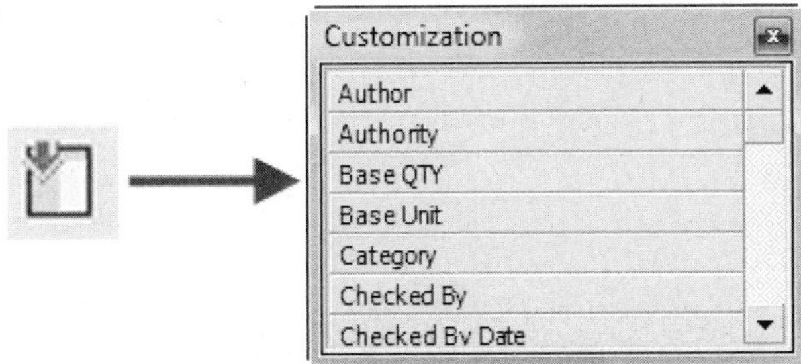

FIGURE 6.126

Changing the View Options of the BOM

Clicking on the View Options button activates a pull-down menu, as shown in Figure 6.127 on the left. You can enable or disable the BOM view based on the current tab. In this example, the current tab is Structured. Clicking on View Properties from the menu will activate the Structured Properties dialog box, as shown in Figure 6.127. Use this dialog to switch the BOM view from First Level to All Levels.

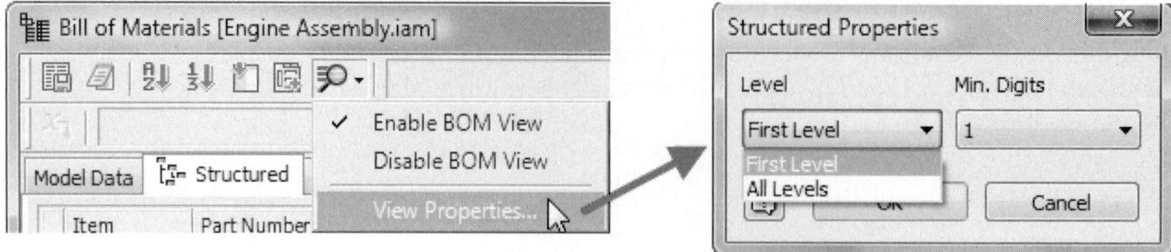

FIGURE 6.127

Editing a BOM Column

Right-clicking on a column in the BOM Editor will display a menu, as shown in Figure 6.128. Use this menu to sort the information in the column by ascending or descending order. Best Fit will resize the column width to fit the information that occupies each cell; you could also double-click on the divider between the column headers. In this case, the column to the left of the divider is resized to fit the contents of the cells in that column. Clicking on Best Fit (all columns) will resize the information in all columns.

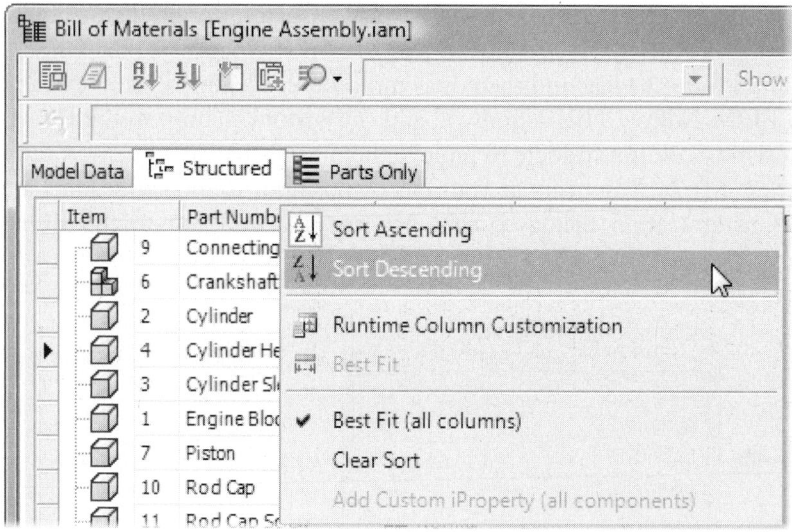

FIGURE 6.128

Highlighting Components in the Browser and BOM

Selecting a cell or group of cells in the BOM editor will highlight all corresponding components located in the assembly browser and in the BOM editor, as shown in Figure 6.129.

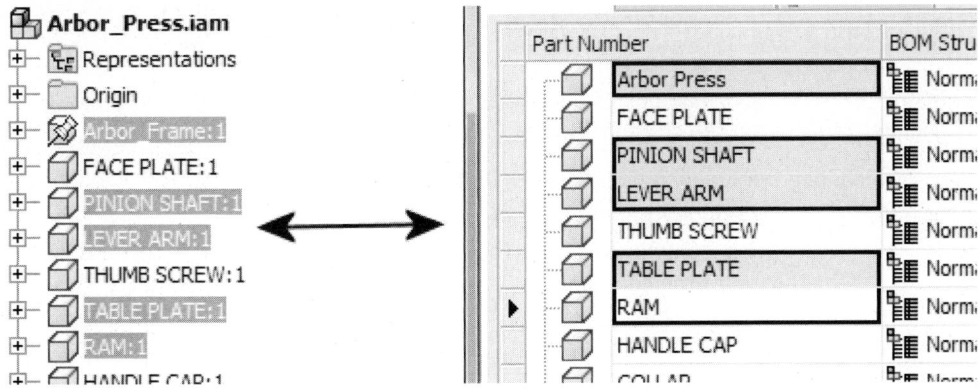

FIGURE 6.129

Opening a Part from the BOM Editor

The BOM editor can be used to open a part or assembly file. To perform this task, open the BOM editor, select the part from the list, right-click and select Open from the menu as shown in Figure 6.130.

FIGURE 6.130

NOTE

While in the BOM editor, the following editing tools work similarly when in Microsoft Excel:

- Use the SHIFT or CTRL keys to make selection sets of information while in the BOM editor.
- Use the grip point located in the lower-right corner of a cell to expand a cell or blocks of cells.
- BOM cells can be copied from the BOM Editor to another BOM view or to an external editor.
- Copy and paste information from the BOM editor into Microsoft Office Excel.
- Additional command options are available through a menu. These options include Open, Copy, Paste, Capitalize, Find, and Replace.

EXERCISE 6-7: EDITING THE BOM

In this exercise, you manipulate a number of items in the BOM Editor and see these changes reflected in the current drawing Parts List.

1. Open the drawing file *ESS_ E06_07.idw*.
2. Examine the existing Parts List, as shown in Figure 6.131. Notice that the Description column is blank. The fields are populated from the iProperties (metadata) of the components.

Parts List			
ITEM	QTY	PART NUMBER	DESCRIPTION
1	2	ESS_E06_07-003	
2	1	ESS_E06_07-001	
3	2	ESS_E06_07-002	
4	2	ESS_E06_07-006	
5	1	ESS_E06_07-007	
6	1	ESS_E06_07-004	
7	6	ESS_E06_07-008	
8	1	ESS_E06_07-005	

FIGURE 6.131

(Courtesy US FIRST Team 342—Robert Bosch Corp., Trident Technical College, Fort Dorchester High School, Summerville High School; Charleston, South Carolina)

3. The individual descriptions of each part can be easily added through the BOM Editor while working in the *.idw* file. Using this method, you will not have to open each individual part file to add a description. Right-click on the Parts List: ESS_E06_07.iam in the browser and, click Bill of Materials from the menu.
4. The BOM Editor will be displayed, as shown in Figure 6.132. Of the three tabs available, (Model Data, Structured, and Parts Only), click the Parts Only tab. Notice the appearance of the Description column.

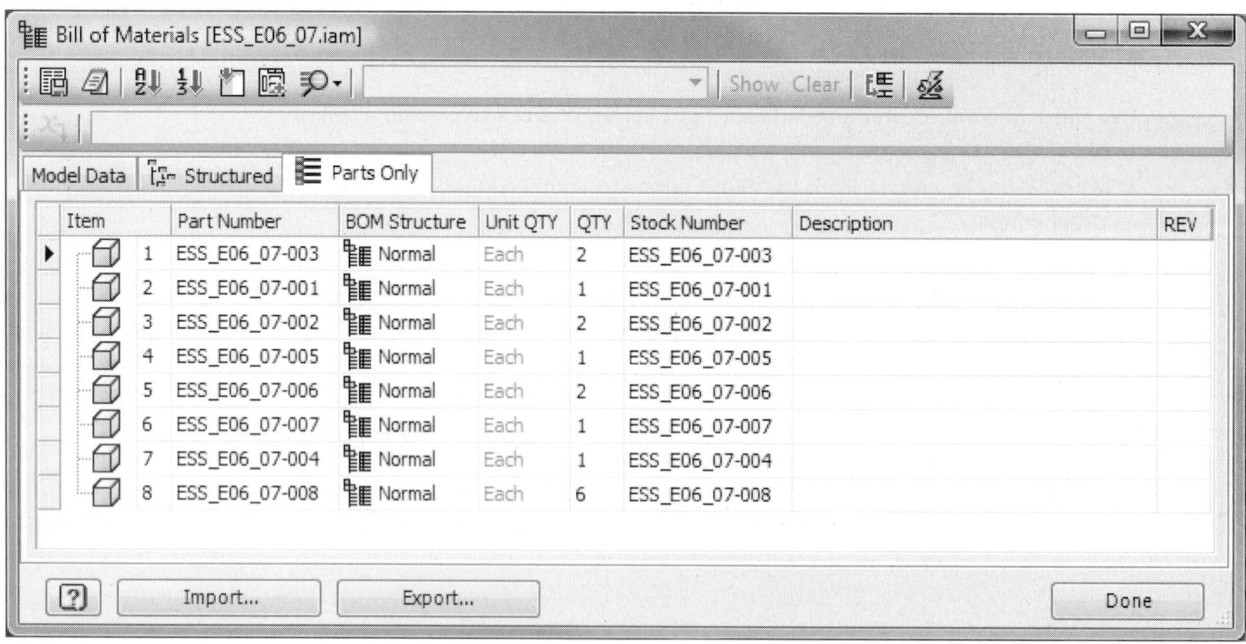

FIGURE 6.132

5. Complete the Description column by clicking in each cell and filling in all needed information as shown in Figure 6.133. When finished, click the Done button to close the BOM Editor.

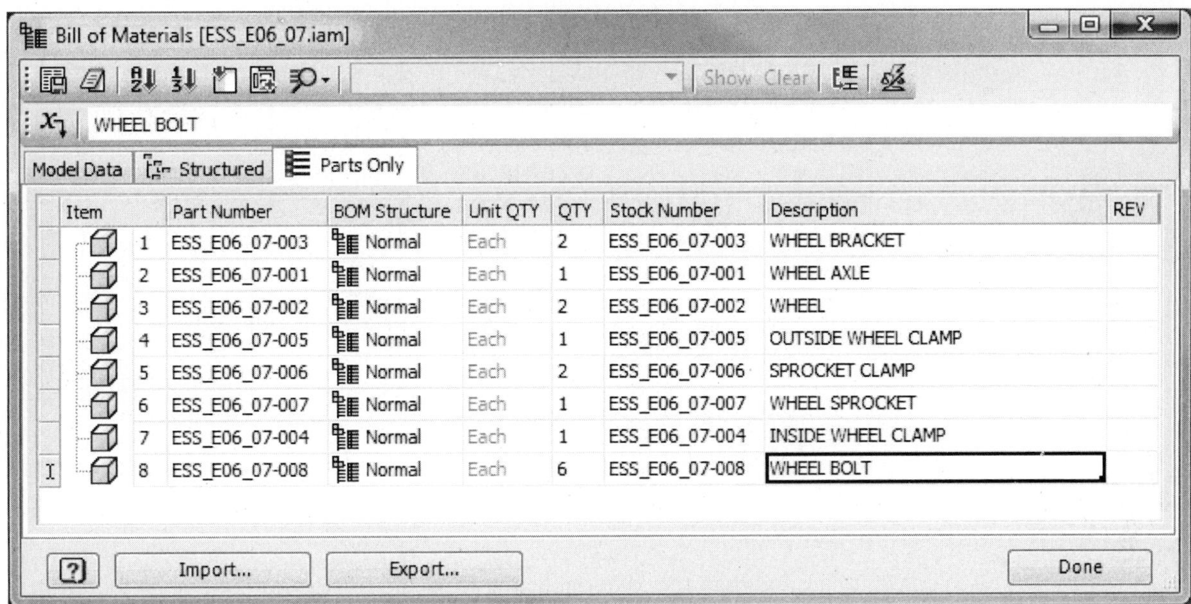

FIGURE 6.133

6. When you return to the drawing file, notice that the Description column of the Parts List updates to reflect the new information, as shown in Figure 6.134.

Parts List			
ITEM	QTY	PART NUMBER	DESCRIPTION
1	2	ESS_E06_07-003	WHEEL BRACKET
2	1	ESS_E06_07-001	WHEEL AXLE
3	2	ESS_E06_07-002	WHEEL
4	2	ESS_E06_07-006	SPROCKET CLAMP
5	1	ESS_E06_07-007	WHEEL SPROCKET
6	1	ESS_E06_07-004	INSIDE WHEEL CLAMP
7	6	ESS_E06_07-008	WHEEL BOLT
8	1	ESS_E06_07-005	OUTSIDE WHEEL CLAMP

FIGURE 6.134

7. Next add a Material column to the BOM Editor to capture the material type for each part. Activate the BOM Editor again from the Parts List in the browser.
8. To add a column called Material, click the Choose Columns button as shown in Figure 6.135. This will display the Customization box of properties in the lower-right corner of the dialog box.
9. Scroll down in the Customization box and locate Material as shown in Figure 6.135.

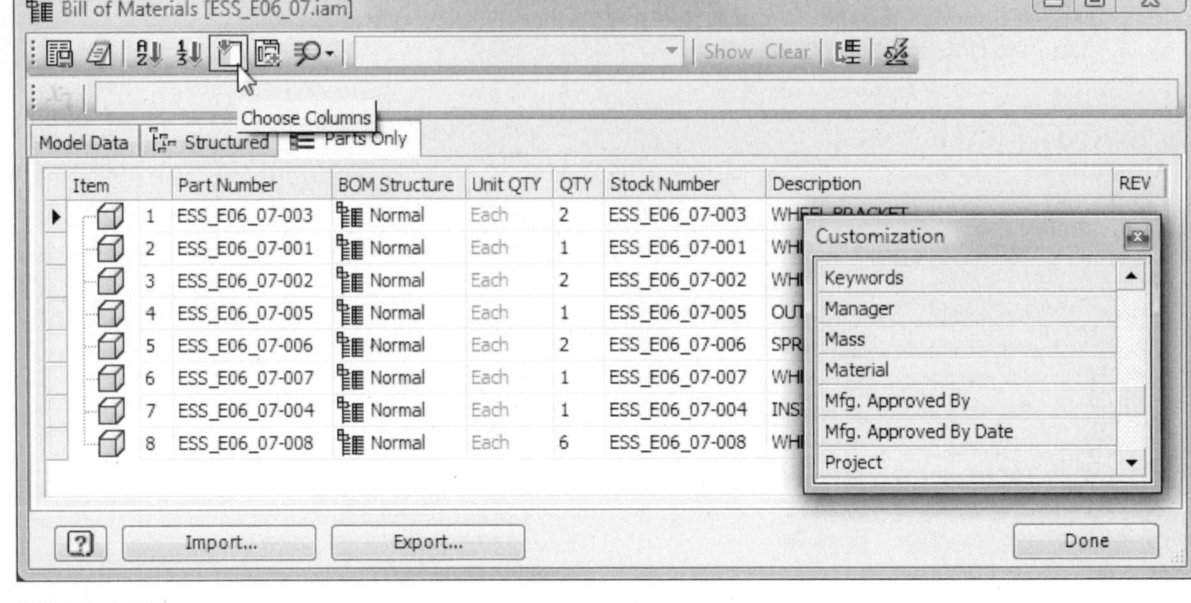

FIGURE 6.135

10. Click and drag the Material listing from the Customization box, and drop it to the right of the Description Column heading as shown in Figure 6.136.

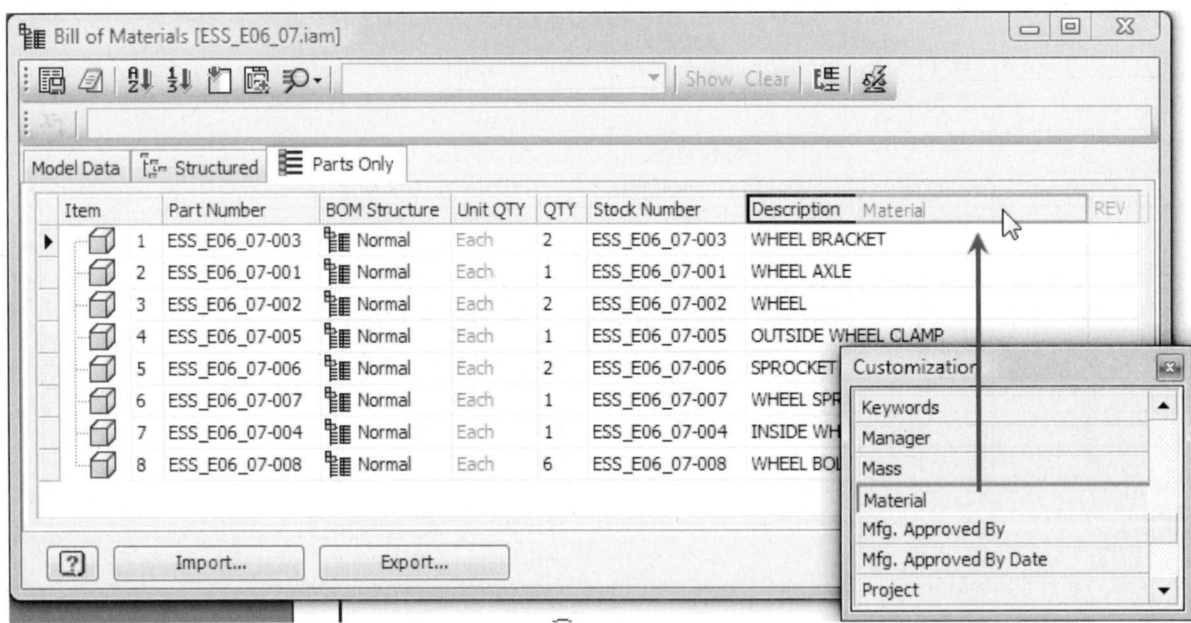

FIGURE 6.136

11. When finished, dismiss the Customization box by clicking on the Close button (X).

12. With the addition of the Material column, your display should appear similar to Figure 6.137. Notice that all Materials are listed as Generic. You will now add a material to each part.

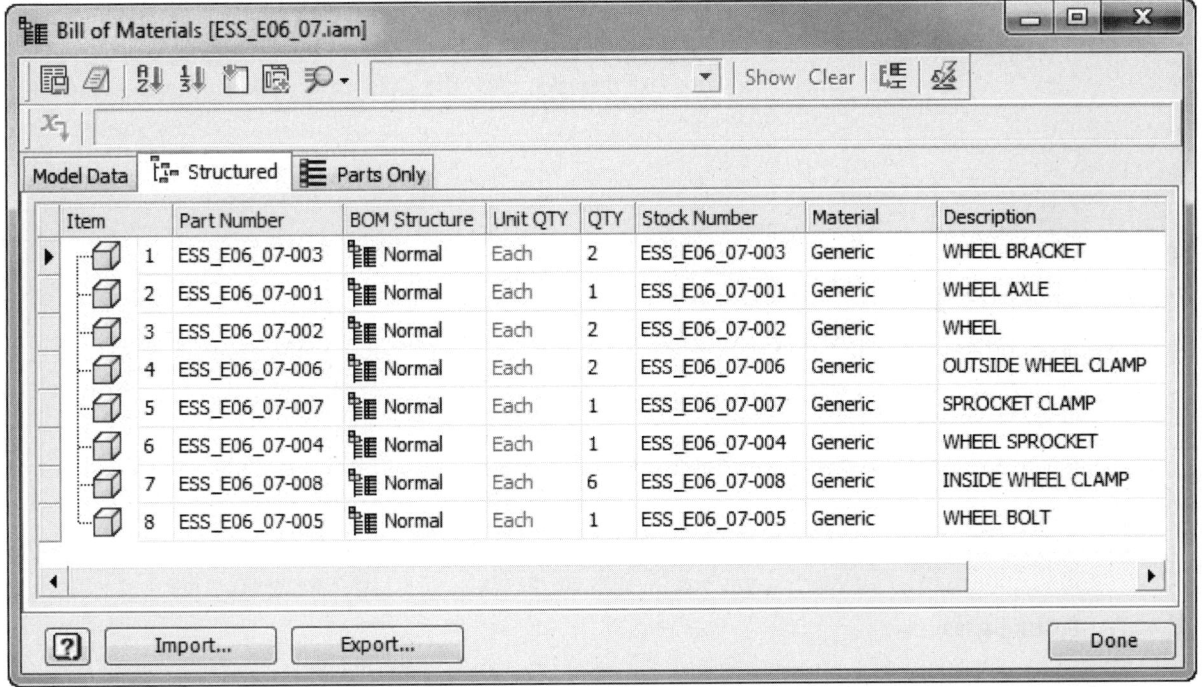

Item		Part Number	BOM Structure	Unit QTY	QTY	Stock Number	Material	Description
	1	ESS_E06_07-003	Normal	Each	2	ESS_E06_07-003	Generic	WHEEL BRACKET
	2	ESS_E06_07-001	Normal	Each	1	ESS_E06_07-001	Generic	WHEEL AXLE
	3	ESS_E06_07-002	Normal	Each	2	ESS_E06_07-002	Generic	WHEEL
	4	ESS_E06_07-006	Normal	Each	2	ESS_E06_07-006	Generic	OUTSIDE WHEEL CLAMP
	5	ESS_E06_07-007	Normal	Each	1	ESS_E06_07-007	Generic	SPROCKET CLAMP
	6	ESS_E06_07-004	Normal	Each	1	ESS_E06_07-004	Generic	WHEEL SPROCKET
	7	ESS_E06_07-008	Normal	Each	6	ESS_E06_07-008	Generic	INSIDE WHEEL CLAMP
	8	ESS_E06_07-005	Normal	Each	1	ESS_E06_07-005	Generic	WHEEL BOLT

FIGURE 6.137

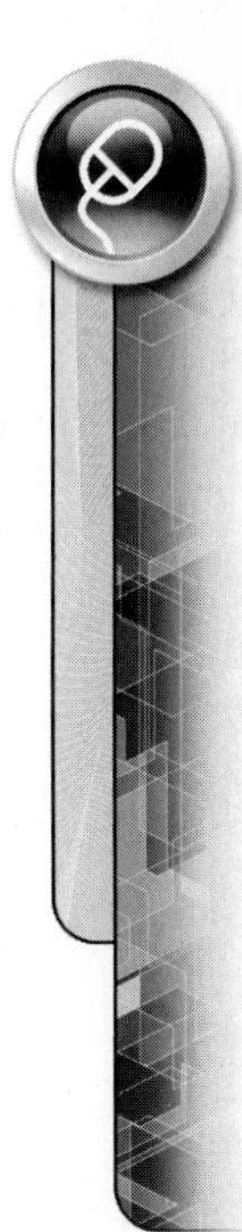

13. Begin filling in the material type by clicking on Generic for Item 1. Double-click the cell to activate the material drop-down list. Select the proper material from the list supplied. Make material assignments that correspond to the part numbers, as shown in Figure 6.138.

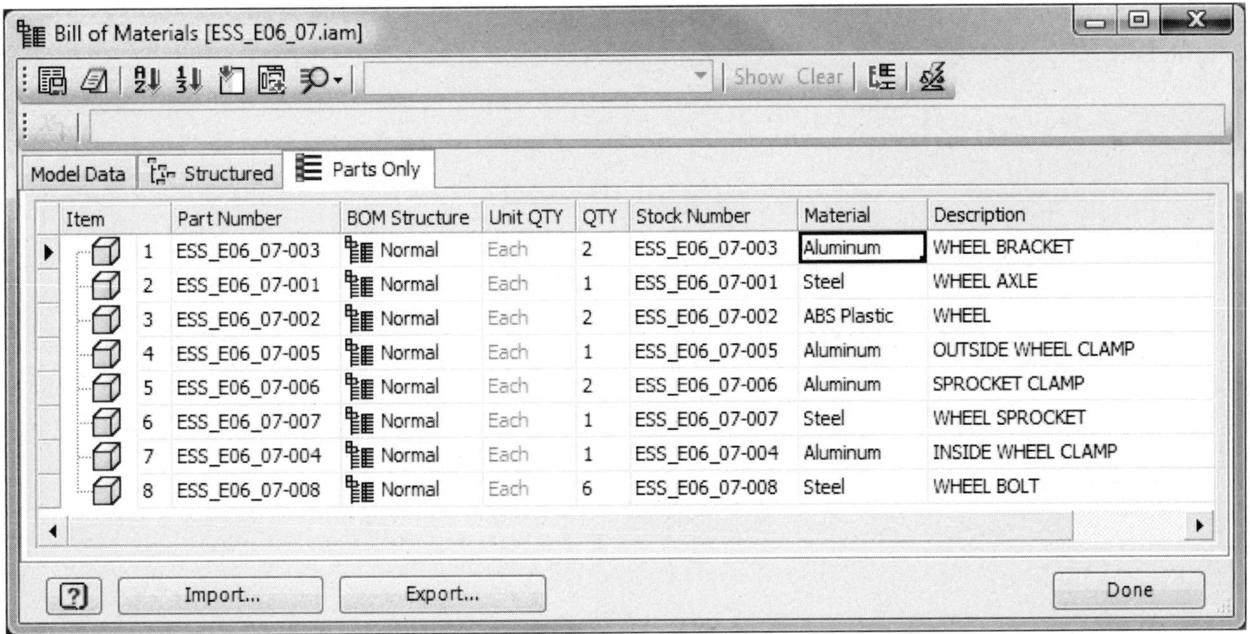

Item		Part Number	BOM Structure	Unit QTY	QTY	Stock Number	Material	Description
	1	ESS_E06_07-003	Normal	Each	2	ESS_E06_07-003	Aluminum	WHEEL BRACKET
	2	ESS_E06_07-001	Normal	Each	1	ESS_E06_07-001	Steel	WHEEL AXLE
	3	ESS_E06_07-002	Normal	Each	2	ESS_E06_07-002	ABS Plastic	WHEEL
	4	ESS_E06_07-005	Normal	Each	1	ESS_E06_07-005	Aluminum	OUTSIDE WHEEL CLAMP
	5	ESS_E06_07-006	Normal	Each	2	ESS_E06_07-006	Aluminum	SPROCKET CLAMP
	6	ESS_E06_07-007	Normal	Each	1	ESS_E06_07-007	Steel	WHEEL SPROCKET
	7	ESS_E06_07-004	Normal	Each	1	ESS_E06_07-004	Aluminum	INSIDE WHEEL CLAMP
	8	ESS_E06_07-008	Normal	Each	6	ESS_E06_07-008	Steel	WHEEL BOLT

FIGURE 6.138

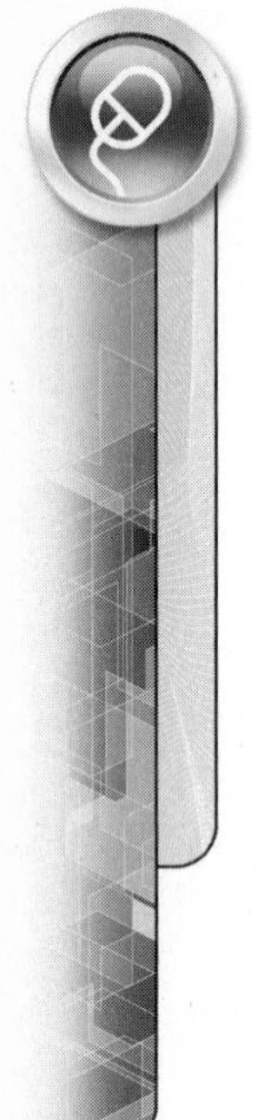

14. When finished, click the Done button to return to the drawing file. Note: Upon leaving the BOM Editor, the iProperty data is written to the individual files.
15. When you return to the drawing, the Material column will not automatically display in the Parts List. To display this column of information, right-click on the Parts List in the browser or in the graphics window, and click Edit Parts List from the menu.
16. When the Parts List displays, click the Column Chooser button, as shown in Figure 6.139.

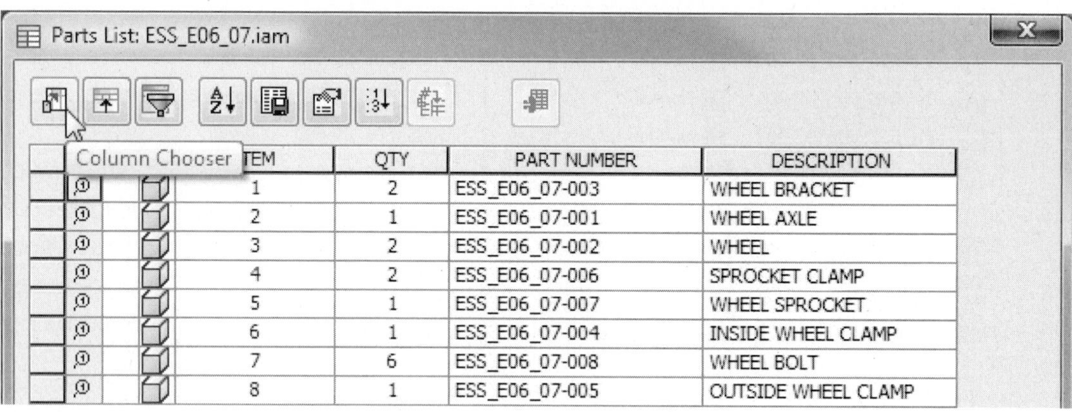

FIGURE 6.139

17. While in the Parts List Column Chooser dialog box, click on the MATERIAL property in the list on the left, and click the Add button to add it under the Selected Properties list, as shown in Figure 6.140 on the left.
18. Select the MATERIAL property in the Selected Properties window, and click the Move Up button to place it above the Description property, as shown in Figure 6.140 on the right. Click the OK button in the Column Chooser and click OK in the Parts List dialog box to return to the drawing.

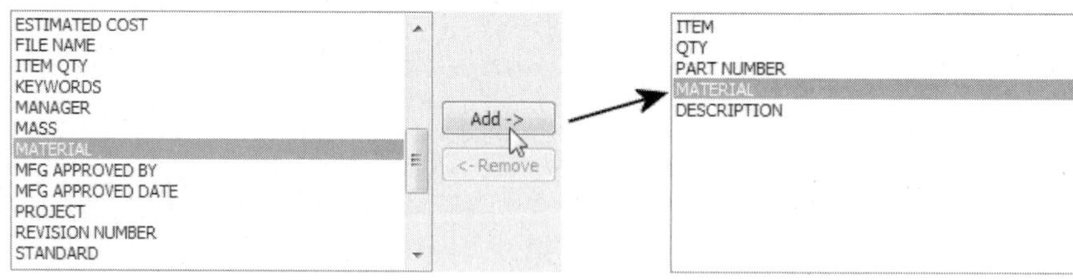

FIGURE 6.140

19. Notice that the Parts List now displays the MATERIAL column and that all information is filled in, as shown in Figure 6.141.

Parts List				
ITEM	QTY	PART NUMBER	MATERIAL	DESCRIPTION
1	2	ESS_E06_07-003	Aluminum	WHEEL BRACKET
2	1	ESS_E06_07-001	Steel	WHEEL AXLE
3	2	ESS_E06_07-002	ABS Plastic	WHEEL
4	2	ESS_E06_07-006	Aluminum	SPROCKET CLAMP
5	1	ESS_E06_07-007	Steel	WHEEL SPROCKET
6	1	ESS_E06_07-004	Aluminum	INSIDE WHEEL CLAMP
7	6	ESS_E06_07-008	Steel	WHEEL BOLT
8	1	ESS_E06_07-005	Aluminum	OUTSIDE WHEEL CLAMP

FIGURE 6.141

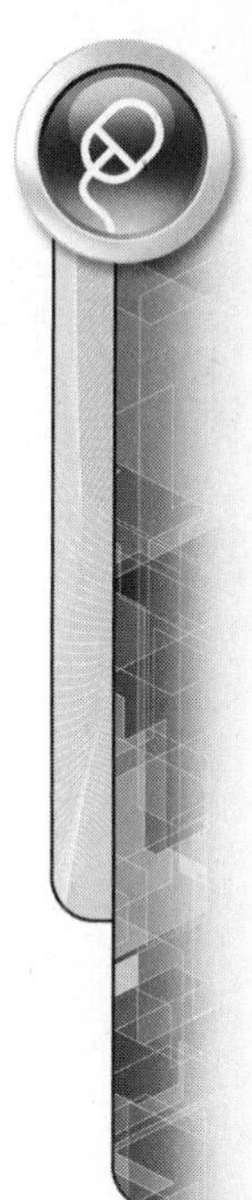

20. Next adjust the width of all columns by the amount of information contained under the heading. To perform this task, edit the parts list by double-clicking on the parts list in the drawing.
21. When the Parts List dialog box displays, right-click the column and pick Column Width from the menu. When the Column Width dialog box appears, enter a value. Change the column headings to the following widths:
    - ITEM = **25**
    - QTY = **20**
    - PART NUMBER = **40**
    - MATERIAL = **30**
    - DESCRIPTION = **55**
22. Click OK to close the Parts List dialog box. The Parts List now reflects the new column widths, as shown in Figure 6.142.

Parts List				
ITEM	QTY	PART NUMBER	MATERIAL	DESCRIPTION
1	2	ESS_E06_07-003	Aluminum	WHEEL BRACKET
2	1	ESS_E06_07-001	Steel	WHEEL AXLE
3	2	ESS_E06_07-002	ABS Plastic	WHEEL
4	2	ESS_E06_07-006	Aluminum	SPROCKET CLAMP
5	1	ESS_E06_07-007	Steel	WHEEL SPROCKET
6	1	ESS_E06_07-004	Aluminum	INSIDE WHEEL CLAMP
7	6	ESS_E06_07-008	Steel	WHEEL BOLT
8	1	ESS_E06_07-005	Aluminum	OUTSIDE WHEEL CLAMP

FIGURE 6.142

23. Close the file. Do not save changes. End of exercise.

CREATING BALLOONS

After you have created a drawing view of an assembly, you can add balloons to the parts and/or subassemblies. You can add balloons individually to components in a drawing, or you can use the Auto Balloon command to automatically add balloons to all components.

To add individual balloons to a drawing, follow these steps:

1. On the Annotate tab > Table panel, click the arrow below the Balloon command, as shown in Figure 6.143. Two icons will appear: the first is used for ballooning components one at a time, and the second is used for automatic ballooning of components in a view via a single operation.
2. To balloon a single component, click the Balloon command or use the hot key B.
3. Select a component to balloon.
4. The BOM Properties dialog box will appear, as shown in Figure 6.143.

NOTE

This dialog box will not appear if a Parts List or balloons already exist in the drawing.

5. Select which Source and BOM Settings to use, as shown in Figure 6.143, and select OK. A preview image of the balloon will appear attached to the cursor.

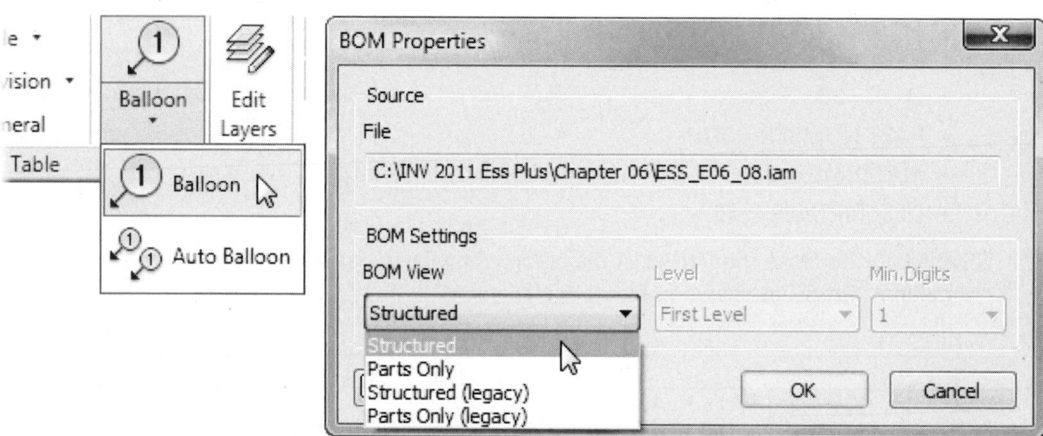

FIGURE 6.143

6. Position the cursor to place the second point for the leader, and then press the left mouse button.
7. Continue to select points to add segments to the balloon's leader, if desired.
8. When finished adding segments to the leader, right-click, and select Continue from the menu to create the balloon.
9. Select Done from the menu to cancel the operation without creating an additional balloon, or select the Back option to undo the last step that was created for the balloon.

NOTE

When placing balloons, the direction of the balloon leader line snaps to 15° increments. The center point of a new balloon can be aligned horizontally or vertically with the center point of a balloon that already exists in a drawing.

BOM Properties Dialog Box Options

Source

Specify the source file on which the BOM will be based.

BOM Settings and BOM View

Structured. A Structured list refers to the top-level components of an assembly in the selected view. Subassemblies and parts that belong to the subassembly will be ballooned, but parts that are in a subassembly will not be.

Parts Only View. Click to balloon only the parts of the assembly in the selected view. Subassemblies will not be ballooned, but the parts in the subassemblies will be.

NOTE

In a Parts Only list, components that are assemblies are not presented in the list unless they are considered inseparable or purchased.

Level

First Level or All Levels determines the level of detail that the Parts List and balloons will display. For example, if an assembly consists of a number of subassemblies and you choose First Level, each subassembly will be considered a single item. If you choose All Levels, the individual parts that make up the subassembly will also be treated as individual components when ballooned.

Min.Digits

Use this command to set the minimum amount of digits displayed for item numbering. The range is fixed from 1 to 6 digits.

Auto Ballooning

In complex assembly drawings, it will become necessary to balloon a number of components. Rather than manually add a balloon to each component, you can use the Auto Balloon command to perform this operation on several components in a single operation automatically. Clicking on the Auto Balloon button under the Annotation tab will display the Auto Balloon dialog box, as shown in Figure 6.144. The areas in this dialog box allow you to control how you place balloons.

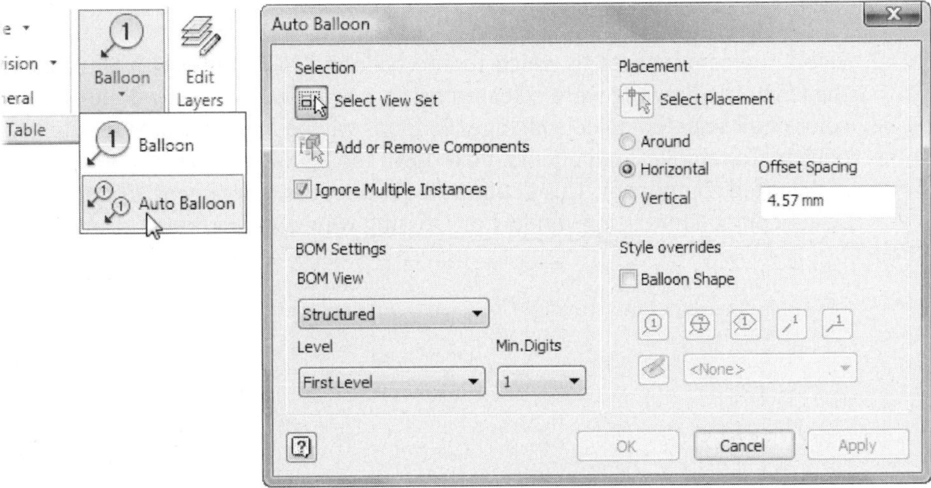

FIGURE 6.144

Selection

This area requires you to select where to apply the balloons. With the view selected, you then add or remove components to the balloon. The Ignore Multiple Instances option appears in this area, and when the box is checked, multiple instances of the same component will not be ballooned. This action greatly reduces the number of balloon callouts in the drawing. If your application requires multiple instances to each have a balloon, remove the check from this box.

Placement

This area allows you to display the balloons along a horizontal axis, along a vertical axis, or around the view being ballooned.

BOM Settings

This area allows you to control if balloons are applied to structured, or first-level components or to parts only.

Style Overrides

The Style overrides area allows you to change the shape of the balloon and assign a user-defined balloon shape. Click Balloon Shape to enable balloon shape style overrides.

To create an auto balloon, follow these steps:

1. Select the view to which to add the balloons, as shown in Figure 6.145.

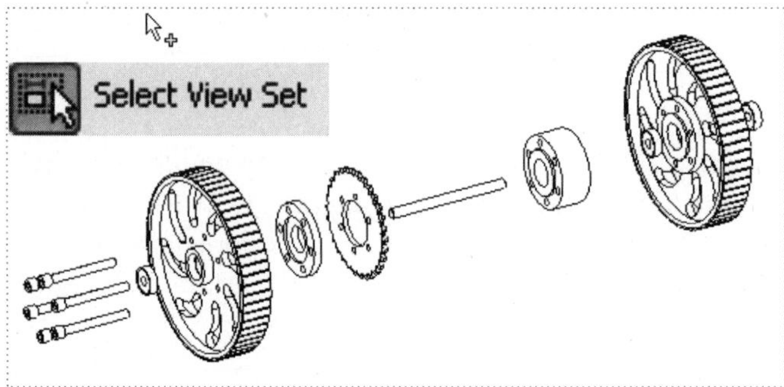

FIGURE 6.145

2. Select the components to which to add balloons. In Figure 6.146, all parts in the isometric drawing were selected using a window box. You will notice the color of all selected objects change. Balloons will be applied to these selected objects. To remove components, hold down the SHIFT or CTRL key, and click a highlighted component. This action will deselect the component. You can also select components using window or crossing window selections.

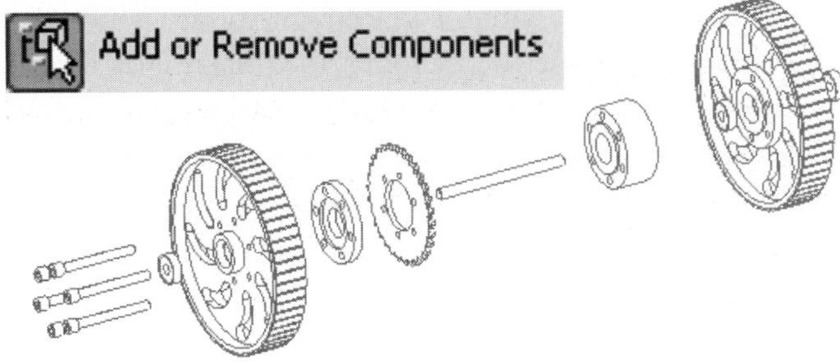

FIGURE 6.146

3. Select one of the three placement modes for the balloons. Figure 6.147 shows an example of using the Horizontal Placement mode. You can further locate this balloon group by moving your cursor around the display screen. This balloon arrangement will retain its horizontal order.

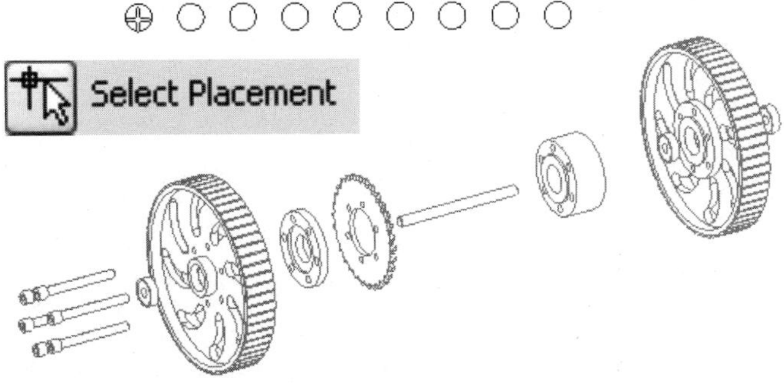

FIGURE 6.147

Figure 6.148 shows an example of using the Vertical Placement mode. You can further locate this balloon group by moving your cursor around the display screen. This balloon arrangement will retain its vertical order.

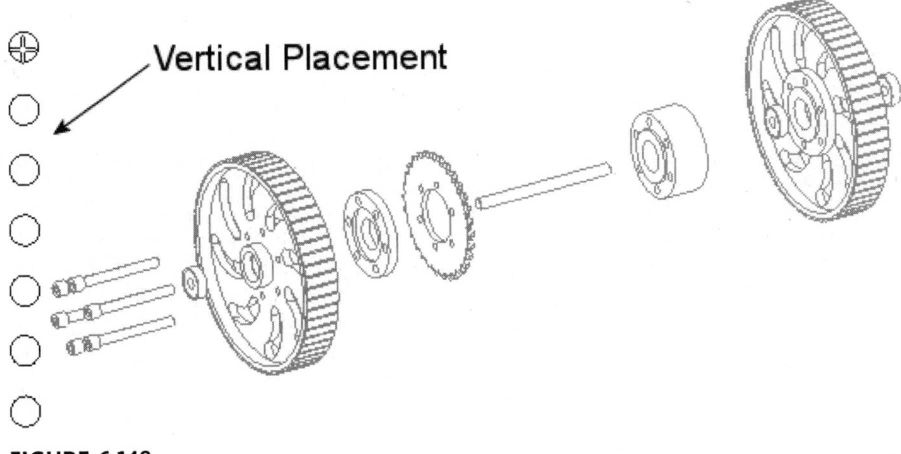

FIGURE 6.148

Figure 6.149 shows an example of using the Around Placement mode. Move the cursor from the center of the view to readjust the balloons to different locations. In this example, the balloons are located in various positions outside of the view box. After you have placed these types of balloons, you may want to drag the balloons and arrow terminators to fine-tune individual balloon locations.

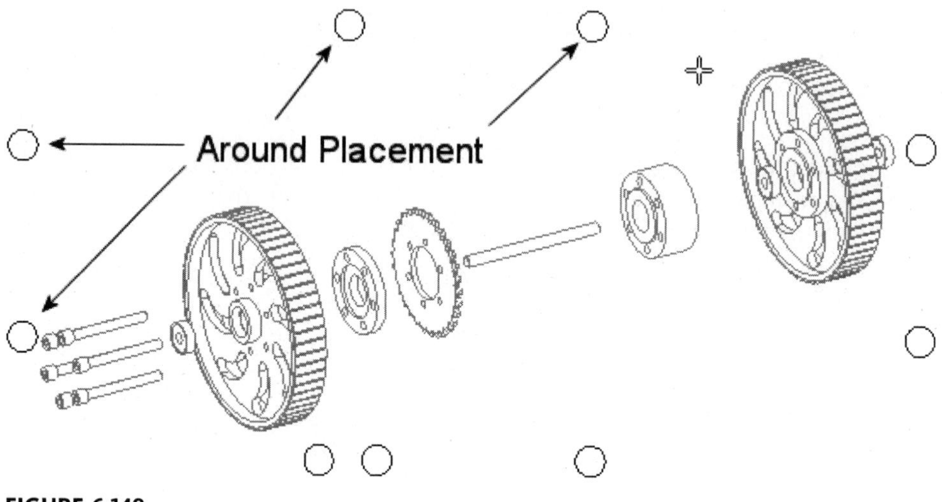

FIGURE 6.149

Editing Balloons

Edit a balloon by right-clicking on it. This action will activate a menu, as shown in Figure 6.150 on the left.

Click Edit Balloon from the menu to activate the Edit Balloon dialog box, as shown in Figure 6.150 on the right.

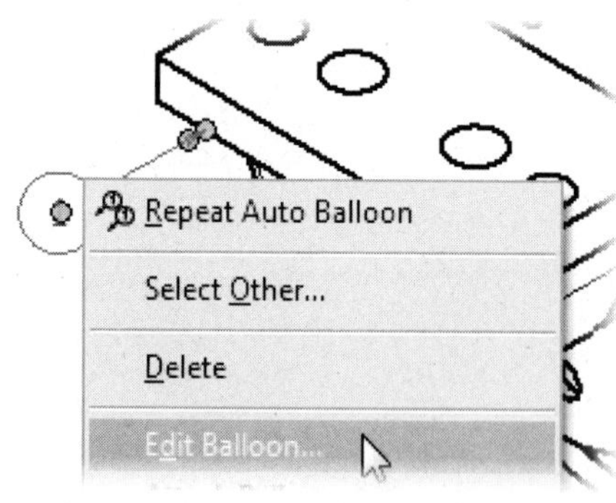

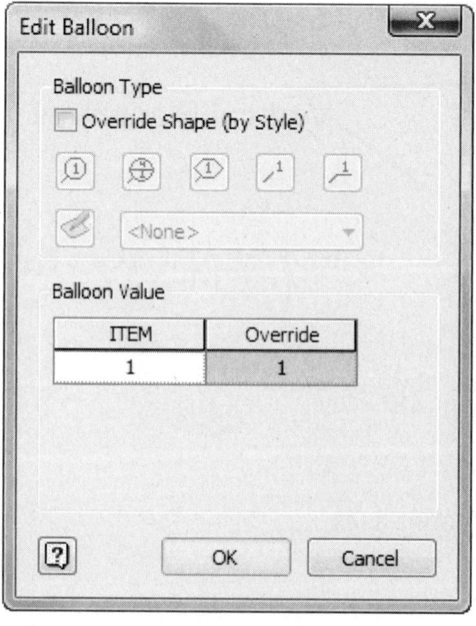

FIGURE 6.150

Use this dialog box to make changes to a balloon on an individual basis. The following features may be edited: Balloon Type and Balloon Value.

Editing the Balloon Value

The columns in the Balloon Value area of the Edit Balloon dialog box in the previous illustration specify Item and Override. When you edit the Item value, changes will be made to the balloon and will be reflected in the Parts List. In Figure 6.151, an Item was changed to 2A. Notice in Figure 6.151, on the left, that both the balloon and the Parts List have updated to the new value.

When you make a change in the Override column, the balloon will update to reflect this change, but the Parts List remains unchanged. Figure 6.151, on the right, shows this. The balloon was overridden with a value of 2A, but the Parts List has the item Rod Cap listed as 2.

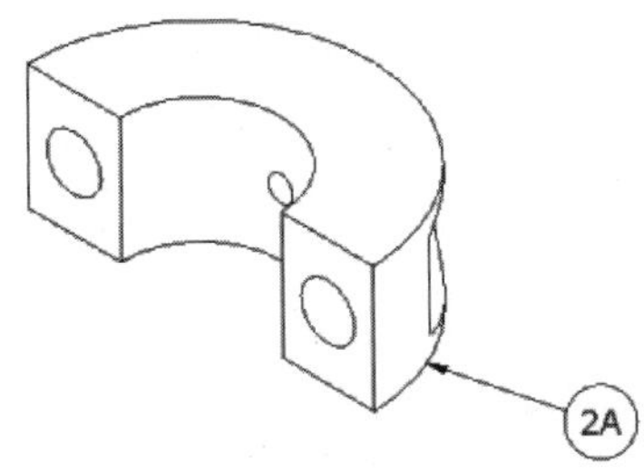

Parts			
ITEM	QTY	PART NUMBER	
1	1	Connecting Rod	
2A	1	Rod Cap	
3	2	Rod Cap Screw	
4	1	Piston	
5	1	Wrist Pin	

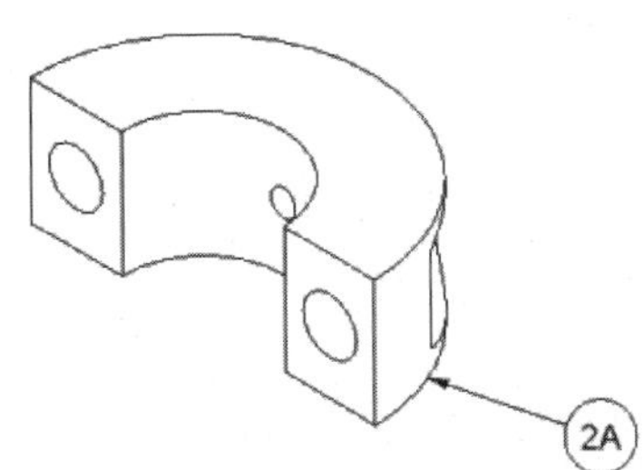

Parts			
ITEM	QTY	PART NUMBER	
1	1	Connecting Rod	
2	1	Rod Cap	
3	2	Rod Cap Screw	
4	1	Piston	
5	1	Wrist Pin	

FIGURE 6.151

NOTE

Any changes performed with the Balloon Edit dialog box will be reflected in all Parts Lists associated with that view.

Attaching a Custom Balloon

If you have predefined a custom part, you can select it from a list of custom parts for custom balloon content. To attach a custom balloon, right-click on the balloon, and select Attach Balloon From List, as shown in Figure 6.152 on the left.

The Attach Balloon dialog box, as shown in Figure 6.152 on the right, will prompt you for your selection of custom balloons. You can select only one custom item at a time.

After selecting the custom part, click the OK button to return to the drawing. Drag the cursor around for placement of the attached balloon. Click to place the balloon in the desired location, as shown in Figure 6.152.

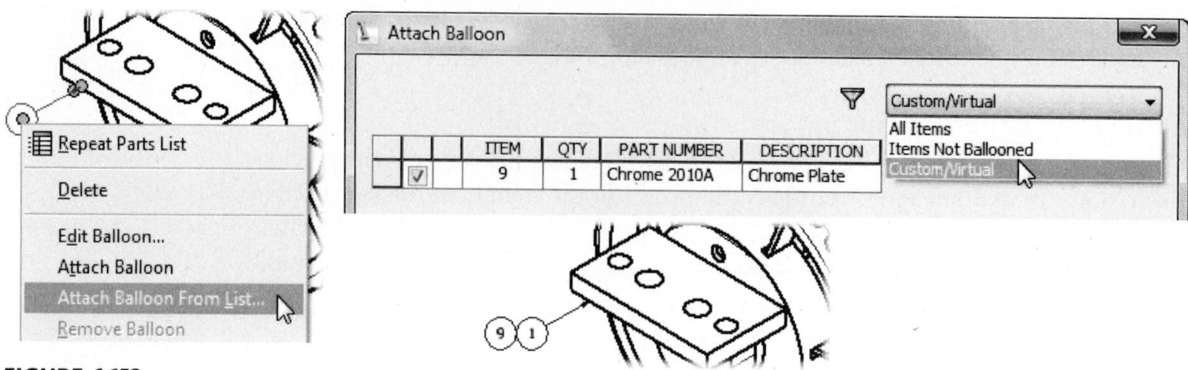

FIGURE 6.152

PARTS LIST

After you have created the drawing views and placed balloons, you can also create a Parts List, as shown in Figure 6.153. As noted earlier in this section, components do not need to be ballooned before creating a Parts List.

Parts List				
ITEM	QTY	PART NUMBER	DESCRIPTION	MATERIAL
1	1	Arbor Press	ARBOR PRESS	Cast Steel
2	1	FACE PLATE	FACE PLATE	Steel, Mild
3	1	PINION SHAFT	PINION SHAFT	Steel, High Strength Low Alloy
4	1	LEVER ARM	LEVER ARM	Steel, Mild
5	1	THUMB SCREW	THUMB SCREW	Steel, Mild
6	1	TABLE PLATE	TABLE PLATE	Steel, High Strength Low Alloy
7	1	RAM	RAM	Steel, High Strength Low Alloy
8	2	HANDLE CAP	HANDLE CAP	Rubber
9	1	COLLAR	COLLAR	Steel, Mild
10	1	GIB PLATE	GIB PLATE	Steel, Mild
11	1	GROOVE PIN	GROOVE PIN	Steel, Mild

FIGURE 6.153

To create a Parts List, select the Annotate tab > Table panel and click the Parts List command, as shown in Figure 6.154 on the left. The Parts List dialog box will appear, as shown in Figure 6.154 on the right. Select a view on which to base the Parts List. After specifying your options, click the OK button, and a Parts List frame preview will appear attached to your cursor. Click a point in the graphics window to place the Parts List. The information in the list is extracted from the properties of each component.

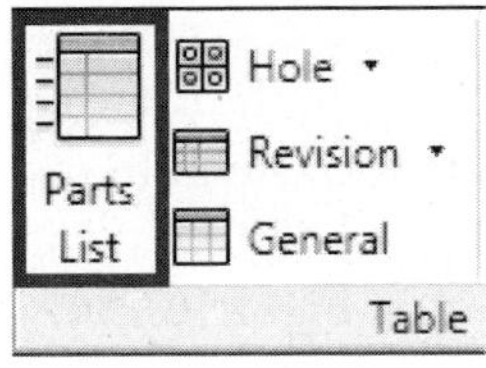

FIGURE 6.154

You can split or wrap a long Parts List into multiple columns through the Parts List dialog box. Notice the Table Wrapping area, as shown in Figure 6.154 on the right. Place a check in the Enable Automatic Wrap box to activate this feature, and then enter the number of sections to break the Parts List into. In this example, 2 is used.

Figure 6.155 shows the results of performing the wrapping operation on a Parts List. Notice the two distinct sections that have been created.

Parts List

ITEM	QTY	PART NUMBER	DESCRIPTION
8	2	HANDLE CAP	HANDLE CAP
9	1	COLLAR	COLLAR
10	1	GIB PLATE	GIB PLATE
11	1	GROOVE PIN	GROOVE PIN
14	4	ANSI B18.3 - 1/4 - 20 UNC - 7/8	Hex Cap Screw
15	4	BS 4168 - M5 x 16	Hex Cap Screws
16	1	ISO 4766 - M5 x 5	Hex Cap Screws

Parts List

ITEM	QTY	PART NUMBER	DESCRIPTION
1	1	Arbor Press	ARBOR PRESS
2	1	FACE PLATE	FACE PLATE
3	1	PINION SHAFT	PINION SHAFT
4	1	LEVER ARM	LEVER ARM
5	1	THUMB SCREW	THUMB SCREW
6	1	TABLE PLATE	TABLE PLATE
7	1	RAM	RAM

FIGURE 6.155

Editing Parts List BOM Data

Once a Parts List is created, you can open the BOM Editor directly from the drawing environment and edit the assembly BOM. All changes are saved in the assembly and the corresponding component files. Right-click on the Parts List in the Drawing browser, and then click Bill of Materials from the menu, as shown in Figure 6.156.

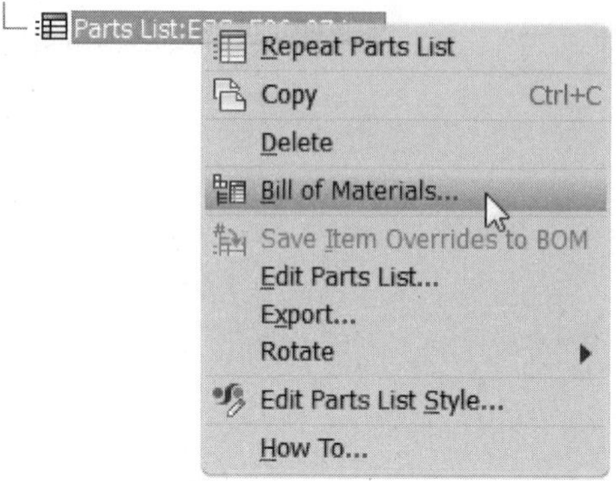

FIGURE 6.156

This action will launch the Bill of Materials editor based on a specific assembly file. The dialog box provides a convenient location to edit the iProperties and Bill of Material properties for all components in the assembly. You can even perform edits on multiple components at once. Notice in Figure 6.157 that the Bill of Materials editor displays the Bill of Materials Structure. In the Bill of Materials editor, perform the desired edits. When finished, click Done to close the dialog box. All changes will be updated in the Parts List and balloons.

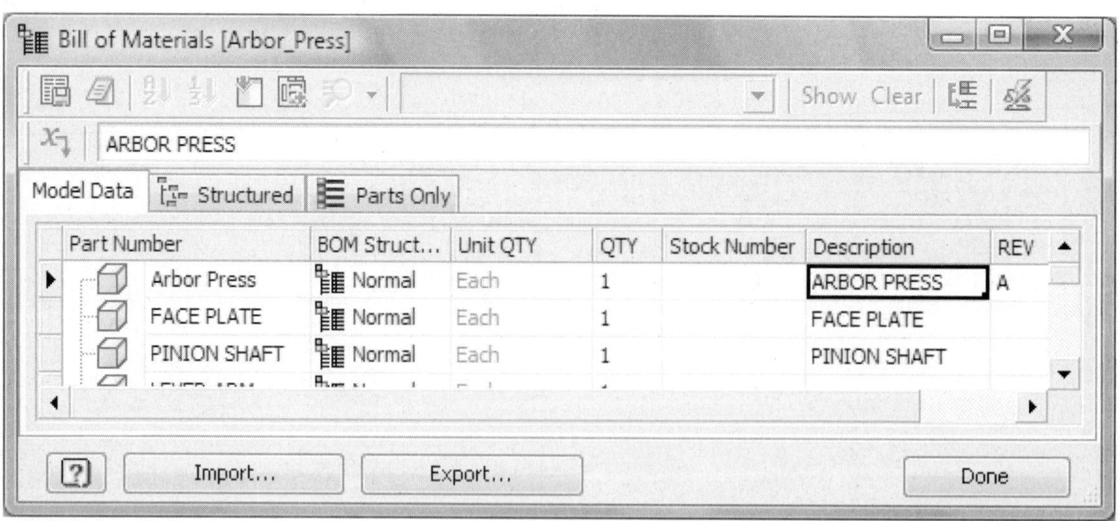

FIGURE 6.157

PARTS LIST COMMANDS

The information supplied earlier in this chapter covered the basics of inserting and manipulating a Parts List. This section will expand on the capabilities of the Parts List and will cover the following topics:

- A detailed account of the Parts List Operations commands available at the top of the Edit Parts List dialog box
- The function of the spreadsheet view
- Additional sections of the Edit Parts List dialog box
- Expanded capabilities of the Edit Parts List dialog box
- Nested Parts Lists

Parts List Operations (Top Icons)

Once you have placed a Parts List in the drawing, controls are available to help you modify the Parts List. Right-click on the Parts List in a drawing and select Edit Parts List from the menu, or double-click on the Parts List in the drawing.

This action displays the Edit Parts List dialog box, as shown in Figure 6.158. The row of icons along the top of the dialog box consist of numerous features that allow you to change the Parts List.

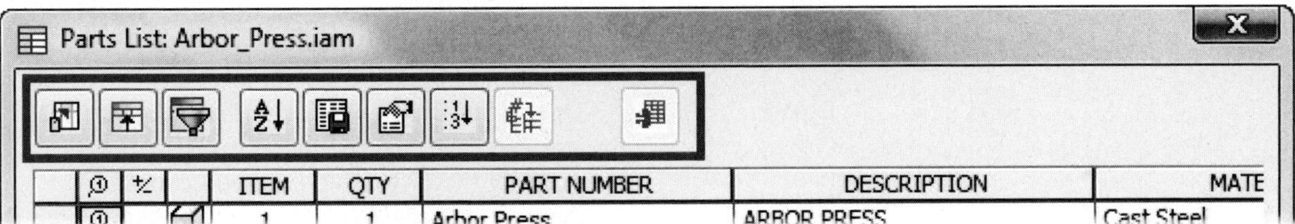

FIGURE 6.158

The Parts List table gives detailed explanations of all Parts List icons:

Parts List Button	Function	Explanation
	Column Chooser	Opens the Parts List Column Chooser dialog box, where you can add, remove, or change the order of the columns for the selected Parts List. Data for these columns is populated from the properties of the component files.
	Group Settings	Opens the Group Settings dialog box, where you select Parts List columns to be used as a grouping key and group different components into one Parts List row. This button is active only when generating a Parts List using the Structured mode.
	Filter Settings	Opens the Filter Settings dialog box, where you can define such filter settings as Assembly View Representations, Ballooned Items Only, Item Number Range, Purchased Items, and Standard Content from which to filter the information. Once a filter is selected, you then filter out rows without changing the data in the Parts List. You can also add Parts List filters to a Parts List style.
	Sort	Opens the Sort Parts List dialog box, where you can change the sort order for items in the selected Parts List.
	Export	Opens the Export Parts List dialog box so that you can save the selected Parts List to an external file of the file type you choose.
	Table Layout	Opens the Parts List Table Layout dialog box where you can change the title text, spacing, or heading location for the selected Parts List.
	Renumber Items	Renumbers item numbers of parts in the Parts List consecutively.
	Save Item Overrides to BOM	Saves item overrides back to the assembly Bill of Materials.
	Member Selection	Used with iAssemblies. Clicking this button opens a dialog box where you can select which members of the iAssembly to include in the Parts List.

Creating Custom Parts

Custom parts are useful for displaying parts in the Parts List that are not components or graphical data, such as paint or a finishing process. To add a custom part, right-click on an existing part in the Edit Parts List dialog box, as shown in Figure 6.159, and click Insert Custom Part. When a custom part no longer needs to be documented in the Parts List, you can right-click in the row of the custom part, and click Remove Custom Part from the menu.

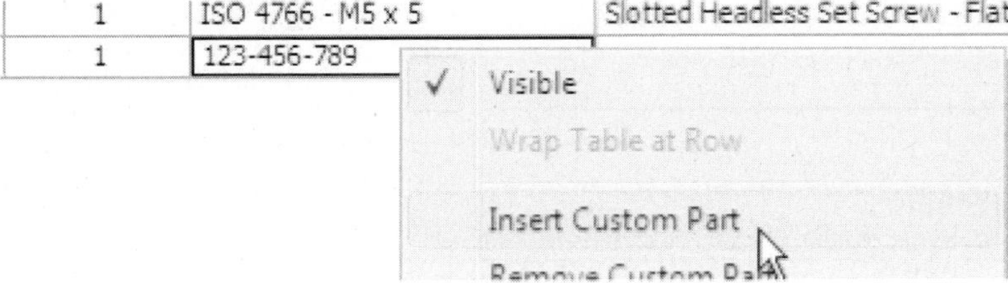

FIGURE 6.159

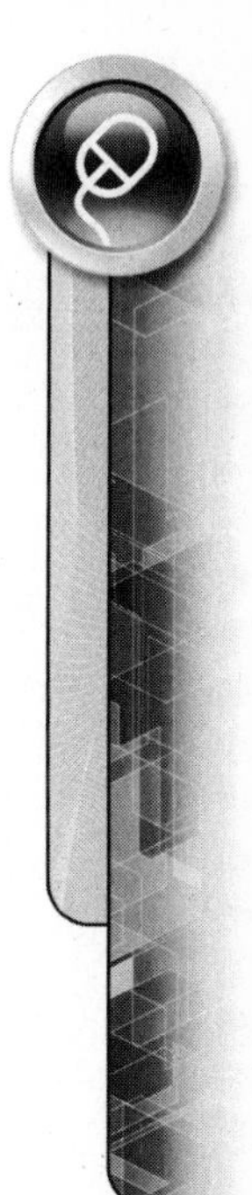

EXERCISE 6-8: CREATING ASSEMBLY DRAWINGS

In this exercise, you create drawing views of an assembly and then add balloons and a Parts List.

1. This exercise begins with formatting a drawing sheet. Open the drawing file *ESS_E06_08.idw* from the Chapter 06 folder. This drawing consists of one blank D sheet with a border and title block.
2. In the browser, right-click Sheet:1 and select Edit Sheet.
3. In the Edit Sheet dialog box, rename the sheet by typing **Assembly** in the Name field.
4. In the Size list, select C. When finished, click the OK button.
5. The next phase of this exercise involves updating the file properties in order to update the title block fields. Select the Inventor Application Menu click iProperties.
6. Click on the Summary tab and make the following changes:
   - In Title, enter **Pump Assembly**.
   - In Author, enter your name or initials.
7. Next click on the Project tab and make the following changes:
   - In Part number, enter **123-456-789**.
   - In Creation Date, click the down arrow, and then select Today's date.
8. When finished, click the OK button. Zoom in to the title block and notice that the entries have updated, as shown in Figure 6.160 on the left.
9. Now create drawing views. Zoom out to view the entire sheet.
10. Click the Base command on the Place Views tab > Create panel to display the Drawing View dialog box.
11. Click the Open an Existing File button next to the File list, select Assembly Files from the Files of Type drop-down list, and double-click *ESS_E06_08.iam* from the Chapter 06 folder as the view source.
12. From the Orientation list, select Top.
13. Set the Style to Hidden Line, not Hidden Line Removed.
14. If needed set the Scale to 1:1
15. Move the view to the upper-left corner of the sheet, and then click to place the view, right-click and select OK. The drawing view should appear similar to Figure 6.160 on the right.

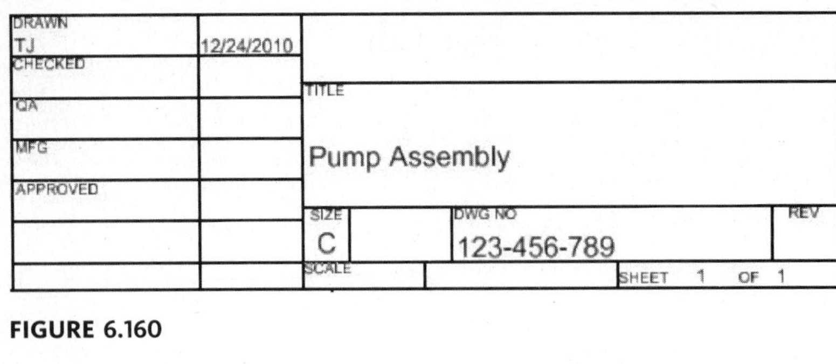

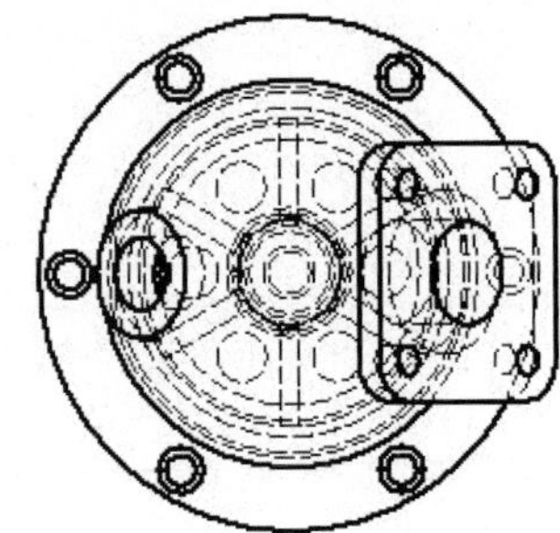

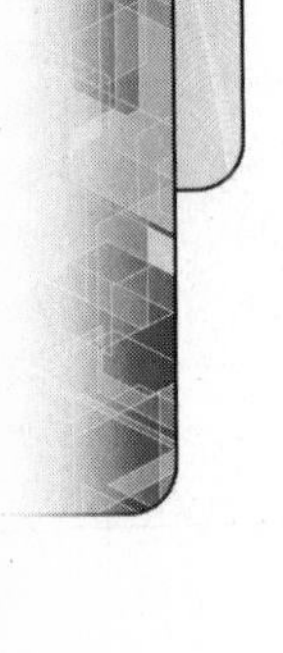

FIGURE 6.160

16. Now begin to generate a section view from the existing top view.
17. Before you create the section view, you will need to turn off the Section property for some components so that they are not sectioned by following the next series of directions:
    - In the browser, expand the Assembly sheet.
    - In the browser, expand View1: *ESS_E06_08.iam*. Your view number may be different.
    - Expand *ESS_E06_08.iam.*
    - Hold down the CTRL key and select *ESS_E06-Pin.ipt, ESS_E06-Spring.ipt, ESS_E06-NutA.ipt*, and *ESS_E06-NutB.ipt.*
    - Right-click any of the highlighted parts in the browser, select Section Participation > None from the menu.

TIP **To select multiple items, hold down the CTRL key while clicking the items from the browser.**

18. Now create the section view by clicking the Section command on the Place Views tab > Create panel.
19. Select the top view of the assembly, and draw a horizontal section line through the center of the top view, as shown in Figure 6.161.
20. Right-click, and select Continue.
21. Move the preview below the top view, and click to place the view.

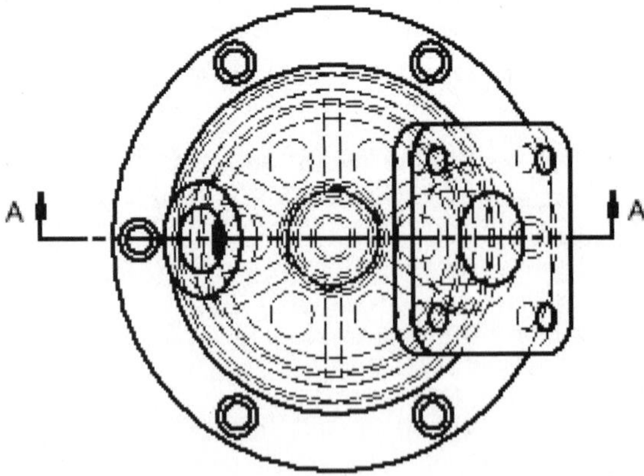

FIGURE 6.161

22. Notice that the nuts, spring, and pin are visible in the view but not sectioned, as shown in Figure 6.162 on the left. Notice also the presence of the section label and drawing view scale.
23. Now edit the section view by double-clicking in the section view, and turn off the label and scale. On the sheet, select the sectioned assembly view, right-click, and select Edit View from the menu to display the Drawing View dialog box.
24. Click the light bulb button (Toggle Label Visibility) to turn off the scale and label in the section view. Click the OK button. The label and scale no longer display at the bottom of the view, as shown in Figure 6.162 on the right.

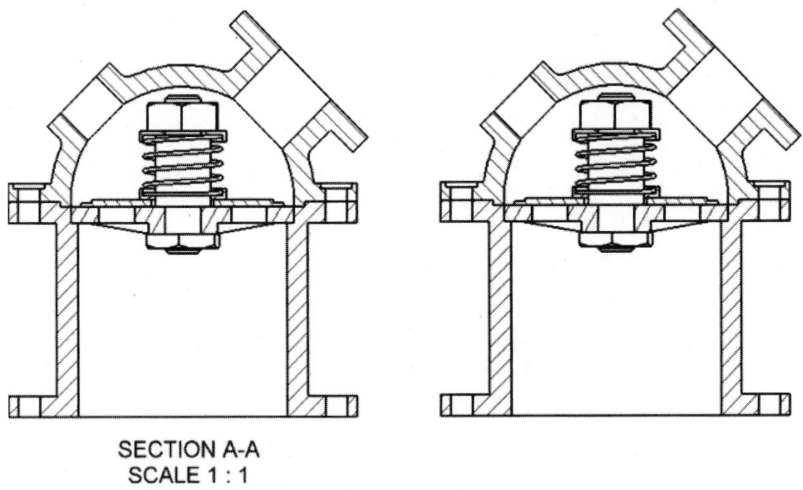

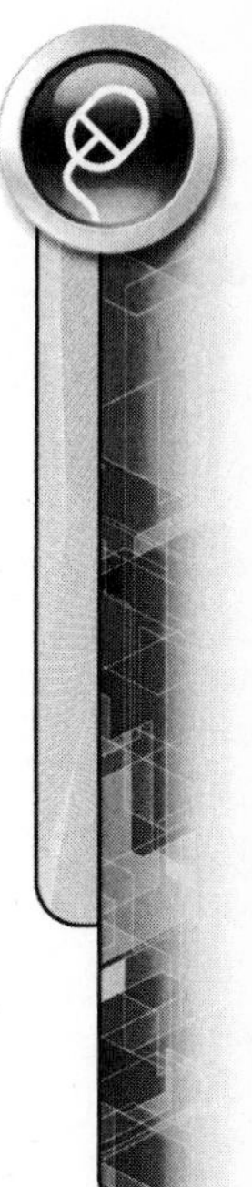

FIGURE 6.162

25. Now hide certain components from displaying in the top view. In the browser, under the expanded component listing for *ESS_E06_08.iam*, select both *ESS_E06-CylHead:1* and *ESS_E06-Spring:1*.
26. Right-click, and clear the checkmark from Visibility. The cylinder head and the spring do not display in the top view, as shown in Figure 6.163.

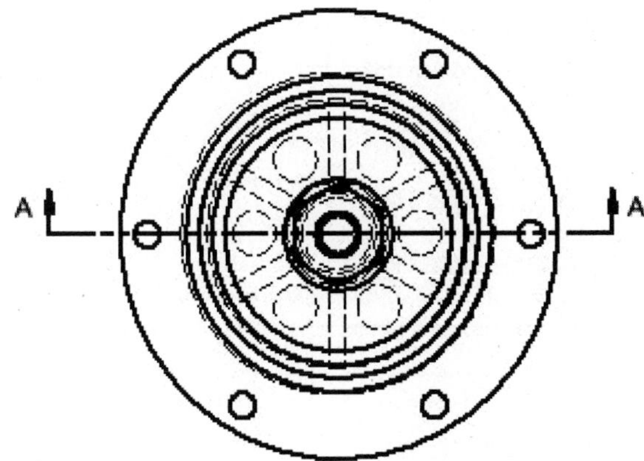

FIGURE 6.163

27. Next add a text note to the top view stating that the cylinder head and spring are not visible. Click the Text command in the Annotate tab > Text panel.
28. Select a point to the left of the top assembly view.

29. In the text entry area, type **CYLINDER HEAD AND**, press ENTER, and type **SPRING REMOVED**. Click OK. Right-click and select Done. The text will be added to the top view, as shown in Figure 6.164. If necessary, select and drag the text to a better location.

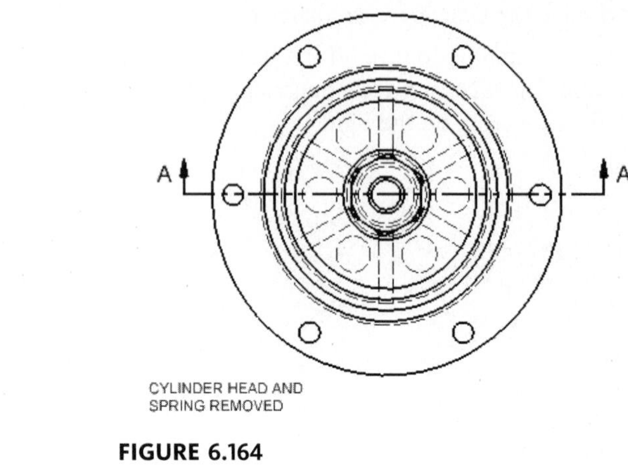

FIGURE 6.164

30. Next, you create a Parts List to identify the item number, quantity, part number, and description. Begin by clicking the Parts List command on the Annotate tab > Table panel.
31. The Parts List dialog box will display. Select the sectioned assembly view.
32. Verify that the BOM View is set to Structured, and click the OK button.
33. Move the Parts List until it joins the title block and border, and then click to place it. Zoom in to display the Parts List, as shown in Figure 6.165.

Parts List			
ITEM	QTY	PART NUMBER	DESCRIPTION
1	1	Cylinder	CYLINDER
2	1	Cyl Head	CYLINDER HEAD
3	1	Valve plate	VALVE PLATE
4	1	Pin	THREADED PIN
5	1	Diaphragm	DIAPHRAGM
6	1	Washer A	RETAINING WASHER
7	1	ESS_E06_05-WasherB	RETAINING WASHER
8	1	Spring	SPRING - Ø1 X Ø22 X 25
9	1	Nut A	FLAT NUT, REG - M16 X 1.5
10	1	Nut B	FLAT NUT, THIN - M16 X 1.5

FIGURE 6.165

34. In the following steps, you will edit the Parts List to reorder the columns and adjust the column widths. Begin by selecting the Parts List in the graphics window or from the browser.
35. Right-click and select Edit Parts List from the menu to display the Parts List dialog box.

36. Click the Column Chooser button to display the Parts List Column Chooser dialog box and make the following changes:
    - In the Available Properties list, select MATERIAL.
    - Click Add.
    - In the Selected Properties list, select DESCRIPTION.
    - Click Move Up two times, and then click OK.
37. Now edit the column widths. While still in the Parts List dialog box, right-click on the DESCRIPTION header, and choose Column Width from the menu. Make the following column width changes:
    - Enter **2.25** in the Column Width dialog box.
    - Change the PART NUMBER field's width to **2**.
    - Verify the ITEM field's width is set to **1**.
    - Change the QTY field's width to **1**.
    - Change the MATERIAL field's width to **1.5**.
38. Click the OK button to close the dialog box.
39. Move the Parts List so that it is flush with the border and the title block. Your display should appear similar to Figure 6.166.

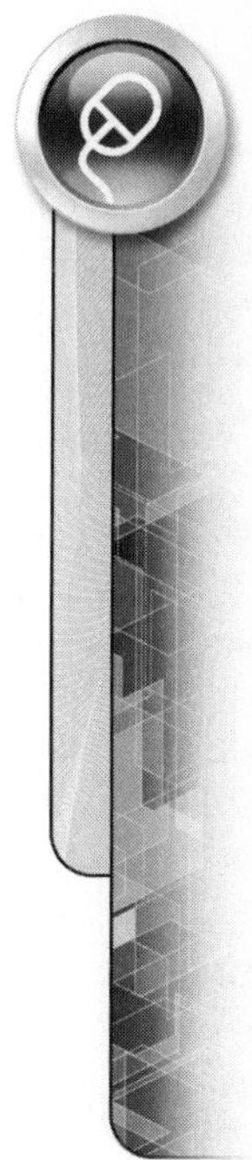

Parts List				
ITEM	DESCRIPTION	QTY	PART NUMBER	MATERIAL
1	CYLINDER	1	Cylinder	Cast Iron
2	CYLINDER HEAD	1	Cyl Head	Cast Iron
3	VALVE PLATE	1	Valve plate	Steel, Mild
4	THREADED PIN	1	Pin	Steel, Mild
5	DIAPHRAGM	1	Diaphragm	Stainless Steel
6	RETAINING WASHER	1	Washer A	Steel, Mild
7	RETAINING WASHER	1	ESS_E06_05-WasherB	Steel, Mild
8	SPRING - Ø1 X Ø22 X 25	1	Spring	Stainless Steel
9	FLAT NUT, REG - M16 X 1.5	1	Nut A	Steel, Mild
10	FLAT NUT, THIN - M16 X 1.5	1	Nut B	Steel, Mild

FIGURE 6.166

40. Now place balloons in the assembly drawing. The item number in the balloon corresponds to the item number in the Parts List.
41. Pan and zoom to display the sectioned assembly view.
42. Click the Balloon command on the Annotate tab > Table panel.
43. Select the edge of the component as the start of the leader as shown in Figure 6.167.
44. Click a point on the sheet to define the end of the first leader segment.
45. Right-click, and select Continue from the menu to place the balloon, as shown in Figure 6.167.

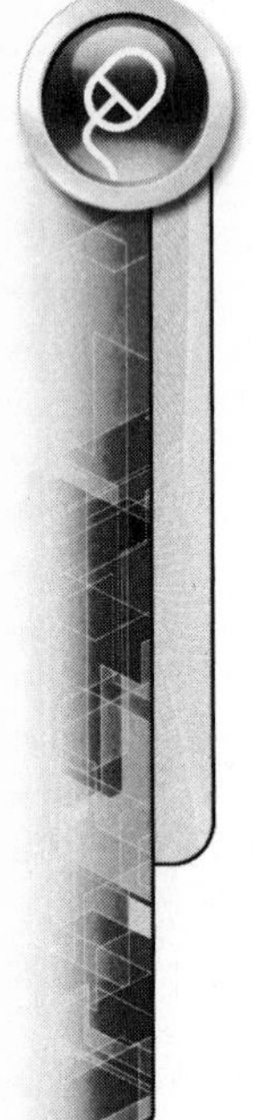

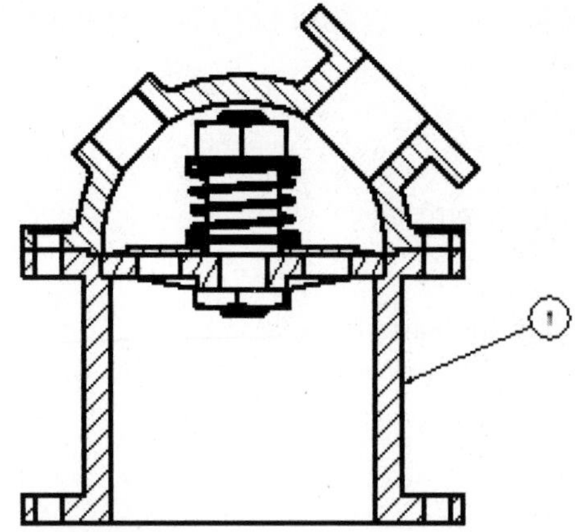

FIGURE 6.167

46. Select the Auto Balloon command from the Annotate tab > Table panel. Select the sectioned assembly view.
47. Window select all the components in the view. The cylinder will not be selected because it has already been ballooned.
48. Click the Select Placement button and select a location above the section view.
49. Select Horizontal for the placement and enter **0.5 in** for the Offset Spacing.
50. Click OK to place the balloons as shown in Figure 6.168.
51. Close all open files. Do not save changes. End of exercise.

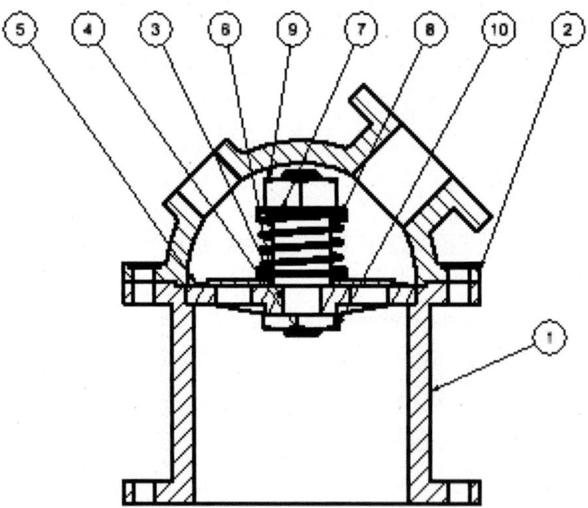

FIGURE 6.168

APPLYING YOUR SKILLS

Skill Exercise 6-1

In this exercise, you create a new component for a charge pump and then assemble the pump.

1. This exercise uses the skills you have learned in previous exercises to assemble a pump and create a new part in place. Open *ESS_E06_09.iam*.

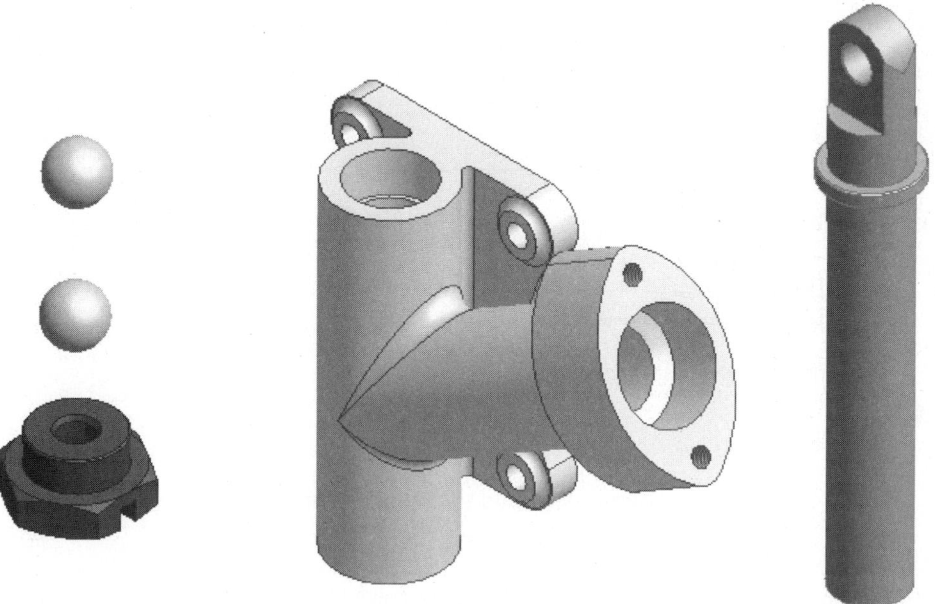

FIGURE 6.169

2. Place the following predefined components into the assembly:
   - 1 occurrence of *ESS_E06_09-Union.ipt*
   - 1 occurrence of *ESS_E06_09-Seal.ipt*
   - 2 occurrences of *ESS_E06_09-M8x30.ipt*
3. Next create the gland in place. Project edges from the pump body to define the flange. See Figure 6.170 for the dimensions needed to complete the gland.

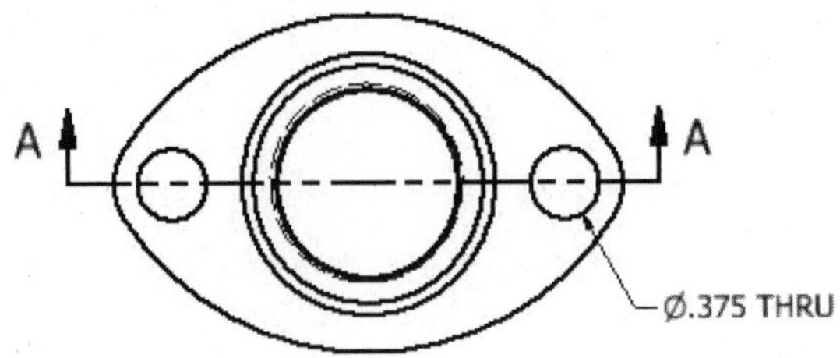

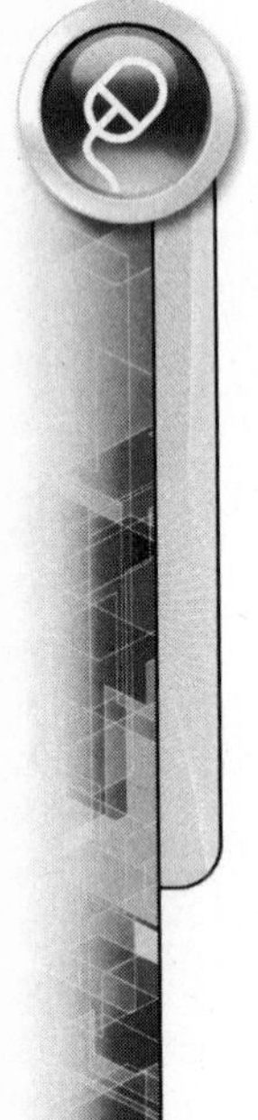

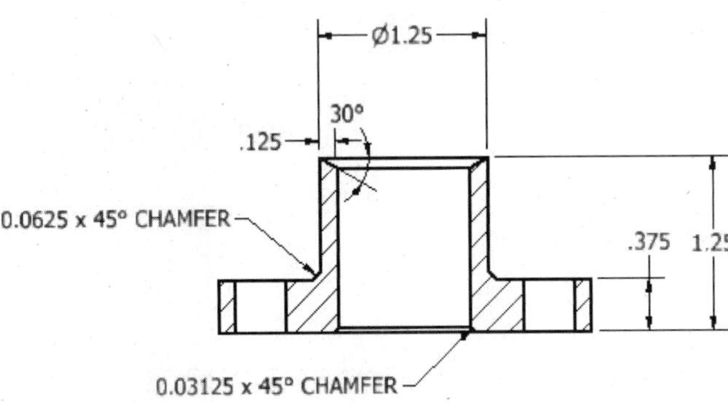

FIGURE 6.170

4. Use assembly constraints to build the model, as shown in Figure 6.171.

TIP

To assemble the balls, mate the center point of the ball with the centerline of the body and then place a tangent constraint between the ball's surface and the sloped surface of the seat.

To assemble the seal into the body, first apply a mate constraint between the conical surface of the seal and the conical surface of the seal's seat in the body. You must use the Select Other command to select the first conical surface. Apply a similar constraint between the gland and the seal.

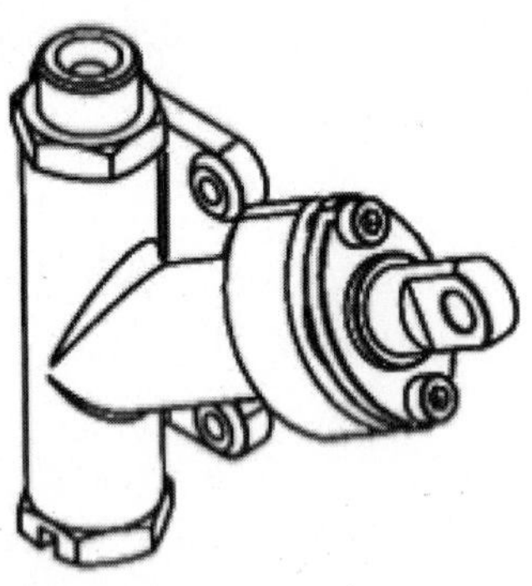

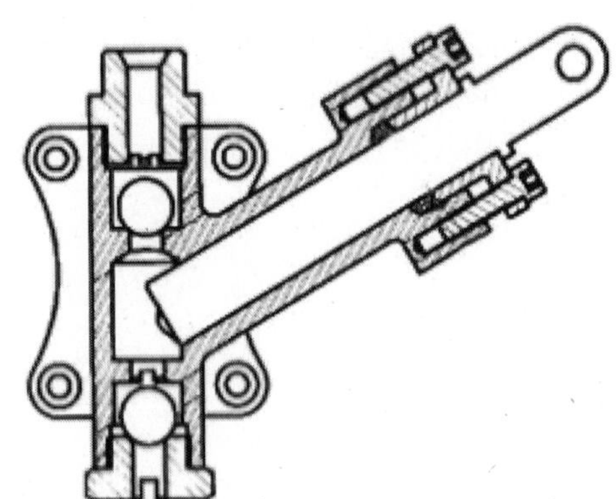

FIGURE 6.171

5. The completed assembly model should appear similar to Figure 6.172.

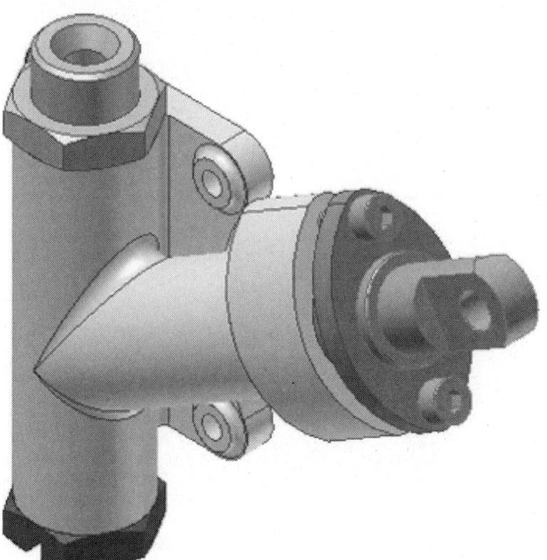

FIGURE 6.172

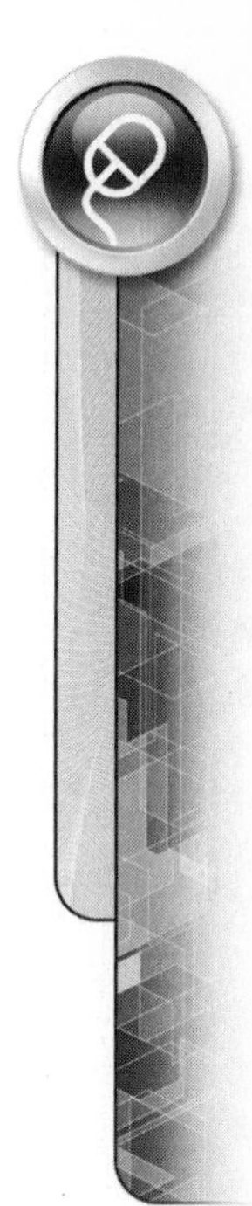

6. Close all open files. Do not save changes. End of exercise.

Skill Exercise 6-2

In this exercise, you create an assembly drawing, a parts list, and balloons for a charge pump. The charge pump assembly is shown in Figure 6.173.

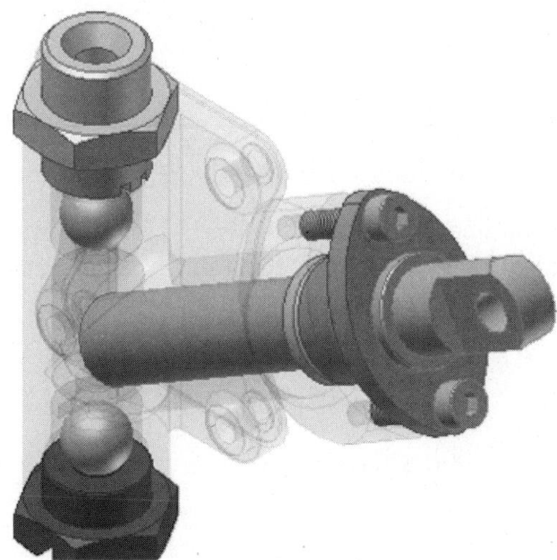

FIGURE 6.173

1. Create a new drawing using the ANSI (in) DWG or IDW template with a single D sheet named Assembly.
2. Insert a top view of *ESS_E06_10.iam* with a scale of 1:1.
3. Using the top view as the base view, create a sectioned front view. Exclude the ram, valve balls, and machine screws from sectioning.
4. Use the Base command and the Change View Orientation button to create an independent top right isometric view of the pump assembly as shown in Figure 6.174.

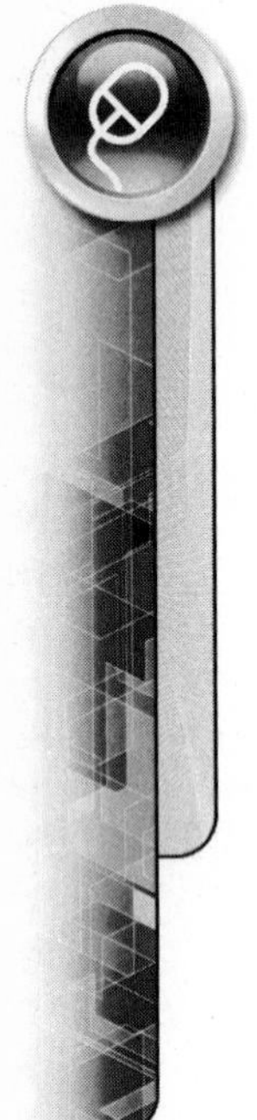

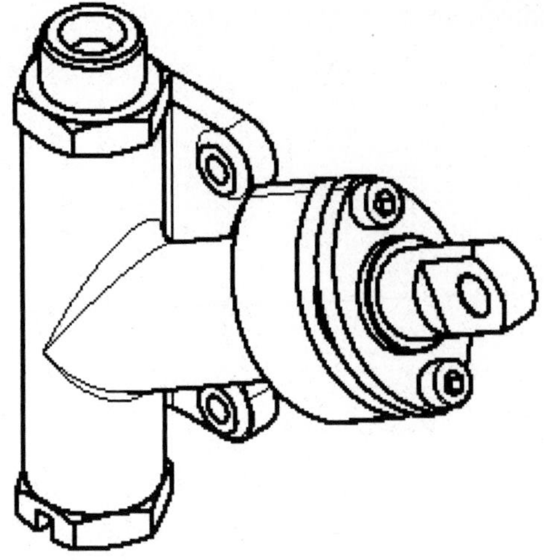

FIGURE 6.174

5. Insert the Parts List, add a Material column and then modify its format according to the column widths shown in the following table.

Column	Width
ITEM	1
QTY	1
DESCRIPTION	1.5
PART NUMBER	2
MATERIAL	2.25

6. Modify iProperties and Parts List parameters so that the Parts List and title block match those shown in Figure 6.175. Note how the item numbers are arranged in ascending order and the header row is at the bottom of the table.

8	1	GLAND	ESS_E06_10-Gland	Brass, Soft Yellow Brass
7	2	PURCHASED	ESS_E06_09-M8x30	Default
6	1	SEAL	ESS_E06_09-Seal	Rubiconium
5	1	UNION	ESS_E06_09-Union	Rubiconium
4	1	SEAT	ESS_E06_09-Seat	Rubiconium
3	1	RAM	ESS_E06_09-Ram	Rubiconium
2	2	BALL	ESS_E06_09-Ball	Rubiconium
1	1	BODY	ESS_E06_09-Body	Rubiconium
ITEM	QTY	DESCRIPTION	PART NUMBER	MATERIAL
PARTS LIST				

DRAWN TJ	12/24/2010			
CHECKED		TITLE		
QA		CHARGE PUMP ASSEMBLY		
MFG				
APPROVED				
		SIZE D	DWG NO	REV
		SCALE		SHEET 1 OF 1

FIGURE 6.175

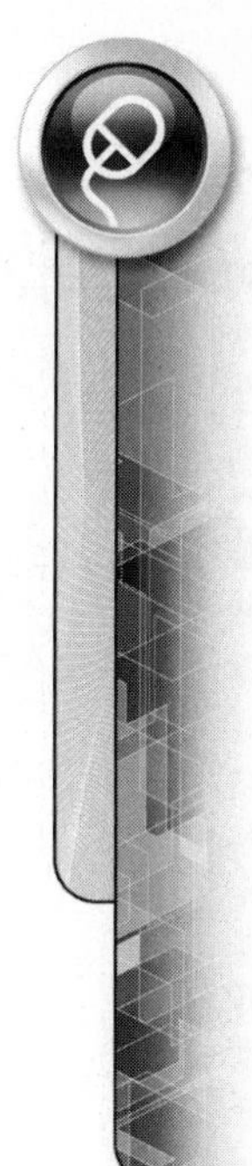

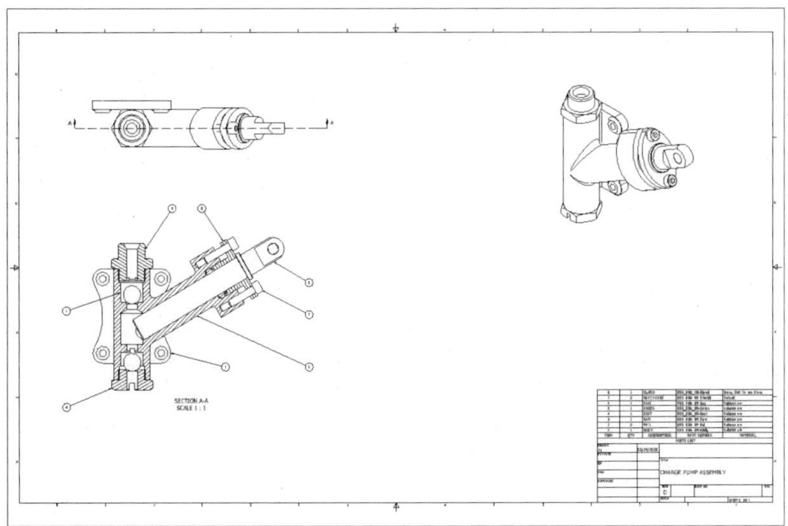

FIGURE 6.176

7. Add balloons to identify the components as shown in Figure 6.176.
8. Close all open files. Do not save changes. End of exercise.

CHECKING YOUR SKILLS

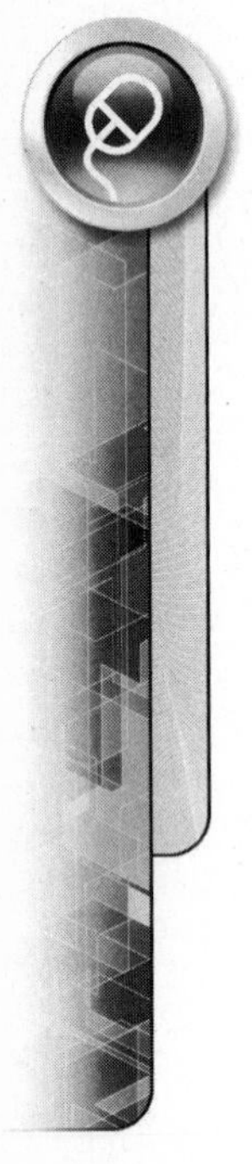

Use these questions to test your knowledge of the material covered in this chapter.

1. True ___ False ___ The only way to create an assembly is by placing existing parts into it.
2. Explain how to create a component in the context of an assembly.
3. True ___ False ___ An occurrence is a new copy of an existing component.
4. True ___ False ___ Only one component can be grounded in an assembly.
5. True ___ False ___ Autodesk Inventor does not require components in an assembly to be fully constrained.
6. True ___ False ___ A sketch must be fully constrained to adapt.
7. What is the purpose of creating a presentation file?
8. True ___ False ___ Balloons can only be placed in a drawing after placing a Parts List.
9. True ___ False ___ When creating drawing views from an assembly, you can create views from multiple presentation views or design views.
10. True ___ False ___ A Bill of Materials only retrieves its data from a Parts List.
11. True ___ False ___ A presentation file can be saved as a Flash animation.
12. True ___ False ___ A BOM Structured view shows all subassemblies and individual parts at the same assembly level.
13. True ___ False ___ An Associated Component Pattern will maintain a relationship to a feature pattern.
14. True ___ False ___ User-defined folders allow you to organize an assembly browser by group assembly constraints in a single folder.
15. True ___ False ___ In a Parts List, you can display custom parts that are not components or graphical data, e.g. paint.

CHAPTER 7

Advanced Modeling Techniques

INTRODUCTION

In this chapter, you will learn how to use advanced modeling techniques. Using advanced modeling techniques, you can create transitions between parts that would otherwise be difficult to create. You can edit advanced features such as other sketched and placed features but typically their creation requires more than one unconsumed sketch. This chapter also introduces you to techniques that will help you be more productive in your modeling.

OBJECTIVES

After completing this chapter, you will be able to:

- Section a part
- Create a design view representation in a part file
- Create ribs, and rib networks
- Emboss text and profiles
- Create sweep features
- Create coil features
- Create loft features
- Split a part and split faces of a part
- Bend a part
- Reorder part features
- Mirror model features
- Suppress features of a part
- Create a derived part based on another part or an assembly
- Shrinkwrap an assembly to create a simplified part file
- Create fillets based on rules
- Simulate stress on a part and an assembly

ADJUSTABLE SECTIONS VIEWS IN A PART FILE

In chapter 6 you learned how to section an assembly to better see inside it. The same functionality is available while working in a part file. As in an assembly, there are four section commands available while working in a part file. To access the section commands, click on the View tab > Appearance panel as shown in Figure 7.1 on the left. Figure 7.1 in the middle shows a part before being sectioned and the view on the right shows the part sectioned with the Three Quarter Section View command. Note that the sections are temporary and sketches cannot be actively placed on a section. As always, you can create a sketch on a plane that was selected to create the section.

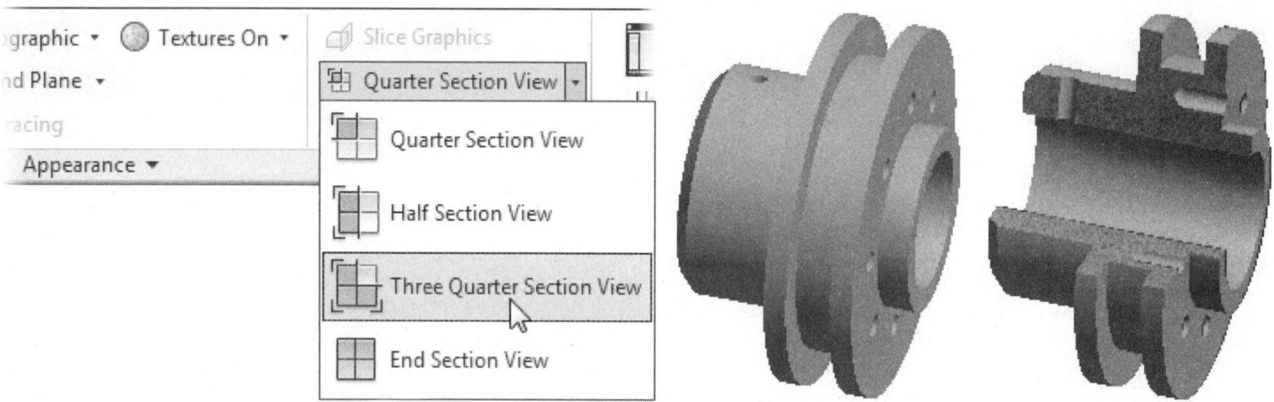

FIGURE 7.1

To section a part, follow these steps.

1. Select the desired section command on the View tab > Appearance panel.
2. Select any existing planar face, work plane, or origin plane.
3. Adjust the section depth by clicking and dragging the visible arrow in the graphics window as shown in Figure 7.2 on the left; enter a value in the Offset cell in the graphics window. Another option to control the offset is to move the cursor over the Offset cell and scroll the wheel on the mouse. You can control the distance that each step takes when you scroll the wheel by right-clicking and click Virtual Movement > Scroll Step Size.
4. To accept the current location click the right faced arrow in the mini-toolbar as shown in Figure 7.2 in the middle, or right-click and click Continue in the marking menu.
5. If you are creating a quarter or three quarter section view, follow the same process to select a second plane.
6. While still in the command, you can specify which side of the section remains visible by right clicking and click the Flip Section or change the section type if Quarter or Three Quarter section was selected.
7. To finish the command, right-click and click OK from the menu.
8. To clear the section, click the End Section View command on the View tab > Appearance panel.

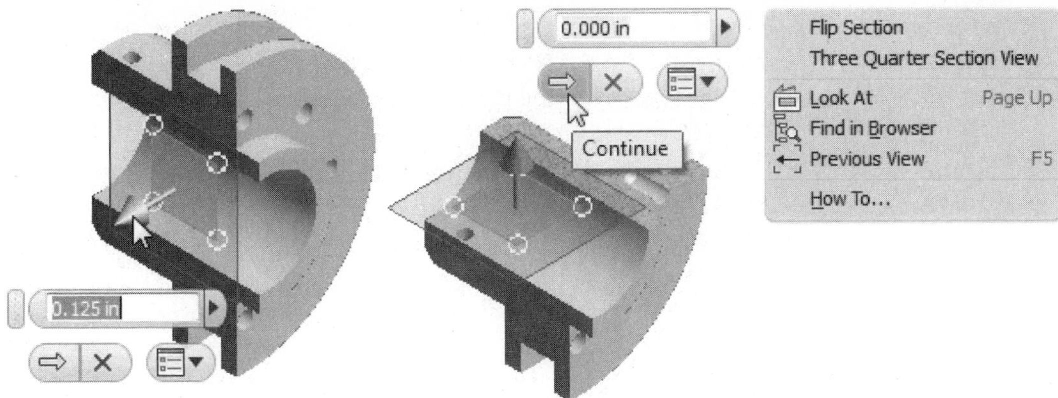

FIGURE 7.2

DESIGN VIEW REPRESENTATION IN A PART FILE

While working on a part file, you can save a design view representation that stores' information related to the current viewpoint, part color, and work feature visibility. You can save as many design view representations as needed, but only one can be current. By default, there is a design view representation named Master, this master design view cannot be changed, deleted, or locked. To create a design view representation in a part file, follow these steps.

1. In the browser, right-click on View: Master and click New from the menu as shown in Figure 7.3 on the left.
2. A design view will be created; you can rename the design view by slowly double-clicking on its name and type in a new name.
3. Once a design view is created, change the view point, section a part, part color, and work feature visibility and those changes are captured in the current design view.
4. Since these changes are actively captured to the current design view, you can lock a design view so any change to the current viewpoint, section a part, part color, or work feature visibility are not saved to the current design view. To lock a design view, right-click on its name in the browser and click Lock from the menu as shown in Figure 7.3 on the right.
5. Create as many design views as needed.
6. Lock and unlock the current design view to capture the settings as needed.
7. To make a design view current, in the browser double-click on its name or right-click on its name and click Activate from the menu. The last settings in the design view or the locked settings in the design view will be displayed in the graphics window.

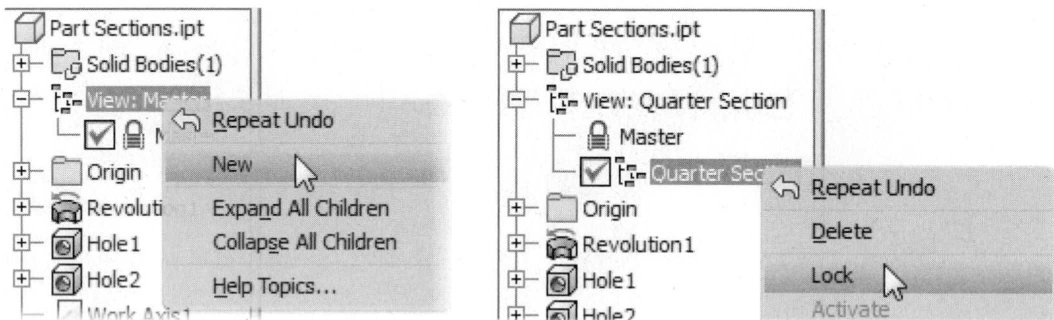

FIGURE 7.3

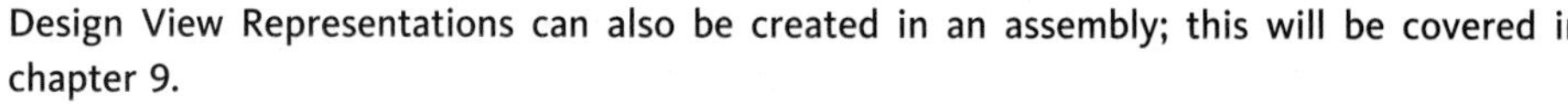

NOTE

Design View Representations can also be created in an assembly; this will be covered in chapter 9.

RIB FEATURES

Ribs are used primarily to reinforce or strengthen features in mold and cast parts, but you can also use them in machined parts and in other cases where additional support and minimal weight are required. Figure 7.4 shows a part with a sketch and with the sketch used to create a rib that uses the To Next and the Finite thickness options.

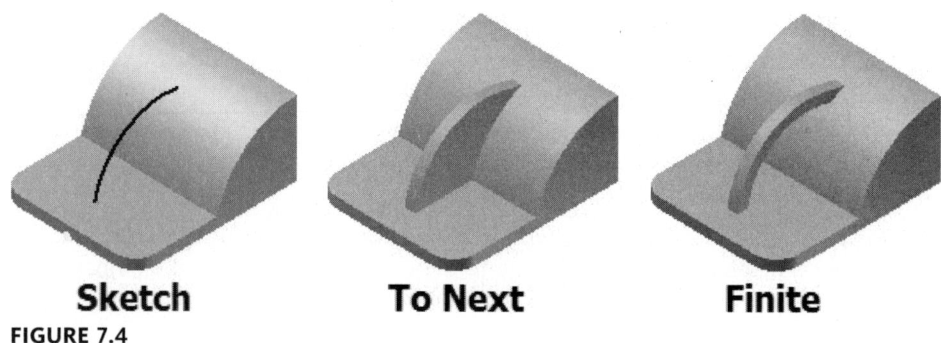

FIGURE 7.4

Using the Rib command on the 3D Model tab > Create panel, as shown in Figure 7.5 on the left, you can create ribs, and rib networks.

- A rib is a thin-walled feature that is typically closed.
- A rib network consists of a series of thin-walled support features.

A rib feature is defined by a single, open, unconsumed profile that is refined using the options in the Rib dialog box. If no unconsumed sketch exists in the part file, Autodesk Inventor will warn you with the message: "No visible unadaptive sketches." After starting the Rib command, the Rib dialog box will appear, as shown in Figure 7.5 on the right. The following sections describe the options in the dialog box.

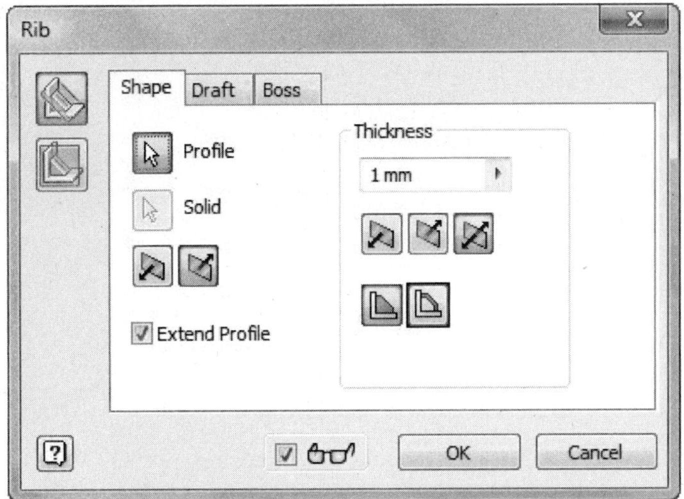

FIGURE 7.5

Shape

Profile

Select this button to select a sketch to create the rib feature from. The sketch can be a single open profile or you can select multiple intersecting or nonintersecting profiles to define a rib or web network.

Rib Type—Normal or Parallel to Sketch Plane

The two options on the left side are Normal to Sketch Plane and Parallel to Sketch Plane. The Normal to Sketch Plane type extrudes the profile normal to the sketch plane and its thickness is parallel to the sketch plane. The Parallel to Sketch Plane type extrudes the profile parallel to the sketch plane and its thickness is normal to the sketch plane. Figures 7.6A and 7.6B show a Normal to Sketch Plane and a Parallel to Sketch Plane type created from the same sketch that is drawn on the visible work plane. Notice how the direction of the thickness changes. Figures 7.6A and 7.6B show an example of a **Normal to Sketch Plane** and a **Parallel to Sketch Plane**.

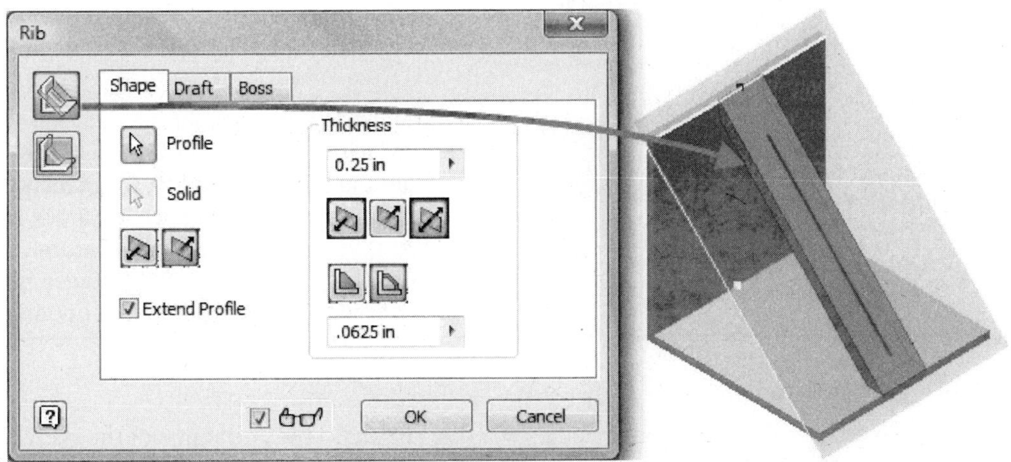

FIGURE 7.6A

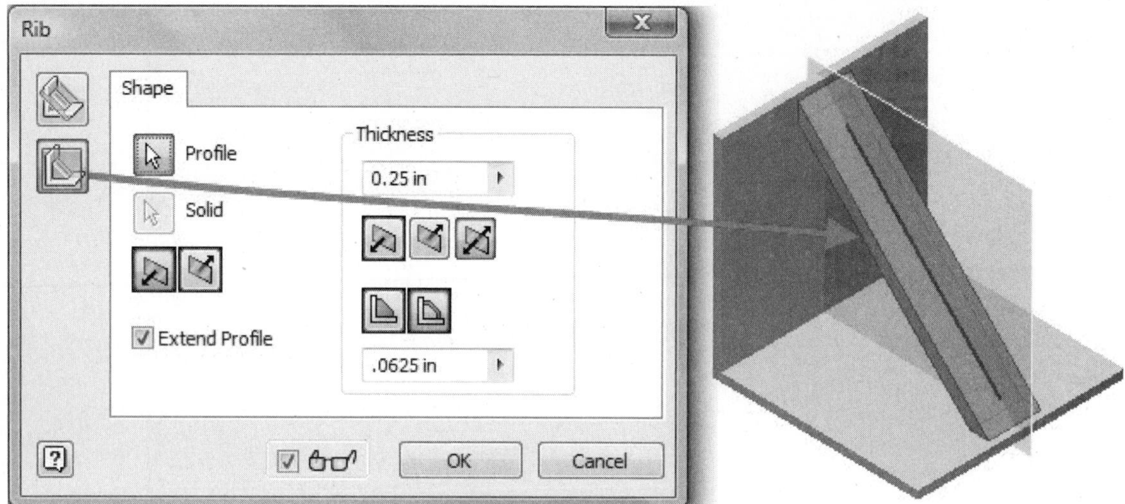

Parallel to Sketch Plane

FIGURE 7.6B

Thickness

In this section, you specify the thickness and thickness direction of the rib.

Value

Enter the width of the rib feature using this edit box.

Flip Buttons

Select the flip buttons to specify which side of the profile to apply the thickness value to or to add the same amount of material to both sides of the profile.

Extents

In this section, you specify a rib (To Next) or a web (Finite).

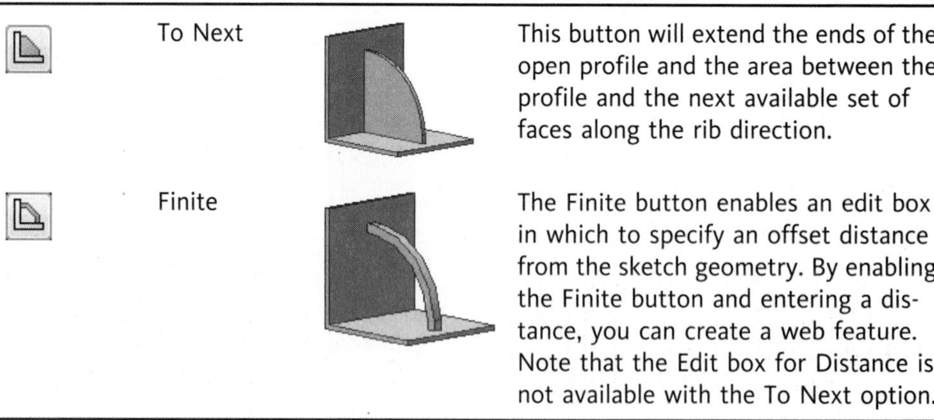

	To Next		This button will extend the ends of the open profile and the area between the profile and the next available set of faces along the rib direction.
	Finite		The Finite button enables an edit box in which to specify an offset distance from the sketch geometry. By enabling the Finite button and entering a distance, you can create a web feature. Note that the Edit box for Distance is not available with the To Next option.

Extend Profile

The Extend Profile checkbox specifies whether to extend the endpoints of the sketch to the next available face or to leave the ends of the open profile as determined by the end of the rib feature. If you click the Extend Profile option, the ends are extended; if you leave the box clear, the ends cap at the end of the sketched profile. The Extend

Profile option is always available for a finite web, but it is only active for the To Next option when the direction and face(s) on the model meet appropriate conditions.

Draft

The Draft tab is only available when the Normal to Sketch Plane type is selected. In the Draft tab you choose which thickness to hold, the top or root, and define the draft angle as shown in Figure 7.7.

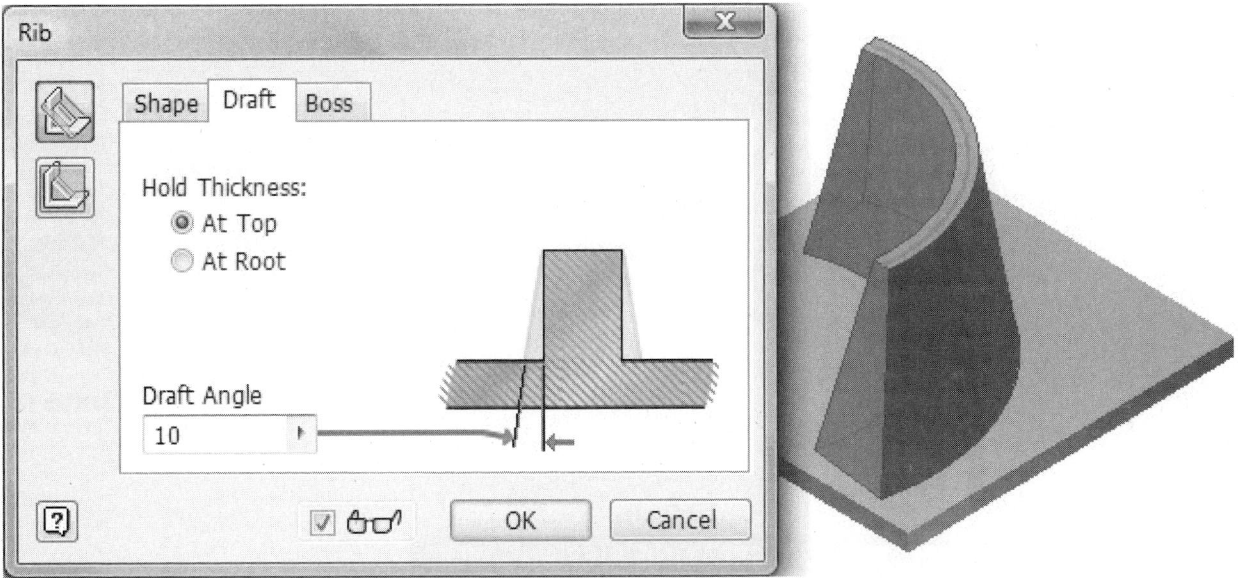

FIGURE 7.7

The Boss tab is only available when the Normal to Sketch Plane type is selected. In the Boss tab you can add a boss to a rib. You place a point, center point(s), in which the boss will be defined. Figure 7.8 shows a boss created on the arc.

Creating Ribs

To create a rib follow these steps:

1. Create an active sketch in the location where you will place the rib.
2. Sketch an open profile that defines the basic shape of one edge of the rib.
3. Add constraints and dimensions to the sketch as needed.
4. Click the Rib command on the 3D Model tab > Create panel.
5. Specify the rib type, normal or parallel to the sketch.
6. Enter a value for the rib or web in the Thickness edit box.
7. Also in the Thickness section, use the Flip buttons to choose which side of the profile to apply the thickness value or to add the same amount of material to both sides of the profile.
8. Specify the depth of the profile by clicking either the To Next or the Finite buttons in the Extents area of the Rib dialog box.
9. If you use the Finite option, enter an offset distance, and click Extend Profile if the endpoints are to be extended to the next available face.
10. If the Normal to Sketch Plane type is selected you can also define draft and a boss(es).
11. To complete the operation, click the OK button.

Rib Network

You can also create a rib network using the Rib command. You can use multiple intersecting or nonintersecting sketch objects within a single profile to create a rib network. The creation process is the same as for creating a single web, except that the thickness is applied to all objects within the profile. When you select the profile objects, you will need to select them individually. If the rib network is to have equal spacing between the objects, use the 2D Rectangular Pattern or the 2D Circular Pattern command in the Sketch tab > Pattern panel to create the profile. Figure 7.8 shows a part with multiple intersecting lines defined in a single profile, the same profile used to create a rib network using the Normal to Sketch Plane type and the thickness set to To Next and Finite.

FIGURE 7.8

EXERCISE 7-1: CREATING RIBS AND WEBS

In this exercise, you sketch an open profile, use the Rib command to create a rib, and then edit the rib feature to change the rib to a web. You complete the exercise by sketching overlapping lines and creating a rib network.

1. Open *ESS_E07_01.ipt* in the Chapter 07 folder.
2. Use the Free Orbit command to examine the part.
3. Create a new sketch on the visible work plane.
4. Turn off the visibility of the work plane by right-clicking on one of the edges of the work plane and click Visibility from the menu.
5. Sketch the line and arc, as shown in Figure 7.9. Make sure that the arc you sketch, if extended to the right, would intersect the top of the part. Constraints and dimensions could be added as required.

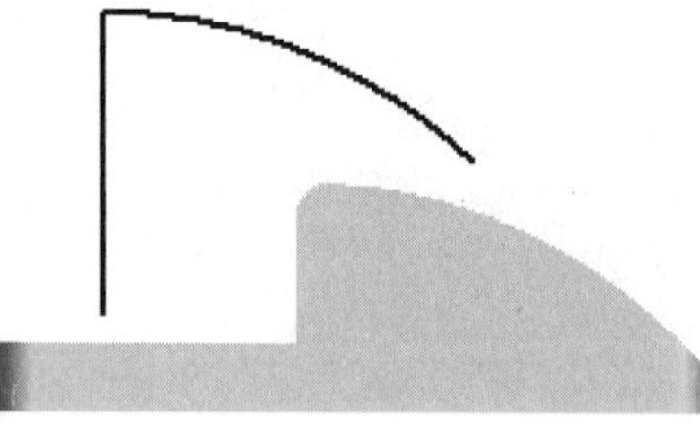

FIGURE 7.9

6. Click the Finish Sketch command on the Sketch tab > Exit panel to finish the sketch.
7. Click the Rib command on the 3D Model tab > Create panel. Make the Normal to Sketch Plane Type current labeled (1) in **Figure 7.10**, the Thickness **.25 in** labeled (2), change the Extents to Finite labeled (3) with a distance of **2 in** labeled (4). If you do not see a preview of the rib, you need to edit the sketch and alter the sketch so it will extend into the part.
8. Click OK to create the rib.

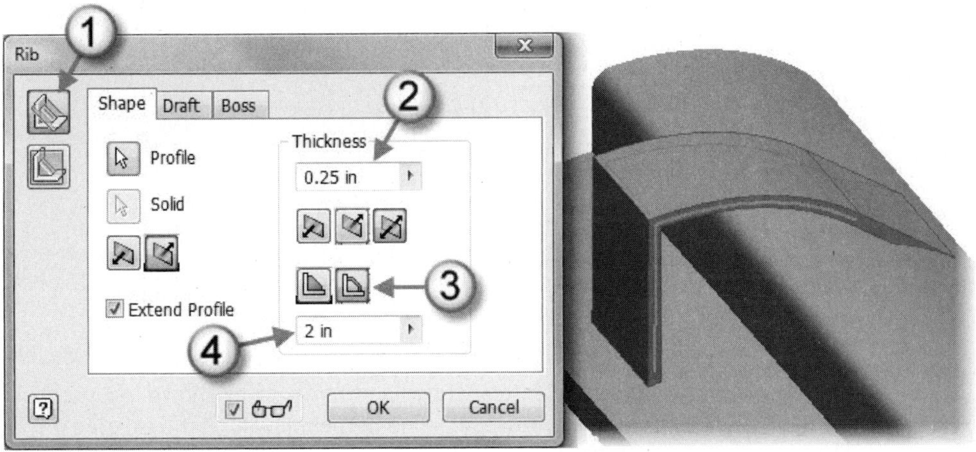

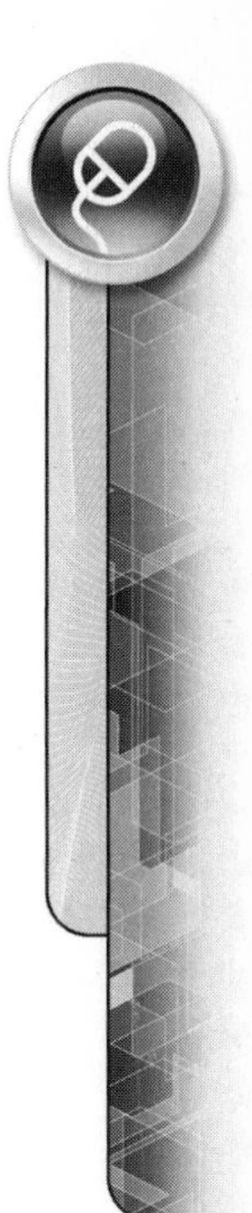

FIGURE 7.10

9. Edit the rib, in the graphics window select a face on the rib and click Edit Rib from the mini-toolbar.
10. In the Rib dialog box, make the Rib Type (Parallel to Sketch Plane) current labeled (1) in Figure 7.11, the Thickness **2 in** labeled (2), ensure the Extents is set to Finite labeled (3) with a distance of **.25 in** labeled (4). Notice how the distances are going in the opposite direction from the Normal to Sketch Plane type.

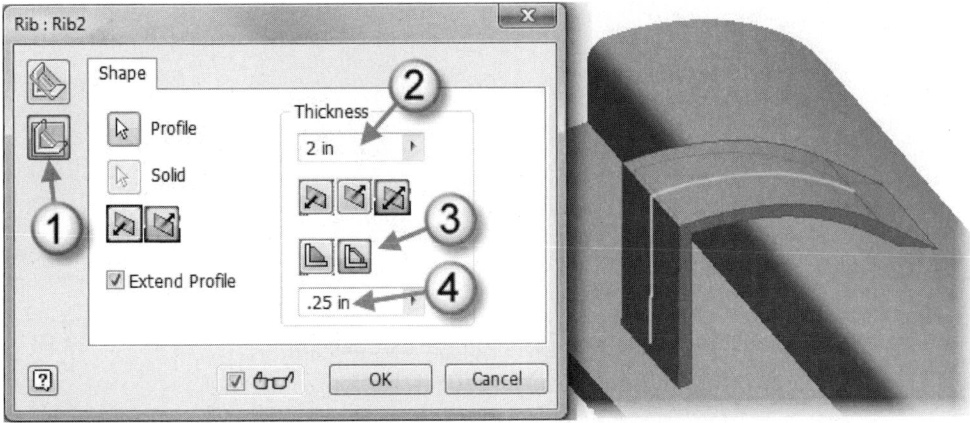

FIGURE 7.11

11. While still editing the rib, change the Extents to To Next labeled (1) as shown in Figure 7.12. Notice how the rib extends to the area between the profile and the next available set of faces. Click OK to complete the edit.

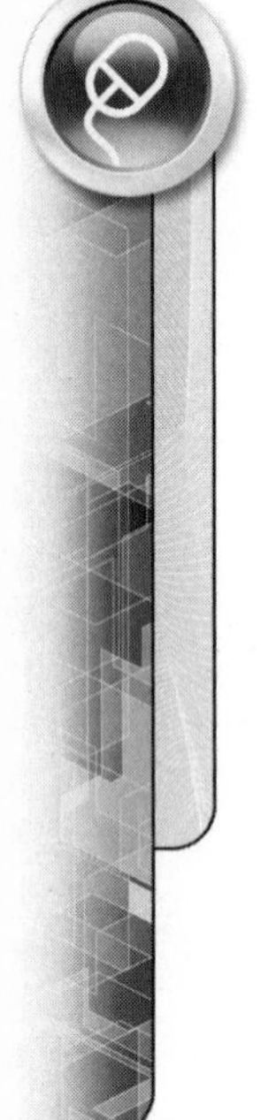

FIGURE 7.12

12. Use the Free Orbit command to change your viewpoint so you can see the bottom of the part, as shown in Figure 7.13 on the left.
13. Create a new sketch on the inside planar face, as shown in Figure 7.13 on the left.
14. Sketch and dimension three, intersecting line segments, as shown in Figure 7.13 on the right. The line should be parallel and perpendicular to the existing edges but not touch the edges. The edges will be extended when the rib feature is created.

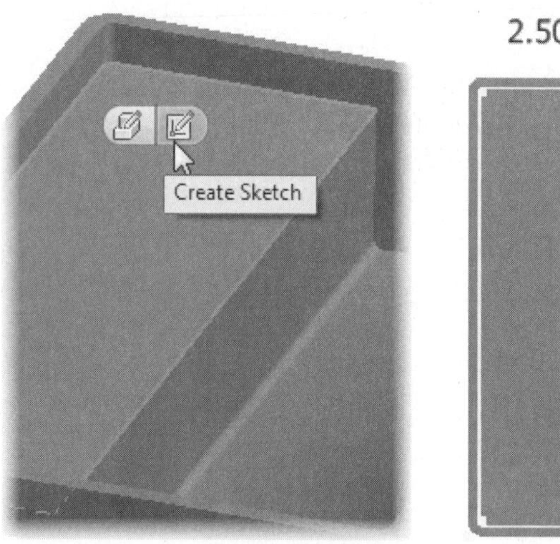

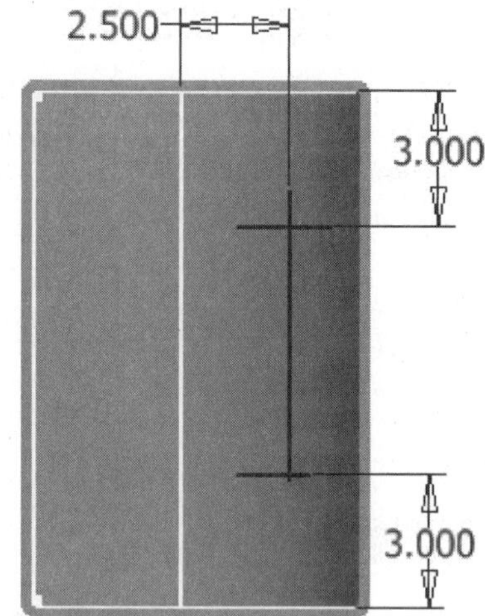

FIGURE 7.13

15. Click the Finish Sketch command on the Exit panel.
16. Click the Rib command on the 3D Model tab > Create panel.
17. Select the three sketched line segments for the profile.
18. In the Rib dialog box, make the Normal to Sketch Plane Type current labeled (1) in Figure 7.14, the Thickness **.5 in** labeled (2), change the Extents to Finite labeled (3) with a distance of **.25 in** labeled (4).

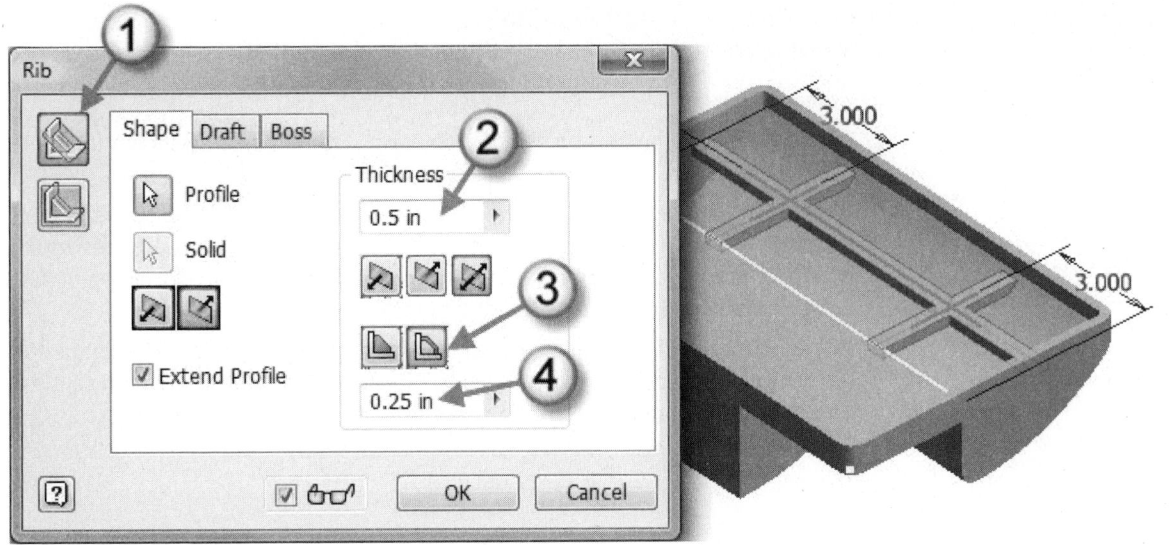

FIGURE 7.14

19. Click OK in the Rib dialog box, and your rib network should resemble Figure 7.15.

FIGURE 7.15

20. Use the Zoom and Free Orbit commands to examine the rib network.
21. Close the file. Do not save changes. End of exercise.

EMBOSSED TEXT AND CLOSED PROFILES

To better define a part, you may need to have a shape or text either embossed (raised) or engraved (cut) into a model. In this section, you will learn how you can emboss or engrave a closed shape or text onto a planar or curved face. You can define a shape using the sketch commands on the Sketch tab > Draw panel. The shape needs to be closed. There are two steps to emboss or engrave: first create the shape or text. Second, you emboss the shape or text onto the part. The text can be oriented in a rectangle or about an arc, circle, or line. The following sections describe the steps.

Step 1: Creating Rectangular Text

To place rectangular text onto a sketch, follow these steps:

1. Create a sketch.
2. Click the Text command on the Sketch tab > Draw panel, as shown in Figure 7.16 on the left.
3. Pick a point or drag a rectangle where you will place the text. If a single point is selected, the text will fit on a single line and can be grip edited to resize the bounding box. If a rectangle is used, it defines the width for text wrapping and can be modified. The Format Text dialog box will appear.
4. In the Format Text dialog box, specify the text font and format style, and enter the text to place on the sketch. Figure 7.16 shows the dialog box with text entered in the bottom pane.

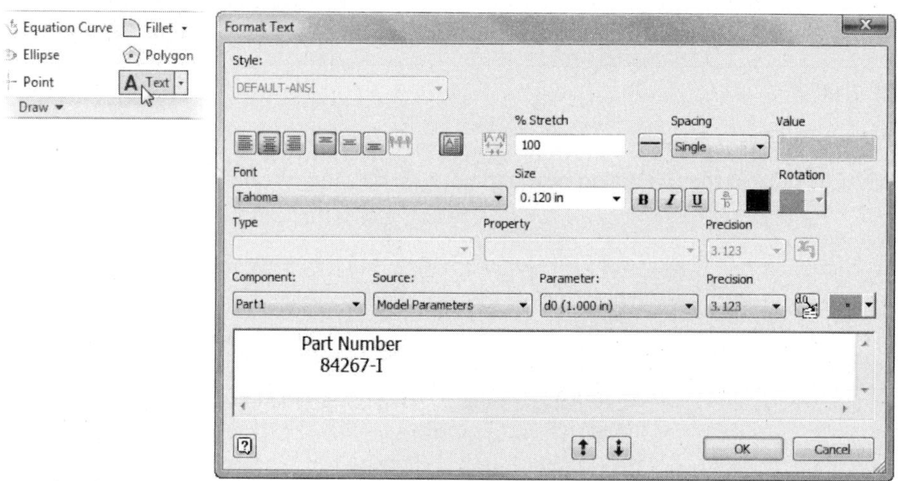

FIGURE 7.16

5. When done typing text, click OK and to complete the command, right-click and click OK in the marking menu.
6. When the text is placed, a rectangular set of construction lines defines the perimeter of the text. You can click and drag a point on the rectangle and change the size of the rectangle. Dimensions or constraints can be added to these construction lines to refine the text's location and orientation, as shown in Figure 7.17.
7. To edit the text, double-click on the text or right-click on the text and click Edit text from the marking menu. The same Format Text dialog box that was used to create the text will appear.

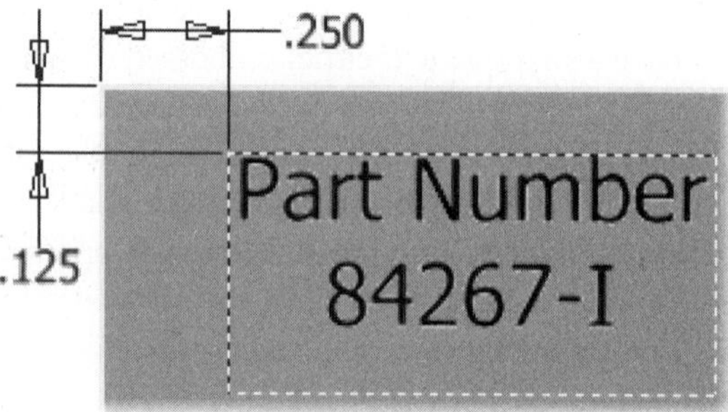

FIGURE 7.17

Step 2: Creating Text about Geometry

To place text about an arc, circle, or line in a sketch, follow these steps:

1. Create a new sketch or make an existing sketch active.
2. Sketch and constrain an arc, circle, or line that the text will follow; it is recommended to make this geometry construction geometry as it will not be used as the geometry for the profile.
3. Click the Geometry Text command on the Sketch tab > Draw panel > Geometry Text command as shown in Figure 7.18 on the left. The command may be under the Text command.
4. In the sketch, select the arc, circle, or line and the Geometry-Text dialog box will appear.
5. In the Geometry-Text dialog box, specify the text font and format style, direction, position, and start angle and enter the text to place on the sketch. The start angle is relative to the left quadrant point of a circle or the start point of the arc. Figure 7.18 on the right shows the dialog box with text entered in the bottom pane.

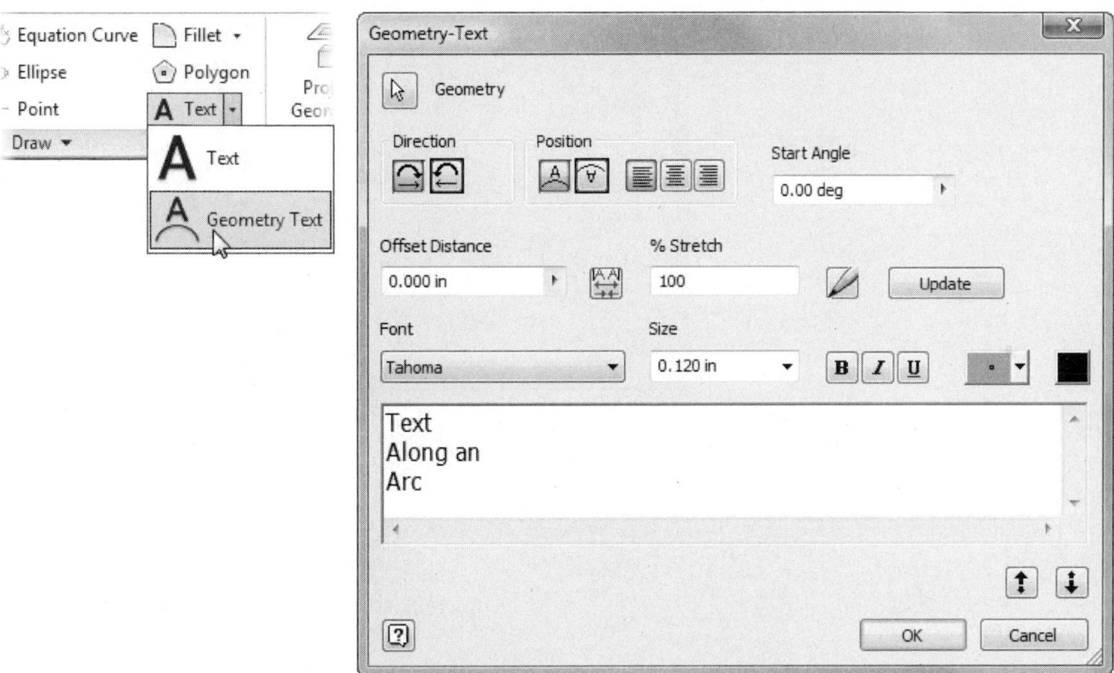

FIGURE 7.18

6. When done entering text, click the Update button in the dialog box to see the preview in the graphics window.
7. Click OK to complete the operation. Figure 7.19 shows text placed along an arc.
8. To edit the text, double-click on the text or right-click on the text and click Edit Geometry Text from the marking menu. The same Format Text dialog box that was used to create the text will appear.

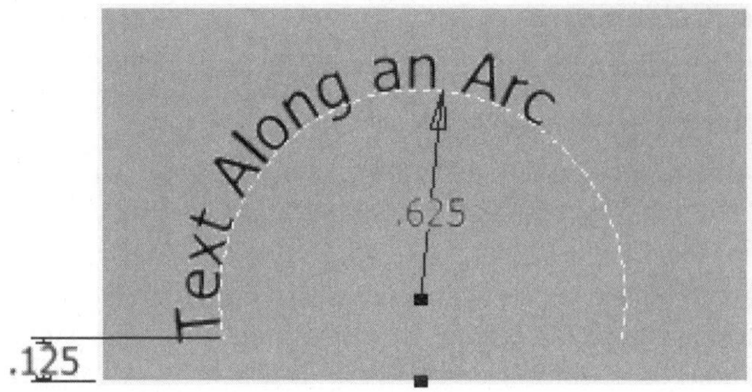

FIGURE 7.19

Step 3: Embossing Text

To emboss text or a closed profile, click the Emboss command on the 3D Model tab > Create panel, as shown in Figure 7.20 on the left. The Emboss dialog box will appear, as shown in Figure 7.20 on the right.

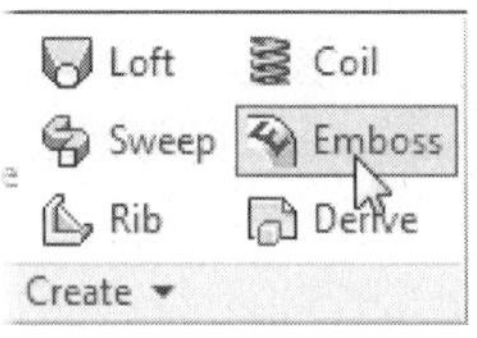

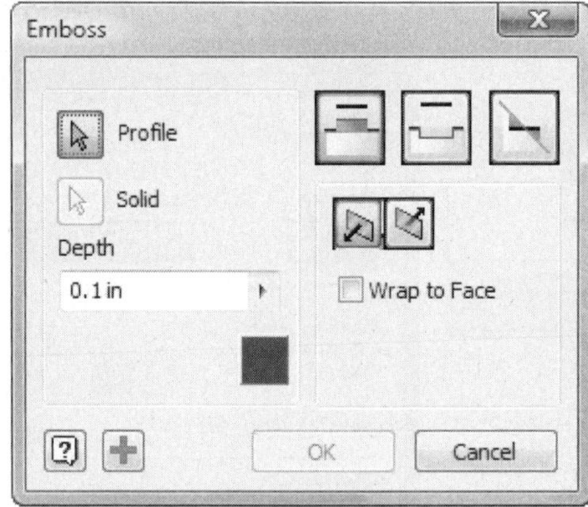

FIGURE 7.20

The Emboss dialog has the following options:

	Profile	Select a profile, meaning closed shape or text, to emboss. You may need to use the Select Other command to select the text.
1 mm	Depth	Enter an offset depth to emboss or engrave the profile if you selected the Emboss from Face or Engrave from Face types.
	Top Face Appearance	Select a color from the drop-down list to define the color of the top face of the embossed area, not its lateral sides.
	Emboss from Face	Select this option to add material to the part.
	Engrave from Face	Select this option to remove material to the part.

(Continued)

	Emboss/ Engrave from Plane	Select this option to add and remove material from the part by extruding both directions from the sketch plane. Direction changes at the tangent point of the profile to a curved face.
	Flip Direction	Select either of these buttons to define the direction of the feature.
	Wrap to Face	Check this box for Emboss from Face or Engrave from Face types to wrap the profile onto a curved face. Only a single cylindrical or conical face can be selected. The profile will be slightly distorted as it is projected onto the face. The wrap stops if a perpendicular face is encountered.

To emboss a closed shape or text, follow these steps:

1. Click the Emboss command on the 3D Model tab > Create panel.
2. Define the profile by selecting a closed shape or text. If needed, use the Select Other—Face Cycling command to select geometry that is not selected immediately when the cursor is moved over it.
3. Select the type of emboss: Emboss from Face, Engrave from Face, or Emboss/ Engrave from Plane.
4. Specify the depth, color, direction, and face as needed. If you selected Emboss/ Engrave from plane option, you can also add a taper angle to the created emboss/engrave feature.

NOTE

You edit the embossed feature like any other feature.

EXERCISE 7-2: CREATING TEXT AND EMBOSS FEATURES

In this exercise, you emboss and engrave sketched profile objects on faces of a razor handle model. You then create a sketch text object and engrave it on the handle.

1. Open *ESS_E07_02.ipt* in the Chapter 07 folder.
2. Use the Free Orbit and Zoom commands to examine the part.
3. From the browser turn on visibility of Sketch11 by right-clicking on Sketch11 and click Visibility from the menu and rotate the view to display the top triangular face, as shown in Figure 7.21.

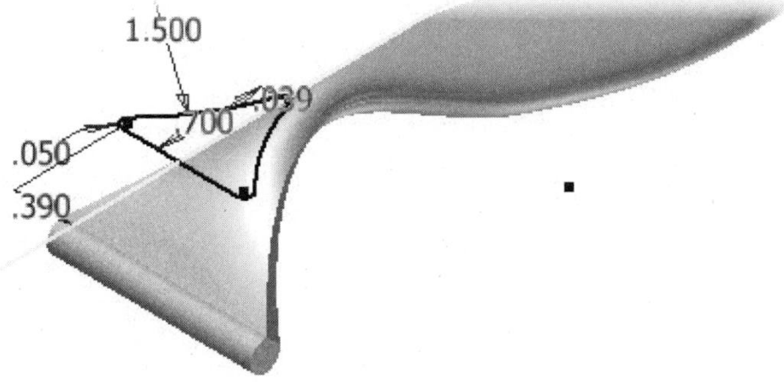

FIGURE 7.21

4. Click the Emboss command on the 3D Model tab > Create panel to engrave a closed sketch profile. Notice that the only visible closed profile is automatically selected.
   - In the Emboss dialog box, click the Engrave from Face option (middle button).
   - Change the Depth to **.03125 in**.
   - Click the Top Face Appearance button, and choose Aluminum (Flat) from the Color dialog box drop-down list, as shown in Figure 7.22 on the left.
5. Click OK twice to close both dialog boxes and create the engraved feature, as shown in Figure 7.22 on the right.

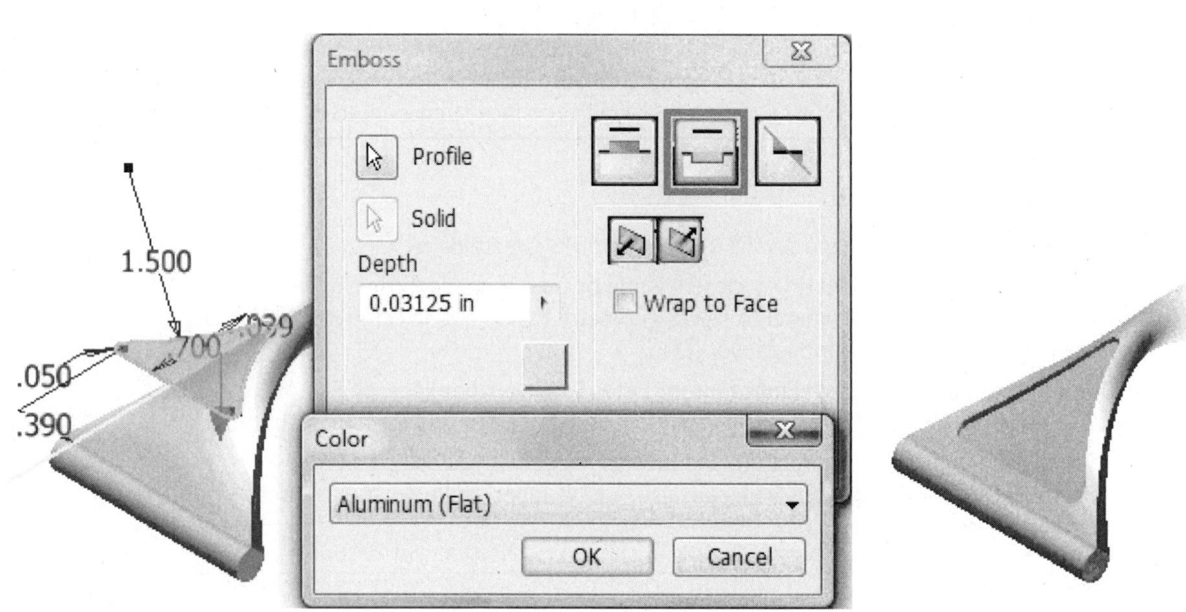

FIGURE 7.22

6. Click the Home View above the ViewCube, and from the browser turn on the visibility of Sketch10.
7. Click the Emboss command from the 3D Model tab > Create panel.
   - For the profile, click the oblong and the six closed profiles. Be sure to select the left and right halves of the herringbone profiles, as shown in Figure 7.23 on the left.
   - In the Emboss dialog box, click the Emboss from Face option (left button).
   - Change the Depth to **.015625 in**.
   - Click the Top Face Appearance button, and select Black from the Color dialog box drop-down list, as shown in Figure 7.23 on the right.

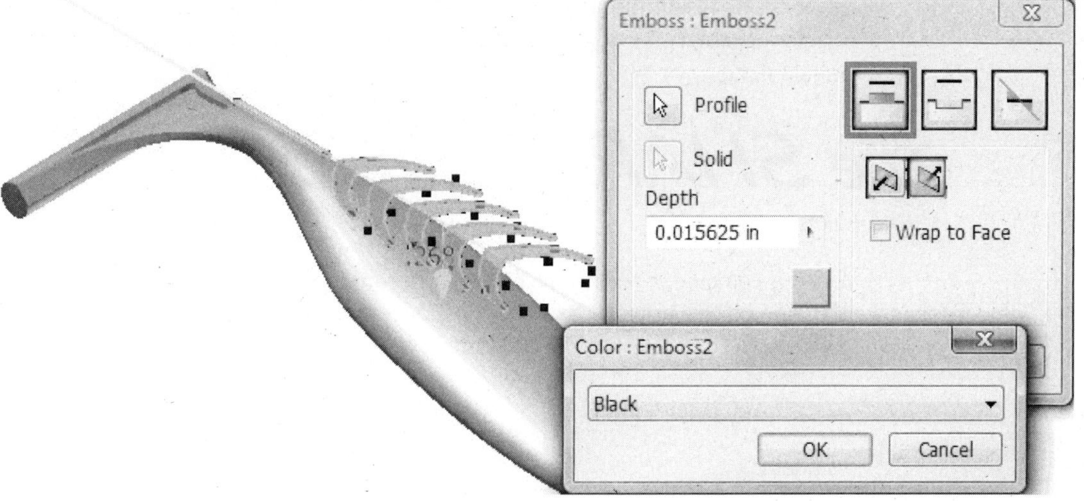

FIGURE 7.23

8. Click OK twice to close both dialog boxes and create the emboss feature.
9. Use the Free Orbit command to examine the engraved and embossed features.
10. Next you create a text sketch object. From the browser turn on the visibility of Sketch9.
11. In the browser, double-click Sketch9 to make it the active sketch.
12. Click the Text command in the Sketch tab > Draw panel.
    - Click in an open area below the part and under the left edge of the construction rectangle in the sketch to specify the insertion point and display the Format Text dialog box.
    - Click the Center and Middle Justification buttons labeled (1) and (2) in Figure 7.24.
    - Click the Italic option labeled (3).
    - Change the % Stretch value to **120** labeled (4).
    - Click in the text field, and type The SHARP EDGE labeled (5).
    - In the text field, double-click on the word "The," and change the text size from 1.20 in to **.09 in** labeled (6).

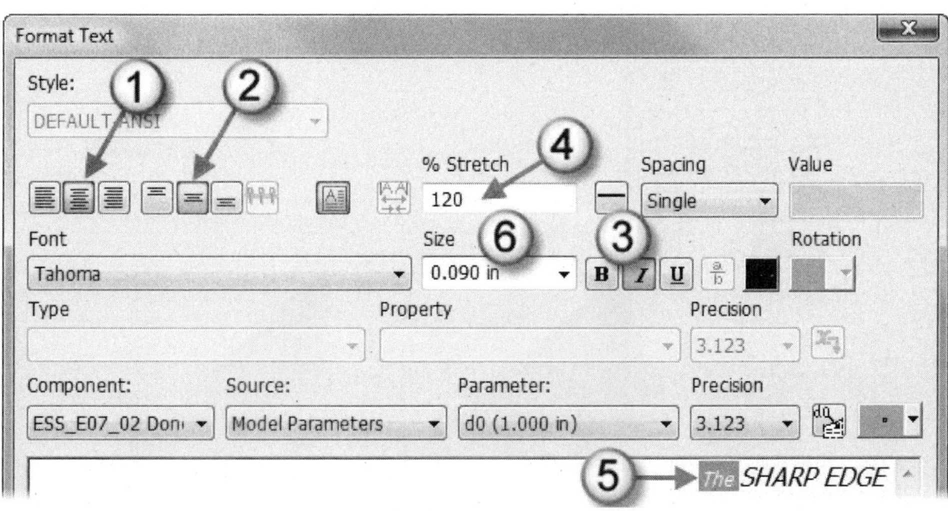

FIGURE 7.24

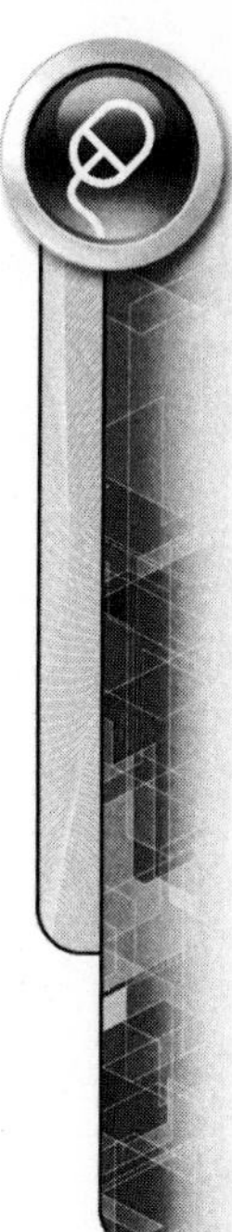

13. Click OK to create the text.
14. Draw a diagonal construction line coincident in the text bounding box corners, as shown in Figure 7.25.

FIGURE 7.25

15. Place a coincident constraint between the midpoint of the diagonal construction line on the text and Sketch9, as shown in Figure 7.26 on the left. Press the ESC key to stop the command. The completed operation is shown in Figure 7.26 on the right.

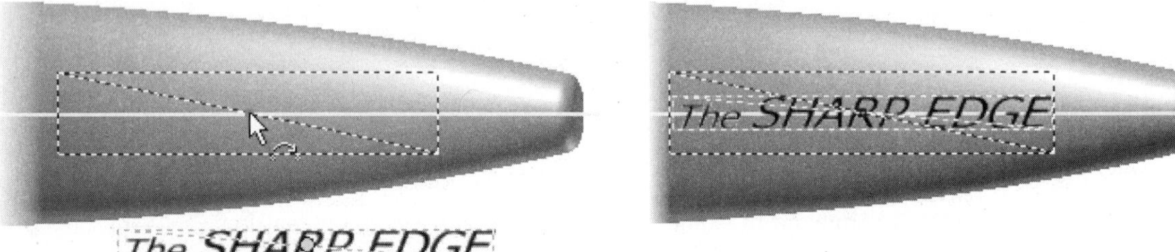

FIGURE 7.26

16. Finish the sketch by right-clicking and click Finish 2D Sketch from the marking menu.
17. Click the Emboss command.
    - For the profile, select the text object.
    - In the Emboss dialog box, click the Engrave from Face option.
    - Change the Depth to **.015625 in**.
    - For the direction, click the right button to change the direction down.
    - Click the Top Face Color button, and click Black from the Color drop-down list as shown in Figure 7.27 on the left.
18. Click OK twice to close both dialog boxes and create the emboss feature.
19. Use the Free Orbit command to examine the engraved text feature as shown in Figure 7.27 on the right.

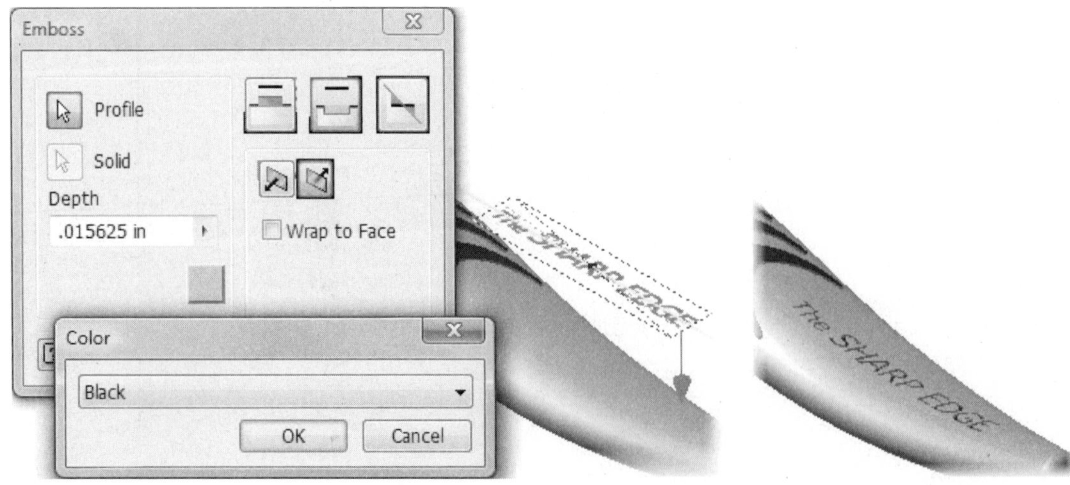

FIGURE 7.27

20. Delete the Emboss3 feature you just created.
21. Edit Sketch9 by double-clicking on it in the browser.
22. In the sketch, delete the existing text, rectangle and angled lines; do NOT delete the yellow horizontal line in the middle of the part (your color may be different depending upon your color scheme).
23. Sketch and dimension a construction circle as shown.

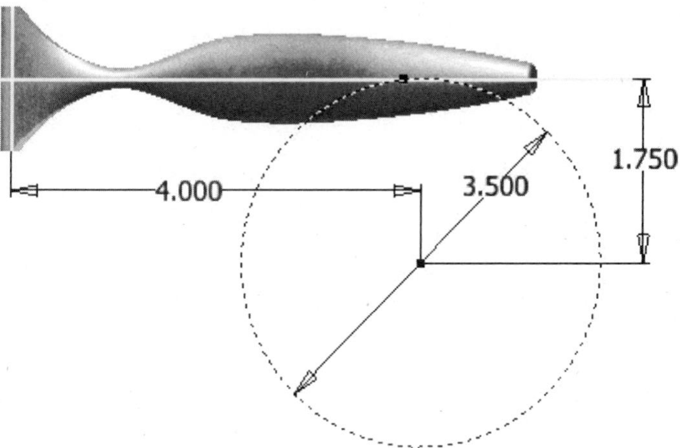

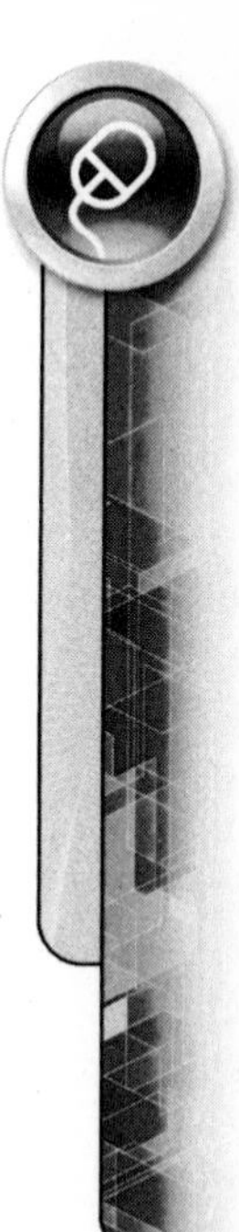

FIGURE 7.28

24. Click the Geometry Text command from the Sketch tab > Draw panel and select the circle for the Geometry selection.
    - Click the Center Justification button (middle button) labeled (1) as shown in **Figure 7.29**.
    - Change the Start Angle to **90.00 deg** labeled (2).
    - Check the Bold button labeled (3).
    - Change the text height to **.125 in** labeled (4).
    - Click in the text field, and enter **The SHARP EDGE** labeled (5) (do not press Enter, the words in your dialog box may be on a single line).
    - To preview the text on the geometry, click the Update button labeled (6) in the Geometry-Text dialog box as shown in Figure 7.29 on the right.

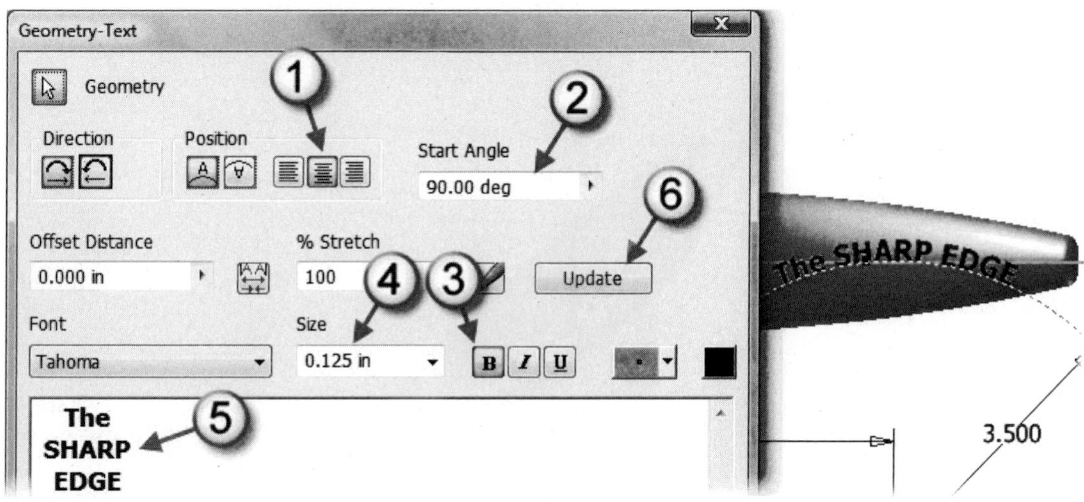

FIGURE 7.29

- Click OK to complete the command.

25. Click the Finish Sketch command on the Sketch tab > Exit panel to finish editing the sketch, and then change to the Home View.
26. Click the Emboss command.
    - For the profile, select the text object.
    - In the Emboss dialog box, click the Engrave from Face option.
    - Flip the direction so it points down.
    - Change the Depth to **.015625 in**.
    - Click the Top Face Appearance button, and click Black from the Color drop-down list.
27. Click OK twice to close both dialog boxes and create the emboss feature.

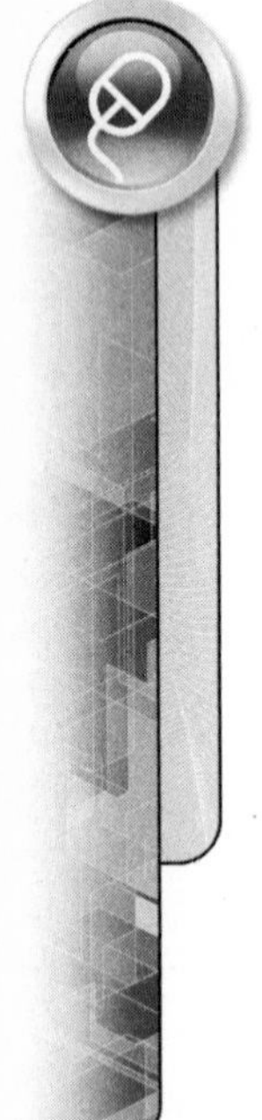

FIGURE 7.30

28. Use the Free Orbit command to examine the emboss features as shown in Figure 7.31.

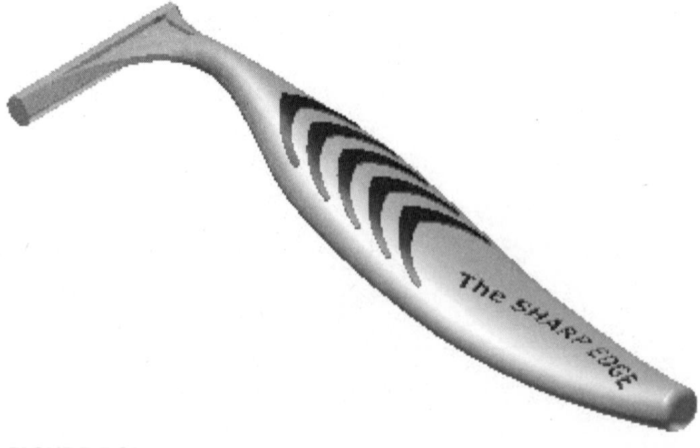

FIGURE 7.31

29. Close the file. Do not save changes. End of exercise.

SWEEP FEATURES

Unlike extrude and revolve features, a sweep feature requires two unconsumed sketches: a profile to be swept and a path that the profile will follow or use edges on a part for the path. Additional profiles can be used as guide rails or a surface can also be used to help shape the feature. The profile sketch and the path sketch cannot lie on the same or parallel planes. The path can be an irregular shape or use part edges by projecting and including the edges onto the active sketch. The path can be either an open or closed profile and can lie in a plane or lie in multiple planes (3D Sketch). Handles, cabling, and piping are examples of sweep features. A sweep feature can be a base or a secondary feature. To create a sweep feature, use the Sweep command on the 3D Model tab > Create panel, as shown in on the left. The Sweep dialog will appear, as shown in Figure 7.32 on the right. The following list includes descriptions of the options in the Sweep dialog box.

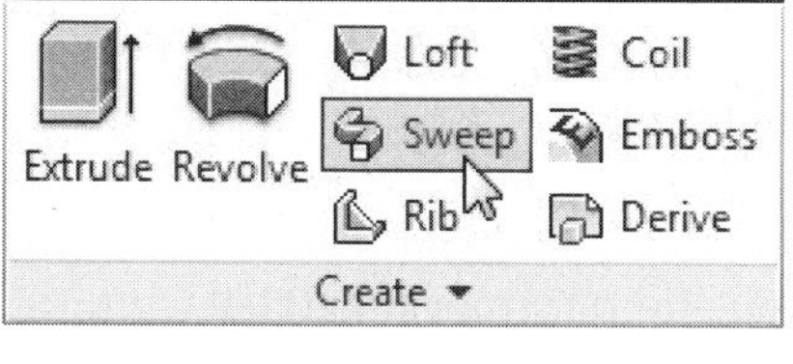

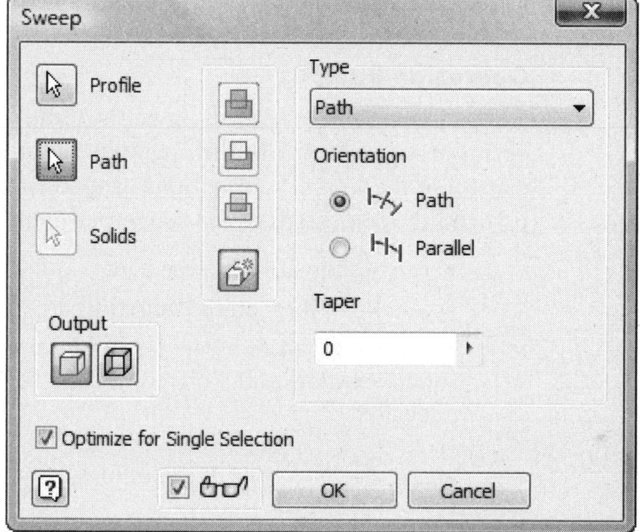

FIGURE 7.32

Profile

Click this button to select the sketch profile to sweep. If the Profile button is depressed and red, it means that you need to select a sketch or sketch area. If there are multiple closed profiles, you will need to select the profile that you want to sweep. If there is only one possible profile, Autodesk Inventor will select it for you and you can skip this step. If you selected the wrong profile or sketch area, depress the Profile button, and deselect the incorrect sketch by clicking it while holding down the CTRL key. Release the CTRL key, and select the desired sketch profile.

Path

Click this button to select the path along which to sweep the profile. The path can be an open or a closed profile on its own sketch but must pierce the sketch that the profile was drawn on. You can also use edge(s) of a part as a path. The profile is typically perpendicular to and intersects with the start point of the path. The start point of the path is often projected into the profile sketch to provide a reference point.

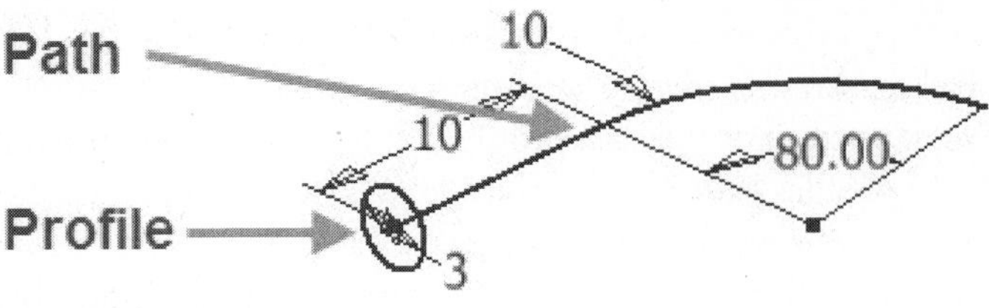

FIGURE 7.33

Solids

If there are multiple solid bodies, click this button to choose the solid body(ies) to participate in the operation.

Output Buttons

In the Output section, click Solid to create a solid feature or Surface to create the feature as a surface.

Operation Buttons

This is the column of buttons down the middle of the dialog box. By default, the Join operation is selected. Use the operation buttons to add or remove material from the part using the Join or Cut options or to keep what is common between the existing part and the completed sweep using the Intersect option.

- *Join:* Adds material to the part.
- *Cut:* Removes material from the part.
- *Intersect:* Creates a new feature from the shared volume of the sweep feature and existing part volume. Material not included in the shared volume is deleted.

Type Box

Path

Creates a sweep feature by sweeping a profile along a path.

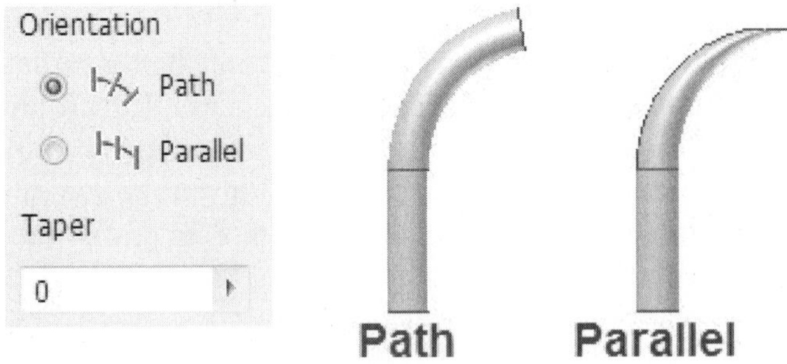

FIGURE 7.34

Orientation

Path

Holds the swept profile constant to the sweep path. All sweep sections maintain the original profile relationship to the path, as shown in Figure 7.34.

Parallel

Holds the swept profile parallel to the original profile as shown in Figure 7.34.

Taper Enter a value for the angle you want the profile to be drafted. By default, the taper angle is 0, as shown in Figure 7.34.

Path & Guide Rail

Creates a sweep feature by sweeping a profile along a path and uses a guide rail to control scale and twist of the swept profile. Figure 7.35 shows the options for the path and guide rail type.

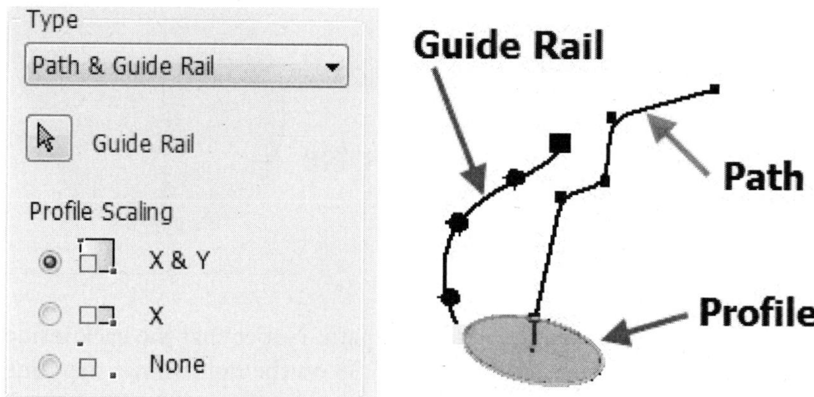

FIGURE 7.35

Guide Rail. Select a guide curve or rail that controls the scaling and twist of the swept profile. The guide rail must touch the profile plane. If you project an edge to position the guide rail and the projected edge is not the path, change the projected edge to a construction line.

Profile Scaling. Specify how the swept section scales to meet the guide rail. Figure 7.36 shows the path and the profile and guide rail from Figure 7.35 with the three different profile scaling options.

X and Y. Scales the profile in both the X and Y directions as the sweep progresses.

X. Scales the profile in the X direction as the sweep progresses. The profile is not scaled in the Y direction.

None. Keeps the profile at a constant shape and size as the sweep progresses. Using this option, the rail controls only the profile twist and is not scaled in the X or Y direction.

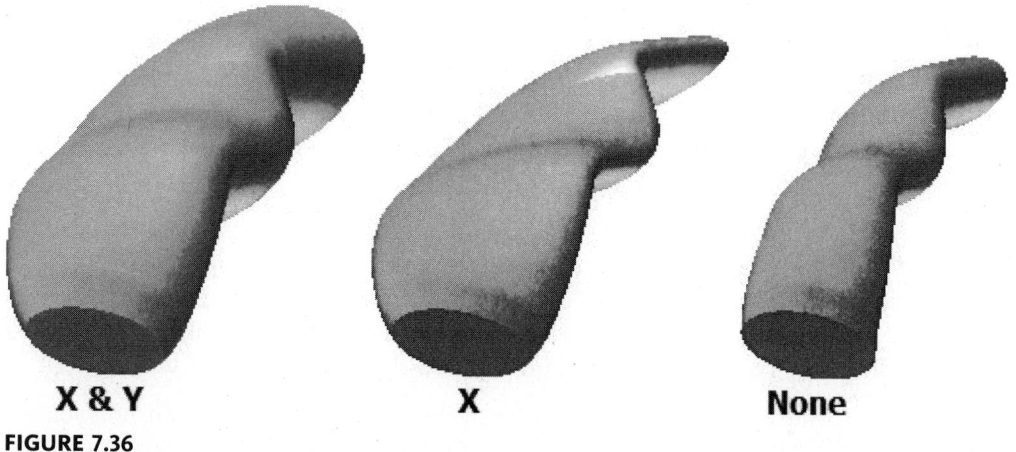

FIGURE 7.36

Path & Guide Surface

Creates a sweep feature by sweeping a profile along a path and a guide surface. The guide surface controls the twist of the swept profile. For best results, the path should touch or be near the guide surface.

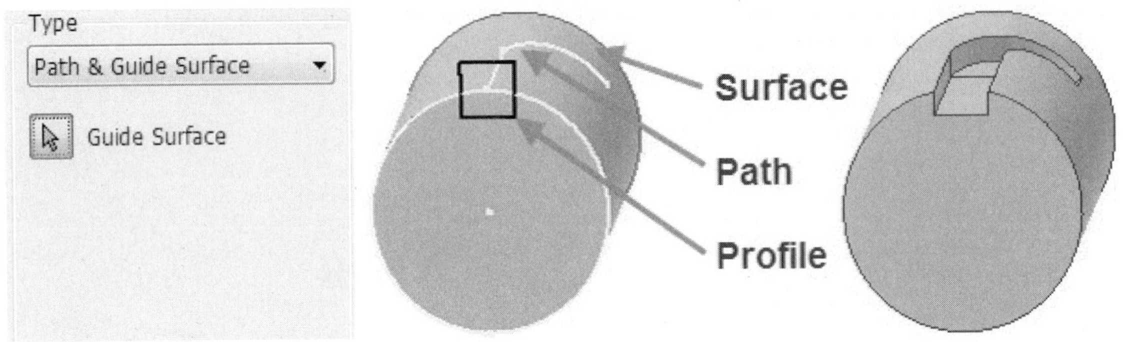

FIGURE 7.37

Figure 7.38 on the left shows a sweep with just the path. Notice that the back inside edge of the cut is angled from the surface. Figure 7.38 on the right shows the same sweep with a guide surface. Notice that the back inside edge is parallel to the top of the surface.

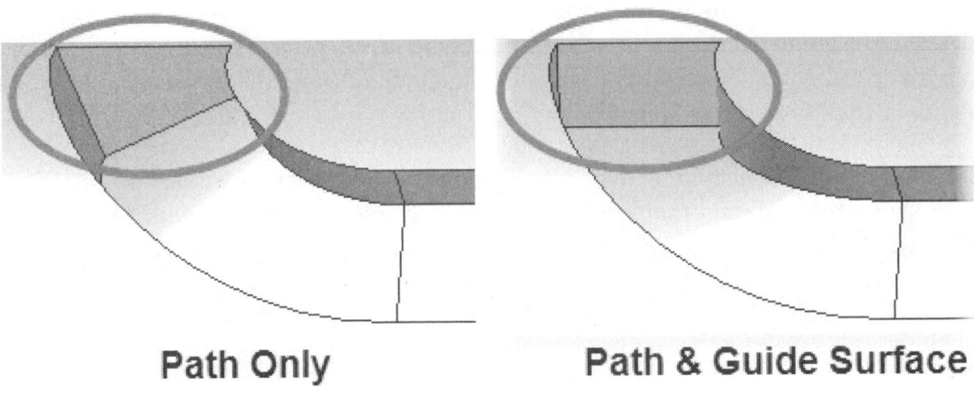

FIGURE 7.38

Optimize for Single Selection

When this option is checked, Autodesk Inventor automatically advances to the next selection option once a single selection is made. Clear the checkbox when multiple selections are required.

Preview

The button with the eyeglasses provides a solid preview of the sweep based on the current selections. If Preview is enabled and no preview appears in the graphics window, then the sweep feature will not be created.

Creating a Sweep Feature

In this section, you will learn how to create sweep features. You first need to have two unconsumed sketches. One sketch will be swept along the second sketch that represents the path. To create a sweep feature, follow these steps:

1. Create an unconsumed sketch for the profile and one for the path, or if you are going to use existing part edges for the path, you don't need to create a second

sketch for the path. The profile and path sketches must lie on separate nonparallel planes. It is recommended that the profile intersect the path. Use work planes to place the location of the sketches, if required. The sketch that you use for the path can be open or closed. Add dimensions and constraints to both sketches as needed. If needed, create a sketch that will be used as a rail, or create a surface to be used as a guide surface.

2. Click the Sweep command on the 3D Model tab > Create panel.
3. The Sweep dialog box will appear. If two unconsumed sketches do not exist, Autodesk Inventor will notify you that two unconsumed sketches are required.
4. Click the Profile button, and then select the sketch that will be swept in the graphics window. If only one closed profile exists, this step is automated for you.
5. If it is not already depressed, click the Path button, and then select the sketch to be used as the path in the graphics window.
6. Select the type of sweep you are creating: Path, Path & Guide Rail, or Path & Guide Surface.
7. Define whether or not the resulting sweep will create a solid or a surface by clicking either the Solid or the Surface button in the Output area. Select the required options, as outlined in the previous descriptions.
8. If this is a secondary feature, click the operation that will specify whether material will be added or removed or if what is common between the existing part and the new sweep feature will be kept.
9. If you want the sweep feature to have a taper, click on the More tab, and enter a value for the taper angle.
10. Click the OK button to complete the operation.

EXERCISE 7-3: CREATING SWEEP FEATURES

In the first part of this exercise, you create a component with the sweep command. Three sketches will be created, one each for the profile, the path, and the guide rail. In the second part of the exercise you sweep a profile using edges on a part as the path.

1. Start a new part file based on the *Standard (in).ipt* template file.
2. Create 2D Sketch on the XZ origin plane.
3. Draw, constrain and dimension the geometry as shown in Figure 7.39. Place the lower left endpoint of the sketch on the origin point. The arcs are tangent to the adjacent lines and apply an equal constraint between the three arcs.

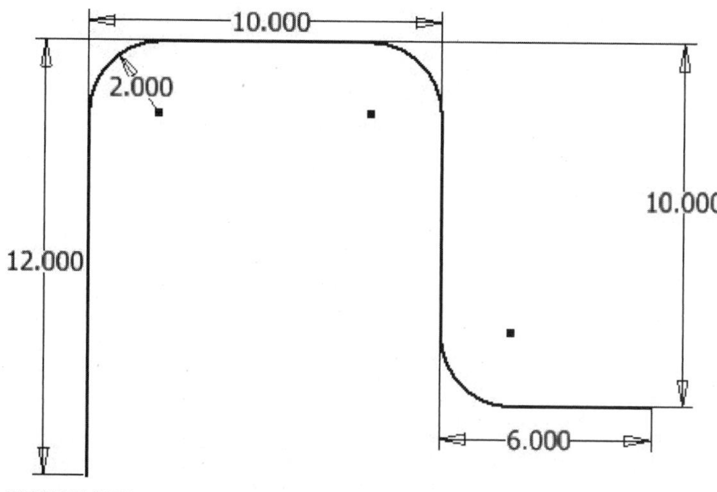

FIGURE 7.39

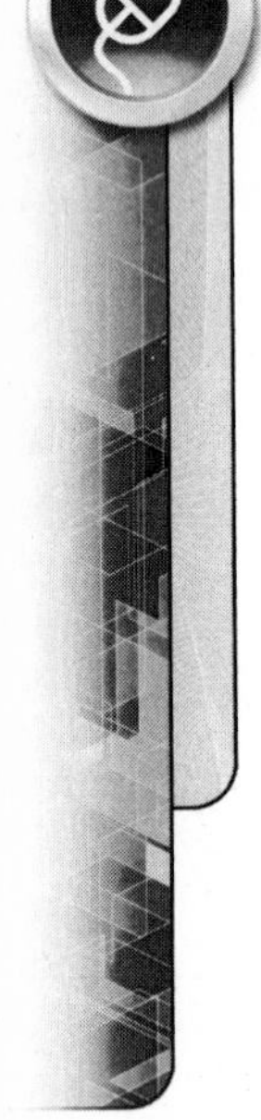

4. Finish the sketch.
5. Create 2D Sketch on the XY origin plane.
6. Draw two concentric circles that are centered on the origin point and add a **1.5 inch** diameter to the outside circle and add a **.1 inch** dimension between the two circles to define the thickness as shown in Figure 7.40 on the left. Note, if the origin point is not on the active sketch, project the origin point or the end-point of the first line. If needed use the Orbit command to change the viewpoint so you can better see the geometry better.
7. Finish the sketch.
8. Click the Sweep command on the 3D Model tab > Create panel.
9. For the profile click a point in between the two circles as shown in Figure 7.41 on the right.

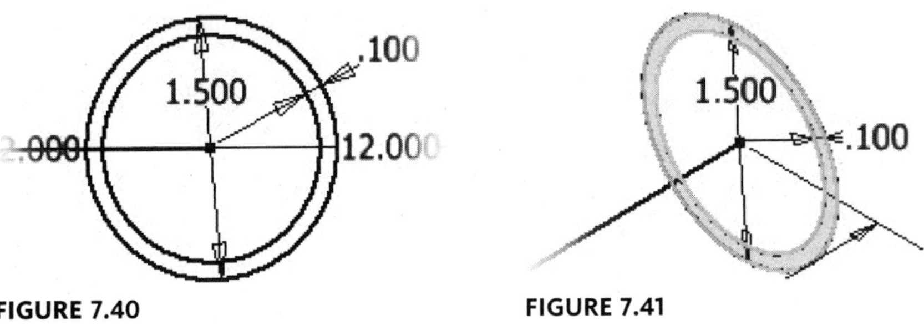

FIGURE 7.40 FIGURE 7.41

10. Next you define the path by clicking the Path button in the Sweep dialog box and then select one of the lines or arcs on the open profile.
11. Click OK to create the sweep feature.
12. Change the material of the part to steel by right-clicking on the part's name in the browser and in the menu click iProperties. From the Physical tab change the Material to Steel as shown in Figure 7.42 on the left and then click OK. You can also change the parts physical material by selecting a material from the Materials list in the Quick Access toolbar.
13. Change your viewpoint so you can see the front of the sweep as shown in Figure 7.42 on the right.

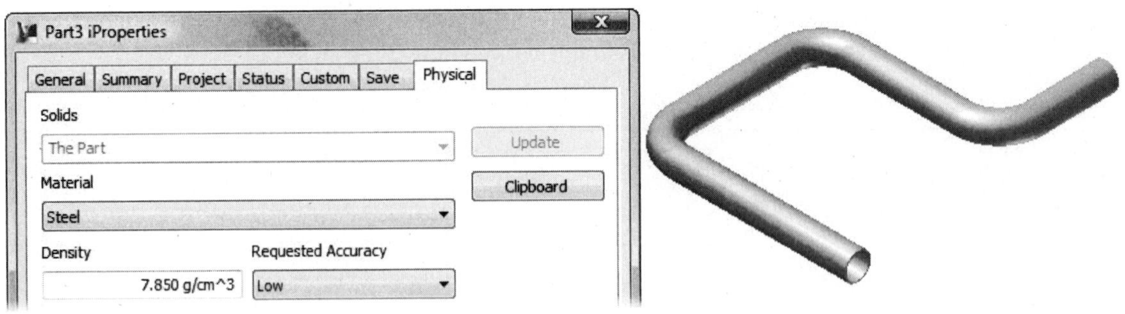

FIGURE 7.42

14. Close the file. Do not save changes.
15. Open *ESS_E07_03-2.ipt* in the Chapter 07 folder.
16. Zoom in on the front right-side of the part, as shown in Figure 7.43 on the left.
17. Click on the front face of the part and click Create Sketch from the mini-toolbar as shown in Figure 7.43 on the left.

18. The viewpoint will change so you are looking directly at the sketch. However, it is difficult to see where you are sketching. To return to your previous view, press the F5 key a few times.
19. Sketch a circle on the projected end point and add a **.0625 in** dimension, as shown in Figure 7.43 on the right. By default, the dimension rounds to three decimal places in the sketch.

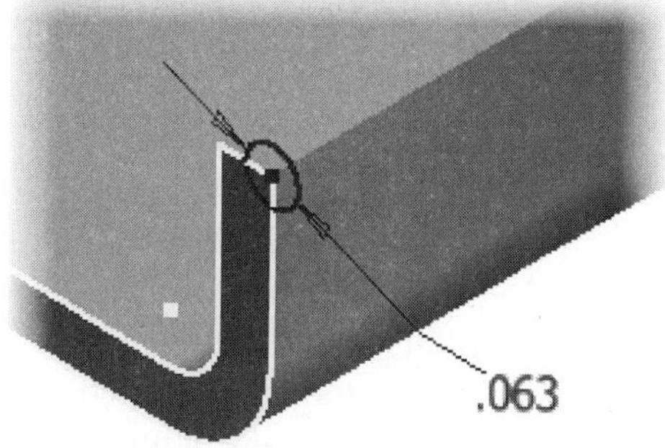

FIGURE 7.43

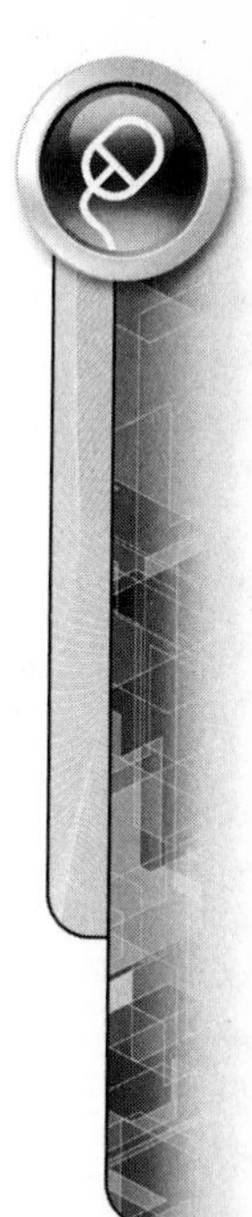

20. Click the Finish Sketch command on the Sketch tab > Exit panel.
21. Next, you sweep the circle along edges of the part using the Sweep command. Click the Sweep command on the Create panel.
22. For the profile select inside the circle and the Path button will then be active. Click on a top-outside tangent edges of the part, Inventor will automatically select all of the top-outside edges that go around the part, as shown in Figure 7.44 on the left.
23. In the Sweep dialog box, click the Cut operation to remove material.
24. Click the OK button in the Sweep dialog box. Figure 7.44 on the right shows the completed part.

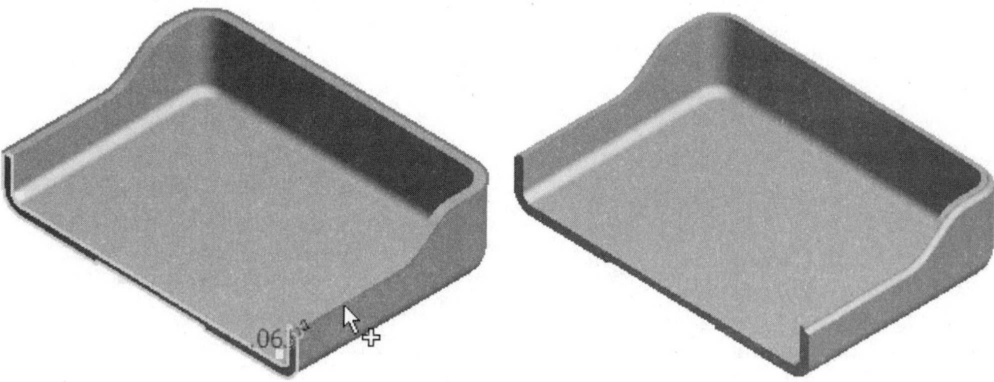

FIGURE 7.44

25. Use the Zoom and Free Orbit commands to examine the Sweep feature. Note that the Lip command can also be used to create rectangular lip features.
26. Close the file. Do not save changes. End of exercise.

3D Sketching

To create a sweep feature whose path does not lie on a single plane, you need to create a 3D sketch that will be used for the path. You can use a 3D sketch to define the path to define the routing path for an assembly component, such as a pipe or duct work that crosses multiple faces on different planes. You need to define a 3D sketch in the part environment, and you can do this within an assembly or in its own part file. You can use the Autodesk Inventor adaptive technology during 3D sketch creation to create a path that updates automatically to reflect changes to referenced assembly components. In this section, you will learn strategies for creating 3D sketches.

3D Sketch Overview

When creating a 3D sketch, you use many of the sketching techniques that you have already learned with the addition of a few commands. 3D sketches use work points and model edges or vertices to define the shape of the 3D sketch by creating line or spline segments between them. You can also create bends between line segments. When creating a 3D sketch, you use a combination of lines, splines, fillet features, work features, constraints, and existing edges and vertices.

3D Sketch Environment

The 3D sketch environment is used to create 3D or a combination of both 2D and 3D curves. Before creating a 3D sketch, change the environment to the 3D sketch environment by clicking the Create 3D Sketch command on the 3D Model tab > Sketch panel beneath the Create 2D Sketch command, as shown in Figure 7.45 on the left. The commands on the ribbon will change to the 3D Sketch commands, as shown in Figure 7.45 on the right. The most common 3D sketch commands are explained in the next section.

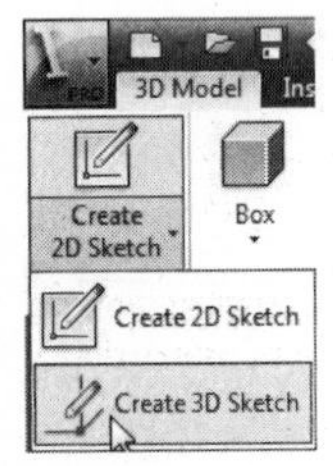

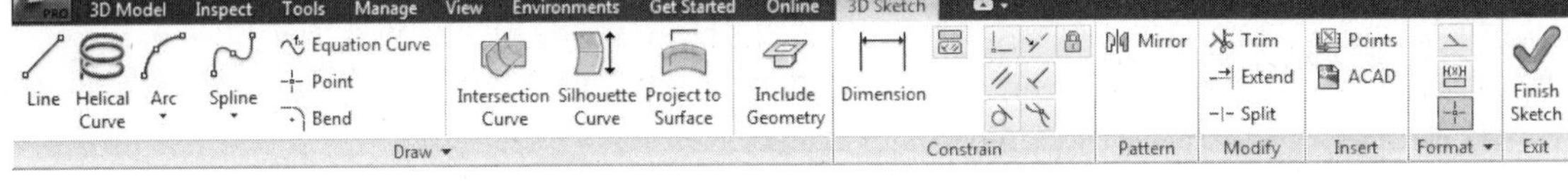

FIGURE 7.45

3D Sketch from Intersection Geometry

Another way to create a 3D path is to use geometry that intersects with the part. If the intersecting geometry defines the 3D path, you can use it. The intersection can be defined by a combination of any of the following: planar or nonplanar part faces, surface faces, a quilt, or work planes. To create a 3D path from an intersection, follow these steps:

1. Create the intersecting features that describe the desired path.
2. Change to the 3D Sketch environment by clicking the Create 3D Sketch command on the Sketch panel under the 2D Sketch command.
3. Click the Intersection Curve command, as shown in Figure 7.46 on the left on the Draw panel.
4. The 3D Intersection Curve dialog box appears, as shown in Figure 7.46 in the middle.

5. Select the two intersecting geometry.
6. To complete the operation, click OK.

Figure 7.46 on the right shows a 3D path being created on a cylindrical face at the edge defined by a work plane that intersects it at an angle.

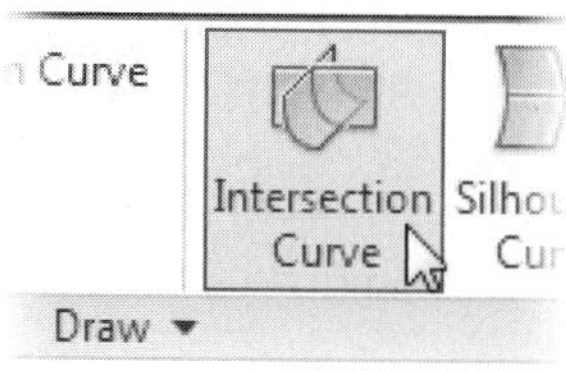

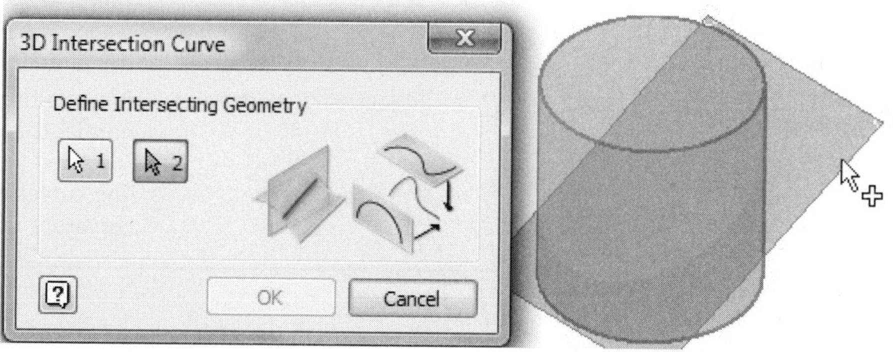

FIGURE 7.46

Project to Surface

While in a 3D sketch, you can project curves, 2D or 3D geometry, part edges, and points onto a face or onto selected faces of a surface or solid. To project curves onto a face, follow these steps:

1. Create the solid or surface onto which the curves will be projected.
2. Create the curves that will be projected onto the face(s).
3. Click the 3D Sketch command on the 3D Model tab > Sketch panel.
4. Click the Project to Surface command on the 3D Sketch tab > Draw panel, as shown in Figure 7.47 on the left.
5. The Project Curve to Surface dialog box will appear, as shown in Figure 7.47 in the middle. The Faces button will be active. In the graphics window, select the face(s) onto which the curves will be projected.
6. Click the Curves Button, and then in the graphics window, select the individual objects to project. Figure 7.47 on the right shows the face and curves selected.

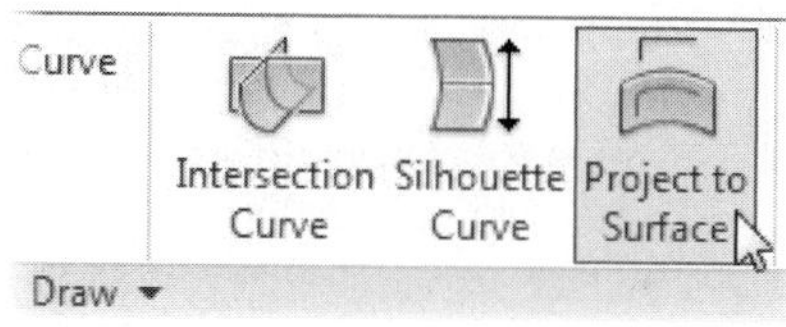

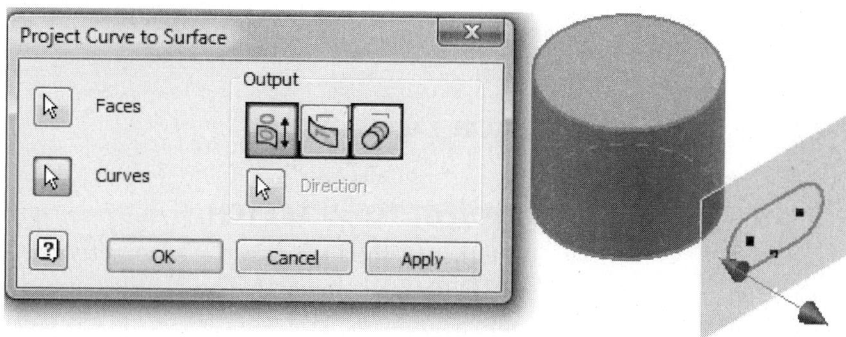

FIGURE 7.47

7. In the Output area, select one of the following options.

	Project along Vector	Specify the vector by clicking the Direction button and selecting a plane, edge, or axis. If a plane is selected, the vector will be normal (90°) from the plane. The curves will be projected in the direction of the vector.
	Project to Closest Point	Projects the curves onto the surface normal to the closest point.
	Wrap to Surface	The curves are wrapped around the curvature of the selected face or faces.

8. Click OK.

By default, the projected curves are linked to the original curve. If the original curves change size, they will be updated. To break the link, move the cursor into the browser over the name of the Projected to Surface entry, right-click, and select Break Link in the menu, as shown in Figure 7.48. You could also display the sketch constraints and delete the reference constraints. You can also change the way the curves were projected by right-clicking on the Project to Surface entry in the browser and click Edit Projection Curve from the menu as shown in Figure 7.48, and the Project Curve to Surface dialog box will appear. Make the changes as required.

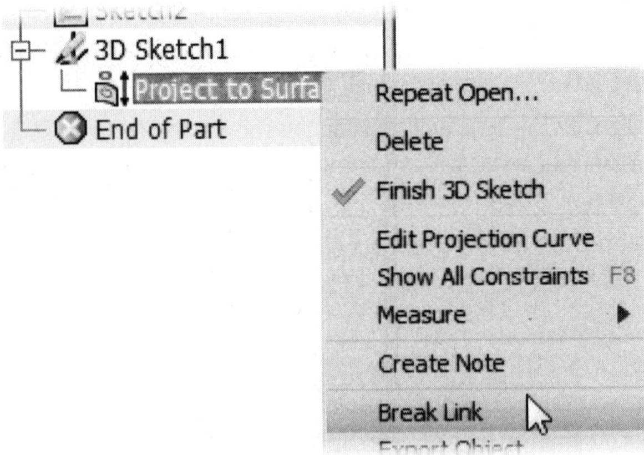

FIGURE 7.48

PROJECT TO 3D SKETCH

Another method to project 2D geometry onto a non-planar face is to use the Project to 3D Sketch command. The 3D Sketch command will project 2D sketch geometry onto a nonplanar face while automatically creating a 3D sketch. To use the Project to 3D Sketch command, follow these steps.

1. To First create a 2D sketch and sketch and constrain geometry as needed or make an existing sketch active that contains the geometry that you want to project.
2. Click the Project to 3D Sketch command from the Sketch tab > Draw panel as shown in Figure 7.49 on the left.

3. In the Project to 3D Sketch dialog box check the Project option.
4. Select a face or faces that the geometry will be projected onto. All of the geometry on the active sketch will preview how it will be projected on the selected face(s) as shown in Figure 7.49 on the right.

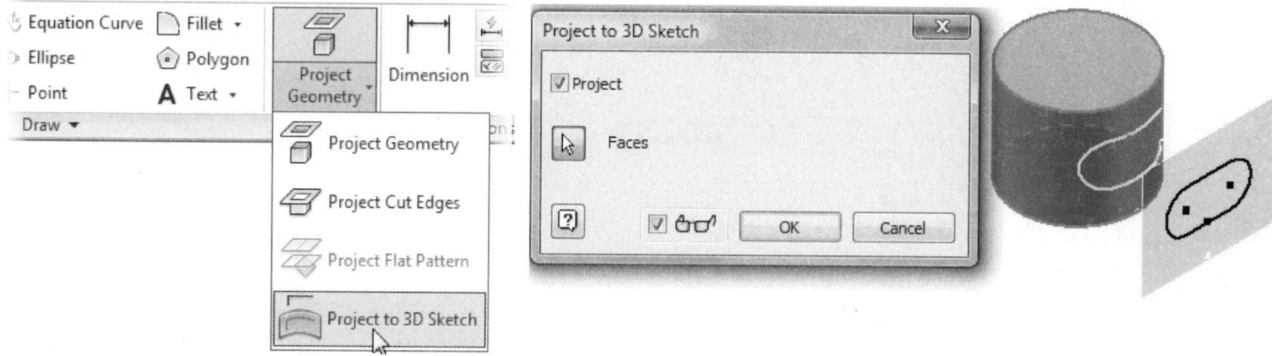

FIGURE 7.49

5. Click OK to create a 3D sketch and project all the geometry on the active sketch to the selected face(s).

3D Lines. An option to create a 3D path is to draw 3D lines. To draw 3D lines follow these steps:

1. While in a 3D Sketch, click the Line command on the 3D Sketch tab > Draw panel as shown in Figure 7.50 on the left. Once in the 3D line command, you can either create lines on the coordinates planes as shown in Figure 7.50 in the middle or enter values in the Precise Input mini-toolbar that is shown in Figure 7.50 on the right. Values can be relative to the last point or absolute to 0, 0, 0. Consult the help system "Create a 3D Line" for more information about using Precise Input.

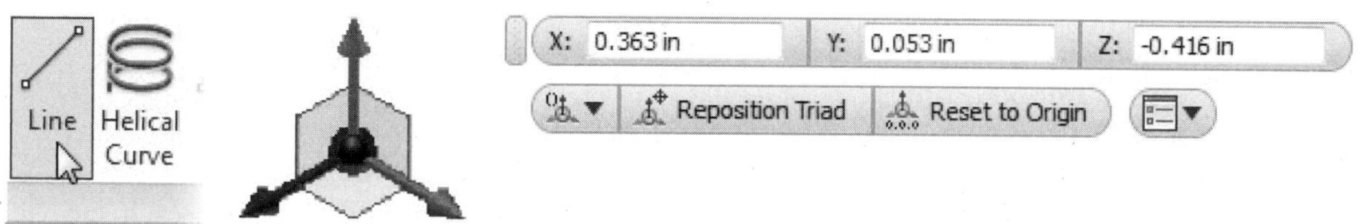

FIGURE 7.50

2. Dimensions and constraints can be applied to the lines.
3. By default, a bend is not applied between 3D line segments, but this option can be toggled on and off by right-clicking while in the 3D line command and selecting or deselecting Auto-Bend on the menu, as shown in Figure 7.51 on the left.
4. To set the default radius of the bend, Click the Document Settings command on Tools tab > Options panel, and then change the 3D Sketch Auto-Bend Radius setting on the Sketch tab.
5. To manually add a bend between two 3D lines, use the Bend command on the 3D Sketch tab > Draw panel, as shown in Figure 7.51 in the middle, or right-click in a blank area in the graphics window and click Bend from the marking menu as shown in Figure 7.51 on the right. The Bend dialog box appears, in the Bend dialog box enter a value for the bend, and then select two 3D lines or the endpoint

where they meet. To complete the command, right-click and click OK from the marking menu. You can edit it by double-clicking on the dimension and entering a new value.

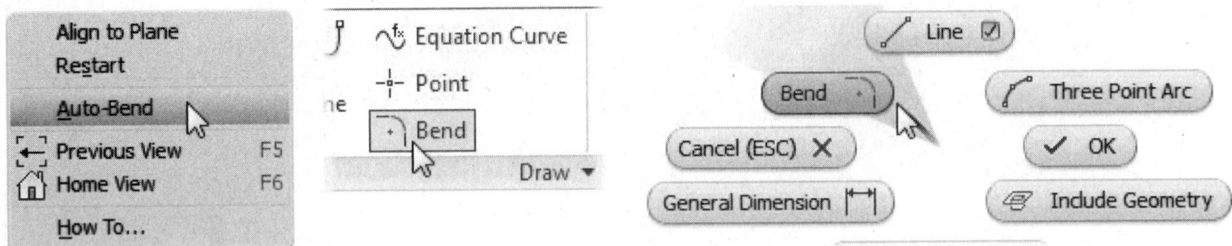

FIGURE 7.51

Create a 3D Sweep. To create a 3D sweep, follow these steps:

1. 3D Model tab > Sketch panel.
2. Use the Precise Input mini-toolbar to create a 3D path.
3. After creating a 3D path, you next create a 2D sketch, draw, and constrain a closed profile. Then start the Sweep command on the 3D Model tab > Create panel and use the closed sketch as the profile and 3D lines as the path.

Figure 7.52 on the left shows a part with 3D lines and dimensions. The image on the right shows the completed sweep.

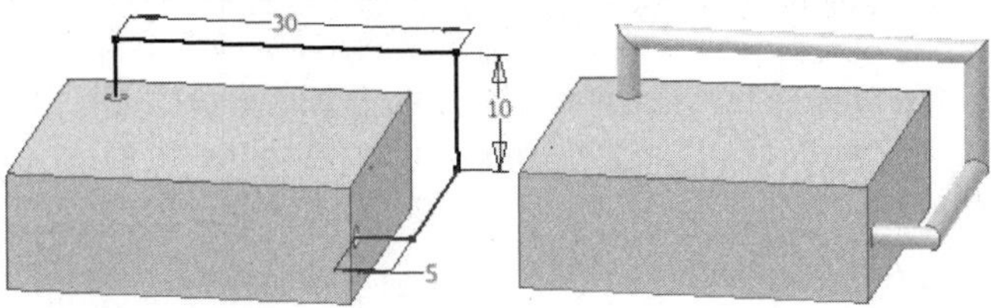

FIGURE 7.52

Import Points

Another option to create geometry is to import X, Y, Z point data from and Excel spreadsheet. The imported points can be connected by lines or splines or left as points. The imported points are not associated back to the spreadsheet: if the data in the spreadsheet changes the imported point data will not update. While in a 2D sketch only X and Y values are imported, while in a 3D sketch X, Y and Z values are imported. The data in the spreadsheet must start in cell A1 and be in the first worksheet. Cell A1 can define a unit; if no unit is defined in cell A1 the document unit will be used. Row be can include X, Y and Z heading if desired but it is not required. The columns must be in the following order, but the C column is not required.

Column A = X value   Column B = Y value   Column C = Z value

Figure 7.53 on the left shows an example of a spreadsheet with all of the data and the image on the right shows an example of only the required fields.

	A	B	C
1	in		
2	X	Y	Z
3	0	0	0
4	5	0	0
5	5	5	0
6	5	5	5
7	10	5	5

	A	B	C
1	0	0	0
2	5	0	0
3	5	5	0
4	5	5	5
5	10	5	5

FIGURE 7.53

To import points from a spreadsheet follow these steps:

1. Create an Excel spreadsheet with X, Y and Z point data. The Z data is optional
2. Create or make a 2D or 3D sketch active.
3. Click the Import Point data command from the 2D Sketch tab > Insert Panel as shown in the following image on the left or 3D Sketch tab > Insert Panel as shown in Figure 7.54 on the right.

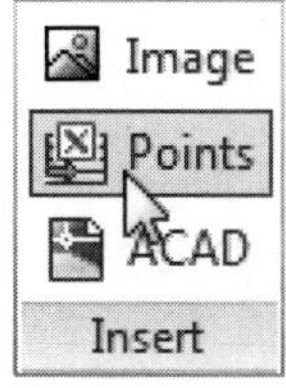

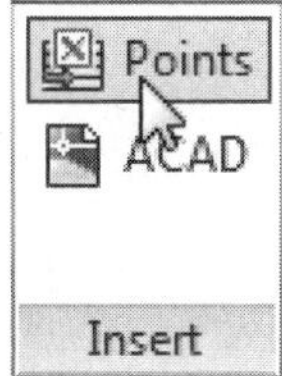

FIGURE 7.54

4. The Open dialog box appears, navigate to and select the Excel spreadsheet.
5. On the bottom-right corner of the Open dialog box click the Options button and the File Open Option dialog box appears as shown in Figure 7.55. Select option to Create points (the default), or connect the point with lines or splines by clicking Create line or Create splines.

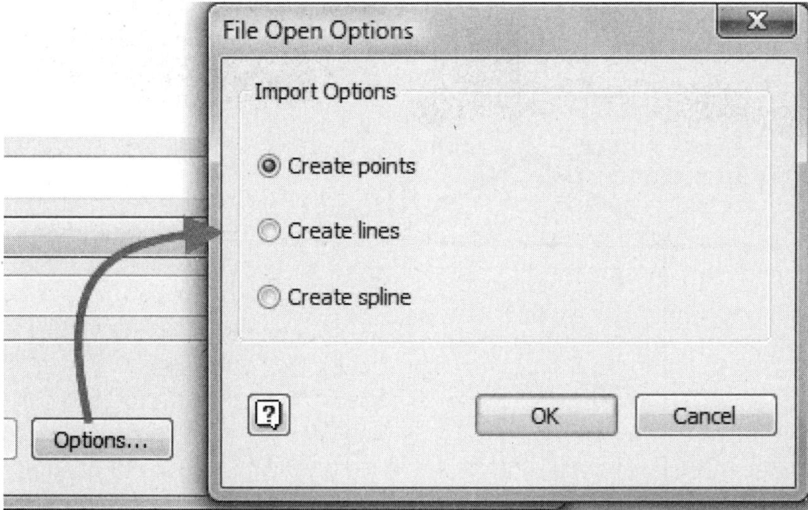

FIGURE 7.55

6. Click OK to accept the options and then click Open to create the data.
7. Add dimensions and constraints as needed.

EXERCISE 7-4: 3D SKETCH—SWEEP FEATURES

In this exercise, you create 3D geometry using different methods.

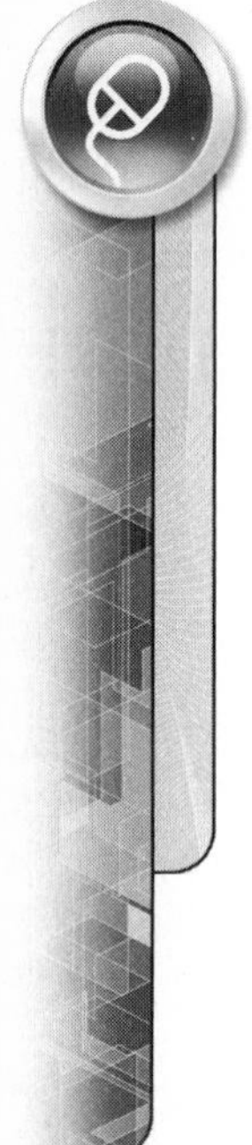

1. Start a new part file based on the *Standard (in).ipt* template file.
2. Create 3D Sketch by clicking Create 3D Sketch on the 3D Model tab > Sketch panel.
3. Press the F6 key to change to the Home view.
4. Click the Line command on the 3D Sketch tab > Draw panel.
5. In the graphics screen select the XY coordinate plane on the 3D coordinator, and in the graphics window draw a line in the X direction and a line in the Y direction as shown in Figure 7.56 on the left. For the first point click a point above the Triad and coordinate planes.
6. In the graphics screen select the YZ coordinate plane on the 3D coordinator, and in the graphics window draw a line in the –Z direction as shown in Figure 7.56 in the middle.
7. In the graphics screen select the XZ coordinate plane on the 3D coordinator, and in the graphics window draw a line in the X direction as shown in Figure 7.56 on the right.

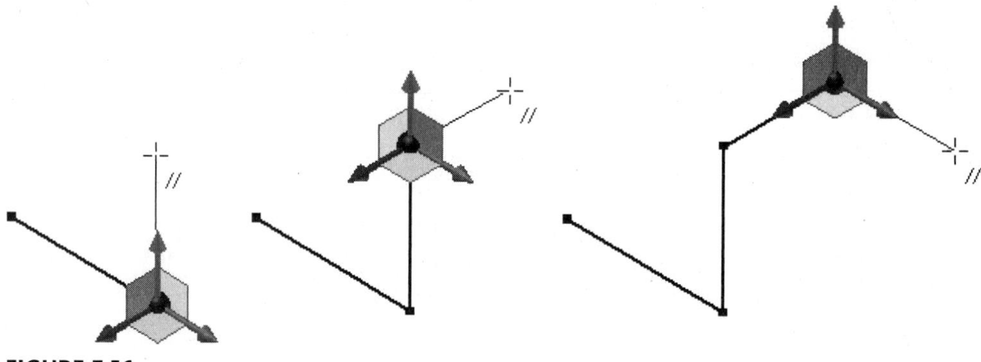

FIGURE 7.56

8. Add dimensions to the lines as shown in Figure 7.57. If needed you can add perpendicular and parallel constraints between the lines.

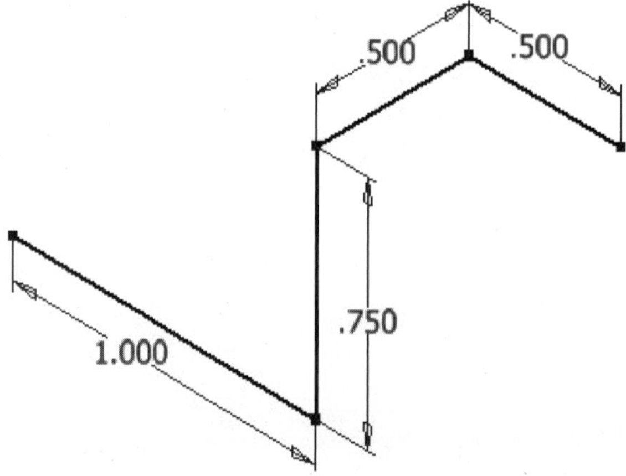

FIGURE 7.57

9. Delete the lines you just created.
10. Click the Line command on the 3D Sketch tab > Draw panel.
11. If the Precise Input mini-toolbar does not appear, click the down arrow on the 3D Sketch tab > Draw panel and click the Precise Input command. Practice creating lines by entering point data in the Precise Input mini-toolbar.

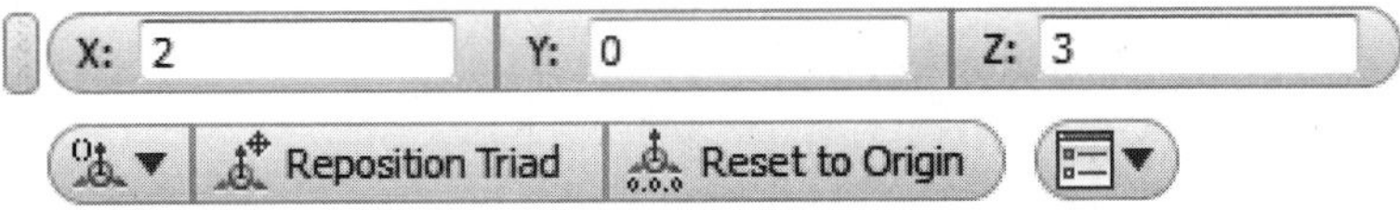

FIGURE 7.58

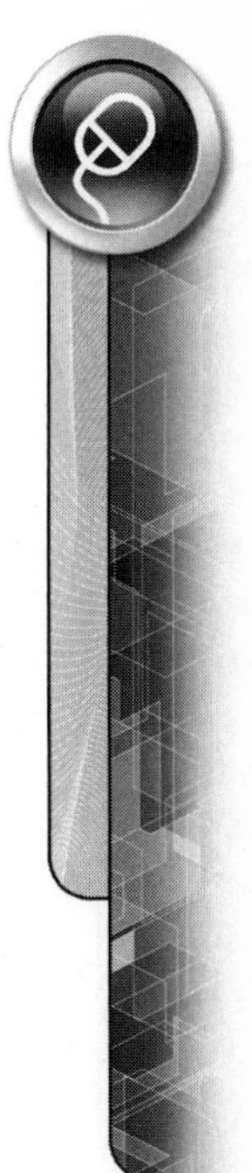

12. Add dimensions and constraints as desired.
13. Delete the lines you just created.
14. Next you import point data from an Excel spreadsheet. Click the Import Point data command from the 3D Sketch tab > Insert Panel.
15. The Open dialog box appears, navigate to and single click on the Excel spreadsheet *XYZ - Point Data.xlsx* in the Chapter 07 folder.
16. On the bottom-right corner of the Open dialog box click the Options button and in the File Open Option dialog box click the Create line option as shown in Figure 7.59.

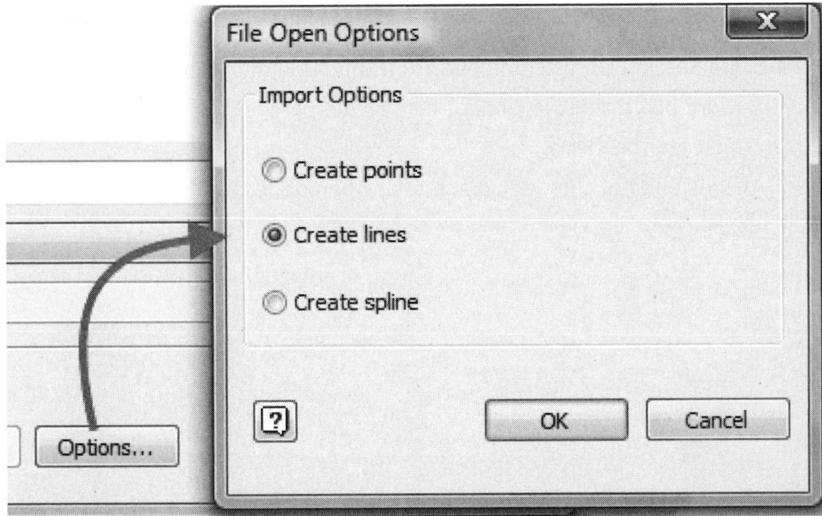

FIGURE 7.59

17. Click OK to accept the options and then click Open to create the data.
18. Zoom out and rotate the viewpoint so you can see the imported points with the connecting lines as shown in Figure 7.60.
19. Add perpendicular and parallel constraints between the lines and add dimensions. Figure 7.61 shows the dimensioned sketch with the constraints visibility on.

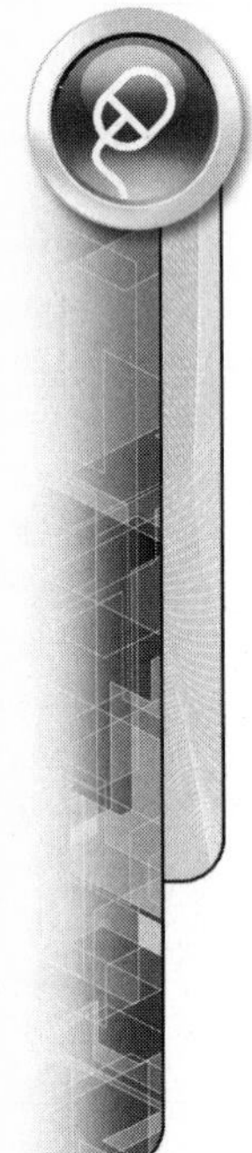

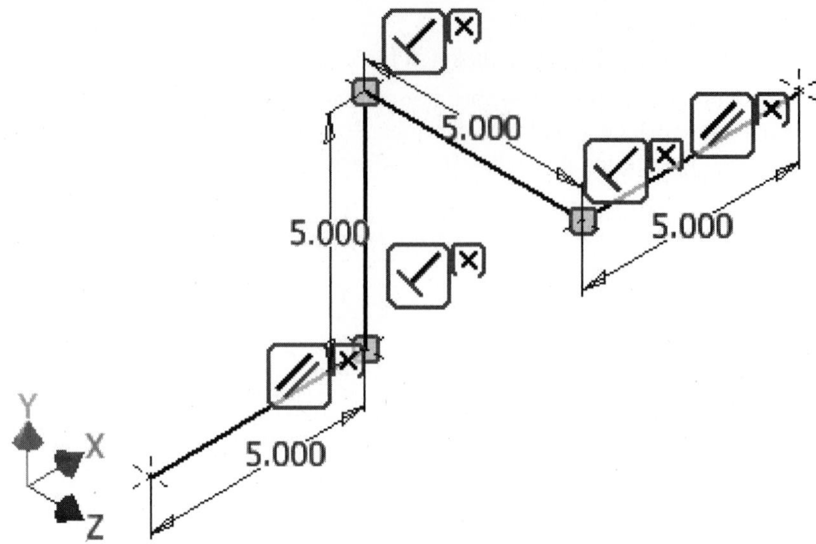

FIGURES 7.60–61

20. Close the file. Do not save changes.
21. In this portion of this exercise you create geometry on a 3D sketch from intersecting geometry. Open *ESS_E07_04-2.ipt* from the Chapter 07 folder. The part contains a cylinder and an angled work plane.
22. Create a 3D Sketch.
23. Click the Intersection Curve command from the 3D Sketch tab > Draw panel.
24. Select the work plane and the circular face of the part.
25. Click OK to create the curve.
26. Turn off the visibility of the work plane and, when done, your screen should resemble Figure 7.62.

FIGURE 7.62

27. Finish the 3D sketch by clicking on Finish Sketch on the Exit panel.
28. Next you create a 2D sketch that the sweep command will use as the profile. Click the Create 2D Sketch command on the Sketch panel under the Create 3D Sketch command, and from the Origin folder in the browser click the YZ Plane, as shown in Figure 7.63 on the left.

29. Change to the Home view.
30. Press the F7 key to slick the graphics.
31. Use the Project Geometry command to project the geometry that was created from the Intersection Geometry command.
32. Change the projected line to a construction line. This will prevent the projected line from being used when you create the sweep.
33. Create a **.5 inch** diameter circle on the bottom left corner of the projected geometry as shown in Figure 7.63 in the middle.
34. Finish the Sketch.
35. Use the sweep command with the following options.
    - Use the circle as the profile.
    - Use the geometry that was created from the intersecting plane and face for the sweep path.
    - Change the operation to cut.
    - Click OK to create the sweep. When done, your screen should resemble Figure 7.63 on the right.

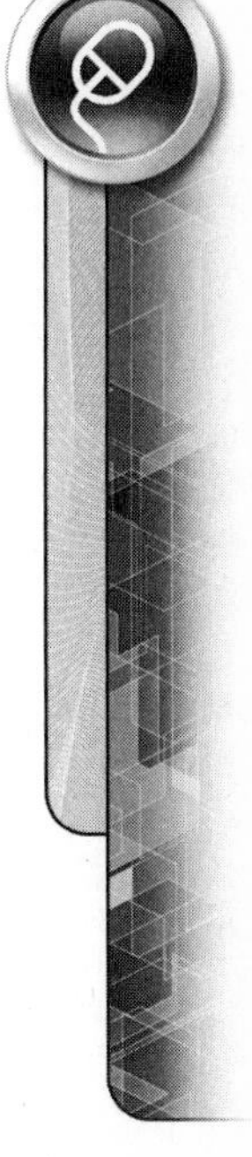

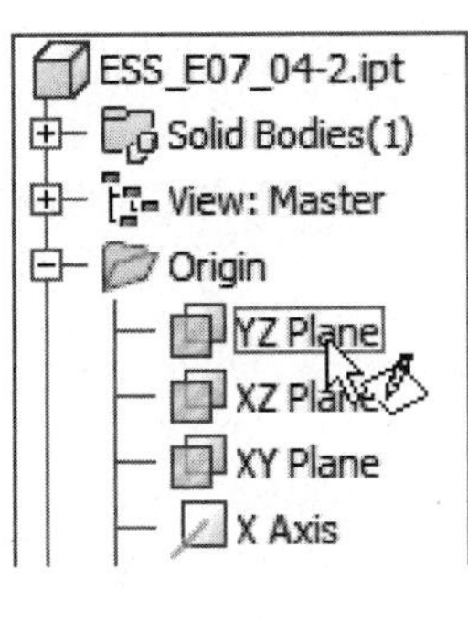

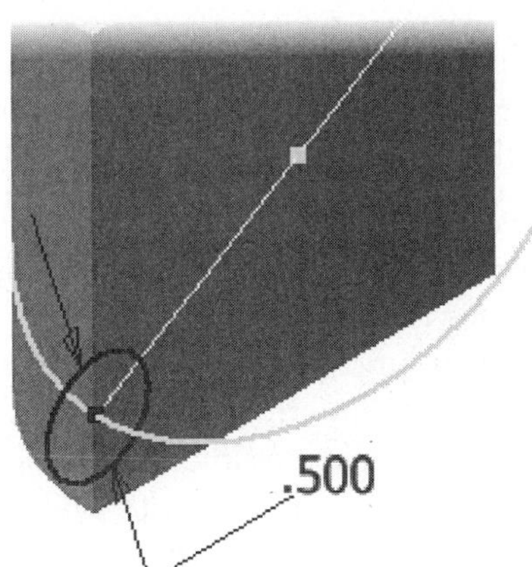

FIGURE 7.63

36. Close the file. Do not save changes.
37. In this portion of this exercise you will use the Project to 3D Sketch command to project geometry onto a cylindrical face. Open *ESS_E07_04-3.ipt* from the Chapter 07 folder. The part contains an extrusion and a constrained sketch.
38. Make Sketch2 active by clicking on one of the arcs or lines in the sketch, and click Edit Sketch from the mini-toolbar or double-clicking on Sketch2 in the browser.
39. To better see the part, change to the Home view.
40. Click the Project to 3D Sketch command on the Sketch tab > Draw panel under the Project Geometry command.
41. In the Project to 3D Sketch dialog box appears, in the graphics window select the inside circular face on the part to project onto as shown in Figure 7.64 on the left.
42. Click OK to complete the operation.
43. Finish the sketch.
44. In the browser, turn off the visibility of Sketch2. When done, your screen should resemble Figure 7.64 on the right.

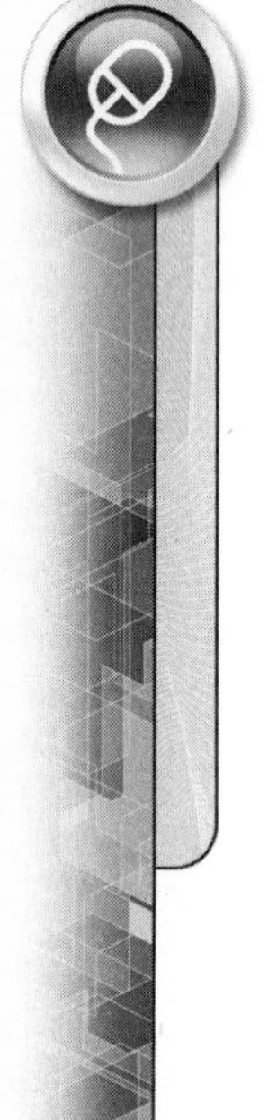

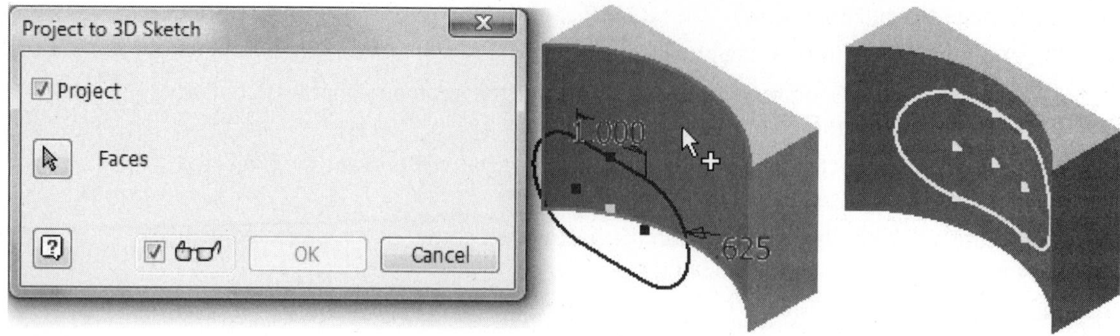

FIGURE 7.64

45. If desired, you can create a profile and sweep it along the projected geometry as you did earlier in this exercise. Note that you could have also used the Project to Surface command in the 3D sketch tab to project the geometry onto the face.
46. Close the file. Do not save changes. End of exercise.

COIL FEATURES

Using the Coil feature, you can easily create many types of helical, coil, or spiral geometry. You can create various types of springs by selecting different settings in the Coil dialog box. You can also use the Coil feature to remove or add a helical shape around the outside of a cylindrical part to represent a thread profile.

To create a coil, you need to have at least one unconsumed or shared sketch available in the part. This sketch describes the profile or shape of the coil feature and can also describe the coil's axis of revolution. If no unconsumed sketch is available, Autodesk Inventor will prompt you with an error message stating, "No unconsumed visible sketches on the part." After an unconsumed sketch is available, you can click the Coil command on the 3D Model tab > Create panel, as shown in Figure 7.65 on the left. The Coil dialog box appears as shown in Figure 7.65 on the right. The following sections explain the tabs.

Coil Shape Tab

The Coil Shape tab allows you to specify the geometry and orientation of the coil.

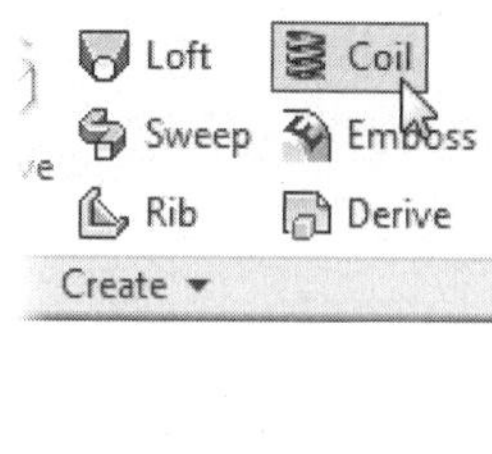

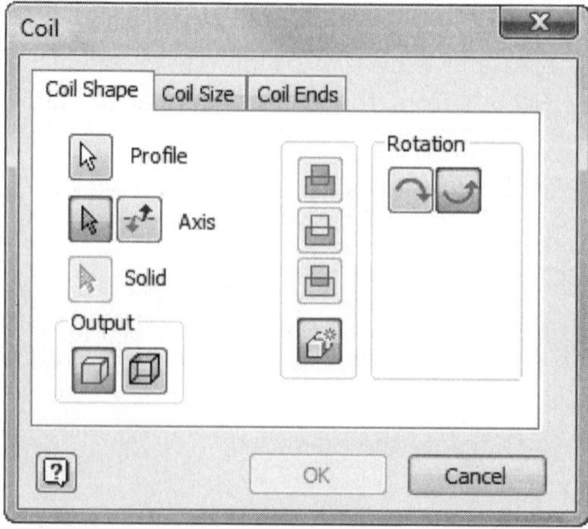

FIGURE 7.65

Profile. Click to select the sketch that you will use as the profile shape of the coil feature. By default, the Profile button is shown depressed; this tells you that you need to select a sketch or sketch area. If there are multiple closed profiles, you will need to select the profile that you want to revolve. If there is only one possible profile, Autodesk Inventor will select it for you, and you can skip this step. If you select the wrong profile or sketch area, click the Profile button and select a new profile or sketch area. You can only use one closed profile to create the coil feature.

Axis. Click to select a sketched line or centerline, a projected straight edge, or an axis about which to revolve the profile sketch. If selecting an edge or sketched centerline, it must be part of the sketch. If selecting a work axis, it cannot intersect the profile.

Flip. Click to change the direction in which the coil will be created along the axis. The direction will be changed on either the positive or negative X or Y axis, depending upon the edge or axis that you selected. You will see a preview of the direction in which the coil will be created.

Solid. If there are multiple solid bodies, click this button to choose the solid body to participate in the operation.

Rotation. Click to specify the direction in which the coil will rotate, either clockwise or counterclockwise.

Operation. The operation buttons are the column of buttons along the center of the dialog box that will appear if a base feature exists, as shown in Figure 7.66. The operation buttons are only available if the Coil feature is not the first feature in the part. By default, the Join operation is selected. You can select the other operations to either add or remove material from the part using the Join or Cut options or to keep what is common between the existing part volume and the completed coil feature using the Intersect option.

- *Join:* Adds material to the part.
- *Cut:* Removes material from the part.
- *Intersect:* Keeps what is common to the part and the coil feature.
- *New solid*: Creates a new solid body. The first solid feature that is created uses this option by default. Select to create a new body in a part file with an existing solid body.

Output. Click to create a solid or a surface.

Coil Size Tab

The Coil Size tab, as shown in Figure 7.66 on the left, allows you to specify how the coil will be created. You have various options for the type of coil that you can create. Based on the type of coil that you select, the other parameters for Pitch, Height, Revolution, and Taper will become active or inactive. Specify two of the three available parameters, and Autodesk Inventor will calculate the last field for you.

Type. Select the parameters that you want to specify: Pitch and Revolution, Revolution and Height, Pitch and Height, or Spiral. If you select Spiral as the Coil Type, only the Pitch and Revolution values are required.

Pitch. Type in the value for the height to which you want the helix to elevate with each revolution.

Revolution. Specify the number of revolutions for the coil. A coil cannot have zero revolutions, and fractions can be used in this field. For example, you can create a coil that contains 2.5 turns. If end conditions are specified, as mentioned in the Coil Ends tab section, the end conditions are included in the number of revolutions.

Height. Specify the height of the coil. This is the total coil height as measured from the center of the profile at the start to the center of the profile at the end.

Taper. Type an angle at which you want the coil to be tapered along its axis.

NOTE

A spiral coil type cannot be tapered.

Coil Ends Tab

The Coil Ends tab, as shown in Figure 7.66 on the right, lets you specify the end conditions for the start and end of the coil. When selecting the Flat option, the helix, not the profile that you selected for the coil, is flattened. The ends of a coil feature can have unique end conditions that are not consistent between the start and the end of the coil.

Start. Select either Natural or Flat for the start of the helix. Click the down arrow to change between the two options.

End. Select either Natural or Flat for the end of the helix. Click the down arrow to change between the two options.

Transition Angle. This is the rotational angle, specified in degrees, in which the coil achieves the coil start or end transition. It normally occurs in less than one revolution.

Flat Angle. This is the rotational angle, specified in degrees, that describes the amount of flat coil that extends after the transition. It specifies the transition from the end of the revolved profile into a flattened end.

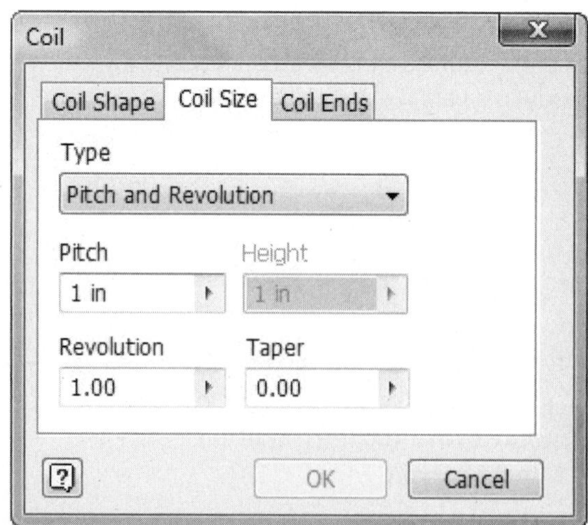

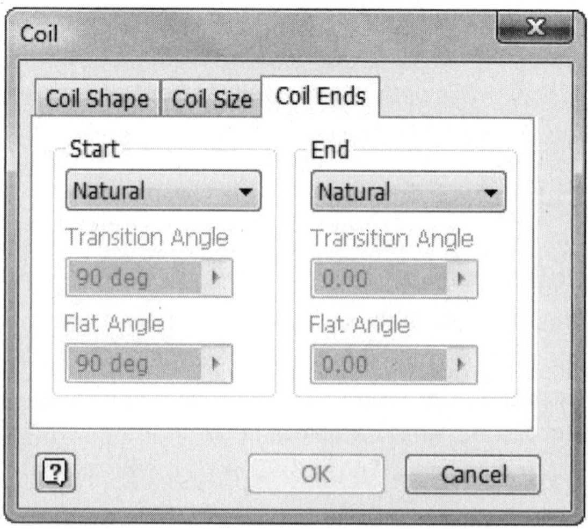

FIGURE 7.66

Figure 7.67 shows a coil created as the base feature. The image on the left shows the coil in its sketch stage; the rectangle will be used as the profile, and the centerline will be used as the axis of rotation. The finished part, as shown on the right, shows the coil with flat ends.

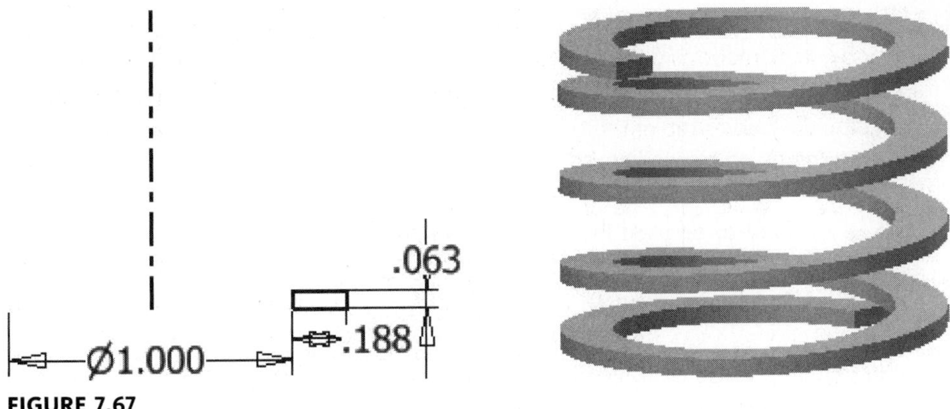

FIGURE 7.67

Figure 7.68 shows a coil created as a secondary feature. The image on the left shows the coil in its sketch stage with the Coil dialog box displayed. The sketch, a rectangle, is drawn tangent to the cylinder. The rectangle will be used as the profile, and the work axis will be used as the axis of rotation. The finished part, as shown on the right, shows the coil with the flat end on the top.

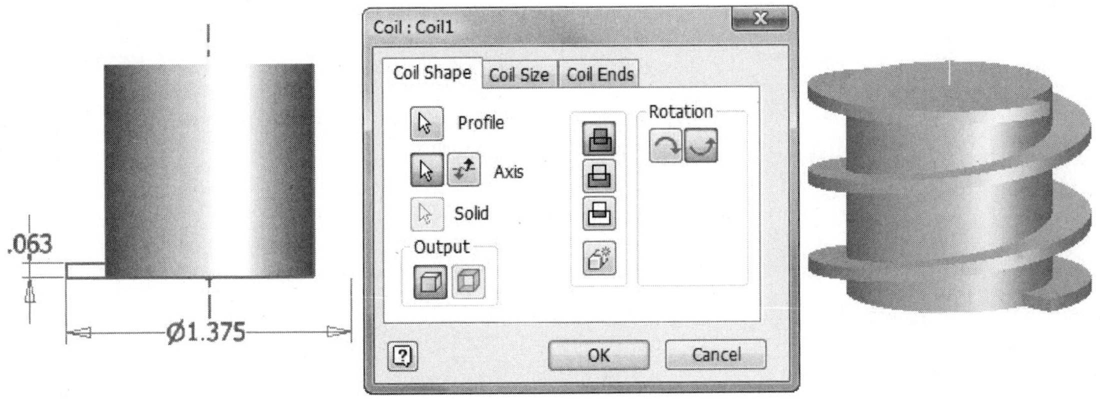

FIGURE 7.68

LOFT FEATURES

The Loft command creates a feature that blends a shape between two or more different sections or profiles. A point can also be used to define the beginning and ending section of the loft. Loft features are used frequently when creating plastic or molded parts. A loft is similar to a sweep, but it can have multiple sections and rails. Many of these types of parts have complex shapes that would be difficult to create using standard modeling techniques. You can create loft features that blend between two or more cross-section profiles that reside on different planes. You can also control the area of a specific section in the loft. You can use a rail, multiple rails, or a centerline to define a path(s) that the loft will follow. There is no limit to the number of sections or rails that you can include in a loft feature. Four types of geometry are used to create a loft: sections, rails, centerlines, and points. The following sections describe these types of geometry.

Create a Loft

To create a loft feature, follow these steps:

1. Create the profiles or points that will be used as the sections to define the loft. If required, use work features, sketches, or projected geometry to position the profiles.

2. Create rails or a centerline that will be used to define the direction or control of the shape between sections.
3. Click the Loft command on the Create panel, as shown in Figure 7.69 on the left.
4. On the Curves tab of the Loft dialog box, the Sections option will be the default. In the graphics window, click the sketches, face loops, or points in the order in which the loft sections will blend.
5. If rails are to be used in the loft, click Click to add in the Rails section of the Curves tab, and then click the rail or rails.
6. If needed, change the options for the loft on the Conditions and Transition tabs.

NOTE

The loft options are also available by right-clicking in a blank area in the graphics window and clicking an option on the menu.

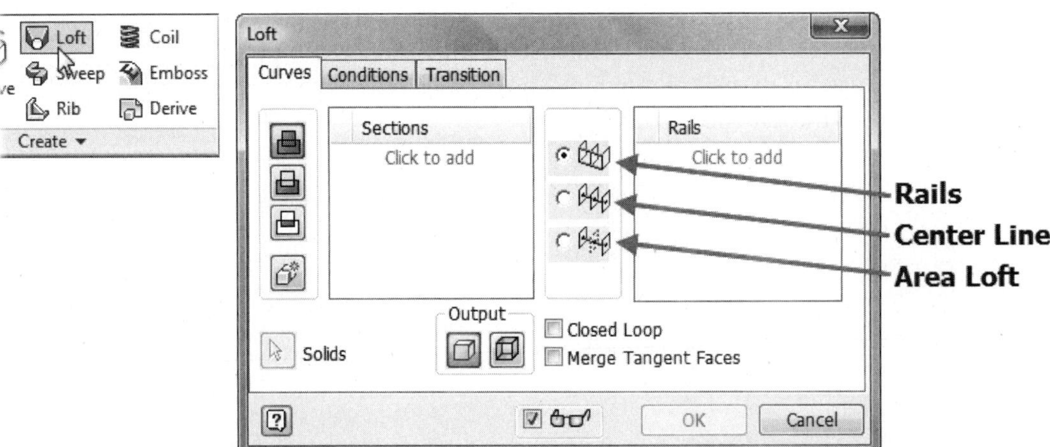

FIGURE 7.69

Figure 7.70 shows a loft created from two sections and a centerline rail. The image on the left shows two sections and a centerline in their sketch stages, shown in the top view. The image on the right shows the completed loft.

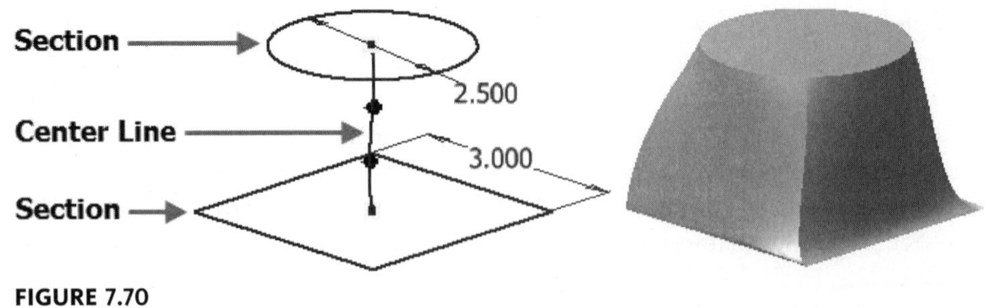

FIGURE 7.70

The following sections explain the options in the Loft dialog box and on its tabs.

Curves Tab

The Curves tab, as shown in Figure 7.71 in this section, allows you to select which sketches, part edges, part faces, or points will be used as sections to select whether or not a rail or centerline will be used and to determine the output condition.

Sections

You can define the shape(s) between which the loft will blend. The following rules apply to sections:

1. There is no limit to the number of sections that you can include in the loft feature.
2. Sections do not have to be sketched on parallel planes.
3. You can define sections with 2D sketches (planar), 3D sketches (nonplanar), and planar or nonplanar faces, edges on a part, or points.
4. All sections must be either open or closed. You cannot mix open and closed profile types within the same loft operation. Open profiles result in a lofted surface.

Points

A point can be used as a section to help define the loft. The following rules apply to points used for a loft profile:

1. A point can be used to define the start or end of the loft.
2. An origin point, sketch point, center point, edge point, or work point can be used.

Rails

You can define rails by the following elements: 2D sketches (planar), 3D sketches (nonplanar), or part faces and edges. The following rules apply to rails:

1. There is no limit to the number of rails that you can create or include in the loft feature.
2. Rails must not cross each other and must not cross mapping curves.
3. Rails affect all of the sections not just faces or sections that they intersect. Section vertices without defined rails are influenced by neighboring rails.
4. All rail curves must be open or closed.
5. Closed rail curves define a closed loft, meaning that the first section is also the last section.
6. No two rails can have identical guide points, even though the curves themselves may be different.
7. Rails can extend beyond the first and last sections. Any part of a rail that comes before the first section or after the last is ignored.
8. You can apply a 2D or 3D sketch tangency or smooth constraint between the rail and the existing geometry on the model.

Centerline

A centerline is treated like a rail, and the loft sections are held normal to the centerline. When a centerline is used, it acts like a path used in the sweep feature, and it maintains a consistent transition between sections. The same rules apply to centerlines as to rails except that the centerline does not need to intersect sections and only one centerline can be used.

Area Loft

Select a sketch to be used as the centerline, and then click on the centerline to define the area of the profile at the selected point. After picking a point, the Section Dimensions dialog box appears, as shown in Figure 7.71. You define its position either as a proportional or absolute distance, and you can define the section's size by area or scale factor related to the area of the original profile.

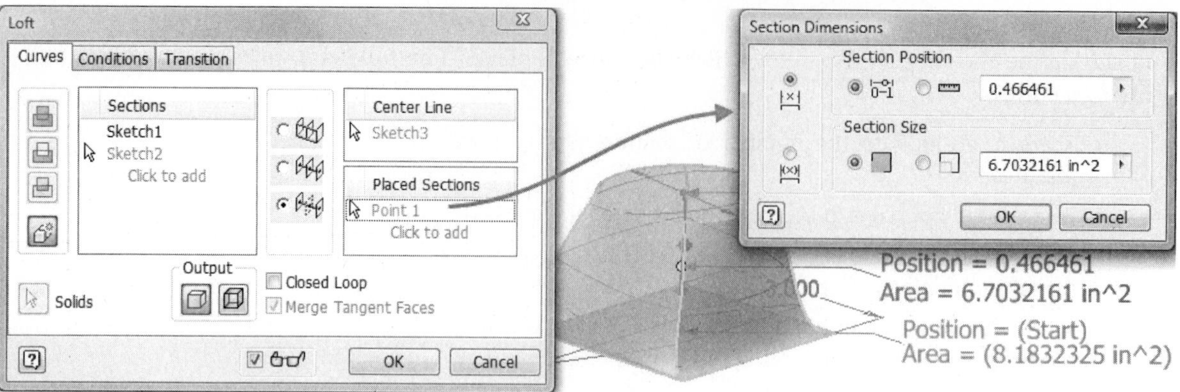

FIGURE 7.71

Output. Select whether the loft feature will be a solid or a surface. Open sketch profiles selected as loft sections will define a lofted surface.

Operation. Select an operations button to add or remove material from the part, using the Join or Cut options, or to keep what is common between the existing part and the completed loft using the Intersect option. By default, the Join operation is selected.

Closed Loop. Click to join the first and last sections of the loft feature to create a closed loop.

Merge Tangent Faces. Click to join tangent faces on the completed loft into a single face.

Conditions Tab

The Conditions tab, as shown in Figure 7.72, allows you to control the tangency condition, boundary angle, and weight condition of the loft feature. These settings affect how the faces on the loft feature relate to geometry at the start and end profiles of the loft. This may be existing part geometry or the plane or work plane containing the loft section sketch.

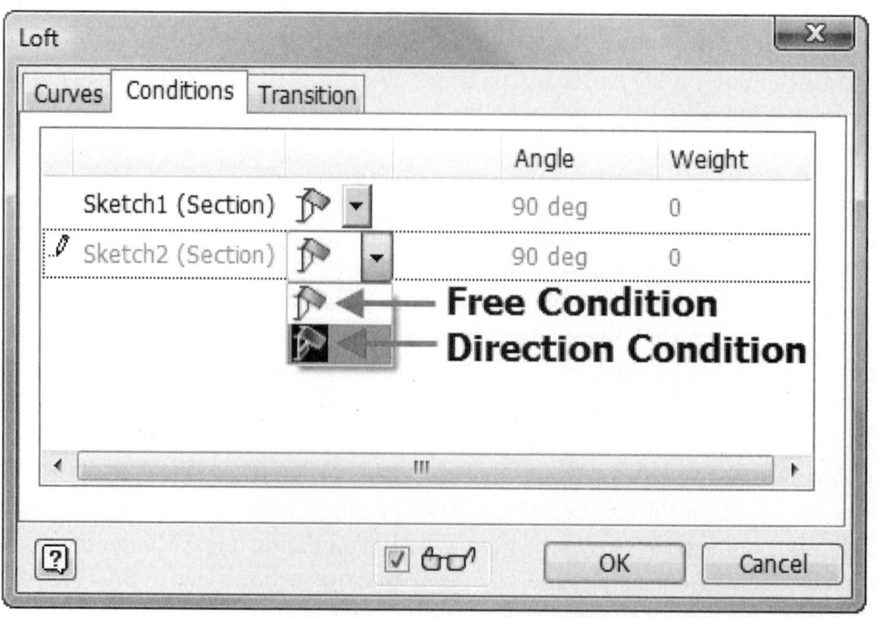

FIGURE 7.72

Conditions. The column on the left lists the sketches and points specified for the sections. To change a sketch's condition, click on its name, and then select a condition option.

Condition Boundaries Sketches

Two boundary conditions are available when the first or last section is a point, as shown in Figure 7.72.

- Free Condition (top button): With this option, there is no boundary condition, and the loft will blend between the sections in the most direct fashion.
- Direction Condition (bottom button): This option is only available when the curve is a 2D sketch. When selected, you specify the angle at which the loft will intersect or transition from the section.

Condition Boundaries Face Loop. When a face loop or edges from a part are used to form a section, as shown in Figure 7.73 on the left, three conditions are available, as shown in Figure 7.73 on the right.

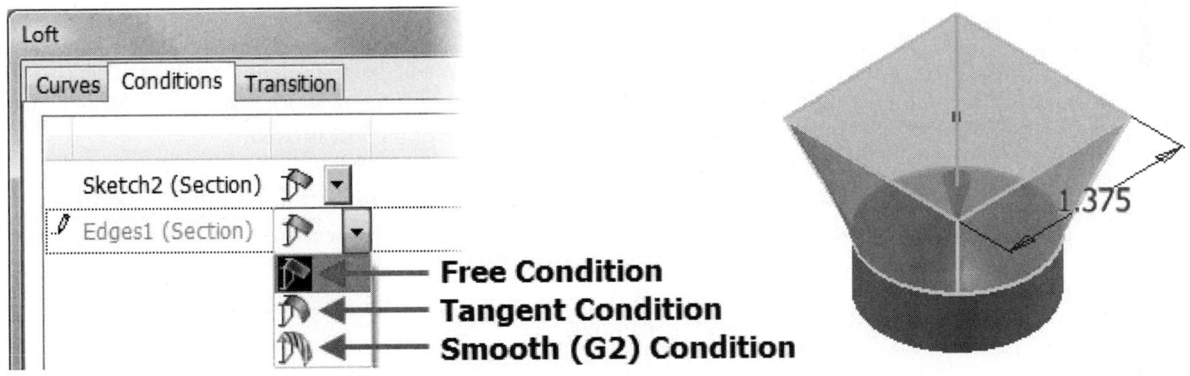

FIGURE 7.73

- Free Condition (top button): With this option, there is no boundary condition, and the loft will blend between the sections in the most direct fashion.
- Tangent Condition (middle button): With this option, the loft will be tangent to the adjacent section of face.
- Smooth (G2) Condition (bottom button): With this option, the loft will have curvature continuity to the adjacent section of face.

Condition Boundaries Point. Three conditions are available when a point is selected for the loft profile, as shown in Figure 7.74.

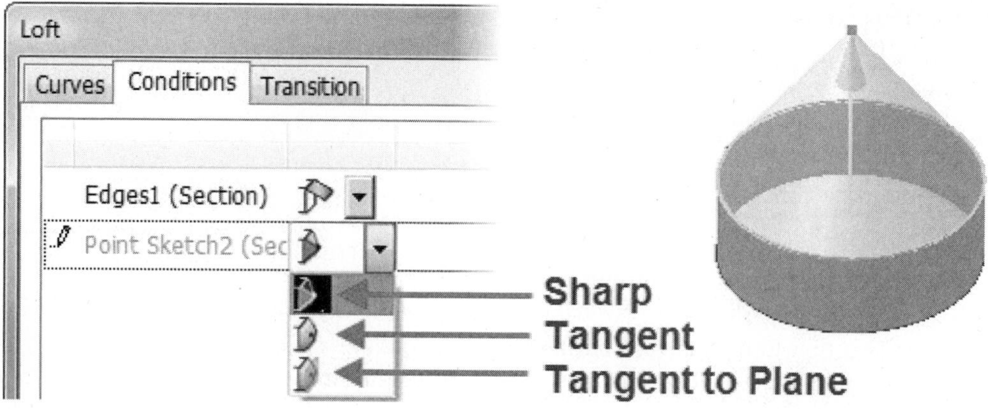

FIGURE 7.74

- Sharp Point (top button): With this option, there is no boundary condition, and the loft will blend from the previous section to the point in the most direct fashion.
- Tangent (middle button): When selected, the loft transitions to a rounded or domed shape at the point.
- Tangent to Plane (bottom button): When selected, the loft transitions to a rounded dome shape. You select a planar face or work plane to be tangent to. This option is not available when a centerline loft is used.

Angle. This option is enabled for a section only when the boundary condition is Tangent or Direction. The default is set to 90° and is measured relative to the profile plane. The option sets the value for an angle formed between the plane that the profile is on and the direction to the next cross-section of the loft feature. Valid entries range from 0.0000001° to 179.99999°.

Weight. The default is set to 0. The weight value controls how much the angle influences the tangency of the loft shape to the normal of the starting and ending profile. A small value will create an abrupt transition, and a large value creates a more gradual transition. High weight values could result in twisting the loft and may cause a self-intersecting shape.

Transition Tab

The Transition tab, as shown in Figure 7.75 with the Automatic mapping box unchecked, allows you to specify point sets. A point set is used to define section point relationships and control how segments blend from one section to the segments of the adjacent sections. Points are reoriented or added on two adjacent sections.

Map Points

You can map points to help define how the sections will blend into each other. The following rules apply to mapping points:

Point Set. The name of the point set appears here.

Map Point. The corresponding sketch for the selected point set appears here.

Position. The location of the selected map point appears here. You can modify the position by entering a new value or by dragging the point to a new location in the graphics window.

To modify the default point sets or to add a point set, follow these steps:

1. Click on the Transition tab, and uncheck the Automatic Mapping box. The dialog box will populate the automatic point data for each section. The list is sorted in the order in which the sections were specified on the Curve tab as shown in Figure 7.75 on the left.
2. To modify a point's position, select the point set in which the point is specified. When you select its name, it will become highlighted in the graphics window.
3. Click in the Position section of the map point that you want to modify, and enter a new value or drag the point to a new location in the graphics window.
4. To add a point set, select Click to add in the point set area.
5. Click a point on the profile of two adjacent sections. As you move the cursor over a valid region of the active section, a green point appears. As the points are placed, they are previewed in the graphics window, as shown in Figure 7.75 on the right.
6. You can modify the new point set in a similar way to the default point set.

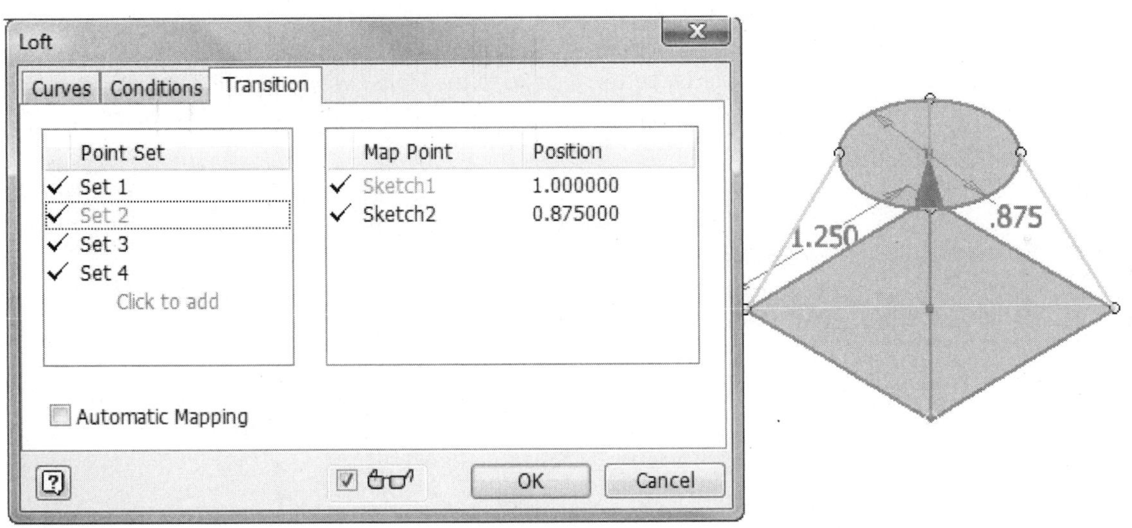

FIGURE 7.75

EXERCISE 7-5: CREATING LOFT FEATURES

In this exercise, you use the loft command to define the shape of a razor handle.

1. Open *ESS_E07_05.ipt* in the Chapter 07 folder.
2. In the browser, double-click Sketch1 to edit the sketch.
3. Click the Point, Center Point command on the Sketch tab > Draw panel.
4. Place the point so it is coincident on the spline near the bottom of the curve, making sure that the coincident glyph is displayed as you place the point, as shown in Figure 7.76.

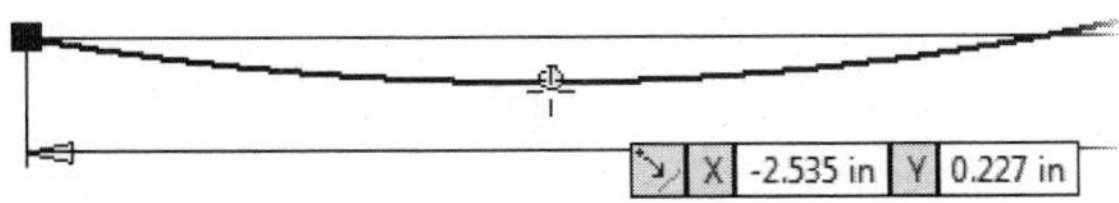

FIGURE 7.76

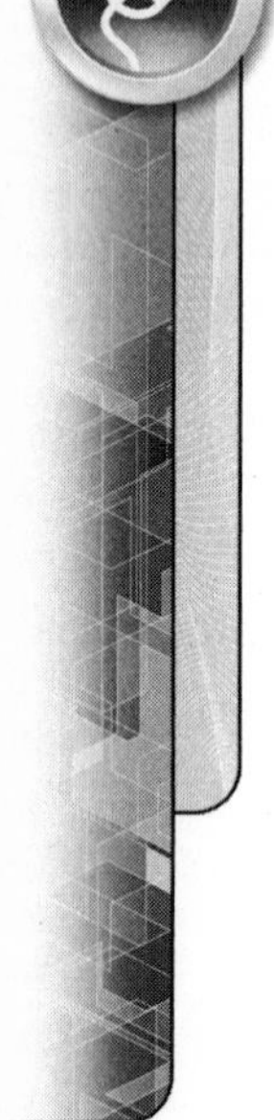

5. Draw two construction lines coincident with the sketched point and the nearest spline points on both sides of the Point that you just created, as shown in Figure 7.77.

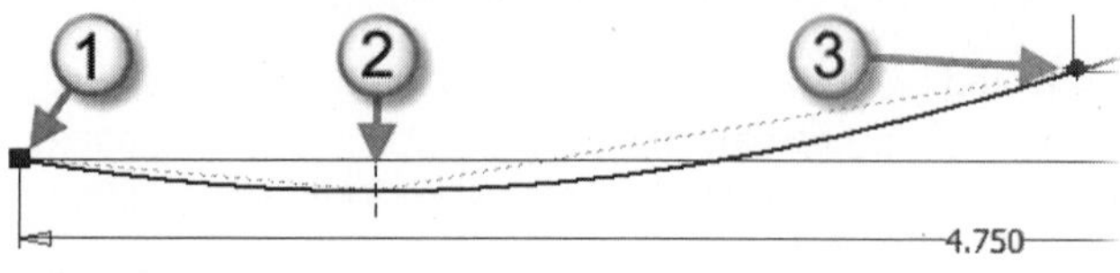

FIGURE 7.77

6. To parametrically position the sketched point midway between the spline points, place an equal constraint between the two construction lines.
7. Change the 4.750 overall horizontal dimension to **5 in**, and verify that the sketched point moves along the spline to maintain its position on the spline.
8. Click Finish Sketch on the Exit panel to finish editing the sketch.
9. Click the Work Plane command on the 3D Model tab > Work Features panel. To create a work plane perpendicular to the spline, click the Center Point you just created, and then select the spline, as shown in Figure 7.78 on the left (do not select the construction line).
10. Next create a profile for a loft section on the new work plane. Create a new sketch on the new work plane.
11. To see the sketch better, change to the Home View.
12. Use the Project Geometry command to project the Point, Center Point that you created in step 4 onto the sketch.
13. Draw an ellipse with its center coincident with the projected point and the second point so the ellipse's axis is horizontally constrained to the Point. Click a third point, as shown in Figure 7.78.
14. Place **.1875 in** and **.375 in** dimensions to control the ellipse size, as shown in Figure 7.78 on the right.

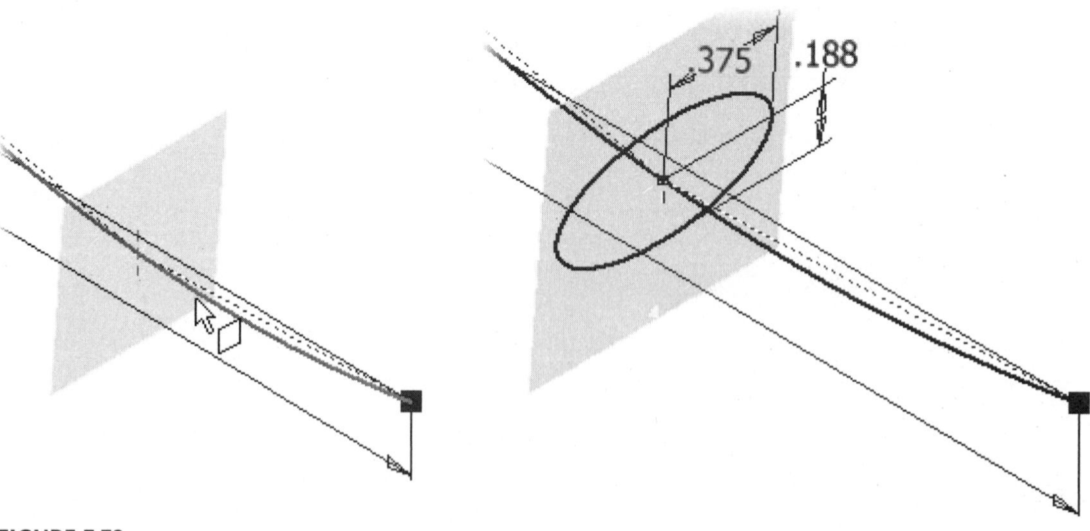

FIGURE 7.78

15. Finish the sketch.
16. Click the Loft command on the 3D Model tab > Create panel.
17. For the first section, click the concave 3D face, as shown in Figure 7.79.

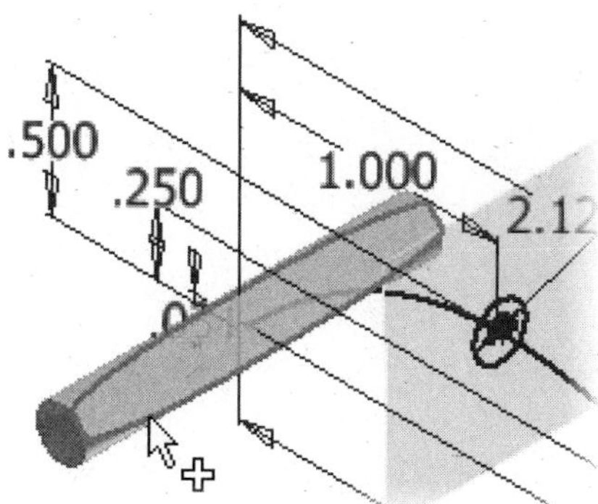

FIGURE 7.79

18. Click the other three profile sections from left to right, and then click the point on the right end of the spline.
19. Click the Center Line option in the Loft dialog box, and then select the spline. When finished, the preview looks like Figure 7.80 on the left.

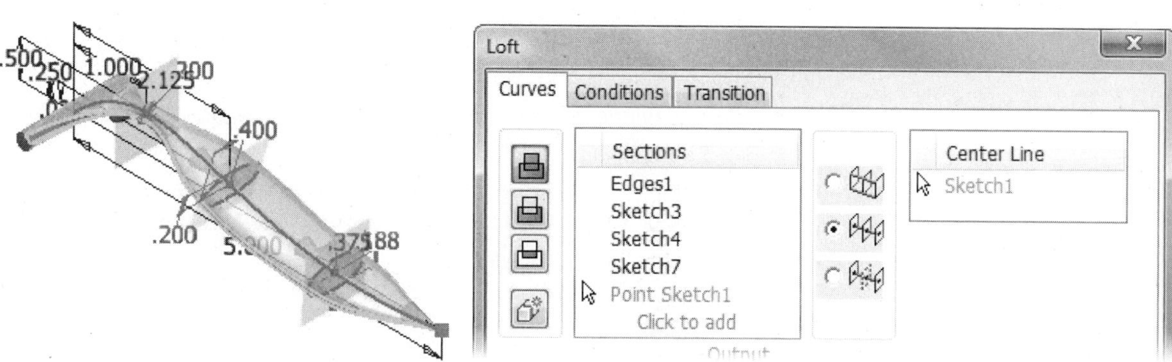

FIGURE 7.80

20. Click OK to create the loft.
21. Turn off the visibility of all the work planes by pressing the ALT and] keys, and then use the Free Orbit command to examine the loft. Notice that the end of the loft on the right side is sharp.

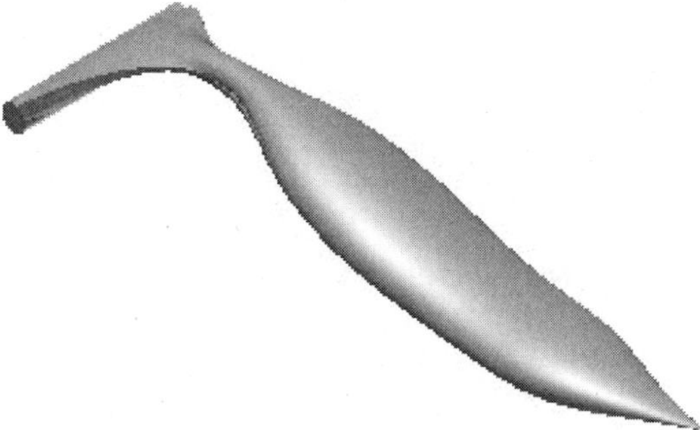

FIGURE 7.81

22. To edit the loft move the cursor over a face on the Loft in the graphics window, and click the Edit Loft from the mini-toolbar that is displayed. Click the Conditions tab, and change the Point entry to a Tangent condition, as shown in Figure 7.82 on the left.

23. Click OK to update the loft. The right side of the part should resemble Figure 7.82 on the right.

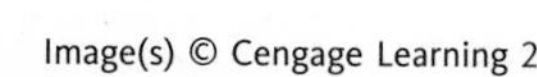

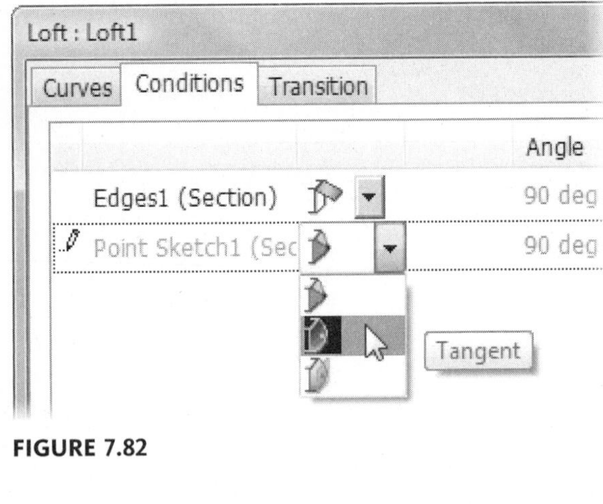

FIGURE 7.82

24. Next edit the loft again and add another section that defines the section's area. In the browser, move the cursor over the entry Loft1 and double-click.
    - From the Curves tab, click Area Loft option labeled (1) in Figure 7.83.
    - In the Placed Sections area, click Click to add labeled (2).
    - In the graphics window, click a point near the middle of the spline labeled (3).

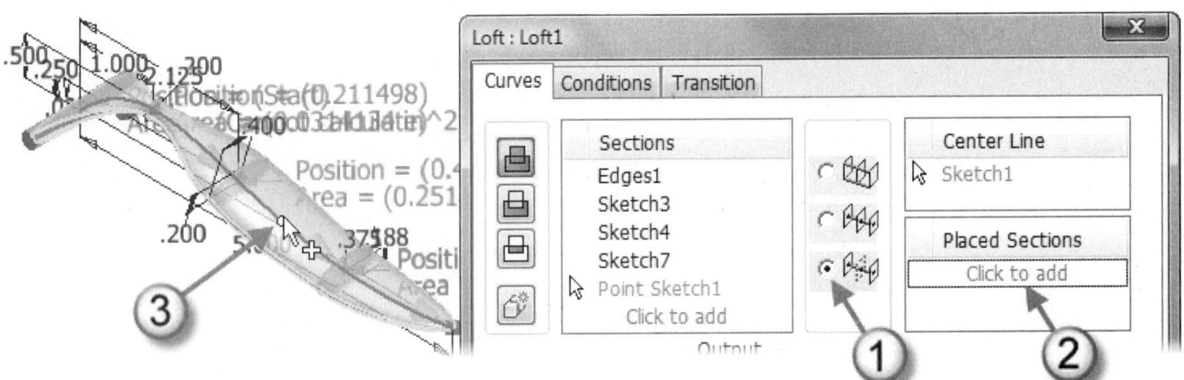

FIGURE 7.83

25. In the Section Dimensions dialog box:
    - Click the Section Position option and change the proportional distance to **0.5** labeled (1) in Figure 7.84.
    - Click the Section Size option and change the area to **0.375 in^2** labeled (2).
    - Click OK.
    - In the graphics window, notice the values of each section and new section half way through the loft.
26. In the Loft dialog box, click OK. Figure 7.84 on the right shows the updated loft.

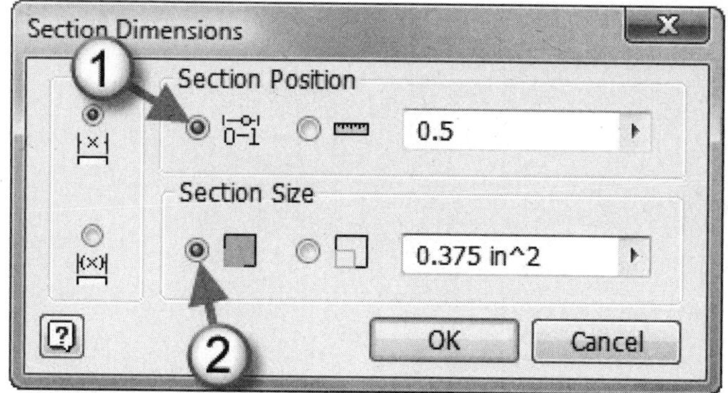

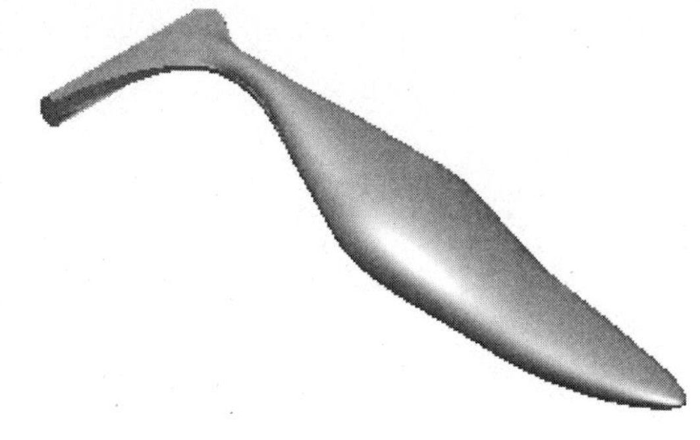

FIGURE 7.84

27. Close the file. Do not save changes. End of exercise.

MULTI-BODY PARTS

As was explained throughout the previous chapters commands such as extrude, revolve, sweep, and loft have a New Solid option that allows you to create a solid body in a part file. There is no limit to the number of solid bodies that can be created in a part file. You can also create a single part and then split it into multiple parts and then, if required, export the solid bodies to their own part files. This modeling method is helpful for designing parts that have complex relationships between the parts.

There are three main methods for creating solid bodies in a part: create within the part, derive an existing part into a part file or split a part into multiple solid bodies.

Create a Solid Body in a Part File

This method creates solid bodies within a part file. To create a solid body within a part file, follow these steps:

1. If needed create a new sketch.
2. Issue the extrude, revolve, sweep, and loft commands.
3. In the dialog box select the New Solid option, as shown in the extrude tool in Figure 7.85 on the left or from the in-canvas operation button, as shown in Figure 7.85 on the right.
4. Define the operation. Only Join and Intersection will create a new solid body.
5. Define the extents of the feature.
6. Create the solid body.

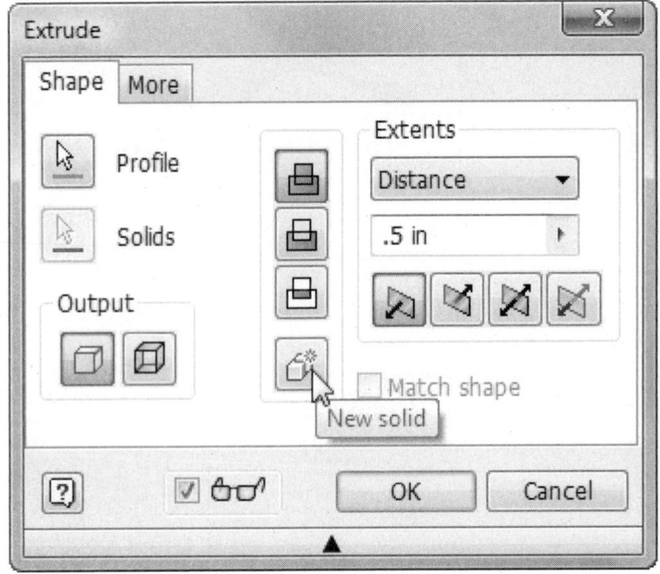

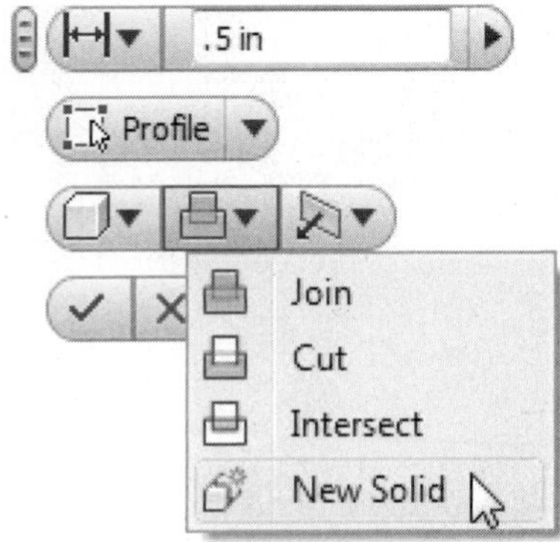

FIGURE 7.85

Derive a Solid Body into a Part File

This method links an existing part file into a part file. To derive a solid body into a part file, follow these steps:

1. While in a part file, start the Derive command from the 3D Model tab > Create panel or from the Manage tab > Insert panel.
2. Select a part or an assembly file.
3. In the Derive Part dialog box, select the geometry that you want to derive.

NOTE The Derive command will be covered later in this chapter.

Split a Part into Solid Bodies

This method will take a solid body and split it into multiple solid bodies. To split a part into multiple solid bodies, follow these steps:

1. Create the part that will be split.
2. Create a work plane or surface that will be used to split the part.
3. Start the split command and select the Split Solid option.

NOTE The Split command will be covered later in this chapter.

Edit Solid Bodies

After a solid body is created it will appear in the browser under the Solid Bodies folder as shown in Figure 7.86 on the left. You can rename the solid bodies by either slowly clicking twice on the solid body's name in the browser and entering a new name or right-clicking on the solid body's name in the browser and clicking Properties from the menu, and in the Body Properties dialog box enter a new name. You can also control the visibility of the solids by right-clicking on each solid in the browser and clicking Visibility or Hide Others as shown in Figure 7.86 on the right.

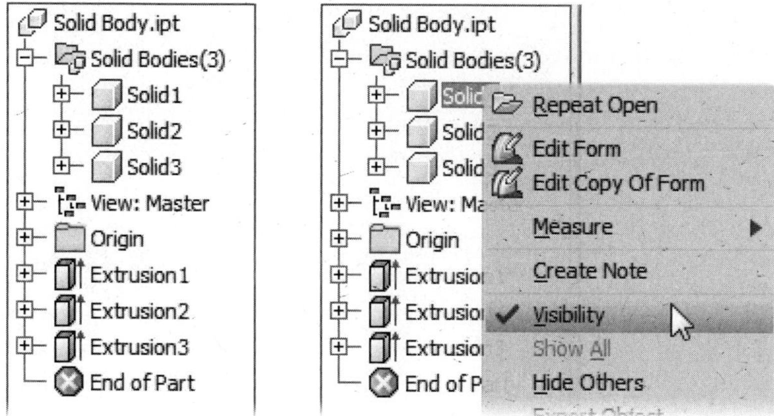

FIGURE 7.86

Autodesk Inventor has tools that can be used to move and combine solid bodies.

Move Bodies

Use the Move Bodies command to move or rotate the location of solid bodies in a part file. Figure 7.87 shows the Move Bodies dialog box. The Move Bodies command is located on the 3D Model tab > Modify panel, as shown in Figure 7.87 on the left.

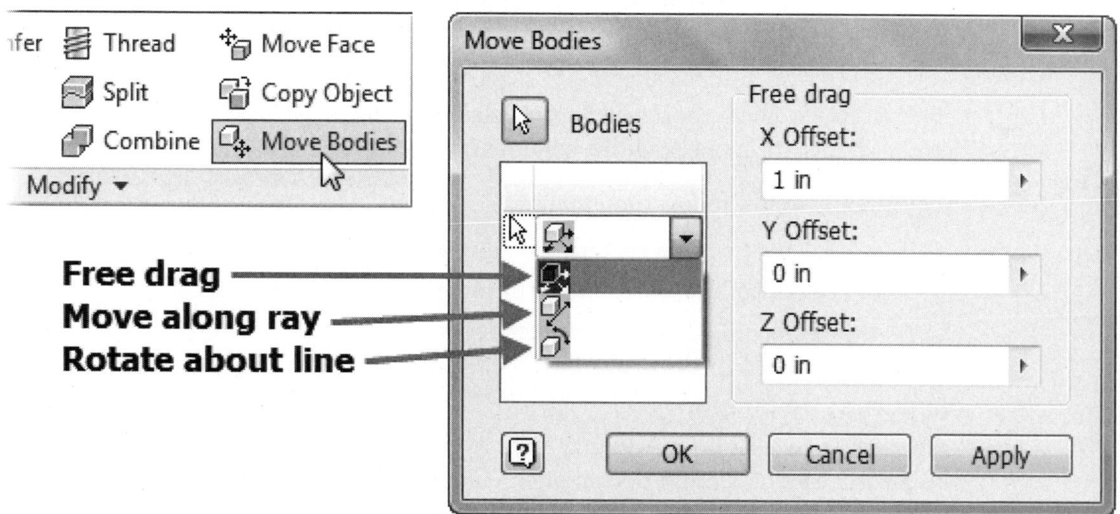

FIGURE 7.87

To move a solid body, you have three Move Type options:

- Free drag: Enter an X, Y, or Z value or select the solid body and free drag the solid by moving the mouse.
- Move along ray: Enter a linear offset value and then select an edge or axis to define the vector.
- Rotate about line: Select an edge or axis to rotate and enter a value for the angle.

To move a solid body(ies), follow these steps:

- Click the Move Bodies command from the Modify panel.
- Select a solid body(ies) to move or rotate.
- Select the Move Type and enter the required information.

Combine

The Combine command will join two or more solids together, remove material from a base solid, or keep the common volume of the selected solids. The Combine command is located on the 3D Model tab > Modify panel, as shown in Figure 7.88 on the left. Figure 7.88 on the right shows the Combine dialog box.

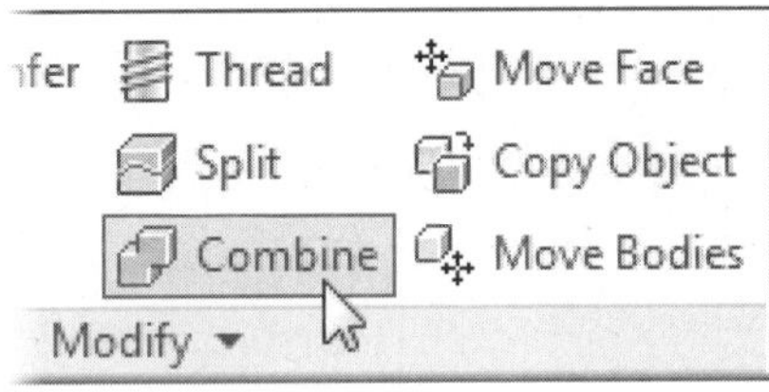

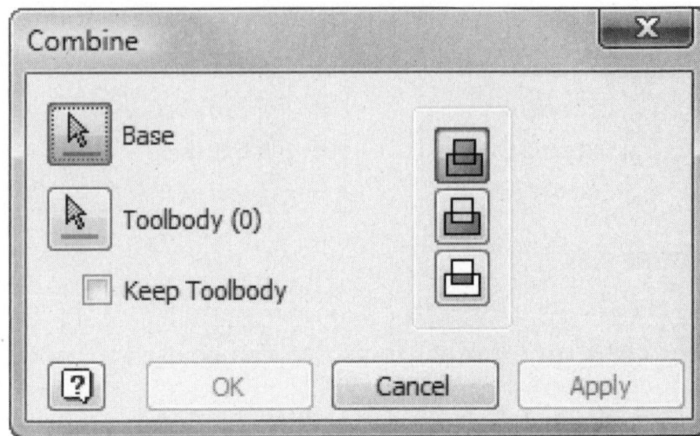

FIGURE 7.88

When combining solids, there are three operations to choose from that are similar to the operations in the feature creation commands.

- *Join:* Combines two or more solid bodies.
- *Cut:* Removes the volume of a solid(s) from the base solid.
- *Intersect:* Keeps the volume that is common between the base and selected solids.

The Keep Toolbody option when checked will keep the solid body in the Solid Bodies folder. If unchecked, the solid body is consumed into the base solid.

To combine solids, follow these steps:

1. Click the Combine command on the Modify panel.
2. Select a base solid from which the other solids will be joined, cut, or intersected.
3. Select the Cut, Join, or Intersect option.

Export Solid Body

After editing a solid body, you may want to export the solid(s) to its own file so it can be used in assemblies and documented. There are two commands to export a solid: Make Part and Make Components. The exported part(s) is linked to the original solid body via the Derived functionality that will be covered later in this chapter.

Make Part

The Make Part command will export a single solid to a new part file. To start the Make Part command, click the Manage tab > Layout panel, as shown in Figure 7.89 on the left and its dialog box is on the right. Via the dialog box, select the objects to be exported in the Status area. Set the scale factor and check the Mirror part option if you want to mirror the part. Then select a template, file name location, and determine if the part should be placed in a new assembly. When done, click OK and the solid will be exported.

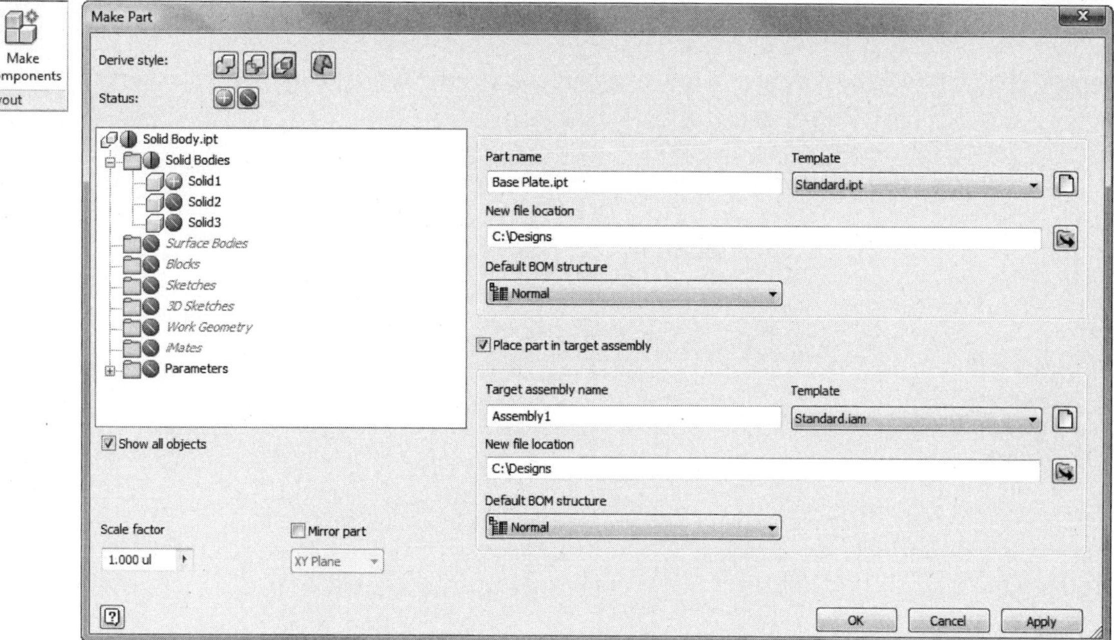

FIGURE 7.89

To Make Part follow these steps.

Make Components

The Make Components command is similar to the Make Part command except it allows multiple solid bodies to be exported to individual files in one operation, to start name of the target assembly, or to uncheck the Insert components in target assembly if you just want the solids to go to a part file. When done, click Next.

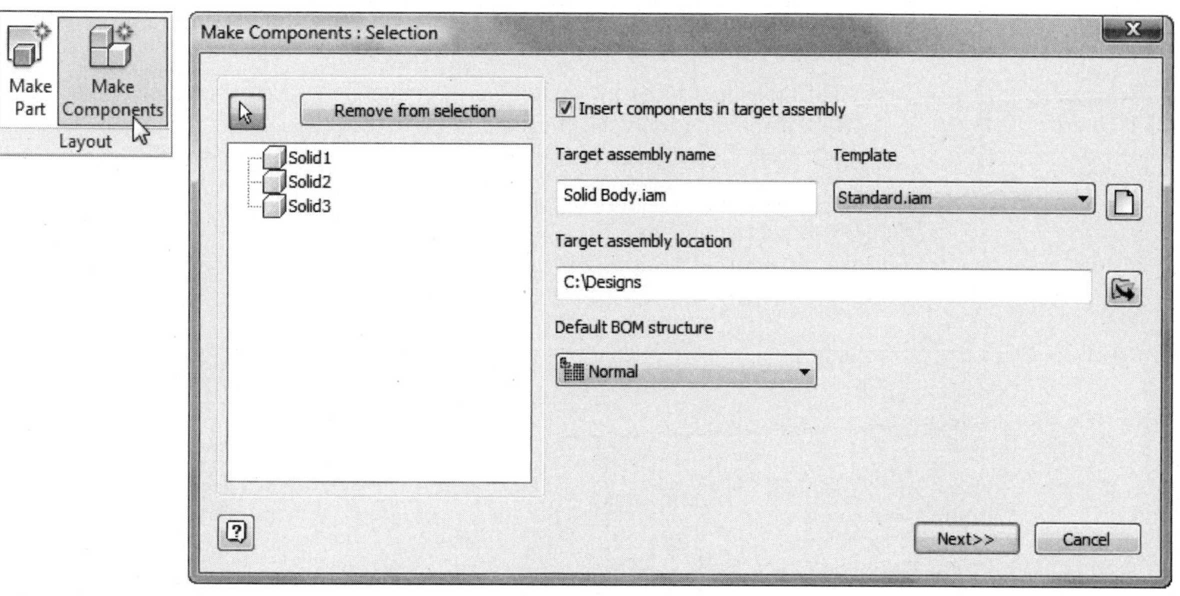

FIGURE 7.90

The Make Components: Bodies dialog box will appear. If needed, change the components name, template, BOM structure, file location, scale factor, determine if the part will be mirrored, as shown in Figure 7.91, and then click OK and the solids will be exported.

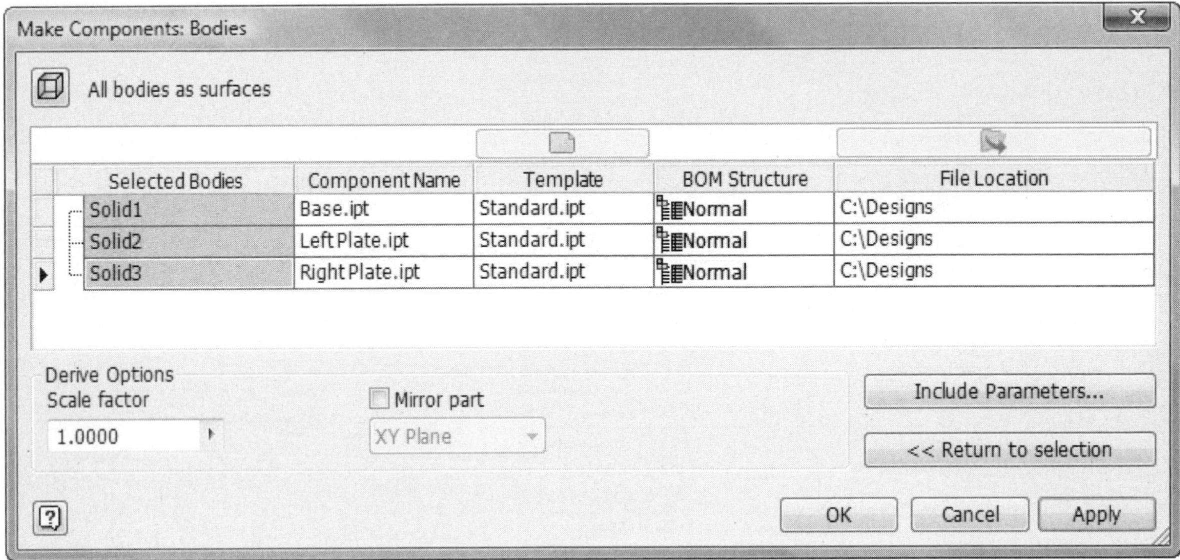

FIGURE 7.91

SPLIT A SOLID, PART, OR FACE

The Split command allows you to split a part into two solids, split the solid by removing one portion of the part, to split individual faces, or to split all faces. A typical application to split a face is to allow the creation of face drafts to the split faces of a part. You can use the Split command to perform the following:

- Split a solid into multiple solid bodies.
- Remove a section of the part by using a surface, planar face, or a work plane and to cut material from the part in the direction you specify. The side that is removed is suppressed rather than deleted. To create a part with the other side removed, edit the split feature, redefine it to keep the other side, save the other half of the part to its own file using the Save Copy As option, or create a derived part.
- Split individual faces by using a surface, sketching a parting line, or placing a work plane, and then selecting the faces to split. You can edit the split feature and modify it to add or remove the desired part faces to be split.

The Split command is located on the 3D Model tab > Modify panel, as shown in Figure 7.92 on the left. Once selected, the Split dialog box will appear, as shown in Figure 7.92 on the right.

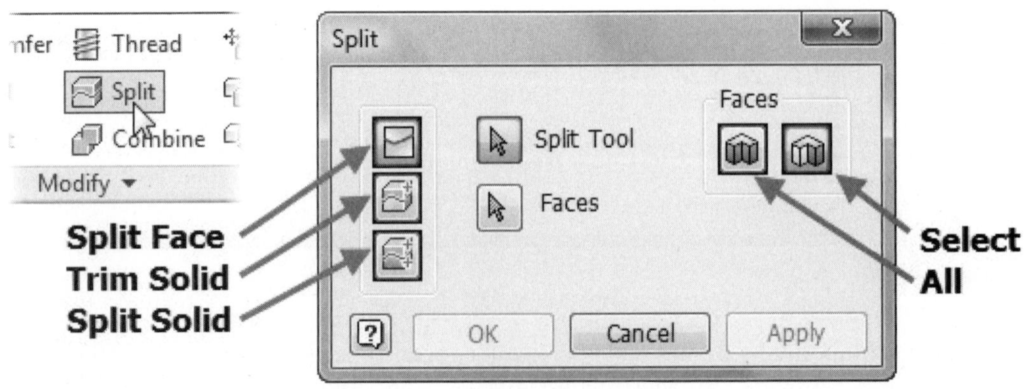

FIGURE 7.92

The Split dialog box contains the following sections:

Method

Split Face. Click this button to split individual faces of a part by selecting a work plane, surface, or sketched geometry and then selecting the faces to split. The split face method can split individual faces or all the faces on the part. When you select the split face method, the Remove area in the dialog box will be replaced with the Faces area, which allows you to select all or individual faces.

Trim Solid. Click this button to split a part using a selected work plane, surface, or sketched geometry to cut or remove material. If you select this option, you are prompted to choose the direction of the material that you want to remove.

Split Solid. Click this button to split a solid into two solid bodies. Use a surface, plane, or work plane that at least touches the outside edges of the solid; it can exceed the exterior faces of the part.

Remove. The option to remove material is only available when you use the Trim Solid method. After splitting a part, you can retrieve the cut material by editing the split feature and clicking to remove the opposite side or by deleting the split feature.

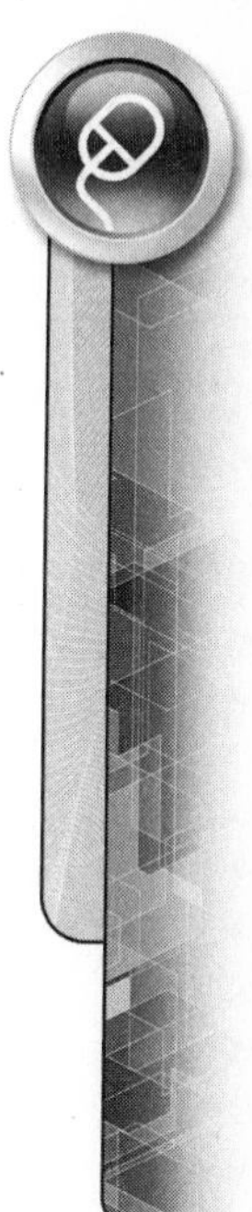

Faces

The Faces option is available only when you select the split face method.

All. Click this button to split all faces of the part that intersect the Split command.

Select. Click this button to enable the selection of specific faces that you want to split. After clicking the Select button, the Faces to Split command becomes active.

Faces. Click this button, and select a surface, work plane, or sketch that you want to use to split the part.

EXERCISE 7-6: SPLITTING A PART INTO MULTIPLE SOLID BODIES

In this exercise, you split a part into multiple solid bodies and export them to an assembly.

1. Open *ESS_E07_06.ipt* in the Chapter 07 folder.
2. In this step you use an origin plane to split the part into two solids. Click the Split command in the 3D Model tab > Modify panel.
    - Click the Split Solid method button, as shown in Figure 7.93.
    - For the Split Tool, click XY Plane from the Origin folder in the browser.

FIGURE 7.93

3. Click OK to split the part into two solids.
4. In the browser, expand the Solid Bodies folder and click on Solid2 and then Solid3 to verify that the solids have been created.
5. Next you turn off the visibility of a solid. In the browser, under the Solid Bodies folder right-click on Solid2 and on the menu click Visibility.
6. Use the Zoom and Free Orbit commands to examine the solid.
7. Next you switch the visibility of the solids. In the browser, under the Solid Bodies folder right-click on Solid2 and on the menu click Hide Others.
8. Use the Zoom and Orbit commands to examine the solid.
9. Turn the visibility of Solid3 back on, under the Solid Bodies folder right-click on Solid3 and on the menu click Visibility or Show All.

NOTE

You can rename a solid by slowly double-clicking on the solid name in the browser or right-clicking on the solid in the browser and clicking Properties from the menu.

10. Next you create a hole feature. From the 3D Model tab > Modify panel click Hole. Add a Through All **.25 in** concentric hole to the top right side of the part, as shown in Figure 7.94.

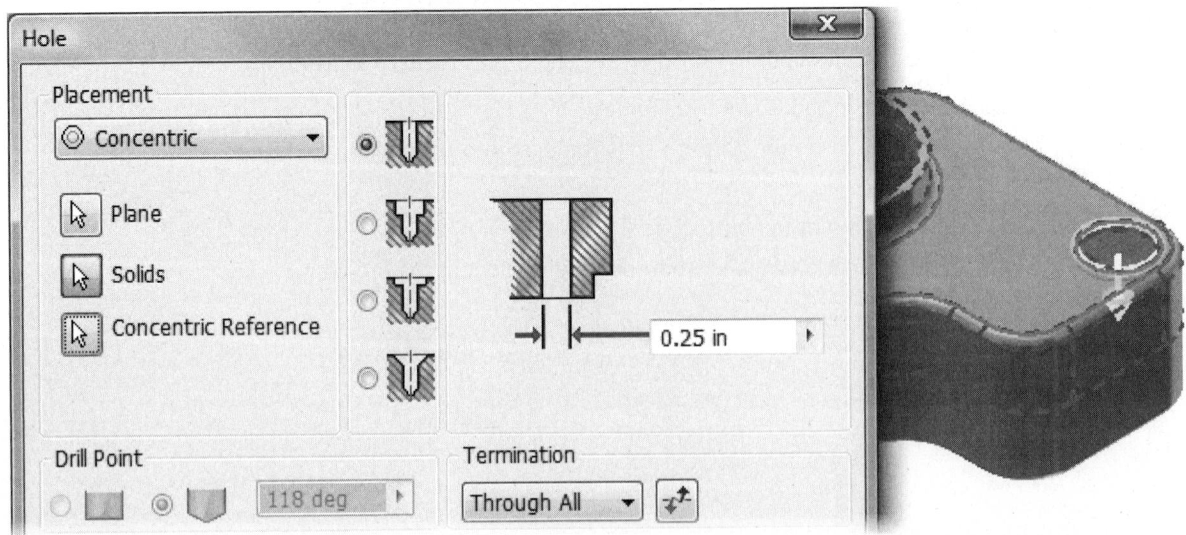

FIGURE 7.94

11. After creating the hole, use the Free Orbit command to examine the bottom of the part. Notice that the hole does not go through the bottom solid.
12. Change to the Home View.
13. In the browser expand the Solid Bodies folder and expand the Solid2 and Solid3 entry. Notice that Hole1 is only under Solid2, as shown in Figure 7.95 on the left.
14. Edit the hole feature and click the Solids button, and then select the lower solid (Solid3) in the graphics window. Click OK to complete the edit.
15. Use the Free Orbit command to ensure that the hole goes through both solids.
16. In the browser also notice that the Hole1 is now under both Solid2 and Solid3 entry.

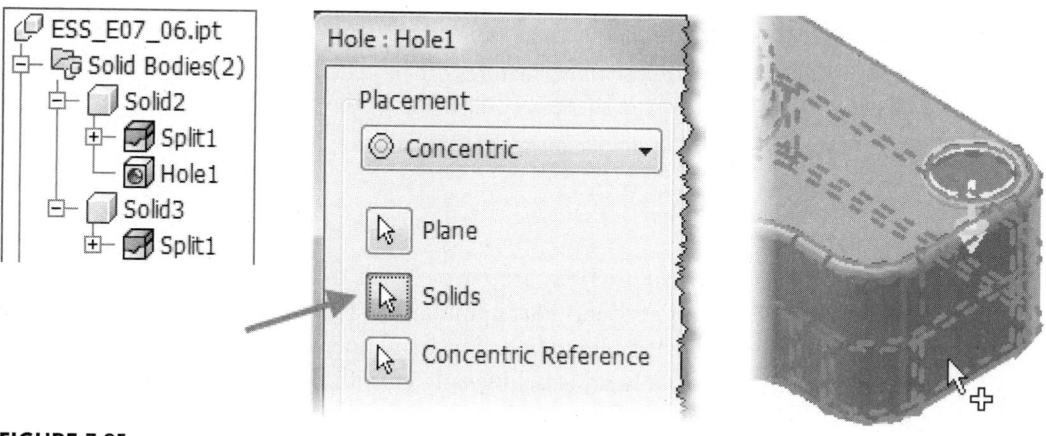

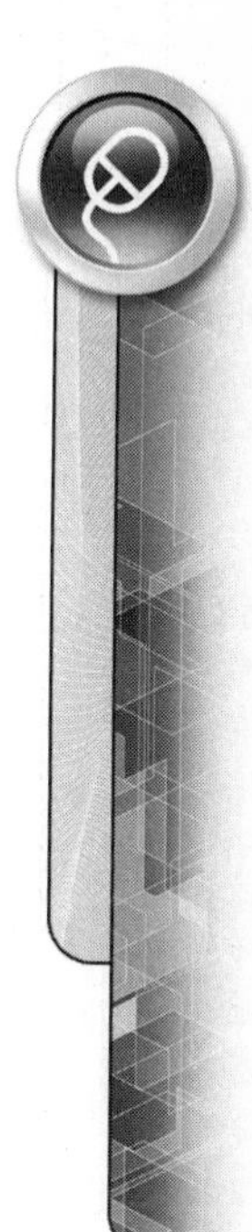

FIGURE 7.95

17. Next you export the solids into an assembly. Click the Make Components command on the Manage tab > Layout panel.
    - Select the two solids.
    - Change the Target assembly location to **C:\Inv 2013 Ess Plus\Chapter 07** as shown in Figure 7.96.

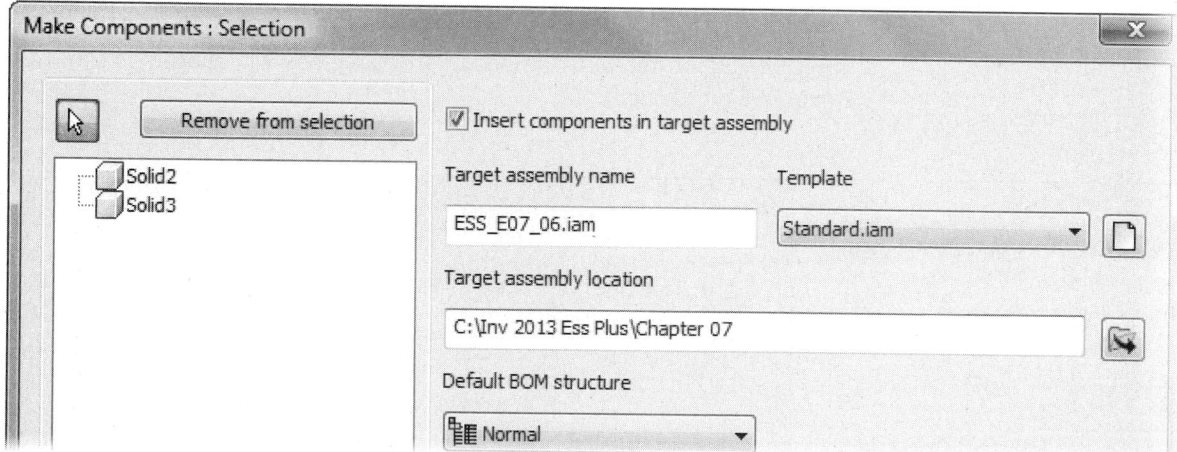

FIGURE 7.96

18. Click Next and click in each of the Component Name Cells and enter new components names, as shown in Figure 7.97.

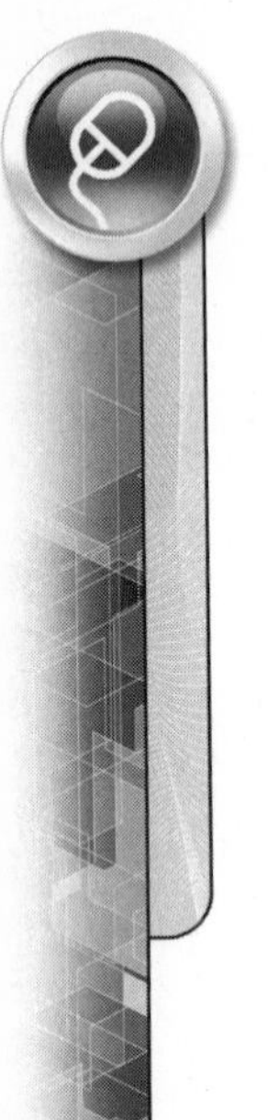

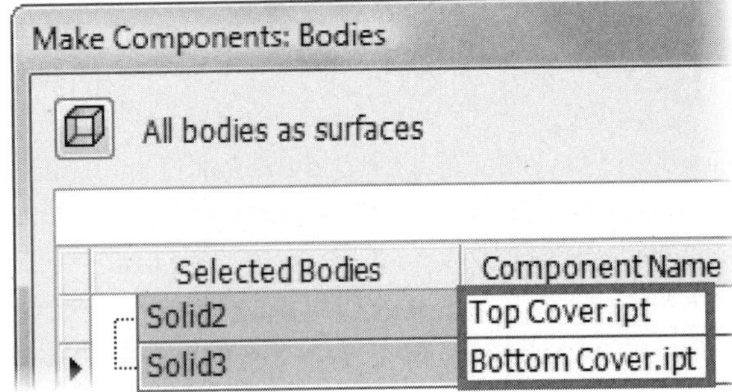

FIGURE 7.97

19. Click OK to create the parts and assembly.
20. Examine the parts. Additional features can be added to the parts as needed but the new part files size and shape are still driven by the original part file.
21. Close the files. Do not save changes. End of exercise.

BEND PART

While designing parts that are bent, it may be easier to create the part in a flattened state and then bend it. The Bend Part command will bend a side or both sides of a part based on the location of a bend line. The bend line will be the location where the bend or the centerline of a bend is bent evenly in both directions. After creating the bend, it will appear in the browser as a feature and can be edited like any other feature. To bend a part, follow these steps.

1. Create or open a part that will be bent.
2. Create a sketch where the bend line will be placed. The plane must intersect the part or lie on a planar face of the part.
3. Draw a line that will be used as the bend line, the line does not need to extend to the limits of the part.
4. Exit the sketch.
5. Click the Bend Part command from the 3D Model tab > Modify panel and click the drop arrow next to Modify to see the Bend Part command, as shown in Figure 7.98 on the left. The Bend Part dialog box will appear, as shown in Figure 7.98 on the right.

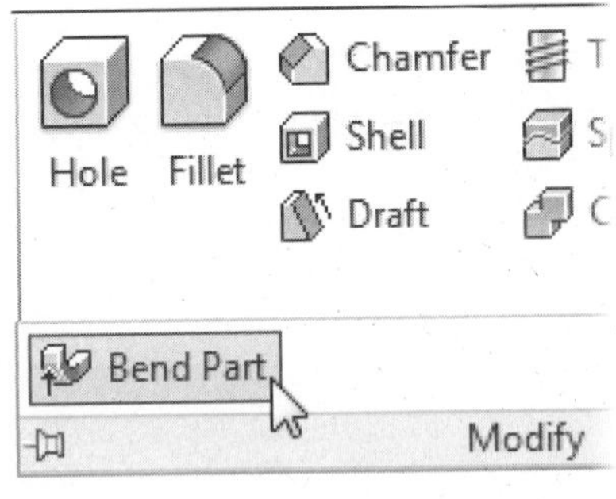

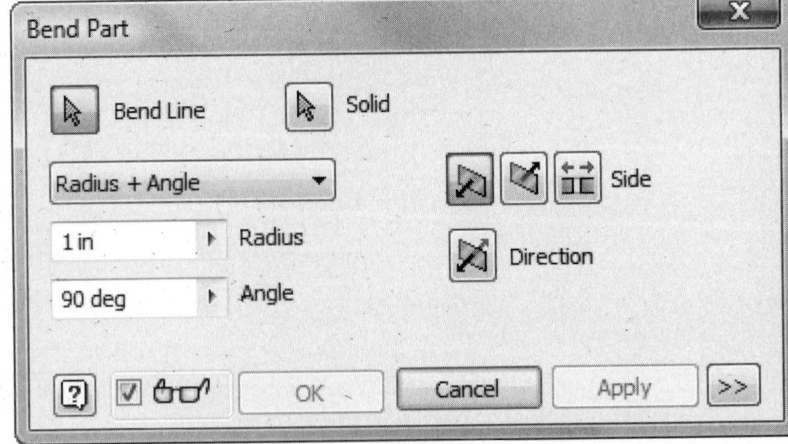

FIGURE 7.98

6. In the dialog box, specify how you want to calculate the bend: Radius + Angle, Radius + Arc Length, or Arc Length = Angle.
7. Enter the data to create the bend.
8. If needed, change the side and direction that the bend occurs.
9. Figure 7.99 on the left shows the part with a bend line in the middle of the inside face and the image on the right shows the bent part.

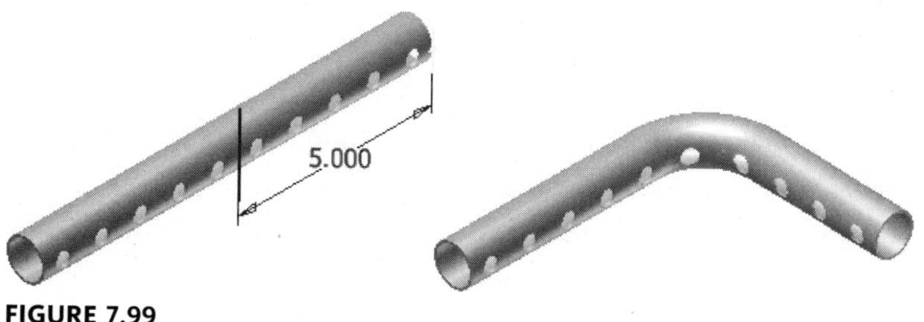

FIGURE 7.99

MIRROR FEATURES

When creating a part that has features that are mirror images of one another, you can use the Mirror Feature command to mirror a feature(s) about a planar face or work plane instead of recreating the features from scratch. Before mirroring a feature, a work plane or planar face that will be used as the mirror plane must exist. The feature(s) will be mirrored about the existing plane; it can be a planar face, a work plane on the part, or an origin plane. The mirrored feature(s) will be dependent on the parent feature—if the parent feature(s) change, the resulting mirror feature will also update to reflect the change. To mirror a feature or features, use the Mirror command on the 3D Model tab > Pattern panel, as shown in Figure 7.100 on the left. The Mirror dialog box will appear, as shown on the right. The following sections explain the options in the Mirror dialog box.

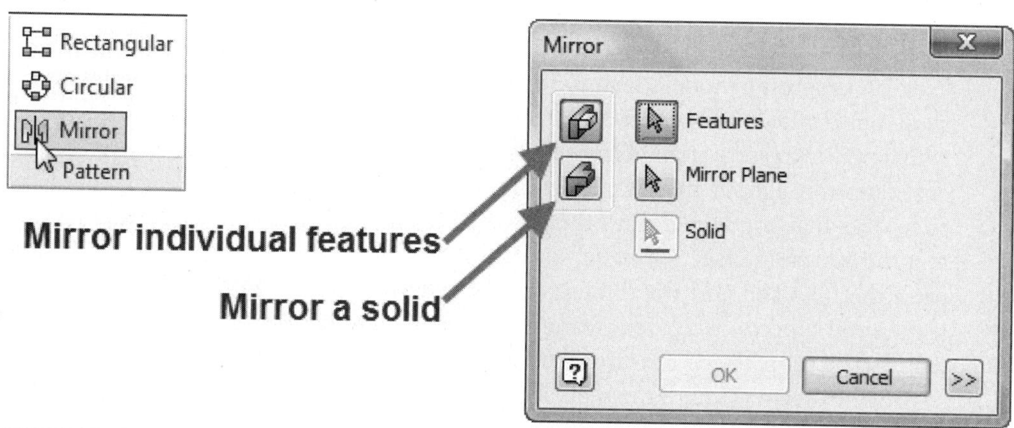

FIGURE 7.100

Mirror Individual Features. Click this option to mirror a feature or features.

Mirror the Entire Solid. Click this option to mirror the solid body.

Mirror Plane. Click and choose a planar face or work plane on which to mirror the feature(s).

Solid. If there are multiple solid bodies, click this button to choose the solid body(ies) to receive the mirrored feature.

To mirror a feature or features, follow these steps:

1. Click the Mirror Feature command on the Pattern panel. The Mirror dialog box will appear.
2. Click the Mirror individual features option or the Mirror the entire solid option.
3. Select the feature(s) or solid body to mirror.
4. Click the Mirror Plane button, and then select the plane on which the feature will be mirrored.
5. Click the OK button to complete the operation.

Figure 7.101 on the left shows a part with the features that will be re-created,the middle image shows the part with a work plane that the features will be mirrored about, and the image on the right shows the features mirrored.

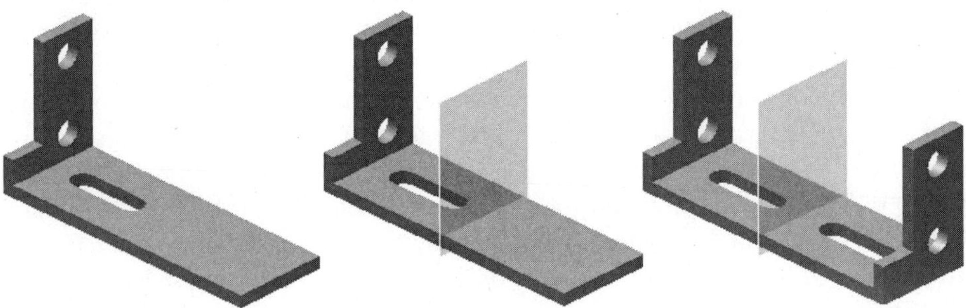

FIGURE 7.101

SUPPRESSING FEATURES

You can suppress a model's feature or features to temporarily turn off their display, as shown in Figure 7.102 on the left. Feature suppression can be used to simplify parts, which may increase system performance. This capability also shows the part in different states throughout the manufacturing process, and it can be used to access faces and edges that you would not otherwise be able to access. If you need to dimension to a theoretical intersection of an edge that was filleted, for example, you could suppress the fillet and add the dimension and then unsuppress the fillet feature. If the feature you suppress is a parent feature for other dependent features, the child features will also be suppressed. Features that are suppressed appear gray in the browser and have a line drawn through them, as shown in Figure 7.102 on the right. A suppressed feature will remain suppressed until it is unsuppressed, which will also return the features browser display to its normal state. Figure 7.102 on the right also shows the suppressed child features that are dependent on the suppressed extrusions.

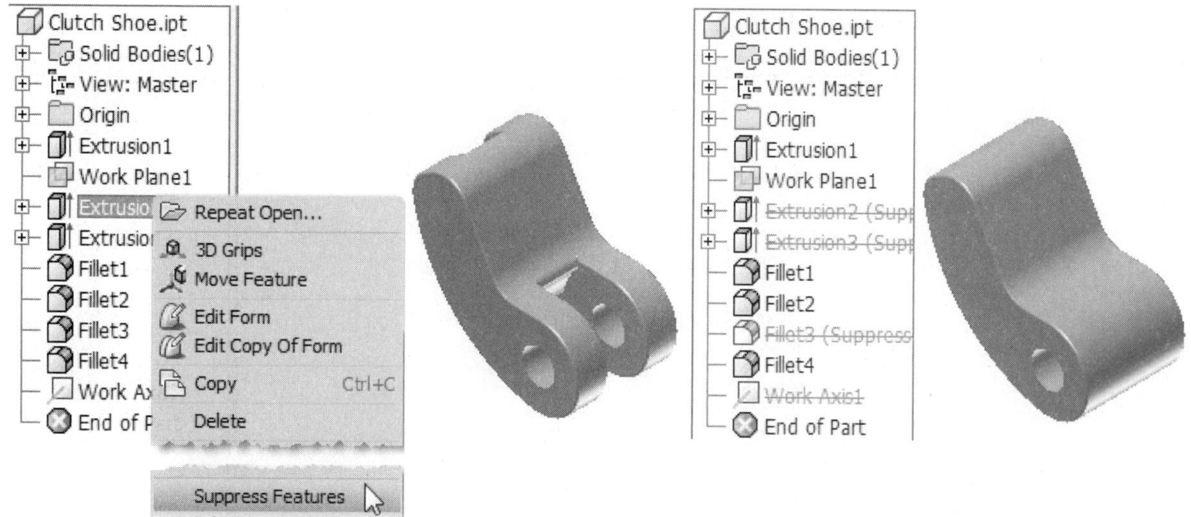

FIGURE 7.102

To suppress a feature on a part, use one of these methods:

1. Right-click on the feature in the browser, and select Suppress Features from the menu, as shown in Figure 7.102 on the left, or
2. Click the Select Feature option from the Quick Access toolbar. This option allows you to select features on the parts in the graphics window.
3. Right-click on the feature in the graphics window, and click the Suppress Features option from the menu.

To unsuppress a feature, right-click on the suppressed feature's name in the browser, and select Unsuppress Features from the menu.

REORDERING A FEATURE

You can reorder a feature in the browser. If you created a fillet feature using the All Fillets or All Rounds option and then created an additional extruded feature, such as a boss, you can move the fillet feature below the boss and include the edges of the new feature in the selected edges of the fillet. To reorder features, follow these steps:

1. Click the feature's name or icon in the browser. Hold down the left mouse button and click and drag the feature to the desired location in the browser. A horizontal line will appear in the browser to show you the feature's relative location while reordering the feature. Figure 7.103 on the left shows a hole feature being reordered in the browser.
2. Release the mouse button and the model features will be recalculated in their new sequence. Figure 7.103 on the right shows the browser and the reordered hole features.

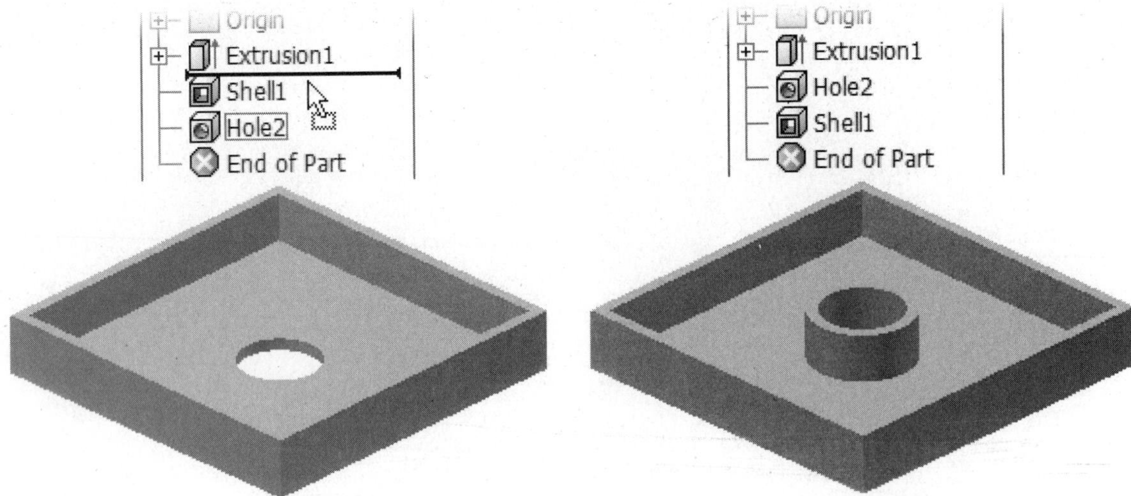

FIGURE 7.103

If you cannot move the feature due to parent-child relationships with other features, Autodesk Inventor will not allow you to drag the feature to the new position. In the browser, the cursor will change to a No symbol instead of a horizontal line, as shown in Figure 7.104.

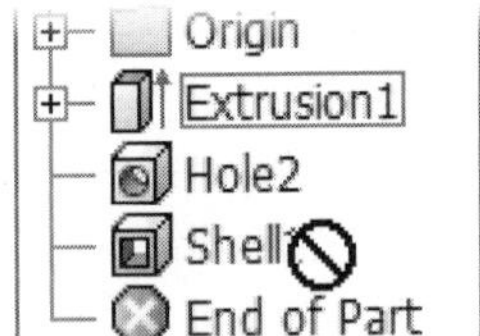

FIGURE 7.104

FEATURE ROLLBACK

While designing, you may not always place features in the order that your design later needs. Earlier in this chapter, you learned how to reorder features, but reordering features will not always allow you to create the desired results. To solve this problem, you can roll back the design to an earlier state and then place the additional new features. To roll back a design, drag the End of Part marker in the browser to the location where the new feature will be placed. To move the End of Part marker, follow these steps:

1. Move the cursor over the End of Part marker in the browser.
2. With the left mouse button depressed, drag the End of Part marker to the new location in the browser. While dragging the marker, a line will appear, as shown in Figure 7.105 on the left.
3. Release the mouse button, and the features below the End of Part marker are removed temporarily from calculation of the part. The End of Part marker will be moved to its new location in the browser as shown in Figure 7.105 on the left. Another method is to right-click on a feature in the browser and click Move EOP Marker as shown in Figure 7.105 in the middle and the End of Part marker will move below the feature.

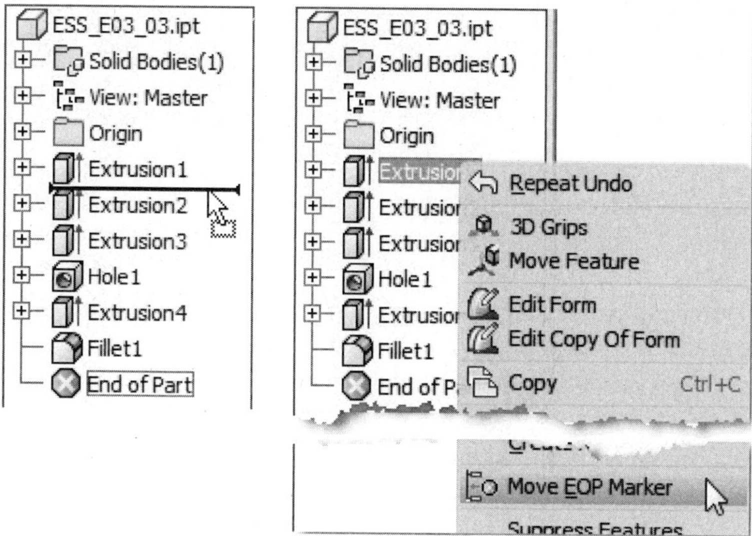

FIGURE 7.105

4. **Figure 7.106** on the left shows the part before the End of Part marker was moved, and the image in the middle shows the part after the End of Part marker was moved. Create new features as needed. The new features will appear in the browser above the End of Part marker.
5. To return the part to its original state, including the new features, drag the End of Part marker below the last feature in the browser or right-click on the End of Part marker and click Move EOP to End as shown in Figure 7.106 on the right.
6. If needed, you can delete all features below the End of Part marker by right-clicking on the End of Part marker. Then click Delete all features below EOP, as shown in Figure 7.106 on the right.

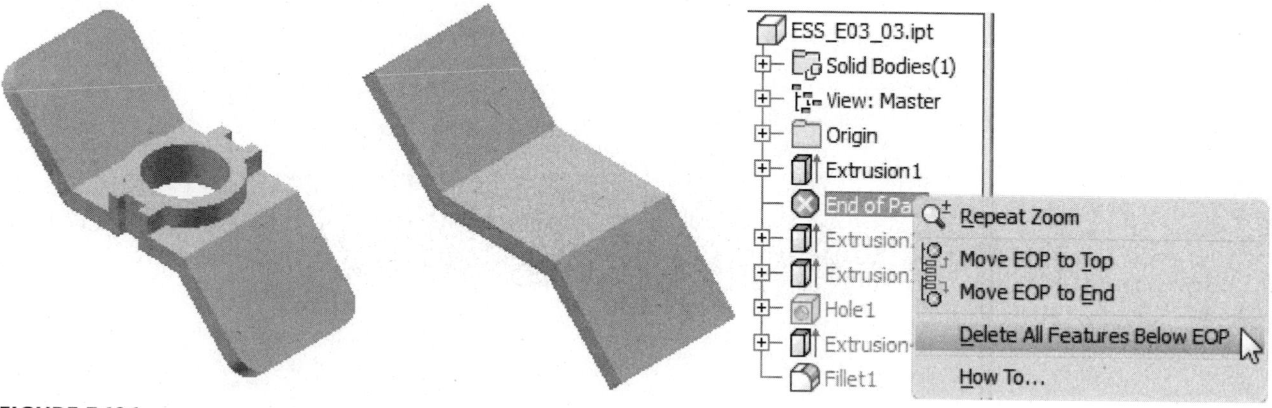

FIGURE 7.106

Rule Fillet

A rule-based fillet is different than the fillet command that was covered in Chapter 4. The fillets created in Chapter 4 were based on selected edges or face. Rule-based fillets are based on a set of rules that dictate when edges touch certain faces or features fillets will automatically be created. When creating a rule fillet, you specify a source and a rule.

Source

Select either a face(s) or feature(s) that when edges touch them fillets will be placed onto them.

Rule

A rule identifies edges to fillet. Following are the rule options:

Feature Rules

- *All Edges:* Fillets are created on all the edges that are created by the feature and on the edges that are intersected by the created features.
- *Against Part:* Edges created by the faces of the features and the faces of the part are filleted.
- *Against Features:* Edges created against a feature are filleted.
- *Free Edges:* Only the edges formed by the faces of the features in the source selection set are filleted.

Face Rules

- *All Edges:* Fillets are created on all the edges that touch the selected face(s).
- *Against Features:* Edges of a feature created against the selected face(s) are filleted.
- *Incident Edges:* Edges that touch the selected face(s) and are parallel to a selected axis are filleted.

NOTE **The Rule Fillet only applies fillets to geometry that exists above it in the browser.**

1. Before creating a Rule Fillet ensure the geometry that you want to fillet exists.
2. Click the Rule Fillet command from the 3D Model tab > Plastic Part panel, as shown in Figure 7.107 on the left. The Rule Fillet dialog box will appear, as shown in Figure 7.107 on the right.

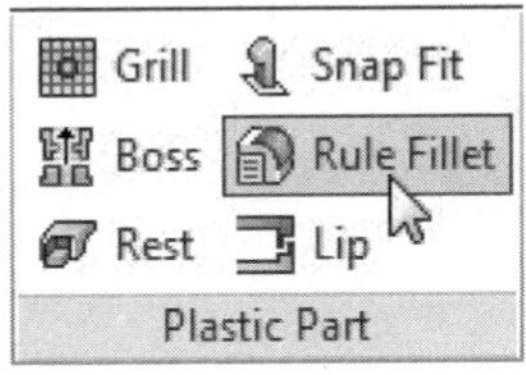

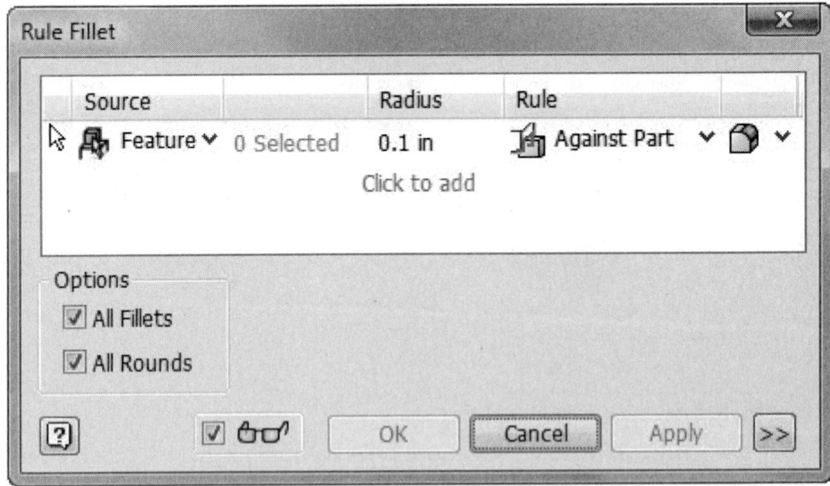

FIGURE 7.107

3. Select the source and apply a rule.
4. Create as many rules as needed.
5. Click OK to create the rule fillets. Figure 7.108 shows an example of edges that were filleted that were incident to the top face.

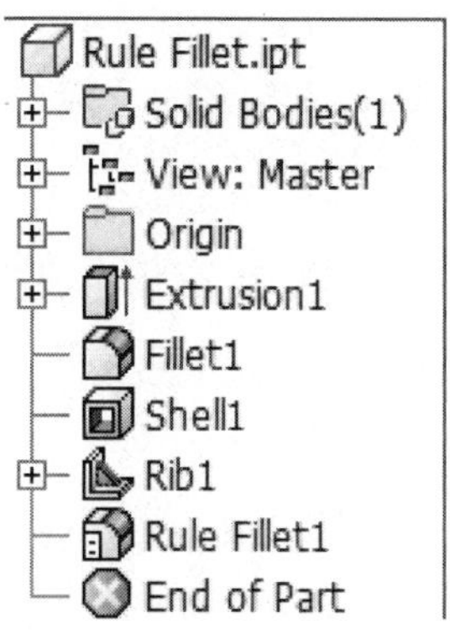

FIGURE 7.108

EXERCISE 7-7: RULE FILLETS

In this exercise you create numerous fillets on a grid using two rule fillets.

1. Open *ESS_E07_07.ipt* in the Chapter 07 folder.
2. Click the Rule Fillet command from the 3D Model tab > Plastic Part panel.
3. In the Rule Fillet dialog box change the Source to Face labeled (1) in Figure 7.109.
4. Select the bottom face of the ribs by moving the cursor over one of the vertical faces on the rib feature and use the Select Other functionality to select the bottom face labeled (2) in Figure 7.109.

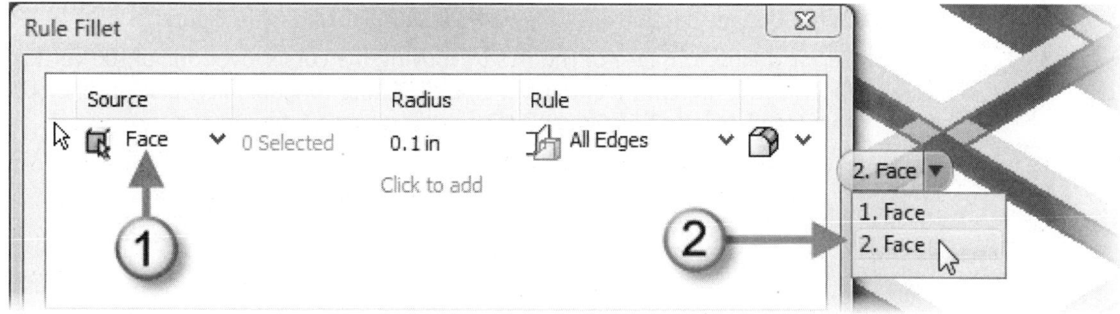

FIGURE 7.109

5. Change the Rule to Incident Edges, labeled (1) in Figure 7.110 on the left.
6. For the Incident Edges Select the top face of the rib feature, labeled (2) in Figure 7.110 on the left. The incident edges that are 90 degrees from the selected face and going in the direction of the arrow labeled (3) in Figure 7.110 on the right are selected.

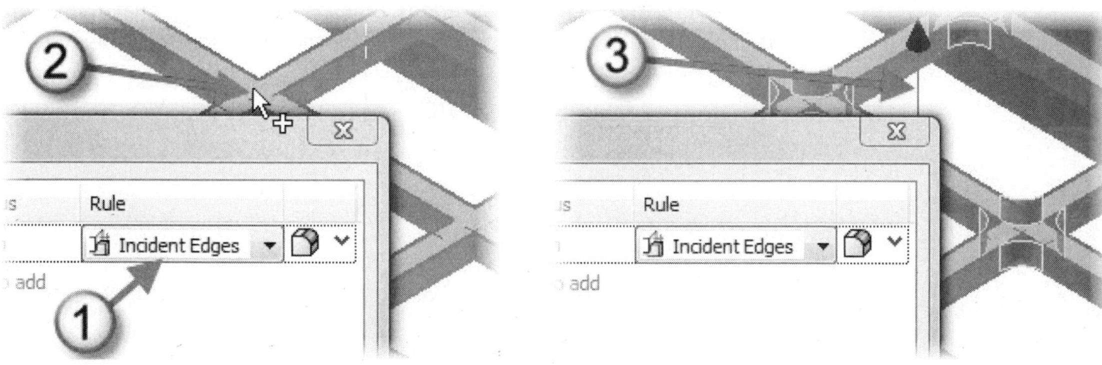

FIGURE 7.110

7. Click Apply to create the rule fillet.
8. In the Rule Fillet dialog box if needed change the Source to Feature labeled (1) in Figure 7.111 on the left.
9. Select the rib feature labeled (2) in Figure 7.111 on the left.
10. If needed change the Rule to Against Part labeled (3) in Figure 7.111 on the right.
11. Change the radius of the Fillet to **0.03** labeled (4) in Figure 7.111 on the right.

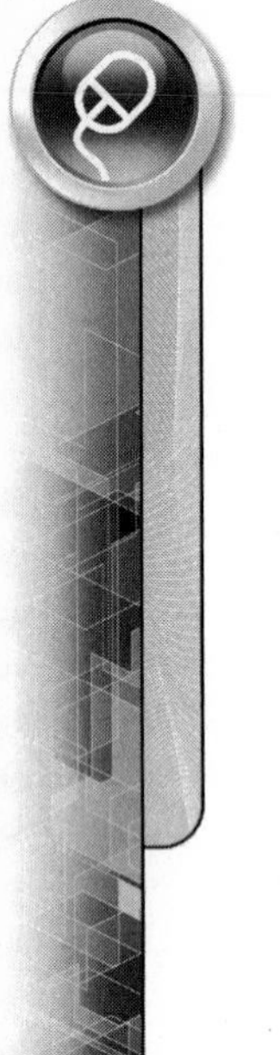

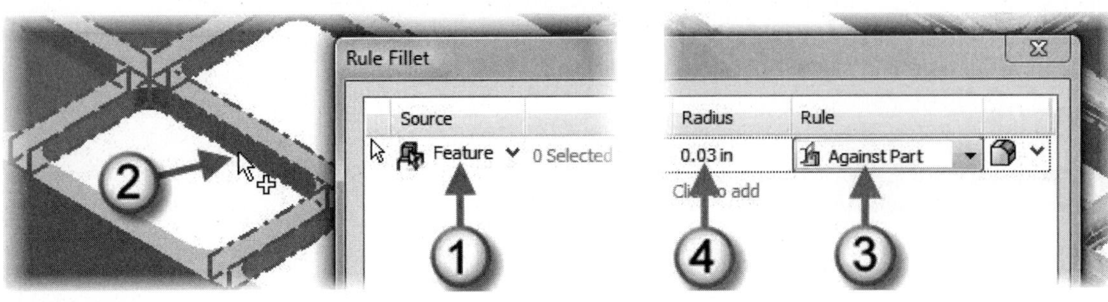

FIGURE 7.111

12. Notice how the top and bottom edges of the rib feature are showing that fillets will be created on them. In this step you remove the edges on the bottom of the face from being filleted.
13. Click the >> button on the lower-right corner of the dialog box to see the additions options on the bottom of the dialog box.
14. In the Exclude section click the Faces button labeled (1) in Figure 7.112,
15. Select the bottom face of the ribs by moving the cursor over one of the vertical faces of the rib feature and use the Select Other functionality to select the bottom face labeled (2) in Figure 7.112.

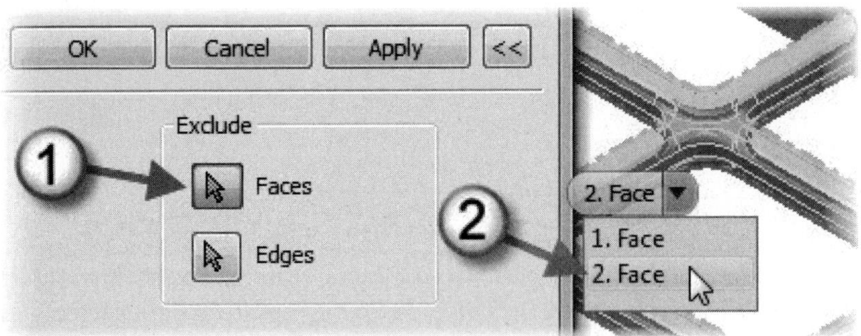

FIGURE 7.112

16. Next you exclude the four outside edges that also previewing that fillets will be created on them. In the Exclude section click the Edges button labeled (1) in Figure 7.113.
17. Select the four outside edges of the part labeled (2), (3), (4), and (5) in Figure 7.113.

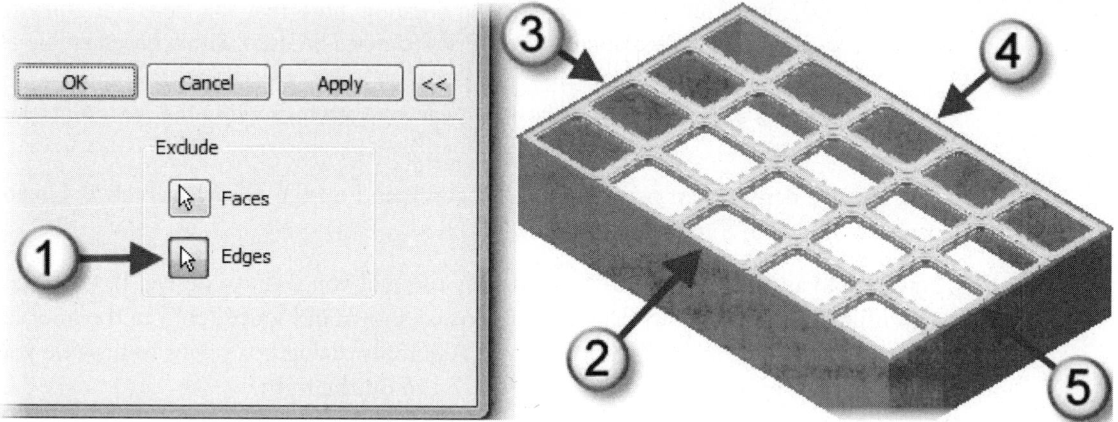

FIGURE 7.113

18. Click OK to create the rule fillet. When done the part should resemble Figure 7.114.

FIGURE 7.114

19. Delete the existing rule fillets and practice creating rule fillets using different rule.
20. Close the file. Do not save changes. End of exercise.

DERIVED PARTS

Derived parts are used to capture the design intent of a part, assembly, specific sketch, work geometry, surface, parameter, or iMate by either creating a new part file or by importing any of the aforementioned selections into a part. A derived part is a part file that links in a selected part, assembly, or sketch from another file. The derived operation is also used in the skeletal modeling technique. Additional features can then be added to the derived part or solid. If the original part changes, the derived part will update to include any changes that are made to the original file. This associativity can be broken if you do not want changes made to the original file to be included in the linked part file. You can derive in as many parts or assemblies into a part as required.

Some possible uses for a derived part or assembly are as follows:

- Use existing parts as solid bodies.
- Create different parts that are machined from a single casting blank.

- Scale or mirror a derived part upon creation: note that this does not apply to derived assemblies. You can derive sketches to be used within an assembly as a layout.
- Derive a part as a work surface that can be used to simplify the representation of a part.

You create a derived part using the Derive command on the 3D Model tab > Create panel, as shown in Figure 7.115 on the left.

Navigate to and select the Part or Assembly file that you want to derive. If you select a part file, the Derived Part dialog box opens, as shown in Figure 7.115 in the middle. If you select an assembly file, the Derived Assembly dialog box opens to provide you with different options, as shown in Figure 7.115 on the right.

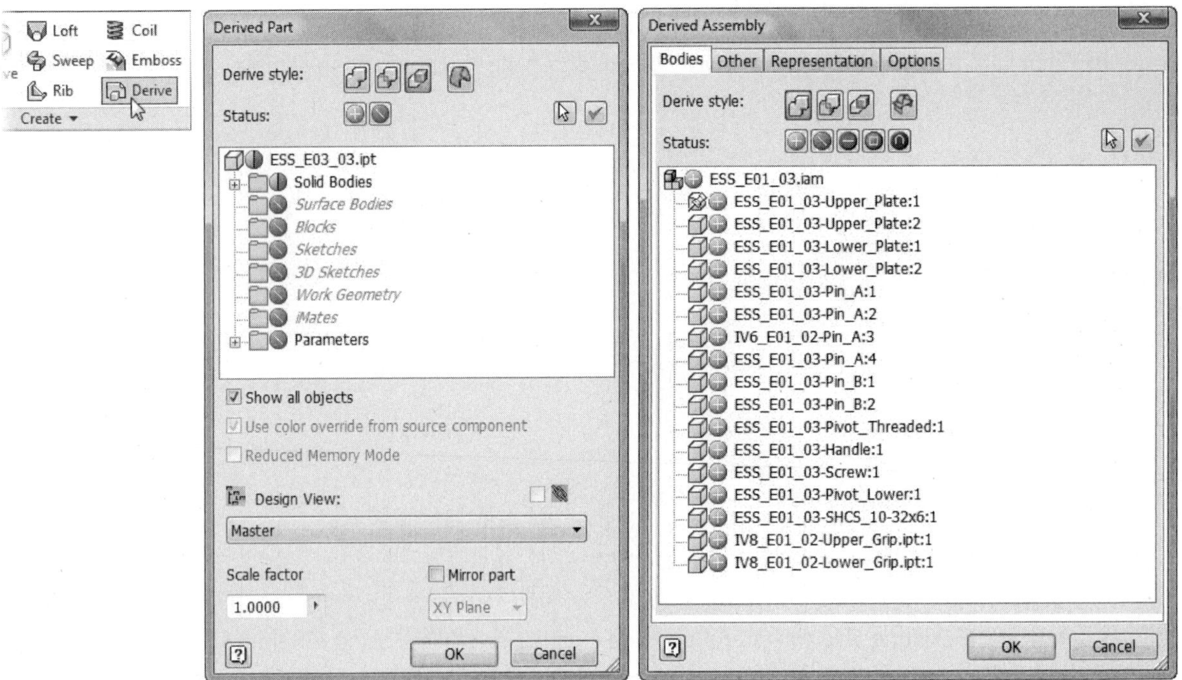

FIGURE 7.115

Derived Part Symbols

The symbols show you how the solid bodies will be handled, the status of geometry that will be contained in the derived part, and whether or not it will be included or excluded from the derived part. The Derived Part Symbols table shows the available symbols.

DERIVED PART SYMBOLS

Symbol	Function	Meaning
	Merge Seams	Single solid body merging out seams between planar faces.
	Retain Seams	Creates a single solid body part that retains the seams between planar faces.
	Solid Bodies	If the source contains a single body, it creates a single body part. If the source contains multiple visible solid bodies, select the required bodies to create a multi-body part. This is the default option.
	Create Surface	Creates a part with the selected bodies as base surfaces.
	Include	Indicates that the selected geometry will be included in the derived part.
	Exclude	Indicates that the selected geometry will not be included in the derived part. If a change is made to this type of geometry in the parent part, it will not be incorporated or updated in the derived part.
	Mixed	Indicates that the folder contains mixed included and excluded objects.
	Select	Switches focus to the base part window so that you can take advantage of the selection commands and keep the Derived Part dialog open.
	Accept	Causes the Derived Part dialog to absorb the selections you made in the base part window, returns you to the Part environment, and highlights your selections in the dialog tree control.

Derived Part Dialog Box

The Derived Part dialog box contains the following options, as shown in the following chart.

Solid Body	Select this option to derive the part as a base solid.
Surface Bodies	Select this option to derive the part body as a work surface. The body is brought in, behaves, and appears as an Autodesk Inventor surface.
Sketches	Select this option to include any unconsumed 2D sketches from the original part.
3D Sketches	Select this option to include any unconsumed 3D sketches from the original part.
Work Geometry	Select this option to include any work features from the original file. They can then be used to create new geometry or to constrain a part in an assembly.
iMates	Select this option to include any iMates that exist in the original part.
Parameters	Select this option to include any parameters designated to be exported parameters in the original file.
Composite Features	Select this option to include any composite features that exist in the original part.

NOTE

If the original file consists only of surfaces, the surface option will be the only available option.

Scale Factor	Select a scale factor in percentage to scale the derived part. The default scale factor is 1.0 or the same size as the original file.
Mirror Part	Select this option to mirror the original part about the XY, XZ, or YZ origin work planes upon derived part creation. You can select the plane about which to mirror the derived part from the drop-down list.

Derived Assembly Symbols

The symbols available when deriving an assembly are different from the symbols available when deriving a part file. The following table shows the derived assembly symbols.

DERIVED ASSEMBLY SYMBOLS

Symbols	Function	Meaning
	Merge Seams	Single solid body merging out seams between planar faces.
	Retain Seams	Creates a single solid body part which retains the seams between planar faces.
	Solid Bodies	If the source contains a single body, it creates a single body part. If the source contains multiple visible solid bodies, select the required bodies to create a multi-body part. This is the default option.
	Composite Surface	Creates a part with the selected bodies as a single composite surface.
	Include	Indicates that the selected component will be included in the derived part.
	Exclude	Indicates that the selected component will not be included in the derived part. Changes to this type of component in the parent assembly will not be incorporated or updated in the derived part.
	Subtract	Indicates that the selected component will be subtracted from the derived part. If the subtracted component intersects another portion of an included part, the result will be a void or cavity in the derived part.
	Bounding Box	Indicates that the selected component will be represented as a bounding box. The bounding box size is determined by the extents of the component. The bounding box is used to represent a component as a placeholder and reduces memory. You can add features to a bounding box, and it will update when changes are made to the original part.

(Continued)

Symbols	Function	Meaning
	Intersect	Intersects the selected component with the derived part. One component must have an Include status. If the component does not intersect the derived part, the result is not a solid.
	Select	Switches focus to the base assembly window so that you can take advantage of the selection commands and keep the Derived Assembly dialog open.
	Accept	Causes the Derived Assembly dialog to absorb the selections you made in the assembly environment, returns to the Part environment, and highlights your selections in the dialog tree control.

Scale Factor	Select a scale factor in percentage to scale the derived part. The default scale factor is 1.0 or the same size as the original file.
Mirror Part	Select this option to mirror the original part about the XY, XZ, or YZ origin work planes upon derived part creation. You can select the plane about which to mirror the derived part from the drop-down list.
Reduced Memory Mode	When checked on the Options Tab, Autodesk Inventor uses less memory by excluding bodies of the parts that are cached in memory. When the link is broken or suppressed the memory savings is removed.

The Other tab of the Derived Assembly dialog box contains options that allow you to select sketches, work geometry, surfaces, and exported parameters from the assembly file or any of the components that are in the assembly. Using the plus and minus icons, you can select which items you want included in the file.

The Representation tab of the Derived Assembly dialog box contains options that allow you to select available design view representation, positional representation, or level of detail representations present in the assembly.

The Options tab of the Derived Assembly dialog box are used to simplify, scale, or mirror an assembly. These options are also available in the Shrinkwrap command that will be covered in the next section.

Because you are using geometry or parts based on another file, any modifications to the original parts are incorporated into the derived component. If you modify a parent file, the update icon next to the derived component in the browser displays an update symbol, as shown in Figure 7.116 on the left. The Local Update command on the Quick Access toolbar also becomes active, as shown in Figure 7.116 in the middle. To update the file, click the Local Update command on the Quick Access toolbar and the update symbol for the derived part/assembly in the browser will change to what is shown in Figure 7.116 on the right.

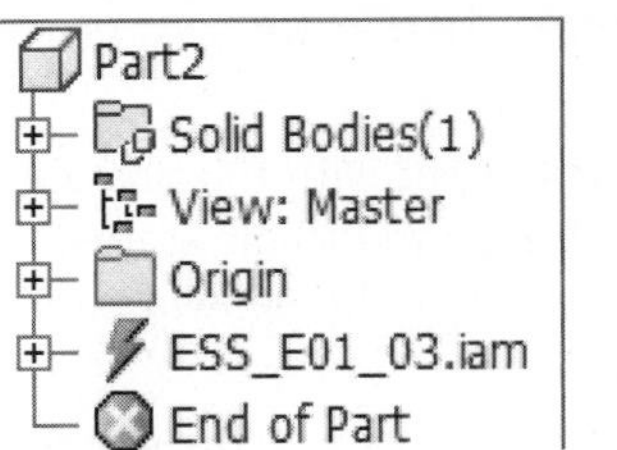

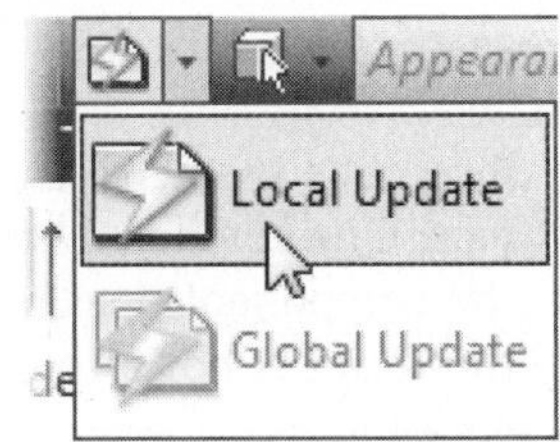

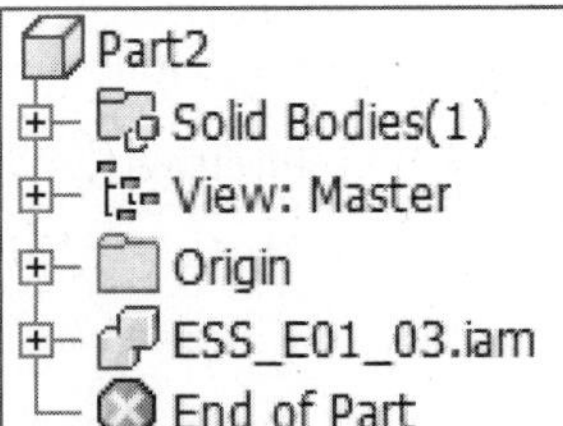

FIGURE 7.116

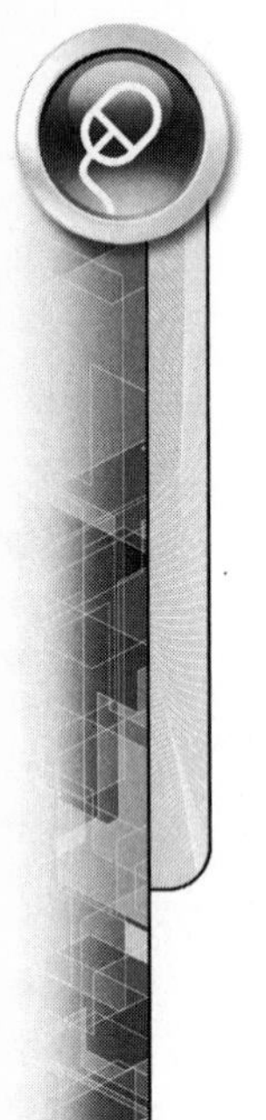

You can edit, break, or suppress the link to the parent file.

To edit the derived part, right-click on the derived parent component in the browser and select either Open Base Component, Edit Derived Assembly or Edit Derived Part from the menu and the same Derived Part dialog box will appear as the derived part was created with.

To break the link with the derived part and its original part, right-click on the derived component in the browser and from the menu click Break Link With Base Component, as shown in Figure 7.117 on the left. Once the link is broken, the derived component icon will display a broken chain link to signify that the link to the parent file no longer exists. Once you have broken the link, you cannot reestablish it.

You can temporarily suppress the changes from the base assembly from affecting the derived part by clicking Suppress Link With Base Component from the menu. To reestablish the link, click Unsuppress Link With Base Component, as shown in Figure 7.117 on the right.

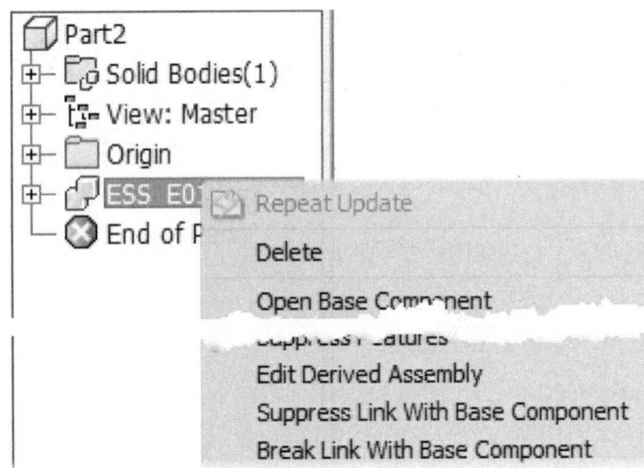

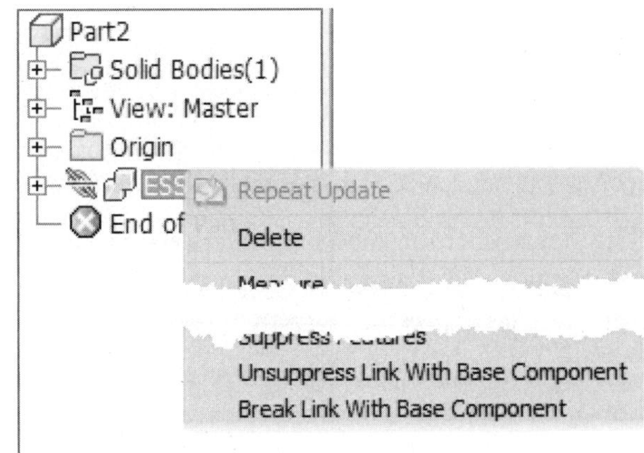

FIGURE 7.117

EXERCISE 7-8: CREATING A DERIVED PART

In this exercise, you create a mold cavity by deriving an assembly consisting of a part that you'll create a mold of and a mold base from which the part cavity will be removed. Assembly constraints have already been applied to the parts in the assembly to center the parts. You then edit the part and update the derived part.

1. Click the New command, click the English folder, and double-click *Standard (in).ipt*.
2. Click the Derive command from the 3D Model tab > Create panel.

3. In the Open dialog box, click the file *ESS_E07_08_Assembly.iam* from the Chapter 07 folder, and then click the Open button in the dialog box.
4. In the Derived Assembly dialog box, click the icon next to the entry *ESS_E07_08-Arm:1* so the – subtract symbol appears, labeled (1) in Figure 7.118 on the left.
5. Click OK to create the mold cavity as shown in Figure 7.118 on the right.

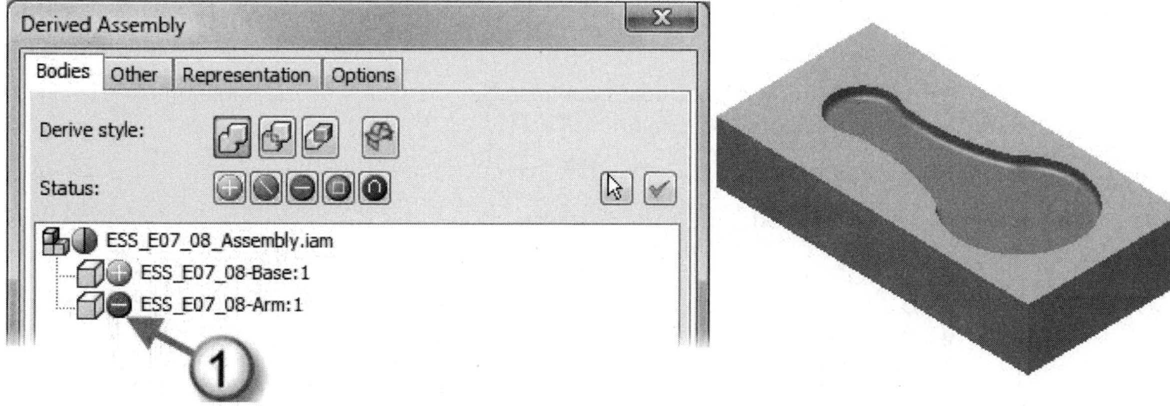

FIGURE 7.118

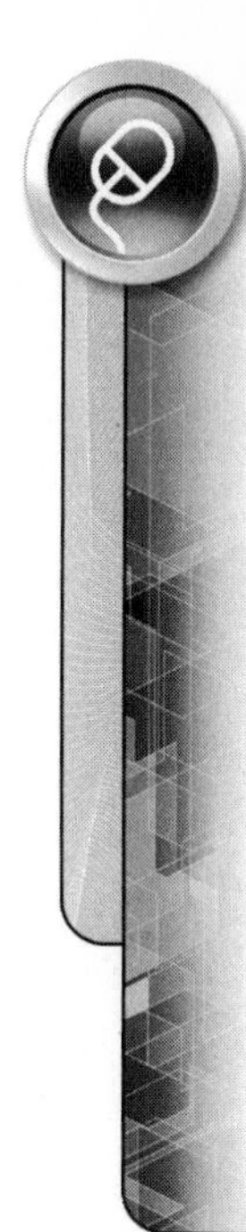

6. Next you create a hole for the material to flow into the cavity. Create a new sketch on the front-left vertical face of the mold base.
7. Place a Point, Center Point on the midpoint of the top edge and then finish the sketch.
8. Start the Hole command, and follow these steps:
   - The point, center point should automatically be selected. If needed change the diameter to **.25 inches**.
   - Change the operation to Distance.
   - Change the distance to 1.5 as shown in Figure 7.119.
   - Click OK to create the hole.

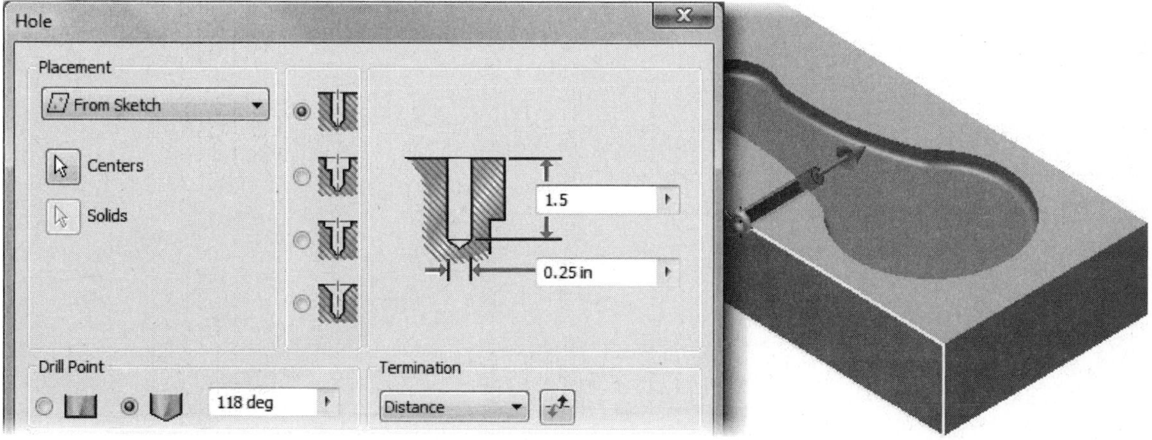

FIGURE 7.119

9. Save the file as *ESS_E07_08_MoldCavity.ipt* in the Chapter 07 folder.
10. Move the cursor in the browser over the entry *ESS_E07_08_Assembly.iam* and click Open Base Component from the menu. Notice that the hole in the derived part does not exist in the original part.

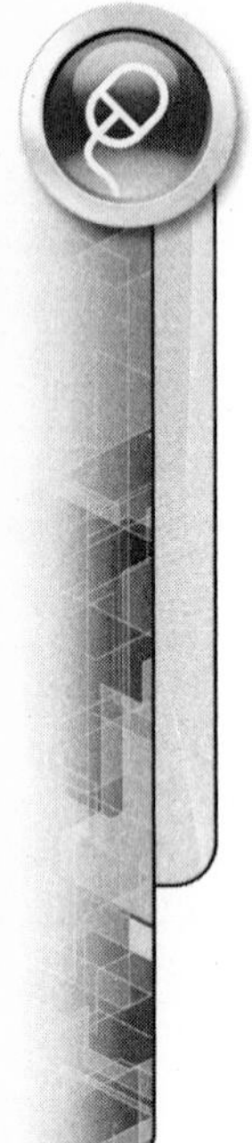

11. In the browser, double-click on *ESS_E07_08-Arm:1* and edit Extrusion1 and Extrusion2 to the distance of **1 inch**.
12. Click the Return command on the 3D Model tab > Return panel and save the assembly and changes to the part files.
13. Close the assembly file and verify that the file *ESS_E07_08_MoldCavity.ipt* is the current file.
14. In the browser, notice the red lightning bolt next to the entry *ESS_E07_08_Assembly.iam*.
15. Click the Local Update command in the Quick Access toolbar, and the cavity should resemble Figure 7.120.

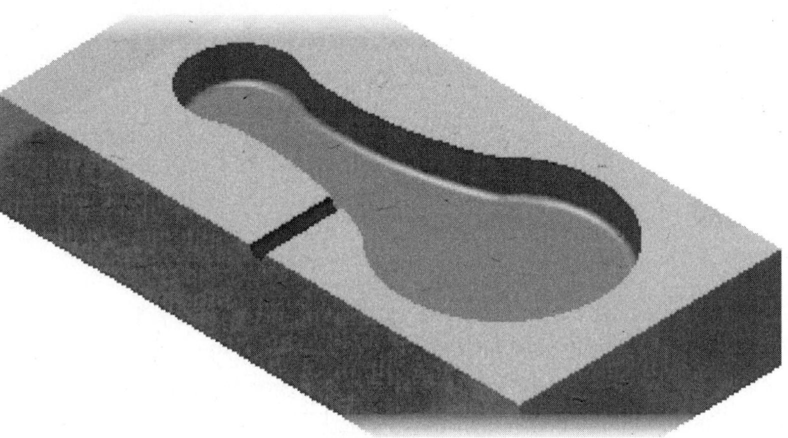

FIGURE 7.120

16. To change the mold cavity to the male portion, move the cursor in the browser over the entry *ESS_E07_08_Assembly.iam* Click Edit Derived Assembly from the menu.
17. In the Derived Assembly dialog box, click the icon next to the entry *ESS_E07_08-Arm:1* so the plus symbol appears, as shown in Figure 7.121 on the left.
18. Click the OK button to update the part. Your part should resemble Figure 7.121 on the right.

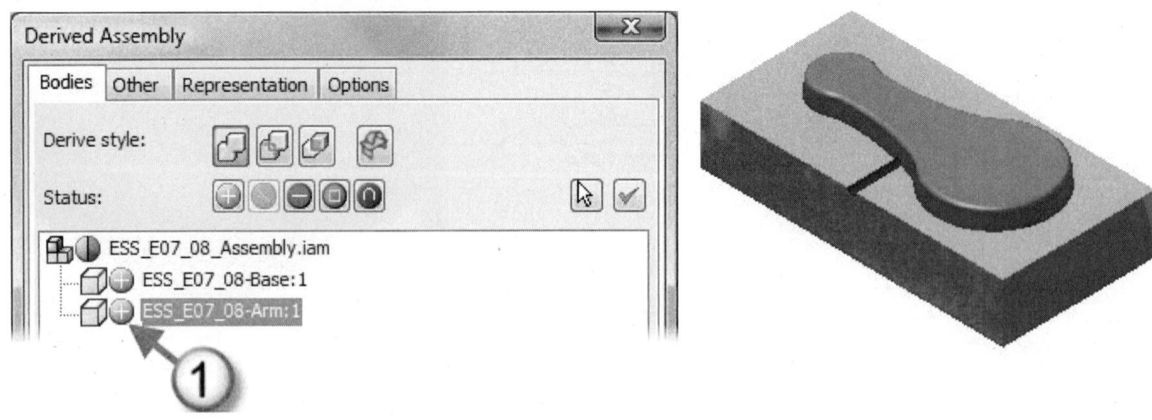

FIGURE 7.121

19. Close the file. Do not save changes. End of exercise.

SHRINKWRAP

Another method to push a derived part is from the Shrinkwrap command. The push technique creates a derived relation by exporting a part and maintains a relationship to this part. The Shrinkwrap command creates a stand-alone, single-part version of a model assembly and simplifies the assembly by removing geometry. Since the part is derived, any changes to the original assembly can be updated in the derived part. The Shrinkwrap command enables you to:

- Reduce detail of data to protect intellectual property and reduce file size.
- Create a substitute part for use with alternative representations.
- Create simplified data for complex purchased assemblies.

To create a shrinkwrap part, follow these steps:

1. Open an assembly that you want to create a simplified part from.
2. Click the Shrinkwrap command from the Assembly tab > Component panel, as shown in Figure 7.122 on the left. You can also use the Shrinkwrap command to create a substitute level of detail, this is covered in Chapter 9.
3. Enter a name, select a template file, and specify a location for the derived part, as shown in Figure 7.122 on the right.

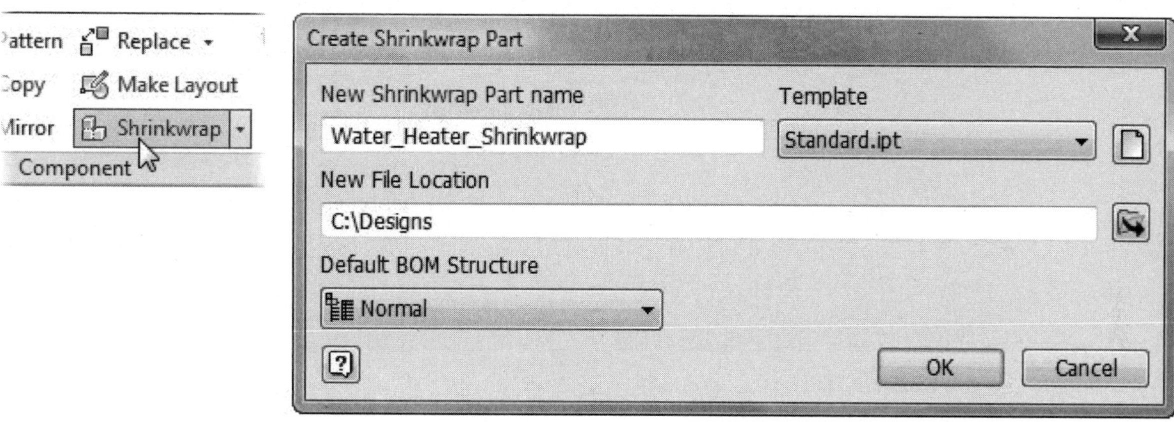

FIGURE 7.122

4. Click OK and the Assembly Shrinkwrap Options dialog box will appear, as shown in Figure 7.123.
5. Select the different options to remove the geometry. Click the Preview button to see the results of the existing options.
6. When done, click OK.

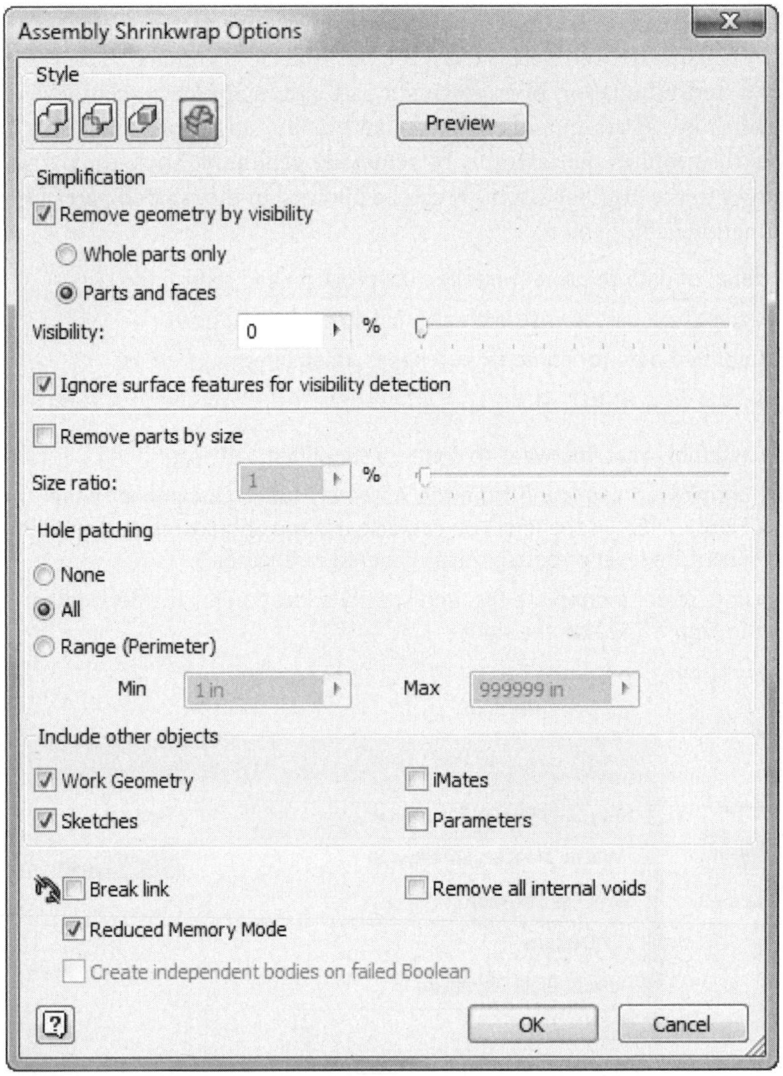

FIGURE 7.123

The options in the dialog box are explained below.

Style

Single solid body merging out seams between planar faces: Select to produce a single solid body without seams between planar faces. When you merge seams between faces, the face assumes a single color.

Solid body keep seams between planar faces: Select to produce a single solid body with seams between planar faces retained.

ASSEMBLY SHRINKWRAP OPTIONS

Symbols	Function	Meaning
	Single Solid	Creates a single solid body merging out seams between planar faces.
	Single Solid with Seams	Creates a single solid body part which retains the seams between planar faces.

(*Continued*)

Symbols	Function	Meaning
	Maintain each solid as a solid body	Creates a multi-body part, each part is created as a unique solid body.
	Single Composite Surface	Creates a part with the selected bodies as a single composite surface.

Simplification

Whole parts only: Whole parts which meet the visibility criteria are removed.

Parts and faces: Removes any face including entire parts which meet the visibility criteria.

Visibility percentage: A value of zero removes all parts or faces that are not visible in any view. Increasing the slider value removes more parts and faces.

Ignore surface features for visibility detection: Available if Remove geometry by visibility is enabled. If enabled, surface features do not impact visibility detection.

Remove parts by size: Check to enable the option to remove parts based on the size ratio. The ratio indicates the difference between the part bounding box and the assembly bounding box.

Hole Patching

None: No holes are removed.

All: Removes all holes that do not cross surface boundaries. Holes do not need to be round to be included; a void in material is seen as a hole.

Range: Specifies the circumference or perimeter of the holes to include or exclude. Holes do not need to be round to be included.

Include other objects

Work Geometry: When checked, any visible work features in the component are exported and can be derived.

Sketches: When checked, any visible and unconsumed 2D or 3D sketches in the component are exported and can be derived.

iMates: When checked, any iMates defined in the source assembly are exported and can be derived.

Parameters: When checked, any parameters in the source assembly are exported and can be derived.

Break link

Permanently disables any updates from the source component.

Reduced Memory Mode

When checked, a part is created using less memory by excluding source bodies from the cache.

Create Independent Bodies on Failed Boolean

Available when Single Solid or the Single Sold with Seams Style is selected. When a single solid body cannot be created individual solid bodies will be created.

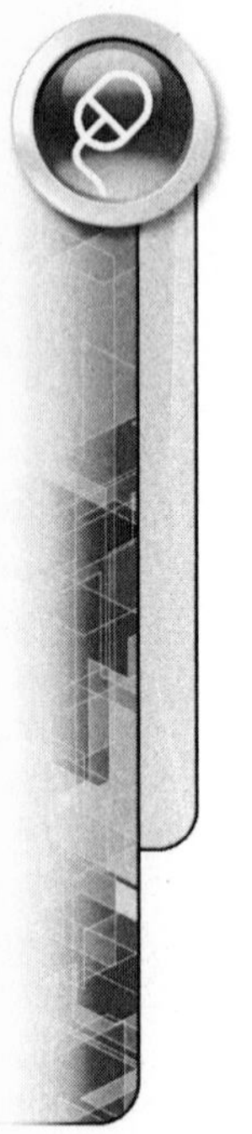

Remove Internal Voids

When checked voids that are internal to the assembly will be filled in with material.

EXERCISE 7-9: SHRINKWRAP

In this exercise, you derive a simplified part from an assembly using the Shrinkwrap command.

1. Open *ESS_E07_09.iam* from the Chapter 07 folder.
2. Use the Free Orbit command to examine the assembly.
3. Return to the Home view.
4. Start the Shrinkwrap command from the Assemble tab > Component panel.
5. Click OK to accept the default name and location.
6. Change the options, as highlighted in Figure 7.124.

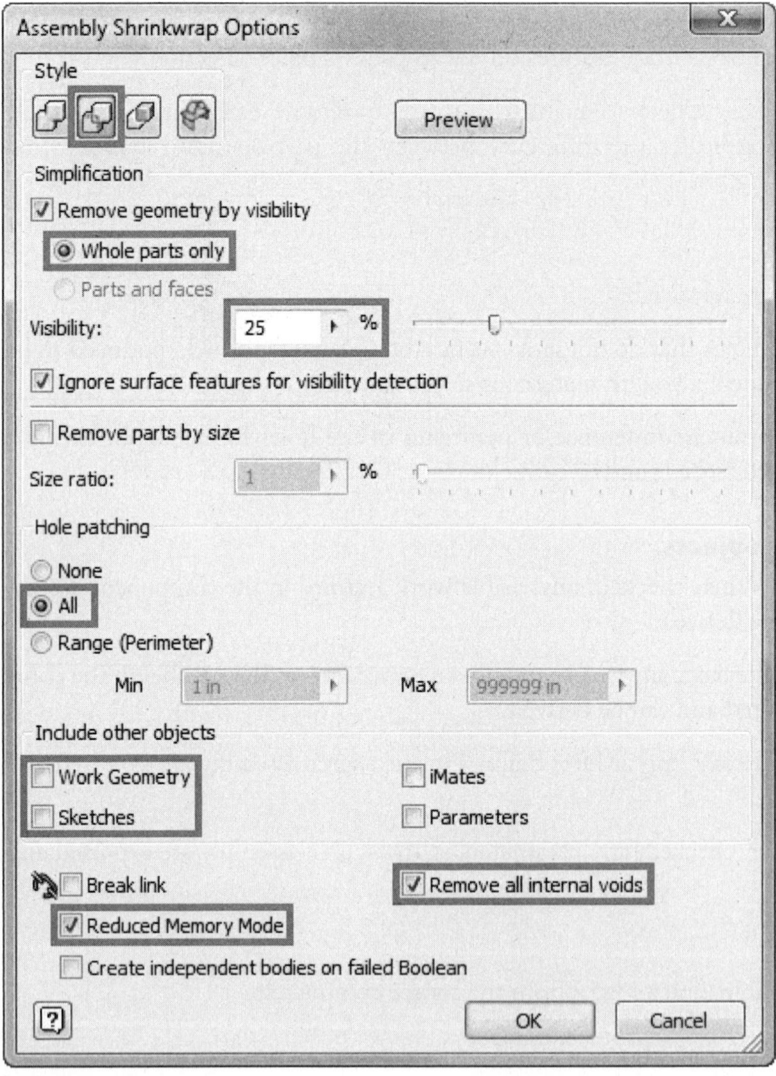

FIGURE 7.124

7. Click the Preview button to see the pending changes.
8. Try other options; click the Preview button to see the affects.
9. Set the options back to the settings in the previous dialog box.
10. Click OK to create the derived part.

11. Next you remove select components from the derived part. In the derived part, right-click in the browser on the entry *ESS_E07_09.iam* and click Edit Derived Assembly from the menu.
12. Change the Derived Status for the parts; ASME B16.18 90 Deg Drain Elbow, ASME B16.4 Reduced Elbow, and ASTM D 1785 Pipe to Excluded, as shown in Figure 7.125 on the left.
13. Click OK to complete the operation; rotate the viewpoint so your screen should resemble Figure 7.125 on the right.

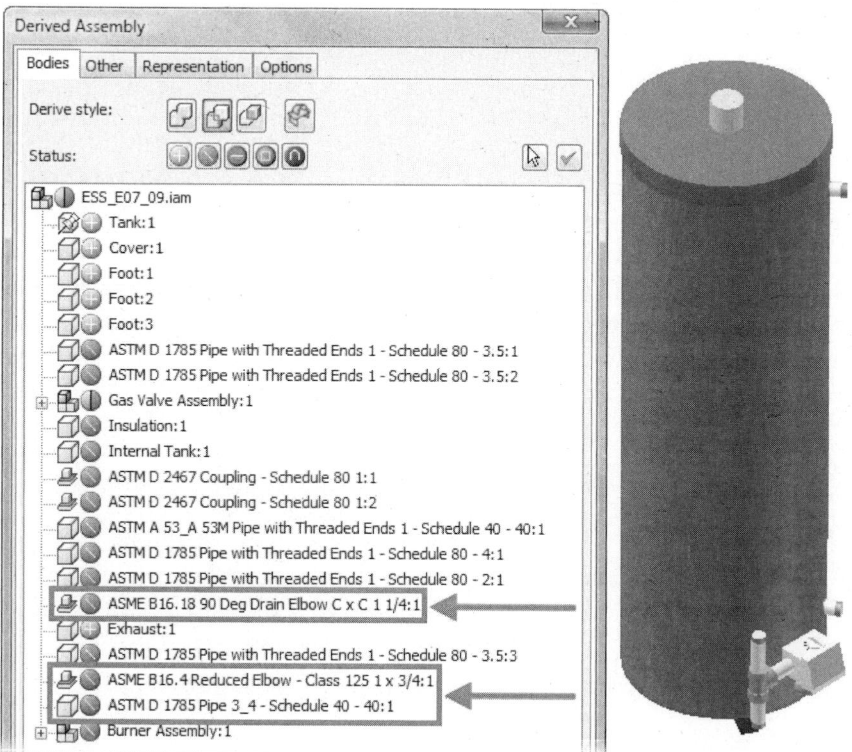

FIGURE 7.125

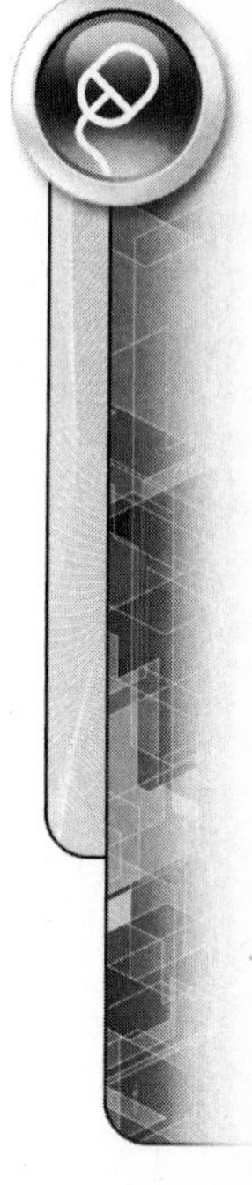

14. Start the Half Section View command from the View tab > Appearance panel and select the XZ origin plane to verify that the internal area has been filled in. When done remove the section by clicking the Exit Section View command from the Appearance panel.
15. Practice changing the Edit Derived Assembly options such as changing the Gas Valve Assembly to the Bounding Box option.
16. Close the file. Do not save changes. End of exercise.

INTRODUCTION TO STRESS ANALYSIS

Autodesk Inventor Professional has the ability to determine the stress on a part, or how the part will react when certain forces are applied. Before a simulation can be performed, a material must be assigned, constraints established, and a load applied as shown in Figure 7.126 on the left. Autodesk Inventor uses Finite Element Analysis (FEA) to create a grid across the surface, called a mesh, as shown in Figure 7.126

in the middle. The mesh divides the object into a continuous set of elements. Using properties of the assigned materials, Inventor calculates the stress in each element as shown in Figure 7.126 on the right.

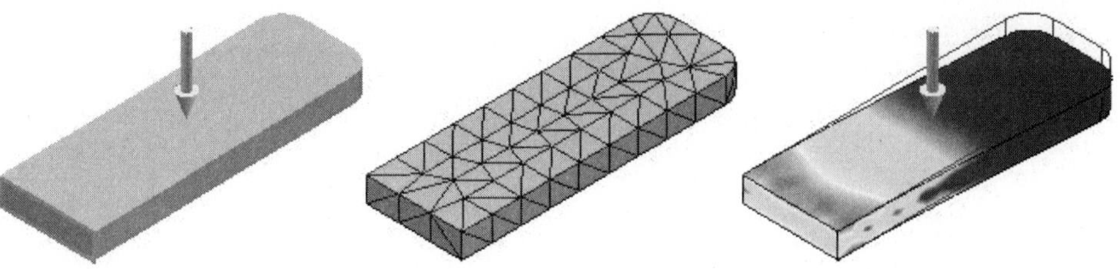

FIGURE 7.126

Create a Simulation

The first step when performing a Stress Analysis is to create a simulation. Inventor allows you to create multiple simulations on the same part or assembly. To begin a simulation, click the Stress Analysis command on the Environments tab > Begin panel, as shown in Figure 7.127 on the left and the Stress Analysis tab will be current. To start a simulation click the Create Simulation command on the Stress Analysis tab > Manage panel as shown in Figure 7.127 in the middle. The Create New Simulation dialog will appear, as shown in Figure 7.127 on the right.

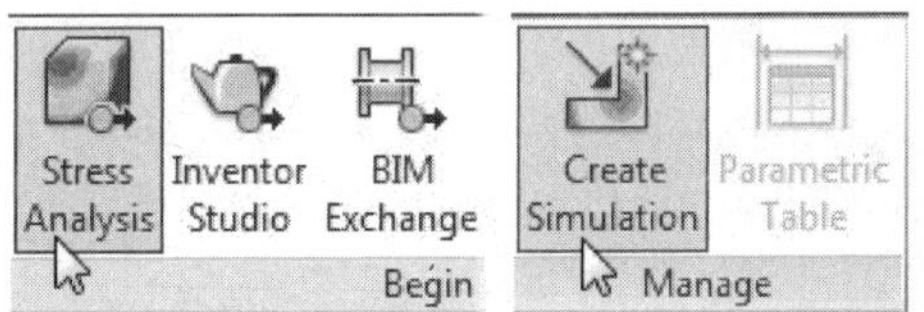

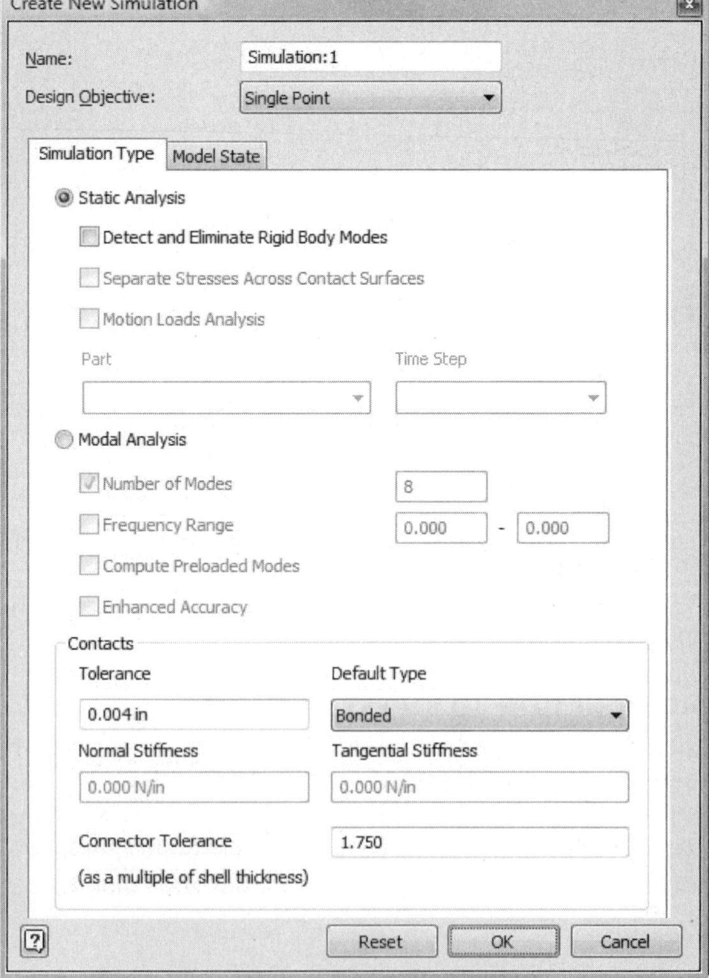

FIGURE 7.127

Before a simulation can be created, you must select the type of simulation that will be created; Static or Modal Analysis. The following description explains the two types of simulations.

Static Analysis

Static analysis performs a simulation that calculates stresses and displacement. The simulation helps you determines the simple structural loading conditions of the part without motion and the load is constant (no impact or changing of the load values).

Modal Analysis

Modal Analysis performs a simulation that calculates the dynamic properties of a model with different frequencies of vibration. Rigid body movements are also a consideration within this type of analysis.

Assign Material

Before an analysis can be performed, a material must be assigned to each part. The material can be applied at the part level or overridden in the simulation. To assign the material at the part level, click the iProperties command on the Inventor Application Menu, as shown in Figure 7.128 on the left. You may also right-click on the file's name in the browser and select iProperties from the menu. In the iProperties dialog box, click the Physical tab. Under the Material drop-down list, select the desired material, as shown in Figure 7.128.

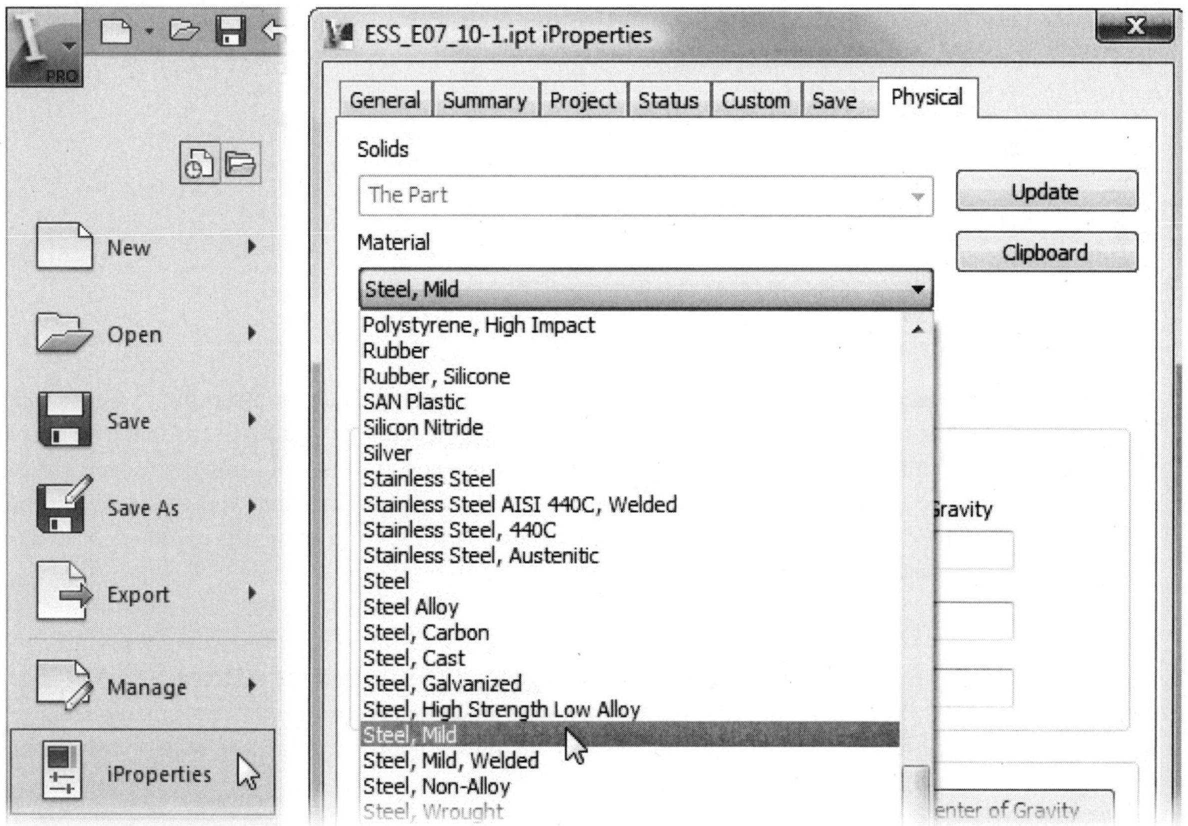

FIGURE 7.128

To assign a material while in a simulation, click the Assign command on the Stress Analysis tab > Material panel, as shown in Figure 7.129 on the left. The Assign Materials dialog will appear, as shown in Figure 7.129 on the right. From the Override

Material drop down menu, select the desired material. The material selected in the override will not change the material set in the iProperties.

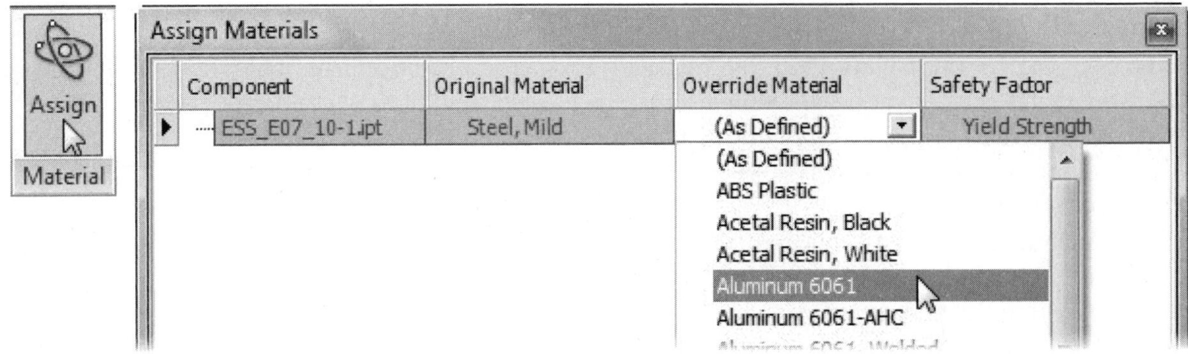

FIGURE 7.129

Applying Constraints

After the material(s) are assigned, you add constraints to the model. These constraints represent the conditions the part or assembly will experience in the real world. Inventor has three constraints within the Stress Analysis environment; fixed, pin and frictionless. These constraints are located in the Stress Analysis tab > Constraints panel, as shown in Figure 7.130 below.

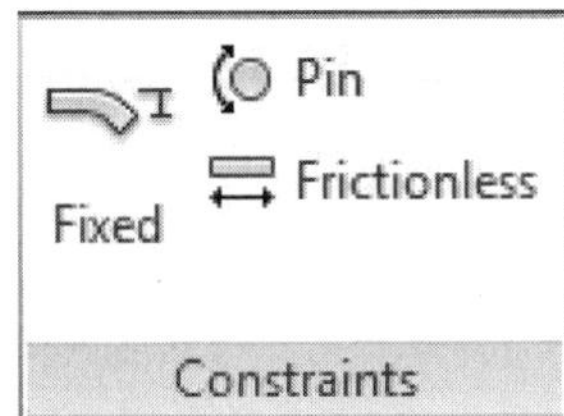

FIGURE 7.130

Fixed, Pin, and Frictionless constraints constrain the part or assembly to assure that the model reacts to an applied force in a realistic manner. Before the proper constraint can be applied, you must determine how the constraints will limit the movement of the model. You need to understand the three constraints that you can apply to the model that emulate the real world scenario that the simulation represents. A brief description of each constraint follows.

Fixed

A fixed constraint removes all the degrees of freedom on the selected face(s) or edge (s). You apply a fix constraint when no translational or rotational movements are permitted. You may apply multiple fix constraints on different faces or edges on the same model.

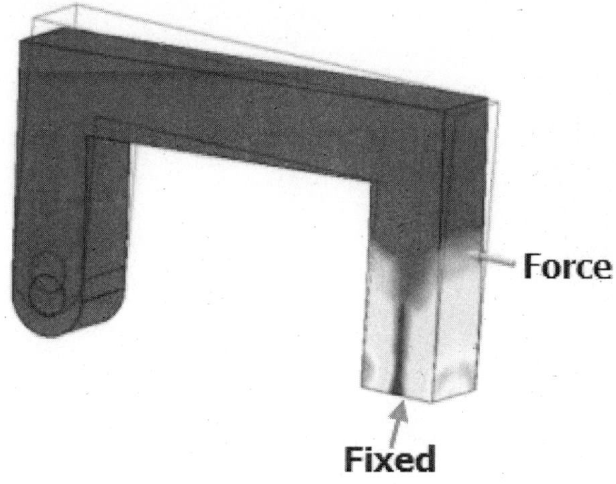

FIGURE 7.131

Pin

A Pin constraint removes translation degrees freedoms along the axes of a cylindrical face (no linear or sliding movement is allowed) but allows rotation. Apply this constraint if rotational movement is desired. A pin constraint can only be applied to a cylindrical face. A Pin constraint can work in conjunction with the other constraints.

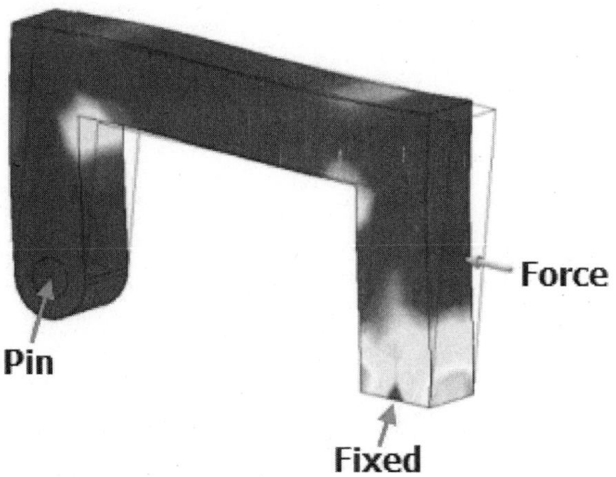

FIGURE 7.132

Frictionless

A Frictionless constraint limits a face to a rotation or translation movement (slide) along a plane. The face is prevented from moving or deforming perpendicular to the plane. A Frictionless constraint can be used in conjunction with the other constraints.

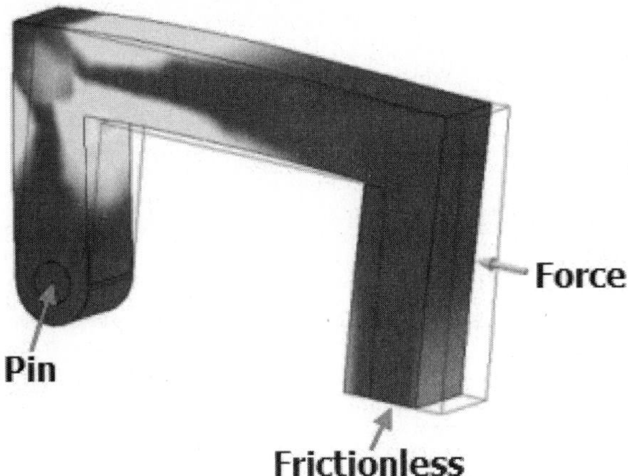

FIGURE 7.133

Applying a Load

The next step to setup a stress analysis is to apply a load to the model. Inventor has different loads that can be applied to the model; force, pressure, bearing, moment and gravity. These constraints are located on the Stress Analysis tab > Loads panel, as shown in Figure 7.134. Review the following descriptions to determine the type of load that closely resembles the desired load you need to apply.

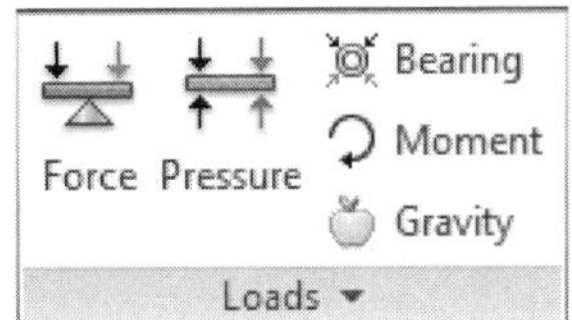

FIGURE 7.134

Force

Applies a user defined load to a face or edge. A Force load is perpendicular to the surface, and parallel to an edge.

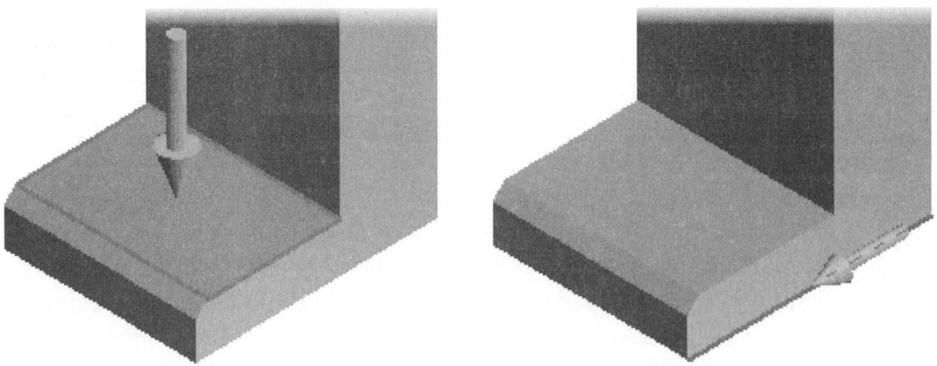

FIGURE 7.135

Pressure

Applies a user defined pressure to a face. Pressure is a uniform load applied perpendicular to the whole surface.

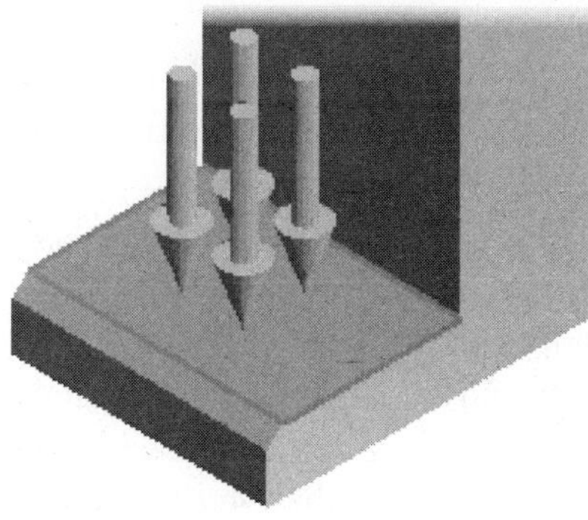

FIGURE 7.136

Bearing

Applies a user defined axial or a radial load to a cylindrical face.

FIGURE 7.137

Moment

Applies a user defined rotational load around an axis. If applied to a face, the momentary load is perpendicular to the surface.

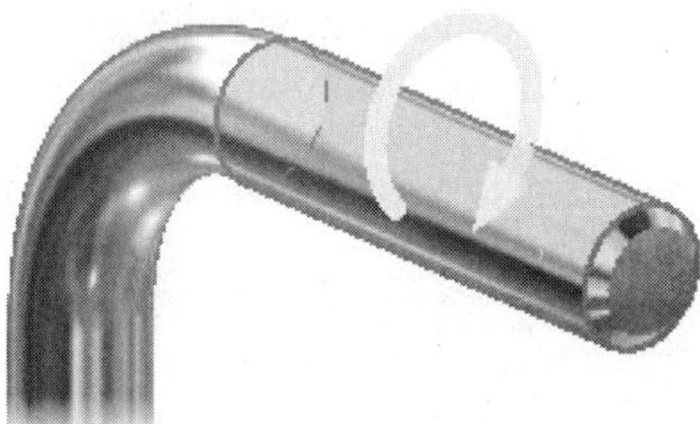

FIGURE 7.138

Gravity

Applies a user defined gravitational pull on a part. The gravitational pull is perpendicular to the surface, or parallel to an edge.

FIGURE 7.139

Mesh Options

As you gain experience performing simulations you may want to adjust the mesh to fine tune the results. There are four mesh commands that you can use from the Stress Analysis tab > Mesh panel, as shown in Figure 7.140.

- *Mesh View*: Use the Mesh View command to compute the mesh the model, if this step is not done before a simulation is performed Inventor will mesh the model automatically before doing the simulation. After the model has been meshed, click the Mesh View command to toggle the display of the mesh on and off.
- *Mesh Settings*: Use the Mesh Settings command to adjust the average element size, minimum element size, grading factor, maximum turn angle and if a curved mesh should be generated.
- *Local Mesh Control*: Use the Local Mesh Control command to set the average element size for a face or an edge.
- *Convergence Settings*: Use the Convergence Settings command to specify the maximum number of refinements, stop criteria, h refinement threshold, select the results to converge (Von Misses Stress, 1st Principal Stress, 3rd principal Stress or Displacement) and select the geometry that will be utilized.

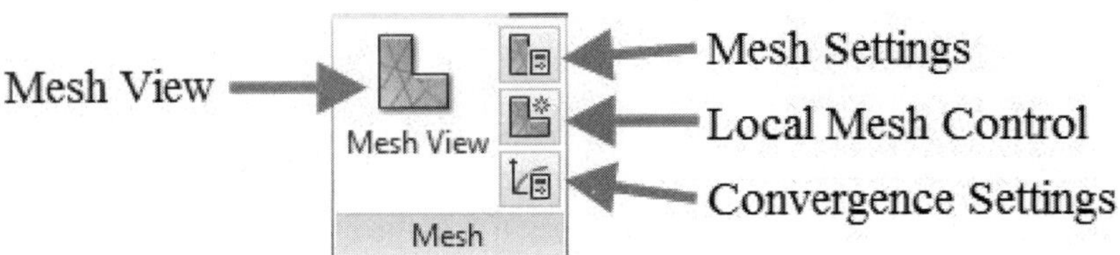

FIGURE 7.140

Simulate

After a material is assigned, loads and constraints are applied; a stress analysis can be simulated. Click the Simulate command from the Stress Analysis tab > Solve panel, as shown in Figure 7.141 on the left. The will appear, then in the Simulate dialog box click the Run button as shown in Figure 7.141 on the right.

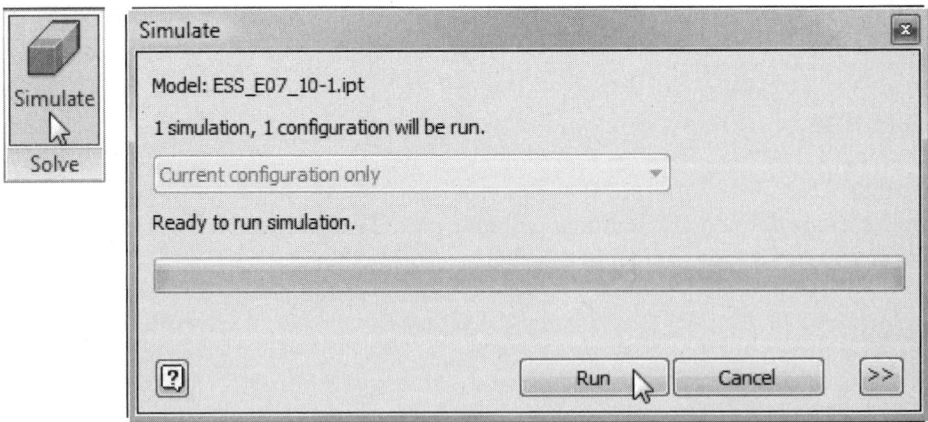

FIGURE 7.141

After a simulation is performed several results can be displayed. To view a different test, check the desired test result in the browser under the results section as seen in Figure 7.142.

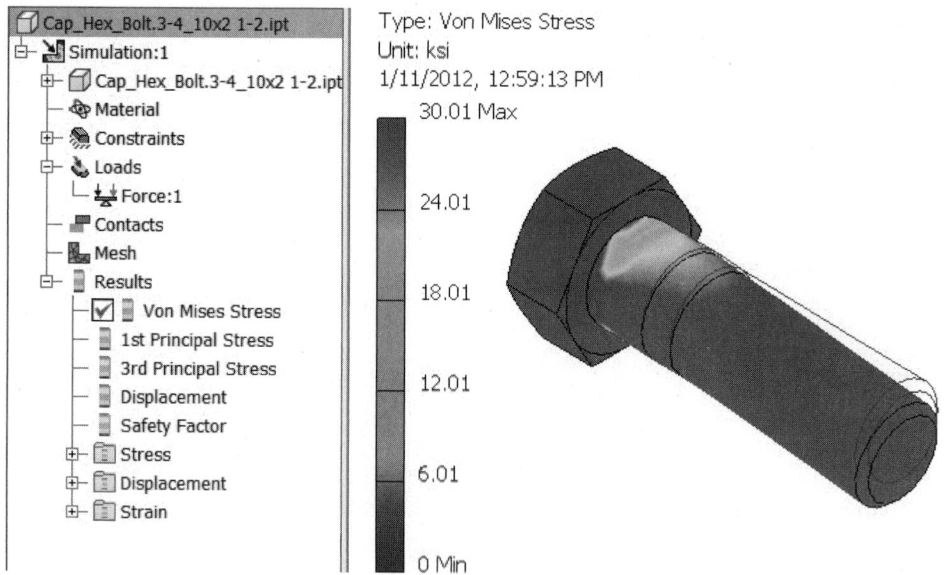

FIGURE 7.142

Display

The results of the simulation can be analyzed by comparing the colors on the part to the color bar on left side of the Graphics Window. How the simulation is displayed can be changed by selecting one of the options under the Stress Analysis tab > Display panel, as shown in Figure 7.143.

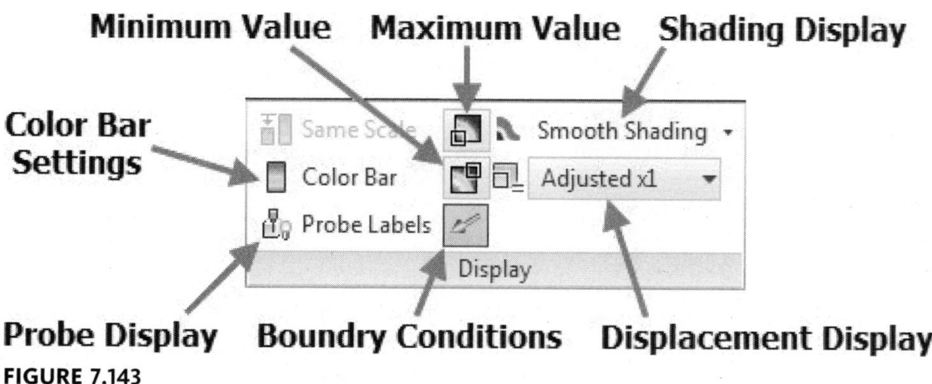

FIGURE 7.143

- *Color Bar Settings:* Opens the Color Bar Settings Menu.
- *Minimum Value:* Displays the location of the minimum value.
- *Maximum Value:* Displays the location of the maximum value.
- *Shading Display:* Allows user to change the quality of the shading.
- *Displacement Display:* Adjusts the level of model deformation.
- *Boundary Conditions:* Toggles the display of glyphs for the loads applied.
- *Probe Display:* Toggles the display of user probes.

Animate or Probe Results

Autodesk Inventor allows you to analyze the results through an animation or probe a specific point(s). These commands are located under the Stress Analysis tab > Results panel, as shown in Figure 7.144.

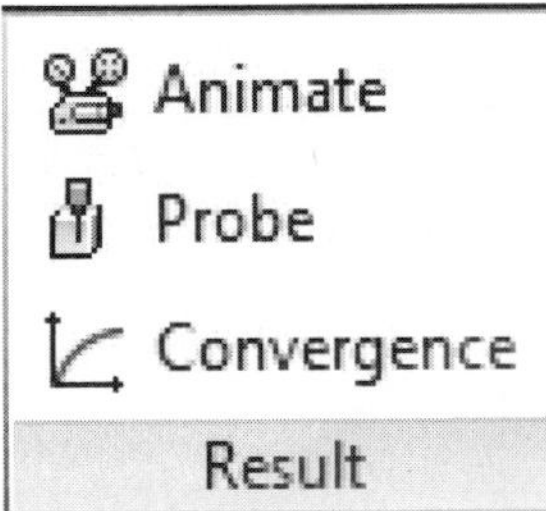

FIGURE 7.144

Animate

The Animate command plays an animation of the results. Once the Animate Results dialog opens click the play button to view animation. If desired, an animation can be recorded.

Probe

The Probe command creates a probe to measure the simulation results. The probe will measure the exact value of the simulation at the selected point. To place a probe, select the area of the model where the exact measurement is to be displayed.

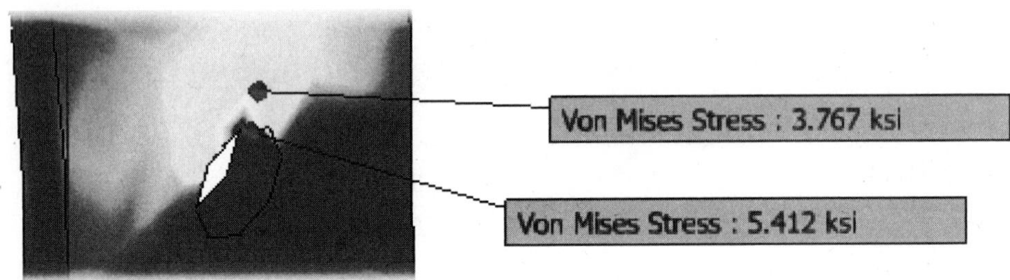

FIGURE 7.145

Report

Once the Simulation is complete a report can be generated and exported. The Report command is located under the Stress Analysis tab > Report panel, as shown in Figure 7.146 on the left. To determine what simulation results to include in the report, click the options desired on the Simulations tab in the report dialog as shown in Figure 7.146 on the right. From the Format tab select the file type to export; html, mhtml or rtf. Click the OK button to create the report.

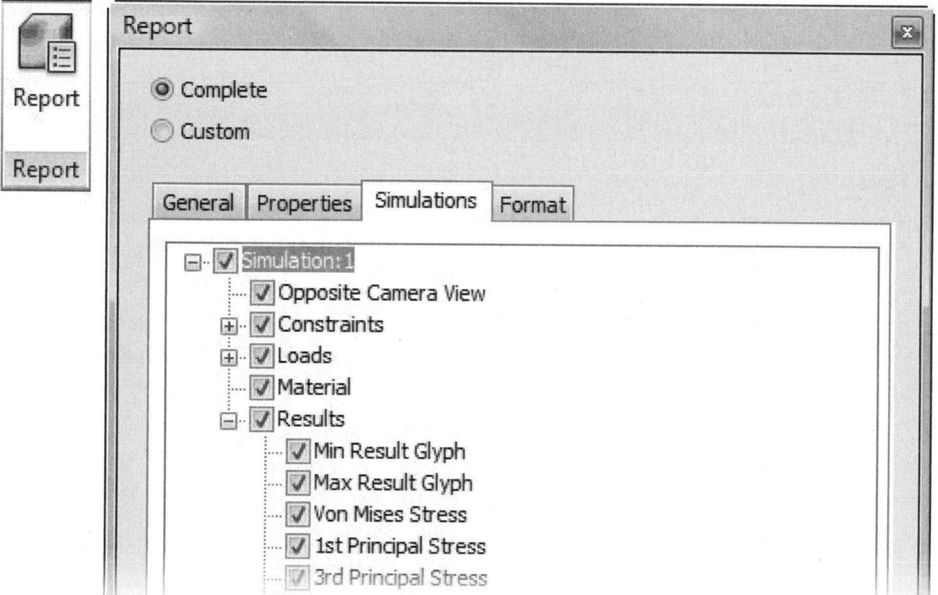

FIGURE 7.146

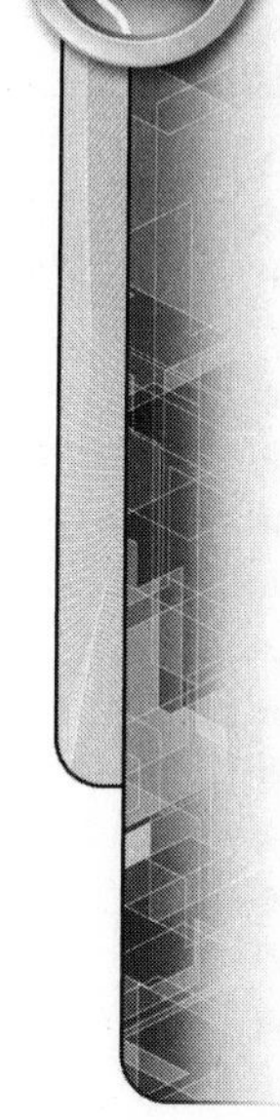

EXERCISE 7-10: RUN A STRESS ANALYSIS ON A PART

In this exercise, you will use the Stress Analysis commands to create a simulation, add constraints, loads, simulate, animate and create a report of a lifting claw that was used in the assembly constraint exercise on chapter 6.

1. Open *ESS_E07_10.ipt* in the Chapter 07 folder.
2. Click the Environments tab.
3. Begin a simulation by clicking the Stress Analysis command on the Environments tab > Begin panel as shown in Figure 7.147 on the left.
4. Create a new simulation by clicking the Create Simulation command on the Environments tab > Manage panel as shown in Figure 7.147 on the right.

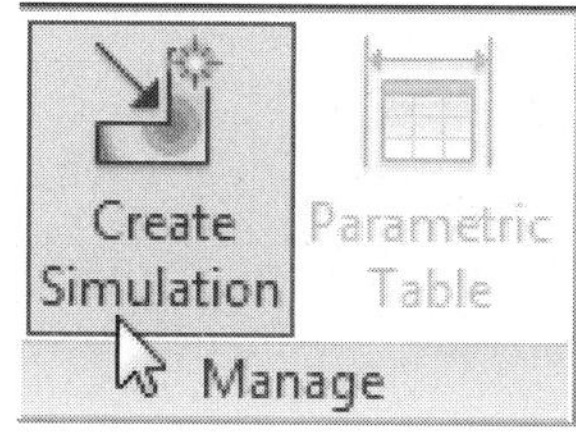

FIGURE 7.147

5. In the Create New Simulation dialog box that appears click OK to create a new static analysis.
6. Verify that a material has been applied to the part, click the Assign Materials command on the Stress Analysis tab > Material panel. Verify that the Original Material assigned to the part via the iProperties is Steel, Mild as shown in Figure 7.148 and click OK.

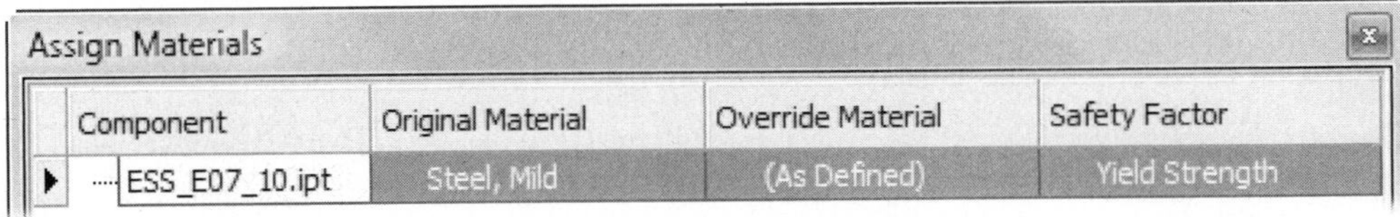

Component	Original Material	Override Material	Safety Factor
ESS_E07_10.ipt	Steel, Mild	(As Defined)	Yield Strength

FIGURE 7.148

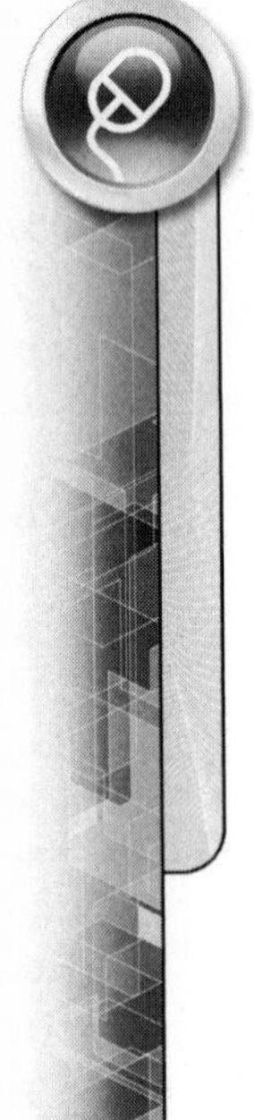

7. Next you constrain the part, click the Pin command from the Stress Analysis tab > Constraints panel.
8. Select the two circular faces of the bottom hole as shown in the following image on the left and click Apply. While still in the Pin Constraint dialog box select the two circular faces of the top hole as shown in Figure 7.149 on the right and click OK.

FIGURE 7.149

9. Next you apply a load to the part, click the Force command from the Stress Analysis tab > Loads panel.
10. Select the bottom horizontal face and change the Magnitude to **200 N** as shown in Figure 7.150. The direction of the force should be directed down. Click OK to create the force.

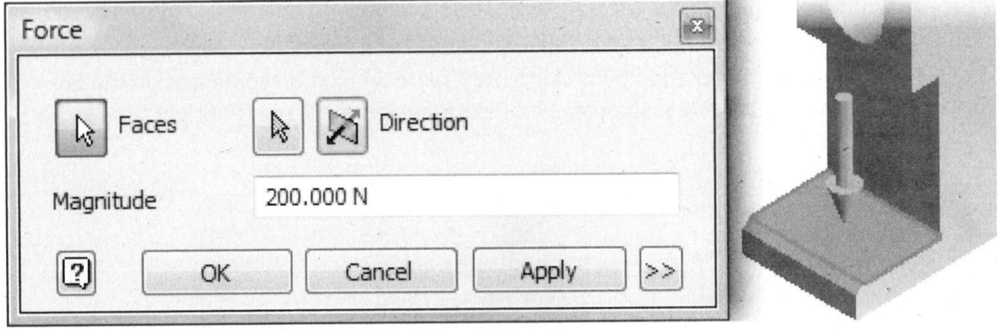

FIGURE 7.150

11. Next you run the simulation. Click the Simulate command from the Stress Analysis tab > Solve panel and then click the Run button in the Simulate dialog box.
12. The result of your simulation should resemble Figure 7.151.

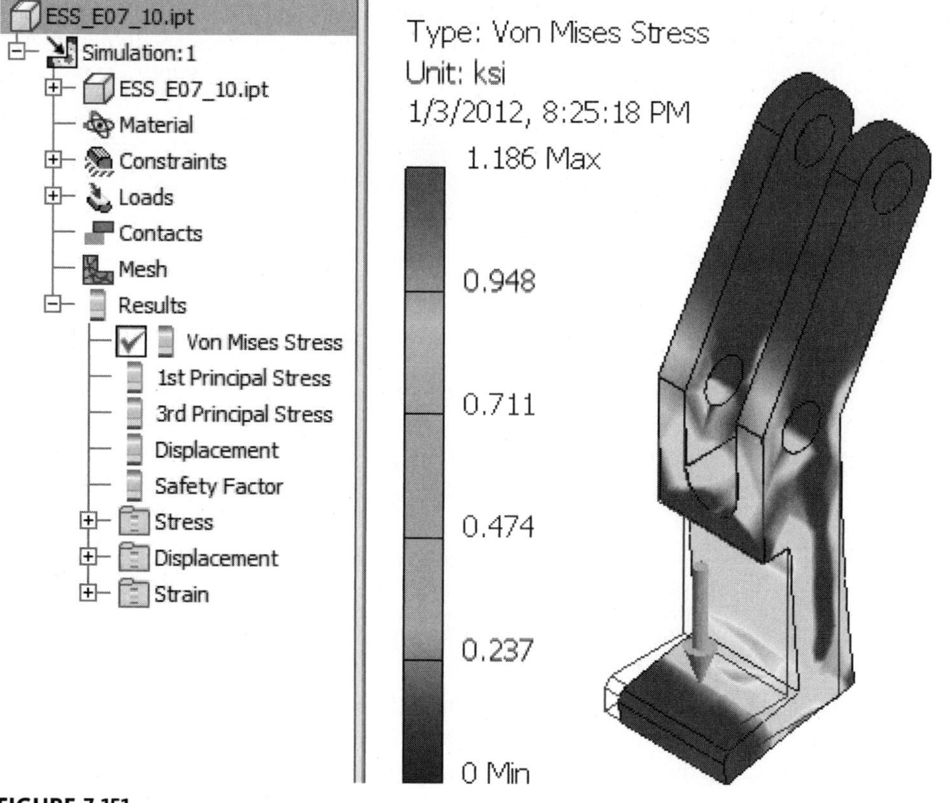

FIGURE 7.151

13. From the browser analyze the results of the simulation by selecting the different results method. When done make the Von Mises Stress results method current.
14. Next you animate the simulation results. Click the Animate command from the Stress Analysis tab > Result panel. To start the animation, click the Play button in the Animate Results dialog. You can adjust the speed by selecting different options from the drop list.
15. Click OK to finish the animation.
16. Next you copy the simulation, override the material of the part and rerun the simulation.
17. Copy the simulation by right-clicking on Simulation:1 in the browser and click Copy Simulation as shown in the following image. The new simulation, Simulation:2 will be the current.

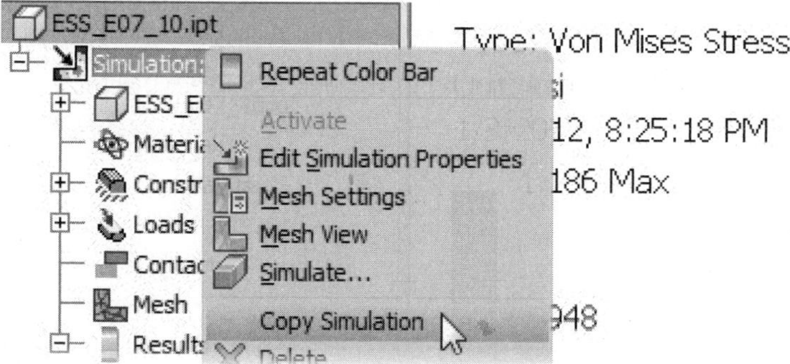

FIGURE 7.152

18. Override the material by clicking the Assign Materials command on the Stress Analysis tab > Material panel. In the Override Material cell click in the cell and from the list of materials click Aluminum 6061 as shown in Figure 7.153 and click OK.

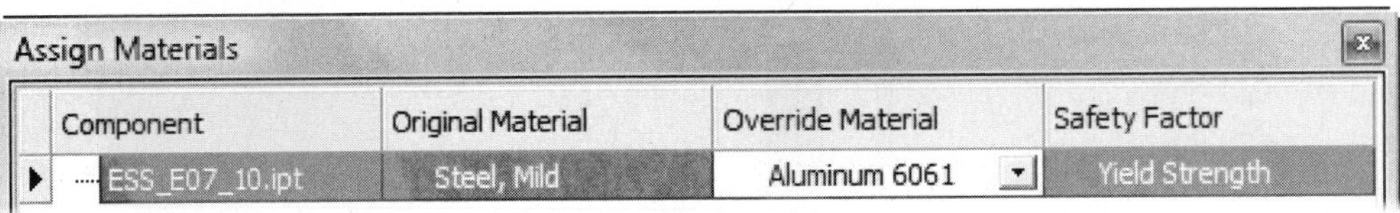

FIGURE 7.153

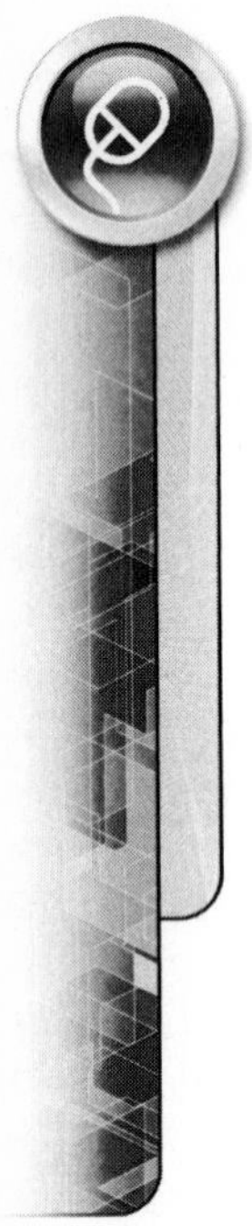

19. The constraints and the load were copied from the first simulation and do not need to be reapplied, but the simulation needs to be rerun to reflect the change in material. Run the simulation by clicking the Simulate command from the Stress Analysis tab > Solve panel and then click the Run button in the Simulate dialog box.
20. From the browser analyze the results of the simulation by selecting the different results method.
21. Lastly you generate a report. Click the Report command from the Stress Analysis tab > Report panel and then click OK.
22. To return to the part file environment click the Finish Stress Analysis command from the Stress Analysis tab > Exit panel.
23. Close the file. Do not save changes. End of exercise.

EXERCISE 7-11: RUN A STRESS ANALYSIS ON AN ASSEMBLY

In this exercise, you will use the Stress Analysis commands to analyze a design of a plant hook to determine how much displacement there will be without a brace and with a brace.

1. Open *ESS_E07_11.iam* in the Chapter 07 folder.
2. Click the Environments tab.
3. Begin a simulation by clicking the Stress Analysis command on the Environments tab > Begin panel.
4. Create a new simulation by clicking the Create Simulation command on the Environments tab > Manage panel.
5. In the Create New Simulation dialog box that appears click OK to create a new static analysis.
6. Verify that a material has been applied to both parts, click the Assign Materials command on the Stress Analysis tab > Material panel. Verify that the Original Material assigned to both parts is Steel, Mild and click OK.
7. For the first simulation, you see how far the plant hook will displace without the brace. In the browser expand ESS_E07_11.iam and exclude the Rib part by right-clicking on Rib:1 in the browser and click Exclude From Simulation from the menu as shown in Figure 7.154.

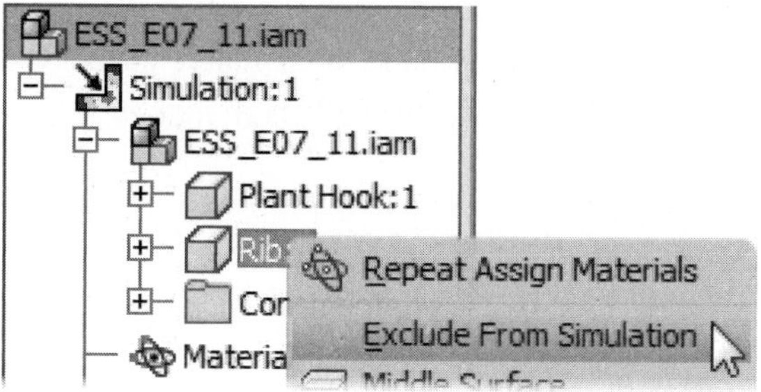

FIGURE 7.154

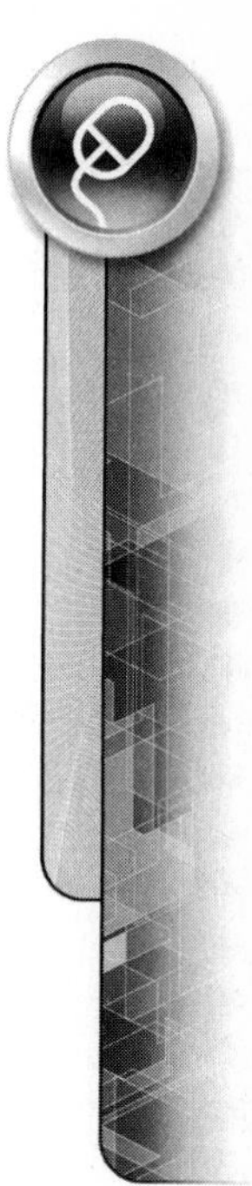

8. Next you constrain the plant hook, click the Fixed command from the Stress Analysis tab > Constraints panel and select the two circular faces of the holes labeled (1) and (2) as shown in Figure 7.155 on the left and click OK.
9. Next you apply a load to the part, click the Force command from the Stress Analysis tab > Loads panel. To create the force, make the following changes.
   - For the location select the inside circular face on the left end of the plant hook labeled (3) in Figure 7.155 on the right.
   - To set the direction of the force based on a vector of the components coordinates click the >> button on the lower right corner of the dialog box and then click the Use Vector Components option labeled (4).
   - In the Fy option (Y vector) enter -**5.000 lbforce** labeled (5) and select in another cell to see the direction of the force going down.
   - Click OK

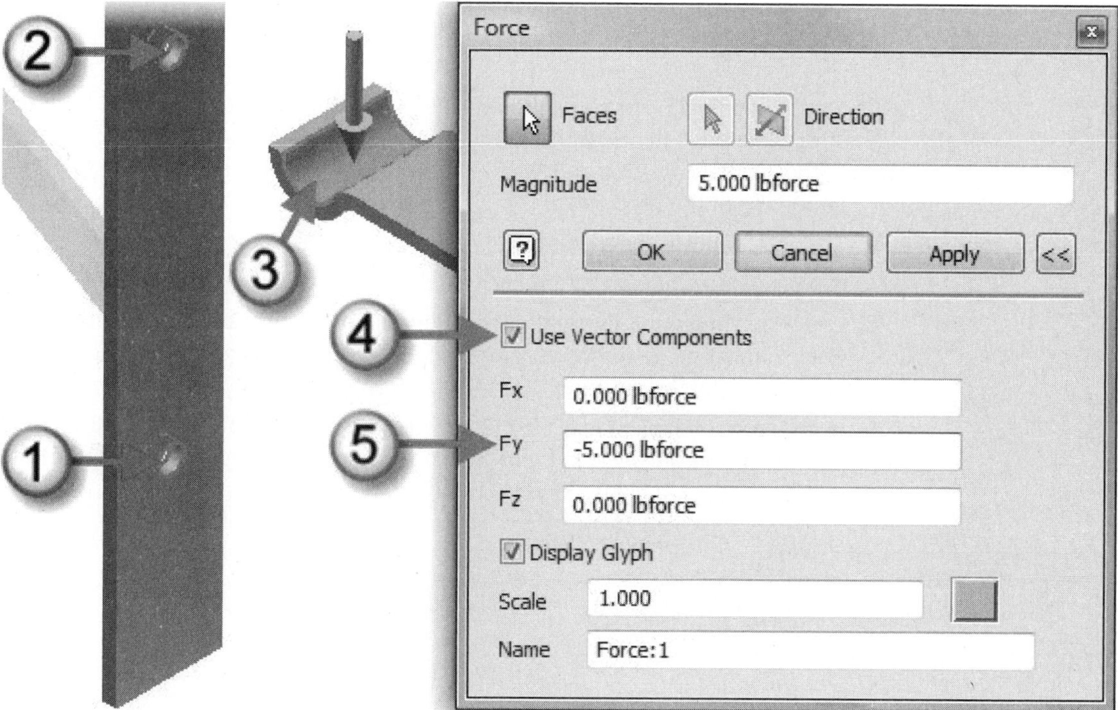

FIGURE 7.155

10. Next you run the simulation. Click the Simulate command from the Stress Analysis tab > Solve panel and then click the Run button in the Simulate dialog box.
11. From the Results entry in the browser click Displacement and your results of your simulation should resemble Figure 7.156.

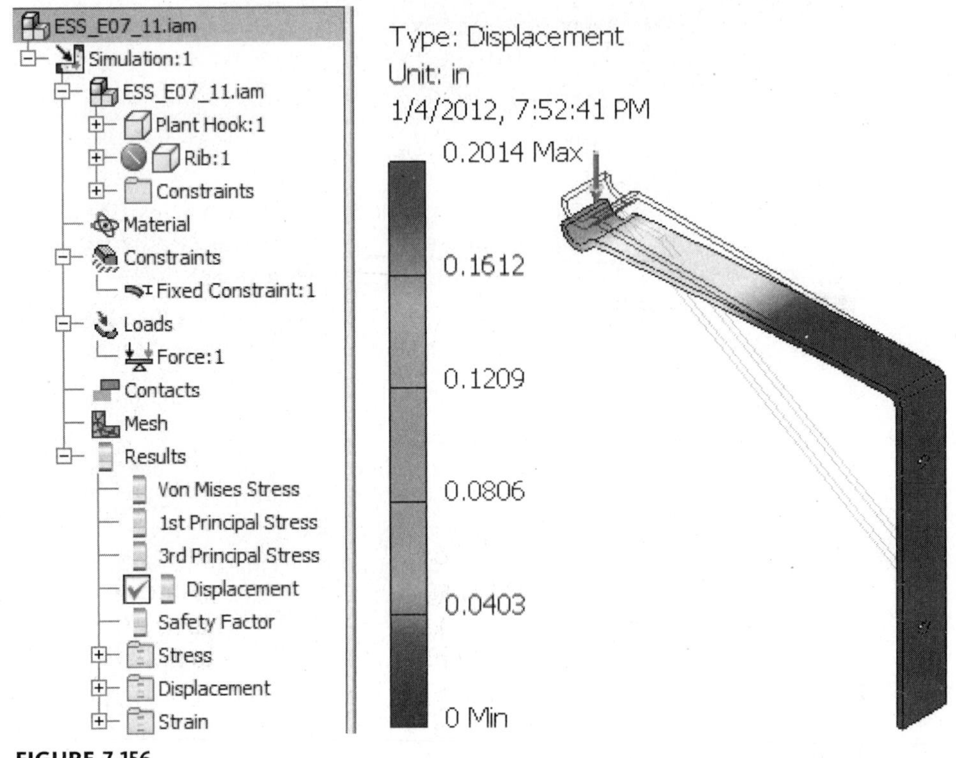

FIGURE 7.156

12. Next you animate the simulation results. Click the Animate command from the Stress Analysis tab > Result panel. To start the animation, click the Play button in the Animate Results dialog. You can adjust the speed by selecting different options from the drop list.
13. Click OK to finish the animation.
14. Next you copy the simulation, include the rib and rerun the simulation.
15. Copy the simulation by right-clicking on Simulation:1 in the browser and click Copy Simulation as shown in Figure 7.157 on the left. The new simulation, Simulation:2 will be the current.
16. In the browser expand ESS_E07_11.iam and include the Rib part by right-clicking on Rib:1 in the browser and click (uncheck) Exclude From Simulation from the menu as shown in Figure 7.157 on the right.

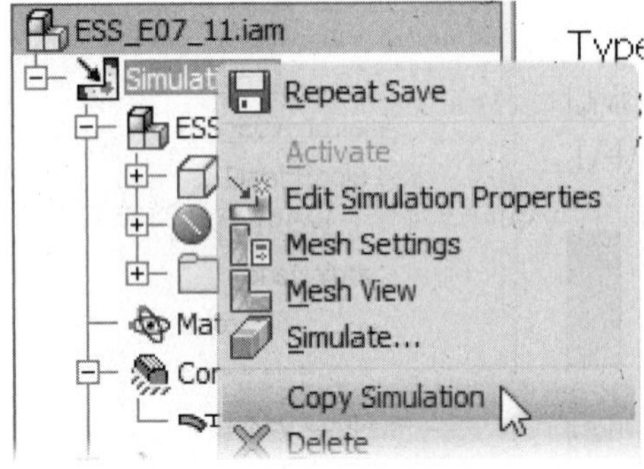

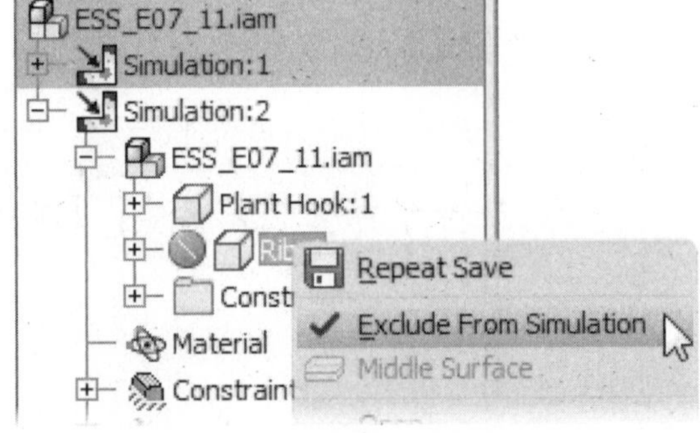

FIGURE 7.157

17. The constraints and the load were copied from the first simulation and do not need to be reapplied, but the simulation needs to be rerun to reflect the change in material. Run the simulation by clicking the Simulate command from the Stress Analysis tab > Solve panel and then click the Run button in the Simulate dialog box.
18. From the Results entry in the browser click Displacement and your results of your simulation should resemble Figure 7.158. As you can see the displacement went from ~.20 inches without the rib to ~.003 inches with the rib.

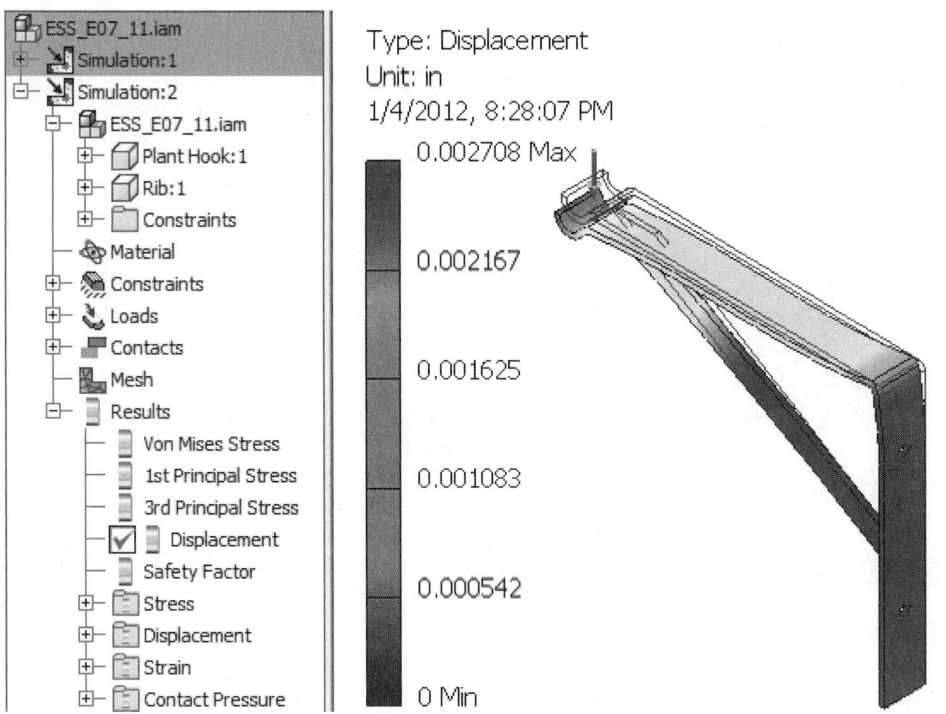

FIGURE 7.158

19. To see where the maximum and minimum value of the active result is on the model click the Maximum Value and or the Minimum Value command from the Stress Analysis tab > Display panel as shown in Figure 7.159 on the left.
20. To see a leader point to the location of the result click and drag the result away from the part. Figure 7.159 on the right shows the maximum and minimum results moved away from the model.

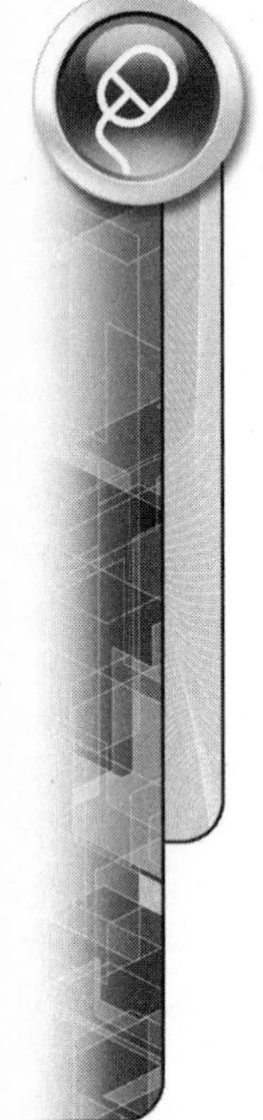

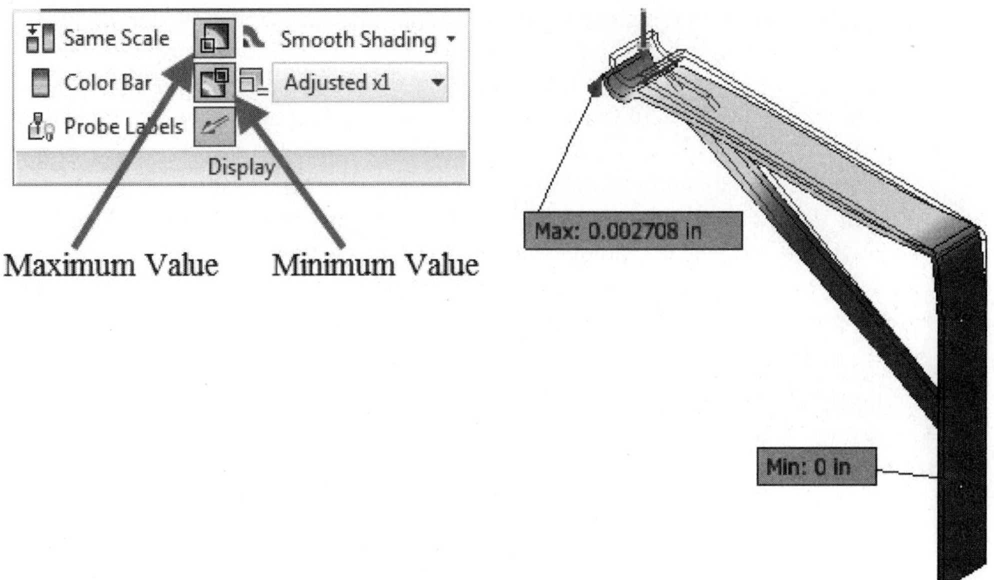

FIGURE 7.159

21. If desired copy the simulation, assign different materials to the parts and edit the location of the holes to see if you can minimize the displacement.
22. To return to the part file environment click the Finish Stress Analysis command from the Stress Analysis tab > Exit panel.
23. Close the file. Do not save changes. End of exercise.

APPLYING YOUR SKILLS

Skill Exercise 7-1

In this exercise, you use the knowledge you gained through this course to create a joystick handle. You will use the loft, split, and a few plastic part commands.

1. Open *ESS_Skills_7-1.ipt* from the Chapter 07 folder.
2. Create a loft, for the sections use the two elliptical shapes and circle in order from top to bottom and for the rails use the splines and your preview should resemble Figure 7.160 on the left.
3. Create a **.1875 inch** fillet around the top edge, as shown in Figure 7.160 on the right.

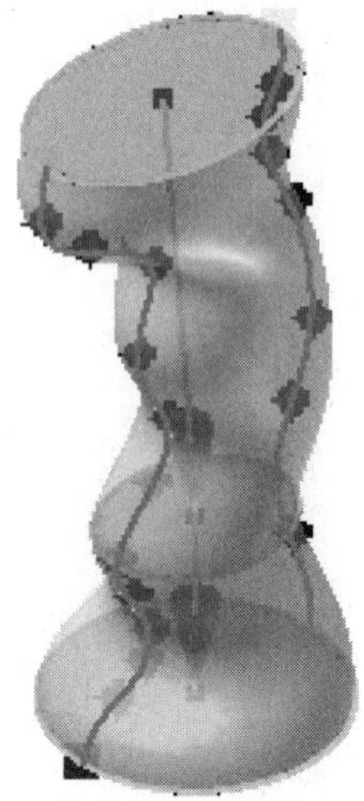

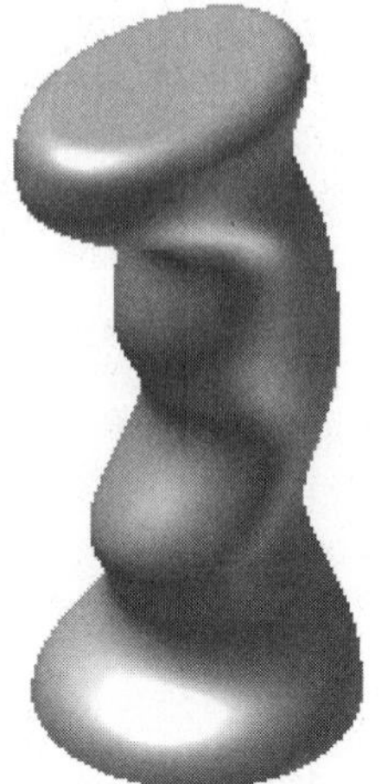

FIGURE 7.160

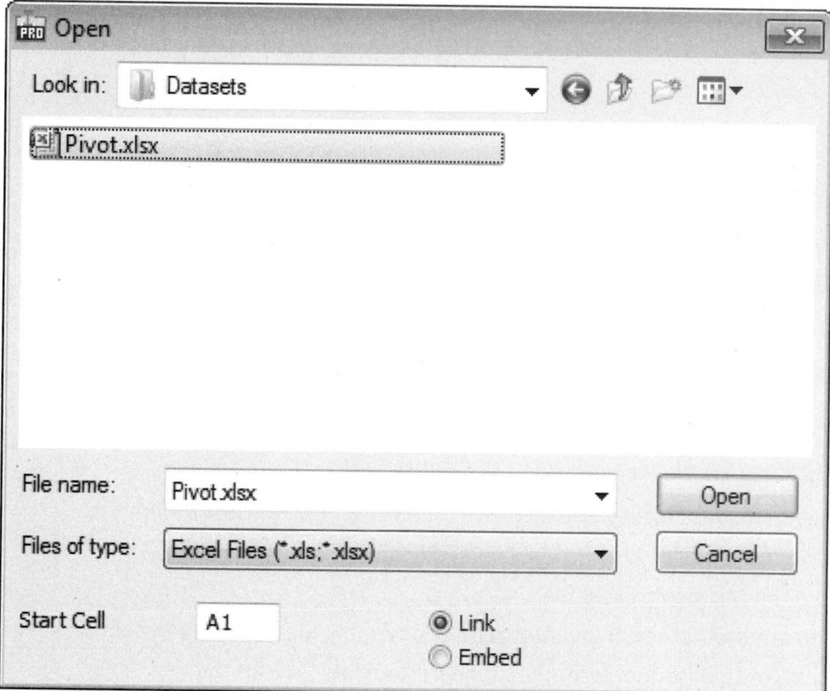

FIGURE 8.39

4. Navigate to and select the Microsoft Excel file to use.
5. In the lower-left corner of the Open dialog box, enter the start cell for the parameter data.
6. Select whether the spreadsheet will be linked or embedded.
7. Click the Open button: a section showing the parameters is added to the Parameters dialog box, as shown in Figure 8.40. If you embedded the spreadsheet, the new section will be titled Embedding #.
8. To complete the operation, click Done.

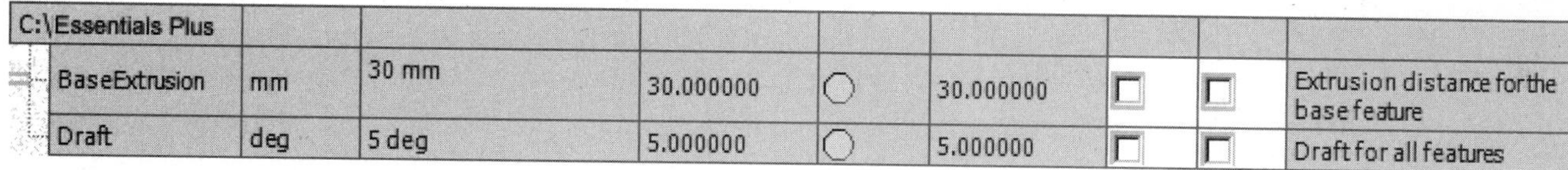

C:\Essentials Plus
BaseExtrusion mm 30 mm 30.000000 30.000000 Extrusion distance for the base feature
Draft deg 5 deg 5.000000 5.000000 Draft for all features

FIGURE 8.40

To edit the parameters that are linked, follow these steps:

1. Open the Microsoft Excel file.
2. Make the required changes.
3. Save the Microsoft Excel file.
4. Open the Autodesk Inventor part or assembly file that uses the spreadsheet.

You can also follow these steps when a spreadsheet has been linked. If you chose to embed the spreadsheet, the following steps must be used to edit the parameters:

1. Open the Autodesk Inventor part or assembly file that uses the spreadsheet.
2. Expand the 3rd Party folder in the browser.

3. Either double-click the spreadsheet or right-click and select Edit from the menu, as shown in Figure 8.41.

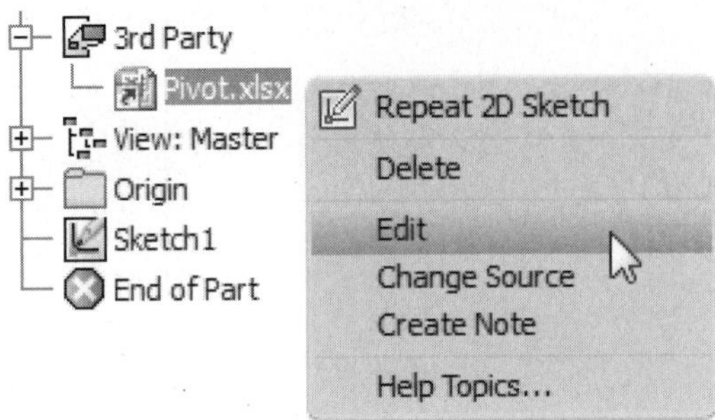

FIGURE 8.41

4. The Microsoft Excel spreadsheet will open in a new window for editing.
5. Make the required changes.
6. Save the Microsoft Excel file.
7. Activate the Autodesk Inventor part or assembly file that uses the spreadsheet.
8. Click the Update tool from the standard toolbar.

NOTE

If you embedded the spreadsheet, the changes will not be saved back to the original file but will only be saved internally to the Autodesk Inventor file.

EXERCISE 8-2: RELATIONSHIPS AND PARAMETERS

In this exercise, you create a sketch and dimension it. You then set up relationships between dimensions, and define User Parameters in both the model and an external spreadsheet.

1. Unless this step has already been performed, click the Application Options command on the Tools tab > Options panel, and then click on the Sketch tab. Check Autoproject Part Origin on Sketch Create to automatically have the origin point projected for the part. Additionally, click on the Part tab and select the Sketch on x-y plane option to have a sketch automatically created on new part creation.
2. Click OK or Close to close the Application Options dialog box.
3. Create a new file based on the default *Standard (in).ipt* file.
4. Create a 2D Sketch on the XY origin plane and then draw the geometry, as shown in Figure 8.42. The lower-left corner should be located at the origin point, and the arc is tangent to both lines. The size should be roughly 25 in wide by 40 in high.

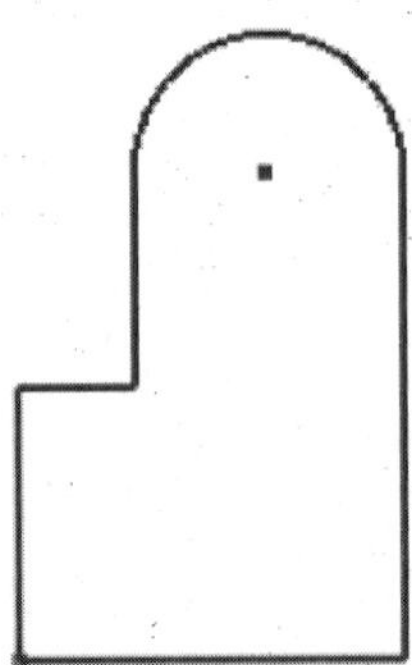

FIGURE 8.42

5. Add the four dimensions, as shown in Figure 8.43.

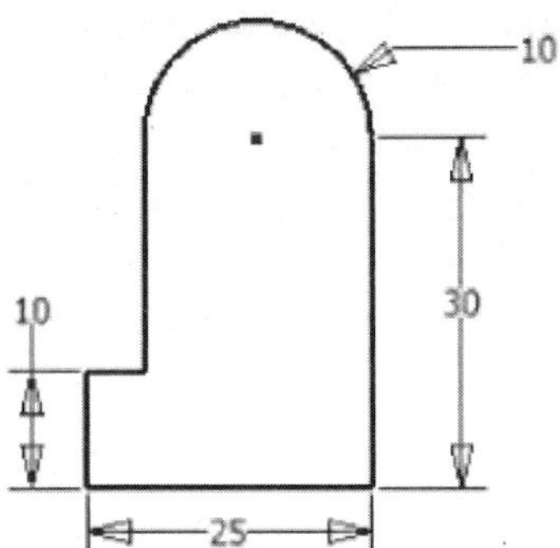

FIGURE 8.43

6. Double-click the 10 in radius dimension of the arc, and for its value, select the 25 in horizontal dimension, as shown in Figure 8.44.

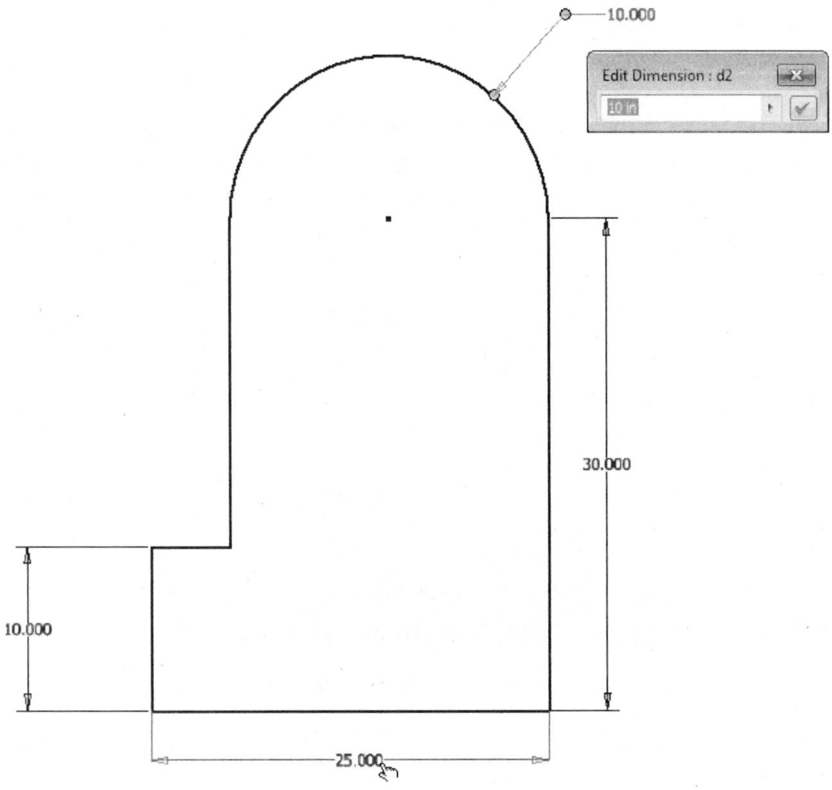

FIGURE 8.44

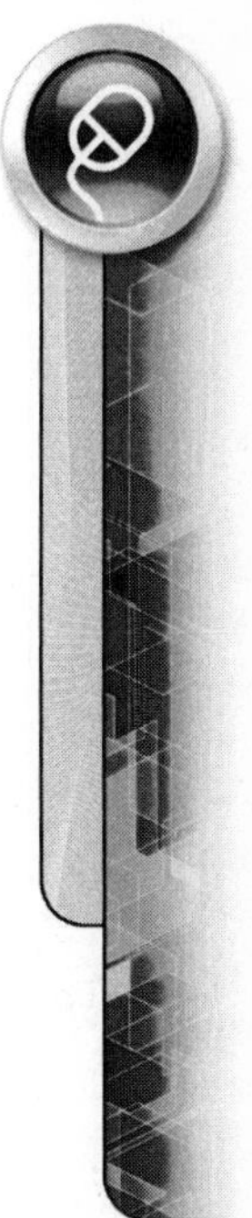

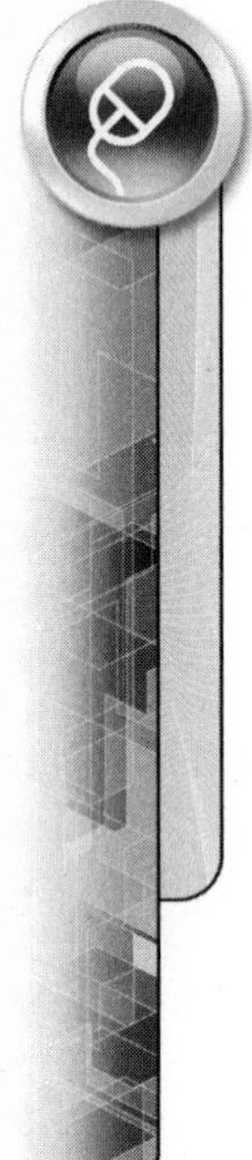

7. In the Edit Dimension dialog box, type **/4** after the d#, creating "d0/4." The 0 may be a different number depending upon the order in which your geometry was dimensioned. Click the green checkmark to accept the dimension value in the Edit Dimension dialog box.
8. Double-click the 10 in vertical dimension, and for its value, select the 30 in vertical dimension. Type **/2** after the d#. Click the green checkmark to accept the dimension value in the Edit Dimension dialog box.
9. Change the value of the 25 in horizontal dimension to **50 in** and the 30 in vertical dimension to **40 in**. When done, your sketch should resemble Figure 8.45. The fx: text that is displayed before the two dimensions denotes that they are equation driven or have a reference to a parameter.

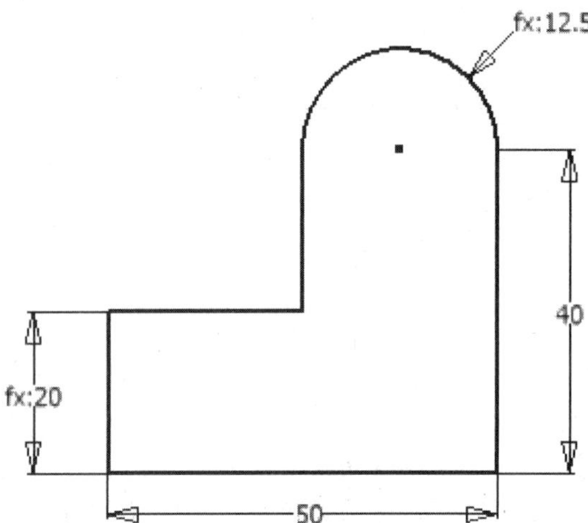

FIGURE 8.45

10. Next create parameters and drive the sketch from them. Click the Parameters command on the Quick Access toolbar or from the Manage tab > Parameters panel.
11. In the Parameter dialog box, create a User Parameter by clicking the Add Numeric button. Type in the following information:
    - Parameter Name = **Length**
    - Units = **in**
    - Equation = **75**
    - Comment = **Bottom length**
12. Create another User Parameter by selecting the Add Numeric button, and type in the following information:
    - Parameter Name = **Height**
    - Units = **in**
    - Equation = **Length/2**
    - Comment = **Height is half of the length**

    When done, the User Parameter area should resemble Figure 8.46.

User Parameters								
Length	in	75 in	75.000000	○	75.000000	☐	☐	Bottom length
Height	in	Length / 2 ul	37.500000	○	37.500000	☐	☐	Height is half of the length

FIGURE 8.46

13. Close the Parameter dialog box by clicking Done.
14. Change the dimension display by clicking the Expression dimension display option on the bottom of the status bar. The Expression option will be under the Tolerance dimension display option. Click OK.
15. Double-click the bottom horizontal dimension, and change its value to the parameter Length as follows: click the arrow; on the right side of the menu, click List Parameters, click Length from the list, and then click the checkmark in the dialog box.
16. Double-click the right vertical dimension, and change its value to Height. When done, your screen should resemble Figure 8.47.

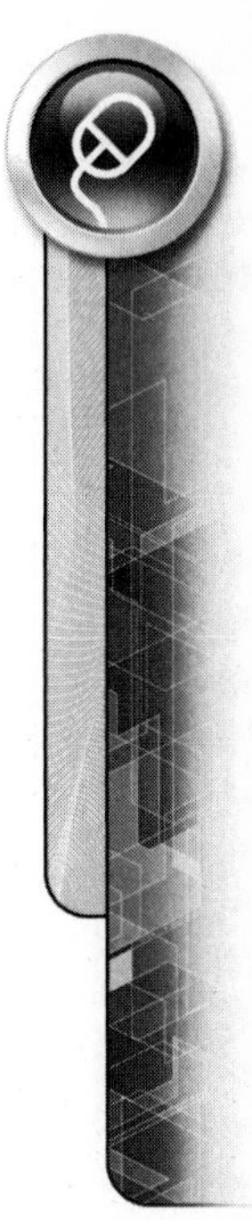

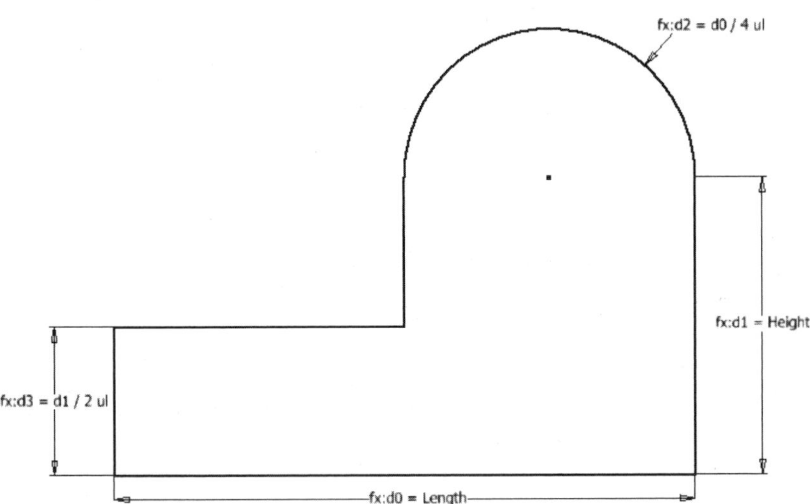

FIGURE 8.47

17. Click the Parameters command on the Quick Access toolbar. From the User Parameters section, click on the equation cell of the parameter name Length, and change its value to **35**.
18. Click in the equation cell of the parameter name Height; change its value to **50** and the comment field to "**Height is half of the right side**." When done making the changes, the User Parameter area should resemble Figure 8.48. The sketch updates automatically when the values are changed in the Parameters dialog box (if the default option is selected). Your sketch should appear similar to Figure 8.49 after the next step of the exercise.

User Parameters								
Length	in	35 in	35.000000	○	35.000000	☐	☐	Bottom length
Height	in	50 in	50.000000	○	50.000000	☐	☐	Height is half of the right side

FIGURE 8.48

19. To complete the changes, click the Done button.

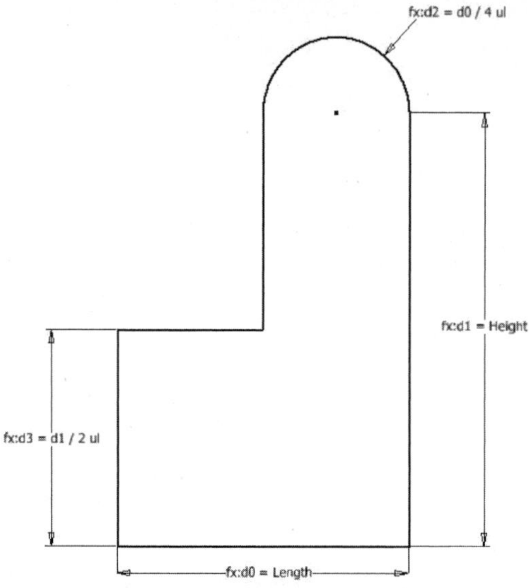

FIGURE 8.49

20. Save the file as *ESS_E08_02.ipt.*
21. Now create a spreadsheet that has two parameters. Create an Excel spreadsheet with the column names Parameter Name, Equation, Unit, and Comment, as shown in Figure 8.50. Enter the following data:
    - Parameter Name = BaseExtrusion
    - Equation = **30**
    - Units = **in**
    - Comment = **Extrusion distance for the base feature**
    - Parameter Name = **Draft**
    - Equation = **5**
    - Units = **deg**
    - Comment = **Draft for all features**

	A	B	C	D
1	Parameter Name	Equation	Unit	Comment
2	BaseExtrusion	30	in	Extrusion distance for the base feature
3	Draft	5	deg	Draft for all features

FIGURE 8.50

22. Save the spreadsheet as *ESS_08_Parameters.xlsx* in the C:\Inv 2013 Ess Plus\Chapter 08 folder. Then close Excel.
23. Make Autodesk Inventor the current application, and then click the Parameters command.
24. In the Parameter dialog box, click the Link button and select the *ESS_08_Parameters.xlsx* file, but do not click Open yet.
25. For the Start Cell, enter **A2**. If you fail to do this, no parameters will be found. The first row contains the names of the columns and will not be imported.
26. Click the Open button, and a new spreadsheet area will appear in the Parameters dialog box, as shown in Figure 8.51. Click the Done button to complete the operation.

C:\Essentials Plus...								
BaseExtrusion	in	30 in	30.000000		30.000000			Extrusion distance for the base feature
Draft	deg	5 deg	5.000000		5.000000			Draft for all features

FIGURE 8.51

27. Change to the home view.
28. Click the Extrude command from the 3D Model tab > Create panel, and click the More tab in the Extrude dialog box, as shown in Figure 8.52. For the Taper's value, enter **Draft** or click the arrow from the menu, click List Parameters, and then click the parameter Draft.

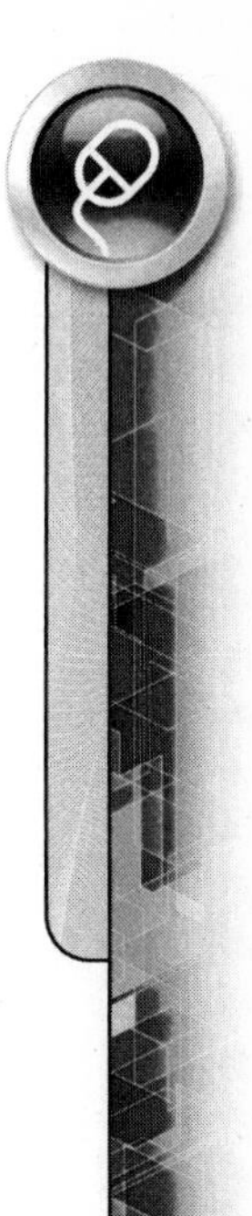

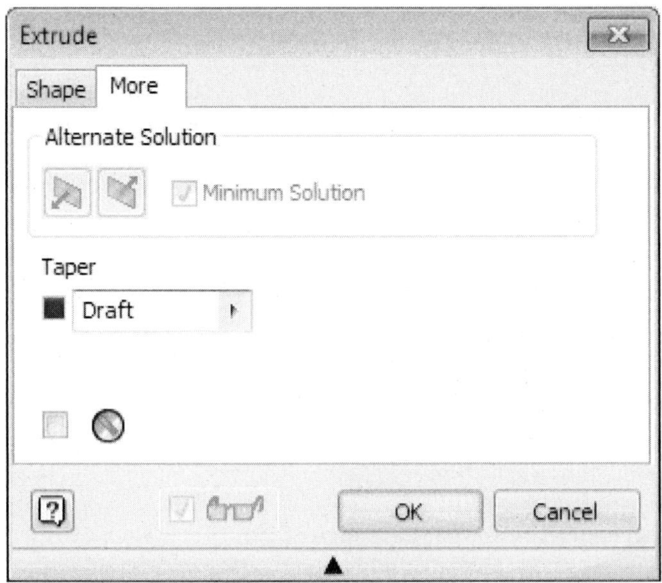

FIGURE 8.52

29. Click the Shape tab. For the Distance value, enter the parameter BaseExtrusion or click the arrow from the menu, click List Parameters, and then click the parameter BaseExtrusion as shown in Figure 8.53.

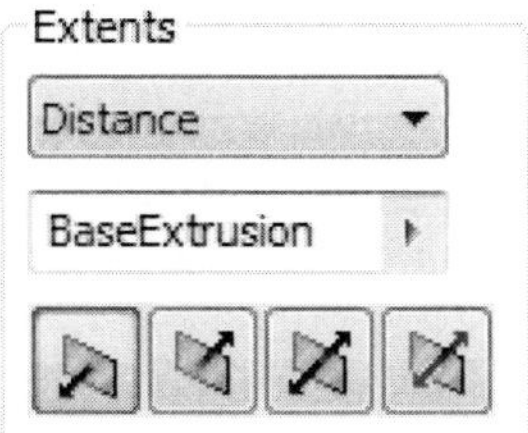

FIGURE 8.53

30. To complete the operation, click the OK button. Orbit your model so that is appears similar to Figure 8.54.

FIGURE 8.54

31. In the browser, expand the 3rd Party icon. Either double-click or right-click on the name *ESS_08_Parameters.xls,* and click Edit from the menu.
32. In the Excel spreadsheet, make the following changes: For the Parameter Name "BaseExtrusion," change the Equation to **50**. For the Parameter Name "Draft," change the Equation to **–3**.
33. Save the spreadsheet, and close Excel.
34. Make Autodesk Inventor the current application, and click the Local Update command from the Quick Access toolbar. When done, your model should resemble Figure 8.55.

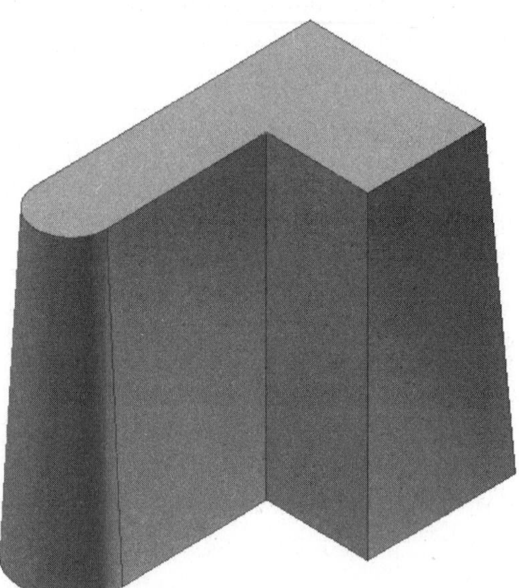

FIGURE 8.55

35. Close the file. Do not save changes. End of exercise.

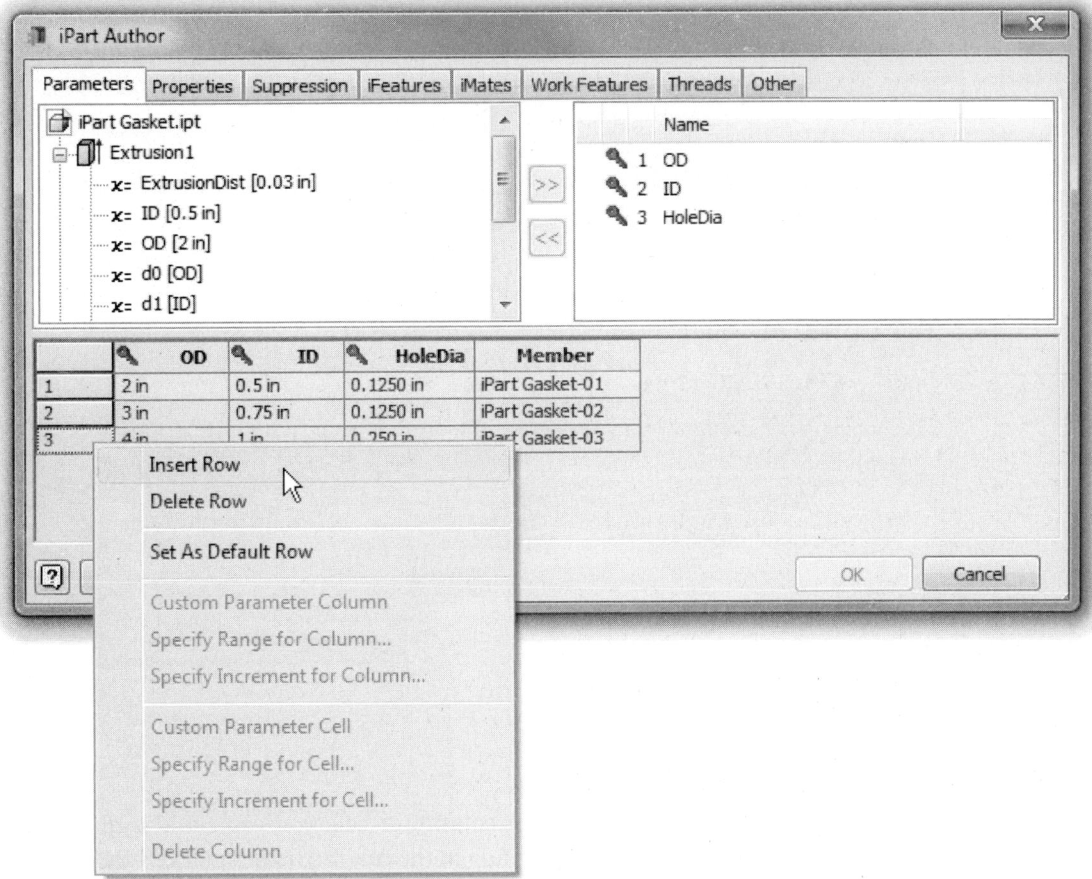

FIGURE 8.58

8. Edit the cell contents by clicking in the cell and typing new values or information as needed. To delete a row, right-click the row number, and select Delete Row from the menu. Each row that you add in the bottom pane of the dialog box represents an additional member within the iPart factory. The table functions similarly to a spreadsheet. In addition to adding or deleting members of the iPart factory, you can also use the table to change members by modifying cell values.

To allow the designer placing the iPart to specify a custom value for a given column, right-click on the column name, and select Custom Parameter Column from the menu, as shown in Figure 8.59. To make a specific cell custom, right-click in the individual cell, and select Custom Parameter Cell from the menu.

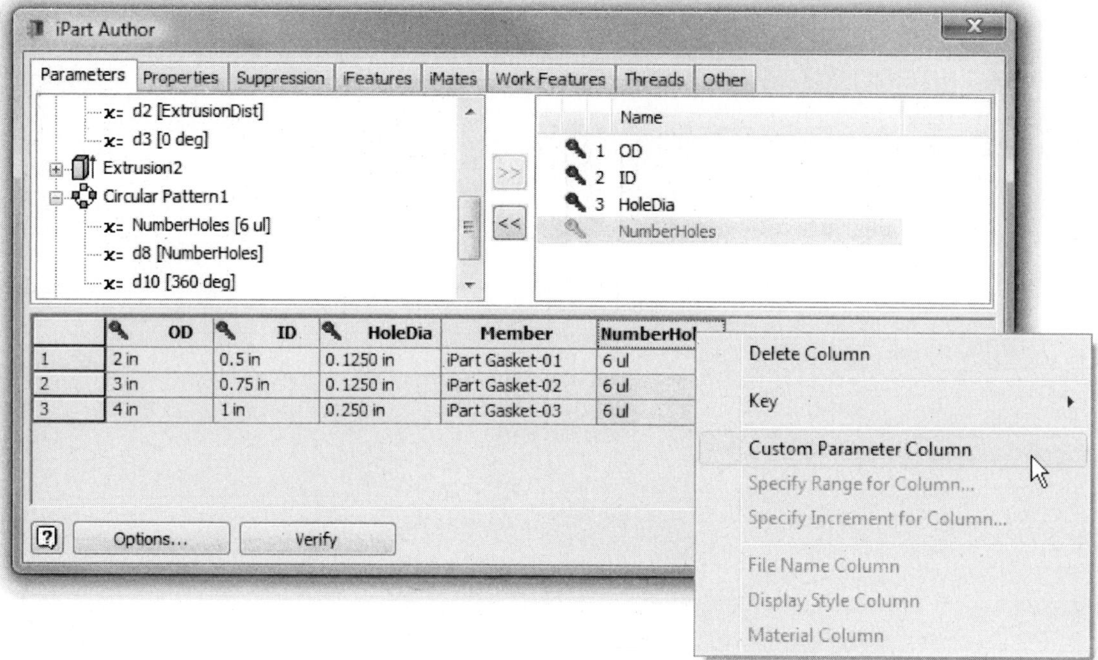

FIGURE 8.59

After designating a custom parameter column or cell, you can set a minimum and maximum range of values by right-clicking in the column heading or cell and selecting Specify Range for Column or Range for Cell from the menu. The Specify Range dialog box appears, as shown in Figure 8.60. Select the options as needed. Custom columns and cells are highlighted with a dark blue background in the table. You can also specify the increment that can be entered for a custom column or cell using the Specify Increment for Column option from the menu.

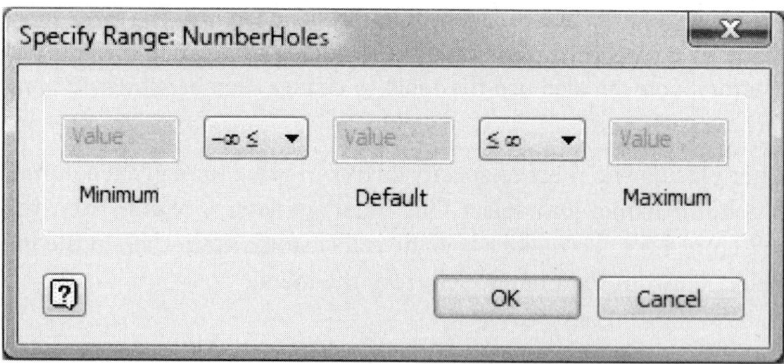

FIGURE 8.60

9. Set the member that will be the default by right-clicking on the row number and selecting Set As Default Row from the menu. You may select the other options as needed.

10. Click the Options button to edit the part number and member naming schemes for the iPart factory, as shown in Figure 8.61.

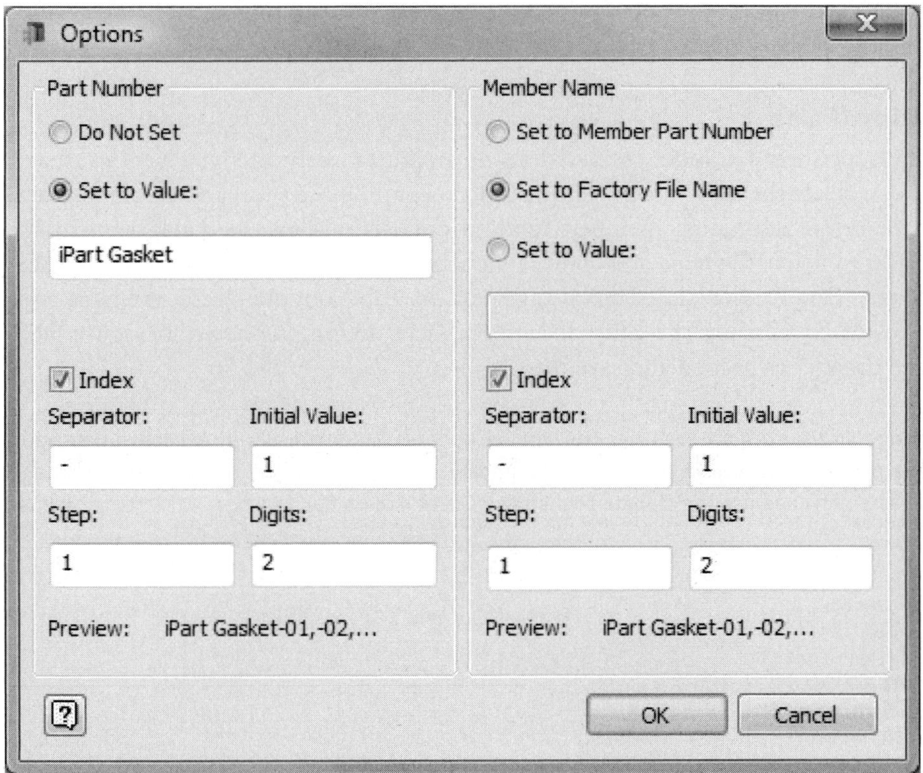

FIGURE 8.61

11. Click OK in the iPart Author dialog box when you have finished defining the contents of the table. The iPart Author dialog box closes, and the part is converted to an iPart factory. The table is saved, and a table icon appears in the browser, as shown in Figure 8.62.

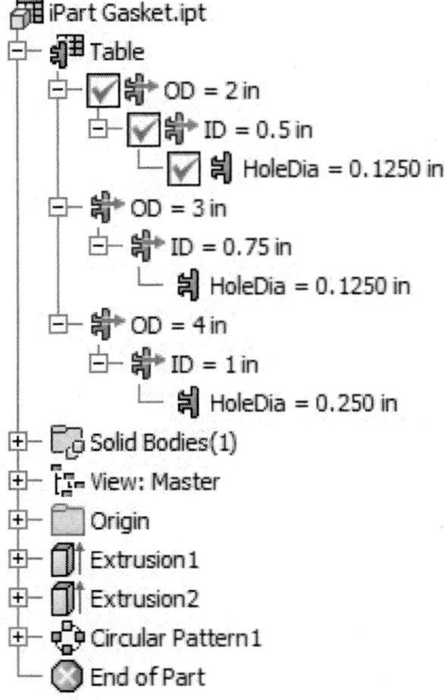

FIGURE 8.62

You can expand the Table icon in the browser to view the iPart members based on the member name or keys and values that you define. The active, or calculated, version appears with a checkmark, as shown in Figure 8.62. You change the browser display to show the member name or keys by right-clicking and selecting either List by Member Name or List by Keys from the menu.

Editing iParts

You can perform a number of operations on an iPart factory after you have created it. You can delete the table, modify the parameters or properties for individual members, add or delete additional members, and so on. Right-click the Table icon in the browser to delete the table and convert the iPart factory back to a part, edit the table with the iPart Author dialog box using the Edit Table option, or edit the table with Microsoft Excel using the Edit via Spread Sheet option, as shown in Figure 8.63. Make changes as needed and save the file.

NOTE

Changes made to iPart factories will not be updated automatically in members that you have previously placed in assemblies. To update the iPart members in an assembly, open the assembly, and use the Update tool on the Quick Access Bar.

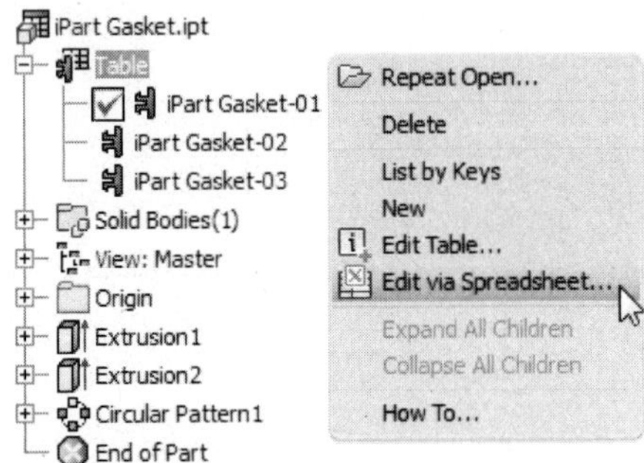

FIGURE 8.63

When editing the spreadsheet using Microsoft Excel, you can incorporate spreadsheet formulas, conditional statements, and multiple sheet data extractions, but you cannot modify spreadsheet formulas and conditional statements from the iPart Author dialog box. These types of cells are inactive and are highlighted in red when displayed in the iPart Author.

iPart Placement

You can place standard iParts in assemblies using the Place Component command. When you select a standard iPart factory, an additional Place Standard iPart dialog box appears. It allows you to use the Keys tab to identify an iPart member by selecting the key values, use the Tree tab to locate a member by expanding the key values, use the Table tab to identify an iPart member by selecting a row in the table, and place multiple instances of different iPart members.

To place a standard iPart into an assembly, follow these steps:

1. Start a new assembly or open an existing assembly in which to place the iPart.
2. Click the Place Component command, and navigate to and select the iPart to place in the assembly in the Place Component dialog box.

3. Click the Open button, and the Place Standard iPart dialog box will appear, as shown in Figure 8.64. If a custom cell(s) or column(s) exists, the Place Custom iPart dialog box appears.

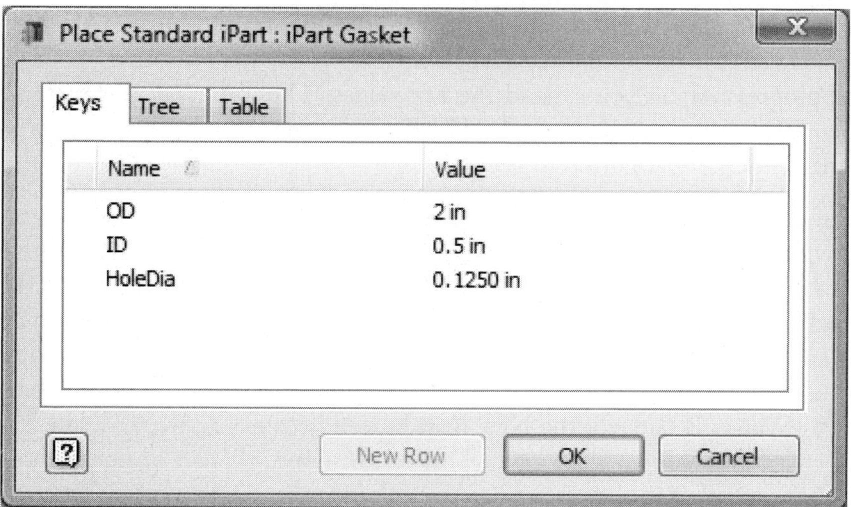

FIGURE 8.64

4. To select from the member list, select the Keys, Tree, or Table tab, and then select the member that defines the part you want to place.
5. If you are placing a custom part, select the Keys, Tree, or Table tab, and enter a value in the right side of the dialog box, as shown in Figure 8.65. The value must fall within the limits set in the iPart factory; otherwise, an alert appears.

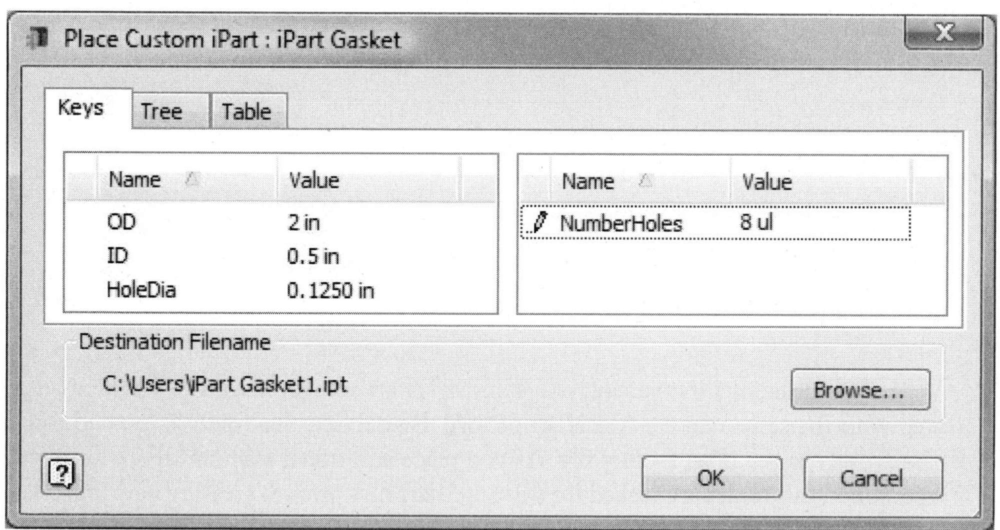

FIGURE 8.65

6. Place the part in the graphics window.
7. Continue placing instances of the iPart as needed.

8. To change an iPart in an assembly to a different configuration, expand the part in the browser, right-click on the table name, and select Change Component. Then select a new member from the Keys, Tree, or Table tab.

The Keys tab displays the primary and any secondary keys defined in the iPart factory. You can select the values of the keys to identify unique members. The Tree tab displays a hierarchical structure of the keys. If secondary keys exist, the values of the keys are filtered progressively as you expand the key values. The Table tab displays the entire table rather than just the keys. To identify a unique iPart member, select a row, and then click the OK button to place the iPart.

You can also dynamically create a new member for the factory when placing a standard iPart with the Table tab, as shown in Figure 8.66. In a row named "New" at the top of the table, you can enter values or select from a drop-down list if appropriate. As you enter values in the row, the table is reduced to display only the rows that match the values entered in the row. As you enter more values, the table continues to be filtered. If no member matches the entered values, a new member is created. When a unique set of values is entered, the New Row button becomes active, and you can select it to create a new row in the table. This can be done whether or not you are placing the new member in the assembly. Any columns that contain values that are not set are set to the same values as the default row.

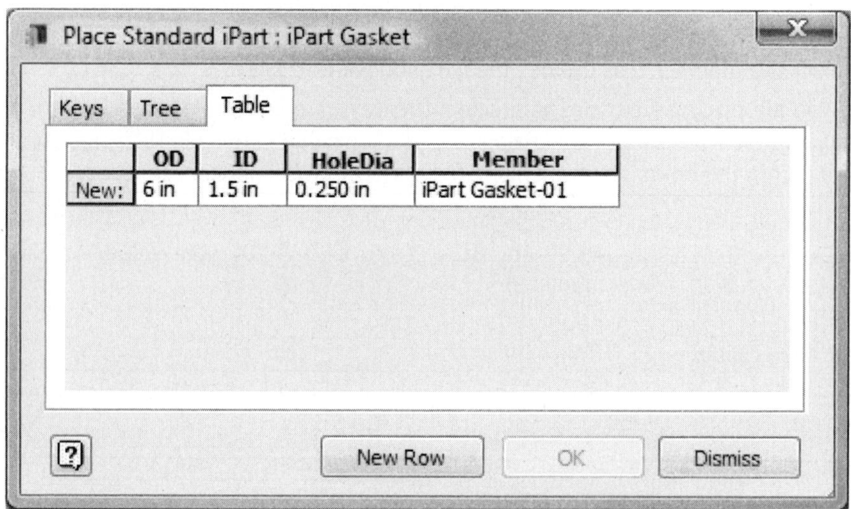

FIGURE 8.66

When you place the first version of a standard iPart into an assembly, a folder is created with the same name as the iPart factory. By default, this folder is created in the same folder as the iPart factory file. As you place additional standard iParts into your assemblies, this folder is checked for existing iPart files prior to creating new iPart files.

Standard iPart Libraries

In a collaborative design environment, the best way to manage iPart factories and the iParts published from those factories is by using libraries. Define a library within your project and place your iPart factories in this library's path. You can also specify that a standard iPart factory publish standard iParts to a different, or proxy, folder. To do this, create another library entry using the same name as the iPart factory library prefixed with an underscore, as shown in Figure 8.67, or right-click on the library's name, and click Add Proxy Path.

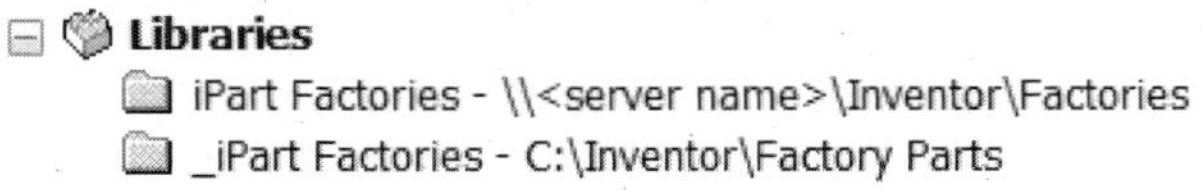

Libraries
iPart Factories - \\<server name>\Inventor\Factories
_iPart Factories - C:\Inventor\Factory Parts

FIGURE 8.67

An example use of the proxy path would be to enable you to store the factory file on a server and then place the created files from the factory locally (or on a different shared drive). This is the reason for the two different library paths. One to find the factory, the other to place and locate the used files.

In the example shown above, you access the iPart Factory part from the server location and when it is placed in the assembly, a folder with the part name is created on your local drive and a member part is created in that local folder. That local member is what is referenced into the assembly.

If you store an iPart factory in a workspace or a workgroup search path, the published iParts are stored in a subdirectory with the same name as the factory.

NOTE

It is recommended that you do not store iParts in the same folders as iPart factories.

Custom iParts

Custom iParts are placed into assemblies in the same way as standard iParts. If you select an existing custom iPart member file, that member of the part is placed into your assembly with no option to define the value of the custom parameters. When you select a custom iPart factory, the Place Custom iPart dialog box appears. As with the Standard iPart Placement dialog box, you can use the Keys tab to identify an iPart member by selecting the key value, the Tree tab to expand key values and select a member, or the Table tab to identify an iPart member by selecting a row in the table. The Place Custom iPart dialog box provides additional options so that you can enter values for custom parameters, define a destination and file name for the custom iPart by selecting Browse in the dialog box, and place multiple instances of different iPart members.

NOTE

Since each custom iPart is unique, you must provide different file names as you place different members.

After you have placed a custom iPart in an assembly, you can add more features to the iPart. The capability to add features to an iPart makes the custom iPart behavior similar to that of a derived part.

Auto-Capture

You can make changes automatically to an iPart or iAssembly, which are covered later in this chapter, using normal tools, without accessing the iPart or iAssembly Author table. You can automatically capture changes to the entire factory or to the active row. This is done using one of the two tools accessed from the Manage tab > Author panel, as shown in Figure 8.68. You can also use the Create iPart or iAssembly tools or edit the table via a spreadsheet.

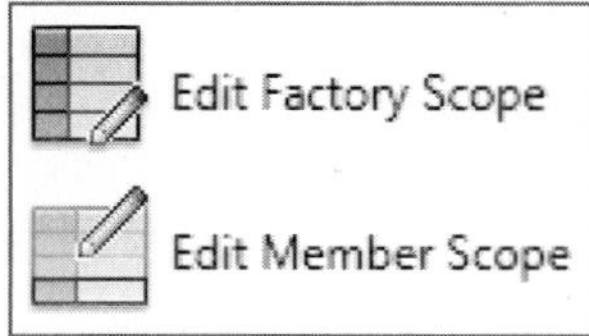

FIGURE 8.68

The available tools are as follows:

Edit Member Scope	Any edit that can be configured is automatically added as one or more new columns or will edit the member's cell value if a column for the change already exists in the table. If a new column is added, the new value of the object is added for the current member's row, and the original value is used for all other members.
Edit Factory Scope	This setting works similarly to Edit Member Scope, except that only columns are modified. New values are set for the entire column. Columns are not added or removed in this mode.
Create iPart/ iAssembly	Opens the iPart or iAssembly Author dialog box depending on which type of file is open.
Edit Using Spread Sheet	Opens the iPart or iAssembly table in a spreadsheet file for modification.

Edit Member Scope Example

For example, say that you have an iPart of a simple rectangle made up of a single extruded rectangle whose sketch has a "Length" and "Width" parameter that define the iPart table, as shown in Figure 8.69.

	Member	Part Number	Length
1	Rectangle-01	Rectangle-01	10 in
2	Rectangle-02	Rectangle-02	20 in
3	Rectangle-03	Rectangle-03	30 in

FIGURE 8.69

The factory is open in Autodesk Inventor, and Rectangle-02 is active. You have selected the Edit Member Scope tool, and you change the "Width" of the extrusion from 5 in to **8 in** by showing the dimensions and modifying them, making the change just as you would to any feature.

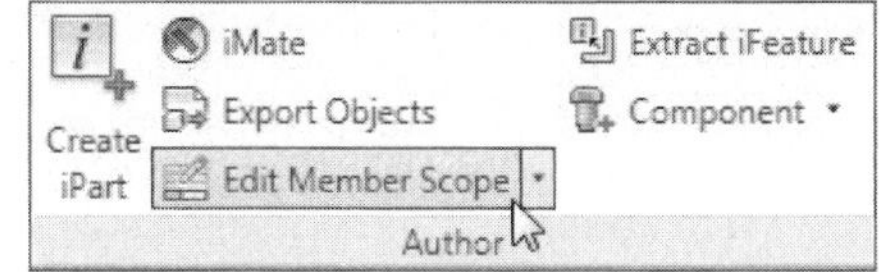

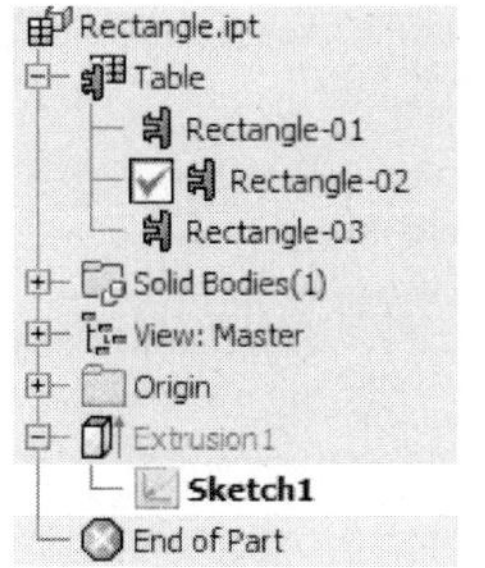

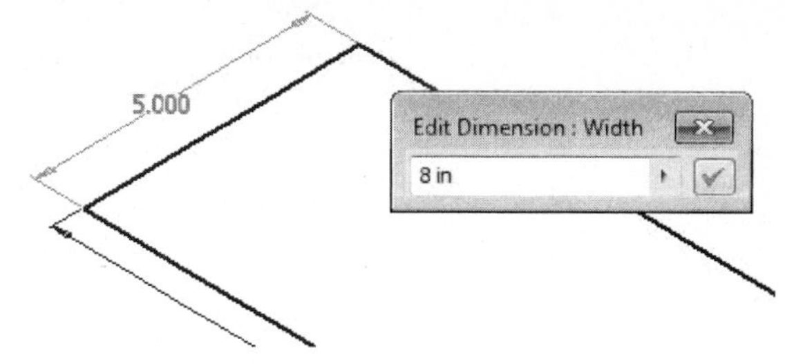

FIGURE 8.70

Because "Width" was not part of the original table, it is added as a new column to the table, and the original value of 5 in is set for all other members in the factory, as shown in Figure 8.71.

	Member	Part Number	Length	Width
1	Rectangle-01	Rectangle-01	10 in	5 in
2	Rectangle-02	Rectangle-02	20 in	8 in
3	Rectangle-03	Rectangle-03	30 in	5 in

FIGURE 8.71

Next you modify "Length" from 20 in to **25 in**. Because the table already has length configured, only the cell for Rectangle-02 is modified. The other members remain as defined.

	Member	Part Number	Length	Width
1	Rectangle-01	Rectangle-01	10 in	5 in
2	Rectangle-02	Rectangle-02	25 in	8 in
3	Rectangle-03	Rectangle-03	30 in	5 in

FIGURE 8.72

Figure 8.72 above show the table in intermediate states. The workflow of changing the width to 8 in and length to 25 in are done in a single edit. Both changes to the table happen simultaneously.

You can access the iPart/iAssembly tools on the Manage tab.

Drawings

When creating a drawing view from an iPart or an iAssembly, you can select the Model State tab of the Drawing View dialog box to access a particular member for the factory, as shown in Figure 8.73.

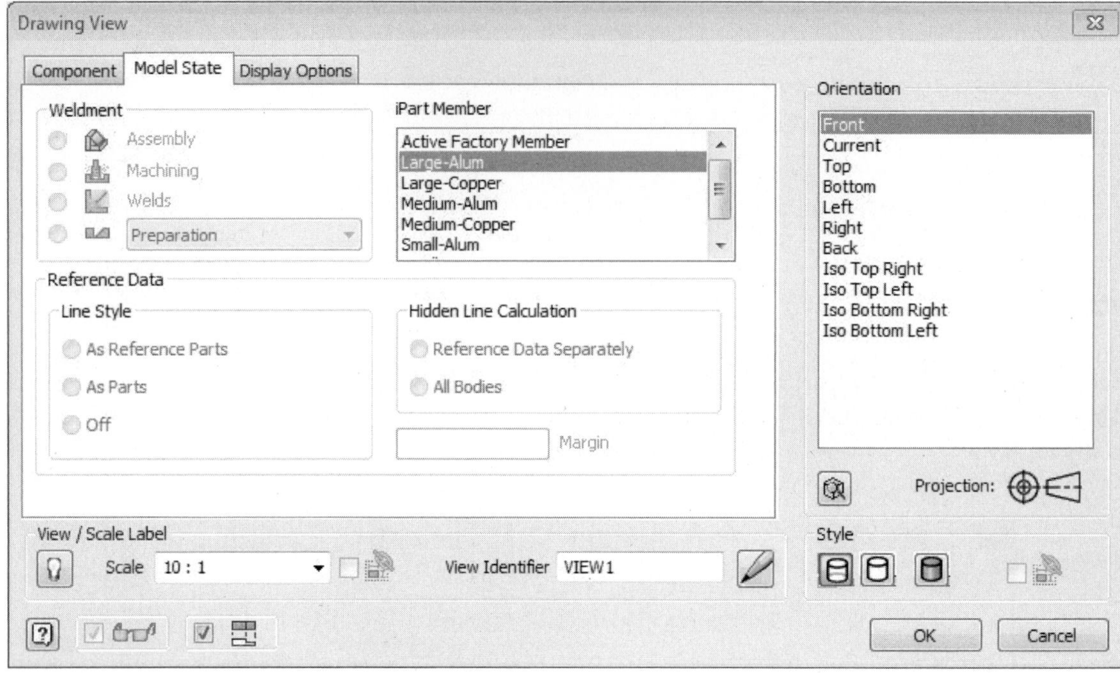

FIGURE 8.73

A General table command, available on the Annotate tab > Table panel, provides access to the Table dialog box, as shown in Figure 8.74. You can use this command to create a configuration table in a drawing. The style of the table is controlled using the Style and Standard Editor.

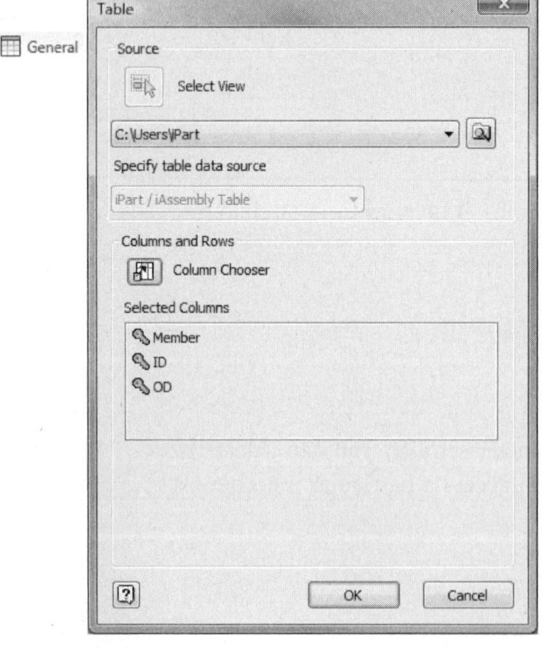

TABLE		
Member	Material	HoleDia
Large-Alum	Aluminum	8.1 mm
Medium-Alum	Aluminum	7.1 mm
Small-Alum	Aluminum	6.1 mm
Large-Copper	Copper	8.1 mm
Medium-Copper	Copper	7.1 mm
Small-Copper	Copper	6.1 mm

FIGURE 8.74

EXERCISE 8-3: CREATING AND PLACING iPARTS

In this exercise, you convert an existing part to a Standard iPart Factory. You then place Standard iParts from that factory into an assembly.

1. Open *ESS_E08_03.ipt* from the Chapter 08 folder.
2. Click the Parameters command on the Quick Access toolbar, and review the parameter names. In this exercise, you will use the parameter HoleDia to drive the iPart factory.
3. Click Done at the bottom of the Parameters dialog box.
4. From the Manage tab > Author panel, click Create iPart.
5. In the iPart Author dialog box, expand the Hole1 feature in the Parameters tab. Notice that the system parameter HoleDia has automatically been added to the list of table-driven items, and a column has been added to the table. User parameters and renamed system parameters are automatically added to the list.
6. HoleDia will be used as the primary key for this iPart Factory in the right side of the dialog box click the key next to HoleDia. It should turn blue, and the number "1" will be placed next to it.
7. You will also control the material from the iPart Factory. To add the Material property to the table, complete the following actions:
   - Click the Properties tab.
   - Collapse Summary.
   - Expand Physical.
   - Double-click Material. Notice that the material Aluminum, defined in the Properties dialog box in the Physical tab, is added to the table.

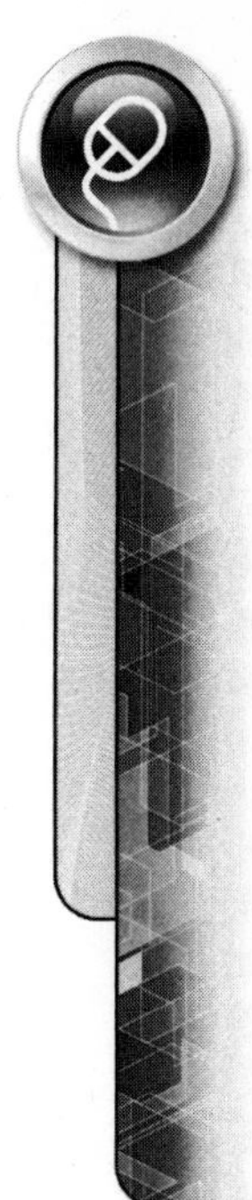

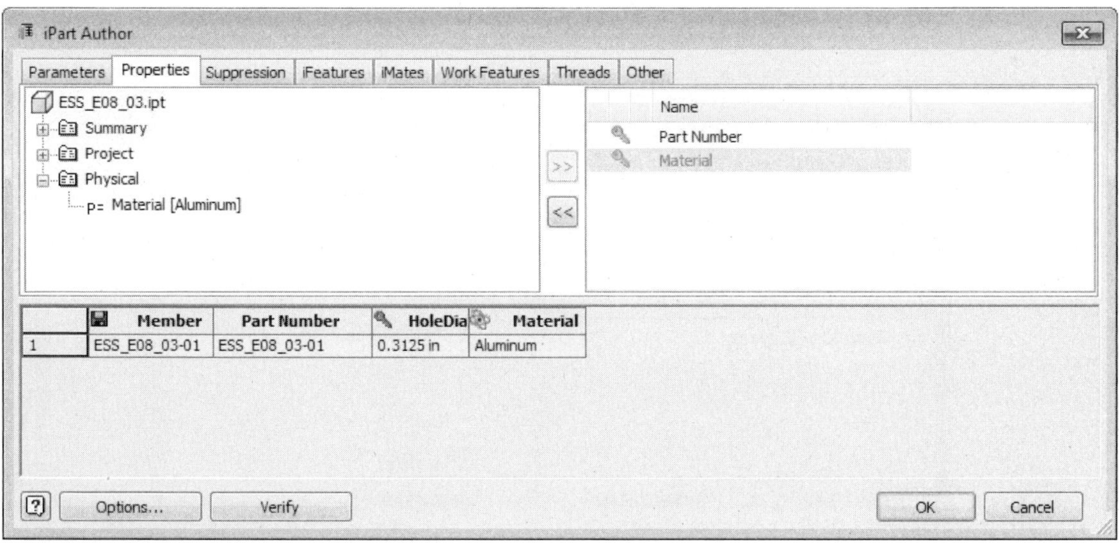

FIGURE 8.75

8. To identify a material when placing this Standard iPart, the Material property will also be defined as a key. To define Material as a secondary key, click the key next to Material. The number next to the key shows the key priority.

9. You can identify which column in the table is used to control the Material for each iPart version. In the table, right-click on the Material column label, and notice the checkmark next to Material Column. If no checkmark appears next to the Material Column option, select it. If one exists, do not deselect it. You'll also notice that a material icon is displayed on the column heading.

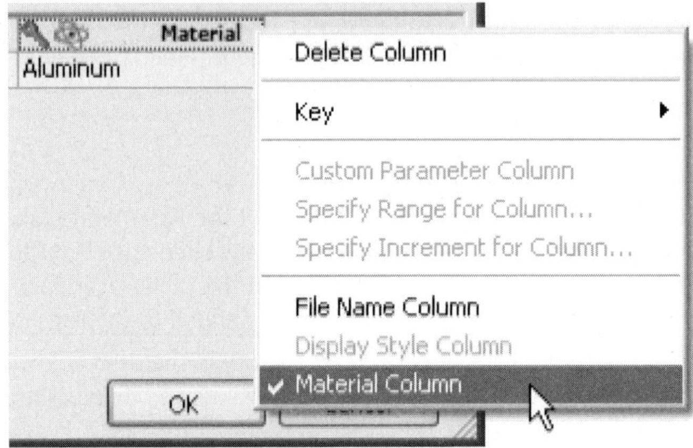

FIGURE 8.76

10. You will also control the Part Number of the Standard iPart when it is published. Click the Options button. Notice that the Set to Value: entry in the Part Number list is selected. Enter **Aluminum Bushing**, and then click OK.

NOTE

You can use the Options button to set rules and values for the Part Number and Member Name. You can further refine the part number by changing the character used for the separator, the initial value, the step increment, and the number of digits. Additionally, you can use the Verify button to ensure that the table contains no syntax errors.

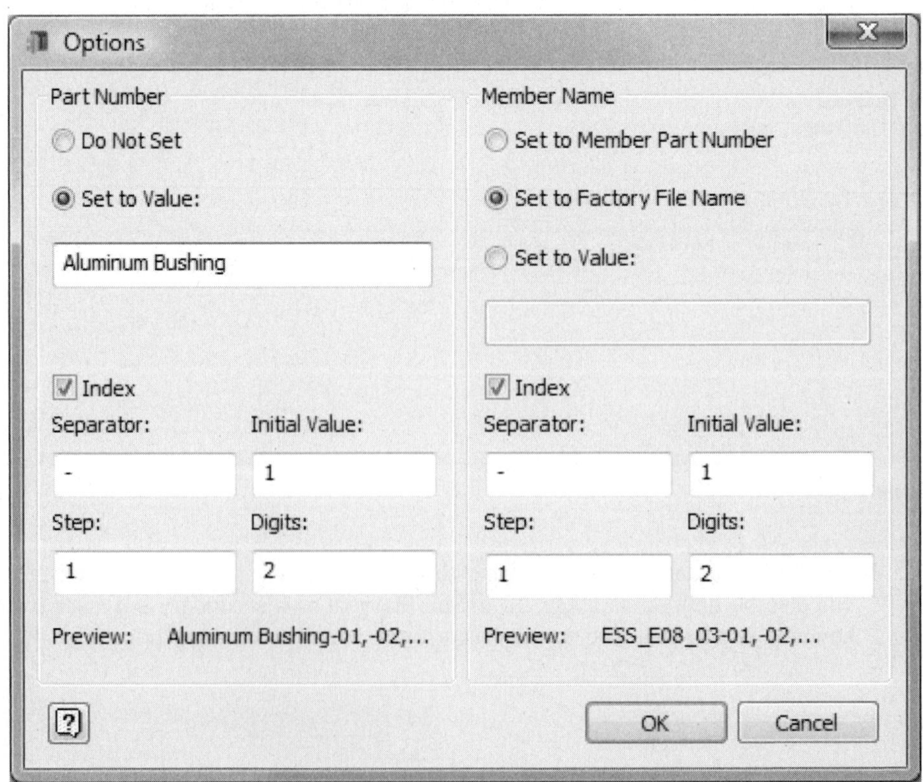

FIGURE 8.77

35. Click Dismiss in the Place Standard iPart dialog box. Notice that the names of the iParts in the browser reflect the names you specified when defining the Member column in the iPart Factory.
36. To change the version of the iPart you just inserted, expand Medium-Alum:4 in the browser, right-click the Table, and click Change Component.

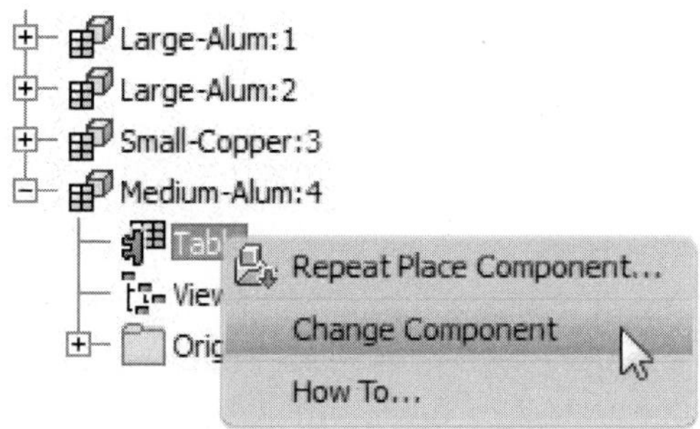

FIGURE 8.87

37. In the Place Standard iPart dialog box, change the HoleDia to **0.125 in** and Material to **Copper**.
38. Click OK in the Place Standard iPart dialog box, and verify that the bushing was replaced in the graphics window.
39. Close all open files. Do not save changes. End of exercise.

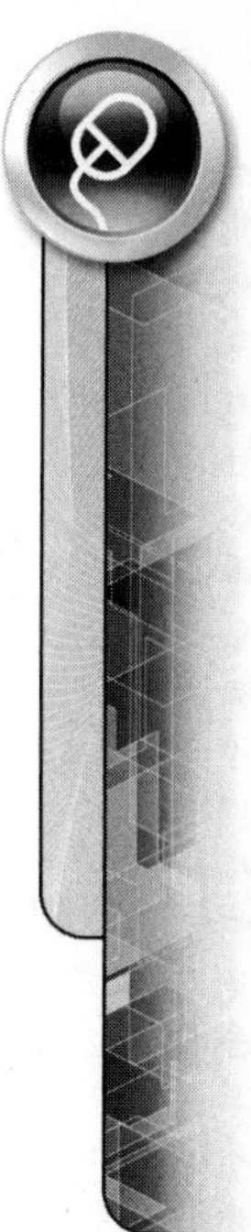

iASSEMBLIES

iAssemblies are used to group a set of similar designs in a table format. These are also commonly referred to as configurations, and they function in a similar method to iParts. An iAssembly factory is the primary definition of the family. The family is defined by a table and is a set of configurations of the same design. Each row of the table defines a unique product definition, which is called a configuration or member. When you use an iAssembly factory in another assembly, you can easily change from one configuration to another.

iAssembly members are stored as separate *.iam* files, similar to iParts. They are named in the table using the File name column designation, reference designation field, or key values. When member files are generated, they are stored in a subfolder with a name identical to that for the factory—similar to iParts, but with the *.iam* file type. You can also create a proxy search path for iAssemblies to designate a set of library locations to search for iAssembly factories and a set of corresponding locations to place populated members. Refer to the iPart section above for an example of a proxy path. You can leverage iMates to assist in assembling configurations and to make sure that the assembly conditions are maintained if you change between members of the iAssembly.

Creating iAssemblies

To create an iAssembly, use the Create iAssembly command on the Assemble tab > iPart/iAssembly panel or on the Manage tab > Author panel, as shown in Figure 8.88.

FIGURE 8.88

After selecting the Create iAssembly command, the iAssembly Author dialog box is displayed, as shown in Figure 8.89. This dialog box displays the configurable items for an assembly and lists the different configurations of the assembly that have been defined—see the rows at the bottom of the dialog box. You can define what controls the iAssembly by using the seven tabs of the dialog box. Each tab has different properties that can be selected and used as configurable items.

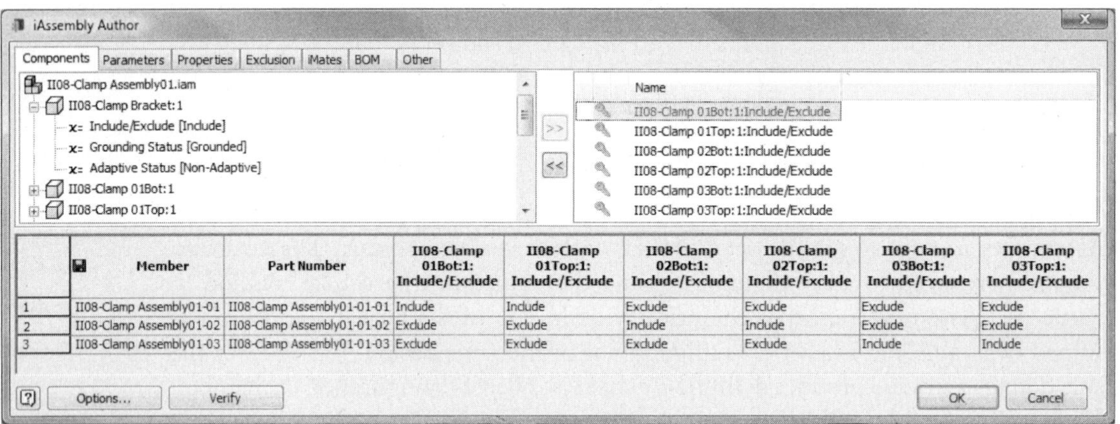

FIGURE 8.89

The iAssembly Author dialog box functions similar to the iPart Author dialog box. You can select a tab across the top to display objects that can be included for configuration in the iAssembly. You use the Add/Remove buttons to include objects from the top-left pane to the top-right pane of the dialog box. The list of configured objects resides in the top-right pane. The lower portion of the dialog box shows the configuration table that displays all of the configured items and defined rows, or members, of the iAssembly. You can use the Options button to set rules and values for the Part Number and Member Name. Additionally, you can use the Verify button to ensure that the table contains no syntax errors. When you have added all objects and defined all members, click OK to complete the authoring process. By right-clicking on a column in the table, you can specify which column you want to use as the file name column, designated by a disk icon in the column header. You can also specify that a column be used as a key column when establishing the iAssembly members.

In addition to creating the iAssembly factory with the iAssembly Author dialog box, you can also use Microsoft Excel or the Autocapture capabilities to create configurations. When using Autocapture, you use familiar tools for editing a design, and changes made to the assembly configuration are applied either to the current configuration or to the entire factory, depending on whether Edit Factory Scope or Edit Member Scope is selected.

When using Edit Factory Scope, the value or changes you make in the assembly are not automatically written to the configuration. If you have a value or settings change that you must click to open the iAssembly Author or to activate a different configuration member, a question dialog box is displayed. This dialog box lists the values in the active row that do not match the current values, and it prompts whether or not you want to update the current member's values with the modified values. Clicking Yes updates the configuration for the active member.

The Edit Member Scope setting will automatically apply changes and edits made to the assembly to the active configuration member. If you edit a property that is not listed as a configurable item in the table, Autocapture will add it to the table definition.

When you create a configuration of an assembly, the components in that assembly can be set to adaptive or flexible. However, a given component in a configuration can only be adaptive in one member of the configuration. All other members must be set to non-adaptive. If you want a component to adjust in size, you must control the size of the component using parameters and formulas or convert the part to an iPart and define the desired sizes. The actual iAssembly can be made flexible when placed in other assembly files, but it cannot be set to adaptive.

The Components tab lists the components of the assembly with the configurable items below. Items listed can be changed in the following ways: Include/Exclude, Grounding Status, Adaptive Status, and Table Replace.

The Parameters tab lists all assembly constraints, assembly features, assembly work features, iMates, component patterns, and other parameters, such as User Parameters, that can be included in the factory.

The Properties tab allows the inclusion and modification of summary, project, physical, and custom properties.

Similar to the Parameters list, the Exclusion tab shows all objects that can be excluded, including components, constraints, assembly features, assembly work features, iMates, Representations, and Component Patterns. Note that these elements can also be set on the Components tab, but the Exclusion list provides a more specific way to view and set the exclusion property.

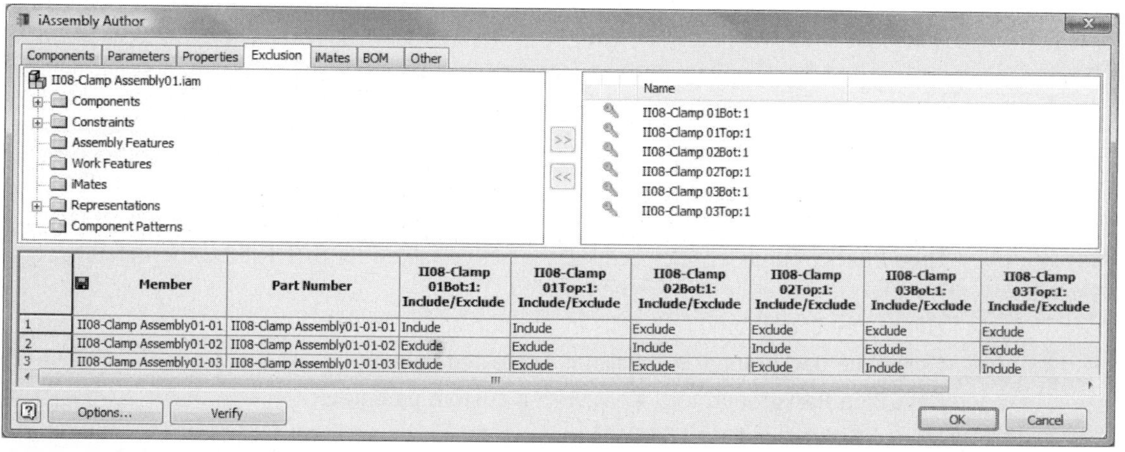

	Member	Part Number	II08-Clamp 01Bot:1: Include/Exclude	II08-Clamp 01Top:1: Include/Exclude	II08-Clamp 02Bot:1: Include/Exclude	II08-Clamp 02Top:1: Include/Exclude	II08-Clamp 03Bot:1: Include/Exclude	II08-Clamp 03Top:1: Include/Exclude
1	II08-Clamp Assembly01-01	II08-Clamp Assembly01-01-01	Include	Include	Exclude	Exclude	Exclude	Exclude
2	II08-Clamp Assembly01-02	II08-Clamp Assembly01-01-02	Exclude	Exclude	Include	Include	Exclude	Exclude
3	II08-Clamp Assembly01-03	II08-Clamp Assembly01-01-03	Exclude	Exclude	Exclude	Exclude	Include	Include

FIGURE 8.90

The iMates tab functions the same as it does for iParts. It lists each iMate with the offset value, include/exclude, matching name, and sequence number available for configuration.

The BOM tab can be used to work with bill of material specific properties, BOM Structure, and BOM Quantity. Two BOM viewing conditions exist for controlling what you see for an iAssembly Bill of Materials. You can view the BOM data from within an assembly that uses a specific iAssembly configuration as an occurrence of the design, or you can view BOM data from within the iAssembly file. In the first scenario, you can view all levels of its components and their details because the item number is a single configuration member.

If viewing the BOM from within the iAssembly, you can view only the top-level structure because the structure will vary based on each configuration member. As you vary the structure of the iAssembly, the item numbers will change from one configuration to another and cause issues for drawing documentation. To view the quantity information in the BOM, you have the option to display the Unit QTY values for a specific configuration member or all members, as shown in Figure 8.91.

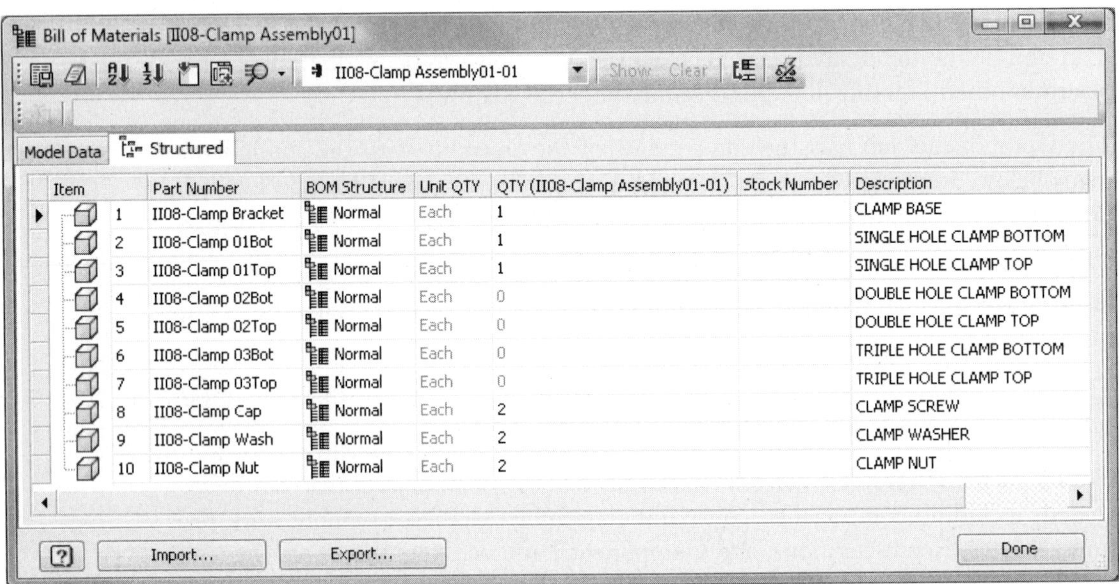

Item	Part Number	BOM Structure	Unit QTY	QTY (II08-Clamp Assembly01-01)	Stock Number	Description
1	II08-Clamp Bracket	Normal	Each	1		CLAMP BASE
2	II08-Clamp 01Bot	Normal	Each	1		SINGLE HOLE CLAMP BOTTOM
3	II08-Clamp 01Top	Normal	Each	1		SINGLE HOLE CLAMP TOP
4	II08-Clamp 02Bot	Normal	Each	0		DOUBLE HOLE CLAMP BOTTOM
5	II08-Clamp 02Top	Normal	Each	0		DOUBLE HOLE CLAMP TOP
6	II08-Clamp 03Bot	Normal	Each	0		TRIPLE HOLE CLAMP BOTTOM
7	II08-Clamp 03Top	Normal	Each	0		TRIPLE HOLE CLAMP TOP
8	II08-Clamp Cap	Normal	Each	2		CLAMP SCREW
9	II08-Clamp Wash	Normal	Each	2		CLAMP WASHER
10	II08-Clamp Nut	Normal	Each	2		CLAMP NUT

FIGURE 8.91

The Other tab functions the same as it does for iParts, and it can be used to specify a custom column that can contain a string value. Custom columns can be designated as keys or as a file name.

As the table is created, you will notice that different cell background colors provide information about the status of the cells as follows:

Green Background	Cells in the active table row that will be the default row when placed
Light Grey Background	Cells in the non-active row
Light Blue Background	Cells in a selected column
Dark Blue Background	Cells with a custom parameter
Mango Background	Cells that are driven by an Excel formula
Yellow Background	Cells that have an error

Placing iAssemblies

iAssemblies are placed into other assemblies using the Place Component command, similar to placing an iPart or any other regular component. When the selected component is an iAssembly file, the Place iAssembly dialog box is displayed, as shown in Figure 8.92.

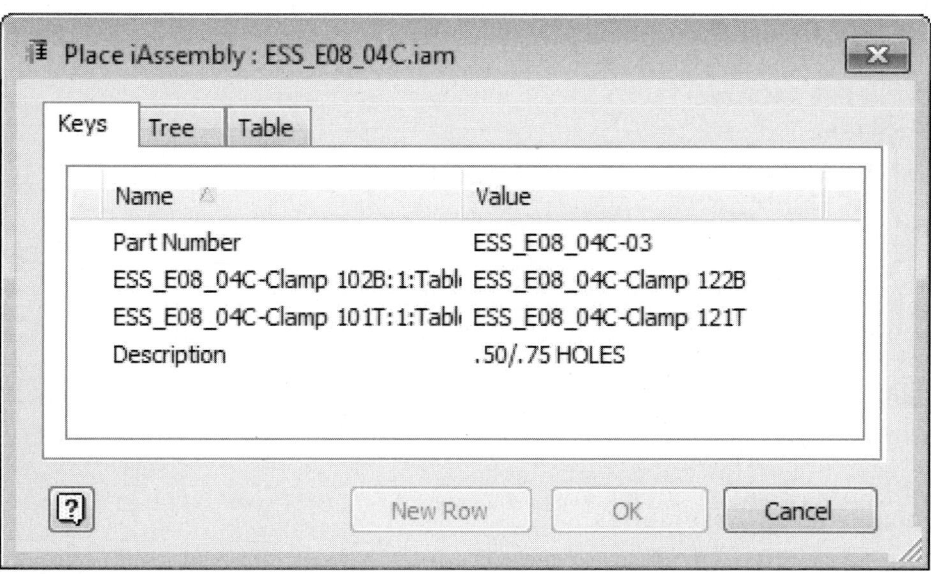

FIGURE 8.92

Before placing an occurrence of the iAssembly, select the configuration member from the Keys, Tree, or Table tabs. Each tab lists the same configurations, but they display the configuration members in different ways. These tabs function the same as when working with iParts. Refer to the iPart Placement section of this chapter for additional information. If iMates are used in the definition of the iAssembly and also in the iAssembly file itself, you can select one of the two iMates options in the Place Component dialog box to speed up the assembly process.

After an iAssembly configuration has been placed, you can change to another configuration by right-clicking the Table node in the browser and selecting Change Component, as shown in Figure 8.93.

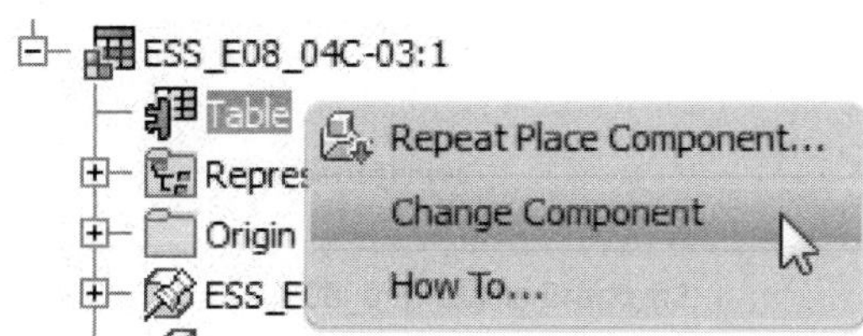

FIGURE 8.93

When a member of an iAssembly configuration is referenced, the member file with the specified and defined properties is generated. This file has either already been created—because it has been previously referenced—or a new member file is created and referenced. Instead of having the member files created when you select to use them in an assembly, you can pre-generate the member files of an iAssembly. This is done by opening the iAssembly and under the Table node, selecting one or more configurations, right-clicking, and selecting Generate Files from the menu, as shown in Figure 8.94. You can use this same method to update member files after changes have been made to an iAssembly, and the same method can be performed for iParts to pre-generate members of the iPart factory.

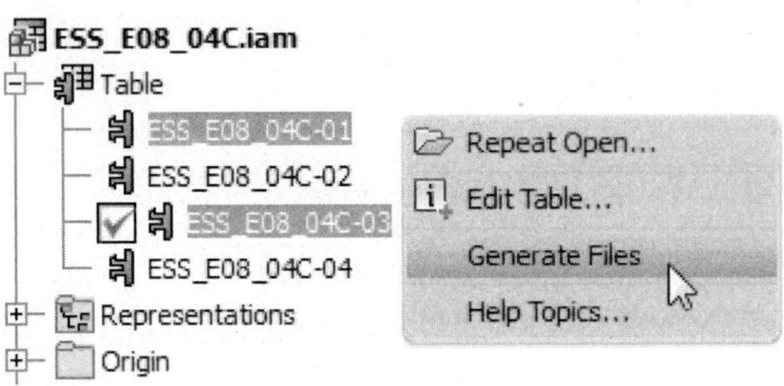

FIGURE 8.94

Documenting iAssemblies

Drawing views of iAssemblies are created using the Base View command. When an iAssembly file is selected, all of the iAssembly members are listed on the Model State tab of the Drawing View dialog box, as shown in Figure 8.95.

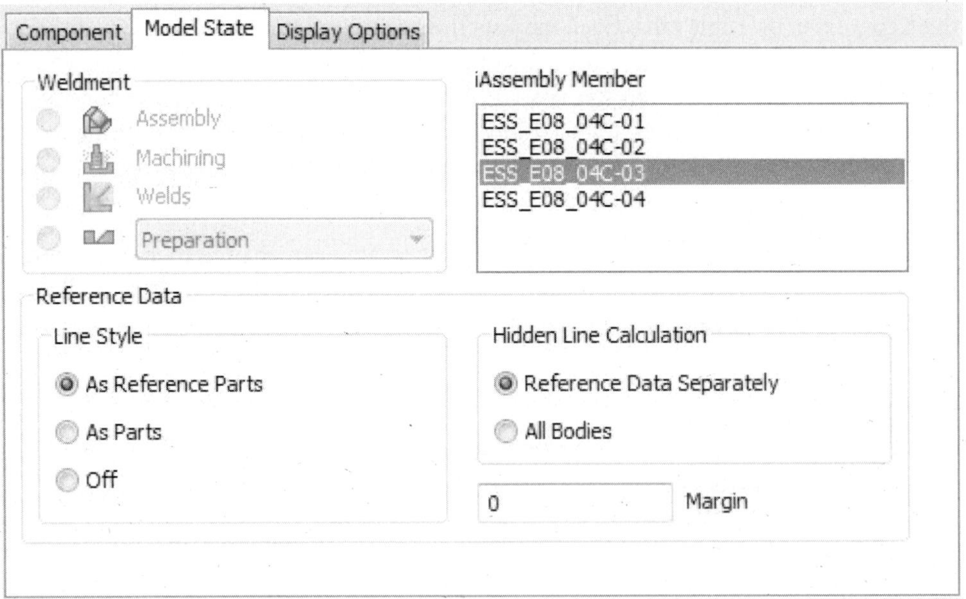

FIGURE 8.95

The created drawing view is based on the selected iAssembly member in that list. After the drawing view has been created, you can change it to a different iAssembly member by editing the view and selecting a different one from the list.

When adding a Parts List to a drawing that is based on an iAssembly, the quantity (QTY) column is based on the method used to create it. If you select an existing drawing view, the configuration member for that view is displayed. If you browse to an iAssembly file, the active configuration member in the file is displayed. Configuration members can be added to a Parts List by editing the existing Parts List and selecting the Member Selection button, as shown in Figure 8.96 on the left. In the Select Member dialog box, shown in Figure 8.96 on the right, you can select the configuration member's checkbox to include it in the Parts List.

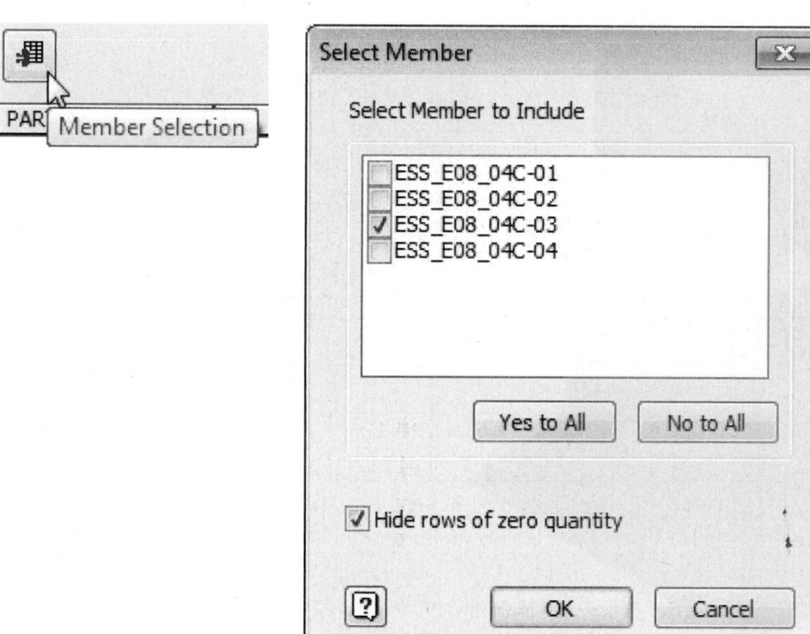

FIGURE 8.96

You can create tables that list iAssembly configuration values using the General table command. After selecting the General table command, you can choose what the table is based on by selecting an iAssembly drawing view or by choosing an iAssembly file. After selecting the source iAssembly, the Table dialog box lists the currently selected columns, as shown in Figure 8.97. By selecting Column Chooser, you can specify the attributes that you want to be included in the table.

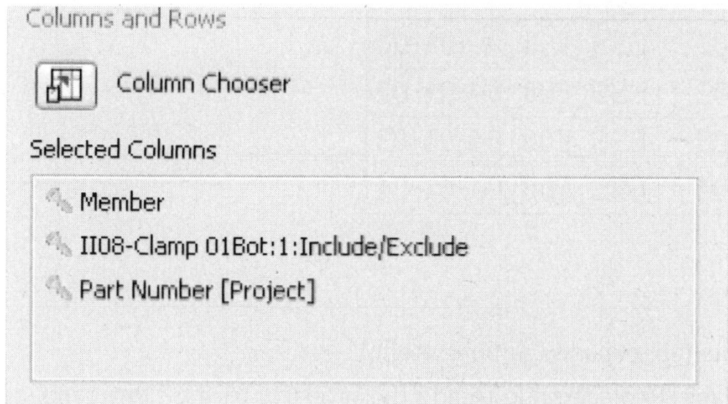

FIGURE 8.97

EXERCISE 8-4: WORKING WITH iASSEMBLIES

In this exercise, you will work with the functionality of iAssemblies. The exercise is meant to familiarize you with the functionality of iAssemblies and enable you to effectively create and work with them. First you will review the iAssembly Author tool.

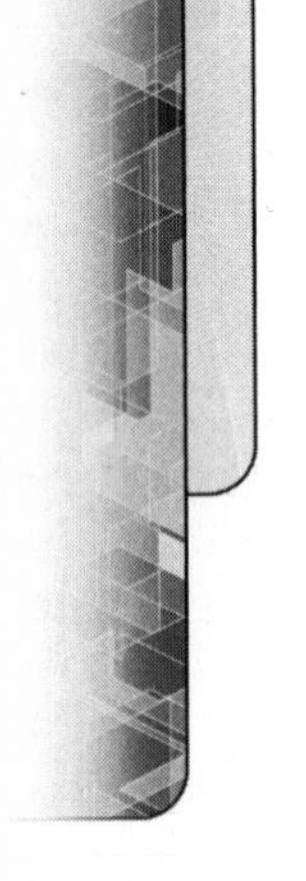

1. Open *ESS_E08_04.iam*. The top and bottom clamp components of this assembly were created as iParts.

FIGURE 8.98

2. On the Assemble tab > iPart/iAssembly panel or on the Manage tab > Author panel, click Create iAssembly.
3. In the iAssembly Author dialog box, right-click row 1 in the table, and click Insert Row from the menu.
4. Add two additional rows to the table so that you have four iAssembly members, as shown in Figure 8.99.

	Member	Part Number
1	ESS_E08_04-01	ESS_E08_04-01-01
2	ESS_E08_04-02	ESS_E08_04-01-02
3	ESS_E08_04-03	ESS_E08_04-01-03
4	ESS_E08_04-04	ESS_E08_04-01-04

FIGURE 8.99

5. On the Components tab, expand Clamp Bottom:1.
6. Double-click the Table Replace node to add the property to the right pane, as shown in Figure 8.100.

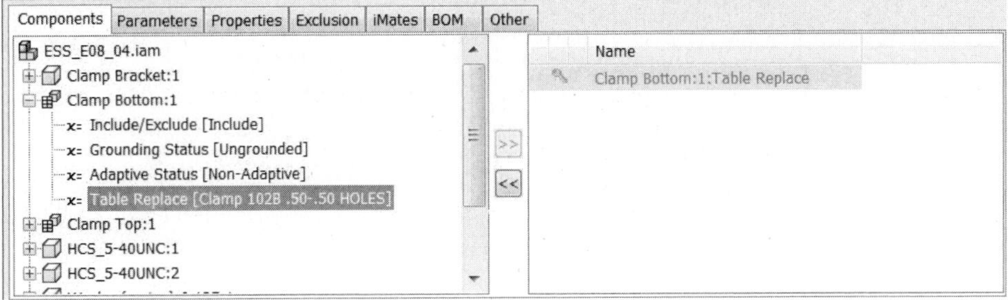

FIGURE 8.100

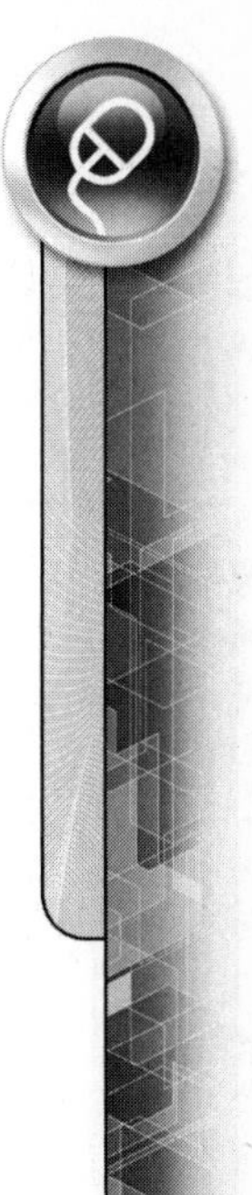

7. Repeat the process to add the Table Replace property for the Clamp Top:1 component.
8. Click the Properties tab. Expand the Project folder in the list and then double-click on Description to add the Description property to the iAssembly definition.
9. Add the description of .50/.50 HOLES to the description of the first member, as shown in Figure 8.101.

	Member	Part Number	Clamp Bottom:1: Table Replace	Clamp Top:1: Table Replace	Description
1	ESS_E08_04-01	ESS_E08_04-01-01	Clamp 102B .50-.50 HOLES	Clamp 101T .50-.50 HOLES	.50/.50 HOLES
2	ESS_E08_04-02	ESS_E08_04-01-02	Clamp 102B .50-.50 HOLES	Clamp 101T .50-.50 HOLES	
3	ESS_E08_04-03	ESS_E08_04-01-03	Clamp 102B .50-.50 HOLES	Clamp 101T .50-.50 HOLES	
4	ESS_E08_04-04	ESS_E08_04-01-04	Clamp 102B .50-.50 HOLES	Clamp 101T .50-.50 HOLES	

FIGURE 8.101

10. Add the remaining three descriptions to the table, as shown in Figure 8.102.

	Member	Part Number	Clamp Bottom:1: Table Replace	Clamp Top:1: Table Replace	Description
1	ESS_E08_04-01	ESS_E08_04-01-01	Clamp 102B .50-.50 HOLES	Clamp 101T .50-.50 HOLES	.50/.50 HOLES
2	ESS_E08_04-02	ESS_E08_04-01-02	Clamp 102B .50-.50 HOLES	Clamp 101T .50-.50 HOLES	.75/.75 HOLES
3	ESS_E08_04-03	ESS_E08_04-01-03	Clamp 102B .50-.50 HOLES	Clamp 101T .50-.50 HOLES	.50/.75 HOLES
4	ESS_E08_04-04	ESS_E08_04-01-04	Clamp 102B .50-.50 HOLES	Clamp 101T .50-.50 HOLES	.75/.50 HOLES

FIGURE 8.102

11. Select the ESS_E08_04-02 member, and replace the following parts in the table, as shown in Figure 8.103:
    - Clamp Bottom:1 with Clamp 112B .75-.75 HOLES
    - Clamp Top:1 with Clamp 111T .75-.75 HOLES

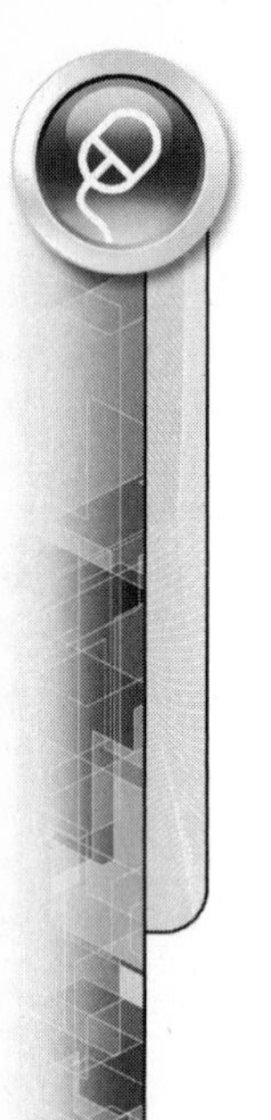

	Member	Part Number	Clamp Bottom:1: Table Replace	Clamp Top:1: Table Replace	Description
1	ESS_E08_04-01	ESS_E08_04-01-01	Clamp 102B .50-.50 HOLES	Clamp 101T .50-.50 HOLES	.50/.50 HOLES
2	ESS_E08_04-02	ESS_E08_04-01-02	Clamp 112B .75-.75 HOLES	p 101T .50-.50 HOLES	.75/.75 HOLES
3	ESS_E08_04-03	ESS_E08_04-01-03	Clamp 102B .50-.50 HOLES	Clamp 101T .50-.50 HOLES	.50/.75 HOLES
4	ESS_E08_04-04	ESS_E08_04-01-04	Clamp 102B .50-.50 HOLES	Clamp 111T .75-.75 HOLES	.75/.50 HOLES
				Clamp 121T .50-.75 HOLES	
				Clamp 131T .75-.50 HOLES	

[?] Options... Verify OK

FIGURE 8.103

12. Repeat the same replacement operations on the ESS_E08_04-03 and ESS_E08_04-04 members, and select the iPart from the list as shown in Figure 8.104.

	Member	Part Number	Clamp Bottom:1: Table Replace	Clamp Top:1: Table Replace	Description
1	ESS_E08_04-01	ESS_E08_04-01-01	Clamp 102B .50-.50 HOLES	Clamp 101T .50-.50 HOLES	.50/.50 HOLES
2	ESS_E08_04-02	ESS_E08_04-01-02	Clamp 112B .75-.75 HOLES	Clamp 111T .75-.75 HOLES	.75/.75 HOLES
3	ESS_E08_04-03	ESS_E08_04-01-03	Clamp 122B .50-.75 HOLES	Clamp 121T .50-.75 HOLES	.50/.75 HOLES
4	ESS_E08_04-04	ESS_E08_04-01-04	Clamp 132B .75-.50 HOLES	Clamp 131T .75-.50 HOLES	.75/.50 HOLES

FIGURE 8.104

13. Click OK to save and exit the iAssembly Author dialog box.
14. A Table node is added to the Assembly browser. Expand the Table entry, and double-click on each assembly configuration to cycle through the different hole arrangements in the clamp, as shown in Figure 8.105.

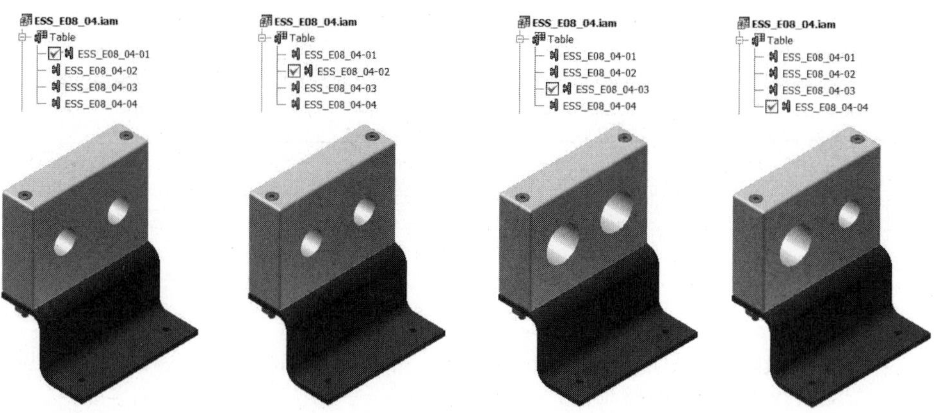

FIGURE 8.105

15. In the browser, double-click ESS_E08_04-03 to ensure this particular configuration is active.

When an iAssembly is created, the Edit Factory Scope and the Edit Member Scope commands on the Assemble tab > iPart/iAssembly panel are now available. You will now work with the Edit Member Scope mode of the iParts/iAssemblies panel, as shown in Figure 8.106. When this mode is active and changes are made to the current iAssembly, the changes are present only in the iAssembly. The remaining iAssembly configurations are unaffected by changes.

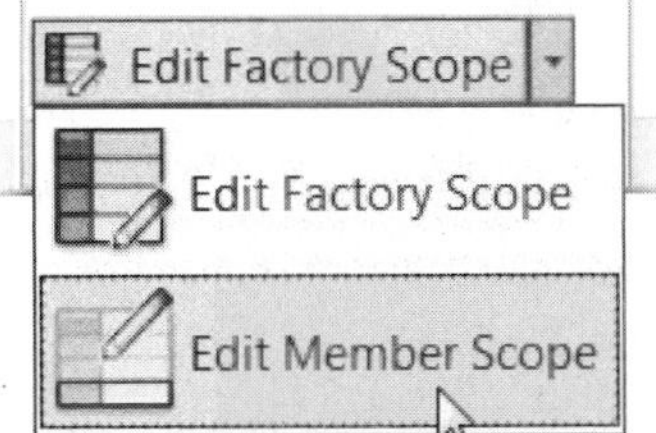

FIGURE 8.106

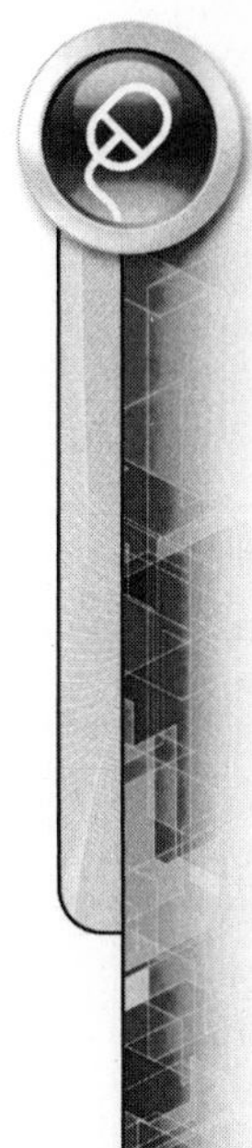

16. Set the current iAssembly to Edit Member Scope from the Assembly tab > iPart/iAssembly panel, as shown in Figure 8.106.
17. On the 3D Model tab > Modify Assembly panel, click Chamfer.
18. Chamfer both edges of both holes on both faces of the iAssembly clamp. Use a chamfer distance of **0.05** units, as shown in Figure 8.107.

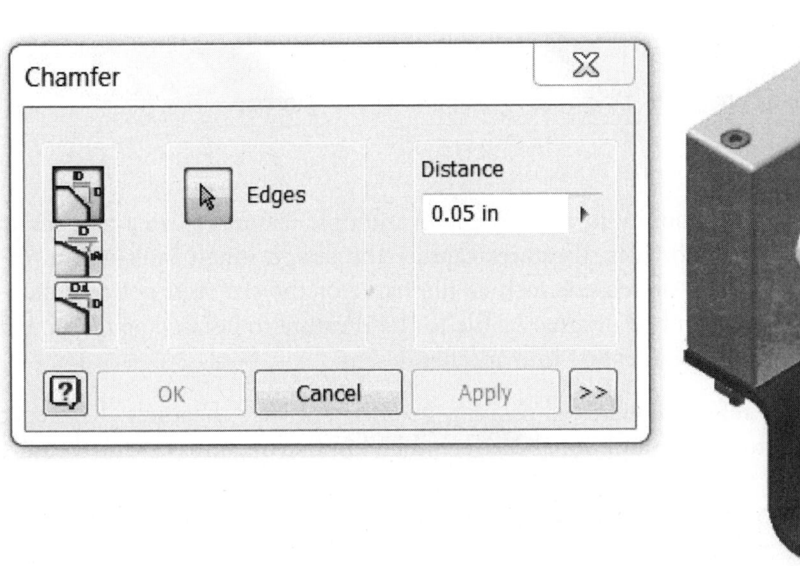

FIGURE 8.107

19. Double-click any of the other configurations of the iAssembly.

    Since Edit Member Scope was active, the chamfers are only available in the configuration that was active when they were created. Notice also that the Chamfer in the browser is excluded from the other configurations, as shown in Figure 8.108. If desired you can edit the table and change the Include/Exclude option for the chamfer feature for each member.

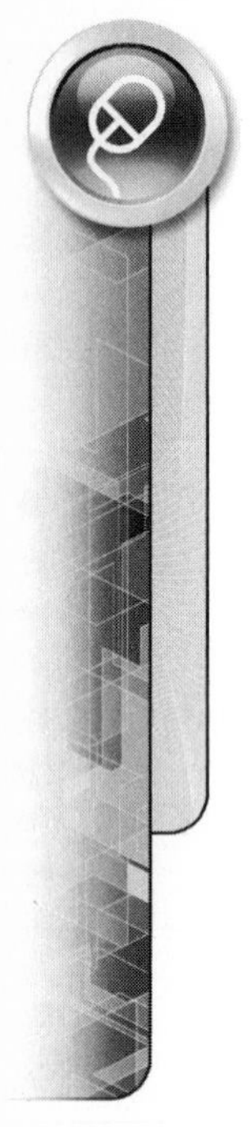

FIGURE 8.108

20. Close all open files. Do not save changes. End of exercise.

iFEATURES

iFeatures give you the ability to reuse single or multiple features from a part file in other Autodesk Inventor files. iFeatures capture the design intent built into a feature(s) that is going to be reused, such as the name or the size and position parameters. You can also embed or attach a file to the iFeature to be used as Placement Help when you place the iFeature into another design.

You can include reference edges as position geometry in your iFeatures. Reference edges allow you to capture additional design intent, but they require that the iFeature be positioned in the same way it was designed originally in every part in which it is placed. iFeatures are saved in their own type of file that has an *.ide* extension and a unique icon, as shown in Figure 8.109.

FIGURE 8.109

iFeatures are stored in a catalog. The catalog is a directory on your computer or on a server that is set to store all of the iFeatures that you create. You can browse the iFeature catalog at any time by clicking the View iFeature Catalog command on the Manage tab > Insert panel, at the bottom of the iFeature browser, as shown in Figure 8.110.

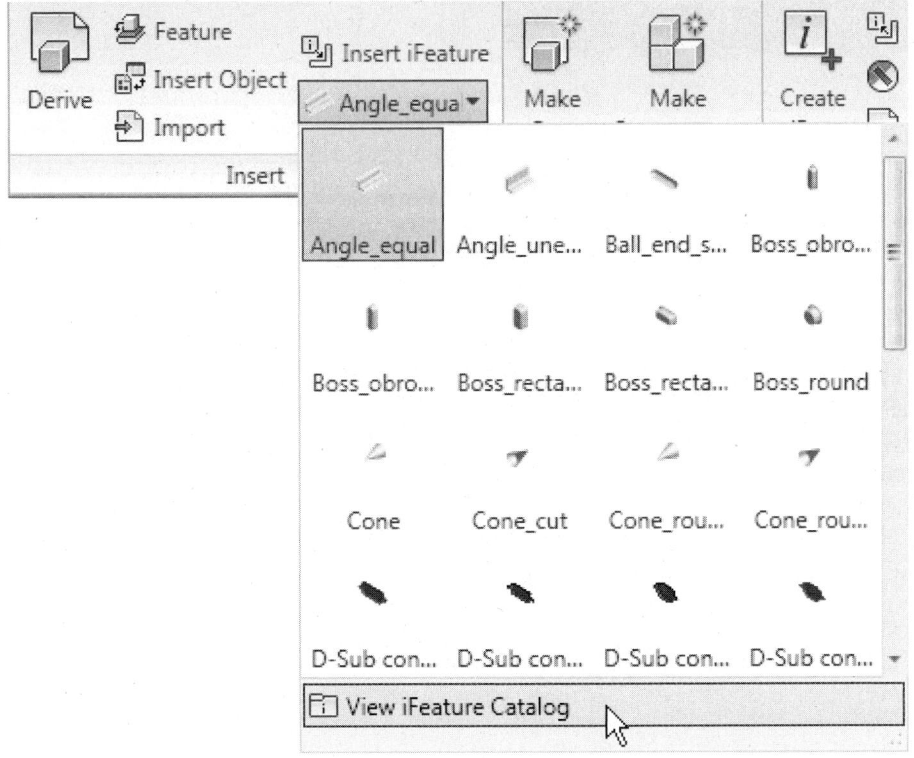

FIGURE 8.110

When the View iFeature Catalog command is selected, Windows Explorer opens to the directory specified in the iFeature root setting on the iFeature tab of the Application Options dialog box. In the iFeature root folder, you can view, copy, edit, or delete iFeatures from your catalog.

Create iFeatures

You create iFeatures using the Extract iFeature command on the Manage tab > Author panel, as shown in Figure 8.111.

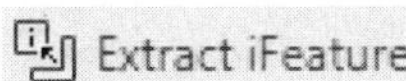

FIGURE 8.111

Once selected, the Extract iFeature dialog box appears, as shown in Figure 8.112.

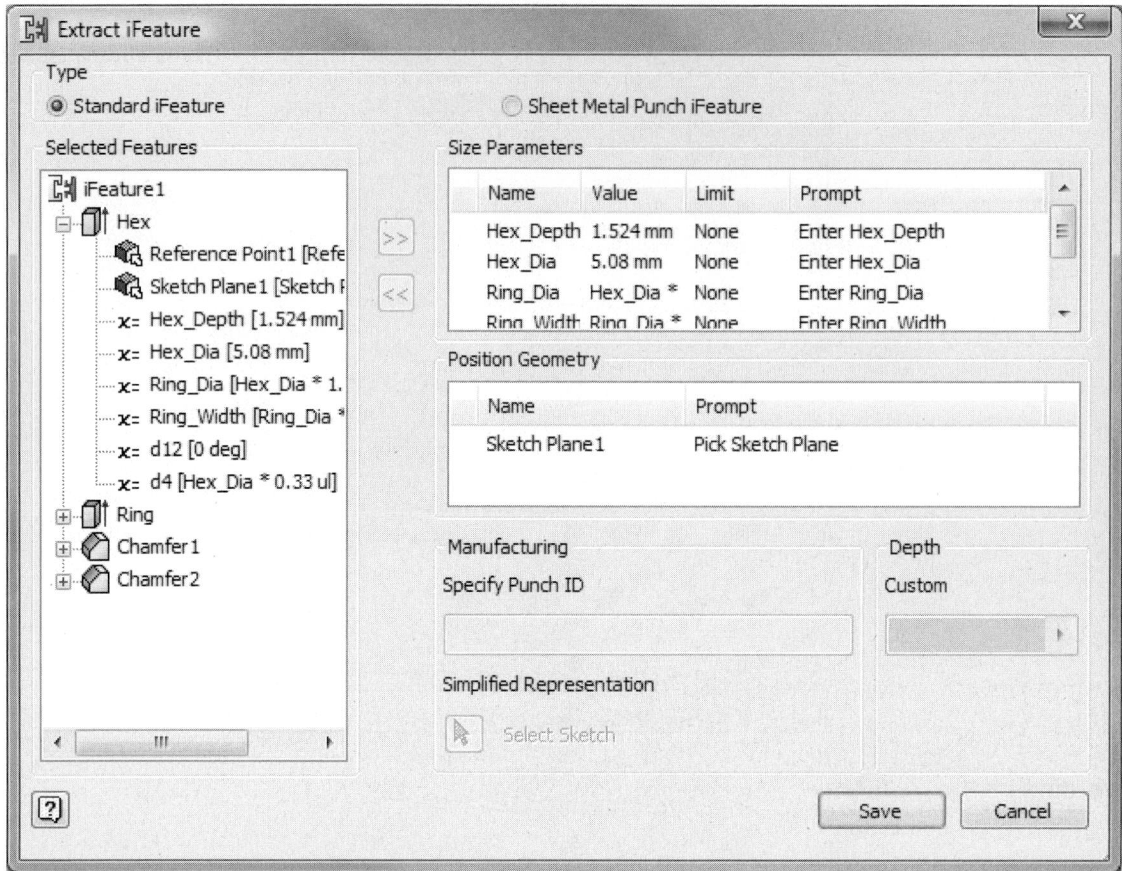

FIGURE 8.112

The Extract iFeature dialog box contains the following sections.

Type

In this area, you can select whether you want to create a Standard iFeature or a Sheet Metal Punch iFeature. In this chapter, we will focus on Standard iFeatures. Refer to Chapter 10 for information regarding the creation and use of the Sheet Metal Punch iFeature.

Selected Features

This area of the dialog box lists the features selected to be included in the iFeature. You can rename features in the list to more descriptive names to assist in working with the iFeature at a later time.

Use the Add parameters (≫) button and the Remove parameters (≪) button to move parameters from the highlighted features in the Selected Features list to the Size Parameters table.

Size Parameters

The Size Parameters table lists all of the parameters that can be modified when the iFeature is placed into another file. You can select the parameters by expanding the features listed in the Selected Features list or directly in the graphics window.

NOTE

Any parameters that have been given a name in the Parameters dialog box appear automatically in the Size Parameters pane of the Create iFeature dialog box upon iFeature creation. Renaming parameters that you want to include in an iFeature can speed up the process of creating them.

Name. Specify a descriptive name for the parameter. Use names that describe the purpose of the parameter.

Value. Place a value to be the default for the parameter when inserting your iFeature into a file. The value is restricted by settings in the Limit column.

Limit. Place restrictions on the values that are available for the parameter by using one of three options, None, Range, or List, as shown in Figure 8.113.

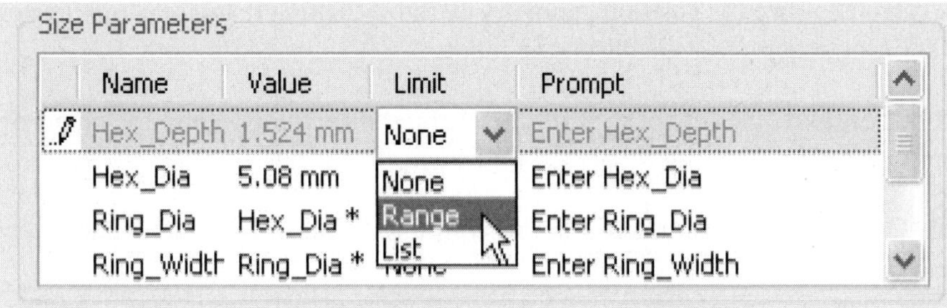

FIGURE 8.113

None specifies that no restrictions be placed on the Value field. Range gives you the ability to specify a minimum and maximum value, including less than, equal to, and infinity, as shown in Figure 8.114 on the left. You can specify the default value to use. The List option, as shown in Figure 8.114 on the right, allows you to predefine a list of values. Upon placement, you can choose from these values for the size parameters.

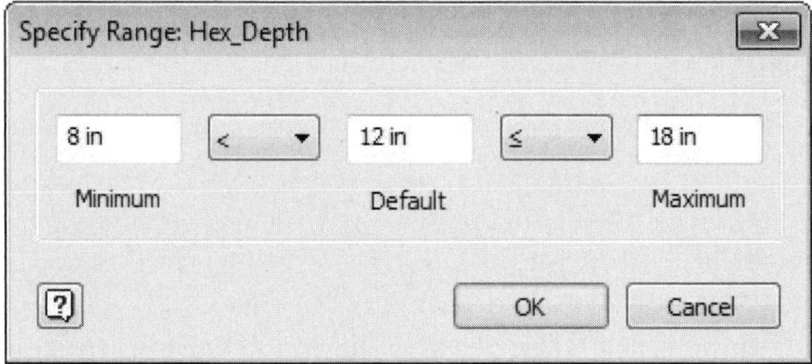

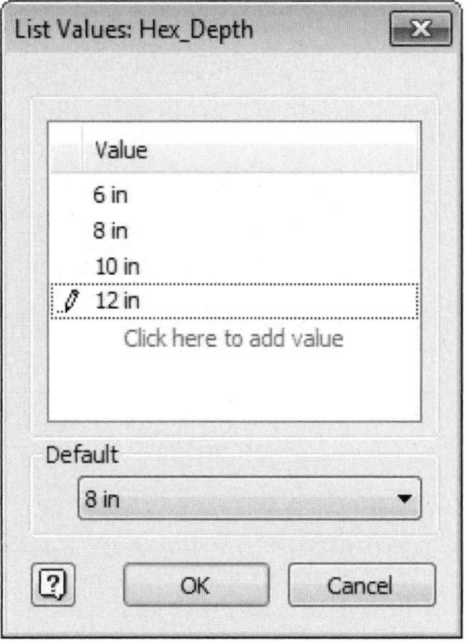

FIGURE 8.114

Prompt. Enter descriptive instructions to explain further why the parameter is used. The text entered in the prompt field appears in a dialog box during the iFeature placement.

Position Geometry

Specify the geometry of the iFeature that is necessary to position it on a part. You can add or remove geometry to or from the Position Geometry list in the Selected Features tree by right-clicking the geometry and selecting Expose Geometry. You remove geometry from the Position Geometry list by right-clicking it and selecting Remove Geometry.

Name. Describes the position geometry.

Prompt. Enter descriptive instructions to prompt a user for position geometry. The text entered in the prompt field appears in a dialog box during the iFeature positioning. You can customize the position geometry by right-clicking it and selecting one of the two additional options available on the menu.

Make Independent. Select this option to separate entries for geometry shared by more than one feature.

Combine Geometry. Select this option to combine listings of geometry shared by more than one feature in a single entry.

Manufacturing

This field is used when creating a Sheet Metal Punch iFeature. Refer to Chapter 10 for more information.

Depth

This field is used when creating a Sheet Metal Punch iFeature. Refer to Chapter 10 for more information.

Insert iFeatures

Insert iFeatures using the Insert iFeature command on the Manage tab > Insert panel, as shown in Figure 8.115.

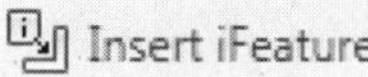

FIGURE 8.115

Once selected, the Insert iFeature dialog box appears, as shown in Figure 8.116. You can select tasks from the tree structure in the left pane of the dialog box or move forward and backward using the Next and Back buttons. Click the Browse button to navigate the iFeatures folder structure and select an iFeature to place.

FIGURE 8.116

The Insert iFeature dialog contains the following sections.

Select

Choose the iFeature that you want to insert using the Browse button.

Position

This feature lists the names of the interface geometries specified during the creation process of the iFeature. After selecting the corresponding geometry in the part where you are placing the iFeature, click the arrowhead on the positioning symbol to either move or rotate the symbol, as shown in Figure 8.117.

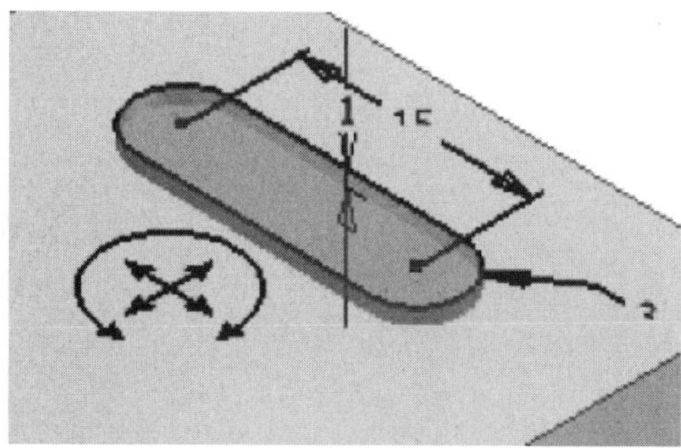

FIGURE 8.117

You can also specify a precise rotation value directly in the angle field of the dialog box. After the requirement has been satisfied, a checkmark will be placed in the left column, as shown in Figure 8.118.

Name. This section lists the named interface geometry.

Angle. This section shows the default angle of the placement geometry on the iFeature.

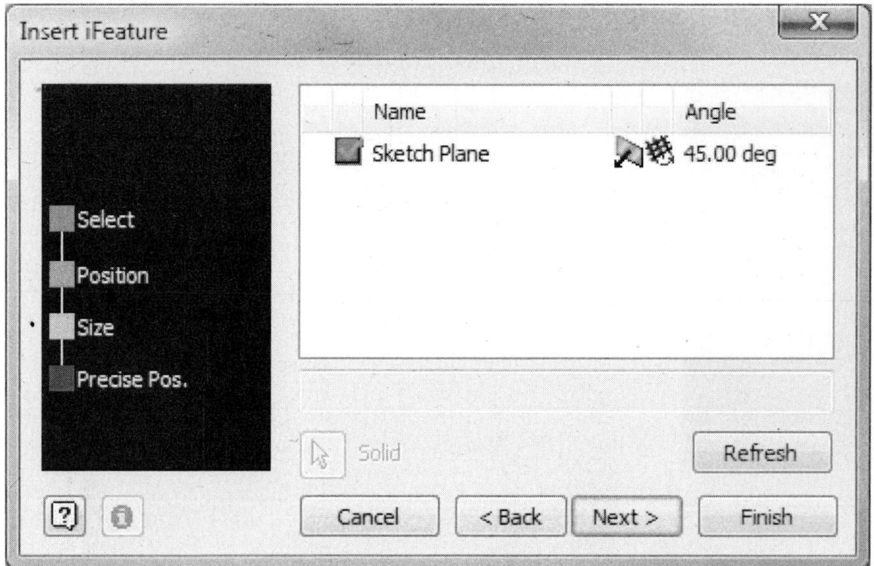

FIGURE 8.118

Move Coordinate System. This section defines horizontal or vertical axes when the iFeature has horizontal or vertical dimensions or constraints included.

Size

This shows the names and default values specified for the iFeature, as shown in Figure 8.119. Click in the row to edit the values, and then click Refresh to preview the changes.

Name. This section lists the name of the parameter.

Value. This section lists the value of the parameter.

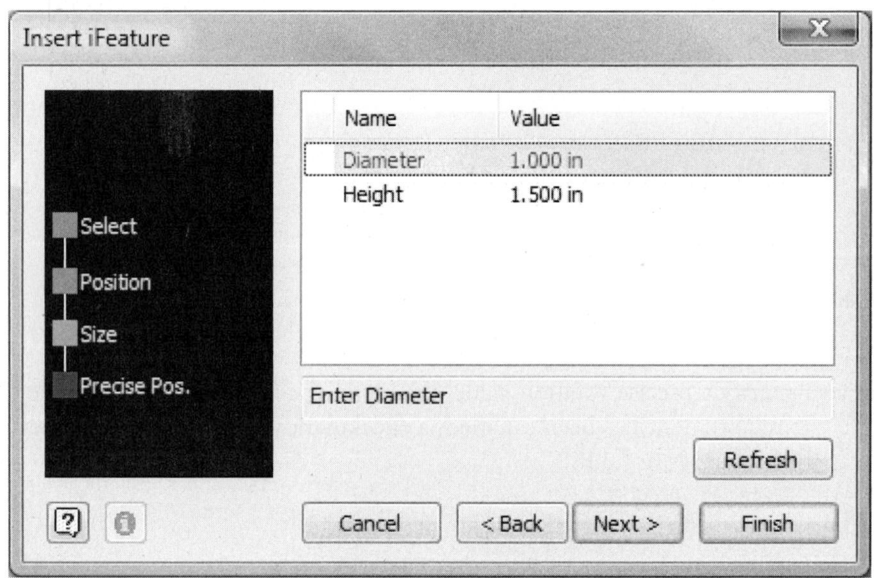

FIGURE 8.119

Precise Position

This feature further refines the position of the iFeature, using either dimensions or constraints after you have placed it.

Activate Sketch Edit Immediately. Click this option to activate the sketch of the iFeature and the 2D Sketch commands. You can then apply additional dimensions and/or constraints to position the iFeature on the part.

Do not Activate Sketch Edit. Click this option, as shown in Figure 8.120, to position the iFeature without applying additional constraints or dimensions.

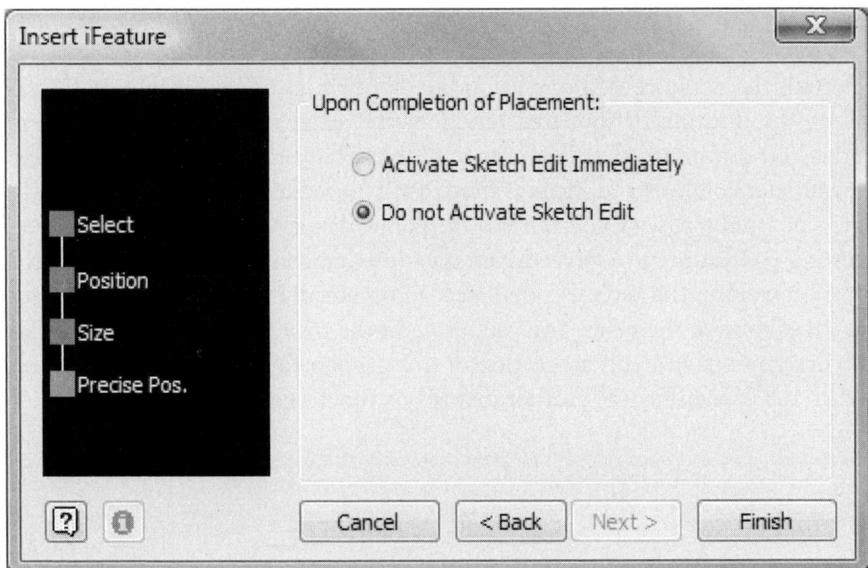

FIGURE 8.120

Editing iFeatures

You can edit an iFeature by opening the *.ide* file in Autodesk Inventor. Five commands are available in the iFeature environment when the file is opened: Edit iFeature, View Catalog, iFeature Author Table, Edit using Spread Sheet, and Change Icon.

The Edit iFeature command, as shown in Figure 8.121, opens the Edit iFeature dialog box. You cannot change which parameters are used to define the iFeature after it has been created, but you can modify the size parameters and position geometry by editing the properties for the name, value, limit, and prompt.

FIGURE 8.121

TIP

If you are building a library of iFeatures, you may want to consider designing them in a simple shape (box, cylinder, etc.) and keeping the part files in an archive folder so you can easily republish the iFeature if necessary. For example, you may want to refine the parameters that are listed or the placement geometry.

The View Catalog command operates the same as noted previously. It opens Windows Explorer to the location specified in the iFeature root field on the iFeature tab of the Application Options dialog box.

The iFeature Author Table command, as shown in Figure 8.122, allows you to create iFeatures that are driven by a table.

FIGURE 8.122

Once selected, the iFeature Author dialog box appears, as shown in Figure 8.123. The table in the iFeature Author tool functions the same as it does when working with iParts. You can insert, delete, and specify the default row. You can also define custom parameter columns and include a range for the value. The main difference is that there is no option to set a column as a file name, display style, or material. If you are converting an iFeature to a table-driven iFeature that contains a parameter with a list of possible values, the table is populated automatically with the different sizes from the list as rows of the table. You can also edit the table via a spreadsheet. Refer to the tab descriptions in the iPart section of this chapter for additional information on how each tab of the iFeature Author dialog box functions.

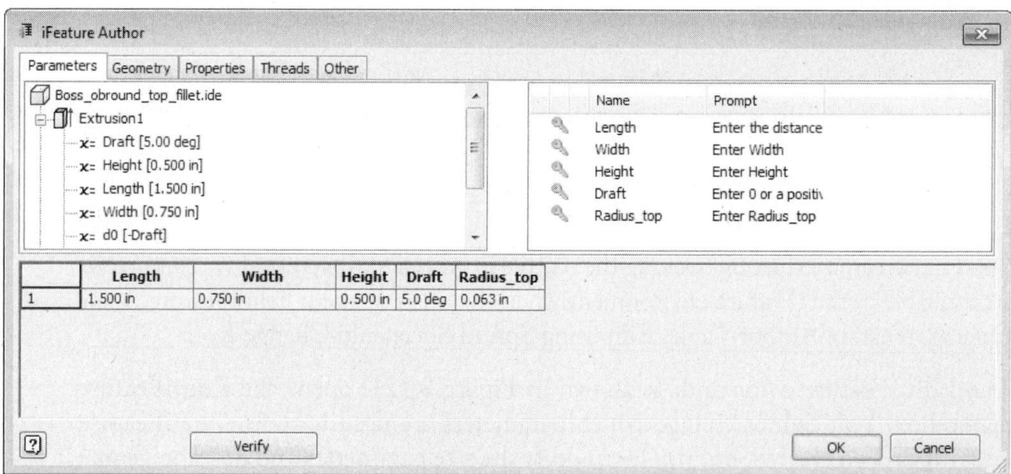

FIGURE 8.123

The Edit Using Spread Sheet command, shown in Figure 8.124, allows you to edit the iFeature table using Microsoft Excel instead of the iFeature Author table. The last command available is Change Icon. The Change Icon command allows you to create a custom icon for the iFeature that is displayed in the browser once it has been included in a part.

FIGURE 8.124

Inserting Table-Driven iFeatures

You insert table-driven iFeatures the same way that you would insert a typical iFeature. The wizard displays the key parameters as a drop-down list while defining the Size parameters, and any custom parameters will have the edit field available for entry.

The browser displays table-driven iFeatures in a similar manner to that used with iParts.

When you place a table-driven iFeature in a part file, a new *.ide* file is not created. The placement is similar to that of an iFeature that is not table driven. The file in which the iFeature is placed contains an independent copy of the iFeature that is not associative to the original *.ide* file. The spreadsheet that contains the table information is not visible in the browser, and you cannot edit it once you place it in the part file.

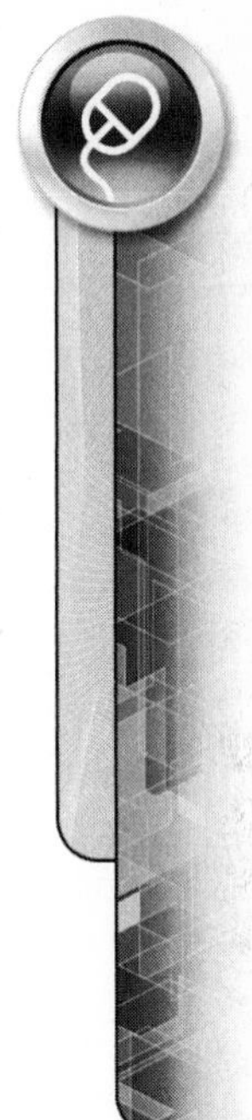

EXERCISE 8-5: CREATING AND PLACING iFEATURES

In this exercise, you will create a hex drive iFeature from an existing component. You will then insert the iFeature into another part file.

1. Open *ESS_E08_05.ipt.*
2. Click the Parameters command on the Quick Access toolbar, and select Filter > Renamed from the lower left corner of the Parameters dialog box, as shown in **Figure 8.125**, to review the list of renamed Model Parameters. The renamed diameter *(Hex_Dia)* and depth *(Hex_Depth)* parameters of the hex drive extrusion are the variables that will be extracted with the iFeature. The geometry of the raised ring around the hex drive is linked to the diameter of the hex drive.

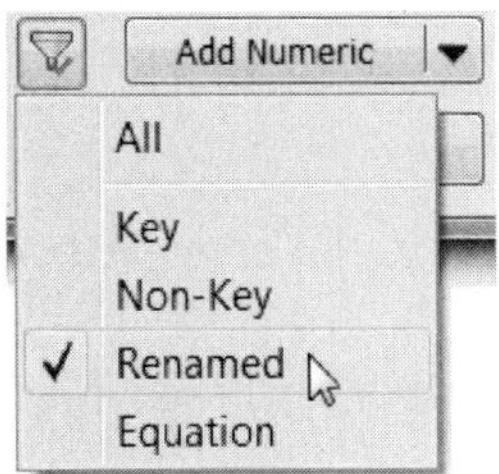

FIGURE 8.125

3. Click the Filter button and change the filter to All.
4. In the Parameters dialog box, click Done.
5. From the Manage tab > Author panel, click Extract iFeature.
6. In the browser click the features named *Hex* and *Ring* in the browser.

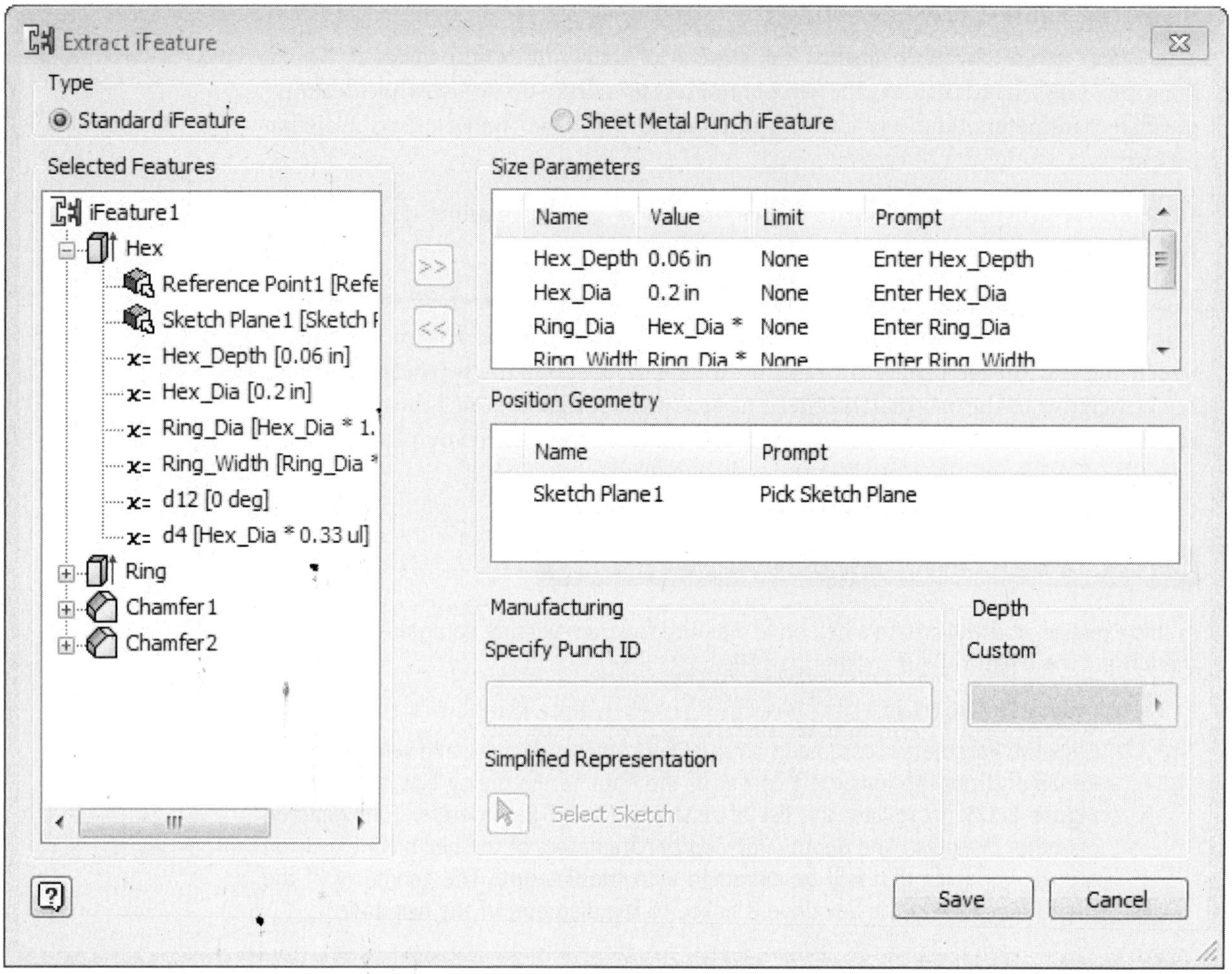

Extract iFeature

Type

Standard iFeature   Sheet Metal Punch iFeature

Selected Features

iFeature1
Hex
Reference Point1 [Refe
Sketch Plane1 [Sketch P
x= Hex_Depth [0.06 in]
x= Hex_Dia [0.2 in]
x= Ring_Dia [Hex_Dia * 1.
x= Ring_Width [Ring_Dia *
x= d12 [0 deg]
x= d4 [Hex_Dia * 0.33 ul]
Ring
Chamfer1
Chamfer2

Size Parameters

Name	Value	Limit	Prompt
Hex_Depth	0.06 in	None	Enter Hex_Depth
Hex_Dia	0.2 in	None	Enter Hex_Dia
Ring_Dia	Hex_Dia *	None	Enter Ring_Dia
Ring_Width	Ring_Dia *	None	Enter Ring_Width

Position Geometry

Name	Prompt
Sketch Plane1	Pick Sketch Plane

Manufacturing   Depth

Specify Punch ID   Custom

Simplified Representation

Select Sketch

Save   Cancel

FIGURE 8.126

7. Click the *Pick Sketch Plane* prompt under Position Geometry, and enter **Select Plane to Position the Hex Drive** as the new prompt.
8. Expand *Hex* under Selected Features, right-click Reference Point1, and click Expose Geometry from the menu, as shown in Figure 8.127.

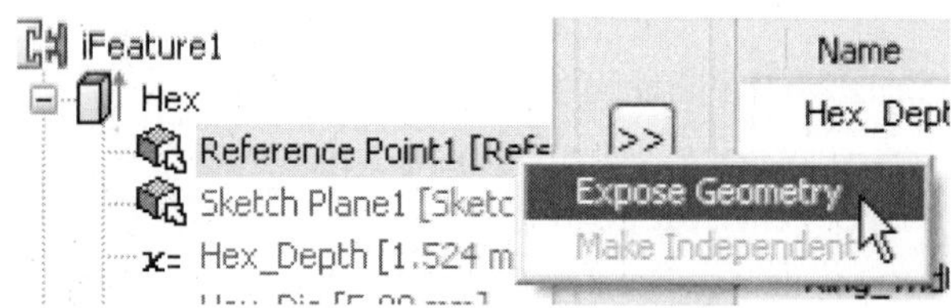

FIGURE 8.127

9. The point is added to the Position Geometry list. This point constrained the hex drive extrusion concentric to the circular face on the end of the cylinder. The point referenced will allow you to select model geometry to position the iFeature during placement in a new part. Click the *Pick Reference Point* prompt under Position Geometry.

- Highlight the text, and replace it with **Select Point to Position Hex Drive Centerline** as the new prompt.
- Click and drag the Reference Point1 position geometry below Sketch Plane1, as shown in Figure 8.128.

Position Geometry

Name	Prompt
Reference Point1	Select Point to Position Hex Drive Centerline
Sketch Plane1	Select Plane to Position the Hex Drive

FIGURE 8.128

- Figure 8.129 shows the completed operation.

Position Geometry

Name	Prompt
Sketch Plane1	Select Plane to Position the Hex Drive
Reference Point1	Select Point to Position Hex Drive Centerline

FIGURE 8.129

10. Three parameters, *Ring_Height, Ring_Width,* and *Ring_Dia,* are linked to the hex drive diameter and are not required in the iFeature. To remove these parameters from the list, perform the following actions:
    - In the Size Parameters area, click *Ring_Height,* and click the Remove parameters button (≪).
    - In the Size Parameters area, click *Ring_Width,* and click the Remove parameters button (≪).
    - In the Size Parameters area, click *Ring_Dia,* and click the Remove parameters button (≪), as shown in Figure 8.130.

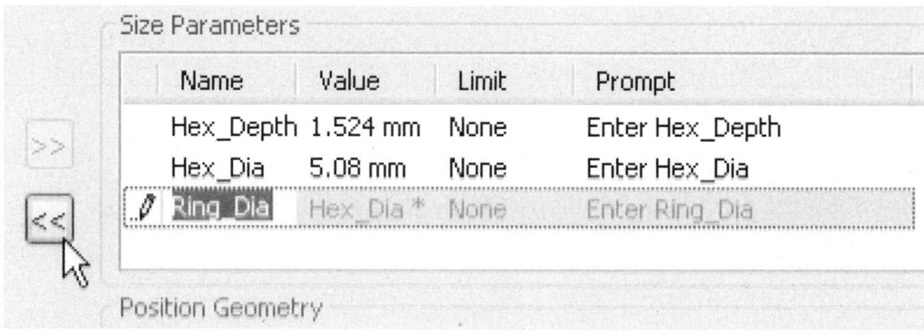
Size Parameters

Name	Value	Limit	Prompt
Hex_Depth	1.524 mm	None	Enter Hex_Depth
Hex_Dia	5.08 mm	None	Enter Hex_Dia
Ring_Dia	Hex_Dia *	None	Enter Ring_Dia

FIGURE 8.130

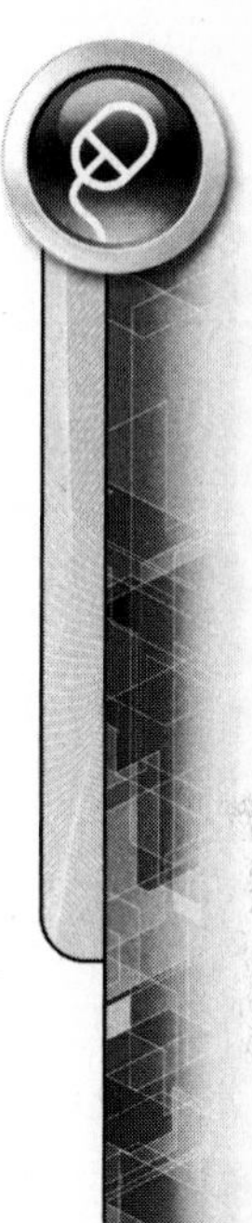

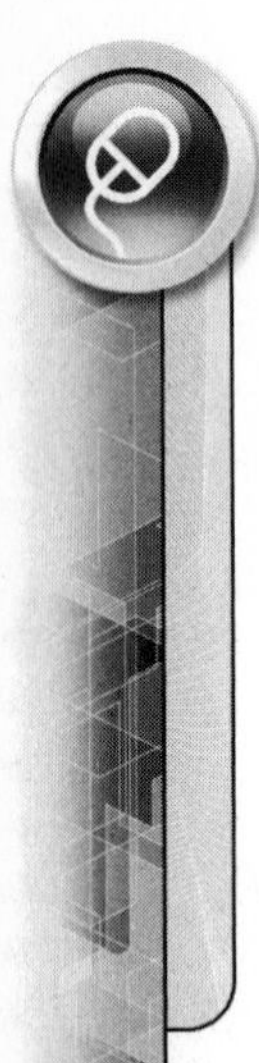

11. Only two parameters should now exist in the Size Parameters area: Hex_Depth and Hex_Dia.
12. To specify a descriptive browser name for the iFeature when placed in a part, at the top of the Selected Features list, right-click iFeature1 and select Rename from the menu. Enter **Hex Drive**. The Save As dialog box will open. **Enter Hex Drive** in the File name field, and place the file in the same location as the exercise files. Click Save, enter **Hex_Drive** in the File name field, and place the file in the Chapter 08 folder where the exercise files are installed.
13. Close the file. Do not save changes.
14. Open *ESS_E08-05-Lock Post.ipt.*
15. Click the Insert iFeature command on the Manage tab > Insert panel.
    - Navigate to the directory where the exercises were installed.
    - Select *Hex Drive.ide.*
    - Click Open.
16. To locate Sketch Plane1, click the front face of the lock post, as shown in Figure 8.131.

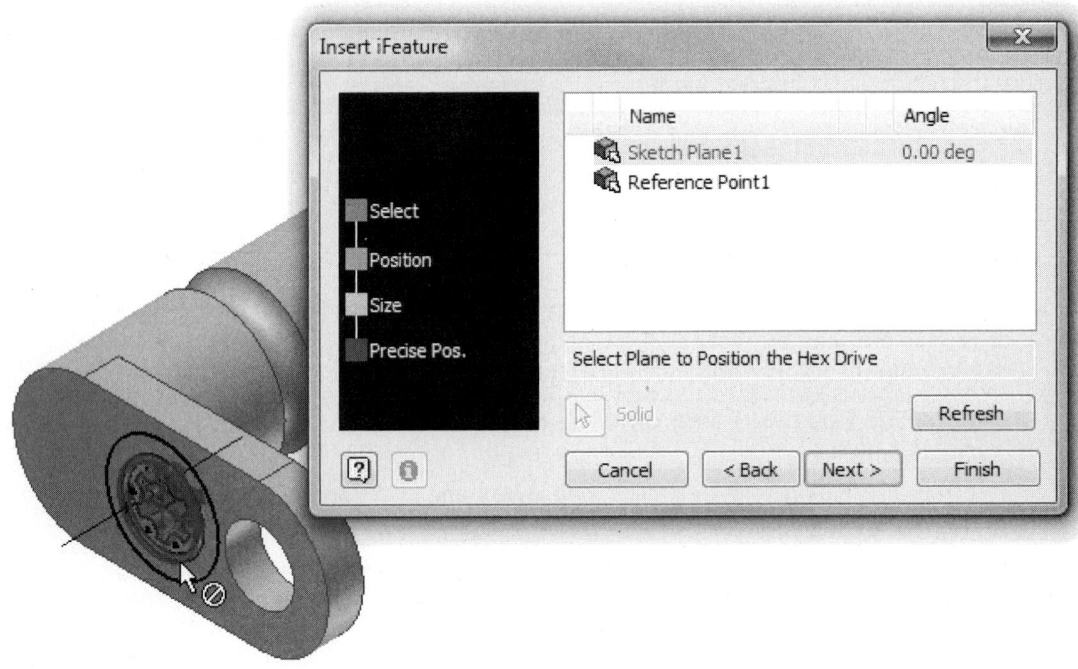

FIGURE 8.131

17. Click Reference Point1 in the Insert iFeature dialog box.
    - Click the left, outer circular edge of the lock post.
    - Click the point that is displayed on the part, as shown in Figure 8.132.

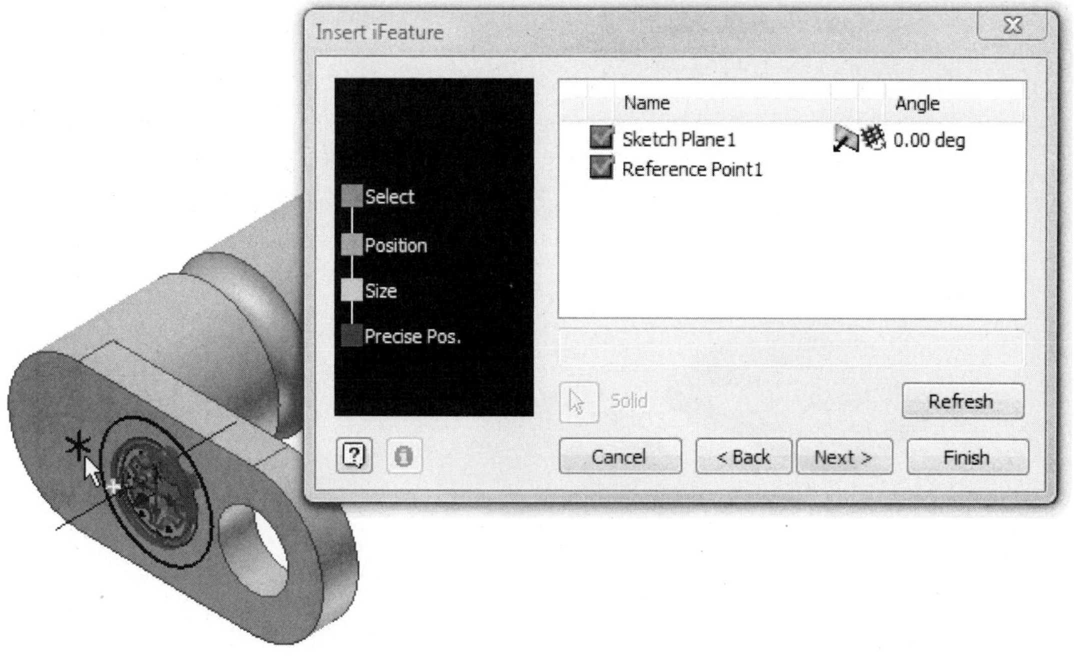

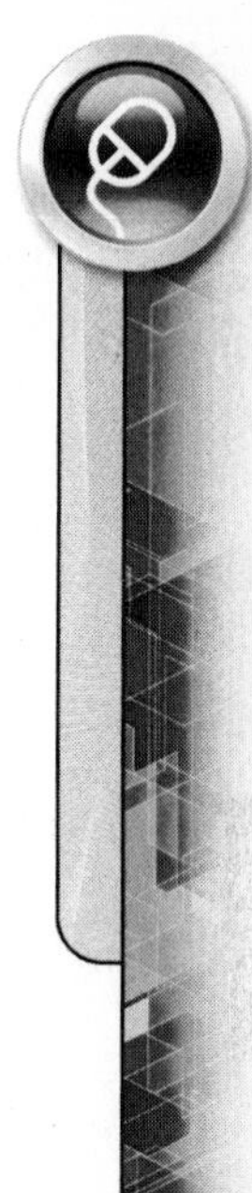

FIGURE 8.132

18. Click Next.
19. Leave the *Hex_Depth* and *Hex_Dia* variables as their default values.
20. Click Next.
21. The Do not Activate Sketch Edit option should be checked, if not check it.
22. Click Finish.
23. The hex drive iFeature is placed with its center point coincident with the selected center point of the arc. Edit the size of the iFeature by double-clicking on the Hex_Drive:1 feature in the browser and enter different values for the feature. Click Finish when done and the feature will update.

FIGURE 8.133

24. Close all open files. Do not save changes. End of exercise.

APPLYING YOUR SKILLS

Skill Exercise 8-1

In this exercise, you automate an iPart factory by controlling two iMates from the table. One version of the iPart has a single button, and the other has two buttons, as shown in Figure 8.134.

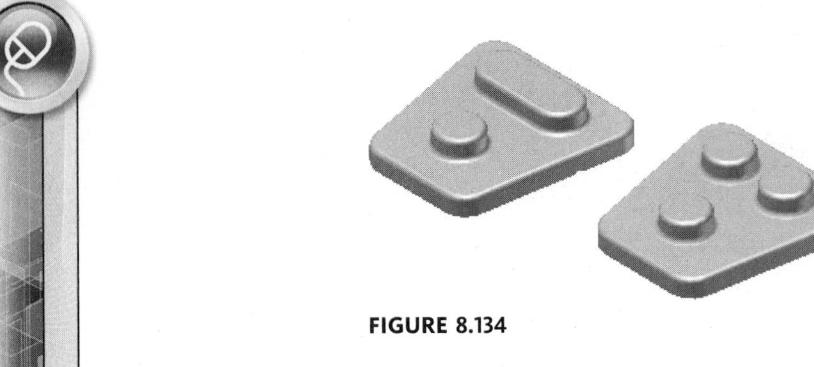

FIGURE 8.134

Place two different versions of the iPart in an assembly. Each version should be automatically positioned to the appropriate mating part using the automated iMates that you create, as shown in Figure 8.135.

FIGURE 8.135

Start with an existing iPart factory of a rubber membrane for a remote keyless entry device.

1. Open ESS_E08_06.ipt and activate the single and double members of the iPart factory. Explore how this iPart factory uses feature suppression from the table to control the number of buttons at the top of the membrane.
2. In the following steps, you will use the button geometry to create a unique iMate for each iPart version.

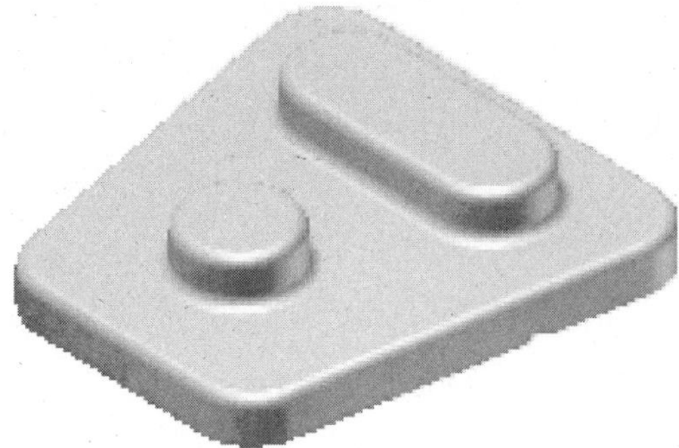

FIGURE 8.136

3. Activate the Single member iPart and then use the Edit Member Scope functionality to add a mate type axial iMate named **Axis1** to the cylindrical face as shown in Figure 8.137.

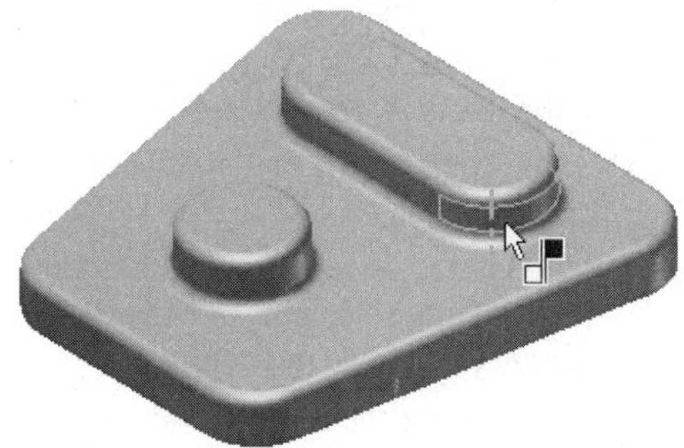

FIGURE 8.137

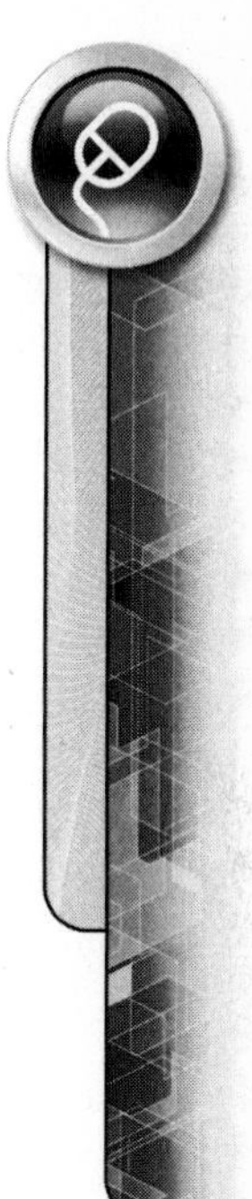

4. Activate the Double member and add a mate type axial iMate named **Axis2**, as shown in Figure 8.138.

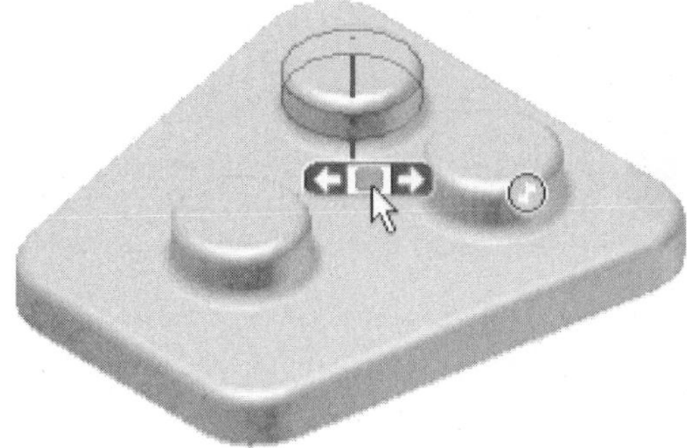

FIGURE 8.138

5. In the iPart Author dialog box, click the iMates tab, and note the Axis1 and Axis2 identifying labels IM0003 and IM0004, as shown in Figure 8.139.

ESS_E08_06.ipt
TopFace:IM0001
MainAxis:IM0002
Axis1:IM0003
Axis2:IM0004

FIGURE 8.139

6. In the table, verify the values for IM0003 and IM0004 with Figure 8.140. These values were automatically added and configured because Edit Member Scope was turned on.

	Member	Part Number	Extrusion3	Fillet3	Extrusion4	Fillet4	IM0003	IM0004
1	Single	Single	Compute	Compute	Suppress	Suppress	Compute	Suppress
2	Double	Double	Suppress	Suppress	Compute	Compute	Suppress	Compute

FIGURE 8.140

7. Verify your changes to the iPart table by activating the two iPart members from the browser.
8. Make the Single member active and save the iPart factory as *Membrane.ipt*.
9. Next open *ESS_E08_06B.iam*, an assembly that contains the mating cases for the Membrane iParts.
10. Use the Place Component command and the Interactively Place with iMates option to place the *Membrane.ipt* file. Use the Generate Remaining iMate Results to place and automatically constrain the previewed single member.
11. Change the member to the Double version and use the Place at All Matching iMates option from the menu to place the component.

FIGURE 8.141

12. Close the file. Do not save changes. End of exercise.

CHECKING YOUR SKILLS

Use these questions to test your knowledge of the material covered in this chapter.

1. True __ False __ iMates are created while inside of a part file.
2. What needs to be selected to have a single component containing iMates be placed and automatically constrained in an assembly?
   a. Automatically Generate iMates on Place
   b. Place Component with iMate
   c. Interactively Place with iMates
   d. Use Composite iMate
3. True __ False __ When creating parameters in a spreadsheet, the data items must be in the following order: parameter name, value or equations, unit of measurement, and, if needed, a comment.
4. What is the difference between a Model Parameter and a User Parameter?
5. True __ False __ Multiple versions of an iPart can be placed in an assembly.
6. True __ False __ When you make changes to an iPart factory, the changes are updated automatically in iParts that have been placed in assemblies.
7. Which tab of the Place iPart or iAssembly dialog box can be used to create a new member of the factory during placement in an assembly?
   a. Keys
   b. Table
   c. Tree
   d. None of the above
8. True __ False __ You can add features to standard iParts after they have been placed in an assembly.
9. True __ False __ Named parameters are added automatically as Size parameters during iFeature creation and cannot be removed.
10. What happens when you create a table-driven iFeature and one of the original parameters contains a list of values for the parameter?
11. True __ False __ iMates must be used as single iMates in order to facilitate easy assembly of components.
12. True __ False __ iMates are automatically assigned an index tag that is a unique value used to identify the iMate properties.
13. Explain what the following three (3) table cell background colors indicate when seen in the iPart author dialog box: Green, Dark Blue, and Red.
14. True __ False __ By defining a Proxy Path in the Libraries section of a project file, you can designate a specific location where you want generated iPart members to be stored on disc.
15. True __ False __ You cannot create an icon that will be displayed in the Browser when defining an iFeature.

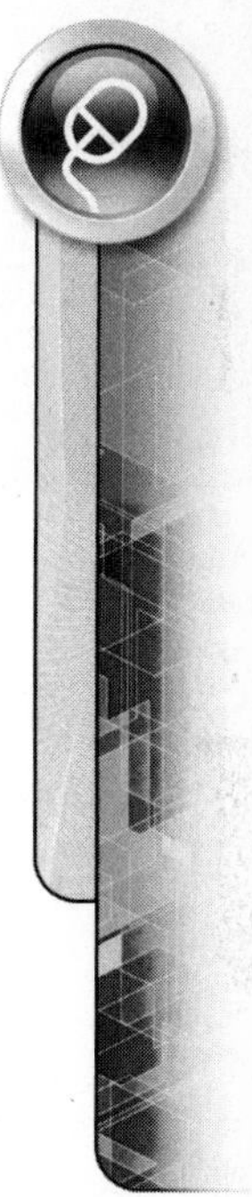

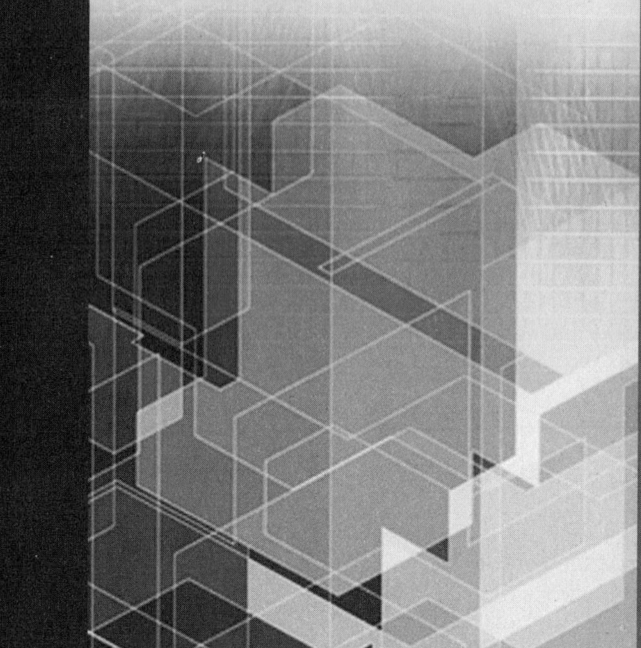

CHAPTER
9
Advanced Assembly Modeling Techniques

INTRODUCTION

In this chapter, you will learn how to use advanced assembly modeling techniques. Using these techniques, you can work more efficiently with assemblies. The techniques discussed will help you manage different views of your assemblies, improve the performance of your system, and analyze assemblies for interference between components. They will also aid in the creation of drawings from those assemblies.

OBJECTIVES

After completing this chapter, you will be able to:

- Create design view representations
- Create an assembly substitution
- Create flexible assemblies
- Create positional representations
- Create overlay drawing views
- Detect contact in assemblies
- Mirror an assembly
- Copy an assembly
- Create assembly work features
- Create assembly features
- Use the Frame Generator
- Locate commands related to the Content Center and Design Accelerator

DESIGN VIEW REPRESENTATIONS

While working in an assembly, you may want to save configurations that show the assembly in different states and from different viewing positions. Design view representations can store the following information:

- Component visibility: visible or not visible
- Sketch visibility: visible or not visible
- Component selection status: enabled or not enabled
- Color settings and style characteristics applied in the assembly
- Zoom magnification
- Viewing angle
- Display of origin work planes, origin work axes, origin work points, user work planes, user work axes, and user work points

You can use design view representations while working on the assembly and when creating presentation views or drawing views. You can also use them to reduce the number of items that display in an assembly, allowing large assembly files to be opened more quickly.

Access design view representations through the Representations folder > View node in the Assembly browser, as shown in Figure 9.1. Each assembly file contains a folder called Representations, as shown in Figure 9.1. Expanding this folder will display all of the representations that can be used. Notice the View, Position, and Level of Detail nodes. Expanding the View node will display all design view representations defined in the assembly model.

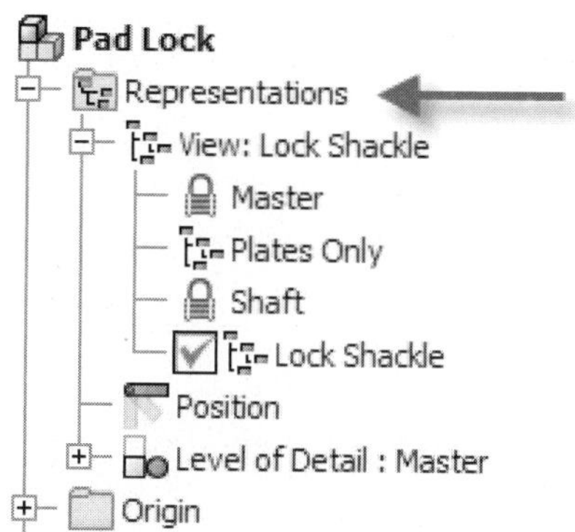

FIGURE 9.1

Creating a New Design View Representation

Begin the process of creating a new design view representation by expanding the Representations folder and the View: Default listing.

To create a design view representation, follow these steps:

1. In the browser, move your cursor over the View listing under the Representations folder and right-click.
2. Select New from the menu, as shown in Figure 9.2 on the left.

Completing these steps will create a new design view representation called View1 and make it current. The current design view representation is identified by the presence of a checkmark next to its name in the browser, as shown in Figure 9.2 on the right. The name of the current design view representation is also shown next to the View node under the Representations folder in the Assembly browser.

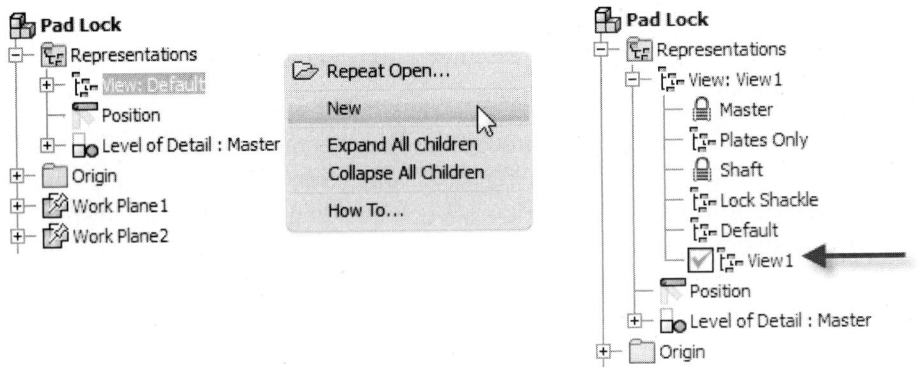

FIGURE 9.2

3. It is considered good practice to change the name of the design view representation to something more meaningful. As shown in Figure 9.3 on the right, View1 has been renamed Alt - Color Blue.

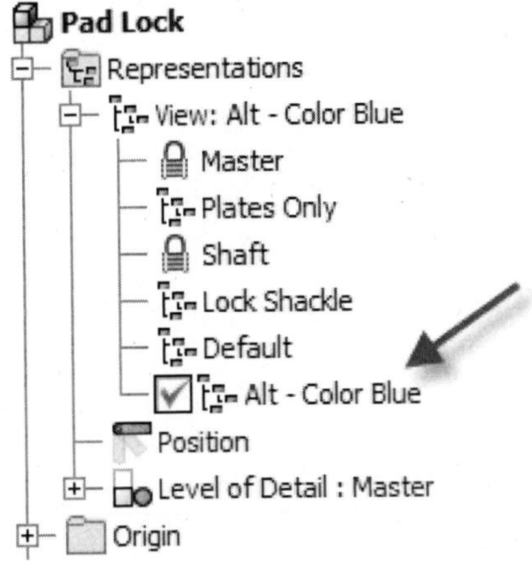

FIGURE 9.3

4. With a current design view representation set, various parts of the assembly can be modified. For this pad lock example, certain components of the assembly would be hidden such as the case, lock shackle, and dial. These changes are saved automatically to the current design view representation. In addition to

component visibility, you can also save the visibility state of sketches, work features, both user and origin, whether or not a component is enabled, color settings of components, the zoom magnification, and the viewing angle.

5. Double-clicking on the Master or any other design view representation name will return the assembly back to its original representation. In Figure 9.4, the default representation of the pad lock is shown on the left, and an alternate representation displaying the combo subassembly is shown on the right.

FIGURE 9.4

6. Create additional design view representations as needed. You may want to create design view representations using alternate color schemes or to assist in the creation of presentation files or drawing views where particular components are disabled or hidden.

Once you are satisfied with all of the changes made to a design view representation, you can prevent any further changes to a representation by locking it. Right-clicking on a design view representation name will display the menu shown in Figure 9.5. Selecting Lock from this menu will add a padlock icon to the design view representation, as shown next to the Master design view representation in Figure 9.5. The Master design view is always locked.

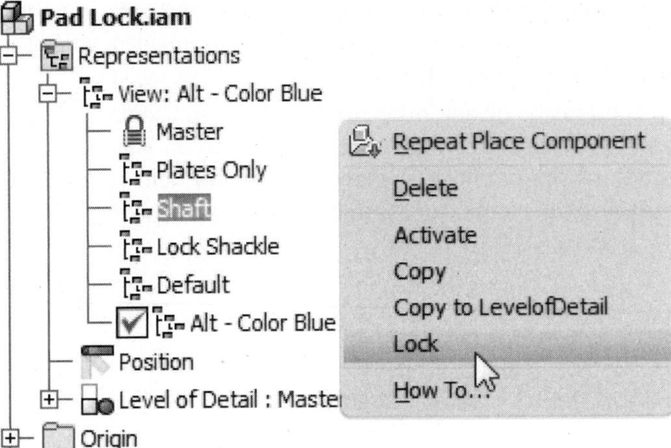

FIGURE 9.5

If you want to make additional changes to a locked design view representation, right-click on the name, and select Unlock from the menu.

Increasing Performance through Design View Representations

Through design view representations, additional control of the visibility of each subassembly is possible, allowing you to turn off the visibility of unimportant components to increase performance. The following sections describe additional design view representation controls, as shown in Figure 9.6.

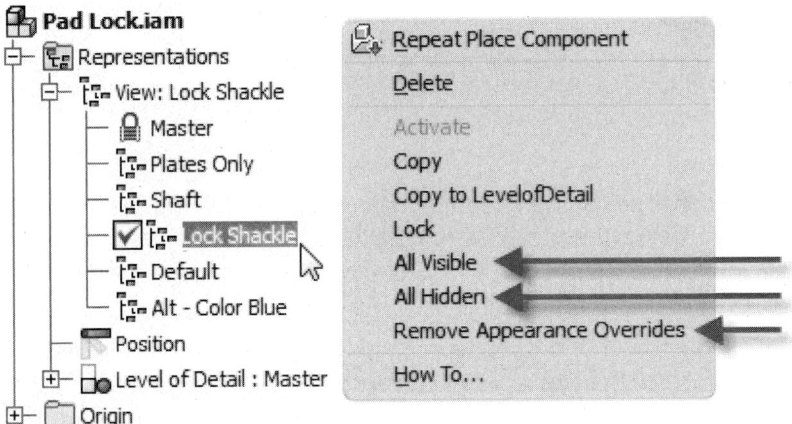

FIGURE 9.6

All Visible. Select this control to make all of the components in the assembly visible.

All Hidden. Select this control so that none of the components in the assembly will be visible. This can be beneficial for managing large assembly files. Opening a large assembly with all components hidden will allow you to manually turn on the visibility for only those components with which you need to work.

Remove Appearance Overrides. Select this control to restore the original appearance settings of an assembly. This works well if you applied an assembly level appearance override to any of the components in an assembly.

NOTE

The All Hidden and All Visible design view representations will override any existing visibility settings in the subassembly. You cannot alter these representations or save new design view representations with the same name.

Creating Drawing Views from Design View Representations

To generate a drawing view based on a design view representation, activate the Drawing View dialog box, and click in the View section of the Representation area to select a valid design view, as shown in Figure 9.7. The view direction and zoom magnification in the selected view representation is ignored during the drawing view creation.

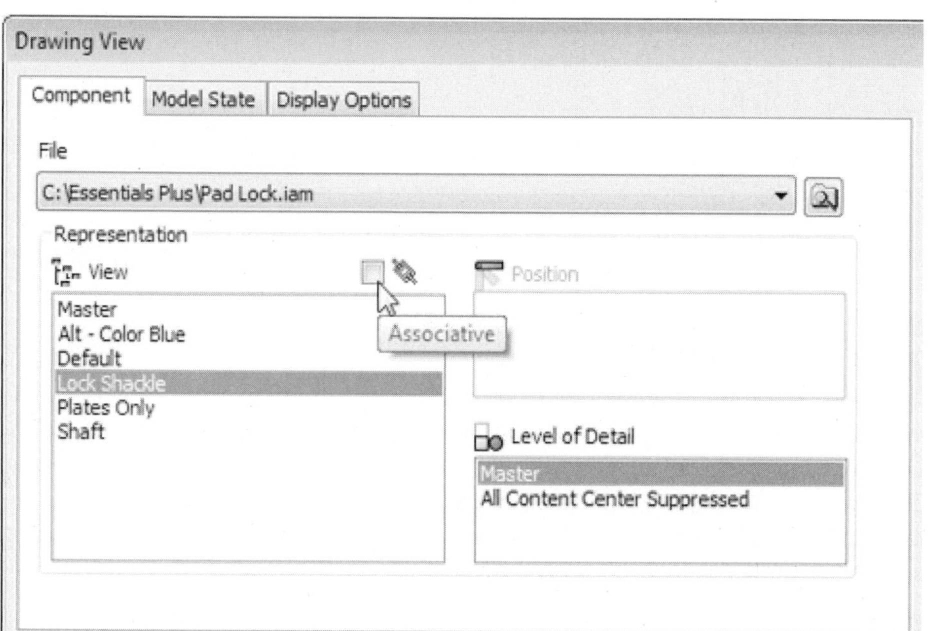

FIGURE 9.7

A projected view will be based on the design view representation from which its parent view was created. You can make drawing views generated from a design view representation associative by selecting the Associative checkbox, as shown in Figure 9.7. This action will make the drawing view generated by the design view representation dependent upon the design view representation found in the assembly file. Any changes made to the design view representation in the assembly will update the drawing view.

Once a drawing view is generated based on a design view representation, you can change the drawing view or views if you select a different design view representation. To perform this operation, first select the drawing view, and then right-click. Select Apply Design View from the menu as shown in Figure 9.8 on the left.

When the Apply Design View Representation dialog box appears, click in the list box, and select a new design view representation, as shown in Figure 9.8 on the right. If multiple views of the assembly appear on the sheet, set the Apply status for each view. Set the Associative status by changing the value from No to Yes.

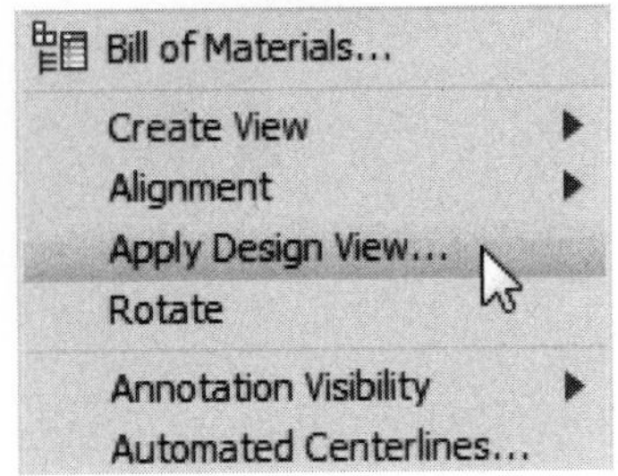

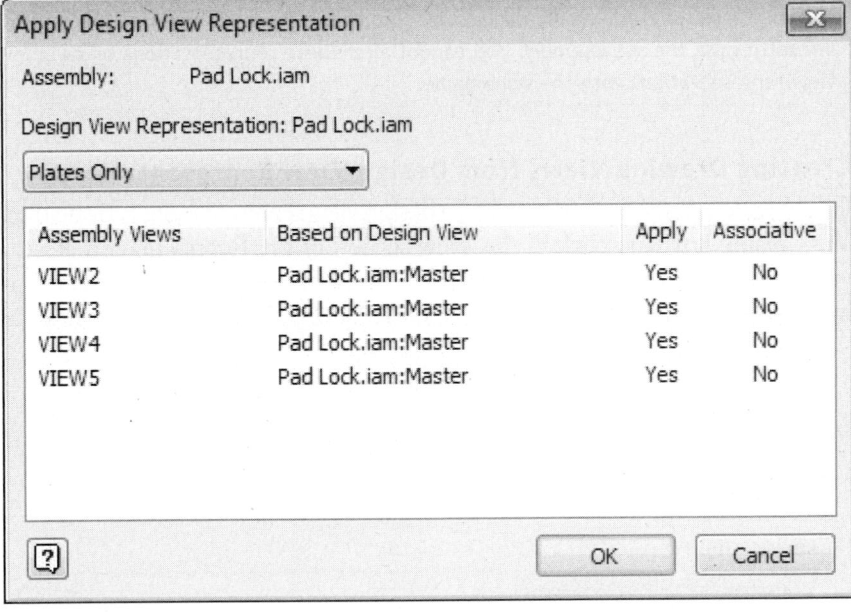

FIGURE 9.8

Levels of Detail Representations

The Levels of Detail Representations command can be used to improve capacity and performance in both the modeling and drawing environments. Levels of detail use component suppression to allow you to selectively choose which models are loaded into memory. The Suppress status of a component, accessed by right-clicking a component in either the browser or the graphics window, unloads the component(s) from memory, as shown in Figure 9.9 on the left.

When a component is suppressed, the component's icon is modified in the browser, the top-level assembly name is modified to notify you which level of detail is active, and the component is not displayed in the canvas. If you place your cursor over or select a suppressed component in the browser, the component bounding box is previewed in the canvas, as shown in Figure 9.9 on the right where two suppressed components have been selected.

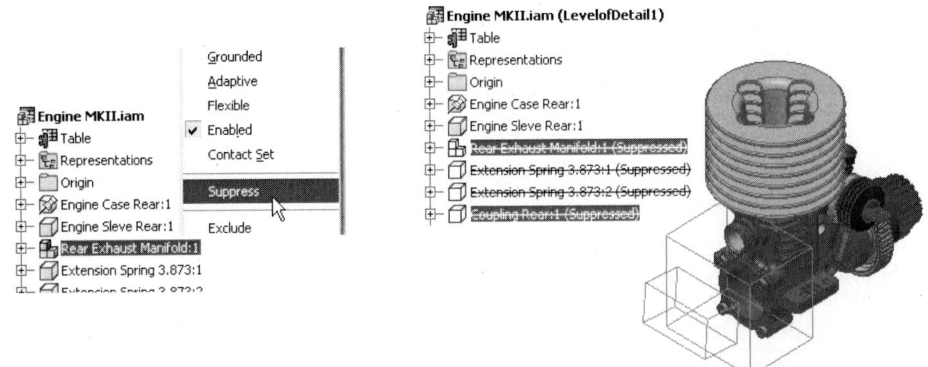

FIGURE 9.9

A Level of Detail node is available under the Representations folder in the browser, as shown in Figure 9.10. By default, four levels of detail are created in each assembly and can be activated to increase performance, if needed. The default levels of detail are Master, which is active by default, All Components Suppressed, All Parts Suppressed, and All Content Center Suppressed. You can activate a level of detail in the same way that you activate a design view representation or a positional representation: double-click on it in the browser.

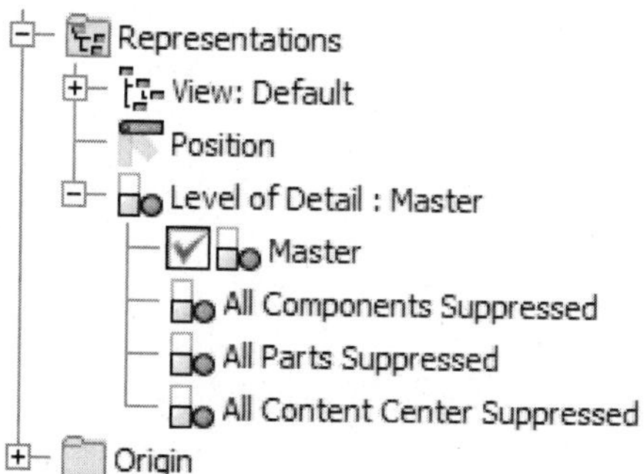

FIGURE 9.10

Similar to Design View Representations, the Master level of detail representation is active by default. You can make modifications to the status of children nodes of an assembly when this representation is active, but the changes are not maintained when the assembly file is saved. In order for the changes to be saved, a new or existing level of detail representation that you create must be active. Refer to the design view representation and positional representation sections of this chapter for more information about creating additional representations.

You can select a level of detail, design view, or positional representation to load when opening a file or placing a component in an assembly. This step is done by selecting the Options button in the File Open or Place Component dialog box and choosing the Level of Detail Representation that you want to load, as shown in Figure 9.11.

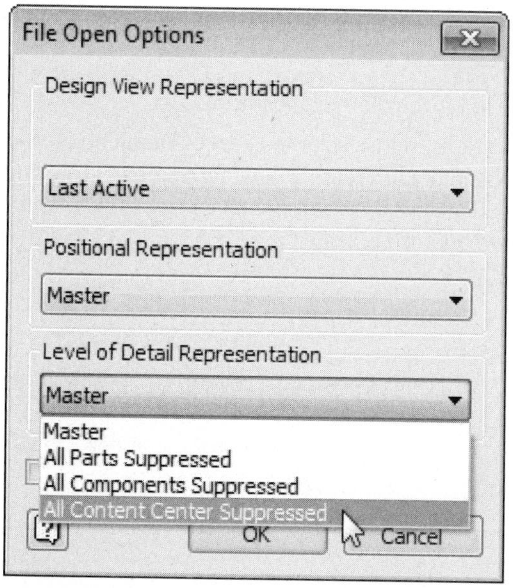

File Open Options
Design View Representation
Last Active
Positional Representation
Master
Level of Detail Representation
Master
Master
All Parts Suppressed
All Components Suppressed
All Content Center Suppressed
OK
Cancel

FIGURE 9.11

In a similar manner, when creating a drawing view of an assembly, you can select a level of detail representation from the Drawing View dialog box, as shown in Figure 9.12.

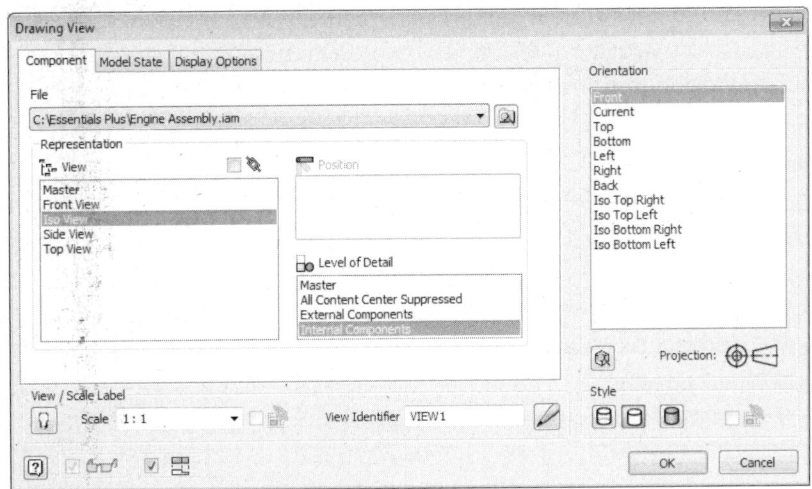

FIGURE 9.12

Assembly Substitution Level of Detail Representation

Another option for a Level of Detail is to create an assembly substitute. While working with an assembly you can use assembly substitution to reduce the memory that the assembly uses and reduce the amount of detail. The Level of Detail needs to be created at the top level of the open assembly. You need to open an assembly to create a substitute; you cannot create a substitute Level of Detail Representation of a subassembly in the active assembly. There is no limit to the number of substitutes that can be created. The iProperty data is maintained from the original assembly when a substitute is created. There are three methods for substituting assemblies; you can create a derived assembly, a shrinkwrap or substitute a subassembly for an existing part file.

Create a Substitute Level of Detail Representation Using Derive Assembly

To create an assembly substitute using the derived method follow these steps. The assembly constraints will be maintained.

1. Open an assembly for which you want to create a substitute.
2. In the browser expand Representations, right-click the Level of Detail node, and click New Substitute > Derive Assembly to create a new derived part from the assembly. The active Level of Detail at the time of creation contains the default body representation for the derive operation.

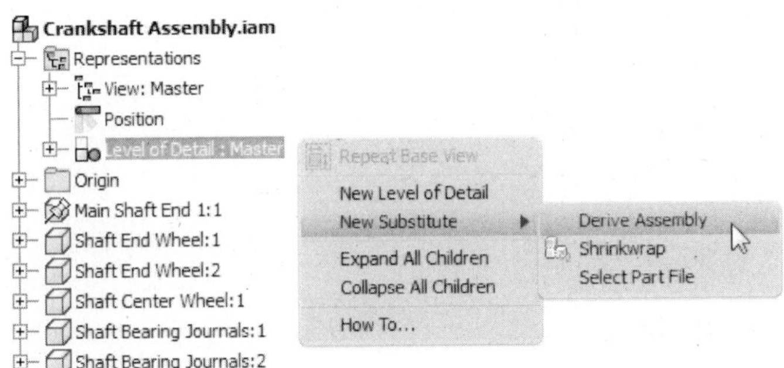

FIGURE 9.13

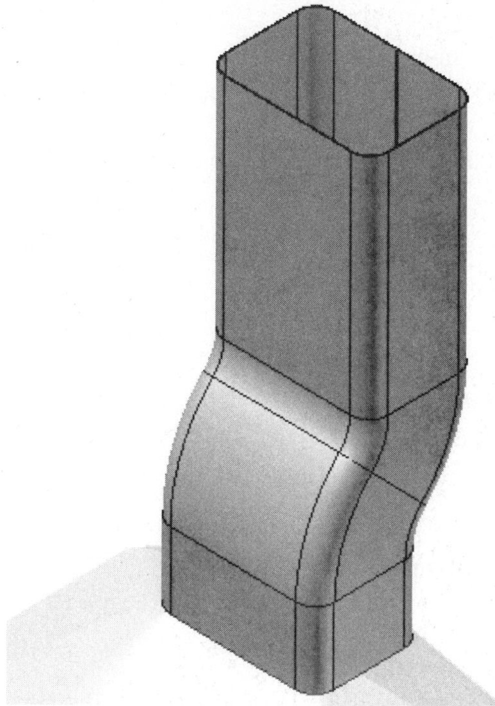

FIGURE 10.70

40. Click the *ESS_E10_02.iam* tab to return to the assembly.
41. In the browser, double-click *ESS_E10_02.iam* to activate the assembly, if it is not already active.

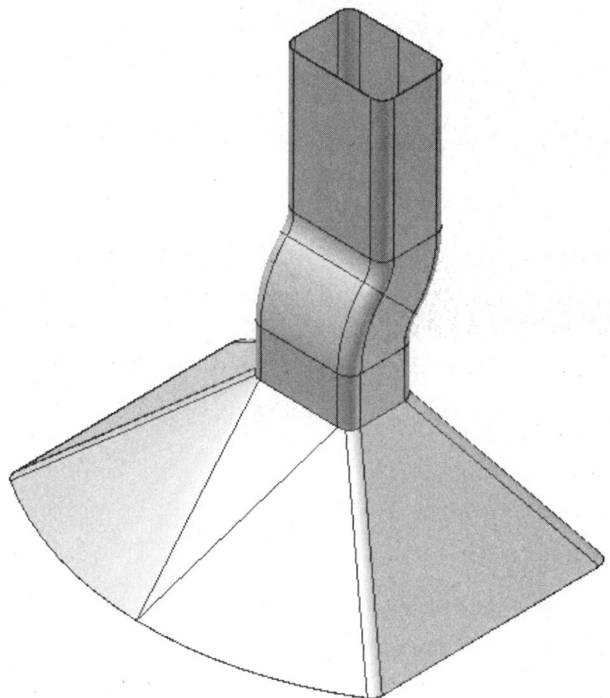

FIGURE 10.71

42. Close the file. Do not save changes. End of exercise.

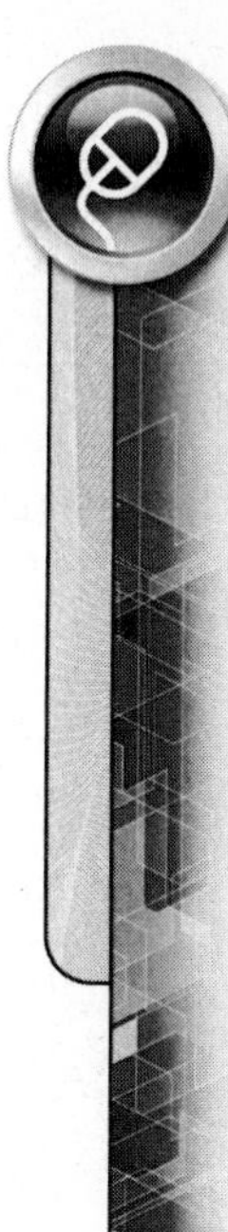

Hem

Hems eliminate sharp edges or strengthen an open edge of a face. Material is folded back over the face with a small gap between the face and the hem. A hem does not change the length of the sheet metal part; the face is trimmed so that the hem is tangent to the original length of the face. Create hems using the Hem command located on the Sheet Metal tab > Create panel, as shown in Figure 10.72.

FIGURE 10.72

NOTE

A minimum of one sheet metal face must exist before creating a hem.

To create a hem, follow these steps:

1. Click the Hem command on the Sheet Metal tab > Create panel. The Hem dialog box appears, as shown in Figure 10.73.

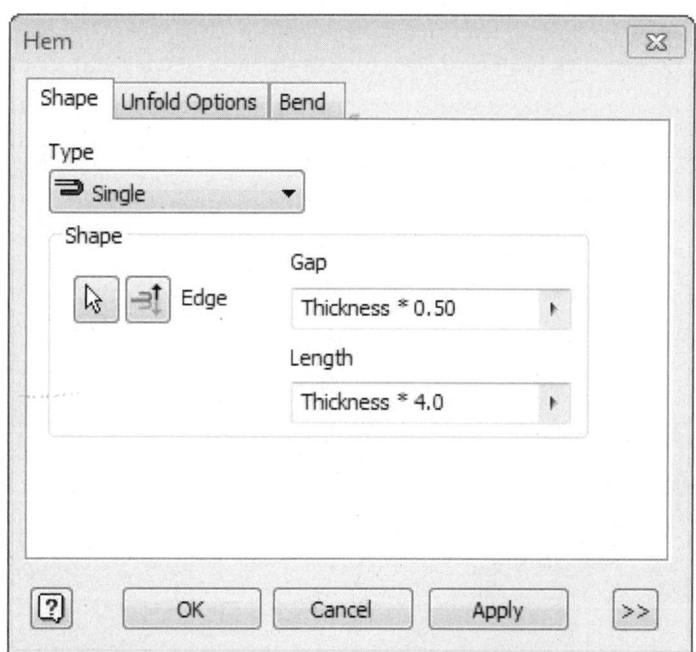

FIGURE 10.73

2. Select an open edge on a sheet metal face.
3. Select the hem type. Examples are shown in Figure 10.74:
   - Single: A 180° flange
   - Teardrop: A single hem in a teardrop shape
   - Rolled: A cylindrical hem
   - Double: Single hem folded 180° resulting in a double-thickness hem

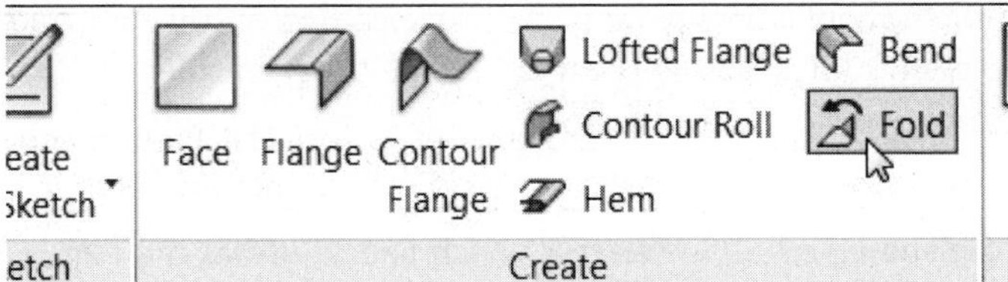

FIGURE 10.102

To create a fold, follow these steps:

1. Create a sketch on an existing face. Sketch a line between two open edges on the face.

The sketched line endpoints must be coincident to the face edge.

2. Click the Fold command on the Sheet Metal tab > Create panel. The Fold dialog box appears, as shown in Figure 10.103.
3. Select the sketch or bend line. The fold direction and angle are previewed in the graphics window. The fold arrows extend from the face that will remain fixed. The face on the other side of the bend line will fold around the bend line.
4. If required, flip the fold direction and side.
5. Enter the angle of the fold.
6. Select the positioning of the fold with respect to the sketched line. The line can define the centerline, start, or end of the bend. The fold preview updates to match the current settings.
7. Make any needed changes on the Unfold Options or Bend tabs.

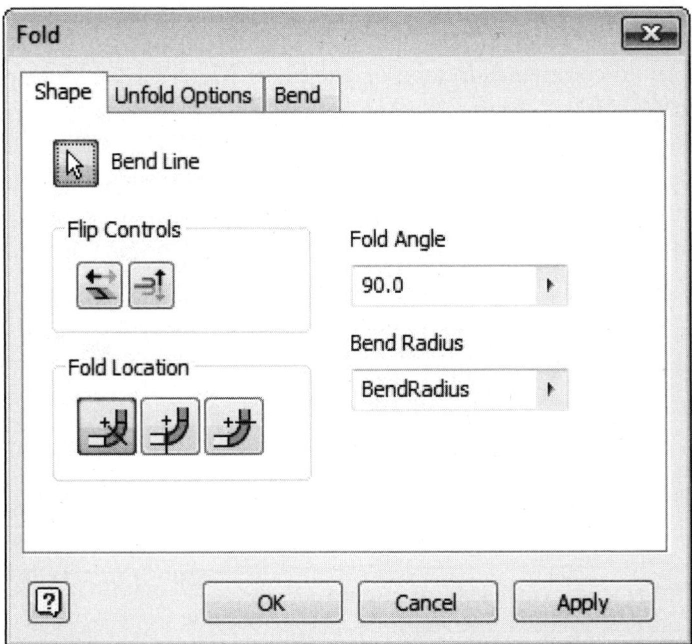

FIGURE 10.103

Shape Tab

The Shape tab allows you to select the bend line for the fold to be created. You can set the location of the selected bend line relative to the fold feature that determines the folded shape. The angle for the fold and direction are also specified on this tab, and you can also override the bend radius that is specified in the sheet metal rule. The following options are available on the Shape tab:

	Bend Line	Select a sketch line to use as the fold line.
	Flip Side	Flip the side used for the angle of the fold.
	Flip Direction	Toggle the direction that the fold will be created.
	Centerline of Bend	Determine the centerline of the fold from the selected sketch line.
	Start of Bend	Determine the start of the fold from the selected sketch line.
	End of Bend	Determine the end of the fold from the selected sketch line.
Fold Angle 90.0	Fold Angle	Specify the angle to apply to the fold.
Radius BendRadius	Bend Radius	Override the default bend radius set by the active sheet metal rule.

The options available on the Unfold Options and Bend tabs were covered in the Face section.

Unfold/Refold

Using the Unfold and Refold commands located on the Sheet Metal tab > Modify panel, as shown in Figure 10.104, you can select bends and rolls to flatten geometry. In many cases, it is easier to create certain sheet metal features while a model is in a flat state. Typically these types of features cross bend lines of a sheet metal part. You can use the Unfold command to flatten a formed state, create the necessary geometry and then use the Refold command to fold the geometry back to the original folded state.

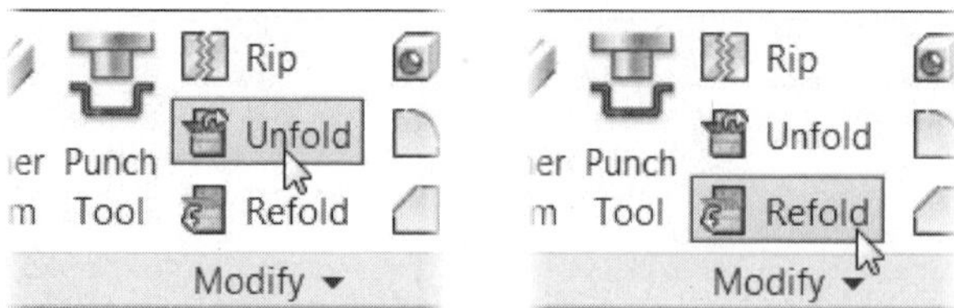

FIGURE 10.104

To unfold sheet metal features, follow these steps:

1. Click the Unfold command on the Sheet Metal tab > Modify panel. The Unfold dialog box appears, as shown in Figure 10.105 on the left.